The Role of Colloidal Systems in Environmental Protection

The Role of Colloidal Systems in Environmental Protection

Edited by

Monzer Fanun

Colloids and Surfaces Research Center,
Al-Quds University,
East Jerusalem,
Palestine

ELSEVIER

AMSTERDAM • BOSTON • HEIDELBERG • LONDON
NEW YORK • OXFORD • PARIS • SAN DIEGO
SAN FRANCISCO • SINGAPORE • SYDNEY • TOKYO

Elsevier
Radarweg 29, PO Box 211, 1000 AE Amsterdam, The Netherlands
The Boulevard, Langford Lane, Kidlington, Oxford OX5 1GB, UK

British Library Cataloguing in Publication Data
A catalogue record for this book is available from the British Library

Library of Congress Cataloging-in-Publication Data
A catalog record for this book is available from the Library of Congress

ISBN: 978-0-444-63283-8

For information on all Elsevier publications
visit our web site at store.elsevier.com

Printed and bound in Poland
14 15 16 17 18   10 9 8 7 6 5 4 3 2 1

Contents

List of Contributors

M.Z. Abdullah
Porous Media Combustion Laboratory, School of Mechanical Engineering, Universiti Sains Malaysia, 14300 Nibong Tebal, Penang, Malaysia

M. Abdul Mujeebu
Department of Building Engineering, College of Architecture and Planning, University of Dammam, 31451 Al-Dammam, Saudi Arabia

Younes Abrouki
Laboratoire de Génie de l'Eau et de l'Environnement, Université Hassan II, Faculté des sciences et techniques de Mohammedia, Maroc

Abdelkader Anouzla
Laboratoire de Génie de l'Eau et de l'Environnement, Université Hassan II, Faculté des sciences et techniques de Mohammedia, Maroc

S. Aytas
Ege University, Institute of Nuclear Sciences, 35100 Bornova-Izmir, Turkey

Vassiliki Belessi
Department of Graphic Arts, Technological Educational Institution of Athens, Agiou Spyridonos Street, 12210 Egaleo, Athens, Greece

J. Beltrán-Heredia
Department of Chemical Engineering and Physical Chemistry, University of Extremadura, 06071 Badajoz, Spain

Leticia A. Bernardez
Ingenium Consultoria em Engenharia Ltda, Avenida Estados Unidos, n. 528, Sala 1213, Salvador, Bahia, CEP 40010-020, Department of Materials Science and Technology, Federal University of Bahia, C.P. 6974, Salvador, BA, Brazil, 41810-971

J.P. Canselier
Université de Toulouse, INPT, UPS; Laboratoire de Génie Chimique, 4 allée Emile Monso, BP 84234, 31432 Toulouse Cedex 4, France

X.H. Chang
School of Chemistry and Chemical Engineering, Luoyang Normal University, Luoyang, 471022, P. R. China

D.R. Chen
School of Chemistry and Chemical Engineering, Shandong University, Jinan, 250100, P. R. China

B. Coco-Rivero
Department of Chemical Engineering and Physical Chemistry, University of Extremadura, 06071 Badajoz, Spain

Afonso Avelino Dantas Neto
Universidade Federal do Rio Grande do Norte, Centro de Tecnologia, Departamento de Engenharia Química, Natal – RN, Brazil

Luiz R.P. de Andrade Lima
Ingenium Consultoria em Engenharia Ltda, Avenida Estados Unidos, n. 528, Sala 1213, Salvador, Bahia, CEP 40010-020, Department of Materials Science and Technology, Federal University of Bahia, C.P. 6974, Salvador, BA, Brazil, 41810-971

Alcides de Oliveira Wanderley Neto
Universidade Federal do Rio Grande do Norte, Centro de Ciências Exatas e da Terra, Instituto de Química, Natal – RN, Brazil

Ali Deriszadeh
Department of Chemical and Petroleum Engineering, University of Calgary, Calgary, Alberta, Canada T2N 1N4

V. Encinas-Sánchez
Department of Chemical Engineering and Physical Chemistry, University of Extremadura, 06071 Badajoz, Spain

Danka Galabova
Department of Microbial Biochemistry, The Stephan Angeloff Institute of Microbiology, Bulgarian Academy of Sciences, Sofia, Bulgaria

Marisol Gallegos-Garcia
Instituto de Metalurgia, Universidad Autónoma de San Luis Potosí, Av. Sierra Leona 550, San Luis Potosí, C. P. 78210, México

O. Gapurova
Institute of Nuclear Physics, Tashkent, Uzbekistan

Maria Gavrilescu
"Gheorghe Asachi" Technical University of Iasi, Faculty of Chemical Engineering and Environmental Protection, Department of Chemical Engineering, Department of Enviromental Engineering and Management, 71 Mangeron Blvd., 700050 Iasi, Romania, Academy of Romanian Scientists, 54 Splaiul Independentei, RO-050094 Bucharest, Romania

C. Gok
Pamukkale University, Denizli Vocational School of Technical Sciences, Department of Electricity and Energy, 20070 Kinikli, Denizli, Turkey

C. Gourdon
Université de Toulouse, INPT, UPS; Laboratoire de Génie Chimique, 4 allée Emile Monso, BP 84234, 31432 Toulouse Cedex 4, France

Alexandre Gurgel
Universidade Federal de Viçosa, Centro de Ciências Exatas e Tecnológicas, Departamento de Química, Viçosa – MG, Brazil

B. Haddou
U. S. T. Oran-MB, Faculté de chimie, Département de génie Chimique BP 1505, M'Nouar, Oran, Algérie

Thomas G. Harding
Department of Chemical and Petroleum Engineering, University of Calgary, Calgary, Alberta, Canada T2N 1N4

Maen M. Husein
Department of Chemical and Petroleum Engineering, University of Calgary, Calgary, Alberta, Canada T2N 1N4

X.L. Jiao
School of Chemistry and Chemical Engineering, Shandong University, Jinan, 250100, P. R. China

A.D. Karathanasis
Department of Plant and Soil Sciences, University of Kentucky, Lexington, KY 40546, USA

Elena Karpenko
Department of Chemistry and Biotechnology, Lviv Department of L.M.Lytvynenko Physical-Organic Chemistry Institute, National Academy of Sciences of Ukraine, Naukova st., 3a, Lviv, Ukraine

Oleksandr Karpenko
Department of Biologically Active Substances, Pharmacy and Biotechnology, Lviv Polytechnic National University, Lviv, Ukraine

Ather Farooq Khan
Interdisciplinary Research Centre in Biomedical Materials, COMSATS Institute of Information Technology, Lahore-54000, Pakistan

Afsar Khan
Department of Chemistry, COMSATS Institute of Information Technology, Abbottabad 22060, Pakistan

Rashid A. Khaydarov
Institute of Nuclear Physics, Tashkent, Uzbekistan

Renat R. Khaydarov
Institute of Nuclear Physics, Tashkent, Uzbekistan

Qaisar Mahmood, (TI)
Department of Environmental Sciences, COMSATS Institute of Information Technology, Abbottabad 22060, Pakistan

J.O. Miller
Department of Plant and Soil Sciences, University of Kentucky, Lexington, KY 40546, USA

A.A. Mohamad
Department of Mechanical and Manufacturing Engineering, CEERE, The University of Calgary, Calgary, Alberta T2N 1N4, Canada

Ana B. Moldes
ETSI, Chemical Engineering Department, University of Vigo, Spain

N. Muthukumar
Corrosion Protection Division, Central Electrochemical Research Institute, Karaikudi—630 006, India; Institute of Advanced Energy, Kyoto University, Gokasho, Uji, Kyoto-611-0011, Japan; Present Address: Water & Process Technologies, GE (General Electric) India Technology Centre, Bangalore - 560066, India

Tereza Neuma de Castro Dantas
Universidade Federal do Rio Grande do Norte, Centro de Ciências Exatas e da Terra, Instituto de Química, Natal – RN, Brazil

N. Nishad Fathima
Chemical Laboratory, Central Leather Research Institute, Council of Scientific and Industrial Research, Adyar, Chennai, India

Dimitris Petridis
Institute of Materials Science, NCSR "Demokritos," 15310 Aghia Paraskevi Attikis, Athens, Greece

J. Raghava Rao
Chemical Laboratory, Central Leather Research Institute, Council of Scientific and Industrial Research, Adyar, Chennai, India

Hicham Rhbal
Laboratoire de Génie de l'Eau et de l'Environnement, Université Hassan II, Faculté des sciences et techniques de Mohammedia, Maroc

Luigi Rizzo
Department of Civil Engineering, University of Salerno, 84084 Fisciano (SA), Italy

Mohamed Safi
Laboratoire de Génie de l'Eau et de l'Environnement, Université Hassan II, Faculté des sciences et techniques de Mohammedia, Maroc

J. Sánchez-Martín
Department of Chemical Engineering and Physical Chemistry, University of Extremadura, 06071 Badajoz, Spain

Sook San Wong
School of Industrial Technology, Universiti Sains Malaysia, 11800 Penang, Malaysia

El-Sayed M. Sherif
Center of Excellence for Research in Engineering Materials (CEREM), College of Engineering, King Saud University, Al-Riyadh, Saudi Arabia, Electrochemistry and Corrosion Laboratory, Department of Physical Chemistry, National Research Centre (NRC), Cairo, Egypt

Alisa Sineva
Faculty of Chemistry, Lomonosov Moscow State University, Moscow

Shaoxian Song
Instituto de Metalurgia, Universidad Autónoma de San Luis Potosí, Av. Sierra Leona 550, San Luis Potosí, C. P. 78210, México

Anna Sotirova
Department of Microbial Biochemistry, The Stephan Angeloff Institute of Microbiology, Bulgarian Academy of Sciences, Sofia, Bulgaria

Salah Souabi
Laboratoire de Génie de l'Eau et de l'Environnement, Université Hassan II,
Faculté des sciences et techniques de Mohammedia, Maroc

Tjoon Tow Teng
School of Industrial Technology, Universiti Sains Malaysia, 11800 Penang, Malaysia

Ling Wei Low
School of Industrial Technology, Universiti Sains Malaysia, 11800 Penang, Malaysia

Daniel C.W. Tsang
Department of Civil and Natural Resources Engineering, University of Canterbury, Christchurch, New Zealand

Balachandran Unni Nair
Chemical Laboratory, Central Leather Research Institute, Council of Scientific and Industrial Research,
Adyar, Chennai, India

Foreword

In recent years, increased anthropogenic inputs of toxic chemicals to terrestrial environments have caused great public concerns relative to contamination of surface and ground water supplies. Early contaminant transport investigations used a two-phase approach in which contaminants were partitioned between an immobile solid phase and a mobile aqueous phase to predict contaminant transport. However, dispersed colloid particles may act as a third mobile phase that can sorb contaminants and carry them through porous media at rates and distances even greater than the soluble phase alone. Organic and mineral colloid particles are abundant in most soil environments in size ranges below 10 μm and possess high specific surface area and chemical reactivity for toxic contaminants. Colloids have been shown to be more mobile in soils with extensive macroporosity, where particle straining is limited and preferential flow may lead to sizeable contaminant load migrations. Under favorable conditions, colloid particles may exceed ordinary transport rates and pose a significant threat to surface and ground water quality. Colloid mobility through porous media is a function of their stability in suspension that is dictated by factors such as their size, pH, ionic strength, composition, and hydrodynamic pore flow conditions. Stable colloidal suspensions are characterized by a net repulsive interaction energy barrier between van der Walls and double-layer electrostatic forces that enhances stability and prevents coagulation and flocculation (DLVO theory). While a significant portion of stable migrating colloids may be filtered by deposition on the pore walls of the soil matrix, substantial amounts may find their way through flow paths and travel to greater than anticipated distances.

A number of investigations involving intact soil columns and field studies have demonstrated strong adsorptive affinities and co-transportabilities of organic and mineral colloid species for a variety of contaminants including metals, phosphorous, bacterial pathogens, radionuclides, and toxic organics. Although preferential sorption to the colloid surface is considered the dominant mechanism by which such contaminants are co-transported through porous media, excessive soluble contaminant loads mobilized in the presence of colloids suggest the involvement of multiple transport mechanisms. Such mechanisms may include complexation with dissolved organic carbon species, co-precipitation, or selective exclusion of soluble species from reactive pore sites blocked by mobilized colloids.

The dramatic worldwide increase in the use of biosolid wastes as cost effective nutrient management alternatives in agricultural and disturbed landscapes also poses significant contamination risks. Many of these wastes may contain high levels of toxic metals, phosphorous, and other pathogens that could threaten surface and groundwater quality. While indigenous mineral soil colloids tend to destabilize and flocculate more readily, organically enriched biocolloids have shown considerably higher stability and mobility through electrostatic and steric repulsive forces even under high ionic strength conditions. Numerous field and soil monolith experiments have demonstrated significant increases in soluble and colloid-bound metal contaminant loads transported in the presence of biocolloids fractionated from a variety of industrial and agricultural sources. Increased biocolloid mobility associated with enhanced metal and phosphorous transportability has also been observed with the application of lime-stabilized biosolids, even though their high pH should have reduced metal solubility. Formation of stable soluble organic complexes and metal-carbonate co-precipitation mechanisms accounted for the increased metal and P migration.

Although the potential role of colloid particles as carriers or facilitators of contaminants has been well documented, a complete physical, chemical, and microbiological framework for the prediction of colloidal transport phenomena is still lacking. In spite of the progress in understanding fundamental colloidal phenomena, most modeling approaches have utilized microcosm experiments with classical column tests that obscure pore scale and interfacial chemical processes. The complexity of natural systems requires the development of reactive transport modeling approaches accounting for spatial and temporal variability in colloid particle transport kinetics and surface chemistry dynamics of the colloid-contaminant interactions. Understanding the susceptibility of colloidal phenomena to hydrological, physicochemical and biological fluctuations of natural environments at the field scale can be regarded as one of the key scientific challenges in the context of colloid facilitated contaminant transport.

A.D. Karathanasis
University of Kentucky

Preface

There is a growing public interest in developing risk assessment framework, environment regulations, and remedial strategies for protecting ecosystems and human from environmental hazards like heavy metals in drinking water, pesticides, food contaminants etc. Colloids are systems made of tiny building elements, and as a consequence, have a high specific surface area, have unique properties, different of those exhibited by ordinary homogeneous or heterogeneous systems, instability due to high free energy, well classified by internal morphology. Colloids are successfully applied in many areas of environmental protection. A number of environmental and energy technologies have already benefited substantially from colloidal technology. The present book describes the role of colloidal systems in environmental protection in areas like drinking water, wastewater treatment, heavy metal remediation, treatment of contaminated soils, abatement of pesticide contamination in food and soil, protection against radioactive materials, corrosion.

The disposal of bio solids through application to agricultural and disturbed lands provides a beneficial soil amendment. Bio solids can increase the organic and nutrient content of soil surfaces, but they often contain inorganic and organic contaminants from industrial and domestic sources. Chapter 1 by Miller and Karathanasis discusses various issues associated with the application to land of bio solid materials particularly in the form of bio colloids, including the types and sources of health risks, the potential generation of bio solid colloids under different environmental settings, their association with industrial and natural environmental contaminants, transport pathways to surface and ground water, mechanisms and fates of transported contaminants, potential risks associated with colloid facilitated transport, as well as pollution prevention and remediation strategies. Natural organic matter is ubiquitous in surface soils and shallow aquifers where most anthropogenic contamination occurs. The influence of natural organic matter, as a dissolved or colloidal phase, on the contaminant removal during contaminated site remediation has been of increasing concern in recent years. Chapter 2 by Tsang provides an overview of the characteristics and geochemical reactions of natural organic matter in the environment, and discusses the effects of natural organic matter on the application of zero-valent iron permeable reactive barrier, which is a widely used technology for groundwater remediation. Bio surfactants are amphiphilic compounds produced by numerous microorganisms in natural and industrial environments. They reduce surface and interfacial tensions by accumulating at the interface of immiscible fluids (liquids, solids or gases) and thus modify interactions between several corpora. They present the same physicochemical properties as surfactants obtained by chemical synthesis (i.e. emulsifying, de-emulsifying, foaming, wetting) but sometimes with lower toxicity and higher biodegradability. As a result, the applications of bio surfactants are many and promising. In Chapter 3 Galabova et al. discuss the potential role and applications of bio surfactants in environment focusing on new data in bioremediation/biodegradation of water and soil pollutants. The bioremediation of hydrocarbon-contaminated sites is limited by the poor availability of these hydrophobic contaminants to microorganisms, which could be improved by using surface-active compounds. Advances on the production of useful metabolites for the environment protection are reported in Chapter 4 by Moldes.

Haddou et al review cloud point extraction that is one of the most promising, environmentally friendly, energy-saving processes for aqueous effluent purification in Chapter 5. Problems of pollution of water resources by synthetic surfactant detergents are considered to be a part of general ecological problems of environmental protection. The increasing use and global scale of production of detergents being pollutants requires the use of new methods of the artificial wastewater treatment based on the principle of nature water auto-purification in ecosystem. Basic process of both water treatment and free-of-waste technologies is an adsorption process, which is connected with surface screening by organic substances and inorganic ions. In Chapter 6, Sineva reports on the adsorption of surfactant detergents on natural adsorbents that are considered to be a component part of colloid chemical base of purification of sewage polluted with synthetic detergents.

Micellar-enhanced ultrafiltration can be applied to removing soluble organics from contaminated waters. It involves addition of a surfactant to the contaminated water at a concentration, which exceeds its critical micelle concentration. Consequently, micelles form and dissolve organic contaminants in their hydrocarbon tails. An ultrafiltration membrane with pores smaller than the micelle size is, then, used to separate the contaminant-containing micelles from the water. Traditional micellar-enhanced ultrafiltration, however, is susceptible to membrane plugging and back contamination, especially at high surfactant dosage. Chapter 7 by Husein et al. sheds some light on amphiphilic naphthenic acid co-contaminants and their role in decreasing the dosage of cetylpyridinium chloride surfactant necessary to achieve appreciable removal of the organic contaminants. The majority of the polluting agents (surfactants, pesticides, heavy Novel tannin-based adsorbents were tested on heavy metals removal and reviewed by Sánchez-Martín et al. in Chapter 8. Coagulation process has been historically used in drinking water treatment to remove colloids from surface water in order to decrease turbidity, typically related to pathogens occurrence. Rizzo reviews advances in coagulation process in terms of prepolymerized, new and natural organic coagulants in Chapter 9. Flocculation remains as the conventional but most reliable mechanism for suspended solids removal in wastewater treatment system. Flocculants come with many forms and they are classified in term of molecular weight, physical form, type of charge and charge density. Coagulation-flocculation method has been used to treat pulp and paper mill wastewater. Processes using single coagulant, single flocculant, and coagulant-flocculant have been reported in Chapter 10 by Teng and Wong. Chapter 11 by Song and Gallegos-Garcia review highlights arsenic removal from drinking water and mining-contaminated water by coagulation process. Textile dyeing processes are among the most environmentally unfriendly industrial processes, because they produce coloured wastewaters that are heavily polluted with synthetic dyes. The investigation presented in Chapter 12 by Anouzla et al. focussed on the steel industrial wastewater FeCl3 rich as an original coagulant to remove the synthetic textile wastewater.

Nanoparticles are currently being developed very rapidly for various applications. Due to their unique physical, chemical and magnetic properties as well as the ability to penetrate through porous media, nanoparticles have promising perspectives for wastewater treatment. There is increasing interest and need to develop a deeper understanding of the nature, fate and behavior of nanoparticles in the wastewater treatment. Literature studies on nanoparticle applications for wastewater treatment primarily focused on employing nanoparticles as reactants for degradation of contaminants, usually organic materials. There are a few studies focused on applying nanoparticles as adsorbents for removal of metal ions from wastewater. Chapter 13 by Belessi and Petridis focuses on the use of modified and non-modified titania that has been found application on wastewater treatment technologies involving adsorption and photo catalysis in aqueous solution. A novel technique of obtaining nanocarbon-polymer composition on the base of nano-carbon colloids and polyethylenimine for the removal of metal ions from contaminated water and remediation of soil has been reported in Chapter 14 by Khaydarov et al. Chapter 15 by Beltrán-Heredia et al. provides a comprehensive picture of the global scenario of applications of Quebracho colorado extract as coagulant for water treatment.

Biosorption can be defined as the removal of substances, such as metal or metalloid species, compounds and particulates from solution by biological material or their products, especially bacteria, algae, yeast and fungi by physicochemical binding. It is an emerging and attractive technology, which involves sorption of dissolved substances by a biomaterial. It is a potential technique for the removal of radionuclides and heavy metals from industrial wastewater. The main advantages of this technique are the reusability of biomaterial, low operating cost, improved selectivity for specific metals of interest, removal of radionuclides or heavy metals from effluent irrespective of toxicity, short operation time, and no production of secondary compounds which might be toxic. In the long run, biopolymers will find their way into industry and everyday life; they are the polymers of the future. The increase in the nuclear industry and other anthropogenic activities has intensified environmental pollution, with the accumulation of radioactive elements as uranium and thorium. Growing attention is being given to the potential health hazard presented by radionuclides to the environment, and the need for economic and effective methods for the removal of radionuclides has resulted in the development of new separation technologies. So, it is very important to identify potential effective and environmentally being sorbents for the remediation/removal of uranium and thorium from aqueous medium in order to protect the environment from this radioactive element and its daughter products. Chapter 16 by Gok and Aytas reviews the state of art of biosorption of uranium and thorium by biopolymers and to compare the results found in the literature and obtained biosorption results of uranium and thorium by calcium-alginate biopolymer beads.

The design and implementation of effective schemes for the remediation of contaminated soils require a clear understanding of the processes controlling the sorption and desorption of the contaminants. This context entails developments on experimental, modeling and large-scale studies on the impact of colloids on contaminant transport in saturated and unsaturated porous media, since stable colloidal particles can travel long distances in subsurface environments and carry particle-reactive contaminants with them. Chapter 17 Gavrilescu analyzes the movements of natural colloids in subsurface environments, which can intensify pollutant transport much more than predicted by the conventional advection-dispersion solute transport equations, mainly due to their association with mobile colloids. Soil characteristics are associated with the presence of the colloid and the fate and transport of contaminants.

Micellar solubilization is an important step for removal of organic contaminants from the soil matrix, especially for low aqueous solubility organic contaminants such as diesel. In addition to fundamental studies, laboratory and field studies on removal of diesel from contaminated soil and mathematical models applied are also reviewed to show the applicability of this technology. The combination of micellar solutions with other remediation techniques such as biodegradation applied to diesel recovery is in Chapter 18 by Bernardez and de Andrade Lima. Corrosion is the deterioration of materials, usually metals that results from the reaction between metal and environment. Chapter 19 by Gurgel et al. aims to provide a general overview on the use of surfactant systems, particularly micellar solutions and microemulsions, as corrosion inhibitors on metallic surfaces. Chapter 20 by Sherif brings together in a general view, the definition, classification, and the of corrosion inhibitors in controlling the corrosion of metals and alloys in harsh environments such as aqueous media and atmosphere. The aim is also extended to shed more light on the most used techniques to obtain the inhibition efficiencies of inhibitors onto surfaces of metals and alloys. During normal pipeline operation, warm petroleum products with small amounts of salt and water travels through the pipeline. Factors viz. nutrients (nitrite; phosphate etc.), oxygen, chloride and bacteria are the causative factors for internal corrosion in petroleum transporting pipeline. In oil transporting pipelines, stagnation of water occurs due to the slopes in the landscape and this acts as a breeding ground for bacteria. Generally degradation starts at the interface between diesel and water. Under this condition good inhibitor is needed for preventing pipeline corrosion. It has also been estimated that 40% of all internal corrosion of pipeline in the gas and oil industry can be attributed to microbial corrosion. Different methods are used for protection of petroleum product pipelines from corrosion. Muthukumar reviews applications of a protective surface coating as efficient corrosion inhibitor in Chapter 21.

Aerogels with open three-dimensional mesoporous structure and high specific surface areas have potential applications as adsorbents, catalysts, thermal insulation and acoustic absorption materials for environmental purposes. Chapter 22 by Chen et al. gives some typical examples to discuss the adsorptions of heavy metals ions and organic pollutants in wastewater and organic molecules in air, the catalytic applications of aerogels for the decomposition of inorganic and organic pollutants. Furthermore, other applications of aerogels used as hydrogen storage, thermal insulation and acoustic absorption materials for environmental purposes are also discussed. Chapter 23 by Rao et al. discusses the effective use of protein wastes from animal hides/skins for the treatment of colored wastewater.

Fast depletion of fossil fuel resources and environmental pollution are two major issues associated with the abundant use of fossil fuels. Porous medium combustion is one of the feasible options to solve these issues to a remarkable extent. It has interesting advantages compared with free flame combustion due to the higher burning rates, the increased power dynamic range, the extension of the lean flammability limits, and the low emissions of pollutants. Though extensive researches have been carried on this topic, more attention is needed to realize its outstanding benefits in practical applications. Chapter 24 by Abdul Mujeebu provides an overview of the widespread applications of Porous medium combustion technology and the updated information on the related research. Based on a thorough review of the available literature, the applications are categorized and presented accordingly. The last chapter by Mahmoud et al. reports on the future perspectives of the use of colloids in environmental protection.

Finally, This book will be of immense use, not only to those working on research and development, over a whole range of different technologies which are concerned with colloids for the environment, but also to academic scientists in the colloid and surface science field. It summarizes recent research in the field of colloids in environmental protection; it eliminates the need to search through stacks of journals for critical information. All the facts needed by the researcher in the laboratory or in the classroom are at his fingertips in this unique book that can be used as an effective guide for planning future research. This book covers recent advances in the formulation, characterization of the properties of colloids for environmental protection. An international community of colloid scientists has come together to create this book. More than 60 individuals from 20 countries contributed to the work. All of them are recognized and respected experts in the areas they wrote about. An important feature of this book is that the author of each chapter has been given the freedom to present, as he/she sees fit, the spectrum of the relevant science, from pure to applied, in his/her particular topic. Any author has his own views on, and approach to, a specific topic, molded by his own experience. I think that this book will familiarize the reader with the technological features of application of colloids in the environmental protection, and to provide experienced researchers, scientists and engineers in academic and industry communities with the latest developments in this field. This book will compliment very well, existing books on environmental protection, which, in general, take the more traditional approach of reviewing systematically the fundamental (pure) aspects of the subject. An important feature of this book is that the author of each chapter has been given the freedom to present, as he/she sees fit, the spectrum of the relevant science, from pure to applied, in his/her particular topic. Any author has his own views on, and approach to, a specific topic, molded by his own experience.

I would like to thank all those who contributed as chapter authors despite their demanding agenda. All of them are prominent and appreciated specialist in the areas they wrote about. None of them is associated with any errors or omissions that remain. I take full responsibility. My sincere gratitude is due to the reviewers for their valuable remarks. My appreciation goes to Anita Koch of Elsevier Science Publishing for her superior handling of this project.

Monzer Fanun PhD.
Colloids and Surfaces Research Center, Al-Quds University, East Jerusalem, Palestine

CHAPTER

Biosolid Colloids as Environmental Contaminant Carriers

1

J.O. Miller, A.D. Karathanasis

Department of Plant and Soil Sciences, University of Kentucky, Lexington, KY 40546, USA

CHAPTER OUTLINE

1 Biosolid materials in the environment

Land application of agricultural and industrial solid wastes as alternative nutrient sources or soil amendments has dramatically increased in the last 30 years. Currently, at least 5.6 million tons of biosolids are disposed of from wastewater plants in the United States, 60% of which is applied to land (NRC, 2002). The EPA has classified sewage sludge as the solid, semisolid, or liquid residue from wastewater treatment, while biosolids are specifically described as the solid waste portion that can be "beneficially recycled." This recycling refers to the application of biosolids to agricultural or reclaimed landscapes for nutrient and organic matter additions, or for general land reclamation purposes (Haering et al., 2000; Brown et al., 2002).

Before biosolids can be applied to land, they must undergo treatment to reduce exposure risks to humans (Brown et al., 2002, NRC, 2002) due to the presence of organic, inorganic, or biological contaminants and pathogens originating from various domestic and industrial sources (NRC, 2002). Besides bacterial and viral pathogens, biosolids are regulated for P and nine other inorganic pollutants (Ar, Cd, Cu, Pb, Hg, Mo, Ni, Se, and Zn). Proponents of land application of biosolid wastes cite a virtual immobilization of these contaminants by the soil matrix through filtration, complexation, and chemisorption processes (Haering et al., 2000; Brown, et al., 2002; Basta et al., 2005). However, substantial shortfalls in metal balances and other organic contaminants have been reported in field and

The Role of Colloidal Systems in Environmental Protection. http://dx.doi.org/10.1016/B978-0-444-63283-8.00001-6

laboratory leaching experiments (McBride et al, 1999; Gove et al, 2001; Kinney et al., 2006), suggesting that alarming levels of these pollutants may be reaching and contaminating our groundwater supplies (Ashworth and Alloway, 2004; Mantovi et al., 2005; Saiers and Ryan, 2006). Recent studies have suggested that the unaccounted for contaminant loads may be associated with mobile biosolid colloid particles (biocolloids) that migrate through soil macropores and thus transport the contaminants to greater soil depths and unexpected distances (Kretzschmar et al., 1995; Gove et al., 2001; Karathanasis and Ming, 2002; Karathanasis et al., 2007). The high organic carbon content and increased surface reactivity of mobile biosolid colloids can enhance chemisorption/complexation interactions with contaminants, thus facilitating pollutant transport by outcompeting soil matrix sorption/filtration forces. Organically coated colloids have also been shown to induce particle stability by acting as steric stabilizers that further enhance their mobility through porous media (Kaplan et al, 1993).

2 Biocolloid properties

There is limited information on biocolloid properties, which may vary according to treatment processes. Biosolids are not purely organic and may contain several mineral components and amorphous oxides (Brown et al., 2002; Basta et al., 2005). Water-dispersible colloids (< 2 μm) generated from three organic wastes, including anaerobically digested biosolids (ADB), lime-stabilized biosolids (LSB), and poultry manure (PMB), varied considerably in composition (Karathanasis and Johnson, 2006). Organic matter (OM) content of the colloids ranged from 20% in LSB to 60% in ADB (Table 1.1). Lime-stabilized biocolloids were characterized by the highest carbonate levels, pH, and EC as well as considerably lower surface area than the other two colloids. The ADB colloid had the significantly highest Fe-oxide and Al-oxide concentrations, comparable to those found in bulk biosolid materials (Brown et al., 2002) and OM content. The PMB colloids were distinguished by the lowest pH and EC and the highest surface area and phyllosilicate mineral content (Table 1.1). Colloid diameter

Table 1.1 Physicochemical and Mineralogical Properties of Suspended Biosolid Colloids (from Karathanasis and Johnson, 2006)

	Colloids		
Properties	**Lime-Stabilized (LSB)**	**Aerobically Digested (ADB)**	**Poultry Manure (PMB)**
pH	11.3	7.2	5.5
EC (μS cm^{-1})	68.2	16.3	12.5
CEC (cmol kg^{-1})	80.4	101.0	93.2
Fe-oxides (g kg^{-1})	0.13	20.1	0.11
Al-oxides (g kg^{-1})	4.1	11.5	0.12
Phyllosilicate minerals (%)	10	18	35
Surface area (m^2 g^{-1})	361	506	578
Colloid Diameter (μm)	0.41	0.46	0.38
OM (%)	20.1	60.6	47.7
Carbonates (%)	60	5	-

was similar for all three colloids, averaging about 0.4 μm (Table 1.1). This size is comparable to that of colloids leached from forest soils with high natural organic matter (NOM) (0.3–0.7 μm), but much smaller than the size of colloid particles eluted from reclaimed soil monoliths receiving biosolid applications (> 2 μm) (Miller et al., 2008, 2011). Generally, organic biosolid colloids tend to have a smaller size diameter than mineral and oxide colloids, thus making them more mobile through heterogeneous pore systems (Pokrvosky and Schott, 2002; Karathanasis, 1999).

The surface area of the biocolloids studied by Karathanasis and Johnson (2006) ranged from 361 m^2 for LSB to 578 m^2 for PMB colloids, with a respective surface charge of 80.4 to 93.2 cmol kg^{-1}. These values compare favorably with the highest reactivity mineral colloid particles. The LSB colloid suspensions had the highest pH (11.3) and electrical conductivity related to their lime stabilization treatment compared to ADB (7.2) and PMB (5.5) suspensions (Table 1.1), providing a wide range of environmental conditions for dispersion and sorption processes.

3 Biocolloid stability and mobilization

The dispersion of soil colloids is controlled by their composition, ionic strength, pH, soil clay content, soil moisture, and soil management (de Jonge et al., 2004). Biosolid colloids, which are a mixture of organic and inorganic particles, are expected to behave similarly. Many studies do not differentiate between dissolved organic carbon (DOC) and colloidal organic carbon (COC) in soils receiving biosolids because an actual separation may be difficult. However, comparisons of OC levels before and after filtration suggested that the type of organic carbon fractions (DOC or COC) may influence biocolloid behavior (Richards et al., 2000; Ashworth and Alloway, 2004). Another important factor that may affect colloid dispersion and mobility is the method used to apply the bulk biosolids to the soil surface (directly or incorporated), which impacts their interaction with infiltrating and percolating water. In naturally structured soils, preferential flow may promote colloid mobilization only through large conduits (de Jonge et al., 2004; Miller et al., 2011) and cause significant filtration of surface-applied biosolids (Bradford and Torkzaban, 2008). In contrast, incorporated biosolid colloids and particularly in-situ colloids can be mobilized from anywhere in the profile (Miller et al., 2011).

Studies of colloids dispersed from biosolid materials have shown that they are largely composed of organic and amorphous minerals (Brown et al., 2002), with their initial suspension being largely controlled by pH and ionic strength. Organically enriched colloids under elevated pH conditions usually exhibit a negatively charged surface that enhances electrostatic repulsion and colloid stability (Figure 1.1) (Kretzschmar and Sticher, 1997; Ashworth and Alloway, 2004; Toribio and Romanya, 2006; Karathanasis et al., 2005). However, extensive biosolid colloid breakthrough from intact soil monoliths receiving a suspended load has also been observed under relatively low pH conditions and has been linked to particles with high NOM and surface charge, low-suspension pH, and increased soil macroporosity (Karathanasis et al., 2005). Recently applied lime-stabilized biosolids may also cause greater dispersion and mobility to soil and biocolloid particles by shifting the pH of the conducting pores to higher levels and overcoming the acidity of the soil matrix (Toribio and Romanya, 2006). Increased stability and mobility have also been observed in iron oxide-dominated biocolloids coated with humic substances (Kretzschmar and Sticher, 1997; Bunn et al., 2002; Pokrovsky and Schott; 2002). Mobilized purely hematitic colloids can easily filtrate in acid porous media, even at low ionic strengths, but show much greater stability in association with organic components (Kretzschmar and

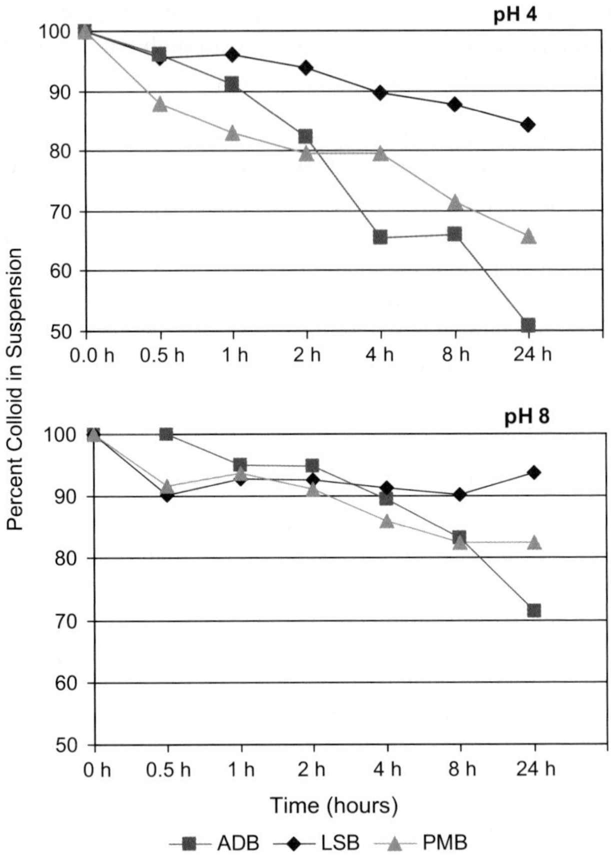

FIGURE 1.1

Settling kinetics (colloid particles remaining in suspension) indicating the stability of suspended biocolloids separated from LSB, ADB, and PMB biosolid wastes as a function of time (Karathanasis and Johnson, 2005). (For color version of this figure, the reader is referred to the online version of this book.)

Sticher, 1997; Pokrovsky and Schott, 2002). The effect of NOM on biocolloid stabilization is particularly prominent in early stages of leaching as the first flush of colloids is usually higher in OM content in natural soils (Laegdsmand et al., 1999). Although increases in ionic strength may inhibit colloid stability and mobility (Kuhnen et al., 2000; Kretzschmar and Sticher, 1997), the presence of organic-coated surfaces in biocolloids may allow them considerable mobility under a wide range of ionic strength fluctuations (Karathanasis et al., 2005). Additionally, biosolid properties and their processing prior to application exert considerable influence on the prevailing environmental conditions and biocolloid behavior. For example, carbonate dissolution in lime-stabilized biosolids may cause increased ionic strengths conducive to biocolloid flocculation, but their inherent high pH may overcome the ionic strength barrier and still cause considerable biocolloid dispersion and mobilization (Karathanasis et al., 2005).

Increased soil macroporosity has also been a factor in the transport of biocolloids and fecal bacteria associated with them, particularly in sites with high clay content and low hydraulic conductivity where bacteria are flushed into pores with greater velocity (Unc and Goss, 2003). Preferential flow through large size pores may accelerate the transport of biocolloids applied as a suspension, but may decrease mobilization of in-situ colloids located further from the flow paths (de Jonge et al., 2004; Miller et al., 2008, 2011). Within undisturbed soils, peak DOC and colloid transport occurred within 1 pore volume (pv), while in soils receiving biosolids, peak DOC elution was delayed after 1 pv and biocolloid release was continuous (Ashworth and Alloway, 2004; Miller et al., 2008, 2011). Biosolids applied in bulk to soil surfaces were also a continuous source for mobilized biocolloids through macropores compared to suspended or in-situ generated soil colloids, which were more evident in early stages of leaching (de Jonge et al., 2004).

A portion of the suspended and mobilized biocolloids along with other mineral and organic particles may be selectively removed from the pore water through filtration, depending on colloid and soil matrix properties (Totsche and Kogel-Knabner, 2004; Saiers and Ryan, 2006). Soil horizons already enriched or saturated with carbon such as spodic and surface horizons are less likely to adsorb mobilized biocolloids (Totsche and Kogel-Knabner, 2004; Karathanasis et al., 2005). The greatest potential for adsorption will be exhibited by soil horizons high in oxides and clay minerals (typically Bt horizons) that usually filter out the more reactive fractions, leaving smaller, less reactive particles in suspension (Totsche and Kogel-Knabner, 2004). Therefore, not only the nature of the biocolloids and the environmental conditions within the porous media, but also the type of the filtering medium (limiting factor) dictate biocolloid mobility and breakthrough in natural and disturbed sites. Generally, biosolid colloids are expected to move easily and extensively through most agricultural soils with high NOM and normal macroporosity, but they experience considerable filtration or sorption by subsoil matrices with high clay content and bulk density unless preferential flow through large channels is possible. Because biosolids are applied to a wide range of agricultural, forest, and reclamation sites, prediction of their mobilization and transport behavior will be highly dependent on detailed assessments of the biosolid and soil characteristics.

4 Biocolloid transport studies

Although many investigations of biosolid leachates have disregarded segregation of the colloidal fraction from DOC, a few studies have evaluated contaminant transport by biocolloid suspensions eluted from bulk biosolid applications (Karathanasis and Ming, 2002; Karathanasis et al., 2005; Miller et al., 2008, 2011). Suspensions of lime-stabilized biosolids (LSB), aerobically digested biosolids (ADB), and poultry manure biosolids (PMB) eluted through two agricultural soils (Woolper clay and Bruno sandy loam) demonstrated significant and diverse biocolloid transport (Karathanasis and Ming, 2002; Karathanasis and Johnson, 2006). Overall, the highest mobility was shown by the ADB colloids, particularly through the sandy soil, and the PMB colloids through the clay soil where high soil NOM content enhanced transport (Figure 1.2) (Karathanasis and Johnson, 2006). The behavior of LSB colloids was more complex, as they were mobile in one study (Karathanasis and Ming, 2002) but demonstrated limited transport in another (Karathanasis et al., 2005) (Figure 1.2). However, even where the initial elution of LSB colloids was limited, their breakthrough increased considerably after the 12th pore volume, suggesting that LSB colloid mobility increased with time as their high pH started buffering the relatively acidic porous media. Different pretreatment (washed and unwashed) of LSB colloid suspensions applied by Karathanasis and Ming (2002) yielded respective pHs of 7 and 11. The pH 7 biocolloids showed a lower breakthrough, but the pH 11 biocolloids were also very stable despite the high ionic strength created by carbonate dissolution (Karathanasis et al., 2005). Thus, high

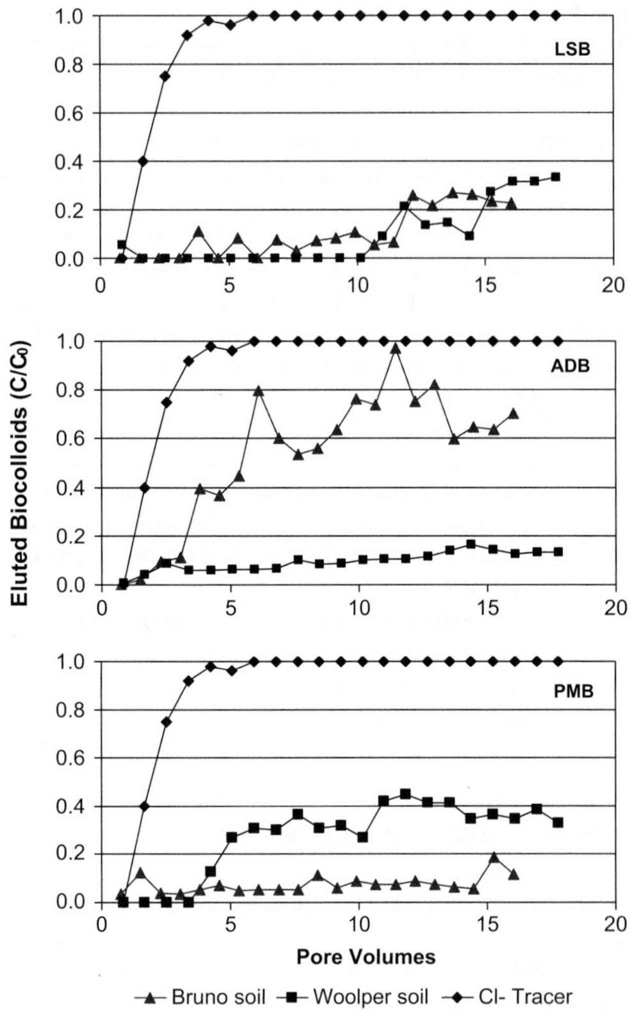

FIGURE 1.2

Breakthrough curves of ADB, LSB, and PMB biocolloids applied on the surface of Woolper (clayey) and Bruno (sandy) soil monoliths (Karathanasis and Johnson, 2006). (For color version of this figure, the reader is referred to the online version of this book.)

pH may play a larger role than high ionic strength in biocolloid dispersion due to organic surface coatings (Kretzschmar and Sticher, 1997; Ashworth and Alloway, 2004; Toribio and Romanya, 2006; Karathanasis et al., 2005).

In mine reclamation, biosolids are applied as a nutrient or organic matter source (Haering et al., 2000) to soils with varying pH, EC, clay content, and water movement patterns (Al and Blowes, 1996; Miller et al., 2011). Biosolids applied to minelands are either incorporated into the soil or simply surface applied. Incorporation into the soil may cause disturbance and more colloid breakthrough (Bunn et al., 2002), or it may decrease colloid elution due to the loss of soil structure (de Jonge et al., 2004). Experimental application of biosolids in bulk to recently reclaimed (5 years ago, T5) and older (30-year-old, T30) mine soils produced variable colloid generation and transport patterns that differed

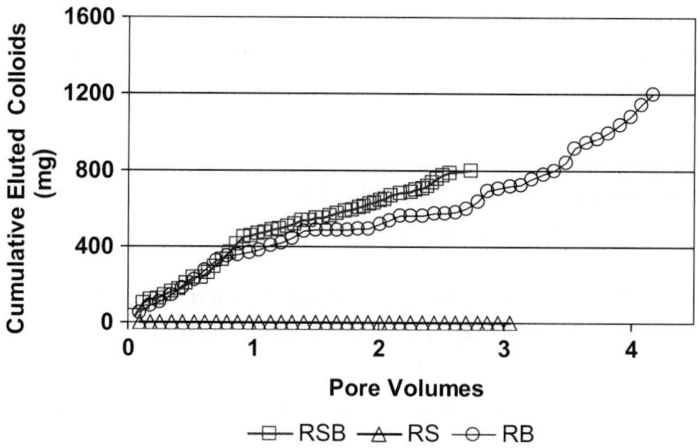

FIGURE 1.3

Biocolloid breakthrough curves from soil monoliths representing 5-year and 30-year old reclaimed sites with and without spoil material following surface application of LSB biosolids after native colloid elution ceased completely (RB = 5-year reclamation site receiving biosolids; RSB = 30-year reclamation site with soil receiving biosolids; RS = 30-year reclamation site with soil without biosolid application) (Miller et al., 2008). (For color version of this figure, the reader is referred to the online version of this book.)

considerably from agricultural soils (Miller et al., 2008, 2011). In both studies, undisturbed forest soils eluted higher DOC, indicating that soluble organics are being leached from organic rich horizons (Totsche and Kogel-Knabner, 2004). Compared to forest soils, disturbed or reclaimed mine soils had lower clay and organic matter contents and lower DOC elution from both T5 and T30 mine soils (Miller et al., 2008, 2011). When T5 monoliths were initially stripped of all in-situ colloids and then received a 20 T/ha LSB application, additional colloids were released (Figure 1.3). In the T30 mine soils, a dense spoil material placed beneath prevented colloid flow, but colloid mobility was detected after the addition of lime-stabilized biosolids (Miller et al., 2011). Biosolid colloid elution patterns were consistent within both T5- and T30-year-old reclaimed soils (Figure 1.3) (Miller et al., 2008, 2011) and were similar to the LSB colloid study of Karathanasis et al. (2005). Although colloid concentrations were not high, colloid elution was continuous and linear in soils receiving biosolids, while in reclaimed soils without biosolid application (control) colloid elution approached a plateau (Miller et al., 2008, 2011). This suggests that applied biosolids may either serve as a constant source of colloids or facilitate colloid release from the soil matrix compared to the reclaimed soils (untreated) where in-situ colloids are slowly being stripped from flow channels by saturated flow conditions (de Jonge et al., 2004). Much like the findings of Karathanasis et al. (2005), LSB applied to reclaimed soils (Miller et al., 2008, 2011) had a consistent colloid release across the entire elution, which indicates that they are finding their way into preferential flow channels.

5 Contaminant transport associated with biocolloids

Biosolid colloids can sorb and carry contaminants through the same mechanisms as in-situ generated colloids. While the environmental behavior of inorganic metal forms is mainly controlled by pH and ionic strength, in the presence of biosolids, metal mobility and reactivity in leachates are dominated by DOC and organic functional groups coating biocolloid surfaces (Ashworth and Alloway, 2004). Therefore, metals such as Ni, Zn, Cu, Cd, Co, and Mn can be sorbed or form complexes on organically coated biosolid colloid surfaces (Pokrovsky and Schott, 2002). Some mechanisms by which biocolloids may facilitate metal transport include direct inner-sphere sorption, surface precipitation, formation of outer-sphere complexes, complexation with DOC, exclusion from sorption onto the soil matrix, and co-precipitation with Fe, Al, or carbonates exposed onto the biocolloid surfaces (Ashworth and Alloway, 2004; Karathanasis and Ming, 2002; Karathanasis et al., 2007; Ippolito et al., 2009). Hence, biosolid type and processing would have significant implications for biocolloid physicochemical characteristics and surface charge behavior (Karathansis et al., 2007). For example, lime-stabilized biosolids have a high pH, which will give the hydroxide and organic colloids abundant negatively charged ligands for metal sorption (McBride, 1994). Also, any processing of biosolids that would cause a loss of functional groups may limit their interaction and association with contaminants.

Few studies have speciated metal transport associated with biosolid colloids or DOC (Karathanasis and Ming, 2002; Karathanasis et al., 2007; Miller et al, 2008). In most studies, colloids have been filtered out (Richards et al., 2000; Ashworth and Alloway, 2004) or *aqua regia* digestions were used prior to metal analysis (Baveye et al., 1999; Gove et al., 2001; Toribio and Romanya, 2006). In cases where biosolid colloids were not separated from DOC, metal transport was exclusively correlated to DOC, while the role of colloids was discounted (Figure 1.4) (Ashworth and Alloway, 2004; Toribio and Romanya, 2006). Those studies discounting biocolloids as vectors of metal transport have shown limited metal mobilization in soils receiving biosolid applications, often below drinking water standards (Gove et al., 2001; Toribio and Romanya, 2006). Even so, the concentrations of leached metals are typically higher compared to soils not receiving biosolids, particularly biosolids having elevated metal concentrations (Brown et al., 1997; Toribio and Romanya, 2006).

In studies considering the transport of only inorganic and DOC-associated metal forms in soils receiving biosolid applications, metal transport varied considerably with metal and biosolid type (Figure 1.4). While Zn mobility was found to be limited in soils receiving anaerobically digested biosolids (Ashworth and Alloway, 2004; Ippolito et al., 2009), substantial breakthroughs occurred with composted fresh or dry sludge applications (Gove et al., 2001). In other investigations, Cu, Ni, Pb, and Zn leaching depended on whether the applied biosolids were digested, limed, or composted (Brown et al., 1997), with pH, DOC, and soil type controlling leaching behavior (Figures 1.5 and 1.6) (Joshua et al., 1998; Baveye et al., 1999; Gove et al., 2001; Karathanasis and Ming, 2002; Toribio and Romanya, 2006; Ippolito et al., 2009). Studies accounting for the biocolloid and/or DOC role in metal transport have shown high Ni and Pb mobility (above drinking water limits) in soils receiving organic amendments (Gove et al., 2001) that correlated well to colloid and DOC breakthrough in leachates (Karathanasis and Ming, 2002; Ashworth and Alloway, 2004; Karathanasis et al., 2007). About 60 to 80% of the water-soluble Ni fraction was associated with DOC and usually flushed from the soil in the first rainfall event (Baveye et al., 1999; Gove et al., 2001; Ashworth and Alloway, 2004). The colloid and DOC fraction was found to be the dominant vector for Pb transport in soils after biosolid application, with up to 97% of the mobile Pb being associated with the biocolloid fraction (Karathanasis and Ming, 2002; Karathanasis et al., 2007; Miller et al., 2008). In moderately acidic soils receiving LSB, significant Cu mobility associated with biocolloid or DOC fractions was induced by high pH and/or co-precipitation on biocolloid carbonatic surfaces (Joshua et al., 1998; Baveye et al., 1999; Ashworth and Alloway, 2004; Karathanasis and Ming, 2002; Karathanasis et al., 2007; Ippolito et al., 2009).

The type and composition of the biosolid colloid and soil may also influence metal mobility. In studies with intact monoliths of three different soils, aerobically digested biocolloids (ADB) carried larger loads of Cd, Zn, Pb, Cu, and Cr than colloids from poultry manure (PMB) or lime-stabilized biosolids (LSB) (Figures 1.5 and 1.6) (Karathanasis et al., 2007). Apparently, the high organic

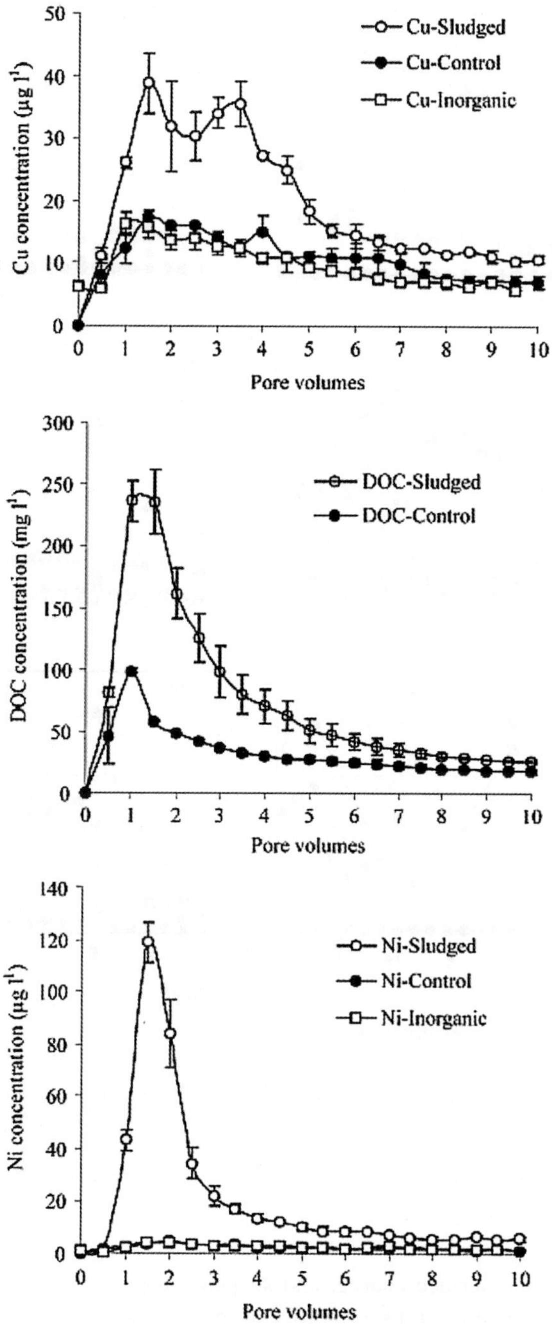

FIGURE 1.4

Elution of Cu, DOC, and Ni from sandy loam columns receiving surface-applied equivalent metal loads in the form of inorganic solutions versus anaerobically digested sewage sludge materials (Ashworth and Alloway, 2004).

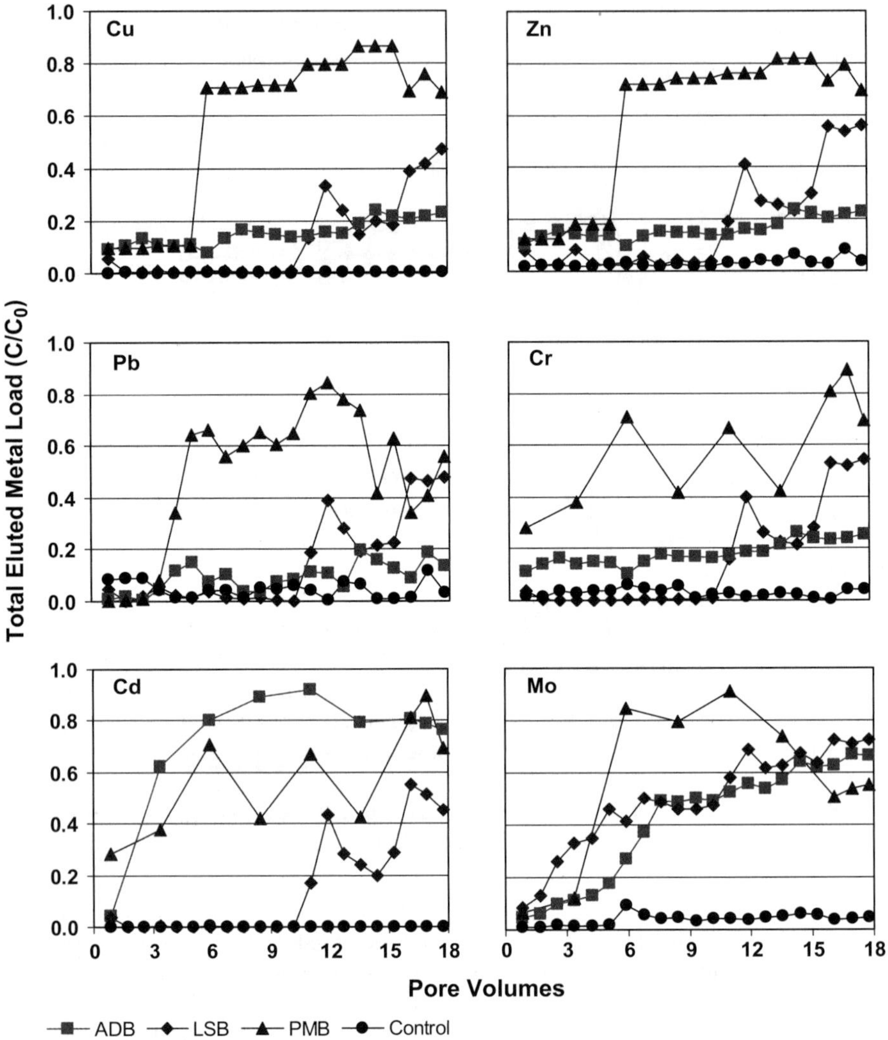

FIGURE 1.5

Metal elution breakthrough curves in the presence and absence (control) of LSB, ADB, and PMB biocolloids following surface application on clayey Woolper soil monoliths (control treatments involved application of equivalent soluble metal loads) (Karathanasis et al., 2007). (For color version of this figure, the reader is referred to the online version of this book.)

carbon content, surface area, surface charge, and suspension pH of the ADB biocolloids enhanced metal transport, while the lower co-transportability of metals associated with PMB and LSB colloids could be related to reduced colloid mobility (Karathanasis et al., 2007). Surprisingly, metal elution patterns in these studies correlated well with breakthrough curves of both the colloid-bound and the soluble fractions (Karathanasis and Ming, 2002; Karathanasis et al., 2007). The higher concentrations

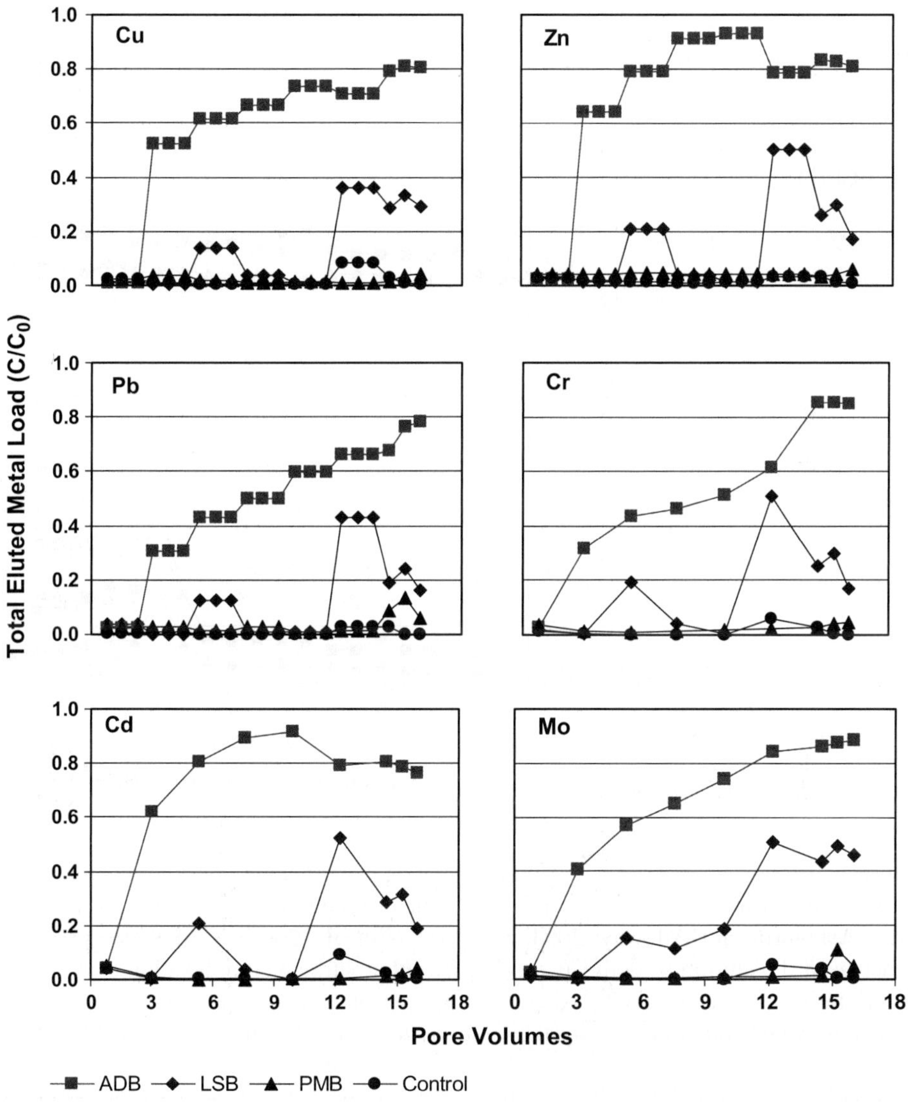

FIGURE 1.6

Metal elution breakthrough curves in the presence and absence (control) of LSB, ADB, and PMB biocolloids following surface application on sandy Bruno soil monoliths (control treatments involved application of equivalent soluble metal loads) (Karathanasis et al., 2007). (For color version of this figure, the reader is referred to the online version of this book.)

of soluble metals eluted in the presence of colloids is probably related to DOC levels, but exclusion phenomena involving blockage of interactive matrix sites by mobile colloids in the pore space are also possible. (Karathanasis and Ming, 2002; Ashworth and Alloway, 2004; Totsche and Kogel-Knabner, 2004; Karathanasis et al., 2007). In experiments with soil monoliths receiving similar metal loads in

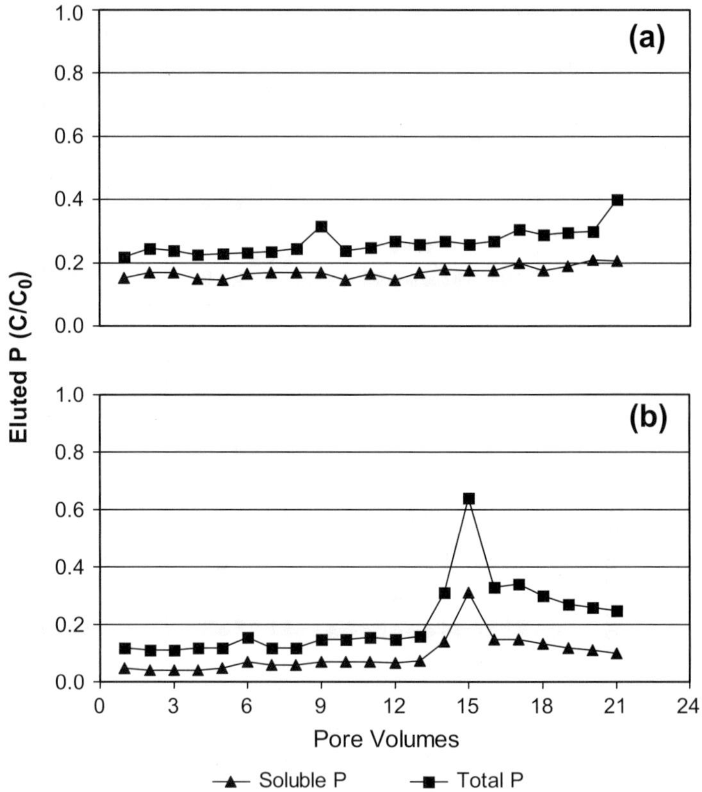

FIGURE 1.7

Elution of total and soluble P loads (colloidal P by difference) from sandy Bruno soil monoliths receiving surface applications of ADB(a) and PMB(b) biosolid colloids. (For color version of this figure, the reader is referred to the online version of this book.)

biocolloid suspensions versus inorganic solutions, metals were detected only in eluents of monoliths receiving biocolloid suspensions, while purely inorganic metal additions were entirely adsorbed by the soil matrix (Ashworth and Alloway, 2004; Karathansis et al., 2007). Carbonates present in LSB colloids increased the elution of both Cd and Mo by either surface complexation or co-precipitation processes (Karathanasis et al., 2007). Also, the high pH of LSB colloids may have contributed to greater dispersivity and enhanced colloid stabilization, thus leading to increased metal breakthrough in the eluents of reclaimed soils receiving biosolid wastes compared to soils receiving similar inorganic metal solution loads (Figure 1.8) (Miller et al., 2008). Although the dissolution of carbonates associated with the LSB may in some soils deter metal solubility and biocolloid mobility, the high pH may occasionally overcome flocculation barriers.

In addition to metals, biosolid wastes may also contain high P concentrations that may degrade water quality by inducing eutrophication in surface and groundwaters of watersheds receiving high biosolid application loads (Pierzynski et al., 2000; Sharpley et al., 2000). Considerable amounts of P are associated with the biocolloid fraction that may upon dispersion be transported through soil macropores and impact water supplies. Most existing studies consider soluble P leached through the soil matrix as the main water-quality culprit, but transported particulate P may be of greater concern for sensitive watersheds and groundwater aquifers. In recent experiments with intact monoliths of a

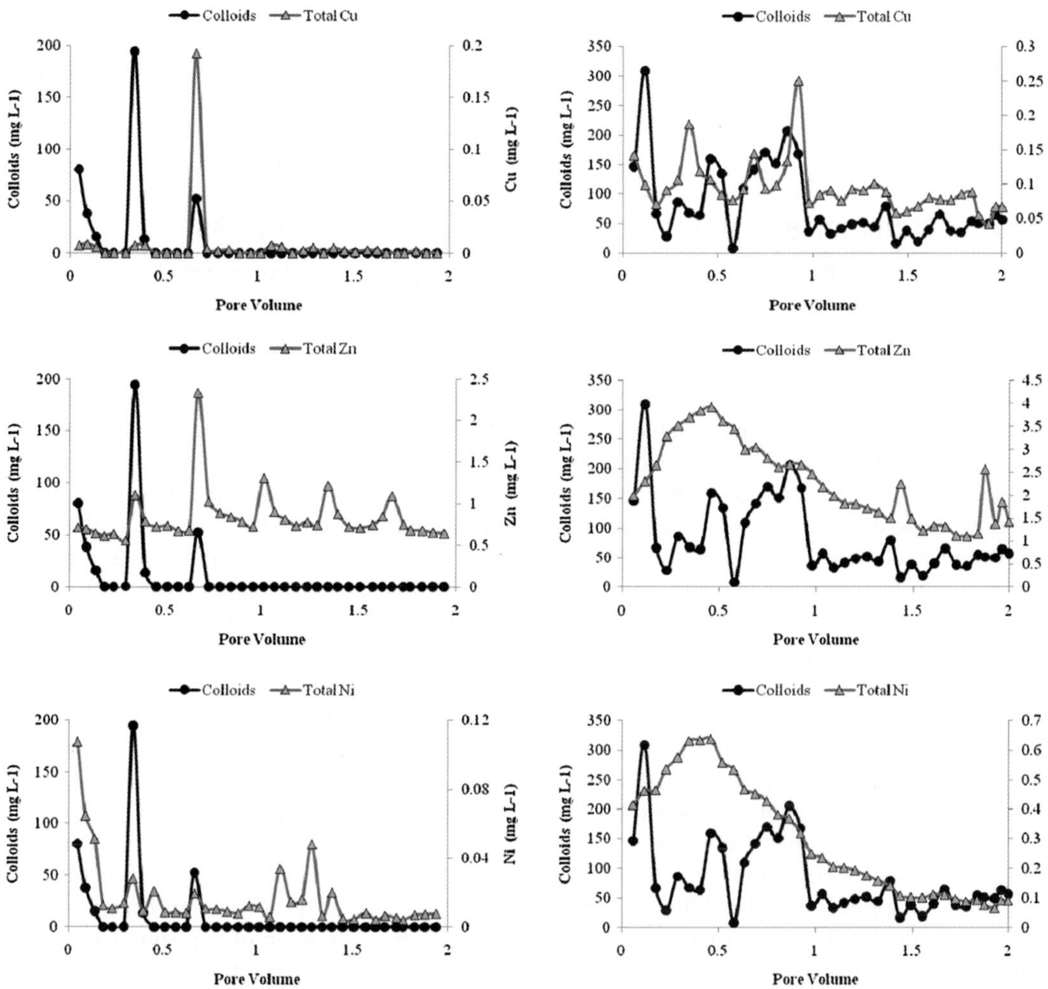

FIGURE 1.8

Elution of colloids, Cu, Zn, and Ni from reclaimed mine soil monoliths receiving surface applied equivalent metal loads in the form of inorganic solutions (left) versus lime stabilized biosolid materials (right) (Miller et al., 2008).

clayey (Woolper) and a sandy (Bruno) soil receiving aerobically digested and poultry manure biocolloids, the eluted colloid particulate P load was more than twice as high that of the soluble P load (Figure 1.7).

Harmful microbes and pathogenic viruses may also be released from applied biosolids in the form of biocolloids or in association with dispersed biocolloids and transported to groundwater through the vadose zone or through seepage from surface waters (Torkzaban et al., 2006; Tufenkji, 2006). Pathogen mobilization and fate depend on infiltration rates, dispersive mixing, and deposition kinetics controlled by pH, electrostatic interactions, and pore water-flow transient phenomena. Surface charge heterogeneity of microbial cells derived from organic functional group variability is particularly important not only for the mobility of microbial pathogens, but also for their capacity to sorb and carry other contaminants through porous soil media (Saiers and Ryan, 2006).

6 Fate of transported contaminants and contamination risks

Untreated biosolids with high levels of metals and P are the most likely sources of water contamination (Brown et al., 1997; Toribio and Romanya, 2006), particularly those with increased OM and dispersion potential that generate biocolloids with greater mobility and surface reactivity (Karathanasis et al., 2005, 2007). Although several studies have compared the metal affinities of soil matrices and eluted colloids in order to assess the capacity of the porous media for contaminant sorption and filtration, predictions have not always been reliable because of the participation of additional transport mechanisms (i.e., complexation, filtration, exclusion, etc.) (Chaney et al., 1996; Karathanasis and Ming, 2002; Karathanasis and Johnson, 2006; Karathanasis et al., 2007). Typical predictions for metal mobilization in soils receiving biosolids suggest that metal migration should not exceed a 30–80 cm depth (Brown et al., 1997; Baveye et al., 1999). However, much greater metal migration in association with biocolloids has been documented, particularly in soils with elevated organic carbon content and macroporosity along preferential flow paths (Totsche and Kogel-Knabner, 2004; Karathanasis et al., 2005, 2007). In contrast, a greater portion of the soluble and sorbed metal fractions transported in the presence of biocolloids is expected to be immobilized along transport pathways in soils having low OM, high clay content, low macroporosity and alkaline pH (Toribio and Romanya, 2006; Karathanasis et al., 2005; Gove et al., 2001; Karathanasis et al., 2007; Totsche and Kogel-Knabner, 2004; Zubillaga et al., 2008). However, transport conditions may change with the type, composition, and physicochemical behavior of the applied biosolid (Gove et al., 2001; Toribio and Romanya, 2006).

Contaminant transport predictions in agricultural soils receiving biosolids is a difficult and complex task because of their variable range in morphological, physical, and chemical properties, the variety of waste materials applied, and the different application methods used. This variability becomes even greater in disturbed soils, such as reclaimed minelands, due to the disruption of the natural structural continuity of the soil matrix imposed by the human intervention (Skousen et al., 2000). This disturbance may also cause changes in pore water chemistry along flow paths associated with heterogeneous acidic and alkaline materials that promote alternating colloid dispersion/flocculation processes (Skousen et al., 2000; Miller et al., 2008, 2011). Although the low OM content and hydraulic conductivity of recently reclaimed mine soils treated with biosolids may initially retard biocolloid-facilitated contaminant transport, as preferential flow paths and OM buildup are gradually established through increased biological activity, the role of biocolloids as contaminant carriers is anticipated to increase dramatically (Al and Blowes, 1996; Skousen et al., 2000; McCarthy and Shevenell, 1998; Seta and Karathanasis, 1996; Miller et al., 2008, 2011). In reclaimed mine soils where initial biocolloid-mediated contaminant transport through the soil may be limited, surface runoff carrying biocolloid-bound contaminants to nearby streams may pose another potential threat to human health (Miller et al., 2008; Meyer et al., 2001; Haering et al., 2000).

Some studies suggest that eluent metal concentrations in soils receiving biosolids are below drinking water standards (Gove et al., 2001; Toribio and Romanya, 2006), but metal loads reaching groundwater resources may be substantially higher when the soluble and the biocolloid fractions are combined (Karathanasis et al., 2007). Similar patterns are anticipated for soluble versus colloidal phosphorous transport processes. Generalized predictions for potential risk and scale of contamination should be avoided because of the variability of geomorphic settings, soil types, and biosolids involved. More accurate risk assessments can be made on a case-by-case basis when considering the most important factors impacting contaminant transport process. However, discounting biocolloid-mediated contaminant transport in agricultural and reclaimed landscape settings would significantly underestimate potential water contamination and human health risks.

7 Prevention and remediation strategies

The greatest risk of contamination is associated with unprocessed biosolids with high metal and P concentrations (Brown et al., 1997; Toribio and Romanya, 2006). Therefore, using sterilized biosolid wastes with low metal and P content would be the first step in reducing potential contamination risks. Unfortunately, in the case of biosolid colloids, reducing dispersion and mobilization is not easily done. Lime stabilization is useful in raising soil pH, precipitating metals, and potentially reducing colloid transport (Karathanasis et al., 2005), but LSB have also been shown to induce colloid stability and transportability (Karathanasis and Ming, 2002). Aerobically digested colloids used in our experiments were very effective in transporting metals (Karathanasis et al., 2007), and their use should be limited to soils with high clay and low OM contents. Poultry manure biocolloids showed variable behavior depending on soil type, and their use should be assessed on a case-by-case basis (Totsche and Kogel-Knabner, 2004; Karathanasis et al., 2005). Finally, biosolids should not be applied to soils with groundwater tables within 30–80 cm of surface applied biosolids, where metal mobility is prevalent (Brown et al., 1997; Baveye et al., 1999) because colloids entering saturated aquifers have a decreased chance of being filtered out by smaller soil pores than under unsaturated conditions (Wan and Tokunaga, 1997; Lenhart and Saiers, 2002).

8 Conclusions

Biosolid wastes are sources of organic and inorganic colloids, both of which have been proven to be mobile within a range of environmental conditions. The high organic carbon makeup of biocolloids induces high surface reactivity and stability that empowers them with increased mobility and cotransportability of contaminants in soil-porous media. Metals and P are likely contaminants in domestic biosolid sources that can be released upon dispersion, be carried through soil macropores to greater soil depths, and even reach groundwater supplies. Our research findings indicate that monitoring only the inorganic or DOC-associated metal and P leached from soils treated with biosolid wastes, while disregarding biocolloids as vectors of metal and P transport, may significantly underestimate risks of water contamination. These risks may be even greater in sites with abundant preferential flow paths to shallow groundwater where migrating biocolloids and contaminants can largely bypass reactive portions of the soil matrix. Using low metal and P content biosolid products and carefully assessing soil morphological, physical and chemical properties before application may alleviate or minimize potential contamination risks.

SYMBOLS AND TERMINOLOGY

PCBs Polychlorinated biphenyls
NRC National Research Council
EPA Environmental Protection Agency
ADB Aerobically digested biosolids
LSB Lime-stabilized biosolids
PMB Poultry manure biosolids
NOM Natural organic matter
DOC Dissolved organic carbon
COC Colloidal organic carbon
OC Organic carbon
PV Pore volume
EC Electrical conductivity
CEC Cation exchange capacity

References

Al, T.A., Blowes, D.W., 1996. Storm water hydrograph separation of run off from a mine-tailings impoundment formed by thickened tailings discharge at Kidd Creek, Ontario. J. Hydrol 180, 55–78.

Ashworth, D.J., Alloway, B.J., 2004. Soil mobility of sewage sludge-derived dissolved organic matter, copper, nickel, and zinc. Environ. Poll. 127, 137–144.

Basta, N.T., Ryan, J.A., Chaney, R.L., 2005. Trace element chemistry in residual-treated soil: Key concepts and metal bioavailability. J. Environ. Qual. 34, 49–63.

Baveye, R., McBride, M.B., Bouldin, D., Hinesly, T.D., Dahdoh, M.S.A., Abdel-sabour, M.F., 1999. Mass balance and distribution of sludge-borne trace elements in a silt loam soil following long-term applications of sewage sludge. Sci. Total Environ. 227, 13–28.

Bradford, S.A., Torkzaban, S., 2008. Colloid transport and retention in unsaturated porous media: A review of interface-, collector-, and pore-scale processes and models. Vadose Zone J 7, 667–681.

Brown, S., Chaney, R., Angle, J.S., 1997. Subsurface liming and metal movement in soils amended with lime-stabilized biosolids. J. Environ. Qual. 26, 724–732.

Brown, S., Chaney, R.L., Sprenger, M., Compton, H., 2002. Soil remediation using biosolids. Part I. Biocycle 43, 41–44.

Bunn, R.A., Magelky, R.D., Ryan, J.P., Elimelech, M., 2002. Mobilization of natural colloids from an iron oxide-coated sand aquifer: Effect of pH and ionic strength. Environ. Sci. Technol. 36, 314–322.

Chaney, R.L., Ryan, J.A., O'Connor, G.A., 1996. Organic contaminants in municipal biosolids: risk assessment, quantitative pathways analysis, and current research priorities. Sci. Total. Environ. 185, 187–216.

de Jonge, L.W., Kjaergaard, C., Moldrup, P., 2004. Colloids and colloid-facilitated transport of contaminants in soils: an introduction. Vadose Zone J. 3, 321–325.

Gove, L., Cooke, C.M., Nicholson, F.A., Beck, A.J., 2001. Movement of water and heavy metals (Zn, Cu, Pb, and Ni) through sand and sandy loam amended with biosolids under steady-state hydrological conditions. Biores. Technol. 78, 171–179.

Haering, K.C., Daniels, W.L., Feagley, S.E., 2000. Reclaiming mined lands with biosolids, manures, and papermill sludges. In: Barnhisel, R.I. (Ed.), Reclamation of drastically disturbed lands. *Agron. Monogr.* 41. ASA, CSSA, and SSSA, Madison, WI, pp. 615–644.

Ippolito, J.A., Barbarick, K.A., Brobst, R.B., 2009. Fate of biosolids Cu and Zn in a semi-arid grassland. Agri. Ecosyst, Environ. 131, 325–332.

Joshua, W.D., Michalk, D.L., Curtis, I.H., Salt, M., Osborne, G.J., 1998. The potential for contamination of soil and surface waters from sewage sludge (biosolids) in a sheep grazing study, Australia. Geoderma 84, 135–156.

Kaplan, D.I., Bertsch, P.M., Adriano, D.C., Miller, W.P., 1993. Soil-borne mobile colloids as influenced by water flow and organic carbon. Environ. Sci. Technol. 27, 1192–1200.

Karathanasis, A.D., 1999. Subsurface migration of copper and zinc mediated by soil colloids. Soil Sci. Soc. Am. J. 63, 830–838.

Karathanasis, A.D., Ming, D.W., 2002. Colloid-mediated transport of metals associated with lime-stabilized biosolids. Devel. Soil Science 28A, 49–62.

Karathanasis, A.D., Johnson, D.M.C., Matocha, C.J., 2005. Biosolid colloid-mediated transport of copper, zinc, and lead in waste-amended soils. J. Environ. Qual. 34, 1153–1164.

Karathanasis, A.D., Johnson, D.M.C., 2006. Stability and transportability of biosolids colloids through undisturbed soil monoliths. Geoderma 130, 334–345.

Karathanasis, A.D., Johnson, C., Matocha, C.J., 2007. Subsurface transport of heavy metals by biosolid colloids in waste-amended soils. In: Frimmel, F.H., von der Kammer, F., Flemming, H. (Eds.), Colloidal transport in porous media. Springer, Berlin, Heidelberg, pp. 175–201.

Kinney, C.A., Furlong, E.T., Zaugg, S.D., Burkhardt, M.R., Werner, S.L., Cahill, J.D., Jorgensen, G.R., 2006. Survey of organic wastewater contaminants in biosolids destined for land application. Environ. Sci. Technol. 40, 7207–7215.

Kretzschmar, R., Robarge, W.P., Amoozegar, A., 1995. Influence of natural organic matter on transport of soil colloids through saprolite. Water Resources Res. 31, 435–445.

Kretzschmar, R., Sticher, H., 1997. Transport of humic-coated iron oxide colloids in a sandy soil:Influence of Ca^{+2} and trace metals. Environ. Sci. Technol. 31, 3497–3504.

Kuhnen, F., Barmettler, K., Bhattacharjee, S., Elimelech, M., Kretzschmar, R., 2000. Transport of iron oxide colloids in packed quartz sand media: Monolayer and multilayer deposition. J. Colloid Interf. Sci. 231, 32–41.

Laegdsmand, M., Villholth, K.G., Ullum, M., Jensen, K.H., 1999. Processes of colloid mobilization and transport in macroporous soil monoliths. Geoderma 93, 33–59.

Lenhart, J.J., Saiers, J.E., 2002. Transport of silica colloids through unsaturated porous media: experimental results and model comparisons. Environ. Sci. Technol 36, 769–777.

Mantovi, T., Baldoni, Z., Toderi, Z., 2005. Reuse of liquid, dewatered, and composted sewage sludge on agricultural land: Effects of long term application on soil and crop. Water Research 39, 289–296.

Meyer, V.F., Redente, E.F., Barbarick, K.A., Brobst, R., 2001. Biosolids application affect runoff water quality following forest fire. J. Environ. Qual. 30, 1528–1532.

Miller, J.O., Karathanasis, A.D., Wendroth, O.O., Matocha, C.J., Barton, C.D., 2008. In-situ colloid mobilization within biosolids amended soils following coal mine reclamation. Proceedings of the NGWA/U.S. EPA Remediation of abandoned mine lands conference (#5019), Denver, CO.

Miller, J.O., Karathanasis, A.D., Matocha, C.J., 2011. In-situ generated colloid transport of Cu and Zn in reclaimed mine soil profiles associated with biosolids application. Appl. Environ. Soil Sci. J. 2011, 1–9.

McBride, M.B., 1994. Environmental chemistry of soils. Oxford University Press, New York.

McBride, M.B., Richards, B.K., Steenhuis, T., Spiers, G., 1999. Long-term leaching of trace elements in a heavily sludge-amended silty clay loam soil. Soil Science 164, 613–623.

McCarthy, J.F., Shevenel, L., 1998. Processes controlling colloid composition in a fractured and karstic aquifer in eastern Tennessee, USA. J. Hydrology 206, 191–218.

National Research Council (NRC), 2002. Bisolids applied to land: Advancing standards and practices. National Academy Press, Washington, DC.

Pierzynski, G.M., Sims, J.T., Vance, G.F., 2000. Soil phosphorous and environmental quality. In: Soils and environmental quality, second ed. Taylor & Francis, Boca Raton, FL, pp. 144–207.

Pokrovsky, O.S., Schott, J., 2002. Iron colloids/organic matter associated transport of major and trace elements in small boreal rivers and their estuaries (NW Russia). Chem. Geol. 190, 141–179.

Richards, B.K., Steenhuis, T.S., Peverly, J.H., McBride, M.B., 2000. Effect of sludge-processing mode, soil texture, and soil pH on metal mobility in undisturbed soil columns under accelerated loading. Environ. Poll. 109, 327–346.

Saiers, J.E., Ryan, J.N., 2006. Introduction to special section on colloid transport in subsurface environments. Water Resources. Res. 42, 4842–4846.

Seta, A.K., Karathanasis, A.D., 1996. Water dispersible colloids and factors influencing their dispersability from soil aggregates. Geoderma 74, 255–266.

Sharpley, A., Foy, B., Withers, P., 2000. Practical and innovative measures for the control of agricultural phosphorous losses to water: an overview. J. Environ. Qual. 29, 1–9.

Skousen, J.G., Sexstone, A., Ziemkiewicz, P.F., 2000. Acid mine drainage control and treatment. In: Barnhisel, R.I. (Ed.), Reclamation of drastically disturbed lands. Agron. Monogr. 41. ASA, CSSA, and SSSA, Madison, WI, pp. 131–168.

Toribio, M., Romanya, J., 2006. Leaching of heavy metals (Cu, Ni, and Zn) and organic matter after sewage sludge application to Mediterranean forest soils. Sci. Total Environ. 363, 11–21.

Torkzaban, S., Hassanizadeh, S.M., Schijven, J.F., van den Berg, H.H.J.L., 2006. Role of air-water interfaces on retention of viruses under unsaturated conditions. Water Resources Res. 42, 4904–4909.

Totsche, K.U., Kogel-Knabner, I., 2004. Mobile organic sorbent affected contaminant transport in soil: Numerical case studies for enhanced and reduced mobility. Vadose Zone J. 3, 352–367.

Tufenkji, N., 2006. Application of dual deposition mode model to evaluate transport of *Escherichia coli* D21 in porous media. Water Resources Res. 42, 4851–4856.

Unc, A., Goss, M.J., 2003. Movement of fecal bacteria through the vadose zone. Water, Air, and Soil Poll. 149, 327–337.

Wan, J., Tokunaga, T.K., 1997. Film straining of colloids in unsaturated porous media: Conceptual model and experimental testing. Environ. Sci. Technol 31, 2413–2420.

Zubillaga, M.S., Bressan, E., Lavado, R.S., 2008. Heavy metal mobility in polluted soils: Effect of different treatments. Am. J. Environ. Sci. 4, 620–624.

CHAPTER 2

Influence of Natural Organic Matter on Contaminant Removal by Permeable Reactive Barrier

Daniel C.W. Tsang

Department of Civil and Natural Resources Engineering, University of Canterbury, Christchurch, New Zealand

CHAPTER OUTLINE

1 Introduction

1.1 Background

Improved understanding of the geochemistry, fate and transport of inorganic and organic colloids is critical for assessing the ecological and human health impacts of contaminants associated with these colloids. A large body of evidence indicates that naturally occurring inorganic and organic colloids play an important role in the contaminant transport, depending on geoenvironmental conditions. Field investigations have revealed that mobile colloids in the subsurface include clay minerals, iron and aluminium hydroxides, silica, carbonates, and natural organic matter. Colloidal particles can be mobilized by changes in geochemical composition, including ionic strength, solution pH, background

The Role of Colloidal Systems in Environmental Protection. http://dx.doi.org/10.1016/B978-0-444-63283-8.00002-8

electrolytes (monovalent/divalent ions), and infiltration of chelating agents or surfactants (Grolimund et al., 1996; Kretzschmar et al., 1999). The resulting dissolution of soil minerals and organic matter, or precipitation of supersaturated mineral phases, generates inorganic or organic colloids. Inorganic colloids such as clay particles and mineral precipitates are capable of sorbing heavy metals and ionic organic contaminants, while organic colloids provide separate moieties in aqueous phase for partitioning of hydrophobic organic contaminants.

As a result, the contaminant transport in the subsurface cannot be accurately predicted and described by the transport models that only take into account the adsorption and retardation of the contaminants. In the presence of colloids that have high-collision efficiency and thus are readily deposited on the soil surface, the associated contaminants are probably less mobile. On the other hand, colloid-facilitated transport of the contaminants (i.e., earlier breakthrough) is likely to be observed if there is a preferential flow of colloids due to size exclusion (from micropore or interior adsorption sites) or electrostatic repulsion (with particle surfaces). The fate and transport of colloidal particles, which depend on their physical size, surface charge, and structural morphology, primarily govern the contaminant transport in the subsurface. Moreover, the bioavailabiliy and fate (e.g., oxidation/reduction and biodegradation) of the contaminants are influenced by the association with colloids.

1.2 Scope and purpose

In recent years, the roles of natural organic matter in contaminated land remediation have received intensive attention. Natural organic matter is ubiquitous in surface soils, shallow aquifers, and sediments, where the most severe anthropogenic contamination is usually encountered. Dissolved or colloidal organic matter, which can exist as macromolecules and small molecular weight organics in aqueous phase, frequently controls contaminant speciation/bioavailability, drives important reactions (e.g., redox and photochemical reactions), and influences interactions of contaminants (e.g., adsorption) and inorganic colloidal particles (e.g., deposition) with mineral surfaces. These are the important factors that need to be considered in the design of engineering systems for groundwater remediation.

Therefore, this chapter aims to provide a critical review of the characteristics and geochemical reactions of natural organic matter in the environment, and discuss their impact (particularly in the aspects of adsorption, complexation, and aggregation) on the application of zero-valent iron permeable reactive barrier which is a widely used technology for groundwater remediation.

2 Natural organic matter in the environment

2.1 Origin, definition, and characteristics

Total organic carbon in the natural environment comprises: (i) carbon in living organisms (biomass); (ii) carbon in undecayed and partially decayed plant and animal tissues (organic residues); and (iii) natural organic matter (humus). In general, organic matter content decreases with increasing soil depth and ranges from less than 1% in sandy and desert soils, 1–5% in the surface horizons (top 15 cm) of typical mineral and agricultural soils, to almost 100% in organic soils. Natural organic matter has an extremely important influence on the chemical and physical properties of soils despite representing just a small percentage of most mineral soils (Sparks, 1996; 1999).

Natural organic matter is classified into nonhumic and humic substances. All the recognizable plant debris and organic compounds, such as polysaccharides, lignin, proteins, and polypeptides, are categorized as nonhumic substances. The remaining high-molecular-weight, highly transformed, brown- to black-colored materials are considered humic substances (Tipping, 2002; Sparks, 2003). The time range that humic substances remain in the environment is wide, ranging from weeks or months in surface waters of lakes, streams, and estuaries to hundreds of years in soils and deep aquifers. The amounts of nonhumic and humic substances in soils differ significantly: Lipids can range from 2% of total organic matter in forest soil to 20% in acid peat soil, protein from 15 to 45%, and carbohydrates from 5 to 25%; humic substances may vary from 33 to 75% of total organic matter.

Humic substances are often described as coiled, long-chain molecules or two- or three-dimensional cross-linked macromolecules with a wide range of functional groups. Recent developments suggest that humic substances are heterogeneous mixtures of molecules self-assembled in supramolecular colloidal aggregates by multiple noncovalent interactions such as hydrogen bonding (Kerner et al., 2003; Sutton and Sposito, 2005). The surface area can be as high as 800–900 $m^2\ g^{-1}$. However, the molecular structure and chemistry of humic substances are still not well understood because of intrinsic chemical heterogeneity, geographical variability (e.g., vegetation, climate, topography), complexity and randomness of formation, and degradation of humic substances. There is not a regularly repeating molecular structure or set of structures that truly defines the configuration of humic substances; structural representations that have been proposed are based on functional groups (the presence of aliphatic and aromatic components) and are meant to convey information about the structural moieties (Sparks, 1999; Tipping, 2002).

Therefore, humic substances are operationally fractionated into fulvic acid (soluble in both acid and alkali), humic acid (soluble in alkali but not in acid), and humin (insoluble in both acid and alkali). Despite the arbitrary nature of this separation scheme, it is mostly used to achieve a segregation of humic substances according to their chemical properties, which chiefly depend on surface functional groups and macromolecular structure.

The major constituents of natural organic matter are C, O, H, and N; other prominent elements are P and S. Average elemental compositions are more distinguishable according to fractionation (fulvic and humic acids) than to sources and geographical areas (e.g., freshwater and marine sources). Fulvic acid has lower contents of C and N but higher contents of O and S than humic acid. Based on elemental compositions, the mole ratios of O/C, H/C, or N/C are calculated to indicate the degree of polarity/aromaticity and unsaturation of carbon chains. The O/C ratios are about 0.5 for humic acid and about 0.7 for fulvic acid. The H/C ratios, which have a larger variation, are about 1.0 for humic acid and about 1.4 for fulvic acid. In general, fulvic acid has greater aliphatic character, while humic acid is of higher aromaticity and hydrophobicity. Insoluble humin is believed to be similar to humic acid, with slightly less aromatic groups and acidity but higher contents of polysaccharide and alcoholic OH (McBride, 1994; Tipping, 2002; Sparks, 2003; Essington, 2004).

It is widely recognized that humic substances are variable-charge materials (i.e., surface charge varies with pH) that contain several major functional groups: carboxyl (R–COOH), phenolic OH (aromatic–OH), alcoholic OH (R–CH_2–OH), and carbonyl (R–C=O–R/H/OR) (Sparks, 2003). Main acidic functional groups are carboxyl and phenolic OH groups; carbonyl and ketonic groups, and the like, are neutral functional groups; amine and amide groups are basic functional groups. The negative charge of humic substances results from deprotonation (ionization) of acidic functional groups. Since the carboxyl (dissociation constant, pK_a of 3–5) and phenolic (pK_a of 5–7) groups can deprotonate at common pH values in many soils, natural organic matter is negatively charged at pH > 3 (Sparks, 1999). Natural organic matter possesses cation exchange capacity ranging from 150 to 300 cmol kg^{-1} and thus is a major contributor of the soil's exchange capacity. Compared with humic acid and humin, fulvic acid contains more carboxyl groups per unit mass and thus possesses a higher total acidity.

2.2 Dissolution and precipitation

Natural organic matter can be adsorbed on almost every type of mineral surfaces, that is, clay minerals and metal oxides; from 52% to 98% of total carbon in soils is estimated to be bound with minerals. Natural organic matter distributed on surfaces and within aggregates of soil minerals is relatively stable against biodegradation, presents a reactive surface coating, and helps bridge soil particles together into stable aggregates. Types of interactions between organic matter and mineral surfaces include van der Waals forces, electrostatic interactions, cation and anion exchange, water and cation bridging, ligand exchange, and hydrogen bonding. Two or more of these mechanisms can simultaneously occur, depending on the nature of the system (Sposito, 1994; Tipping, 2002; Sparks, 2003).

Therefore, natural organic matter can be mobilized as organic colloids along with proton-promoted dissolution of iron/aluminium oxides and kaolinite under acidic conditions (pH < 5). On the other hand, natural organic matter can be directly dissolved in solution at high pH. The solubility of organic matter is observed to dramatically increase at pH above 7 or 8, resulting in a high content of dissolved organic matter. Because at higher pH the particle surfaces become more deprotonated, leading to higher negative charges on both organic matter and mineral surfaces, the electrostatic repulsion causes dispersion (intermolecular repulsion) and dissolution (intramolecular repulsion) of organic matter from solid surfaces (Tipping, 2002). The ratio of dissolved humic acid to dissolved fulvic acid in solution also increases with increasing pH because the extent of increase of humic acid dissolution is greater than that of fulvic acid dissolution due to the difference of their chemical nature (You et al., 1999; Sauve et al., 2000; Yin et al., 2002).

The molecular weight of dissolved fulvic acid may range from 500 to 5000 Da, while that of dissolved humic acid may range from 3000 to 1,000,000 Da. Fulvic acid usually possesses a more flexible conformation than humic acid. The structure and molecular size of dissolved organic matter are determined by inter- and intramolecular interactions that stabilize (minimize) the electrostatic energy of the structures. The interactions include hydrogen bonding, metal complexation, and electrostatic interactions (Tipping, 2002; Sparks, 2003). Thus, the morphology and molecular weight (size) of dissolved organic matter are affected by solution pH, ionic strength, background cations, and organic matter concentrations.

In general, dissolved or colloidal organic matter has a more compact structure and tends to aggregate and precipitate with decreasing solution pH, increasing ionic strength, increasing organic matter concentration, and a change of background cations from monovalent to divalent or polyvalent cations (Kretzschmar et al., 1999; Tipping, 2002; Lu and Pignatello, 2004; Sutton and Sposito, 2005). At low pH there is little or no deprotonation of acidic functional groups, thus, inter- and intramolecular electrostatic repulsion is minimal. At a higher ionic strength, the thickness of the electrical double layers of particles is reduced, resulting in a net attractive force due to the predominance of van der Waals force over electrostatic repulsion force. The aqueous solubility of natural organic matter is not high, and therefore high organic matter concentration leads to precipitation. The presence of divalent (e.g., calcium and magnesium in hard water) or polyvalent cations (e.g., iron and aluminium due to mineral dissolution) allows metal complexation with surface functional groups that reduces electrostatic repulsion and/or promotes inter- and intramolecular cation bridging.

Furthermore, it should be noted that natural organic matter exhibit more than one type of macromolecular structure (e.g., globular and net-, ring-, and sheetlike structures) in aqueous solutions (Maurice and Namjesnik-Dejanovic, 1999; Myneni et al., 1999). Moreover, the conformation of natural organic matter is probably altered upon association with particle surfaces (Wershaw, 1999; Namjesnik-Dejanovic and Maurice, 2000; Gorham et al., 2007).

3 Influence on permeable reactive barrier

3.1 Overview of zero-valent iron permeable reactive barrier

Over the last decade, risk-based remediation has captured intensive attention. This approach puts the emphasis on reducing the potential risk to human health and the environment to an acceptable level rather than to technologically feasible levels. As illustrated in Figure 2.1, managerial and/or remedial actions on contaminated land can be taken to restrict the access of receptors to the site, block the exposure pathways, and/or remove the source contaminants. The shift of remediation focus to managing site-specific risk allows setting a more realistic treatment objective, reducing the treatment cost and duration, and better reserving scarce natural resources (e.g., energy, chemicals, and construction materials) for other priorities.

Traditional pump-and-treat technology for remediation of contaminated groundwater has been proven difficult and costly. The removal of contaminant source by groundwater extraction has achieved limited success, in terms of attaining the treatment goals and completing the remediation in a timely manner, due to rate-limited contaminant desorption from soils, while generating a substantial volume of extracted groundwater that requires above-ground wastewater treatment. Therefore, the application of permeable reactive barriers has gained increasing popularity in North America and Europe since 1995. This in-situ remediation technology aims to block the exposure pathway of contaminants to downstream drinking water wells by installing a permeable reactive barrier that intercepts the contaminated plume and immobilizes and/or degrades the contaminants in the groundwater. Figure 2.2 outlines the design approach for permeable reactive barriers. Figure 2.3 presents schematic diagrams of three commonly used configurations of permeable reactive barriers.

Permeable reactive barriers have been considered an established remediation technology and are applied at field scale for treating chlorinated solvents, metals and inorganics, fuel hydrocarbons, and so on (Figure 2.4). Among a range of reactive materials, zero-valent iron (cast iron, a low-cost commercial product) has been most widely applied (Figure 2.5) because a variety of organic and inorganic contaminants can be treated as a result of oxidative corrosion of zero-valent iron. Chlorinated solvents

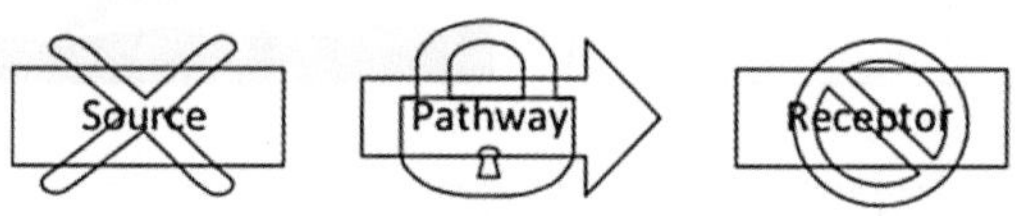

FIGURE 2.1

Contaminated land management approach.

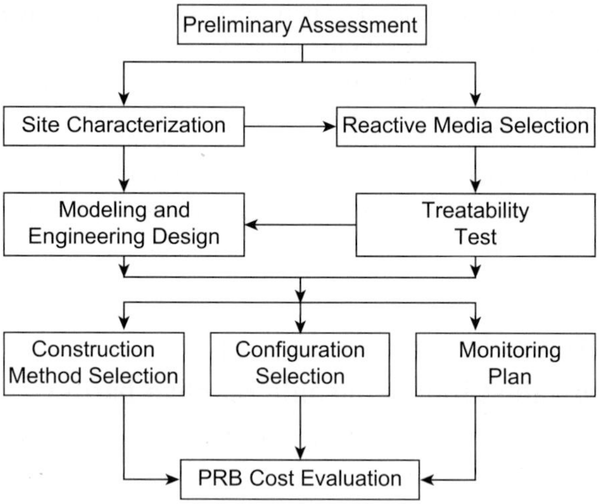

FIGURE 2.2

Design approach for permeable reactive barriers.

(adapted from Lo et al., 2006).

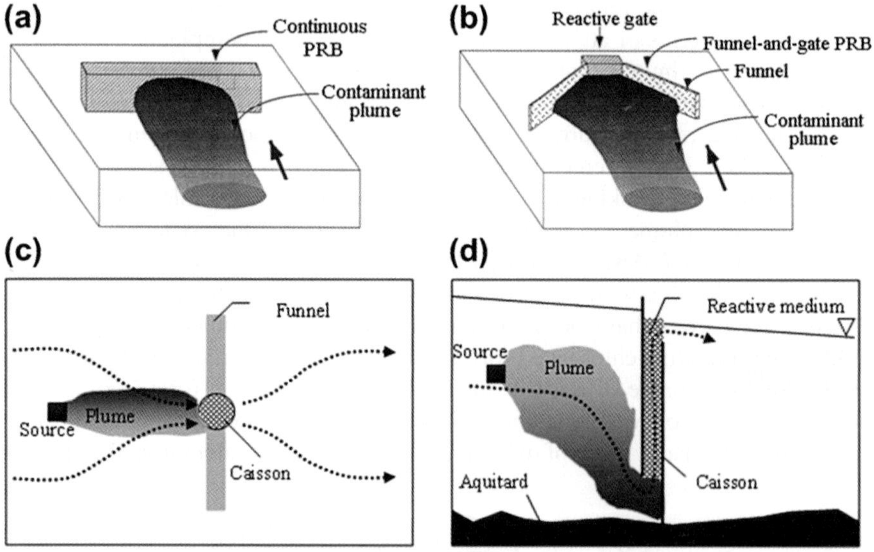

FIGURE 2.3

Schematic configurations of: (a) continuous permeable reactive barriers; (b) funnel-and-gate permeable reactive barriers; and (c) plan view and (d) side view of cassion permeable reactive barriers.

(adapted from U.S. EPA, 2002).

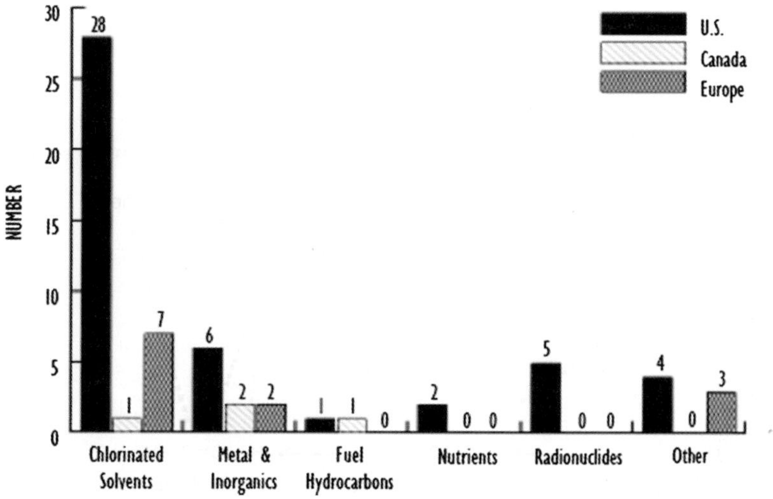

FIGURE 2.4

Contaminants treated by permeable reactive barriers.

(adapted from U.S. EPA, 2002).

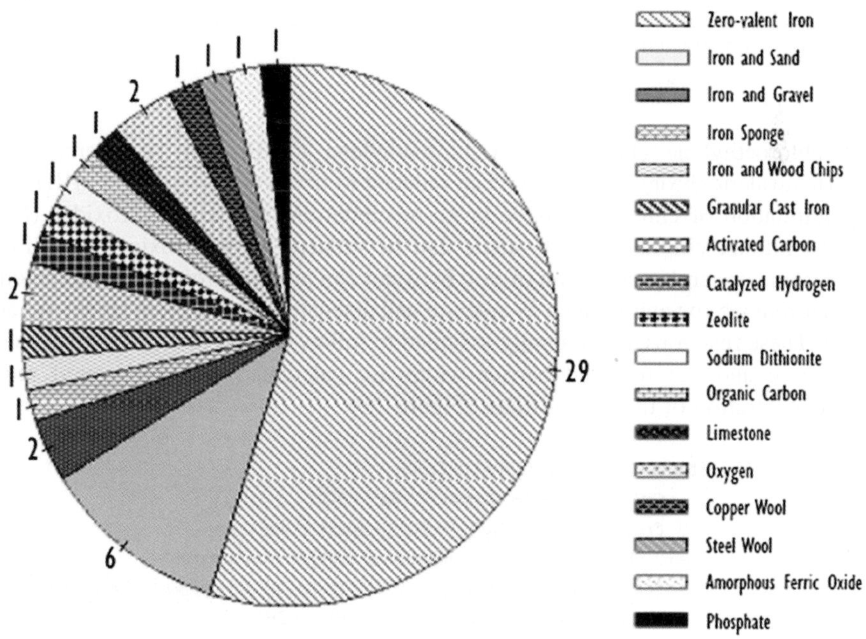

FIGURE 2.5

Reactive materials used for permeable reactive barriers.

(adapted from U.S. EPA, 2002).

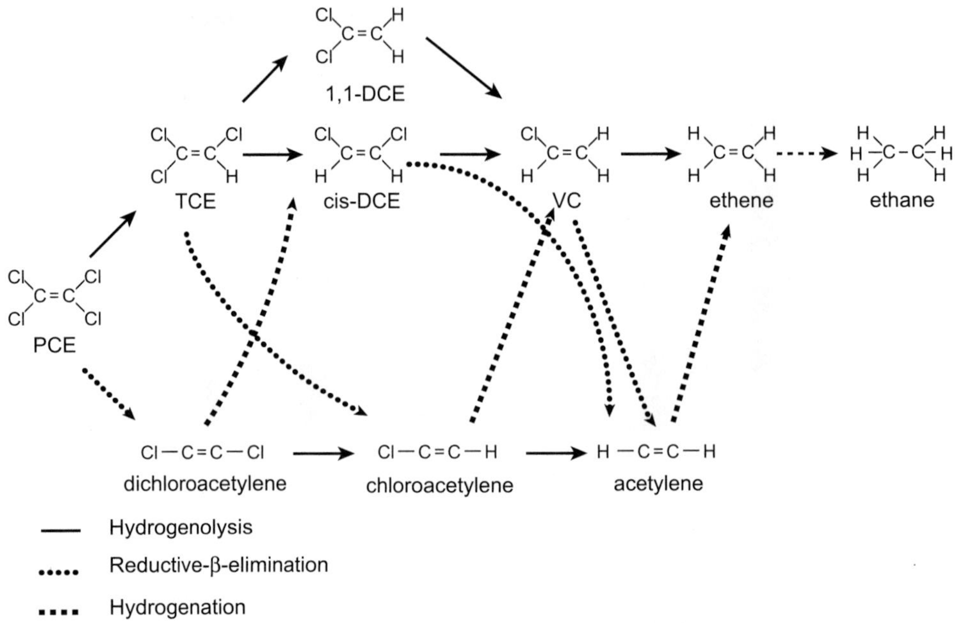

FIGURE 2.6

Reduction pathways of chlorinated solvents.

(adapted from Lo et al., 2006).

such as tetrachloroethylene (PCE) and trichloroethylene (TCE) can be chemically reduced (Figure 2.6). Hexavalent chromium (Cr(VI)) can reduced to trivalent chromium (Cr(III)) and immobilized by precipitation of chromium oxide or co-precipitation with iron hydroxides (Figure 2.7). Anions such as nitrate, chlorate, and bromate can be reduced to less toxic forms (Figure 2.8). Moreover, anaerobic or aerobic corrosion of iron by water can produce iron hydroxide precipitates on the surfaces that remove oxyanions such as arsenite and arsenate by adsorption (inner-sphere surface complexation). These reactions are generally assumed to follow pseudo first-order kinetics, by which the residence time and the required barrier width are calculated for the desired reduction in contaminant concentration of the groundwater (Figure 2.9).

Step 1 – Reduction of Cr(VI) to Cr(III)

$$CrO_4^{2-} + Fe^0_{(s)} + 4H_2O_{(l)} \Rightarrow Cr^{3+} + Fe^{3+} + 8OH^-$$

$$Cr_2O_7^{2-} + 2Fe^0_{(s)} + 7H_2O_{(l)} \Rightarrow 2Cr^{3+} + 14OH^- + 2Fe^{3+}$$

Step 2 – Precipitation/Co-precipitation of Cr(III) on iron surfaces

$$2Cr^{3+} + 6OH^- \Leftrightarrow 2Cr(OH)_{3(s)}$$

$$(1\text{-}x)Fe^{3+} + xCr^{3+} + 3H_2O_{(l)} \Leftrightarrow Cr_xFe_{1-x}(OH)_{3(s)} + 3H^+$$

$$(1\text{-}x)Fe^{3+} + xCr^{3+} + 2H_2O_{(l)} \Leftrightarrow Cr_xFe_{1-x}OOH_{(s)} + 3H^+$$

FIGURE 2.7

Reduction and immobilization of chromate.

Nitrate

$$4Fe^0_{(s)} + NO_3^- + 10H^+ \Rightarrow 4Fe^{2+} + NH_4^+ + 3H_2O_{(l)}$$

Chlorate

$$3Fe^0_{(s)} + ClO_3^- + 6H^+ \Rightarrow 3Fe^{2+} + Cl^- + 3H_2O_{(l)}$$

Bromate

$$3Fe^0_{(s)} + BrO_3^- + 6H^+ \Rightarrow 3Fe^{2+} + Br^- + 3H_2O_{(l)}$$

FIGURE 2.8

Reduction of nitrate, chlorate, and bromated.

- Pseudo first-order reaction

$$C_e = C_i e^{-kt}$$

where C_i is influent concentration, C_e is effluent concentration, and k is the observed first-order rate coefficient

- Residence time (t_{res}) to achieve the desired reduction in contaminant concentration

$$t_{res} = \frac{1}{k}\ln\left(\frac{C_i}{C_e}\right)$$

- Required barrier width (W) based on groundwater velocity (v)

$$W = vt_{res}$$

FIGURE 2.9

Design calculations for permeable reactive barriers.

(adapted from U.S. EPA, 2002).

3.2 Adsorption of natural organic matter on iron surfaces

As discussed in previous sections, a fraction of natural organic matter is adsorbed as coatings on mineral surfaces, which may completely block or kinetically reduce the accessibility of the underlying reactive sites for adsorption or reduction (Zachara et al., 1992; Yin et al., 1997). Moreover, even at low surface coverage, natural organic matter can significantly modify the properties of particle surfaces such as neutralizing, or even reversing, the electrostatic charges (Saito et al., 2004), which in turn alters the interactions between the surfaces and chemicals/colloids in solution. A number of studies have demonstrated the significant roles of natural organic matter in controlling the fate of contaminant in the subsurface through statistical analysis (Lee et al., 1996; McBride et al., 1997; Gray et al., 1999; Sauve et al., 2000; Yin et al., 2002), addition or removal of indigenous organic matter (Yin et al., 1997; Nachtegaal and Sparks, 2003), and chemical alteration of the functional groups and structure of organic matter (Schilling and Cooper, 2004).

As many iron hydroxides phases (magnetite, maghemite, ferrihydrite, lepidocrosite, green rust, ferrous hydroxide, etc.) are typically present on the zero-valent iron surfaces (Furukawa et al., 2002; Ritter et al., 2002; Kohn et al., 2005), natural organic matter tends to readily adsorb onto the iron surfaces. Under the alkaline pH range (due to iron corrosion by water) observed in permeable reactive barriers, both organic matter and iron hydroxides present in the interfacial region are negatively charged; thus, specific interactions such as inner-sphere surface complexation should dominate natural organic matter adsorption. Chlorinated and nitroaromatic hydrocarbons, for example, that are expected to form weakly bonded surface complexes may therefore be excluded from the iron surface due to direct competition for adsorption or steric hindrance from the reactive sites. As a result, it has been observed that the contaminant reduction by zero-valent iron, which is known to be surface-mediated, is inhibited in the presence of natural organic matter (Tratnyek et al., 2001; Klausen et al., 2003; Marconetto et al., 2005; Doong and Lai, 2006).

Furthermore, the sources and characteristics of natural organic matter appear to result in a varying extent of inhibitory effects. The effects of humic acids were found to be greater than that of fulvic acids, because humic acids are known to have a greater aromatic content and adsorb onto iron hydroxide surfaces more strongly than fulvic acids (Tratnyek et al., 2001; Klausen et al., 2003). Along the same line, natural organic matter from soil sources exerted a more significant effect than that from aquatic sources, due to stronger adsorption of the soil sources (Tratnyek et al., 2001). Besides, the relatively smaller molecules of organic matter may be able to access surface sites located in small pores that are precluded for the larger molecules, leading to competition for a larger number of reactive sites (Marconetto et al., 2005). These results suggest that the physical and chemical properties of natural organic matter are important factors.

In addition to the steric effect that is often predominant, adsorption of natural organic matter may alter the reduction potential of neighboring reactive sites on the iron surfaces and mediate contaminant reduction through a minor quantity of redox-active components such as quinone moieties. However, the extent and kinetics are relatively minor based on previous findings (Jardine et al.,1999; Xie and Shang, 2005).

3.3 Complexation of released iron with natural organic matter

It is well recognized that natural organic matter has high complex stability with divalent and trivalent metals. Soluble metals may predominantly exist in the form of organo-metal complexes, especially at pH above 7 or 8 due to the enhanced solubility of organic matter and the greater dissociation of surface functional groups (McBride et al., 1997; You et al., 1999; Sauve et al., 2000; Yin et al., 2002). The complexation with dissolved organic matter can increase the concentration of dissolved metals by more than two order-of-magnitudes according to direct measurements of metal complexes (Weng et al., 2002). Mobilization (facilitated transport) of heavy metals is therefore significant even at low concentrations of dissolved organic matter in an alkaline region (Buerge-Weirich et al., 2002). In addition, the low-molecular-weight fraction of dissolved organic matter forms labile metal complexes, while the high-molecular-weight fraction largely forms nonlabile soluble metal complexes, probably because of a high degree of aromaticity or strongly sorbing thiol and amine functional groups (Martinez et al., 2001; Karlsson et al., 2005). These findings demonstrate the important roles of natural organic matter in controlling the speciation, toxicity, mobility, and bioavailability of metals in the environment.

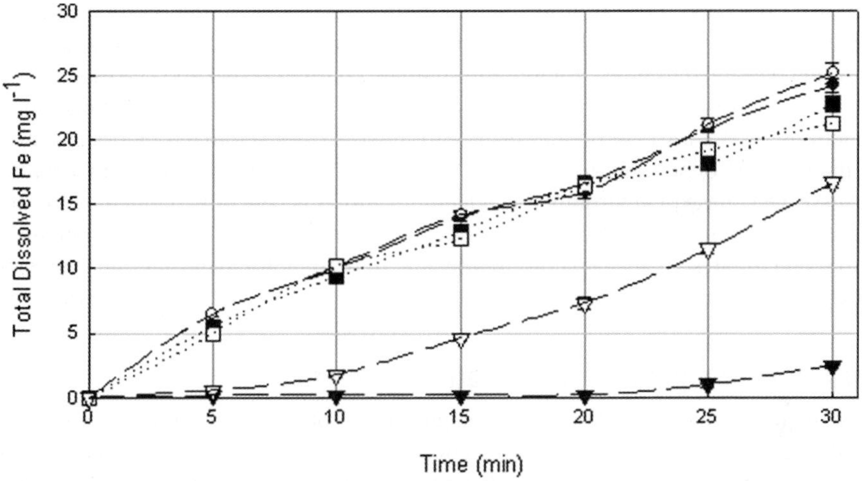

FIGURE 2.10 Dissolved iron concentration in batch kinetics experiments

(humic acid alone at initial pH 6 ● and pH 9.5 ○; humic acid and Cr(VI) at initial pH 6 ■ and pH 9.5 □; Cr(VI) alone at initial pH 6 ▼ and pH 9.5 ▽).

(adapted from Liu et al., 2008).

In zero-valent iron systems, dissolved ferrous and ferric iron is released from the surfaces as a result of iron corrosion by water and contaminant reduction. It is likely that most dissolved iron expeditiously forms precipitates such as iron hydroxides that can exist as colloidal particles in solution or deposits on the iron surfaces, depending on geochemical conditions. The presence of dissolved humic acid has been found to increase the amount of dissolved iron (Figure 2.10), probably due to the

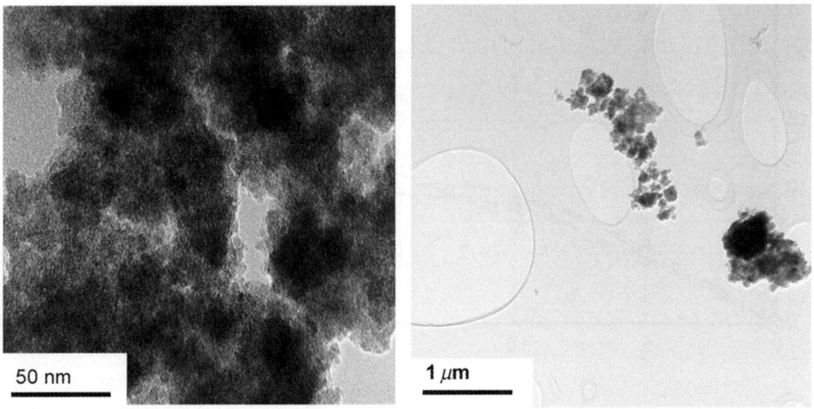

FIGURE 2.11

Transmission electron microscopic images of dissolved and colloidal iron hydroxides and iron-humate complexes.

(adapted from Liu et al., 2008).

high affinity of ferrous and ferric iron for binding with humic acid. Based on the centrifugation and filtration results and transmission electron microscopic image (Figure 2.11), it was estimated that about 40% of the dissolved iron was soluble iron-humate complexes and about 50% was colloids (Liu et al., 2008). Furthermore, another recent study showed that at high iron/organic matter ratios, colloidal fractions (< 0.2 μm) are predominantly iron hydroxides stabilized by organic matter, but at low ratios, they consist of soluble iron attached to dissolved/colloidal organic matter (Gaffney et al., 2008). The dissolved phase of environmental aqueous samples is generally defined by filtration at 0.45 μm (or 0.2 μm) (Jackson et al., 2005). Thus, both iron complexes with dissolved organic matter and colloidal iron precipitates in the aqueous phase are important for determining iron mobility and bioavailability.

3.4 Aggregation of natural organic matter in solution

Nevertheless, in divalent cation (calcium or magnesium) solutions, iron-humate colloids gradually agglomerated to large size clusters along with reaction time, as reflected by the subsequent decrease in the amount of dissolved iron (Figure 2.12). By analyzing the solution using high-performance size-exclusion chromatography (Figure 2.13), initial humic acid solution showed a broad peak centered at about 1.5 kDa, while humic acid solution after reaction with zero-valent iron displayed a major sharp peak centered at about 25 kDa. After longer reaction time, 97% of humic acid was aggregated and filtered out, and the small fraction that remained dissolved was composed of molecular weight less than 1 kDa. Such an increase of molecular weight was similarly observed for various humic acids (Figure 2.14), indicating the continuous aggregation of dissolved humic acid in solution (Tsang et al., 2009a).

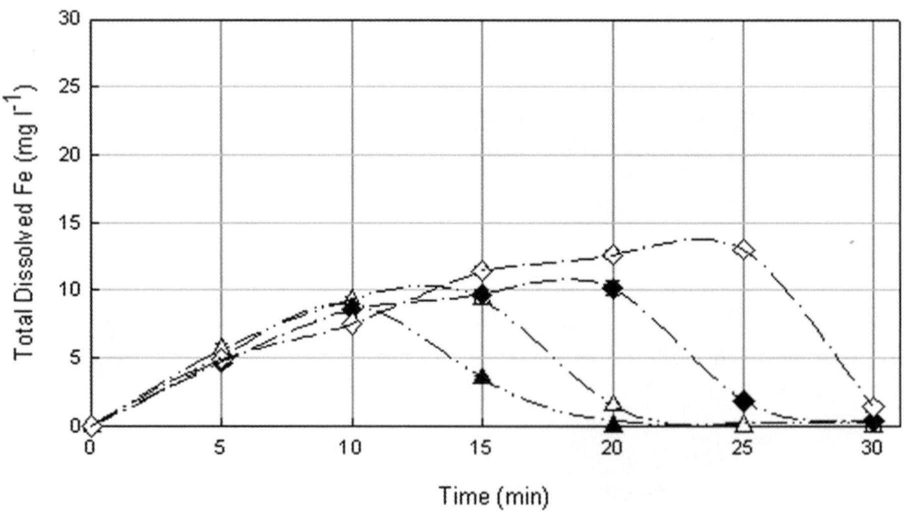

FIGURE 2.12

Dissolved iron concentration in the presence of calcium/magnesium in batch kinetics experiments (humic acid and Cr(VI) in Ca-solution at initial pH 6 ▲ and pH 9.5 △; humic acid and Cr(VI) in Mg-solution at initial pH 6 ◆ and pH 9.5 ◇).

(adapted from Liu et al., 2008).

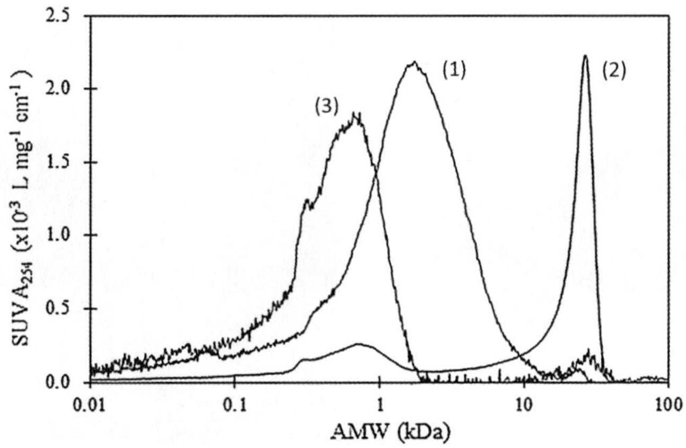

FIGURE 2.13 High-performance size exclusion chromatograph of dissolved humic acid

(initial solution (1); after 70% of dissolved humic acid was removed due to reaction with zero-valent iron (2); after 97% of dissolved humic acid was removed due to reaction with zero-valent iron (3); AMW: apparent molecular weight; $SUVA_{254}$: specific UV absorbance at 254 nm).

(adapted from Tsang et al., 2009a).

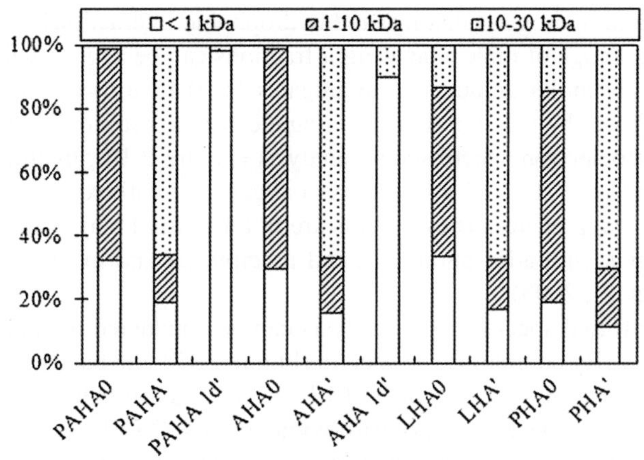

FIGURE 2.14 Apparent molecular weight fractions of various types of dissolved humic acids

(HA0: initial solution; HA': after 70% of dissolved humic acid was removed due to reaction with zero-valent iron; HA 1d': after 97% of dissolved humic acid was removed due to reaction with zero-valent iron).

(adapted from Tsang et al., 2009a).

The molecular weights of humic acids in calcium solution alone (without reaction with zero-valent iron) increased slightly but contained negligible amounts of humic acid aggregates even after aging for one month. Therefore, the continuous release of dissolved iron into solution due to iron corrosion is necessary for humic acid aggregation. Humic acid aggregation can be promoted through charge neutralization, co-precipitation, and inter- and intramolecular bridging by background calcium and dissolved iron. Low-molecular-weight fractions contain more abundant acidic functional groups,

FIGURE 2.15

Transmission and scanning electron microscopic images of iron-humate aggregates.

(adapted from Liu et al., 2008).

while high-molecular-weight fractions are subject to stronger hydrophobic interactions and smaller loss of translational entropy upon adsorption. Since both acidic functional group-related processes (e.g., ligand exchange and hydrogen bonding) and hydrophobic interactions may contribute to humic acid aggregation, a wide range of molecular-weight fractions can be aggregated (Tsang et al., 2009a).

The formation of substantial amounts of aggregates ($> 0.45\ \mu m$) of natural organic matter in solution is clearly illustrated by transmission/scanning electron microscopic images (Figure 2.15). As indicated by X-ray photoelectron spectroscopic analysis (Figure 2.16), hydrolyzed species and precipitates of calcium and iron, as well as hydroxyl oxygen and carboxyl carbon of natural organic matter, are present in the aggregates. These results are in line with the hypothesis that natural organic matter aggregation begins with adsorption on metal precipitates and hydrolyzed species in solution (Randtke, 1988; Jung et al., 2005).

Fourier-transform infrared spectroscopic analysis can be conducted to study the bonding characteristics of the aggregates. By comparing the spectra (Figure 2.17), the bands at ~ 2917 and $2849\ cm^{-1}$ (C-H stretching of aliphatic CH_2 or CH_3), at $\sim 1617\ cm^{-1}$ ($C = C$ stretching of aromatic ring), and at $\sim 1386\ cm^{-1}$ (symmetric $C = O$ stretching of carboxylic groups, C-OH stretching of phenolic groups) were found to be unchanged. Nevertheless, the bands at $\sim 1716\ cm^{-1}$ ($C = O$ stretching of carboxylic or carbonyl groups) and at $\sim 1249\ cm^{-1}$ (C-O stretching of esters, ethers, and phenols) disappeared, while the bands at $\sim 1039\ cm^{-1}$ (C-O stretching of carbohydrate or aliphatic alcohol) were shifted in the spectra of humic acid aggregates. Such disappearance and shift of bands were, however, not illustrated in the spectra of humic acid aggregates that had been repeatedly acid-washed. These observations evidence that, within the aggregates, inner-sphere complexation was formed between hydrolyzed metal species and humic acids involving different oxygen-containing functional groups such as carboxylic groups. In addition, the bond stretching of oxygen-containing functional groups and aliphatic CH_2 or CH_3 significantly diminished after a long reaction time but reappeared after acid-washing, reflecting a greater extent of complexation (which constrained the bond stretching) with increasing reaction time (Tsang et al., 2009a).

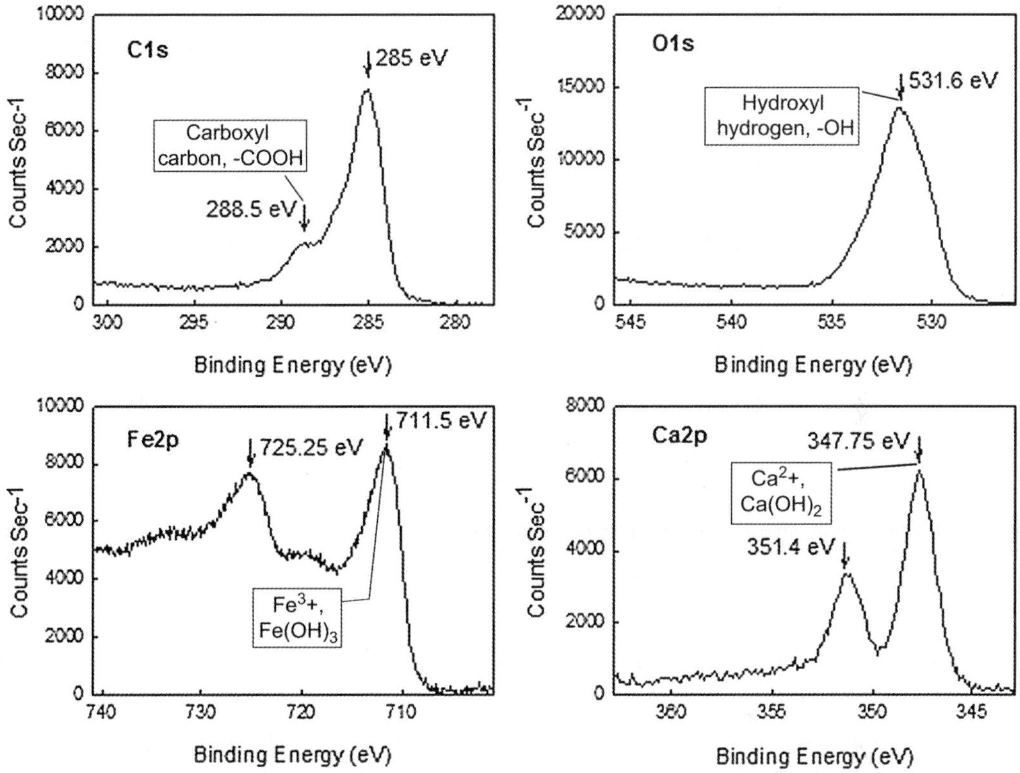

FIGURE 2.16

X-ray photoelectron spectra of iron-humate aggregates: (a) C1s; (b) O1s; (c) Fe2p; and (d) Ca2p.

(adapted from Liu et al., 2008).

In addition to inner-sphere complexation, the contribution of hydrophobic interactions appears to be significant as well (Hur and Schlautman, 2003), which can be reflected by the degrees of aromaticity and hydrophobicity of natural organic matter (Weber et al., 2006). Because a variety of underlying processes (e.g., complexation and hydrophobic interactions) occur simultaneously, multiple cross-correlated physicochemical properties of natural organic matter (molecular weight, functional group density, aromaticity and hydrophobicity) govern the extent of aggregation. Thus, little quantitative interpretation or estimation can be made based on only one of these properties of natural organic matter.

3.5 Effects on removal of chlorinated solvents and chromate

As discussed in previous sections, there are multiple interactions (adsorption, complexation, aggregation, etc.) of natural organic matter with zero-valent iron. Not surprisingly, mixed effects of natural organic matter have been observed on the contaminant removal by zero-valent iron permeable reactive barriers. Inhibition of contaminant removal due to natural organic matter was reported in some studies (Tratnyek et al., 2001; Marconetto et al., 2005; Doong and Lai, 2006), while it was also observed that

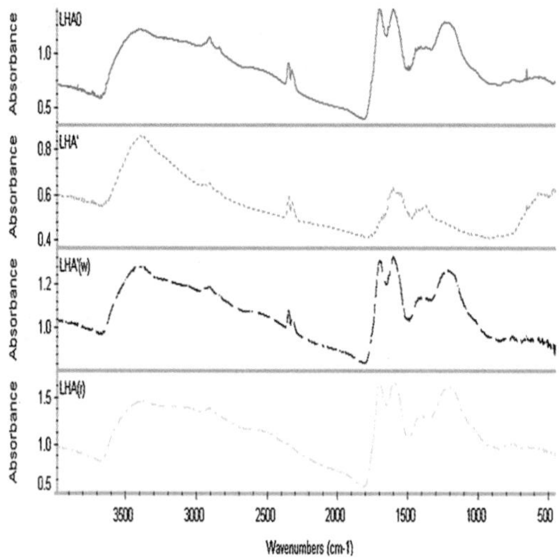

FIGURE 2.17 Fourier-transform infrared spectra of humic acid

LHA0: standard Leonardite humic acid; LHA': iron-humate aggregates collected after reaction with zero-valent iron; LHA'(w): iron-humate aggregates washed with 1 M HCl repeatedly; LHA(r): humic acid that remained dissolved in solution after reaction with zero-valent iron. (For color version of this figure, the reader is referred to the online version of this book.)

(adapted from Tsang et al., 2009a).

contaminant removal was influenced to an insignificant extent (Dries et al., 2005; Liu et al., 2008) or enhanced through contaminant adsorption onto organic matter (which can be adsorbed, aggregated, or indigenously present on the iron surfaces) (Burris et al., 1998; Dries et al., 2004; Cho and Park, 2006; Tsang et al., 2009a). The following discussion exemplifies the potential effects of natural organic matter aggregation on the removal of chlorinated solvents and chromate.

The removal of chlorinated solvents, due to hydrophobic character, results from both reduction at reactive sites and adsorption on nonreactive sites on the iron surfaces (Burris et al., 1998; Dries et al., 2004). Their removal may increase with increasing concentrations of natural organic matter through enhanced partitioning into adsorbed natural organic matter (Cho and Park, 2006). Newly formed natural organic matter aggregates, which constitute more inter- and intramolecular hydrophobic domains for hydrophobic partitioning than dissolved organic matter (Tsang et al., 2009a). This may result in two distinct effects, that is, enhanced solubilization or enhanced adsorption, depending on whether these aggregates are readily deposited on the iron surfaces.

In view of increasing molecular size and high collision efficiency in calcium solution, natural organic matter aggregates are likely to deposit on the iron surfaces through interception and gravitational sedimentation (Kretzschmar et al., 1999; Logan, 1999). Adsorption of chlorinated solvents may thus be enhanced. However, the deposited aggregates may at the same time result in steric blocking of neighboring reactive sites on the iron surfaces and inhibit reduction of chlorinated solvents. The enhancement in adsorption and the suppression of reduction, which offset each other to some extent, probably account for the observations of mixed results (Tsang et al., 2009a,b), that is, an apparent increase of trichloroethylene removal (Figure 2.18) but statistically insignificant effects on tetrachloroethylene removal (Figure 2.19).

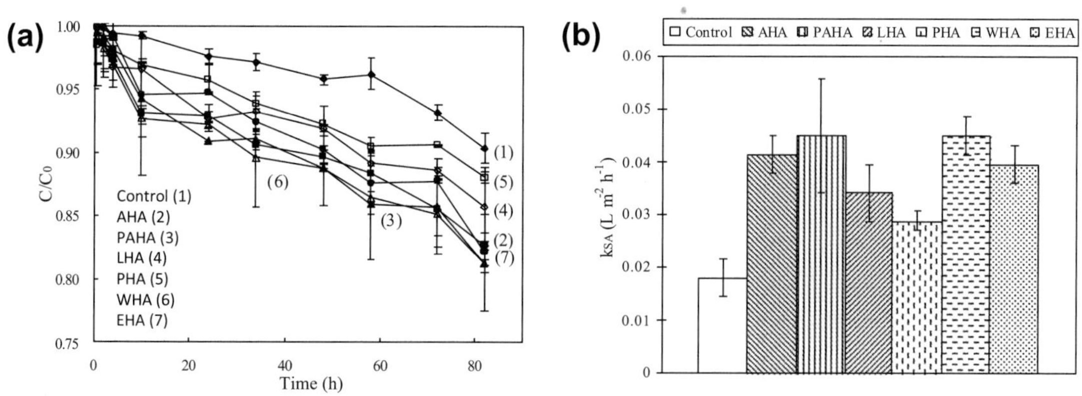

FIGURE 2.18

Trichloroethylene removal by zero-valent iron in the presence of various humic acids: (a) removal kinetics; and (b) surface area-normalized first-order rate coefficients.

(adapted from Tsang et al., 2009a).

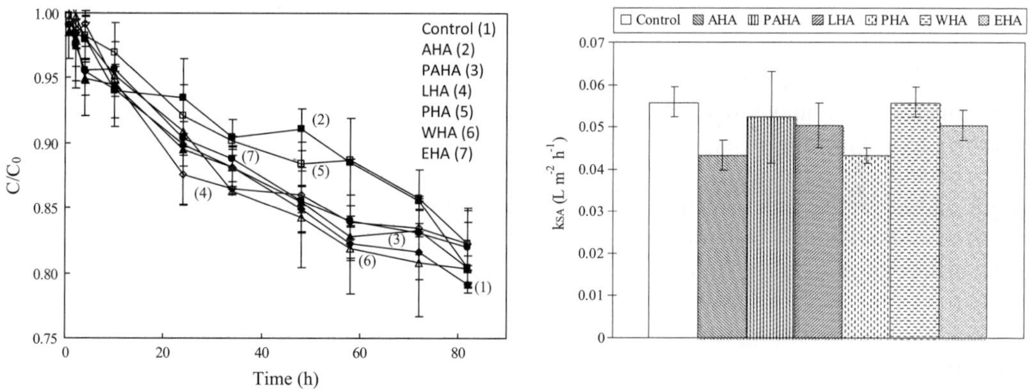

FIGURE 2.19

Tetrachloroethylene removal by zero-valent iron in the presence of various humic acids: (a) removal kinetics; and (b) surface area-normalized first-order rate coefficients.

(adapted from Tsang et al., 2009b).

On the other hand, chromate removal is less likely to be affected by natural organic matter because chromate does not significantly complex with dissolved organic matter or adsorb onto natural organic matter aggregates. Figure 2.20 shows that the chromate removal was marginally influenced by natural organic matter, with a 7–9% reduction in the removal rate constants. This is in agreement with previous findings that, on the iron surfaces, adsorption sites (metal precipitates and graphite impurities) for natural organic matter are different from reactive sites (zero-valent iron) for contaminant reduction (Burris et al., 1998; Dries et al., 2004), and that the minor inhibitory effect results from steric blocking of neighboring reactive sites (Dries et al., 2005; Liu et al., 2008).

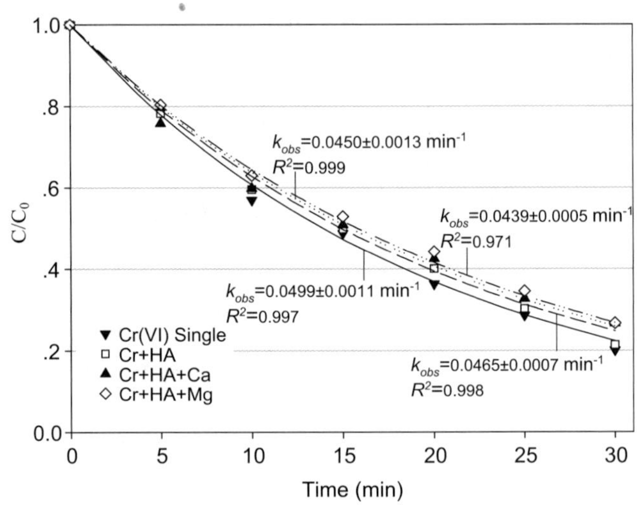

FIGURE 2.20

Chromate removal kinetics by zero-valent iron in the presence of humic acid.

(adapted from Liu et al., 2008).

Nevertheless, continuous deposition of aggregates on the surfaces probably results in a more noticeable steric blocking of neighboring reactive sites and clogging of flow paths, which in turn inhibit contaminant removal in the long term. Therefore, the factors such as solution pH, ionic strength, and respective index ion that are crucial for controlling the deposition and transport colloids (as well as colloid-bound contaminants) also play key roles in determining the effects of natural organic matter on zero-valent iron permeable reactive barriers for groundwater remediation.

4 Summary

Natural organic matter is primarily a heterogeneous mixture of organic molecules that are self-assembled in supramolecular colloidal aggregates or macromolecules. These vaguely defined, transformed organic molecules possess cross-correlated physicochemical properties (molecular weight, functional group density, aromaticity, and hydrophobicity) and thus are able to simultaneously undergo multiple interactions (partitioning, complexation, redox reactions, etc.) with mineral surfaces and dissolved contaminants. Natural organic matter is dissolved and/or mobilized as organic colloids at alkaline pH values or along with mineral dissolution. In the aqueous phase, adsorption, complexation, or aggregation of natural organic matter takes place to a varying extent depending on geochemical conditions, which in turn result in complex effects on groundwater remediation such as zero-valent iron permeable reactive barriers.

A fraction of natural organic matter tends to be adsorbed onto the iron surfaces and cause steric blocking of neighboring reactive sites, which may affect removal of chlorinated solvents to a greater extent than removal of chromate. At the same time, released iron resulting from continuous corrosion of zero-valent iron surfaces forms colloidal particles and complexes with dissolved organic matter in solution. Elevated concentrations of dissolved iron and inorganic and organic colloids in groundwater may arouse aesthetic and safety concerns. Furthermore, if there is calcium or magnesium in the aqueous phase (e.g., hard water), natural organic matter aggregates with hydrolyzed species and precipitates of iron and calcium/magnesium. These aggregates provide additional partitioning domains for chlorinated solvents. But in view of their increasing molecular weight with reaction time, they also tend to deposit on the iron surfaces, resulting in steric blocking of reactive sites and clogging of flow path that eventually inhibit the removal of both chlorinated solvents and chromate in the long term. Therefore, it is important to take into account multiple interactions in different aspects when assessing the effects of natural organic matter.

SYMBOLS AND TERMINOLOGY

AMW apparent molecular weight of dissolved organic matter
C contaminant concentration in aqueous phase
C_i influent contaminant concentration of permeable reactive barriers
C_e effluent contaminant concentration of permeable reactive barriers
K observed first-order reaction rate coefficient
k_{SA} surface area-normalized first-order reaction rate coefficient
$SUVA_{254}$ specific UV absorbance at a wavelength of 254 nm normalized to total organic carbon of aqueous samples
V average linear velocity of groundwater flow
W width of permeable reactive barriers
t_{res} fluid residence time in permeable reactive barriers
Cr(VI) hexavalent chromium (chromate)
Fe^0 zero-valent iron
HA humic acid
NOM natural organic matter
PCE tetrachloroethylene
PRB permeable reactive barriers
TCE trichloroethylene

Glossary

Colloids inorganic or organic particles with size smaller than 0.1 μm (or 1 μm)
Dissolved phase pass through 0.45-μm (or 0.2-μm) membrane filtration, including ions, organic molecules, soluble and colloidal organic matter
Fourier-transform infrared spectroscopy used to identify surface functional groups and investigate bonding characteristics such as inner-sphere or outer-sphere, monodentate or bidentate, surface complexation
Humic acid an operationally defined fraction of humic substances that is soluble in alkali but not in acid
Humic substances natural organic matter except recognizable plant debris and organic compounds such as polysaccharides, lignin, proteins and polypeptides (no-humic substances)
High-performance size exclusion chromatography used to investigate molecular-weight distribution of dissolved organic matter

Iron-humate aggregates clusters of colloidal precipitates of iron and humic acid that are filtered out by 0.45-μm membrane filtration

Natural organic matter total organic carbon in the natural environment except carbon in living organisms (biomass) and carbon in undecayed and partially decayed plant and animal tissues (organic residues)

Permeable reactive barriers a passive groundwater remediation technology, in which reactive materials are installed to intercept contaminated plume, and immobilize and/or degrade the contaminants in groundwater

Scanning/Transmission electron microscopy used to provide microscopic image of particle surface

X-ray photoelectron spectroscopy used to analyze chemical oxidation states and compositions of particle surface

Zero-valent iron a low-cost commercial product, mostly cast iron and low alloy steels, that is widely used for contaminant reduction and immobilization in permeable reactive barriers

References

Buerge-Weirich, D., Hari, R., Xue, H., Behra, P., Sigg, L., 2002. Adsorption of Cu, Cd, and Ni on goethite in the presence of natural groundwater ligands. Environ. Sci. Technol. 36, 328–336.

Burris, D.R., Allen-King, R.M., Manoranjan, V.S., Campbell, T.J., Loraine, G.A., Deng, B.J., 1998. Chlorinated ethene reduction by cast iron: Sorption and mass transfer. J. Environ. Eng. 124, 1012–1019.

Cho, H., Park, J., 2006. Sorption and reduction of tetrachloroethylene with zero valent iron and amphiphilic molecules. Chemosphere 64, 1047–1052.

Doong, R., Lai, Y., 2006. Effect of metal ions and humic acid on the dechlorination of tetrachloroethylene by zerovalent iron. Chemosphere 64, 371–378.

Dries, J., Bastiaens, L., Springael, D., Agathos, S.N., Diels, L., 2004. Competition for sorption and degradation of chlorinated ethenes in batch zero-valent iron systems. Environ. Sci. Technol. 38, 2879–2884.

Dries, J., Bastiaens, L., Springael, D., Kuypers, S., Agathos, S.N., Diels, L., 2005. Effect of humic acids on heavy metal removal by zero-valent iron in batch and continuous flow column systems. Water Res. 39, 3531–3540.

Essington, M.E., 2004. Soil and water chemistry: An integrative approach. CRC Press, New York.

Furukawa, Y., Kim, J.W., Watkins, J., Wilkin, R.T., 2002. Formation of ferrihydrite and associated iron corrosion products in permeable reactive barriers of zero-valent iron. Environ. Sci. Technol. 36, 5469–5475.

Gaffney, J.W., White, K.N., Boult, S., 2008. Oxidation state and size of Fe controlled by organic matter in natural waters. Environ. Sci. Technol. 42, 3575–3581.

Gorham, J.M., Wnuk, J.D., Shin, M., Fairbrother, H., 2007. Adsorption of natural organic matter onto carbonaceous surfaces: Atomic force microscopy study. Environ. Sci. Technol. 41, 1238–1244.

Gray, C.W., McLaren, R.G., Roberts, A.H.C., Condron, L.M., 1999. Solubility, sorption and desorption of native and added cadmium in relation to properties of soils in New Zealand. J. Soil Sci. 50, 127–137.

Grolimund, D., Borkovec, M., Barmettler, K., Sticher, H., 1996. Colloid-facilitated transport of strongly sorbing contaminants in natural porous media: A laboratory column study. Environ. Sci. Technol. 30, 3118–3123.

Hur, J., Schlautman, M.A., 2003. Molecular weight fractionation of humic substances by adsorption onto minerals. J. Colloid Interf. Sci. 264, 313–321.

Jardine, P.M., Fendorf, S.E., Mayes, M.A., Larsen, I.L., Brooks, S.C., Bailey, W.B., 1999. Fate and transport of hexavalent chromium in undisturbed heterogeneous soil. Environ. Sci. Technol. 33, 2939–2944.

Jung, A.V., Chanudet, V., Ghanbaja, J., Lartiges, B.S., Bersillon, J.L., 2005. Coagulation of humic substances and dissolved organic with a ferric salt: An electron energy loss spectroscopy investigation. Water Res. 39, 3849–3862.

Karlsson, T., Persson, P., Skyllberg, U., 2005. Extended X-ray absorption fine structure spectroscopy evidence for the complexation of cadmium by reduced sulfur groups in natural organic matter. Environ. Sci. Technol. 39, 3048–3055.

Kerner, M., Hohenberg, H., Ertl, S., Reckermann, M., Spitzy, A., 2003. Self-organization of dissolved organic matter to micelle-like microparticles in river water. Nature 422, 150–154.

Klausen, J., Vilesland, P.J., Kohn, T., Burris, D.R., Ball, W.P., Roberts, A.L., 2003. Longevity of granular iron in groundwater treatment processes: solution composition effects on reduction of organohalides and nitroaromatic compounds. Environ. Sci. Technol. 37, 1208–1218.

Kohn, T., Livi, K.J.T., Roberts, A.L., Vikesland, P.J., 2005. Longevity of granular iron in groundwater treatment processes: Corrosion product development. Environ. Sci. Technol. 39, 2867–2879.

Kretzschmar, R., Borkovec, M., Grolimund, D., Elimelech, M., 1999. Mobile subsurface colloids and their role in contaminant transport. Advances in Agronomy 66, 121–193.

Lee, S.Z., Allen, H.E., Huang, C.P., Sparks, D.L., Sanders, P.F., Peijnenburg, W.J.G.M., 1996. Predicting soil-water partition coefficients for cadmium. Environ. Sci. Technol. 30, 3418–3424.

Liu, T., Tsang, D.C.W., Lo, I.M.C., 2008. Chromium(VI) reduction kinetics by zero-valent iron in moderately hard water with humic acid: Iron dissolution and humic acid adsorption. Environ. Sci. Technol. 42, 2092–2098.

Logan, B.E., 1999. Environmental transport processes. John Wiley and Sons, New York.

Lo, I.M.C., Surampalli, R.Y., Lai, K.C.K. (Eds.), 2006. Zero-valent iron reactive materials for hazardous waste and inorganics removal. Reston, Virginia, American Society of Civil Engineers.

Marconetto, S., Gui, L., Gillham, R., 2005. Adsorption of natural organic matter and its effects on TCE degradation by iron PRBs. In: Bringing Groundwater Quality Research to the Watershed Scale, Proceedings of GQ 2004, 4th International Groundwater Quality Conference. Waterloo, Canada, July, 2004. IAHS Publ. 297, 389–397.

Lu, Y., Pignatello, J.J., 2004. Sorption of apolar aromatic compounds to soil humic acid particles affected by aluminum(III) ion cross-linking. J. Environ. Qual 33, 1314–1321.

Maurice, P.A., Namjesnik-Dejanovic, K., 1999. Aggregate structures of sorbed humic substances observed in aqueous solutions. Environ. Sci. Technol. 33, 1538–1541.

McBride, M.B., 1994. Environmental chemistry of soils. Oxford University Press, New York.

McBride, M.B., Sauve, S., Hendershot, W., 1997. Solubility control of Cu, Zn, Cd and Pb in contaminated soils. J. Soil Sci. 48, 337–346.

Myneni, S.C.B., Brown, J.T., Martinez, G.A., Meyer-Ilse, W., 1999. Imaging of humic substance macromolecular structures in water. Science 286, 1335–1337.

Nachtegaal, M., Sparks, D.L., 2003. Nickel sequestration in a kaolinite-humic acid complex. Environ. Sci. Technol. 37, 529–534.

Namjesnik-Dejanovic, K., Maurice, P.A., 2000. Conformations and aggregate structures of sorbed natural organic matter on muscovite and hematite. Geochim. Cosmochim. Acta 65, 1047–1057.

Randtke, S.J., 1988. Organic contaminant removal by coagulation and related process combinations. J. Am. Water Works Assoc. 80 (5), 40–56.

Ritter, K., Odziemkowski, M.S., Gillham, R.W., 2002. An in situ study of the role of surface films on granular iron in the permeable iron wall technology. J. Contam. Hydrol. 55, 87–111.

Saito, T., Koopal, L.K., van Riemsdijk, W.H., Nagasaki, S., Tanaka, S., 2004. Adsorption of humic acid on goethite: Isotherms, charge adjustments, and potential profiles. Langmuir 20, 689–700.

Sauve, S., Hendershot, W., Allen, H.E., 2000. Solid-solution partitioning of metals in contaminated soils: Dependence on pH, total metal burden, and organic matter. Environ. Sci. Technol. 34, 1125–1131.

Schilling, M., Cooper, W.T., 2004. Identification of copper binding sites in soil organic matter through chemical modifications and ^{13}C CP-MAS NMR spectroscopy. Environ. Sci. Technol. 38, 5059–5063.

Sparks, D.L., 1999. Soil physical chemistry, 2nd ed. CRC Press, New York.

Sparks, D.L., 2003. Environmental soil chemistry, 2nd ed. Academic Press, Amsterdam.
Sposito, G., 1994. Chemical equilibria and kinetics in soils. Oxford University Press, New York.
Sutton, R., Sposito, G., 2005. Molecular structure in soil humic substances: The new view. Environ. Sci. Technol. 39, 9009–9015.
Tipping, E., 2002. Cation binding by humic substances. Cambridge University Press, London.
Tratnyek, P.G., Scherer, M.M., Deng, B., Hu, S., 2001. Effects of natural organic matter, anthropogenic surfactants, and model quinones on the reduction of contaminants by zero-valent iron. Water Res. 35, 4435–4443.
Tsang, D.C.W., Graham, N.J.D., Lo, I.M.C., 2009a. Humic acid aggregation in zero-valent iron systems and its effects on trichloroethylene removal. Chemosphere 75, 1338–1343.
Tsang, D.C.W., Graham, N.J.D., Lo, I.M.C., 2009b. Chlorinated hydrocarbons removal by zero-valent iron: Effects of humic acid aggregation. Third International Contaminated Site Remediation Conference, 27–30. September, Adelaide, Australia.
U.S. EPA, 2002. Field applications of in situ remediation technologies: Permeable reactive barriers. EPA/68/W-00/084, Technology Innovation Office, Office of Solid Waste and Emergency Response. United States Environmental Protection Agency, Washington, DC, 20460, United States.
Weng, L., Temminghoff, E.J.M., Lofts, S., Tipping, E., van Riemsdijk, W.H., 2002. Complexation with dissolved organic matter and solubility control of heavy metals in a sandy soil. Environ. Sci. Technol. 36, 4804–4810.
Weber Jr., W.J., Tang, J., Huang, Q., 2006. Development of engineered natural organic sorbents for environmental applications: 1. Materials, approaches, and characterizations. Environ. Sci. Technol. 40, 1650–1656.
Wershaw, R.L., 1999. Molecular aggregation of humic substances. Soil Sci. 164, 803–813.
Xie, L., Shang, C., 2005. Role of humic acid and quinone model compounds in bromate reduction by zerovalent iron. Environ. Sci. Technol. 39, 1092–1100.
Yin, Y., Allen, H.E., Huang, C.P., Sparks, D.L., Sanders, P.F., 1997. Kinetics of mercury (II) adsorption and desorption on soil. Environ. Sci. Technol. 31, 496–503.
Yin, Y., Impellitteri, C.A., You, S.J., Allen, H.E., 2002. The importance of organic matter distribution and extract soil:solution ratio on the desorption of heavy metals from soils. Sci. Total Environ. 287, 107–119.
You, S.J., Yin, Y., Allen, H.E., 1999. Partitioning of organic matter in soils: effects of pH and water/soil ratio. Sci. Total Environ. 227, 155–160.
Zachara, J.M., Smith, S.C., Resch, C.T., Cowan, C.E., 1992. Cadmium sorption to soil separates containing layer silicates and iron and aluminum oxides. Soil Sci. Soc. Am. J. 56, 1074–1084.

CHAPTER

Role of Microbial Surface-Active Compounds in Environmental Protection

3

Danka Galabova*, Anna Sotirova*, Elena Karpenko[†], Oleksandr Karpenko**

** Department of Microbial Biochemistry, The Stephan Angeloff Institute of Microbiology, Bulgarian Academy of Sciences, Sofia, Bulgaria, † Department of Chemistry and Biotechnology, Lviv Department of L.M.Lytvynenko Physical-Organic Chemistry Institute, National Academy of Sciences of Ukraine, Naukova st., 3a, Lviv, Ukraine, ** Department of Biologically Active Substances, Pharmacy and Biotechnology, Lviv Polytechnic National University, Lviv, Ukraine*

CHAPTER OUTLINE

The Role of Colloidal Systems in Environmental Protection. http://dx.doi.org/10.1016/B978-0-444-63283-8.00003-X

1 Introduction

In our modern world, the threat of global warming remains a major topic of discussion and activities for many people and organizations. But we must deal with yet another threat to human health: the threat of pollution of our environment. Our quality of life is linked to the quality of the environment. Years ago, we believed that we had unlimited abundance of and control over land and resources; today, the problems associated with contaminated sites reveal how careless we have been with our resources. More and more marine, freshwater, and soil ecosystems are contaminated with products of human activities. The problem is worldwide, and the estimated number of contaminated sites is enormous. Contamination generally results from past industrial activities when there was generally little awareness of the dangers to human health posed by the production, use, and disposal of hazardous materials. Today most soil and water contamination is the result of accidental oil and organic chemical waste spills; of improperly managed landfills; and of unregulated disposal of household wastes, including pesticides, cleaning agents, and dyes.

The conventional techniques used for remediation have been to dig up contaminated soil and move it to landfill. In reality, however, all that does is move the contaminants from one place to another, thereby creating significant risks in the excavation, handling, and transport of hazardous material.

A better approach than the conventional methods is to completely destroy pollutants if possible, or at least to transform them into innocuous substances. These technologies are based on physical and chemical methods such as high-temperature incineration and various types of chemical decomposition (e.g., UV oxidation and base-catalyzed dechlorination). They can be effective in reducing levels of contaminants, but they do have some drawbacks—such as technological complexity, high cost—and most importantly, they may cause an increase in pollution in the event of human faults and mistakes.

Different techniques are employed, depending on an area's degree of saturation and aeration. In-situ techniques are defined as those that are applied to soil and groundwater at the site with minimal disturbance. Ex-situ techniques are those that are applied to soil and groundwater at the site which have been removed from the site through excavation (soil) or pumping (water) (Joshi et al., 2013). In-situ bioremediation by indigenous microbial population is an increasingly popular, ecofriendly option for cleanup of contaminated sites (Zeyaullah et al., 2009).

The techniques applied for elimination of pollutants have to be cost effective and at the same time not harm the environment. Bioremediation of organic compounds by microorganisms represents one of the primary mechanisms for eliminating pollutants from the environment. Bioremediation offers the possibility of destroying contaminants in aquatic and terrestrial ecosystems using natural biological activities. The use of bioremediation technologies for removing these contaminants provides a safe and economic alternative to commonly used physical-chemical treatments (Leahy and Colwell, 1990; Atlas and Bartha, 1992; Allard and Neilson, 1997; Margesin and Shinner, 1997; Zeyaullah et al. 2009). By definition, bioremediation is the use of living organisms, mainly microorganisms, to degrade environmental contaminants into less toxic forms. It uses naturally occurring bacteria, yeasts, fungi, and plants to degrade or detoxify substances hazardous to human health. The microorganisms may be indigenous to a contaminated area or they may be isolated from other places and brought to contaminated site. The process of importing microorganisms to contaminated sites to enhance degradation is known as bioaugmentation.

The control and optimization of the bioremediation process is a complex of many factors: the existence of microbial populations capable of degrading the pollutants; the availability of contaminants to the microbial population; environment factors (temperature, pH, presence of oxygen or other electron acceptors, and nutrients). The benefits of the bioremediation of polluted environment are as follows:

- It treats contamination in place.
- It harnesses natural processes.
- It reduces environmental stress.

A great deal of research has been devoted to finding the organisms, usually bacteria, capable of altering or degrading such pollutants to environmentally tolerable forms. Bacteria are particularly suitable for biodegradation applications because of the wide variety of carbon sources or electron acceptors used by various strains. In most environments, enrichment of petroleum-degrading microbial communities occurs soon after contamination (Atlas and Bartha, 1992; Margesin and Shinner, 1997).

Single strains of bacteria are often insufficient to degrade certain pollutants; complete catabolism may require consortia or communities composed of two or more taxa (Wagner-Dobler et al., 1998; Hubert et al., 1999; Piehler et al., 1999). In part, the efforts are directed toward isolating individual strains of bacteria from indigenous polluted sites with desirable biodegradative abilities, and cooperation of appropriate strains forming selected communities, which could be used in bioremediation of toxic sites, accelerating the naturally occurring biodegradation. On the other hand, hydrolysis of the organic matter in polluted sites is largely dependent on bacterial activity (Deming and Baross, 1993). Extracellular enzyme activity is a key step in the degradation and utilization of organic polymers, since only compounds with molecular mass lower than 600 daltons can pass through cell pores (Hoppe, 1991; Meyer-Reil, 1991). The key factor in processes of bioremediation/biodegradation, however, is the solubility of the contaminants. Contaminants such as aliphatic and polycyclic aromatic hydrocarbons (PAHs) and polychlorinated biphenyls (PCBs) are poorly soluble. Bioremediation requires the use of microorganisms, which must be capable of developing in the presence of pollutants and either metabolizing or accumulating organic compounds. In both cases, the pollutants have to cross the microbial cell envelope (i.e., cell walls and the membranes) this transport represents a major step of the bioremediation process. In view of these considerations, use of bacteria capable of recycling organic polymers by producing extracellular enzymes, with hydrocarbon-degrading microorganisms, is a significant addition to bioaugmentation in hydrocarbon pollution control.

Surface-active compounds (surfactants) play an important role in bioremediation in facilitating the solubilization of hydrophobic pollutants and enhancing their penetration inside microbial cells (Lawniczak et al., 2013). The action of surfactants also modifies the cell envelope and changes the fluidity of membranes. Furthermore, the surfactants which are tools for remediation cannot be harmful to the environment; they have to be biodegradable. The use of biosurfactants presents a real advantage over synthetic derivatives, as they are biodegradable after a certain period.

In this chapter we discuss the potential role and applications of biosurfactants in the environment, focusing on new data on the bioremedation/biodegradation of water and soil pollutants.

2 Biosurfactants

2.1 Nature

Surfactants (surface-active compounds) are amphiphilic compounds with hydrophobic and hydrophilic domains, which allow them to partion at the interface between fluid phases with different degrees of polarity. They reduce surface and interfacial tension and form microemulsions, thus solubilizing the organic compounds in water. By reducing the interfacial tension and forming micelles, surfactants or biosurfactants have led to many environmental applications, including enhanced oil recovery, removal of heavy metals from contaminated soil, and remediation of hydrophobic organic compounds from soil (Mulligan and Wang, 2006; Urum et al., 2003; Li, Y.-Y. et al., 2006; Whang et al., 2008; Fracchia et al., 2012; Kaczorek et al., 2013).

Typical desirable properties include solubility enhancement, surface tension reduction, and low critical micelle concentrations (CMC).The surface tension is a measure of the surface free energy per unit area required to move a molecule from the bulk phase to the surface (Rosen, 1978).

The surface tension correlates with the concentration of the surfactant until the CMC is reached. The CMC is the minimal concentration of surfactant necessary to initiate micelle formation or the maximal concentration of surfactant monomers in water. Efficient surfactants have low CMC.

Surfactants currently used in industry, agriculture, and medicine are mainly products of chemical synthesis. Their application for bioremediation of industrial pollutants, crude oil recovery, and the food processing industry is limited by their toxicity and low biodegradability.

In recent years, significant interest has been focused on surfactants, synthesized by microorganisms (biosurfactants). It is due to their environmental compatibility, as well as to the possibility of their production from cheap and waste materials, thus producing additional environmental impact. Biosurfactants are a structurally diverse group of amphipathic surface-active molecules synthesized extracellularly or as part of cell membranes by bacteria, yeast, and fungi.

The complex chemical structure and physical properties of biosurfactants result in properties that are equal to, or that exceed, many synthetic surfactants. Biosurfactants (BSAC) have many advantages compared to their chemically synthesized counterparts. The main features of biosurfactants and a brief description of each property are as follows:

- Highly effective, environmentally friendly, nontoxic, nonallergic, biodegradable, inexpensive products
- Universal emulsification (solubilization) of various substances, activity at a wide range of pHs and temperatures
- Regulation of transfer of hydrophobic substances into living cells
- Modification of various surfaces, regulation of sorption–desorption processes
- Safe, inexpensive, renewed, natural feedstock (e.g., industrial wastes) for their production
- Synergetic activity when used together with synthetic surfactants

However, further investigation is required to assess the potential of these compounds in many applications. Research on biosurfactant application focuses mainly on their use in environmental protection, and despite their numerous advantages only a few studies have been done on their potential application in medicine.

2.2 Physiological roles of biosurfactants

The main physiological role of biosurfactants is to permit microorganisms to grow on water-immiscible substrates by reducing the surface tension at the phase boundary, therefore making the substrate more readily available for uptake and metabolism.

However, the molecular mechanisms of the uptake process of these substrates (e.g., alkanes) are still unclear. In addition to emulsification of the carbon source, they are also involved in the adhesion of microbial cells to the hydrocarbon. The cellular adsorption of the hydrocarbon-degrading microbes to water-immiscible substrates and the excretion of surface-active compounds together allow growth on such carbon sources. In general, negatively charged biosurfactants inhibit, whereas positively charged biosurfactants promote microbial adhesion to hydrophobic phases. Furthermore, biosurfactants have been shown to have antagonistic effects on other microbes in the environment.

2.3 Classification of biosurfactants

Unlike chemically synthesized surfactants, which are usually classified according to the nature of their polar grouping, biosurfactants are generally categorized by their chemical composition and microbial origin.

Rosenberg and Ron (1999) suggested that biosurfactants can be divided into low-molecular-mass molecules, which efficiently lower surface and interfacial tension, and high-molecular-mass polymers, which are more effective as emulsion-stabilizing agents. The major classes of low-mass surfactants include glycolipids, lipopeptides and phospholipids, whereas high-mass surfactants include polymeric and particulate surfactants (Table 3.1). Most biosurfactants are either anionic or neutral, and the hydrophobic moiety is based on long-chain fatty acids or fatty acid derivatives, whereas the hydrophilic portion can be a carbohydrate, amino acid, phosphate, or cyclic peptide (Table 3.1). A brief discussion of each class of biosurfactant follows.

Glycolipids: Most known biosurfactants are glycolipids. They are carbohydrates in combination with long-chain aliphatic acids or hydroxyaliphatic acids. The linkage is realized by means of either ether or an ester group. Among the glycolipids, the best known are rhamnolipids, trehalose lipids, and sophorolipids. Synthesis of surface-active substances by three actinobacterial strains, *Rhodococcus erythropolis, Gordonia rubripertincta,* and *Acinetobacter calcoaceticus,* was studied by Karpenko

Table 3.1 Major Biosurfactant Classes and Microbial Producers

Biosurfactant Class	*Microorganism/Genus*
Glycolipids	
Rhamnolipids	*Pseudomonas aeruginosa*
Trehalose lipids	*Rhodococcus erithropolis, Arthobacter* sp.
Sophorolipids	
Mannosylerythritol lipids	*Candida bombicola, C. apicola, C. antarctica*
Lipopeptides	
Surfactin/iturin/fengycin	*Bacillus subtilis*
Viscosin	*Pseudomonas fluorescens*
Lichenysin	*Bacillus licheniformis*
Serrawettin	*Serratia marcescens*
Phospholipids	*Acinetobacter* sp., *Corynebacterium lepus*
Fatty acids/neutral lipids	
Corynomicolic acids	*Corynebacterium insidibasseosum*
Polymeric surfactants	
Emulsan	*Acinetobacter calcoaceticus, Gordonia* sp.
Alasan	*Acinetobacter radioresistens*
Liposan	*Candida lipolytica*
Lipomanan	*Candida tropicalis*

Source: Karanth et al., 1999; Rosenberg and Ron, 1999; Franzetti et al., 2008.

et al. (2006; Prystay et al., 2006; Pirog et al., 2009). Through the method of thin-layer chromatography (marker analysis) they determined that the fractions, which were identified as trehalosemycolate and trehalosedimycolate, have prevailed in the content of nonpolar lipids as well as of polar lipids. Mycolic acids, cetyl alcohol, and ethers of fatty acids were identified among the nonpolar lipids.

Rhamnolipids: These glycolipids, in which one or two molecules of rhamnose are linked to one or two molecules of β-hydroxydecanoic acid, are the most extensively researched biosurfactants. While the OH group of one of the acids is involved in glycosidic linkage with the reducing end of the rhamnose disaccharide, the OH group of the second acid is involved in ester formation. L-rhamnosyl-β-hydroxydecanoyl-β-hydrocydecanoate (Figure 3.1) and L-Rhamnosyl-L-rhamnosyl-β-hydroxydecanoyl-β-hydroxydecanoate, referred to as rhamnolipids 1 and 2, respectively, are the principal glycolipids produced by *P. aeruginosa* (Edward et al., 1965). Up to seven homologues have now been identified (Abalos et al., 2001). Because rhamnolipids are widely applicable, they will be discussed in more detail later in this chapter.

Trehalose lipids: Several structural types of microbial trehalose lipid biosurfactants have been reported. Disaccharide trehalose linked at C-6 and C-6′ to mycolic acid is associated with most species of *Mycobacterium, Nocardia,* and *Corynebacterium*. Mycolic acids are long-chain, α-branched-β-hydroxy fatty acids. Trehalose lipids from different organisms differ in the size and structure of

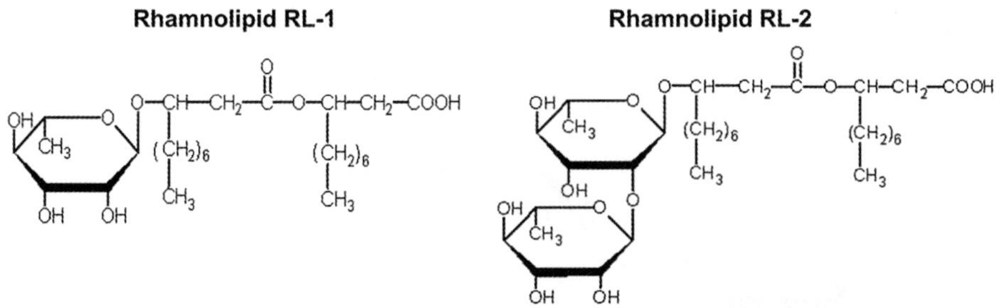

FIGURE 3.1

Extracellular surface-active substances from the bacterial strain *Pseudomonas* sp. PS-17.

mycolic acid, the number of carbon atoms, and the degree of unsaturation (Asselineau et al, 1978) (Figure 3.2). Trehalose lipids from *Rhodococcus erythropolis* and *Arthrobacter* sp. lowered the surface and interfacial tension in culture broth from 25 to 40 and 1 to 5 mN/m, respectively (Shulga et al., 1990; Prystay et al., 2006; Pirog et al., 2004).

Sophorolipids: These glycolipids are produced mainly by yeast such as *Torulopsis bombicola, T. petrophilum, T. apicola, and C. lipolytica;* they consist of a dimeric carbohydrate sophorose linked to a long-chain hydroxyl fatty acid by glycosidic linkage (Cavalero and Cooper, 2003; Cooper and Paddock, 1984). The sophorolipids reduce surface tensions between individual molecules at the surface, although they are effective emulsifying agents (Bognolo, 1999). These biosurfactants are a mixture of at least six to nine different hydrophobic sophorolipids.

Lipopeptides and lipoproteins A large number of cyclic lipopetides, including decapeptide antibiotics (gramicidins), and lipopeptide antibiotics (polymyxins), are produced. These consist of a lipid attached to a polypeptide chain.

Surfactin: The cyclic lipopeptide surfactin produced by *Bacillus subtilis* (Whang et al., 2008, 2009; Shulga et al., 1990) is one of the most powerful biosurfactants. It is composed of a seven amino-acid ring structure coupled to a fatty-acid chain via lactone linkage. It can lower the surface tension from 72 to 27.9 mN/m at concentrations as low as 0.005% (Sen, 1997).

Lichenysin: Bacillus licheniformis produces several biosurfactants that act synergistically and exhibit excellent temperature, pH, and salt stability. These are also similar in structural and physicochemical properties to the surfactin (Grangemard, I., 2001). The surfactants produced by *B.licheniformis* are capable of lowering the surface tension of water to 27 mN/m and the interfacial tension between water and *n*-hexadecane to 0.36 mN/m (Joshi et al., 2007).

Fatty acids, phospholipids, and neutral lipids Several bacteria and yeast produce large quantities of fatty acids and phospholipid surfactants during growth on *n*-alkanes (Cirigliano and Carman, 1985). The hydrophilic and lipophilic balance (HLB) is directly related to the length of the hydrocarbon chain in their structures. In *Acinetobacter* sp. strain HO1-N, phosphatidylethanolamine-rich vesicles are produced (Kappeli et al., 1979), which form optically clear microemulsions of alkanes in water. Phosphatidylethanolamine produced by *R. erythropolis* grown on *n*-alkane causes a lowering of interfacial tension between water and hexadecane to less than 1 mN/m and a critical micelle concentration (CMC) of 30 mg/l (Kretschmer et al., 1982).

Polymeric biosurfactants The best-studied polymeric biosurfactants are emulsan, liposan, alasan, lipomanan, and other polysaccharide–protein complexes (Rosenberg and Ron, 1999). Emulsan

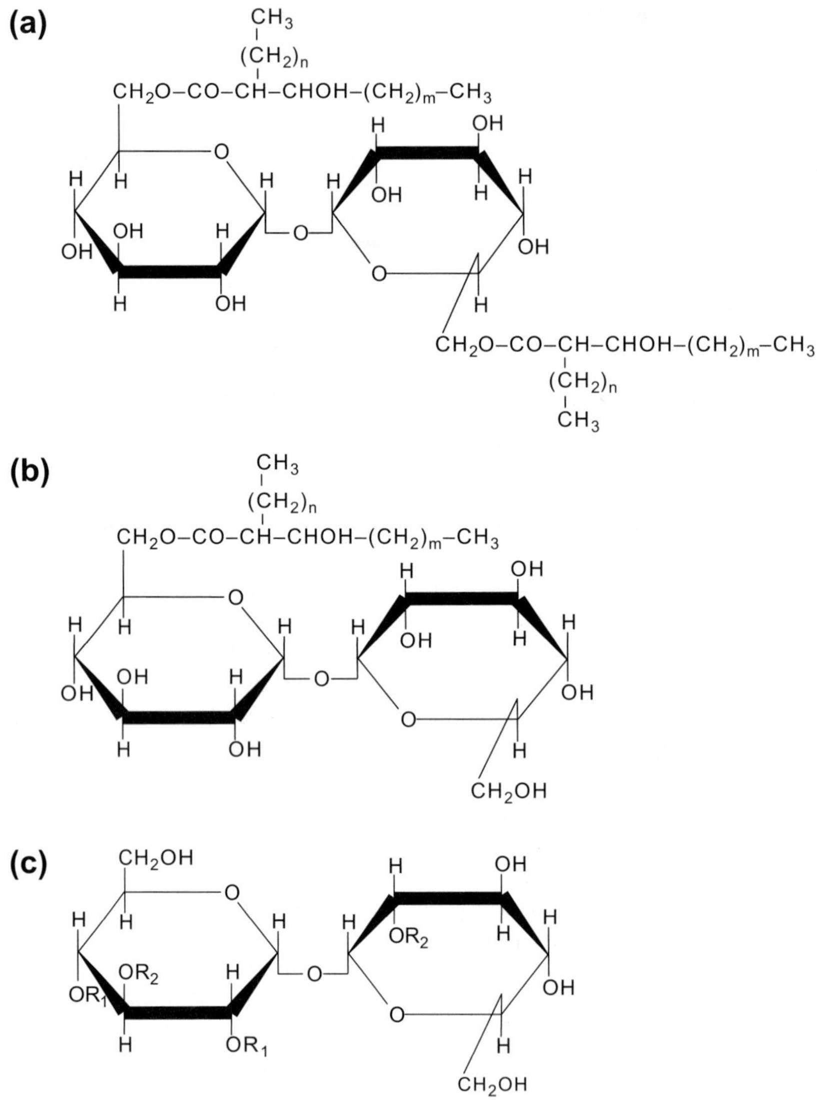

FIGURE 3.2

Structural formulas of trehalosolipids (a) trehaloso-6,6-dimicolate; (b) trehaloso-6-monomicolate (c) trehaloso-2, 3, 4, 2-tetraether (R_1=$OC(CH_2)_6CH_3$; R_2=$OC(CH_2)_8CH_3$) *Rhodococcus erythropolis DSM* 43215.

is an effective emulsifying agent for hydrocarbons in water (Zosim et al., 1982; Amiriyan et al., 2004; Su, 2009), even at a concentration as low as 0.001 to 0.01%. Liposan is an extracellular water-soluble emulsifier synthesized by *Candida lipolytica* and is composed of 83% carbohydrate and 17% protein (Cirigliano and Carman, 1985).

2.4 Mode of action (properties of biosurfactants)

Biosurfactants are of increasing interest for commercial use because of the continually growing spectrum of available substances. Biosurfactants have many advantages over their chemically synthesized counterpart. In a recent review Kitamoto et al. (2009) underlined the following properties of biosurfactants when compared with chemical surfactants: (i) One or more functional groups and chiral centers, (ii) bulky but sophisticated structures, (iii) higher biodegradability and lower toxicity, (iv) lower critical micelle concentration and higher surface activity, (v) gradual adsorption and continuing activity, (vi) superior ability to form molecular assembly and liquid crystal, and (vii) versatile biological activity (antimicrobial and antitumor actions, etc.).

The main distinctive features of biosurfactants and a brief description of each property are given next.

Surface and interface activity A good surfactant can lower the surface tension of water from 72 to 35 mN/m and the interfacial tension of water/hexadecane from 40 to 1 mN/m (Mulligan, 2005). Surfactin from *B. subtilis* can reduce the surface tension of water to 25 mN/m and interfacial tension of water/hexadecane to < 1 mN/m (Cooper et al., 1981). Rhamnolipids from *P. aeruginosa* decrease the surface tension of water to 26 mN/m and the interfacial tension of water/hexadecane to < 1 mN/m (Finnerty, 1994). The sophorolipids from *T. bombicola* have been reported to reduce the surface tension to 33 mN/m and the interfacial tension to 5 mN/m (Cavalero and Cooper, 2003) In general, biosurfactants are more effective and efficient and their CMC is lower than that of chemical surfactants; that is, less surfactant is necessary to get a maximum decrease in surface tension (Medrzycka et al., 2005; Grabovska et al., 2008).

Temperature, pH and ionic strength tolerance Many biosurfactants and their surface activities are not affected by environmental conditions such as temperature and pH. McInerney et al. (1990) reported that lichenysin from *B. licheniformis* JF-2 was not affected by temperature (up to 50°C), pH (4.5–9.0) and by NaCl and Ca concentrations up to 50 and 25 g/l, respectively. A lipopeptide from *B. subtilis* LB5a was stable after autoclaving (121°C/20 min) and after 6 months at –18°C; the surface activity did not change from pH 5 to 11 and NaCl concentrations up to 20% (Nitschke et al., 2006).

Emulsion forming and emulsion breaking Biosurfactants may stabilize (emulsifiers) or destabilize (de-emulsifiers) the emulsion. High-molecular-mass biosurfactants are in general better emulsifiers than low-molecular-mass biosurfactants. Sophorolipids from *T. bombicola* have been shown to reduce surface and interfacial tension but are not good emulsifiers (Cavalero and Cooper, 2003). By contrast, liposan does not reduce surface tension but has been used successfully to emulsify oils (Cirigliano and Carman, 1985).

The chief advantages of biosurfactants can be summarized as follows:

- Highly effective, environmentally friendly, nontoxic, nonallergic, biodegradable, inexpensive products
- Universal emulsification (solubilization) of various substances, activity at the wide range of pH and temperature
- Regulation of transfer of hydrophobic substances into microbial cells
- Modification of various surfaces, regulation of sorption-desorption processes
- Safe, inexpensive, renewed, natural feedstock (e.g., industrial wastes) for their production
- Synergetic activity when used together with synthetic surfactants

The likelihood that products derived from renewable resources will be successful increases if they can be shown to have a quality and price comparable to that of their synthetic counterparts. Several features make biosurfactants commercially promising: their effectiveness at high salinity and within a wide pH range and the fact that they offer new possibilities for industrial applications. Their most important advantage is probably their ecological applicability; they are biodegradable, and they are produced by a variety of microorganisms that occur naturally in the environment.

2.5 Rhamnolipids as most applicable in the bioremediation technologies group of biosurfactants

As mentioned earlier, rhamnolipid biosurfactants are the most intensively studied as they are the most applicable biosurfactants in bioremediation technologies (Rosenberg and Ron, 1999; Cameotra and Makkar, 2004). Rhamnolipids from *Pseudomonas aeruginosa* strains, Gram-negative bacteria, are between biosurfactants that have been the focus of many researchers (Beal and Betts, 2000; Lang and Wullbtandt, 1999; Mulligan, 2005; Sotirova et al., 2008). These strains are capable of utilizing hydrocarbons and producing rhamnolipid-type biosurfactants across aa wide range of temperatures (30–45 C) and salinity (0–6%) (Kumar et al., 2008).

Although there are many types of rhamnolipids, all of them possess similar chemical structures. Generally, rhamnolipids contain a hydrophilic head formed by one or two rhamnose molecules and a hydrophobic tail composed of one or two fatty acid chains, as mentioned above. L-rhamnosyl-β-hydroxydecanoyl-β-hydrocydecanoate and L-Rhamnosyl-L-rhamnosyl-β-hydroxydecanoyl-β-hydrocydecanoate, referred to as rhamnolipids 1 and 2, respectively (Figure 3.1), are the principal glycolipids produced by *P. aeruginosa* (Edward and Hayashi, 1965).

Most of the biosurfactants produced by *P. aeruginosa* strains, however, have been reported to be dirhamnolipid (Sebba, 1987); only a few reports have shown monorhamnolipid to be the predominant component (Sim et al., 1997). The difference in the type and proportion of rhamnolipid compounds in the excreted biosurfactant has been proposed to be governed by the age of the culture, the selected bacterial strain (Déziel et al., 1999), substrate composition, and specific culture conditions (Costa et al., 2006).

From the perspective of surfactant properties, rhamnolipids are one of the most interesting biosurfactants (Cameotra and Makkar, 2004). Rhamnolipids can reduce the surface tension of pure water from 72 to below 30 mN/m with a critical micelle concentration (CMC) in the range of 5–200 mg/l, depending on the rhamnolipid components in the excreted biosurfactant (Finnerty, 1994). Rhamnolipids can maintain their surface activities even under extreme conditions of temperature and pH (Pornsunthorntawee 2008). In addition, rhamnolipids also possess distinguishing biological activities, including antiproliferative activity against a human breast cancer cell line (MCF-7) (Thanomsub, 2006) and antimicrobial activity against both bacteria and phytopathogenic fungi species (Benincasa, 2004). Like other synthetic surfactants, biosurfactants form a variety of microstructures such as spherical, globular, and cylindrical micelles; spherical and irregular vesicles; tubular and irregular bilayers; lamellar sheets; and lyotropic liquid crystalline phases with lamellar, hexagonal, and cubic symmetries (Vinson, et al., 1989). The morphology of these surfactant aggregates is affected by surfactant concentration (Sánchez, et al., 2007), pH (Champion, et al., 1995), temperature (Bakshi et al., 2006), ionic strength (Mata, et al., 2004), and co-solutes or contaminants such as alcohols and metals (Champion et al., 1995).

In the case of rhamnolipids, it has been demonstrated that the aggregate structures of a monorhamnolipid can be affected by pH, cadmium, and octadecane (Champion et al., 1995), while those of a dirhamnolipid are affected by biosurfactant concentration (Sánchez et al., 2007). One of the most important functions of these molecular aggregates involves the solubilization capacity. The mechanism and technique of solubilization (Saeki et al., 2000) are used in many products, such as cleaning products, pharmaceuticals, cosmetics, and food. However, the solubilization capacity of rhamnolipid aggregates has rarely been studied. Further knowledge about the aggregation behavior of rhamnolipid biosurfactants could perhaps be used to introduce these surface-active molecules into high-value-added applications in the near future, enlarging their potential applications (Pinzon et al., 2013). Like chemical surfactants, these compounds can be used in many processes involving emulsification, foaming, detergency, wetting, and dispersing or solubilizing.

It has been found that the strain *Pseudomonas* sp. PS-17 is the effective producer of extracellular rhamnolipids and biopolymer (Karpenko et al., 2004). Rhamnolipid PS-17 represents a unique natural biocomplex. It consists of two homologous extracellular biosurfactants glycolipids RL-1 and RL-2 (Figure 3.1). It has been found that they are rhamnolipids and are capable of causing a considerable decrease in the surface and interfacial tensions to values of 27.5–29.8 and 0.04–0.07 mN/m, respectively. The parameters for their surface and interfacial tensions, as well as their critical concentrations for micelle formation CMC (20–80 mg/l) and emulsifying index, indicate their high surface activity (Table 3.2). The extracellular biopolymer is also synthesized by the strain *Pseudomonas* sp. *PS-17*. It is a polysaccharide with alginate nature and molecular weight in the range 3-4 x 10^5 (Figure 3.3). The biopolymer forms a surface-active complex with rhamnolipids.

The biosurfactant/biopolymer complex has specific functional properties, including effect on microbial cells and washing activity, in comparison with rhamnolipids (Table 3.2).

As Table 3.2 shows, surfactants and complex can cause a considerable decrease of surface and interfacial tensions to values of 29.0 and 0.1–0.2 mN/m, respectively, and form stable, highly dispersed emulsions. The critical micelle dilutions were determined based on the adsorption isotherms at temperature 25°C and pH 7 and to amounted 131.8 mg/l for RL1 and RL2. The surface tension of the surfactants were 26 mN/m and 28 mN/m, respectively. The speed of effective reduction of surface tension is characterized by the pC_{20} parameter, which was 4.67 for RL1 and 10.05 for RL2.

The hydrophilic-lipophilic balances of groups, which belong to structural formulas of surfactants, were determined on the basis of table data on the indexes of hydrophilicity and lipophilicity. The hydrophilic/lipophilic balance (HLB) of monorhamnolipid equaled 13 and dirhamnolipid – 21, which indicates their ability to form stable emulsions of the "oil in water" kind; monorhamnolipid can be used as an effective detergent and washing agent. The surface activity was performed by two main components: less polar monorhamnolipid RL1 and more polar dirhamnolipid RL2 (Medrzycka et al., 2005; Grabovska et al., 2008).

Table 3.2 Properties of Rhamnolipids and Biocomplex from the Strain *Pseudomonas* sp. PS-17

Substance	Surface Tension, mN/m	CMC, mg/l	Index of Emulsifying, E_{24}, %
Complex PS	29.7	85	65
Lipid extract	29.8	80	50
Rhamnolipids	28.0	20–50	55

Source: Karpenko et al., 2007.

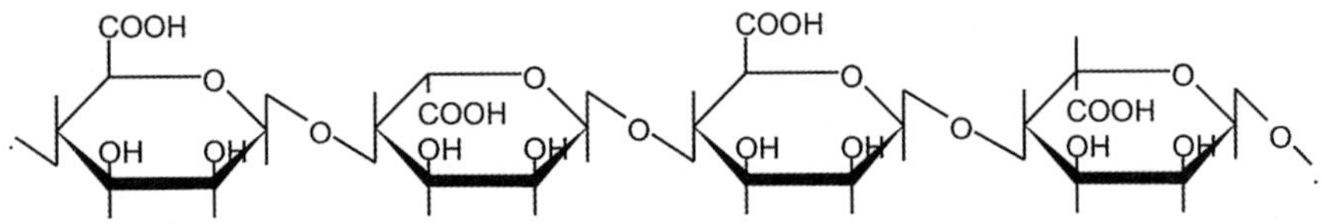

FIGURE 3.3

Extracellular biopolymer from the bacterial strain *Pseudomonas* sp. PS-17.

The foam fractionation technique in batch mode has been applied successfully to recover and concentrate rhamnolipid biosurfactant from a free-cell culture medium of *Pseudomonas aeruginosa* SP4 (Sarachat et al., 2010). The results suggest that the foam fractionation method, which is an economical and environmentally friendly technique, can potentially be used for a primary purification and concentration of the biosurfactant products. This technique allows the recycling of microbial cells and thus may enhance the commercial exploitation of rhamnolipids by reducing the cost of biosynthesis and downstream processing.

Generally, rhamnolipids have many potential industrial and environmental applications, including the production of fine chemicals, the characterization of surfaces and surface coatings, as additives for environmental remediation, and as a biological control agent (Lang and Wullbtandt, 1999; Maier and Sobéron-Chávez, 2000; Ron and Rosenberg, 2001, 2002; Nitschke et al., 2005). Realization of this wide variety of applications requires economical commercial-scale production of rhamnolipids.

3 Bioremediation/Biodegradation of polluted compounds

Pollutants and waste products are widespread in nature. The techniques applied for elimination of pollutants cannot be too costly and at the same time must not harm the environment. For this purpose, the bioremediation of organic compounds by microorganisms represents one of the primary mechanisms by which pollutants are eliminated from the environment. In the simplest terms, bioremediation is the use of microorganisms (yeast, fungi, or bacteria) to decompose toxic pollutants, transforming them into less harmful compounds. To fully understand bioremediation, we have to discuss the term *biodegradation*. Biodegradation is a natural process by which microbes alter and break down pollutants into other substances. The resulting products can be carbon dioxide, water, and simpler compounds that do not affect the environment. Bioremediation is the optimization of biodegradation. This acceleration can be accomplished by two forms of technology: (1) fertilizing (adding nutrients) and/or (2) seeding (adding microbes). These additions are necessary to overcome certain environmental factors that may limit or prevent biodegradation. Knowledge on bioremediation allows us to manipulate environmental factors to enhance natural biodegradation. This knowledge includes:

- Identification of microbes capable of degrading petroleum hydrocarbons.
- Nutrient requirements of these microbes, such as carbon, nitrogen and phosphorus.
- Environmental requirements such as oxygen, water, and temperature.
- Metabolic pathways of decomposition for hydrocarbon fractions. These fractions include the aromatics aliphatics and asphaltenes.

Bioremediation has potential to help clean up marine and land oil spills, pesticides, and hazardous waste. It can also save money, being ecologically sound, destroying contaminates (rather than just moving them form one place to another), and treating waste on site.

By using the science of microbiology and chemistry, scientists have been able to recognize the processes and factors that allow biodegradation to take place. This same knowledge allows efforts to increase these natural processes (bioremediation) and to achieve results faster than can occur naturally. Unfortunately, no formula has been devised that will allow bioremediation to work every single time. Each situation is unique and requires a separate study to determine its particular needs.

Microbial surfactants find potential applications within physical-chemical technologies for remediation of both organic and metal contaminations, such as in-situ soil flushing and ex-situ soil washing for remediation of the unsaturated zone and "pump and treat" for aquifer remediation. They also improve the biodegradation rate of organic compounds in bioremediation technologies (Kitamoto et al., 2009; Christofi and Ivshina, 2002; Cameotra and Bollag, 2003). Biosurfactants have attracted considerable interest among researchers working on environment remediation techniques because they have some distinct advantages over their synthetic counterparts, such as lower toxicity, higher biodegradability, and high selectivity. The low-molecular-weight microbial surface-active compounds or biosurfactants lower the surface tension at air–water interfaces and the interfacial tension at oil–water interfaces, whereas the high-molecular-weight compounds, called bioemulsifiers, are more effective in stabilizing oil-in-water emulsions (Banat et al., 2010). In the literature, results of surfactant addition in bioremediation experiments are very contrasting. Rhamnolipids are known to be effective as enhancers in the biodegradation of both aliphatic and aromatic hydrocarbons (Zeng et al., 2011; Sponza and Gok, 2012; Chrzanowski et al., 2011).

Several studies, however, have demonstrated that the addition of rhamnolipids did not result in enhancing biodegradation for most contaminants (Noordman and Janssen, 2002; Mata-Sandoval et al., 2001; Shin et al.,2005). The waste treatment in situ or ex situ, with application of chemical or microbial surfactants, is a dual-action process used on different kinds of contaminants and on microbes—producers of surfactants and degraders of pollutants. Many reviews and research articles show the effect of surfactants on pollutants and examine recent advances in the environmental application of biosurfactants (Chrzanowski et al., 2011; Mulligan, 2009; Lu et al., 2011; Hua and Wang, 2012; Shcheglova et al., 2007). Thus, in the next sections we will focus on aspects that have not been discussed widely as well as very new applications concerning the relationship between surfactants and microbial strains and consortia participating in bioremediation and degradation of pollutants. Some very recent findings on the use of biosurfactants in the biodegradation of aliphatic and polycyclic aromatic hydrocarbons when removing heavy metals will also be presented.

3.1 Biosurfactants, microbial strains, and enzyme activities

Bioremediation implies the use of microorganisms that can be developed in the presence of pollutants and that either metabolize or accumulate organic compounds. In both cases, the pollutants have to penetrate the microbial cell envelope, that is, the cell wall and the membrane; this transport is a major step in the bioremediation process. Microorganisms must also be able to excrete several enzymes that can transform or degrade the pollutants. These enzymes often are hydrolases, oxidases, esterases, and so on. Vasileva-Tonkova and Galabova (2003) characterized the bacterial isolates from lubricant-contaminated waters capable to produced glycolipids and enzymes. All members of the selected microbial consortium were grown on hexadecane as a sole carbon source. These researchers found a correlation between the secretion of surface-active compounds and bacterial growth on hydrocarbons: Isolates with the best growth showed higher levels of glycolipids (Vasileva-Tonkova et al., 2008). The results of this study revealed the ability of isolates to produce a wide range of hydrolytic enzymes. Microorganisms capable of secreting extracellular enzymes and degrading hydrocarbons appear to be infrequent in the environment (Tzintzun-Camacho et al., 2012, Covino et al., 2010). The bacterial strains studied in this work were isolated from a relatively closed ecosystem, the settler of "mineral oil fraction" in the Electric Power Station. The results showed a well-balanced microbial consortium in which different participants have their own roles and support each other. Some of the releasing biosurfactants relieve the assimilation of hydrocarbons from the other isolates, which alone cannot grow on hydrocarbons.

On the other hand, the group of bacteria possessing enzyme systems for degradation of basic organic polymers provides degrading products for the other members. Bioaugmentation of polluted sites only with producers of surface-active compounds or only with producers of hydrolytic enzymes may cause a disturbance in relationships within microbial communities and delay the remediation process. As contaminated sites (wastewaters) usually contain a variety of hydrocarbons and organic polymers, in the case of bioaugmentation of these sites a mixed culture of at least three indigenous bacteria groups should be done: (1) isolates with the best growth on hydrocarbons, (2) biosurfactant producers, and (3), the most active producers of hydrolytic enzymes. This consortium model may find application for treatment of wastewaters contaminated with different hydrocarbons and organic polymers (Tzintzun-Camacho et al., 2012).

Because of their potential for bioremediation of oil spills, microorganisms capable of growth on hydrocarbons have been extensively studied. Petroleum is an extremely complex mixture of homologous series, including alkanes, isoalkanes, alkyl aromatic, and heteroatom compounds. Lower-molecular-weight *n*-alkanes are normally biodegraded more rapidly than multiply branched and long-chain alkanes (Whyte et al., 1998). The rapid depletion of *n*-alkanes from petroleum mixtures by microorganisms may be driven by two factors: the activity of metabolic enzymes for oxidation of *n*-alkanes relative to other compounds and the transport of alkanes into the cells. Karpenko et al. (2006, 2008) studied the effect of surface-active complex PS, consisting of biosurfactant rhamnolipid and biopolymer polysaccharides on microbial enzyme activities (hydrolases and oxidases). Enzymes that play an important role in the degradation of petroleum hydrocarbons—urease, catalase, and dehydrogenase—were selected for the experiments. In most cases biosurfactant induces the activation of enzymes. A stimulating effect of biosurfactants on the dehydrogenase activity of microorganisms-destructors was established as well.

3.2 Role of biosurfactants in the bioremediation of polluted waters and soil

3.2.1 Decay of benzo(a)pyrene in the presence of rhamnolipid biosurfactants

Rhamnolipid microbial surfactants were studied for their potential in enhancing the bioavailability and biodegradation of benzo(a)pyrene (Malachovska-Yutsz et al., 2007). Polycyclic aromatic hydrocarbons (PAHs), which consist of more than four rings, belong to hardly degradable compounds. The rhamnolipid biosurfactant used in this study was extracted from culture supernatants of *Pseudomonas* sp. PS-17. Biodegradation experiments conducted with an initial benzo(a)pyrene concentration of 100 mg/liter showed that 82% of the benzo(a)pyrene was mineralized in 60 days in the presence of 500 mg of rhamnolipid per liter, compared with only 64% benzo(a)pyrene mineralization when no surfactant was present (Figure. 3.3). These results indicate that rhamnolipids may facilitate the bioremediation of sites contaminated with hydrocarbons that have limited water solubility. The soil particles can affect the local activity of microorganisms. The fractions of soil that have the greatest impact on the biological activity of microorganisms and the excretion of enzymes are colloids, silt minerals, and humic substances (Johnsen et al., 2005). Relatively long benzo(a)pyrene decay time probably is connected with the bioavailability of soil fractions formed by substances that are dissolved in the aqueous phase, as well as those that are weakly bound to the solid phase of soil and can quickly diffuse and/or be exposed to desorption into the soil extract. Concentrations of bioavailable fractions are not constant but may change over time, as evidenced by various biotic and abiotic processes subjected to pollution in the soil environment. Another reason for slow biodegradation could be a reduction in activity of extracellular enzymes (through their decomposition, adsorption, or denaturation). Nevertheless, the positive effect of rhamnolipid biosurfactants is obvious: The addition of rhamnolipids to samples contributed an 18% increase in the degree of hydrocarbons decomposition compared with the control.

3.2.2 *Influence of rhamnolipids from* Pseudomonas *sp. PS-17 on coal tar and petroleum residue biodegradation*

Rhamnolipid surfactants of the bacteria *Pseudomonas* sp. PS-17 stimulate biodegradation of coal tar waste obtained from former gas work and petroleum residues from atmospheric distillation of light oil. The coal tars and heavy petroleum fractions are highly hydrophobic and toxic pollutants. They are multicomponent mixtures of paraffinic, cycloparaffinic, mono- and polycyclic aromatic hydrocarbons, and heteroorganic compounds with a chemical nature. Liquid products of coal contain a number of PAH, phenols, and heterocyclic compounds. These substances have been shown to have a low susceptibility to biodegradation. Sliwka et al. (2009) found that addition of rhamnolipids from *Pseudomonas* sp. PS-17 accelerated the biodegradation of coal tar components. The highest efficiency of the biodegradation process was observed at a rhamnolipids concentration of 125 mg in 1 dm^3 of the cultivation medium. For the petroleum residue, the maximum efficiency was 55% at a rhamnolipid concentration of 250 mg/dm^3.

The coal tar components were not susceptible to biological decay. No signs of biodestruction of tar by using bacteria were observed unless a surfactant was added. The introduction of the rhamnolipid initiated the biodegradation process. Rhamnolipids (as high as 125 mg/dm^3) stimulated the biodegradation of coal tar. The loss of the gas-work tar in this case reached 27.7% wt. after 14 days of incubation.

The biodegradation efficiency was twice as high (34.4%) as the control when minimal concentration of rhamnolipid, 15.6 mg/dm^3, was used, and it reached its peak (54.7%) in variants with an addition of 250 mg/dm^3 of rhamnolipid.

3.2.3 *Effect of biosurfactants and synthetic surfactants on the biodegragation of petroleum hydrocarbons*

The comparative effects of microbial and synthetic surfactants on the biodegradation of petroleum hydrocarbons has been studied by Kolwzan et al. (2003). The research illustrates that rhamnolipid biosurfactants from the *Pseudomonas* sp. PS-17 strain culture affects the degree of biodegradation of middle-diesel oil (Figure 3.4). Diesel oil has been obtained as a result of raw-atmospheric distillation

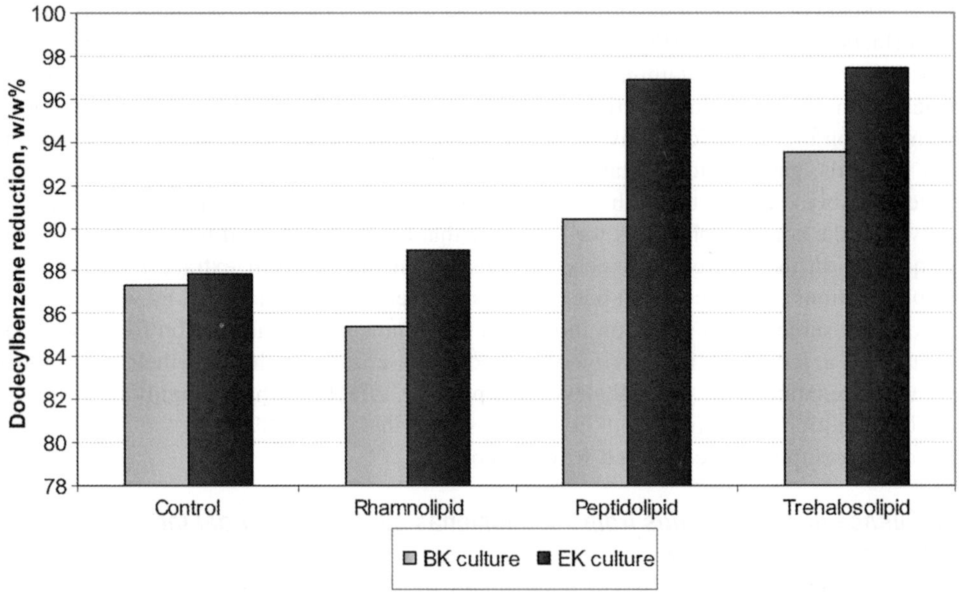

FIGURE 3.4

Reduction of dodecylbenzene concentration in the process of biodegradation influenced by biosurfactants. BK, EK - bacterial associations (Karpenko et al., 2008).

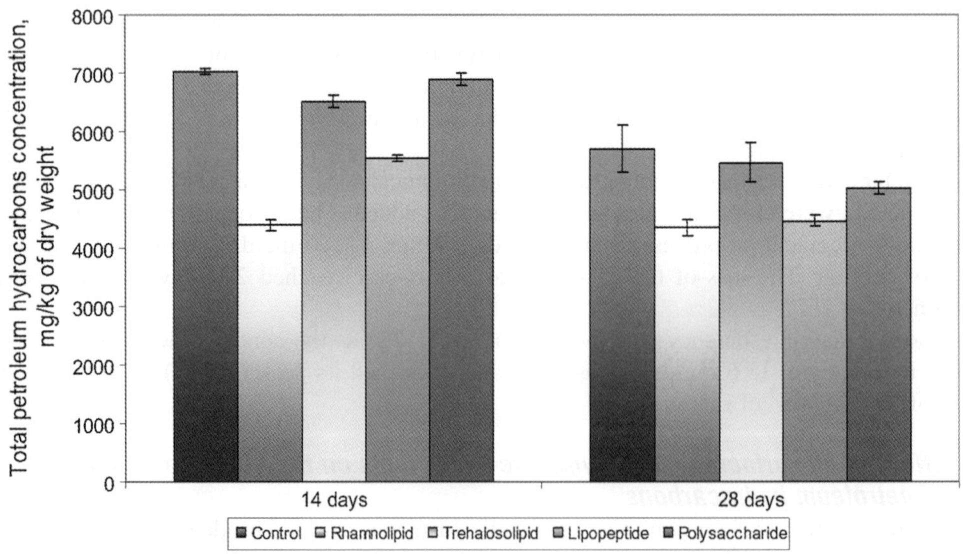

FIGURE 3.5

The content of the residual hydrocarbons in soil after 14 and 28 days of biodegradation (Karpenko et al., 2008).

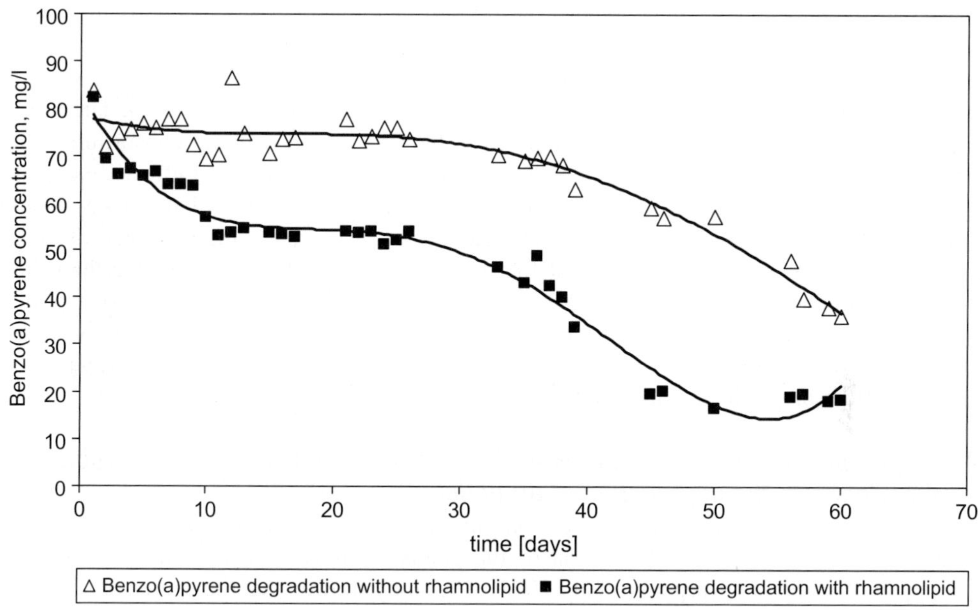

FIGURE 3.6

Kinetics of benzo(a)pyrene degradation (Malachovska-Yutsz et al., 2007).

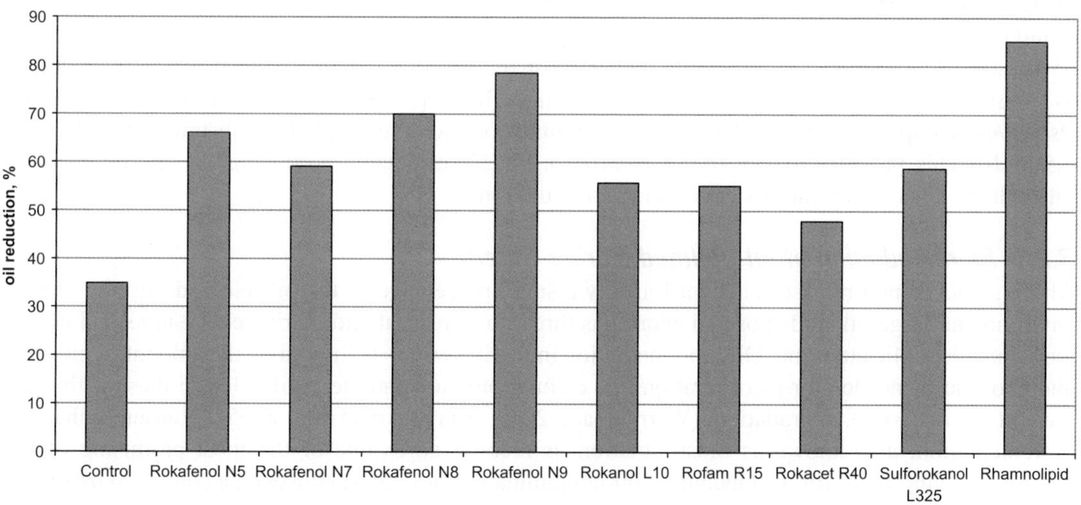

FIGURE 3.7

Influence of rhamnolipid and synthetic surfactants (rokafenol, rokanol, rofam, rokacet, sulforokanol) upon the process of middle-diesel oil biodegradalion (Malachovska-Yutsz A. et al., 2007).

of Russian petroleum from the URAL deposit (with a boiling range of 295–310°C). By using GC-MS analysis, it has been proven that the oil contains hydrocarbons C_9-C_{29} (paraffins, naphthenes, and aromatics). n-Paraffins C_{17}-C_{25} and isoparaffins were the predominant components: pristane (2, 6, 10, 14- tetramethylpentadecane) and phytane (2, 6, 10, 14-tetramethylheptadecane). The biodegradation process of middle-diesel oil was conducted utilizing a mixed-bacteria consortium. The hydrocarbons were characterized by a low availability to biodegradation. The effectiveness of fuel biodegradation in the presence of biosurfactant of the *Pseudomonas* sp. PS-17 strain was as high as 85% (after 21 days of incubation) for the concentrations of surfactants ranging from 0.3 to 0.7 g/l (Figure 3.4). Synthetic surfactants on diesel fuel biodegradation were less effective than the biosurfactant. Introduction of Rokaphenol N9 into the substrate increased the efficiency of fuel breakdown from 36 to 79%–yielding the best results (Figure 3.4). Thus, the stimulative effect of the biosurfactant of the *Pseudomonas* sp. PS-17 strain in treating oil-contaminated soils significantly exceeded the effect of synthetic surfactants.

3.2.4 The prospects of using the bacteria of the genus *Rodococcus* and microbial surfactants for the degradation of oil pollutants

The possibility of accelerating oil degradation through an enrichment culture of oil-oxidizing microorganisms in the presence of bacteria of the genus *Rhodococcus* and microbial surfactants was the subject of a study by Karpenko et al. (2006). The study showed that the degree of consumption of crude oil (2% vol.) after 192 h of enrichment culture growth reached 84%. Inoculation of the active hydrocarbon-oxidizing strain *Rhodococcus erythropolis* EK-1 and exogenous surfactants produced by *Pseudomonas* sp. PS-27 increased this degree to 70–90%.

Emulsification of organic substances (in particular, oil and petroleum products) with surfactants is crucial for a fast biotechnological purification of water and soil. Studies of the emulsifying properties of the culture liquid obtained by growing *Pseudomonas* sp. PS-27 with hexadecane showed that the liquid efficiently emulsified various hydrophobic substrates with emulsification indexes varying from 65 to 80%. The feature of surfactants that is most important for their practical use is their ability to emulsify hydrocarbons over a broad pH range. It was shown that surfactants produced by *Pseudomonas* sp. PS-27 could form fine stable emulsions with hydrophobic substrates at a pH from 5.5 to 10. This property makes *Pseudomonas* sp. PS-27 surfactants promising for the treatment of oil-polluted sites under various environmental conditions.

3.2.5 Bioremediation of oil-polluted marine waters

Although accidental releases account for only a small percentage of the oil released into the marine environment, large oil spills pose an enormous threat to terrestrial and marine ecosystems and hence, attract much public attention. Oil bioremediation in marine waters is limited by the amount of nitrogen and phosphorus needed for incorporation into cellular biomass and the availability of these within the area of hydrocarbon degradation (Warr et al., 2009; Karpenko et al., 2005). Nutrient-enhanced bioremediation, also known as biostimulation, is recognized as one of the most promising of the emerging technologies for the cleanup of oil-contaminated shorelines. However, a combined treatment with a fertilizer and biosurfactant results in total utilization of most of the crude oil fractions after 7 days of treatment (Nikolopoulou and Kalogerakis, 2008). The application of biosurfactants in the solutions also enhanced the growth of the hydrocarbon degraders (Pantsyrnaya et al., 2011). Biostimulation with nutrients and biosurfactants enables naturally occurring microbes to adapt better and faster to the oil spill environment, resulting in a shorter lag phase and faster crude oil degradation. Thus it is an effective tool for combating oil spills (Cheng et al., 2013).

3.2.6 Biodegradation of diesel-contaminated water and soil

Very hydrophobic contaminants tend to bind tightly to soil particles in a manner that renders them inaccessible to microorganism degraders. Traditionally, chemical surfactants are used in soil washing (Lee et al., 2005). Biosurfactants have also been found useful for remediation of oil spills, for dispersal of oil slicks into fine droplets, and for conversion of mousse oil into oil-in-water emulsion. Shulga et al. (2000) examined the use of biosurfactants in the cleanup of sea sand as well as bird feathers from oil. Whang et al. (2008, 2009), studying the potential application of two biosurfactants, surfactin and rhamnolipid, for enhanced biodegradation in diesel-contaminated water and soil, found that both biosurfactants are capable of reducing surface tension and increasing diesel solubility with increased biosurfactant addition. The following major outcomes can be drawn from this study. In diesel/water batch experiments, the addition of rhamnolipid to diesel/water systems from 0 to 80 mg/L significantly increased biomass growth and diesel biodegradation percentage from 40 to 100%, respectively. Addition of rhamnolipid in a concentration of 160 mg/L provided similar results to those from a 80 mg/L concentration. Similar trends have also been observed for specific biomass growth rate and specific diesel degradation rate constant as the addition of rhamnolipid increased from 0 to 160 mg/L. The addition of 40 mg/L of surfactin to diesel/water systems substantially enhances biomass growth and diesel biodegradation to 94%, compared to 40% biodegradation in controls. Addition of more than 40 mg/L of surfactin, however, decreases both biomass growth and diesel biodegradation efficiency, with a worse diesel biodegradation percentage (0%) at 400 mg/L of surfactin addition. Their superior enhancing capability with regard to both the efficiency and rate of diesel biodegradation in diesel/soil systems confirms the potential application of surfactin and rhamnolipid in stimulating indigenous microorganisms for enhanced bioremediation of diesel-contaminated soil. Kaczorek et al. (2013) found that rhamnolipid and saponin surfactants increased the biodegradation of diesel oil by *Achromobacter* sp. and also modified the cell surface. Cell-surface hydrophobicity in the presence of rhamnolipids remarkably increased, whereas, following the addition of saponins, it remained unchanged. These results constitute a significant increase in knowledge of this species' ability to degrade petroleum and are highly important in improving future bioremediation applications.

Karpenko et al. (2006) also studied the possibility of accelerating oil degradation by an enrichment culture of oil-oxidizing microorganisms in the presence of bacteria of the genus *Rhodococcus* and microbial surfactants. It was shown that the degree of consumption of crude oil (2% vol.) after 192 h of enrichment culture growth reached 84%. Inoculation of the active hydrocarbon-oxidizing strain *Rhodococcus erythropolis* EK-1 and exogenous surfactants produced by *Pseudomonas* sp. PS-27 increased this degree to 70–90%.

Emulsification of organic substances (in particular, oil and petroleum products) with surfactants is crucial for fast biotechnological purification of water and soil. Studies of the emulsifying properties of the culture liquid obtained by growing *Pseudomonas* sp. PS-27 with hexadecane showed that the liquid efficiently emulsified various hydrophobic substrates, with emulsification indexes varying from 65 to 80%. One feature of surfactants that is most important for their practical use is their ability to emulsify hydrocarbons over a broad pH range. It has been shown by Karpenko et al. (2006) that surfactants produced by *Pseudomonas* sp. PS-27 can form fine, stable emulsions with hydrophobic substrates at a pH from 5.5 to 10. This property makes *Pseudomonas* sp. PS-27 surfactants promising for the treatment of oil-polluted sites under various environmental conditions.

3.2.7 Bioremediation of metal-polluted sites

The common use of heavy metals in many industrial applications has led to their wide distribution in wastewaters, sediments and soils. Today, due to the high toxicity and nonbiodegradable nature of heavy metals, metal-polluted soils have become a serious environmental problem. Remediation technologies developed for metal-contaminated soil are based on: (1) allowing heavy metals to remain in the polluted site after decreasing their availability by solidification/stabilization processes, or (2) removing heavy metals from soil by, for example, phytoremediation or soil extraction (flushing/washing). Techniques such as ex-situ soil washing and in-situ soil flushing transfer heavy metals to a liquid phase through desorption and solubilization. To support the removal of heavy metals from the soil matrix, the washing water used in these methods is usually augmented with acids, bases, chelating agents, or surfactants. Sorption and sequestration of xenobiotic compounds within the soil matrix are critical processes affecting contaminant mobility, toxicity, and persistence. Slow desorption and release from the soil matrix to the aqueous phase represents a long-term contaminant source and hinders remediation efforts. The prospect for use of microbial surfactants in hydrocarbon-contaminated soil washing depends on the capacity of these compounds to enhance the desorption and dissolution of the organic pollutants, and to increase their transport rate in soils. The mechanisms involved in the hydrocarbon removal from soils are related to the mechanisms that increase the bioavailability in bioremediation. The properties of stabilizing oil/water emulsions and increasing hydrocarbon solubility may enhance, on one hand, the biodegradation rate and, on the other hand, the hydrocarbon removal from soils (Ron and Rosenberg, 2001). The use of chemical as well as microbial surfactants has been reported to be efficient in removing hydrocarbons from soils (Khalladi et al., 2009; Mulligan and Wang, 2006). The removal efficiency is in fact determined by the physical-chemical properties of the biosurfactants and not by their effects on the cell-surface properties and microbial metabolism.

Both synthetic and biological surfactants have been shown to enhance the apparent aqueous solubility of nonpolar organic compounds, resulting in increased bioavailability and biodegradation. However, there are also reports suggesting that some synthetic surfactants inhibit biodegradation. This inhibition is generally attributed to toxicity or reductions in bioavailability due to partitioning of contaminants into surfactant micelles.

Biosurfactants, as already mentioned, have gained considerable interest as an environmental remediation technique owing to some of its distinct advantages over its synthetic counterparts, such as lower toxicity, higher biodegradability, and high selectivity.

The modes of interaction between surfactants and metals are not completely understood. It is known that surfactants can remove metals from surfaces through different mechanisms. Non-ionic metals can form complexes with biosurfactants, enhancing their removal from porous media (Wang and Mulligan, 2009). Anionic surfactants interact with cationic metals, leading to their desorption from surfaces (Christofi and Ivshina, 2002). Nevertheless, cationic surfactants can also play a role through a competitive binding to negatively charged binding sites. Recently, Mulligan and coworkers extensively studied the potential of rhamnolipids, sophorolipids, and surfactins in washing metal-contaminated soils and sediments (Mulligan and Wang, 2006; Mulligan, 2009; Dahrazma and Mulligan, 2007).

4 New aspects in the process of bioremediation of low-soluble contaminants: biosurfactant effects on microbial cell-surface characteristics and on microbial cell-surface structure

Pollutants and waste products are widespread throughout the world. Different techniques to remove them are employed depending on an area's degree of saturation and aeration. In-situ techniques are defined as those that are applied to soil and groundwater at the site with minimal disturbance. Ex-situ techniques are those that are applied to soil and groundwater at the site, which have been removed from the site via excavation (soil) or pumping (water). In-situ bioremediation by an indigenous microbial population is an increasingly popular, ecofriendly option for cleanup of contaminated sites (Zeyaullah et al., 2009).

Specifically, in-situ treatments are well known and cost-effective methods that usually eliminate the need for expensive excavation and emission control (Watanabe, 2001). Thus, microbiological decontamination of petroleum hydrocarbons in polluted environments is claimed to be an efficient, economic, and versatile alternative, or complement, to physicochemical treatments (Iwamoto et al, 2001). As noted earlier, the techniques applied for elimination of pollution have to be both relatively inexpensive and not harmful to the environment. For this purpose, the bioremediation of organic compounds by microorganisms represents one of the primary mechanisms by which pollutants are eliminated from the environment.

A great deal of research has been devoted to finding the organisms, usually bacteria, that are capable of altering or degrading such pollutants to environmentally tolerable forms. Bacteria are particularly suitable for application of biodegradation because of the wide variety of carbon sources or electron acceptors used by various strains. Single strains of bacteria are often insufficient to degrade certain pollutants; complete catabolism may require the development of consortia or communities made up of two or more taxa (Wagner-Dobler et al., 1998; Hubert et al., 1999). In part, the efforts are directed toward isolating individual strains of bacteria from indigenous polluted sites with desirable biodegradative abilities and cooperation of appropriate strains forming selected communities, which can be used in the bioremediation of toxic sites, accelerating the naturally occurring biodegradation. On the other hand, hydrolysis of organic matter in polluted sites is largely dependent on bacterial activity (Deming and Baross, 1993). Extracellular enzyme activity is a key step in the degradation and utilization of organic polymers, since only compounds with molecular mass lower than 600 daltons can pass through cell pores (Hoppe, 1991; Meyer-Reil, 1991). The enzymes often are hydrolases, oxidases, esterases, and so on.

Bioremediation requires the use of microorganisms that must be capableof development in the presence of the pollutants and that either metabolize or accumulate organic compounds. In both cases, the pollutants have to cross the microbial cell envelope, that is, the cell wall and the membrane, and, as noted earlier, this transport represents a major step in the bioremediation process.

Surfactants play an important role in bioremediation because they facilitate the solubilization of hydrophobic pollutants and enhance their penetration within the microbial cells. The action of surfactants also modifies the cell envelope and changes the fluidity of membranes. Furthermore, the surfactants which are tools for remediation cannot be harmful to the environment; they have to be biodegradable. Biosurfactants possess a real advantage over synthetic derivatives as they are biodegradable after a certain period. The role of biosurfactants in the bioremediation of polluted environments (soils and waters) is discussed in many research studies and reviews (Li J-L et al., 2009; Beal and Betts, 2000; Mulligan, 2009; etc.). Because of their chemicophysical properties, surfactants are involved in interactions between microbial cells and immiscible hydrocarbons by different mechanisms, including: (i) emulsification, (ii) micellarization, (iii) adhesion–de-adhesion of microorganisms to and from hydrocarbons, and (iv) desorption of contaminants (Ron and Rosenberg, 2001; Singh et al., 2007).

Despite the intensive studies on the application of biosurfactants in various fields, their effects on microbial cells—specifically on membrane permeability, cell-surface characteristics, and cell structure—are still obscure (Willumsen et al., 1998). The next section discusses some new findings on the influence of biosurfactants on microbial cells.

4.1 Biosurfactant effects on microbial cell-surface characteristics

4.1.1 Effects on hydrophobic interactions

Hydrocarbon-degrading bacteria have developed different adaptation mechanisms for utilizing poorly soluble substrates. One major mechanism employed by hydrocarbon degrading organisms to utilize such substrates is the production of biosurfactants.

The mode of action of these compounds is in modifying cell-surface hydrophobicity and/or in promoting emulsification and/or solubilization of substrates. Two modes of hydrocarbon uptake are generally considered: (1) interfacial uptake (direct contact of cells with hydrocarbon droplets), and (2) biosurfactant-mediated hydrocarbon transfer (cell contact with emulsified or solubilized hydrocarbons) (Goswami and Singh, 1991; Hommel, 1994). Usually, both mechanisms take place, but the domination of one or the other depends on the strain.

The probable modes of hydrocarbon uptake in *Candida* cells were studied by Coimbra et al. (2009) based on data for cell hydrophobicity, emulsifying activity, surface tension, and interfacial tension of the cell-free culture medium. Six *Candida* strains were cultivated in insoluble and soluble substrates for 144 h, including in these studies, soybean oil, ground-nut oil refinery residue, corn steep liquor, and glucose. The results obtained showed the potential use of yeasts in the removal of hydrophobic compounds. Depending on the strain and substrate used, the adhesion ability of the yeast cells and the production of biosurfactants can vary and occur simultaneously, thus increasing the efficiency of the bioremediation treatment of petroleum pollution. The application of crude biosurfactants separated from the yeast cells was also demonstrated by tests in which petroleum and the derivate motor oil adsorbed in sand samples were removed. Biosurfactants produced in a low-cost medium were able to remove 90% of the hydrophobic contaminants. Besides the possible application of yeast *Candida* in the bioremediation of polluted sites, the results obtained can help clarify the natural relationships between members of the ecosystem studied. The occurrence of both mechanisms, promoting the solubilization or emulsification of substrate or modifying cell hydrophobicity, has also been proposed. These observations suggest a broader role for biosurfactants than just hydrocarbon uptake. A likely possibility is the more general participation in adhesion and de-adhesion interactions between microorganisms and interfaces.

Vasileva-Tonkova et al. (2008, 2011) characterized bacterial strains isolated from hydrocarbon-contaminated wastewater samples according to probable modes of hexadecane uptake and their relative distribution. They attempted to analyze the natural adaptive mechanisms of bacterial strains, helping them to grow on hydrophobic compounds. In this case cell hydrophobicity, emulsifying activity, glycoside content, and surface tension of cell-free culture broths were followed during growth of isolates on hexadecane. Hexadecane is one of the water-immiscible hydrocarbon substrates that forms an important constituent of oil.

The bacterial isolates used in these studies belong to seven genera–five Gram-positive and two Gram-negative. *Pseudomonas* was the most frequent isolate (five strains, 33%). It was found that strains differing in their initial cell hydrophobicity were able to produce effective emulsifying agents. The strains that showed the highest emulsifying activity, *Pseudomonas fluorescens* HW-6 and *Arthrobacter* sp. HW-8, were very different in their initial cell hydrophobicity (27.8% and 66.5%, respectively). Similar observations have been reported by other authors showing that both hydrophilic and hydrophobic bacteria produce surface-active compounds when grown on glucose or hexadecane (Bouchez-Nautili et al., 1999).

Hexadecane-grown strains showed rather higher emulsifying activity than that of glucose-grown strains, and this finding is most probably connected with utilization of the water-insoluble substrate.

Based on data obtained for growth, cell hydrophobicity, emulsifying activity, glycoside production, and surface tension of cell-free culture broths, could be proposed as the most probable mechanism for hexadecane uptake by bacterial strains tested. The increase in cell hydrophobicity and glycoside production by all the strains during growth on hexadecane indicated that the strains could employ biosurfactant-mediated alkane uptake. Surfactants produced by the strains make their cell surfaces more hydrophobic (above 42%), thus increasing the attachment of bacterial cells to the large alkane droplets. Based on high hydrophobicity and the relatively low emulsifying activity of the strains *Staphylococcus sp.* HW-2 and *Streptococcus sp.* HW-9, the mechanism of surfactant-enhanced interfacial uptake of hexadecane could be suggested as predominant for these isolates. Also included in this group could be the strain *Bacillus sp.* HW-4, which grew slightly at the stage tested. Relatively high values for cell hydrophobicity, glycoside content, and emulsifying activity of *Arthrobacter* sp. HW-8 isolate suggested that this strain could employ both direct interfacial uptake and biosurfactant-enhanced alkane emulsification.

The mode of action of biosurfactants is usually considered to promote both substrate emulsification and solubilization. In the case where biosurfactants act mainly by solubilization of the alkane, the formation of micelles with hydrophilic outer layer is observed. This mode of micellar transfer appears to be well suited to hydrophilic cells, allowing efficient contact with the alkane-degrading bacteria. Of the strains studied, such a mechanism of solubilization rather than emulsification appears to be appropriate only for the most hydrophilic strain *Streptococcus* sp. Most of the strains (67%) exhibited both high or medium hydrophobicity and high or medium emulsifying activity. This suggested the existence of both mechanisms of the biosurfactant action, thereby promoting interfacial hexadecane uptake and emulsification to a similar degree. Production of biosurfactants also occurs during growth on soluble carbon sources, which suggests a broader role of biosurfactants than just hydrocarbon uptake—for example, their participation in adhesion and deadhesion interactions between microorganisms and interfaces (Neu, 1996). Biosurfactant-enhanced alkane uptake could be one aspect of this general role.

The role of the biosurfactant does not end here. The hydrocarbon droplet enclosed within a surfactant layer is the form in which cells seem to take up the hydrocarbon (Rathledge, 1992). Another factor could be utilization of the internalized biosurfactant for the formation of membranes around the hydrocarbon droplet and its further partitioning into smaller compartments for increased surface area for metabolism. The entire process, however, is a combination of predominantly two modes of uptake proposed earlier: uptake by direct contact and pseudosolubilization of hydrocarbon to microdroplets, both facilitated by biosurfactants and occurring simultaneously to enhance and facilitate hexadecane uptake.

4.1.2 Effects of biosurfactants on the microbial cell-surface structure

As mentioned elsewhere, the biosurfactants that are studied most extensively in terms of their industrial and environmental application are the rhamnolipids. Nevertheless, little is known about their interaction with bacterial cells, although their surface and membrane active properties likely play an important role (Lang and Wullbtandt, 1999; Singh and Cameotra, 2004). A number of studies suggest that it is possible to modify the outer membrane of Gram-negative bacteria by mutation (Nikaido, 2003) or by addition of chemical agents that can act as membrane-permeabilizing agents (Denyer and Maillard, 2002). These processes result in changes in cell-surface properties as well as in increased permeability and hydrophobicity of the bacterial membrane.

Sotirova et al. (2009) studying the effect of rhamnolipid-biosurfactant produced by *Pseudomonas* sp. on the cell-surface structures of the *Pseudomonas aeruginosa* strain demonstrated that the rhamnolipid at concentrations below and above CMC provoked a multicomponent response of the bacterial cells without affecting their growth and viability. It was found that in the presence of rhamnolipid-biosurfactant at concentrations above CMC the cell surface became more hydrophobic. The results suggest that the enhanced cell hydrophobicity is associated with a reduction in the total cellular lipopolysaccharide (LPS) content. LPSs are the main component of the outer membrane of Gram-negative bacteria and are responsible for the cell impermeability characteristics (Denyer and Maillard, 2002). The quantity and type of LPS have a profound effect on the interactions between the microbial cell and its environment (Makin and Beveridge, 1996). Al-Tahhan et al. (2000) observed a loss of LPS in *P. aeruginosa* strains treated with rhamnolipid at low concentrations, and this resulted in increased cell surface hydrophobicity. Sotirova et al. (2009) observed similar changes with *P. aeruginosa* strain grown in the presence of rhamnolipid-biosurfactant but in concentrations above CMC. Above CMC, the rhamnolipid caused a reduction in total cellular LPS content of 22%, which can be associated with an increase in cell hydrophobicity to 31% adherence. The release of LPS from the cell surface may be due to solubilization of the outer membrane by binding the aggregated biosurfactant to the membrane, followed by removal of the LPS component.

The rhamnolipid-biosurfactant at concentrations below CMC did not affect the LPS component of the bacterial outer membrane but caused changes in the outer membrane protein (OMP) composition of *P. aeruginosa*. Examination of the OMP profiles revealed that the amount of major proteins (OprF, OprD, OprJ, and OprM) markedly decreased. The results of this research demonstrate that rhamnolipid-biosurfactant in concentrations below CMC caused a decrease in the amount of important OMPs, which may lead to alterations in the pathogenic properties and antibiotic sensitivity of the *P.aeruginosa* strain. The observed effect of the rhamnolipid-biosurfactant on the protein composition was probably due to the binding of biosurfactant monomers to the outer membrane, causing changes in the membrane organization. The reduction in the amount of OMPs may decrease the compactness of the outer membrane, leading to it becoming smoother, thinner, and indistinct as was demonstrated by transmission electron microscopy (TEM). The disruption of the outer membrane observed in certain zones allows for increased passage of hydrophobic compounds. Still, these changes did not affect cell viability. This is the first report that shows changes in the bacterial OMPs in the presence of a rhamnolipid-biosurfactant. Understanding the biosurfactant effects on the cell surfaces can contribute to the use of biosurfactants in bioremediation technologies and other applications.

Recently, it was shown that the outer membrane protein A (OmpA) in several members of the *Acinetobacter* family has emulsifying properties (Walzer et al., 2006). These properties depend on the amino acid composition of four putative extramembrane loops, which in various strains of *Acinetobacter*, but not in *E. coli*, are highly hydrophobic. As many *Acinetobacter* strains can utilize hydrophobic carbon sources, such as oil, the emulsifying activity of their OmpA may be important for the utilization and uptake of hydrocarbons. If outer membrane proteins with emulsifying activity are physiologically important, they may exist in additional oil degrading bacteria. In order to identify such proteins, Walzer et al. (2009) describe a method for using protein sequence data to predict the hydrophobic properties of the extramembrane loops of outer membrane proteins.

Several authors reported on changes in lipid and fatty acid composition of bacterial cell surface as a consequence of the effects of rhamnolipids (Sotirova et al., 2012, Kaczorek et al., 2013, Bharali et al., 2013). The predominant anionic phospholipids in bacterial membranes of many Gram-positive bacteria are phosphatidylglycerol and cardiolipin. The physical properties of cardiolipin determine interactions that are crucial for the structural organization of biological membranes. Study of the effect of rhamnolipid on the lipid composition of the *Bacillus subtilis* membrane demonstrated a remarkable increase of cardiolipin, which probably results in altered barrier functions of the bacterial surface envelope (Sotirova et al., 2012). The influence of rhamnolipids on the membrane's fatty acid composition and cell-surface hydrophobicity of *Stenotrophomonas maltophilia* was investigated by Kaczorek et al. (2013). These authors observed an increase of the branched acids amount in the outer membrane and showed that lower hydrophobicity corresponds to the increasing amount of hydroxy fatty acids.

The infrared spectroscopy investigation on the interaction of rhamnolipid with cell surfaces of *Staphylococcus aureus* and *Klebsiella pneumoniae* reflected the alteration in cellular constituents such as proteins, polysaccharides, fatty acids, and nucleic acids and indicated clearly cellular damages (Bharali et al., 2013).

5 Permeabilizing ability of biosurfactants: from activation to inhibition

5.1 Biosurfactant effects on permeabilization of growing and resting microbial cells

As already mentioned, despite the intensive studies on biosurfactant application in various fields, the data on their effects on microbial cells remain obscure (Al-Tahhan et al., 2000).

Even less clear are the data on their application as permeabilizing agents, although the permeabilizing ability of biosurfactants may play an important role in many biotechnological processes. Their use in different fields of industries and especially in environmental bioremediation processes leads to direct contact with microbial cells, undoubtedly resulting in an increase of cell permeability. Sotirova et al. (2008) studied the potential of biosurfactant PS to permeabilize the bacterial cells of *Pseudomonas aeruginosa*, *Escherichia coli*, and *Bacillus subtilis* on growing (in vivo) and resting (in vitro) cells. Biosurfactant PS has been isolated from the bacterial strain *Pseudomonas sp*. PS-17 and is a biocomplex of biosurfactant–glycolipid and biopolymer–alginate with high-surface and emulsification activities (Karpenko et al., 2004). It was shown that the presence of biosurfactant PS in culture media was detrimental to the growth of tested bacteria or neutral. The determining factor in the first case was surfactant concentration. Lower concentrations, close to CMC of biosurfactant PS, do not influence the growth of Gram-positive and Gram-negative strains. Concentrations higher than CMC, however, exhibit toxic conditions for Gram-positive *B. subtilis* cells but not for Gram-negative P. *aeruginosa*. Biosurfactant-enhanced levels of extracellular protein in *B. subtilis* 168 cells were compared with those of *P. aeruginosa*, which confirmed the higher susceptibility of Gram-positive cells to the effect of the studied biosurfactant. Scanning-electron microscopy (SEM) observations showed that the biosurfactant PS did not exert a disruptive action on resting cells opposite to the detrimental effect on growing cells of *B. subtilis*.

This is a good grounding for application of biosurfactant PS in many biotechnological processes. Low CMC and interfacial tension values, mild action on nongrowing cells, and neutral effects on the growth of microbial strains at low surfactant concentrations make biosurfactant PS a potential candidate for application in environmental and industrial processes. Based on the results of this work it appears, however, that high concentrations of biosurfactant PS must be applied carefully in the development of environmental cleaning technologies, despite the widely accepted concept of microbial surfactant biocompatibility.

Similar results were obtained on the effect of rhamnolipid biosurfactant produced by *P. fluorescens* on laboratory bacterial strains and isolates from industrial wastewater (Vasileva-Tonkova et al., 2011). For all treated strains, the authors observed an increase in the amount of released protein, suggesting increased cell permeability. In preliminary studies, Magalhães and Nitschke (2013) also observed that rhamnolipids promoted an increase in cell permeabilization. All these results suggest the efficiency of biosurfactants in modifying the cell-surface properties, making them potentially applicable in different industrial fields.

5.2 Effect of chemical surfactant and biosurfactant on PAH mineralization

Avramova et al. (2008) studied the effect of one synthetic surfactant, Triton X-100, and one microbial surfactant, rhamnolipid biosurfactant PS-17, on phenanthrene mineralization by *Pseudomonas* sp. growing cells (in vivo) and resting cells (in vitro). Previous studies have shown that Triton X-100 has potential as a permeabilazing agent (Galabova et al., 1996; Christova et al., 1996). In spite of the fact that Triton X-100 and rhamnolipid biosurfactant are the most studied surfactants effective in bioremediation/biodegradation of PAHs (Shin et al., 2004), the results obtained are quite contradictory. Both surfactants showed minimal effect on the growing culture. The addition of Triton X-100 to the reaction medium of the resting cells inhibited the phenanthrene mineralization, while rhamnolipid PS-17 had no effect on the process. The resting cells permeabilized with Triton X-100 accelerated the initial rate of the biodegradation, while the cells permeabilized with rhamnolipid degraded phenanthrene more slowly than with the untreated resting cells. The reduced degrading capability of permeabilized resting cells may be due to the direct contact between the rhamnolipid biosurfactant and *Pseudomonas* sp. cells leading to changes in the membrane organization that result in its decreased compactness (Sotirova et al., 2009). The inhibitory effect of non-ionic surfactant Triton X-100 on the dynamics of phenanthrene degradation by growing cells was not surprising, as many researchers reported about such an effect, giving several explanations for it (Willumsen et al., 1998; Yang et al., 2003). Bramwell and Laha (2000), working with *Pseudomonas aeruginosa,* indicated that the presence of solubilized phenanthrene increased the toxicity of the surfactant used through the experiments.

The increased toxicity of the solubilized substrate also has to be considered when surfactants are applied in the biodegradation processes.

The preliminary permeabilization reduced the toxic effect of the prolonged incubation with the surfactant and the toxicity of solubilized phenanthrene, thus leading to an increase in the initial phenanthrene mineralization rate. Perhaps the longer permeabilization rate facilitated the incorporation of surfactants in cell membranes and increased cell-surface hydrophobicity. Such changes in the cell surface have beneficial effects on the phenanthrene mineralization rate (Brown and Jaffe, 2006; Tecon and van der Meer 2010).

The literature describes the possibility of resting microbial cells to produce biosurfactants with greater efficiency compared to growing culture (Kitamoto et al., 2002; Desai and Banat, 1997) Based on these results, these authors hypothesized that the resting *Pseudomonas sp.* cells have the potential to produce their own biosurfactant as a mechanism of adaptation to the reaction conditions.

Thus far, the role of preliminary permeabilization with regard to the surfactants of bacterial cells in the biodegradation of phenanthrene has not been studied intensively. Obviously, resting and permeabilized microbial cells are applicable in in-situ as well as ex-situ technologies for converting pollutants into benign environmental substances. This study reconfirmed the complicated mechanism of polycyclic aromatic hydrocarbon mineralization in the presence of surfactants.

The literature results on the addition of surfactants in bioremediation experiments are very contrasting. Rhamnolipids are known to be effective enhancers in the biodegradation of both aliphatic and aromatic hydrocarbons (Zhang and Miller, 1995; Rahman et al., 2003) as well as in soil washing of hydrocarbons (Urum et al., 2003). However, several studies have demonstrated that the addition of rhamnolipids did not enhance biodegradation for most of the contaminants (Mata-Sandoval et al., 2001; Noordman and Janssen, 2002; Ward, 2008). Franzetti et al. (2008) isolated from a diesel-contaminated soil a strain of *Gordonia* sp., producing two different types of surface-active compounds: extracellular bioemulsions and cell-bound biosurfactants that extensively degrade aliphatic hydrocarbons. Bioremediation results showed that the bioemulsans are able to slightly enhance the biodegradation of recalcitrant branched hydrocarbons but are effective in removing crude oil and PAHs from soil. Also, it was shown that the addition of rhamnolipids lowered the biodegradation rates for *Pseudomonas stutzeri* strain 9 isolated from petroleum-contaminated soil, able to degrade aromatic alkyl derivatives (butylbenzene, sec-butylbenzene, tert-butylbenzene and isobutylbenzene) (Kaczorek et al., 2013). However, the rhamnolipids did not succeed in increasing either the rate or the extent of biodegradation for most of the individual contaminants studied.

In PAH experiments, the final concentration of pyrene in rhamnolipid-treated microcosms was double that in the controls. The authors' attempts to promote biodegradation by exogenous (bio) surfactant additions in soil bioremediation failed. Cases of inhibition of biodegradation by biosurfactants have been reported from other authors as well, and the reasons of these failures have been proposed. The surfactant micelles can provide a barrier between bacteria and the contaminants (Yuan et al., 2000). The capability of microorganisms to access the micelle-contaminant complex depends on the hydrophobicity of the cell surface (Bouchez-Nautili et al., 1999). Furthermore, the surfactants can inhibit hydrocarbon degradation due to competitive biodegradation (Goudar et al., 1999).

Our increasing ability to analyze the microbial diversity in natural environments is expected to expand our knowledge of microbial surfactants with respect to their exploitation for commercial applications and their roles in the physiology of the producing microorganisms.

6 Industrial perspectives on biosurfactants application

Advantages such as biodegradability and production of renewable-resource substrates may be a green alternative in a variety of industrial fields. The potential applications of biosurfactants relate to the food, cosmetics, petroleum, and pharmaceutical industries. Emulsification and de-emulsification, micellarization, dispersion, foaming, and regulation of the adhesion-deadhesion of microorganisms are some of the numerous activities that make the biosurfactants potential candidates for active exploitation in industry. The literature contains a number of reports on different aspects of the applications of biosurfactants. The ability of biosurfactants to enhance the pseudosolubilization and emulsification of the immiscible fractions of the contaminants, thus enhancing their bioavailability to degrading microorganisms, allows the use of biosurfactants in the bioremediation of soil and groundwater contaminated by insoluble organic pollutants (Franzetti et al., 2010).

6.1 Biosurfactants in medicine and the pharmaceutical industry

Some properties of biosurfactants have therapeutic and biomedical importance. Biosurfactants have potential for applications in biomedicine as they act as biologically active compounds. They have antibacterial, antifungal (Kitamoto et al., 2002; Joshi et al., 2007; Sotirova et al., 2008; Saravanakumari and Mani, 2010; Bharali et al., 2013), and antiviral activities (Desai and Banat, 1997; Remichkova et al., 2008; Muthusamy et al., 2008); they have antiadhesive action against several pathogenic microorganisms (Singh and Cameotra, 2004; Rodrigues et al., 2004, 2006; Mulligan, 2005; Sotirova et al., 2009; Banat et al., 2010; Janek et al., 2012). Biosurfactants are capable of inhibiting pathogen adhesion to solid surfaces. Rhamnolipids significantly reduced the initial attachments of the Gram-negative *Escherichia coli*, *Pseudomonas putida,* and *P. aeruginosa,* and the Gram-positive *B. subtilis* on hydrophilic glass and hydrophobic modified surfaces (Sodagari et al., 2013). The application of these compounds to surfaces may be a novel and efficient method of preventing the colonization of pathogens (Singh and Cameotra, 2004; Araujo Lívia et al., 2011; Dusane et al., 2011). Many authors reported for the anticarcinogenic effects of biosurfactants on lung, cervical, breast, and brain (Isoda et al., 1997; Cao et al., 2009; Chiewpattanakul et al., 2010). It was found that the extracellular glycolipids induce cell differentiation instead of cell proliferation in the human promyelocytic leukemia cell line, making them novel reagents for the treatment of cancer cells (Muthusamy et al., 2008). The lipopeptide biosurfactant pseudofactin II has specific activity toward melanoma cells and induces the apoptosis of these cells (Janek et al., 2013).

Some biosurfactants have been reported as effective alternatives to synthetic pharmaceuticals for oral and dermal application (Lang et al., 1989; Rodrigues et al. 2006; Saravanakumari and Mani, 2010). Glycolipids and their derivatives have a variety of biological activities such as antibacterial, antifungal, antiviral, anticarcinogenic, and immune-modulating and are appropriate for therapeutic uses. The glycolipids become a viable alternative to be used in the pharmaceutical, biomedical, food, and health industries (Cortés-Sánchez et al., 2013). The combination of rhamnolipid biosurfactant and antimicrobial agents has a strong synergistic effect, leading to decreasing the bactericidal and fungicidal concentrations of antimicrobials (Sotirova et al., 2010, 2012; Magalhães and Nitschke, 2013). This allows creating a new generation of antimicrobials. Biosurfactants also act as anti-inflammatory, immune-modulator, and cellular differentiation agents (Banat et al., 2010). These observations indicate the potential use of biosurfactants in medicine and the pharmaceutical industry.

6.2 Application of biosurfactants in the cosmetic industry

The major and most widely used biosurfactants in cosmetics and personal care products are glycolipids (sophorolipids, rhamnolipids, and mannosylerythritol lipids). Their physicochemical properties, biological activities, biocompatibility, and biodegradability make them suitable ingredients for cosmetics. Sophorolipids are used as emulsifiers, foaming agents, solubilizers, wetting agents, and active ingredients in cosmetics. They have good skin compatibility and excellent moisturizing properties (Lourith and Kanlayavattanakul, 2009). Sorophollipids have antibacterial activity and are used in the treatment of acne, dandruff, and body odors (Mager et al., 1987). They also exhibit antioxidant properties. Sophorolipids are produced commercially and are present in cosmetics for skin applications. The sophorolipids are also present in dishwashing products, lotions, hair products, pencil-shaped lip rouge, lip cream, and eye shadow as well as in compressed powder cosmetics (Lourith and Kanlayavattanakul, 2009; Bhadoriya et al., 2013). As reported by Lourith and Kanlayavattanakul (2009), rhamnolipids are used in health care products in several different formulations—insect repellents, antacids, acne pads, antidandruff products, deodorants, nail care products and toothpastes. Cosmetics containing rhamnolipids were patented and used as anti-wrinkle and anti-aging products.

Mannosylerythritol lipids have a high hydrophilicity and low critical aggregation concentration (Imura et al., 2006) and are suitable as emulsifiers, dispersants, and detergents. They also have antimicrobial activities and are ingredients in skin smoothing, skin care, and anti-wrinkle products (Bhadoriya et al., 2013).

6.3 Application of biosurfactants in the food industry

Biosurfactants have potential applications in the food industry as food-formulation ingredients and antiadhesive agents. Biosurfactants decrease surface and interfacial tension, thus promoting the formation and stabilization of emulsions. Other functions of biosurfactants in food processing include: controlling the agglomeration of fat globules, stabilizing aerated systems, improving the texture and shelf life of starch-containing products, and improving the consistency and texture of fat-based products (Muthusamy et al., 2008). In bakery and ice cream manufactures, biosurfactants are used to: control consistency, extend freshness, and solubilize flavor oils. They are also utilized as fat stabilizers and antispattering agents during the cooking of oil and fats. The addition of rhamnolipid biosurfactant improves the stability in dough; the texture, volume, and conservation of bakery products; and the properties of butter cream, croissants, and frozen confectionery products (Muthusamy et al., 2008).

The new strategy of controlling the adherence of microorganisms to food contact surfaces and thus preventing biofilm formation is to involve biosurfactants. Bacterial biofilms are potential sources of contamination, which may lead to food spoilage and disease transmission (Muthusamy et al., 2008).

Besides the listed applications, biosurfactants can be useful for food processing industries as antimicrobial agents. Magalhães and Nitschke (2013) established the notion that the rhamnolipids showed antimicrobial activity against *Listeria monocytogenes*—an important foodborne pathogen that can cause serious illness.

The activity demonstrated by biosurfactants makes them potential new tools in developing appropriate technologies to prevention and preparation of foods.

6.4 Application of biosurfactants in the petroleum industry

The high-quality light crude oils that are easily extracted are limited in availability, with an ultimate recoverable resource being estimated between 2 and 4 trillion barrels (Hall et al., 2003).

Only less than half of the crude oil content of the reservoirs can be extracted by the current techniques (Banat, 1995; Simayev et al., 1999). These require optimizing processing efficiency throughout all stages in the petroleum industry. This opportunity offers modern biotechnology using biosurfactants.

Biosurfactants can potentially be used throughout all stages of petroleum processing, including extraction, transportation, upgrading and refining, and petrochemical manufacturing (Hayes et al., 1990; Cameotra and Singh, 2008; Piotrowska-Cyplik et al., 2012).

Perfumo et al. (2010) reported that biosurfactant ares able to exert oil displacement and movement through oil-bearing rocks by means of three main mechanisms:

1. Reduction of interfacial tension between oil–rocks and oil–brine
2. Modification of the wettability of porous media
3. Emulsification of crude oil.

Perfumo et al. (2010) also reported that biosurfactants produced ex-situ enhanced oil recovery when biosurfactants were injected into oil resources. Another successful strategy for enhancing oil recovery is to inject selected biosurfactant-producing microorganisms into the oil reservoir (Magot, 2005).

The appropriate approach is to produce the biosurfactant in the reservoir itself (in situ) by providing nutrients to indigenous microbial populations in the reservoir. Technologies involved injection of nutrient solutions into the oil well to stimulate the resident microbial communities (Perfumo et al., 2010).

Biosurfactants can find application in the next stage as well: crude oil transportation in the pipeline. Oil viscosity slows the flow in the pipeline when the crude oil is transported over long distances from the extraction fields to the refineries. A promising technology designed to reduce oil viscosity is to produce a stable oil-in-water emulsion that facilitates oil motility. Suitable for this application are high-molecular-weight biosurfactants (e.g., emulsan) (Hayes et al., 1990). They are not effective in reducing interfacial tensions but have the capability to stabilize oil-in-water emulsions.

Another problem for the petroleum industry is the cleanup of oil containers and storage tanks used in oil transportation. For the first time in 1981 the use of biosurfactants was proposed for cleaning tanks (Gutnick and Rosenberg, 1981) as an alternative to traditional methods. Ten years later Banat et al. (1991) described the application of microbial biosurfactants for the cleanup of oil storage tanks. A field trial carried out at the Kuwait Oil Company demonstrated that the addition of rhamnolipid effectively lifted and mobilized oil sludge from the bottom of the tank and solubilized it within the emulsion formed. The treatment recovered 91% of hydrocarbons in the sludge (Perfumo et al., 2010). An improved form of this technology was patented in 2004 by Idrabel Italia (Italy) and Jeneil Biosurfactant Company (United States). As a result of the implementation of the proposed process, recovery of oil is generally $> 90\%$ and reduction of material to be disposed is $<5\%$. Also, the use of natural biosurfactant products is with high environmental compatibility (Perfumo et al., 2010).

One problem associated with the petroleum industry (production, distribution, and use of petroleum) is the extremely damaging effects on the environment: pollution of water resources and contamination of the soil. Bioremediation techniques have become a promising approach for removing oil residues from polluted areas (Mulligan and Yong, 1997). One reliable technique designed to enhance natural biodegradation by indigenous microorganisms is the use of biosurfactants (Helmy et al., 2010; Kaczorek, 2012; Joo and Kim., 2013). The examination of long-term contact of bacterial cells with diesel oil in the presence of rhamnolipids revealed that the rates of biodegradation were higher. It was observed that the addition of rhamnolipids led to a significant increase in the efficiency of diesel oil removal (Kaczorek et al., 2013).

Although the biosurfactants and bioemulsifiers have been extensively investigated and there is a large number of patents in the petroleum industry, only a few had successful commercialization. The problem is high production costs.

7 Risk assessment aspects of biosurfactant application

According to the U.S. EPA definition, risk assessment characterizes the nature and magnitude of health risks to humans and ecological receptors from chemical contaminants and other stressors, which may be present in the environment (http://epa.gov/riskassessment/basicinformation.htm#arisk, 2012). So, in general, the risk assessment process can be divided into two major directions—human health risk assessment and ecological risk assessment. A human health risk assessment estimates the nature and probability of adverse health effects in humans who may be exposed to some certain agent or agents (http://www.epa.gov/risk_assessment/health-risk.htm, 2012). An ecological risk assessment evaluates the probability of the impact on the environment as a result of exposure to one or more agents (http://www.epa.gov/risk_assessment/ecological-risk.htm, 2012). Due to their general biodegradability, low toxicity, mild action as well as low distribution of biosurfactants, in the past the problems of their risk assessment were barely studied and reviewed. Unlike biogenous surfactants, the risk assessment for synthetic surfactants was conducted and reviewed in numerous papers (Lewis, 1991; Wibbertmann et al., 2011). With regard to biosurfactants, two main risks can be connected with their wider distribution: (1) certain toxicity levels of the specific biosurfactants; and (2) the possibility of their biodegradation in the environment.

7.1 Biodegradability of biosurfactants

High biodegradability of biosurfactants is declared in numerous reviews (Mulligan, 2005; Rodrigues et al., 2006; Muthusamy et al., 2008). At the same time, their actual biodegradation in water or soil media is seldom studied. The biosurfactants surfactin, iturin, and fengycin produced by *Bacillus* sp., surfactin produced by *B. subtilis*, glycolipid produced by *Dietzia maris*, arthrofactin produced by *Arthrobacter oxydans,* and flavolipids produced by *Flavobacterium* sp. were used as carbon sources in liquid medium and soil microcosms by a mixed culture composed of three bacterial strains representatives of the genera *Acinetobacter* and *Pseudomonas*. The biosurfactants did not differ in their ability to stimulate the respiratory activity of the mixed culture during seven days of incubation. The biodegradability of all biosurfactants by mixed bacterial culture in soil microcosms did not differ significantly and ranged from 42.5% up to 73.4%. In a comparative analysis, the biological surfactants had a higher degradation potential than the synthetic surfactant SDS, which was also mineralized by these microorganisms, but at a lower rate—24.8% (Lima et al., 2011). Hirata et al., (2009) have shown that *Candida bombicola* sophorolipids can be classified as "readily" biodegradable chemicals, which are defined as chemicals that are degraded 60% within 28 days under specified test methods using the OECD Guidelines for Testing of Chemicals (301C, Modified MITI Test) (Hirata et al., 2009). The rhamnolipid biodegradation in black loamy soil and red sandy soil after 7 days amounted to 92%; the initial biosurfactant concentration in soil made 240 mg/kg (Pei et al., 2009).

According to OECD 301D "Ready Biodegradability," the biosurfactant showed ready biodegradability by demonstrating a biodegradability rate of 68.4% on day 10 of the 28-day test cycle. This is an excellent test result, clearly demonstrating that JBR 425 is readily biodegradable (Uysal and Türkman, 2005). The research by Mohan et al. indicated the 74% biodegradability of rhamnolipid after 10 days under aerobic conditions and 47% after 6 days under anaerobic conditions, while synthetic surfactant Triton X-100 was only slightly biodegradable under aerobic conditions (Mohan et al., 2006). The investigation of Frank et al. (2010) on the degradation of selected biosurfactants by mixed bacterial cultures demonstrated that the cultures were able to use all of the tested biosurfactants (rhamnolipid, sophorolipid, and trehalose tetraester) as the sole carbon source, but only the rhamnolipid could be degraded completely. Only a primary degradation was observed for the other biosurfactants (Frank et al., 2010).

In conclusion, the biodegradation of biosurfactants under environmental conditions allows their wide application as bioremediation stimulators, but conduction of biodegradability tests is recommended before their field use.

7.2 Toxicity of biosurfactants

During recent years, several studies have been devoted to the assessment of toxicity of biosurfactants to various test organisms. Kuyukina et al. have shown that the biosurfactant glycolipid complex synthesized by *Rhodococcus ruber* is not toxic and exhibits no appreciable effect on the proliferative activity of peripheral blood leukocytes (Kuyukina et al., 2007). Hwang et al. (2009) have reported that when studying the effect of *Bacillus subtilis* surfactin on Sprague-Dawley rats, there were no surfactin C-related toxicities in survival, clinical signs, and hematological parameters in the experimental period. The highest dose of surfactin C showed the decrease in body weight gain despite normal food and water consumption and the increase in relative liver weight (Hwang et al., 2009). Certain toxicity levels of a mixture of mono- and di-rhamnolipids—EC50 45-167 mg/L studied via exposure to luminescent bacteria—were reported (Bondarenko et al., 2010). Glycolipids produced by *Rhodococcus* sp. Strain 413A exhibited 50% less toxicity than Tween 80 in naphthalene solubilization tests (Kanga et al., 1997). Ivshina et al. have determined that the toxicity (IC50) of R.ruber biosurfactants was 10-100 times lower than that of several commercial synthetic surfactants (Ivshina et al., 1998). It was shown that biosurfactants from *Pseudomonas* sp. PS-17 possess low toxicity and that the biocomplex PS (a natural mixture of rhamnolipid and biopolymer polysaccharide) was less toxic than the rhamnolipid alone (Kolwzan et al., 2008). Therefore, it can be assumed that despite revealing certain toxicity levels biosurfactants are more acceptable for use and are considerably safer to living organisms than their synthetic counterparts.

Another aspect of biosurfactant risk assessment is the hazard, which can be caused by the biosurfactant-producing microorganisms since several species are opportunistic pathogens (Lyczak et al., 2000). This demands providing special conditions and complying with safety requirements during their production.

8 Conclusions

Microbial surfactants are receiving increasing attention due to their potential commercial applications as substitutes for synthetic surfactants. Similar to synthetic surfactants, they exhibit high surface and emulsifying activities, and they are stable under extreme chemicophysical conditions. Their most important advantage is their ecological applicability, their biodegradability, and their production y a variety of microorganisms that occur naturally in the environment. Thanks to these features, public acceptance is higher for microbial surfactants than for synthetic surfactants (Banat et al., 2000; Marchant and Banat, 2012). Microbial surfactants have been used to enhance contaminant removal in soil and water (Ron and Rosenberg, 2002; Franzetti et al., 2009). They have also been used in chemicophysical processes design to remediate hydrocarbon or heavy metal contaminated sites, such as in in-situ soil flushing and ex-situ soil washing for remediation of unsaturated zone or pump and treat for aquifer remediation (Mulligan, 2005; Franzetti et al., 2010). Due to their heterogeneity, microbial surfactants display a broad range of potential applications in the oil industry, as well as in agriculture, medicine, and the cosmetic and food industries (Banat et al., 2000; Nitschke et al, 2005; Karpenko et al, 2005; Marchant and Banat, 2012). Thus, the screening for new surface-active compounds-producing microorganisms is timely.

TERMINOLOGY

Surfactants surface-active compounds,; amphiphilic compounds that can reduce surface and interfacial tensions by accumulating at the interface of immiscible fluids and increase the solubility and mobility of hydrophobic or insoluble organic compounds

Biosurfactants surfactants produced by microbes;

CMC critical micelle concentration, defined as the minimum concentration necessary to initiate micelle formation.

Abbreviations

LPS lipopolysaccharides

OMP outer microbial membrane proteins

PAH polycyclic aromatic hydrocarbons

References

Abalos, A., Pinaso, A., Infante, M.R., et al., 2001. Physicochemical and antimicrobial properties of new rhamnolipids by *Pseudomonas aeruginosa* AT10 from soybean oil refinery wastes. Langmuir 17, 1367–1371.

Allard, A.S., Neilson, A.H., 1997. Bioremediation of organic waste sites: a critical review of microbiological aspects. Int Biodeterior Biodegrad 39, 253–285.

Al-Tahhan, R.A., Sandrin, T.R., Bodour, A.A., Maier, R.M., 2000. Rhamnolipid–induced removal of lipopolysaccharide from *Pseudomonas aeruginosa*: effect on cell surface properties and interaction with hydrophobic substrates. Appl. Environ. Microbiol. 66, 3262–3268.

Amiriyan, A., Assadi, M.M., Saggadian, V.A., Noohi, A., 2004. Bioemulsan production by Iranian oil reservoirs microorganisms. Iran J. Environ. Health Sci. Eng. 1, 28–35.

Araujo, L.V., Abreu, F., Lins, U., Santa Anna, L.M.M., Nitschke, M., Freire, D.M.G., 2011. Rhamnolipid and surfactin inhibit *Listeria monocytogenes* adhesion. Food Res. Int. 44, 481–488.

Asselineau, C., Asselineau, J., 1978. Trehalose containing glycolipids. Prog. Chem. Fats Lipids 16, 59–99.

Atlas, R.M., Bartha, R., 1992. Hydrocarbon biodegradation and oil spill bioremediation. Adv. Microbial. Ecol. 12, 287–338.

Avramova, T., Sotirova, A., Galabova, D., et al., 2008. Effect of Triton X-100 and rhamnolipid PS-17 on the mineralization of phenanthrene by *Pseudomonas* sp. cells. Int. Biodeterioration & Biodegradation 62, 415–420.

Bakshi, M.S., Sachar, S., 2006. Influence of temperature on the mixed micelles of Pluronic F127 and P103 with dimethylene-bis-(dodecyldimethylammonium bromide). J. Colloid Interface Sci. 296, 309–315.

Banat, I.M., Makkar, R.S., Cameotra, S.S., 2000. Potential commercial applications of microbial surfactants. Appl. Microbiol. Biotechnol. 53, 495–508.

Banat, I., Samarath, N., Murad, M., et al., 1991. Biosurfactant production and use in oil tank clean-up. World J. Microbiol. Biotechnol. 7, 80–88.

Banat, I.M., 1995. Biosurfactant production and possible uses in microbial enhanced oil recovery and oil pollution remediation—a review. Bioresour. Technol. 51, 1–12.

Banat, I.M., Franzetti, A., Gandolfi, I., et al., 2010. Microbial biosurfactants production, applications and future potential. Appl. Microbiol. Biotechnol. 7, 427–444.

Beal, R., Betts, W.B., 2000. Role of rhamnolipid biosurfactants in the uptake and mineralization of hexadecane in *Pseudomonas aeruginosa*. J. Appl. Microbiol. 89, 158–168.

Benincasa, M., Abalos, A., Oliveira, I., Manresa, A., 2004. Chemical structure, surface properties and biological activities of the biosurfactant produced by *Pseudomonas aeruginosa* LBI from soapstock. Antonie van Leeuwenhoek. Int. J. Gen. Mol. Microbiol. 85, 1–8.

Bhadoriya, S., Madoriya, N., Shukla, K., Parihar, M.S., 2013. Biosurfactants: A new pharmaceutical additive for solubility enhancement and pharmaceutical development. Biochem Pharmacol. 2, 2.

Bharali, P., Saikia, J.P., Ray, A., Konwar, B.K., 2013. Rhamnolipid (RL) from *Pseudomonas aeruginosa* OBP1: A novel chemotaxis and antibacterial agent. Colloids and Surfaces B: Biointerfaces 103, 502–509.

Bognolo, G., 1999. Biosurfactants as emulsifying agents for hydrocarbons. Colloids and Surfaces, A 152, 41–52.

Bondarenko, O., Rahman, P., Rahman, T., Kahru, A., Ivask, A., 2010. Effects of rhamnolipids from *Pseudomonas aeruginosa* DS10-129 on luminescent bacteria: Toxicity and modulation of cadmium bioavailability. Microb. Ecol. 59, 588–600.

Bouchez-Nautili, M., Rakatozafy, H., Marchal, R., Leveau, J.-Y., Vandecasteele, J.-P., 1999. Diversity of bacterial strains degrading hexadecane in relation to the mode of substrate uptake. J. Appl. Microbiol. 86, 421–428.

Bramwell, D.P., Laha, S., 2000. Effects of surfactant addition on the biomineralization and microbial toxicity of phenanthrene. Biodegradation 11, 263–277.

Brown, D.G., Jaffe, P.R., 2006. Effects of nonionic surfactants on cell surface hydrophobicity and apparent Hamaker constant of a *Sphingomonas* sp. Env. Sci. Technol. 40, 195–201.

Cameotra, S.S., Bollag, J.M., 2003. Biosurfactant-enhanced bioremediation of polycyclic aromatic hydrocarbons. Crit. Rev. Env. Sci. Tec. 30, 111–126.

Cameotra, S.S., Singh, P., 2008. Bioremediation of oil sludge using crude biosurfactants. Int. Biodeterioration & Biodegradation 62, 274–280.

Cameotra, S.S., Makkar, R.S., 2004. Recent applications of biosurfactants as biological and immunological molecules. Curr. Opin. Microbiol. 7, 262–266.

Cao, X.H., Liao, Z.Y., Wang, C.L., Yang, W.Y., Lu, M.F., 2009. Evaluation of a lipopeptide biosurfactant from *Bacillus natto* TK-1 as a potential source of anti-adhesive, antimicrobial and antitumor activities. Braz. J. Microbiol. 40, 373–379.

Cavalero, D.A., Cooper, D.G., 2003. The effect of medium composition on the structure and physical state of sophorolipids produced by *Candida bombicola* ATCC22214. J. Biotechnol. 103, 31–41.

Champion, J.T., Gilkey, J.C., Lamparski, H., Retterer, J., Miller, R.M., 1995. Electron microscopy of rhamnolipid (biosurfactant) morphology: Effects of pH, cadmium, and octadecane. J. Colloid Interface Sci. 170, 569–574.

Cheng, F.Y., Tang, C., Yang, H., Yu, H., Chen, Y., Shen, Z.Y., 2013. Characterization of a Blend-Biosurfactant of Glycolipid and Lipopeptide Produced by *Bacillus subtilis* TU2 Isolated from Underground Oil-Extraction Wastewater. J. Microbiol. Biotechnol. 23, 390–396.

Chiewpattanakul, P., Phonnok, S., Durand, A., Marie, E., Thanomsub, B.W., 2010. Bioproduction and anticancer activity of biosurfactant produced by the dematiaceous fungus *Exophiala dermatitidis* SK80. J. Microbiol. Biotechnol. 20, 1664–1671.

Christofi, N., Ivshina, I.B., 2002. Microbial surfactants and their use in field studies of soil remediation. J. Appl. Microbiol. 93, 915–929.

Christova, N., Tuleva, B., Galabova, D., 1996. Dependance of yeast permeabilization with Triton X-100 on the cell age. Biotechnol. Tech. 10, 77–79.

Chrzanowski, L., Owsianiak, M., Szulc, A., et al., 2011. Interactions between rhamnolipid biosurfactants and toxic chlorinated phenols enhance biodegradation of a model hydrocarbon-rich effluent. Int. Biodeterioration & Biodegradation 65, 605–611.

Cirigliano, M.C., Carman, G.M., 1985. Purification and characterization of liposan, a bioemulsifier from Candida lipolytica. Appl. Environ. Microbiol. 50, 846–850.

Coimbra, C.D., Rufino, R.D., Luna, J.M., Sarubbo, L.A., 2009. Studies of the cell surface properties of *Candida* species and relation to the production of biosurfactants for environmental applications. Curr. Microbiol. 58, 245–251.

Cooper, D.G., Macdonald, C.R., Duff, S.J.B., Kosaric, N., 1981. Enhanced production of surfactin from Bacillus subtilis by continuous product removal and metal cation additions. Appl. Environ. Microbiol. 42, 408–412.

Cooper, D.G., Paddock, D.A., 1984. Production of a biosurfactant from Torulopsis bombicola. Appl. Environ. Microbiol. 47, 173–176.

Cortés-Sánchez, A., Hernández-Sánchez, H., Jaramillo-Flores, M.E., 2013. Biological activity of glycolipids produced by microorganisms: New trends and possible therapeutic alternatives. Microbiol. Res. 168, 22–32.

Costa, S.G.V.A.O., Nitschke, M., Haddad, R., Eberlin, M.N., Contiero, J., 2006. Production of *Pseudomonas aeruginosa* LBI rhamnolipids following growth on Brazilian native oils. Process Biochem. 41, 483–488.

Covino, S., Svobodová, K., Křesinová, Z., et al., 2010. In vivo and in vitro polycyclic aromatic hydrocarbons degradation by *Lentinus (Panus) tigrinus* CBS 577.79. Bioresour. Technol. 101, 3004–3012.

Dahrazma, B., Mulligan, C., 2007. Investigation of the removal of heavy metals from sediments using rhamnolipid in a continuous flow configuration. Chemosphere 69, 705–711.

Deming, J.W., Baross, J.A., 1993. The early diagenesis of organic matter: Bacterial activity. In: Engel, M.H., Macks, S.A. (Eds.), Organic Geochemistry: Principles and Applications. Plenum Press, New York, pp. 119–144.

Denyer, S.P., Maillard, J.-Y., 2002. Cellular impermeability and uptake of biocides and antibiotics in Gram-negative bacteria. J. Appl. Microbiol. 92, 35–45.

Desai, J.D., Banat, I.M., 1997. Microbial production of surfactants and their commercial potential. Microbiol. Mol. Biol. Rev. 61, 47–64.

Déziel, E., Lépine, F., Dennie, D., et al., 1999. Liquid chromatography/mass spectrometry analysis of mixtures of rhamnolipids produced by *Pseudomonas aeruginosa* strain 57RP grown on mannitol or naphthalene. Biochimica et Biophysica Acta—Molecular and Cell Biology of Lipids 1440, 244–252.

Dusane, D.H., Pawar, V.C., Nancharaiah, Y.V., et al., 2011. Anti-biofilm potential of a glycolipid surfactant produced by a tropical marine strain of Serratia marcescens. Biofouling 27, 645–654.

Edward, J.R., Hayashi, J.A., 1965. Structure of rhamnolipid from *Pseudomonas aeruginosa*. Arch. Biochem. Biophys 111, 415–421.

Finnerty, W.R., 1994. Biosurfactants in environmental biotechnology. Current Opinion in Biotechnology 5, 291–295.

Fracchia, L., Cavallo, M., Martinotti, M.G., Banat, I., 2012. Biosurfactants and bioemulsifiers biomedical and related applications—Present status and future potentials. In: Biomedical science, engineering and technology, Chapter 14, 325–370, Ghista, D.N. (Ed.), InTech.

Frank, N., Lißner, A., Winkelmann, M., Hu, R., et al., 2010. Degradation of selected (bio-)surfactants by bacterial cultures monitored by calorimetric methods. Biodegradation 21, 179–191.

Franzetti, A., Bestetti, G., Caredda, P., La Colla, P., Tamburini, E., 2008. Surface-active compounds and their role in the access to hydrocarbons in *Gordonia* strains. FEMS Microbiol. Ecol 63, 238–248.

Franzetti, A., Caredda, P., Ruggeri, C., et al., 2009. Potential applications of surface active compounds by *Gordonia* sp. strain BS29 in soil remediation technologies. Chemosphere 75, 801–807.

Franzetti, A., Gandolfi, I., Bestetti, G., Banat, I., 2010. (Bio)surfactant and bioremediation, Successes and failures. In: Plaza, G. (Ed.), Trends in bioremediation and phytoremediation. Research Signpost, Kerala, pp. 145–156.

Galabova, D., Tuleva, B., Spasova, D., 1996. Permeabilization of *Yarrowia lipolytica* cells by Triton X-100. Enzyme Microbial Technol. 18, 18–22.

Goswami, P., Singh, H.D., 1991. Different modes of hydrocarbon uptake by two *Pseudomonas* species. Biotechnol. Bioeng. 37, 1–11.

Goudar, C., Strevett, K., Grego, J., 1999. Competitive substrate biodegradation during surfactant-enhanced remediation. J. Environ. Eng. 125, 1142–1148.

Grabovska, A.P., Voloshinets, V.A., Semenyuk, I.V., Karpenko, E.V., 2008. Colloid properties of synthetic emulsifiers and natural rhamnolipids. Questions Chem. Chem. Technol. 6, 149–152.

Grangemard, I., Wallach, J., Maget-Dana, R., Peypoux, F., 2001. Lichenysin: A more efficient cation chelator than surfactin. Appl. Biochem. Biotechnol.—Part A Enzyme Eng. Biotechnol. 90, 199–210.

Gutnick, D., Rosenberg, E., 1981. Cleaning oil-contaminated vessels with α-emulsans. U.S. Patent, 4,276,094.

Hall, C., Tharakan, P., Hallock, J., Cleveland, C., Jefferson, M., 2003. Hydrocarbons and the evolution of human culture. Nature 426, 318–322.

Hayes, M., Hrebenar, K., Murphy, P., et al., 1990. Bioemulsifier-stabilized hydrocarbosols. U.S. Patent 4 (943), 390.

Helmy, Q., Kardena, E., Nurachman, Z., Wisjnuprapto, 2010. Application of biosurfactant produced by *Azotobacter vinelandii* AV01 for enhanced oil recovery and biodegradation of oil sludge. Int. J. Civil Environ. Eng. IJCEE 10, 7–14.

Hirata, Y., Ryu, M., Oda, Y., Igarashi, K., Nagatsuka, A., Furuta, T., Sugiura, M., 2009. Novel characteristics of sophorolipids, yeast glycolipid biosurfactants, as biodegradable low-foaming surfactants. J. Biosci. Bioeng. 108, 142–146.

Hommel, R.K., 1994. Formation and function of biosurfactants for degradation of water-insoluble substrates. In: Ratledge, C. (Ed.), Biochemistry of microbial biodegradation. Kluwer Academic Publishers, Dordrecht, pp. 63–87.

Hoppe, H.G., 1991. Microbial extracellular enzyme activity: a new key parameter in aquatic ecology. In: Chrost, R.G. (Ed.), Microbial enzymes in aquatic environment. Springer-Publ, Berlin, pp. 60–80.

Hua, F., Wang, H., 2012. Uptake modes of octadecane by *Pseudomonas* sp. DG17 and synthesis of biosurfactant. J. Appl. Microbiol. 112, 25–37.

Hubert, C., Shen, Y., Voordouw, G., 1999. Composition of toluene-degrading microbial communities from soil at different concentrations of toluene. Appl. Environ. Microbiol. 65, 3064–3070.

Hwang, Y.H., Kim, M.S., Song, I.B., Park, B.K., Lim, J.H., Park, S.C., Yun, H.I., 2009. Subacute (28 day) toxicity of surfactin C, a lipopeptide produced by *Bacillus subtilis*, in rats. J. Health Sci. 55, 351–355.

Imura, T., Ohta, N., Inoue, K., et al., 2006. Naturally engineered glycolipid biosurfactant leading to distinctive self-assembled structures. Chem. Eur. J. 12, 2434–2440.

Isoda, H., Kitamoto, D., Shinmoto, H., et al., 1997. Microbial extracellular glycolipid induction of differentiation and inhibition of protein kinase C activity of human promyelocytic leukaemia cell line HL60. Biosci. Biotechnol. Biochem. 61, 609–614.

Ivshina, I.B., Kuyukina, M.S., Philp, J.C., Christofi, N., 1998. Oil desorption from mineral and organic materials using biosurfactant complexes produced by *Rhodococcus* species. World J. Microbiol. Biotechnol. 14, 711–717.

Iwamoto, T., Nasu, M., 2001. Current bioremediation practice and perspective. J. Biosci. Bioeng. 92, 1–8.

Janek, T., Krasowska, A., Radwańska, A., Łukaszewicz, M., 2013. Lipopeptide biosurfactant pseudofactin II induced apoptosis of melanoma A 375 cells by specific interaction with the plasma membrane. PLoS ONE 8, e57991.

Janek, T., Łukaszewicz, M., Krasowska, A., 2012. Antiadhesive activity of the biosurfactant pseudofactin II secreted by the Arctic bacterium *Pseudomonas fluorescens* BD5. BMC Microbiol. 12, 24.

Johnsen, A.R., Wick, L.Y., Harms, H., 2005. Principles of microbial PAH-degradation in soil. Environ. Pollut. 133, 71–84.

Joo, M.H., Kim, J.Y., 2013. Characteristics of crude oil biodegradation by biosurfactant-producing bacterium *Bacillus subtilis* JK-1. J. Korean Soc. Appl. Biological Chem. 56, 193–200.

Joshi, S., Yadav, S., Nerurkar, A., Desai, A.J., 2007. Statistical optimization of medium components for the production of biosurfactant by *Bacillus licheniformis* K51. J. Microbiol. Biotechnol. 17, 313–319.

Joshi, S.J., Desai, A.J., 2013. Bench-Scale production of biosurfactants and their potential in *ex-situ* MEOR application. Soil & Sediment Contam. 22, 701–715.

Kaczorek, E., 2012. Effect of external addition of rhamnolipids biosurfactant on the modification of Gram positive and Gram negative bacteria cell surfaces during biodegradation of hydrocarbon fuel contamination. Pol. J. Environ. Stud. 21, 901–909.

Kaczorek, E., Sałek, K., Guzik, U., Dudzińska-Bajorek, B., 2013. Cell surface properties and fatty acids composition of *Stenotrophomonas maltophilia* under the influence of hydrophobic compounds and surfactants. New Biotechnol. 30 (2), 173–182.

Kaczorek, E., Sałek, K., Guzik, U., Jesionowski, T., Cybulski, Z., 2013. Biodegradation of alkyl derivatives of aromatic hydrocarbons and cell surface properties of a strain of *Pseudomonas aeruginosa*. Chemosphere 90, 471–478.

Kaczorek, E., Sałek, K., Guzik, U., Dudzińska-Bajorek, B., Olszanowski, A., 2013. The impact of long-term contact of Achromobacter sp. 4(2010) with diesel oil—Changes in biodegradation, surface properties and hexadecane monooxygenase activity. Int. Biodeterioration and Biodegradation 78, 7–16.

Kanga, S.H., Bonner, J.S., Page, C.A., Mills, M.A., Autenrieth, R.L., 1997. Solubilization of naphthalene from crude oil using biosurfactants. Environ. Sci. Technol. 31, 556–561.

Kappeli, O., Finnerty, W.R., 1979. Partition of alkane by an extracellular vesicle derived from hexadecane-grown *Acinetobacter*. J. Bacteriol. 140, 707–712.

Karanth, N., Deo, P., Veenanadig, N., 1999. Microbial production of biosurfactants and their importance. Curr. Sci. 77, 116–123.

Karpenko, E., Kolwzan, B., Sliwka, E., et al., 2008. Influence of microbial surfactants on biodegradation of oil hydrocarbons. Ekotoksykologia w ochronie srodowiska 884, 179–184.

Karpenko, E., Kolwzan, B., Vildanova, R., et al., 2005. Method of remediation of soils from petroleum-products - enhanced cleanning by biosurfactants. Proc. of the Int. Conf. on Bioremediation Soil and Groundwater—EBD, Silesian Univ. Technol., 79–88.

Karpenko, E., Lisova, N., Scheglova, N., et al., 2005. The perspectives of using ecologically safe surfactants for agriculture. In: Pawelczyk, A. (Ed.), Chemistry for Agriculture, 6. Czech-Pol Trade, Prague–Bruxelles, pp. 786–793.

Karpenko, E., Scheglova, N., Vildanova, R., Kolwzan, B., 2007. Application of microbial surfactants in new formulation for plant protection. In: Pawelczyk, A. (Ed.), Chemicals in Agriculture and Environment, 8 Czech-Pol Trade, Prague–Bruxelles, pp. 113–117.

Karpenko, E.V., Martynyuk, N.B., Shulga, A.N., Vildanova-Martsishin, R.I., 2004. Surface active biopreparation. Patent of Ukraine. N 71792, Bull.N 12.

Karpenko, E.V., Vildanova, R.I., Shcheglova, N.S., 2006. The prospect of using bacteria of genus *Rhodococcus* and microbial surfactants for degradation of oil pollutants. Appl.Biochem. and Microbiol. MAIK, Nauka/ Interperiodica 42, 156–159.

Khalladi, R., Benhabiles, O., Bentahar, F., Moulai-Mostefa, N., 2009. Surfactant remediation of diesel fuel polluted soil. J. Hazard. Mater. 164, 1179–1184.

Kitamato, D., Isoda, H., Hara, T.N., 2002. Functions and potential applications of glycolipid biosurfactants from energy-saving materials to gene delivery carriers. J. Bio. Sci. Bio. Eng. 94, 187–201.

Kitamoto, D., Morita, T., Fukuoka, T., et al., 2009. Self-assembling properties of glycolipid biosurfactants and their potential applications. Curr. Opin. Colloid Interface Sci. 14, 315–328.

Kolwzan, B., Karpenko, E., Vildanova-Martsishin, R., Sliwka, E., Grabas, K., 2003. Influence of biosurfactants on biodegradation of petroleum hydrocarbons. Proc. 2nd Sci. conf. “Surfaktanty i Układy Zdyspergowane w Teorii i Praktyce,”, 383–388.

Kolwzan, B., Biazik, J., Czarny, A., Zaczyńska, E., Karpenko, E., 2008. Assessment of toxicity of biosurfactants produced by *Pseudomonas* sp ps-17. In: Ecotoxicology in Environmental protection management (Pol). PZITS Oddz. Dolnosl., Wroclaw.

Kretschmer, A., Bock, H., Wagner, F., 1982. Chemical and physical characterization of interfacial-active lipids from *Rhodococcus erythropolis* grown on n-alkanes. Appl. Environ. Microbiol. 44, 864–870.

Kumar, M., Leon, V., De Sisto Materano, A., Ilzins, O.A., Luis, L., 2008. Biosurfactant production and hydrocarbon-degradation by halotolerant and thermotolerant *Pseudomonas* sp. World. J. Microbiol. Biotechnol. 24, 1047–1057.

Kuyukina, M.S., Ivshina, I.B., Gein, S.V., Baeva, T.A., Chereshnev, V.A., 2007. In vitro immunomodulating activity of biosurfactant glycolipid complex from *Rhodococcus ruber.* Bull Exp. Biol. Med. 144, 326–330.

Lang, S., Katsiwela, E., Wagner, F., 1989. Antimicrobial effects of biosurfactants. Fat Sci. Technol. 91, 363–366.

Lang, S., Philp, J.C., 1998. Surface-active lipids in rhodococci. Antonie Van Leeuwenhoek. 74, 59–70.

Lang, S., Wullbtandt, D., 1999. Rhamnoselipids—biosynthesis, microbial production and application potential. Appl. Microbiol. Biotechnol. 51, 22–32.

Lawniczak, L., Marecik, R., Chrzanowski, L., 2013. Contributions of biosurfactants to natural or induced bioremediation. Appl. Microbiol. Biotechnol. 97, 2327–2329.

Leahy, J.G., Colwell, R.R., 1990. Microbial degradation of hydrocarbons in the environment. Microbiol. Rev. 54, 305–331.

Lee, M., Kang, H., Do, W., 2005. Application of nonionic surfactant-enhanced in situ flushing to a diesel contaminated site. Water. Res. 39, 139–146.

Lewis, M.A., 1991. Chronic and sublethal toxicities of surfactants to aquatic animals: A review and risk assessment. Water Res. 25, 101–113.

Li, J.-L., Chen, B.-H., 2009. Surfactant-mediated biodegradation of polycyclic aromatic hydrocarbons. Materials 2, 76–94.

Li, Y.Y., Zheng, X.L., Li, B., 2006. Influence of biosurfactant on the diesel oil remediation in soil-water system. J. Environ. Sci. 18, 587–590.

Lima, T.M.S., Procópio, L.C., Brandão, F.D., Carvalho, A.M.X., Tótola, M.R., Borges, A.C., 2011. Biodegradability of bacterial surfactants. Biodegradation 22, 585–592.

Lourith, N., Kanlayavattanakul, M., 2009. Natural surfactants used in cosmetics: Glycolipids. Int. J. Cosmet Sci. 31, 225–261.

Lu, X.-Y., Zhang, T., Fang, H.H.-P., 2011. Bacteria-mediated PAH degradation in soil and sediment. Appl. Microbiol. Biotechnol. 89, 1357–1371.

Lyczak, J.B., Cannon, C.L., Pier, G.B., 2000. Establishment of *Pseudomonas aeruginosa* infection: lessons from a versatile opportunist. Microbes Infect 2, 1051–1060.

Magalhães, L., Nitschke, M., 2013. Antimicrobial activity of rhamnolipids against *Listeria monocytogenes* and their synergistic interaction with nisin. Food Control 29, 138–142.

Mager, H., Rothlisberger, R., Wagner, F., 1987. Use of sophorose-lipid lactone for the treatment of dandruffs and body odor. European Patent 0 209 783. Institut Francais du Petrole, Malmaison.

Magot, M., 2005. Indigenous microbial communities in oil fields. In: Ollivier, B., Magot, M. (Eds.), Petroleum microbiology. ASM Press, Washington, DC, pp. 21–33.

Maier, R.M., Soberon-Chavez, G., 2000. *Pseudomonas aeruginosa* rhamnolipids: Biosynthesis and potential applications. Appl. Microbiol. Biotechnol. 54, 625–633.

Makin, S.A., Beveridge, T.J., 1996. The influence of A-band and B-band lipopolysaccharide on the surface characteristics and adhesion of *Pseudomonas aeruginosa* to surfaces. Microbiology 142, 299–307.

Malachovska-Yutsz, A., Pokyn'broda, T., Karpenko, E., 2007. Degradation of Benzpyrene by Soil Microorganisms in the Presence of Glycolipids Produced by Strain *Pseudomonas* sp. PS-17. Biotechnology 3, 68–73.

Marchant, R., Banat, I., 2012. Biosurfactants: a sustainable replacement for chemical surfactants? Biotech. Lett. 34, 1597–1605.

Margesin, R., Schinner, F., 1997. Bioremediation of diesel-oil-contaminated alpine soils at low temperatures. Appl. Microbiol. Biotechnol. 47, 462–468.

Mata, J., Joshi, T., Varade, D., Ghosh, G., Bahadur, P., 2004. Aggregation behavior of a PEO-PPO-PEO block copolymer + ionic surfactants mixed systems in water and aqueous salt solutions. Colloids Surfaces A: Physicochem. Eng. Aspects 247, 1–7.

Mata-Sandoval, J.C., Karns, J., Torrents, A., 2001. Influence of rhamnolipids and triton X-100 on the biodegradation of three pesticides in aqueous phase and soil slurries. J. Agri. Food. Chem. 49, 3296–3303.

McInerney, M.J., Javaheri, M., Nagle Jr, D.P., 1990. Properties of the biosurfactant produced by *Bacillus licheniformis* strain JF-2. J. Ind. Microbiol. 5, 95–102.

Medrzycka, K., Hallmann, E., Karpenko, E., Pastewski, S., 2005. Adsorption properties of triton X-100 and rhamnolipid biosurfactant in the cortex of soil remediation. Proc. of 4-th Int. Conf. "Oils Environ.", 138–143.

Meyer-Reil, L.A., 1991. Ecological aspects of enzymatic activity in marine sediments. In: Chrost, R.G. (Ed.), Microbial enzymes in aquatic environment. Springer-Publ, Berlin, pp. 84–95.

Miller, R.M., 1995. Biosurfactant-facilitated remediation of metal-contaminated soils. Environ. Health Perspect. 103, 59–62.

Mohan, P.K., Nakhla, G., Yanful, E.K., 2006. Biokinetics of biodegradation of surfactants under aerobic, anoxic and anaerobic conditions. Water Res. 40, 533–540.

Mulligan, C.N., 2005. Environmental applications for biosurfactants. Environ. Pollut. 133, 183–198.

Mulligan, C.N., 2009. Recent advances in the environmental applications of biosurfactants. Curr. Opin. Colloid Interface Sci. 14, 372–378.

Mulligan, C.N., Wang, S., 2006. Remediation of a heavy metal contaminated soil by a rhamnolipid foam. Eng. Geol. 85, 75–81.

Mulligan, C.N., Yong, R.N., 1997. The use of biosurfactants in the removal of metals from oil-contaminated soil. In: Yong, R.N. (Ed.), Contaminated ground: Fate of pollutants and remediation. Thomas Telford, London, pp. 461–466.

Muthusamy, K., Gopalakrishnan, S., Kochupappy, R.T., et al., 2008. Biosurfactants: Properties, commercial production and application. Curr. Sci. 94, 736–747.

Neu, T., 1996. Significance of bacterial surface-active compounds in interaction of bacteria with interfaces. Microbiol. Rev. 60, 151–166.

Nikaido, H., 2003. Molecular basis of bacterial outer membrane permeability. Microbiol. Mol. Biol. Rev. 67, 593–656.

Nikolopoulou, M., Kalogerakis, N., 2008. Enhanced bioremediation of crude oil utilizing lipophilic fertilizers combined with biosurfactants and molasses. Mar. Pollut. Bull. 56, 1855–1861.

Nitschke, M., Costa, S.G.V.A.O., Contiero, J., 2005. Rhamnolipid surfactants: An update on the general aspects of these remarkable biomolecules. Biotechnol. Progress 21, 1593–1600.

Nitschke, M., Pastore, G.M., 2006. Production and properties of a surfactant obtained from *Bacillus subtilis* grown on cassava wastewater. Bioresour. Technol. 97, 336–341.

Noordman, W.H., Janssen, D.B., 2002. Rhamnolipid stimulates uptake of hydrophobic compounds by *Pseudomonas aeruginosa*. Appl. Environ. Microb. 68, 4502–4508.

Pantsyrnaya, T., Blanchard, F., Delaunay, S., et al., 2011. Effect of surfactants, dispersion and temperature on solubility and biodegradation of phenanthrene in aqueous media. Chemosphere 83, 29–33.

Pei, X., Zhan, X., Zhou, L., 2009. Effect of biosurfactant on the sorption of phenanthrene onto original and H2O2-treated soils. J. Env. Sci. 21, 1378–1385.

Perfumo, A., Rancich, I., Banat, I., 2010. Possibilities and challenges for biosurfactants use in petroleum industry. Adv. Exp. Med. Biol. 672, 135–145.

Piehler, M.F., Swistak, J.G., Pinckney, J.L., Paerl, H.W., 1999. Stimulation of diesel fuel biodegradation by indigenous nitrogen fixing bacterial consortia. Microb. Ecol. 38, 69–78.

Pinzon, N.M., Cook, A.G., Ju, L.K., 2013. Continuous rhamnolipid production using denitrifying *Pseudomonas aeruginosa* cells in hollow-fiber bioreactor. J. Chem. Technol. Biotechnol. 88, 719–726.

Piotrowska-Cyplik, A., Cyplik, P., Marecik, R., et al., 2012. Genetic and chemical analyses of transformations in compost compounds during biodegradation of oiled bleaching earth with waste sludge. Bioresour. Technol. 114, 75–83.

Pirog, T.P., Antoniuk, S.I., Karpenko, E.V., Shevchuk, T.A., 2009. Effect of cultivation of strain *Acinetobacter calcoaceticus* K-4 on the synthesis of surface-active agents. Appl. Biochem. Microbiol. 45, 304–310.

Pirog, T.P., Shevchuk, T.A., Voloshina, I.N., Karpenko, E.V., 2004. Production of surfactants by *Rhodococcus erythropolis* Strain EK-1. Appl. Biochem. Microbiol. 5, 544–550.

Pornsunthorntawee, O., Wongpanit, P., Chavadej, S., Abe, M., Rujiravanit, R., 2008. Structural and physicochemical characterization of crude biosurfactant produced by *Pseudomonas aeruginosa* SP4 isolated from petroleum-contaminated soil. Bioresour. Technol. 99, 1589–1595.

Prystay, M.V., Nogina, T.M., Karpenko, E.V., 2006. Selection of new effective biosurfactant producers from bacteria of genus *Rhodococcus* and *Gordonia*. Bull. Nat. Univ. "Lviv polytechnic," "Chem. technol. subst. appl." 590, 175–178.

Rahman, K.S.M., Rahman, T.J., Kourkoutoas, Y., et al., 2003. Enhanced bioremediation of n-alkane in petroleum sludge using bacterial consortium amended with rhamnolipid and micronutrients. Bioresour. Technol. 90, 159–168.

Ratledge, C., 1992. Chemistry of aliphatic hydrocarbons and degradation. In: Latin American Biodeterioration Symposium. Campos do Jordoa, SP, Brasil, pp. 236–250.

Remichkova, M., Galabova, D., Roeva, I., et al., 2008. Anti-herpesvirus activities of *Pseudomonas sp. S-17* Rhamnolipid and its complex with alginate. Z. Naturforsch. 63, 75–81.

Rodrigues, L., Banat, I.M., Teixeira, J., Oliveira, R., 2006. Biosurfactants: Potential applications in medicine. J. Antimicrob. Chemother. 57, 609–618.

Rodrigues, L., Mei, H.C.V., Teixeira, J., Oliveira, R., 2004. Influence of biosurfactants from probiotic bacteria on formation of biofilms on voice prostheses. Appl. Environ. Microbiol. 70, 4408–4410.

Ron, E.Z., Rosenberg, E., 2001. Natural roles of biosurfactants. Environ. Microbiol. 3, 229–236.

Ron, E.Z., Rosenberg, E., 2002. Biosurfactants and oil bioremediation. Curr. Opin. Biotechnol. 13, 249–252.

Rosen, M.J., 1978. Surfactants and interfacial phenomena. John Wiley and Sons, New York.

Rosenberg, E., Ron, E.Z., 1999. High- and low-molecular-mass microbial surfactants. Appl. Microbiol. Biotechnol. 52, 154–162.

Saeki, A., Sakai, H., Kamogawa, K., et al., 2000. Solubilization of oily compounds into fluorocarbon-hydrocarbon hybrid surfactant admicelles formed on alumina surfaces. Langmuir 16, 9991–9995.

Sánchez, M., Aranda, F.J., Espuny, M.J., et al., 2007. Aggregation behaviour of a dirhamnolipid biosurfactant secreted by *Pseudomonas aeruginosa* in aqueous media. J. Colloid Interface Sci. 307, 246–253.

Sarachat, T., Pornsunthorntawee, O., Chavade, S., et al., 2010. Purification and concentration of a rhamnolipid biosurfactant produced by *Pseudomonas aeruginosa* SP4 using foam fractionation. Bioresour. Technol. 101, 324–330.

Saravanakumari, P., Mani, K., 2010. Structural characterization of a novel xylolipid biosurfactant from *Lactococcus lactis* and analysis of antibacterial activity against multi-drug resistant pathogens. Bioresour. Technol. 101, 8851–8854.

Sebba, F., 1987. Foams and Bilquid Foams-Aphrons. John Wiley and Sons, Toronto, 236 pages.

Sen, R., 1997. Response surface optimization of the critical media components for the production of surfactin. J. Chem. Technol. Biotechnol. 68, 263–270.

Shcheglova, N., Vildanova, R., Karpenko, O., 2007. Remediation and Protection of Soil by Microbial Preparations and Natural Sorbent. In: Pawielczyk, A., et al. (Eds.), Chemicals in Agriculture and Environment series: Chemistry for Agriculture. Czech-Pol Trade, Prague–Bruxelles, pp. 211–216.

Shin, K.H., Ahn, Y., Kim, K.W., 2005. Toxic effect of biosurfactant addition on the biodegradation of phenanthrene. Environ. Toxicol. Chem. 24, 2768–2774.

Shin, K.-H., Kim, K.-W., Seagren, E.A., 2004. Combined effects of pH and biosurfactant addition on solubilization and biodegradation of phenanthrene. Appl. Microbiol. Biotechnol. 65, 336–343.

Shulga, A., Karpenko, E., Vildanova-Martishin, R., Turovsky, A., Soltys, M., 2000. Biosurfactant-enhanced remediation of oil contaminated environments. Adsorp. Sci. Technol. 18, 171–176.

Shulga, A.N., Karpenko, E.V., Eliseev, S.A., Koronelly, T.V., 1990. Extracellular lipids and surface-active properties of bacteria. R. erythropolis. Microbiologia. 59, 443–447.

Sim, L., Ward, O.P., Li, Z.-Y., 1997. Production and characterisation of a biosurfactant isolated from *Pseudomonas aeruginosa* UW-1. J. Ind. Microbiol. Biotechnol. 19, 232–238.

Simayev, Yu.M., Bazekina, L.V., Timirkhanov, N.Sh., Karpenko, E.V., Shulga, A.N., 1999. Composition for oil recovery. Patent of Russia. No. 2125152.

Singh, A., Van Hamme, J.D., Ward, O.P., 2007. Surfactants in microbiology and biotechnology. Part 2. Application aspects. Biotechnol. Adv. 25, 99–121.

Singh, P., Cameotra, S.S., 2004. Potential applications of microbial surfactants in biomedical sciences. Trends Biotechnol. 22, 142–146.

Sliwka, E., Kolwzan, B., Grabas, K., Karpenko, E., Rutkowski, P., 2009. Influence of rhamnolipids from *Pseudomonas* PS-17 on coal tar and petroleum residue biodegradation. Environ. Prot. Eng. 35, 139–150.

Sodagari, M., Wang, H., Newby, B.M., Ju, L.K., 2013. Effect of rhamnolipids on initial attachment of bacteria on glass and octadecyltrichlorosilane-modified glass. Colloids Surf. B: Biointerfaces 103, 121–128.

Sotirova, A., Avramova, T., Stoitsova, S., Lazarkevich, I., Lubenets, V., Karpenko, E., Galabova, D., 2012. The importance of rhamnolipid-biosurfactant-induced changes in bacterial membrane lipids of *Bacillus subtilis* for the antimicrobial activity of thiosulfonates. Curr Microbiol. 65 (5), 534–541.

Sotirova, A., Avramova, T., Lazarkevich, I., Lubenets, V., Karpenko, E., Galabova, D., 2010. Antimicrobial potential of selected thiosulfonates-based biocides and biosurfactants against bacteria and fungi. C. R. Acad. Bulg. Sci. 63, 1307–1314.

Sotirova, A., Spasova, D., Galabova, D., Karpenko, E., Shulga, A., 2008. Rhamnolipid-biosurfactant permeabilizing effects on Gram-positive and Gram-negative bacterial strains. Curr. Micrrobiol. 56, 639–644.

Sotirova, A., Spasova, D., Vasileva-Tonkova, E., Galabova, D., 2009. Effects of rhamnolipid-biosurfactant on cell surface of *Pseudomonas aeruginosa*. Microbiol. Res. 164, 297–303.

Sponza, D.T., Gok, O., 2012. Aerobic biodegradation and inhibition kinetics of poly-aromatic hydrocarbones (PAHs) in a petrochemical industry wastewater in the presence of biosurfactants. J. Chem. Technol. Biotechnol. 87, 658–672.

Su, W.-T., Chen, W.-J., Lin, Y.-F., 2009. Optimizing emulsan production of *A. venetianus* RAG-1 using response surface methodology. Appl. Microbiol. Biotechnol. 84, 271–279.

Tecon, R., van der Meer, J.R., 2010. Effect of two types of biosurfactants on phenanthrene availability to the bacterial bioreporter *Burkholderia sartisoli* strain RP037. Appl Microbiol Biotechnol. 85, 1131–1139.

Thanomsub, B., Pumeechockchai, W., Limtrakul, A., Arunrattiyakorn, P., Petchleelaha, W., Nitoda, T., Kanzaki, H., 2006. Chemical structures and biological activities of rhamnolipids produced by *Pseudomonas aeruginosa* B189 isolated from milk factory waste. Bioresour. Technol. 97, 2457–2461.

Tzintzun-Camacho, O., Loera, O., Ramírez-Saad, H.C., et al., 2012. Comparison of mechanisms of hexadecane uptake among pure and mixed cultures derived from a bacterial consortium. Int. Biodeterioration & Biodegradation 70, 1–7.

Urum, K., Pekdemir, T., Gopur, M., 2003. Optimum conditions for washing of crude oil contaminated soil with biosurfactant solutions. Trans. Inst. Chem. Eng. 81B, 203–209.

Uysal, A., Turkman, A., 2005. Effect of biosurfactant on 2,4-dichlorophenol biodegradation in an activated sludge bioreactor. Process Biochem. 40, 2745–2749.

Vasileva-Tonkova, E., Sotirova, A., Galabova, D., 2011. The effect of rhamnolipid biosurfactant produced by *Pseudomonas fluorescens* on model bacterial strains and isolates from industrial wastewater. Curr Microbiol 62, 427–433.

Vasileva-Tonkova, E., Galabova, D., 2003. Hydrolytic enzymes and surfactants of bacterial isolates from lubricant-contaminated wastewater. Z. Naturforsch. 58c, 87–92.

Vasileva-Tonkova, E., Galabova, D., Stoimenova, E., Lalchev, Z., 2008. Characterization of bacterial isolates from industrial wastewater according to probable modes of hexadecane uptake. Microbiol. Res. 163, 481–486.

Vasileva-Tonkova, E., Sotirova, A., Galabova, D., 2011. The effect of rhamnolipid biosurfactant produced by *Pseudomonas fluorescens* on model bacterial strains and isolates from industrial wastewater. Curr. Micrrobiol. 62, 427–433.

Vinson, P.K., Talmon, Y., Walter, A., 1989. Vesicle-micelle transition of phosphatidyl-choline and octyl glucoside elucidated by cryo-transmission electron microscopy. Biophys. J. 56, 669–681.

Wagner-Dobler, I., Bennasar, A., Vancanneyt, M., et al., 1998. Microcosm enrichment of biphenyl degrading microbial communities from soils and sediments. Appl. Environ. Microbiol. 64, 3014–3022.

Walzer, G., Rosenberg, E., Ron, E.Z., 2006. The *Acinetobacter* outer membrane protein A(OmpA) is a secreted emulsifier. Environ. Microbiol. 8, 1026–1032.

Walzer, G., Rosenberg, E., Ron, E.Z., 2009. Identification of outer membrane proteins with emulsifying activity by prediction of β-barrel regions. J. Microbiol. Methods 76, 52–57.

Wang, S., Mulligan, C.N., 2009. Rhamnolipid biosurfactant-enhanced soil flushing for the removal of arsenic and heavy metals from mine tailings. Process Biochem. 44, 296–301.

Ward, O.P., 2008. Microbial biosurfactants and biodegradation. In: Sen, R. (Ed.), Biosurfactants. Landes Bioscience, Austin, TX, 2008.

Warr, L.N., Perdrial, J.N., Lett, M.-C., Heinrich-Salmeron, A., Khodja, M., 2009. Clay mineral-enhanced bioremediation of marine oil pollution. Appl. Clay Sci. 46, 337–345.

Watanabe, K., 2001. Microorganisms relevant to bioremediation. Curr. Opin. Biotechnol. 12, 237–241.

Whang, L.-M., Liu, P.-W.G., Ma, C.-C., Cheng, S.-S., 2009. Application of rhamnolipid and surfactin for enhanced diesel biodegradation—Effects of pH and ammonium addition. J. Hazard. Mater. 164, 1045–1050.

Whang, L.-M., Liu, P.-W.G., Ma, C.-C., Cheng, S.-S., 2008. Application of biosurfactants, rhamnolipid, and surfactin, for enhanced biodegradation of diesel-contaminated water and soil. J. Hazard. Mater. 151, 155–163.

Whyte, L.G., Hawari, J., Zhou, E., et al., 1998. Biodegradation of variable-chain-length alkanes at low temperatures by a psychrotrophic *Rhodococcus* sp. Appl. Environ. Microbiol. 64, 2578–2584.

Wibbertmann, A., Mangelsdorf, I., Gamon, K., Sedlak, R., 2011. Toxicological properties and risk assessment of the anionic surfactants category: Alkyl sulfates, primary alkane sulfonates, and α-olefin sulfonates. Ecotoxicol. Environ. Saf. 74, 1089–1106.

Willumsen, P.A., Karlson, U., Pritchard, P.H., 1998. Response of fluoranthrenedegrading bacteria to surfactants. Appl. Microbiol. Biotechnol. 50, 475–483.

Yang, J.G., Liu, X., Long, T., Yu, G., Peng, S., Zheng, L., 2003. Influence of nonionic surfactant on the solubilization and biodegradation of phenanthrene. J. Environ. Sci. (China) 15, 859–862.

Yuan, S.Y., Wei, S.H., Chang, B.V., 2000. Biodegradation of polycyclic aromatic hydrocarbons by a mixed culture. Chemosphere 41, 1463–1468.

Zhang, Y., Miller, R.M., 1995. Effect of Rhamnolipid (Biosurfactant) Structure on Solubilization and Biodegradation of n-Alkanes. Appl. Environ. Microbiol. 61, 2247–2251.

Zeng, G., Liu, Z., Zhong, H., et al., 2011. Effect of monorhamnolipid on the degradation of n-hexadecane by *Candida tropicalis* and the association with cell surface properties. Appl. Microbiol. Biotechnol. 90, 1155–1161.

Zeyaullah, M., Atif, M., Islam, B., et al., 2009. Bioremediation: A tool for environmental cleaning. Afr. J. Microbiol. Res. 3, 310–314.

Zosim, Z., Guntick, D.L., Rosenberg, E., 1982. Properties of hydrocarbon in water emulsion. Biotechnol Bioeng, 281–292.

CHAPTER

4 Review on the Utilization of Low-cost Substrates to Obtain Metabolites for Protection of the Environment

Ana B. Moldes
ETSI, Chemical Engineering Department, University of Vigo, Spain

CHAPTER OUTLINE

1 Introduction

The bioremediation of contaminated sites is limited by the poor availability of hydrophobic contaminants such as pesticides, hydrocarbons, or chemical industrial wastes to microorganisms, which could be improved by using surface-active compounds. The main natural hydrocarbon removal mechanisms are photo-oxidation, evaporation, and microbial degradation, which may take years to stabilize and, consequently, years to remediate contaminated sites (Bognolo, 1998). Some bacteria, yeast, and fungi are able to produce metabolites useful for protection of the environment such as biosurfactants or bioemulsifiers, which are receiving increasing attention. These surface-active compounds increase the surface area of hydrophobic contaminants in soil or water, increasing their aqueous solubility and their microbial degradation. Biosurfactants are compounds produced by microorganisms that reduce the surface tension at the air–water interface, whereas bioemulsifiers are surface-active compounds that reduce the interfacial tension between immiscible liquids or at the solid–liquid interface (Batista et al., 2006). A criterion cited to identify bioemulsifiers is their ability to maintain at least 50% of the original emulsion volume 24 h after formation (Willumsen & Karlson, 1997), whereas a substance is

The Role of Colloidal Systems in Environmental Protection. http://dx.doi.org/10.1016/B978-0-444-63283-8.00004-1

considered a biosurfactant if it is able to reduce the surface tension of the media more than 8 mN m^{-1} (Van der Vegt et al., 1991).

For the bioremediation of contaminated sites, in using microbial metabolites it is important to focus on two questions. The microorganisms employed for obtaining the biosurfactants and/or bioemulsifier must be safe, and the large-scale production of these metabolites must be cost competitive with the synthesis of chemical surfactants and/or chemical emulsifiers. Thus use of low-cost agroindustrial wastes as a carbon source can lead to advances in the industrial use of these metabolites in the coming decade (Das & Mukherje, 2007).

2 Microorganisms that produce surface-active compounds

Pseudomona, Candida, Bacillus, and *Acinectobater* are the best studied microbial genera for obtaining biosurfactants. Biosurfactants obtained by these strains are well characterized; for example, *Pseudomona* is able to produce rhamnose-containing surfactants, also known as rhamnolipids (Mercade et al., 1993; Mercade & Manresa, 1994; Abalos et al., 2001; Mahnaz et al., 2004) as well as lipopeptides and lipoproteins with surface-active properties (Robert et al., 1989; Neu et al., 1990). On the other hand, *Candida bombicola and Candida apicola* produce sophorolipids (García Ochoa & Casas, 1997a; García Ochoa & Casas, 1997b); whereas some authors have reported that *Candida lipolytica* produces *a* polymeric biosurfactant named Liposan (Cirigliano & Carman, 1985). Another well-characterized biosurfactant is surfactin, which is produced by *Bacillus subtilis* (Cooper et al., 1981; Ahimou et al., 2000).

Recently, novel bacteria strains isolated from different hydrocarbon-contaminated environments have been referenced in the literature. For instance succinoyl trehalose lipids (STLs) are promising glycolipid biosurfactants produced from n-alkanes by *Rhodococcus* sp. These compounds show not only unique interfacial properties but also versatile biochemical actions (Tokumoto et al., 2009). Thus, Ruggeri et al. (2009) isolated several bacteria able to produce surface-active compounds from hydrocarbon-contaminated environments. These authors also found that *Cupriavidus* sp produces extracellular biosurfactants that reduce the surface tension into the culture medium up to 37.1 mN m^{-1}, whereas *Bradyrhizobium* sp and *Bosea* sp show emulsification activities comparable to those of synthetic surfactants. On the other hand, Calvo et al. (2008) have demonstrated that *Ochrobactrum anthropi* produces bioemulsifiers with potential applications in hydrocarbon bioremediation, whereas Bodour et al. (2004) have reported a new class of biosurfactant, a flavolipid, produced by *Flavobacterium sp.*

Other kinds of biosurfactants are the mannosylerythritol lipids (MELS), surface-active compounds that belong to the glycolipid class of biosurfactants abundantly produced by *Basidiomycetous* yeasts such as strains of *Pseudozyma.* Although MELs have been known for over five decades, they recently regained attention because of their environmental compatibility, mild production conditions, structural diversity, self-assembling properties, and versatile biochemical functions; they also exhibit not only excellent interfacial properties, but also remarkable differentiation-inducing activities against human leukemia cells (Morita et al., 2009). Another microorganism that is able to accumulate large amounts of MELs in the culture medium is the smut fungus *Ustilago scitaminea* (Morita et al., 2009).

It is important, however, to consider that biosurfactants are produced by opportunistic pathogen strains and that the surface-active compounds obtained with these microorganisms have to overcome a lot of problems associated with the complex environmental regulation affecting their biosynthesis. From this point of view, *Lactobacillus* strains are considered safe industrial strains, and several authors proposed that they produced biosurfactants instead of opportunistic pathogen strains like *P. aeruginosa* (Moldes et al., 2007; Portilla et al., 2007, 2008a, 2008b). For this reason in the next section we pay special attention to the production of biosurfactants from lactic acid bacteria.

3 Biosurfactants from lactic acid bacteria

Lactic acid bacteria can produce biosurfactants extracellularly or cell bounded to its plasmatic membrane. Velraeds et al. (1996) found that four strains of *Lactobacillus* (*Lactobacillus casei* subsp, *rhamnosus* 36 and ATCC 7469, *Lactobacillus fermentum* B54, and *Lactobacillus acidophilus* RC14) produced biosurfactants in their mid-exponential (4-5 h) and stationary growth phases (18 h). During the mid-exponential and stationary phase, biosurfactants were similar in their surface activities properties, but the latter one had a lower and better defined critical micelle concentration. In particular, biosurfactants from the stationary phase from *L. acidophilus* RC14 reached a low liquid surface tension phosphate buffered saline, with a critical micelle concentration of 1.0 mg ml^{-1}. These biosurfactants were characterized as protein and polysaccharides that possibly contained bound phosphate groups. On the other hand, Rodrigues et al. (2006a and 2006b) found that *L. pentosus; L. rhamnosus; L. casei; and L. coryniformis* produce cell-bound biosurfactants, whereas *Lactococcus lactis 53* and *Streptococcus thermophilus A* produce extracellular biosurfactants. Most recently lactic acid bacteria also were subjected to screening by Golek et al. (2009), and they found that the most favorable synthesis of biosurfactants was observed in the culture of *Lactobacillus casei*. The biosurfactant biosynthesized by this microorganism is constituted by protein fractions and polysaccharide fractions, with a typical structure of glycoproteins, which is affected by the medium composition and the phase of growth of the biosurfactant-synthesizing strain; this fact is in agreement with the information reported by Velraeds (1996).

4 Low-cost substrates as a carbon source for obtaining biosurfactants

Biosurfactants have several advantages over chemical surfactants, including lower toxicity and higher biodegradability and effectiveness at extreme temperatures or pH values. In spite of these advantages, biosurfactants must be cost competitive with chemical surfactants. From this point of view, utilization of agricultural residues as substrates can reduce the cost of the biotechnological production of biosurfactants. Table 4.1 summarizes most of the renewable raw materials reported in the literature that can currently be employed for obtaining biosurfactants; most of them consist of agricultural residues.

5 Lignocellulosic residues

The cost of biosurfactant production can be decreased by using renewable raw materials such as lignocellulosic residues (Moldes et al., 2007). Various agricultural residues can be employed as a carbon source for obtaining biosurfactants, and the utilization of these lignocellulosic residues for biotechnological purposes, like biosurfactant production, have various benefits. First, it prevents the

Table 4.1 Agricultural Residues Employed as Carbon Source for Producing Biosurfactants

Carbon Source	Strain	Localization	Reference
Barley bran husks	*Lactobacillus pentosus*	Cell bound	Moldes et al., 2007
Trimming vine shoots	*Lactobacillus pentosus*	Cell bound	Moldes et al., 2007
Corn cobs	*Lactobacillus pentosus*	Cell bound	Moldes et al., 2007
Eucalyptus *globulus* chips	*Lactobacillus pentosus*	Cell bound	Moldes et al., 2007
Grape marc	*Lactobacillus pentosus*	Cell bound	Portilla et al., 2007
Hazelnut shells	*Lactobacillus pentosus*	Cell bound	Portilla et al., 2008a
Walnut shells	*Lactobacillus pentosus*	Cell bound	Portilla et al., 2008a
Cassava waste	*Bacillus subtilis*	Extracellular	Nitschke & Pastore, 2006
Canola oil and glucose	*Candida lipolytica*	Extracellular	Sarubbo et al., 2007
Babassu oil	*Candida lipolytica*	Extracellular	Sarubbo et al., 1999
Sunflower-oil	*Pseudomonas aeruginosa*	Extracellular	Benincasa & Acorsini, 2008
Acidified soybean oil	*Pseudomonas aeruginosa*	Extracellular	Cha et al., 2008
Olive oil mill effluent	*Pseudomonas* sp	Extracellular	Mercade et al., 1993
Turkish corn oil and honey	*Candida Bombicola*	Extracellular	Pekin et al., 2005
Whey	*Lactococcus lactis*	Extracellular	Rodrigues et al., 2006b
Whey	*Streptococcus thermophilus A*	Extracellular	Rodrigues et al., 2006b

field burning of agricultural residues and prevents certain gases such as CO_2, CH_4, and N_2O from being produced in the tropospheric atmosphere, absorbing infrared radiation that brings out the greenhouse effect. Moreover, lignin is one of the main contributors of the total carbon of agricultural wastes. When lignocellulosic residues are field-burned polycyclic aromatic hydrocarbon components such as benzopyrene, catechol, hydroquinone, phenanthrene, and naphthalene are obtained (Kjällstrand et al., 1998). Some authors (Tsutsui et al., 1997) have reported that all these compounds can inhibit DNA synthesis and induce cancerous tumors in animals and humans. Therefore, utilization of these agricultural residues as raw material to obtain useful metabolites reduces the environmental impact produced by this kind of residue.

Among the lignocellulosic residues assayed, for obtaining biosurfactants, trimming vine shoots and grape marc hydrolyzates showed greater effectiveness than other agricultural residues such as corn cobs or barley bran husks (Moldes et al., 2007). However, before employing these lignocellulosic residues as a carbon source for biosurfactants, it is necessary to obtain solubilization of the lignocellulosic sugars by acid hydrolysis. The hydrolysis of these lignocellulosic residues can be carried out using diluted H_2SO_4, obtaining a liquid phase rich in glucose and xylose that, following neutralization with $CaCO_3$ and nutrient supplementation, can be sterilized and then employed as fermentation media for biosurfactant production (Moldes et al., 2007). Moreover, during the hydrolysis, described previously, a solid phase rich in cellulose and lignin is also obtained, which can be delignified with NaOH or acetic acid to obtain useful metabolites (Cruz et al., 1997). Therefore, once the lignin is removed from the cellulosic residue, this cellulose fraction can be saccharified by acid or enzymatic hydrolysis in order to obtain glucose-rich solutions that can be employed as a carbon source to produce

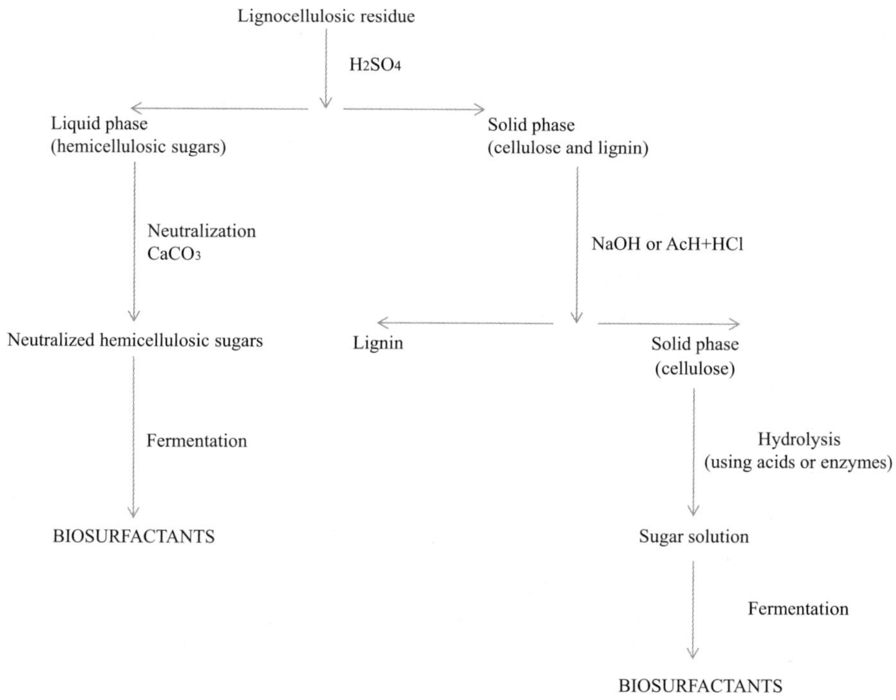

FIGURE 4.1

Flow diagram of the fractionation of lignocellulosic residues in order to obtain sugar solutions.

biosurfactants (Cruz et al., 1997). Figure 4.1 shows the flow diagram for obtaining sugar solutions from lignocellulosic residues that can be fermented by microorganisms to produce biosurfactants.

It is important to point out that biosurfactants are not the main metabolites, and at the same time that biosurfactants are synthesized other primary metabolites are produced in the fermentation media. For example, when biosurfactants are obtained from lactobacillus strains, lactic acid is also obtained at high concentrations; this fact could improve the profits associated with industrial production of biosurfactants because in this case not only biosurfactants but also lactic acid are obtained.

Table 4.2 shows the composition of some lignocellulosic materials proposed in the literature to obtain biosurfactants, and Table 4.3 shows the conditions employed for the diluted-acid hydrolysis of these lignocellulosic residues as well as the hemicellulosic sugars released.

On the other hand, Table 4.4 shows the mineral composition of hydrolyzates from grape marc, as well as the composition of the general medium of *Lactobacillus* reported by Portilla et al. (2009). It can be observed that hemicellulosic hydrolyzates from grape marc are rich in K, Ca, and Mg. It is important to point out that, among the mentioned minerals, Ca and Mg are found in much higher concentrations in grape marc hydrolyzates than in the general medium of *Lactobacillus* proposed by Mercier et al. (1992). Moreover, Portilla et al. (2009) observed that grape marc hydrolyzates are composed of glucose, arabinose, and xylose as well as other sugars consisting in galactose, mannose, and fructose. In some cases, it was observed that the presence of glucose in the fermentation media, at

Table 4.2 Composition (%) of Lignocellulosic Residues. The Results are Expressed as Percentage of the Initial Weight of Sample on a Dry Basis

Residue	Cellulose	Xilan	Araban	Acetyl Groups	Lignin
Barley bran husks	23.0	26.6	6.1	1.6	21.4
Trimming vine shoots	34.1	12.8	0.9	5.3	27.1
Corn cobs	31.7	30.9	3.8	3.4	20.3
Eucalyptus globulus chips	46.3	16.6	0.5	3.6	22.9
Hazelnut shells	23.7	19.6	0.7	4.1	40.8
Walnut shells	23.0	16.0	1.2	3.8	37.4
Grape marc	10.8	7.5	2.2	1.6	50.9

Data from Moldes et al. (2007) and Portilla et al. (2008a)

Table 4.3 Conditions Selected for the Acidic Hydrolysis of the Different Lignocellulosic Residues and Concentrations of Sugars Released During the Treatments[a]

Residue	H_2SO_4 (%)	Time (min)	Glucose (g L^{-1})	Xylose (g L^{-1})	Arabinose (g L^{-1})
Barley bran husks	3	15	5.82	40.7	7.47
Trimming vine shoots	3	15	9.18	19.1	2.81
Corn cobs	2	15	2.79	36.9	4.10
Eucalyptus globulus chips	3	60	2.24	19.5	1.25
Hazelnut shells	3	15	0.6	20.0	0.6
Walnut shells	3	15	1.3	18.4	1.6
Grape marc	3	30	2.2	8.2	2.1

[a]*Temperature = 130 °C; Solid/liquid ratio = 8 g g^{-1}.*
Data from Moldes et al. (2007) and Portilla et al. (2008a)

Table 4.4 Micronutrients Found in the General Medium of *Lactobacillus* in Comparison with the Micronutrients Found in Grape Marc Hydrolyzates (units are in mg Kg^{-1})

Medium	Mn	K	Na	Ca	Mg	Al	Ni	Cu	Zn	Fe
Grape marc hydrolyzates	10.5	2707	16.3	3681	198.5	1.2	<1.0	<1.0	<1.0	<1.0
Mercier medium[a]	<0.5	1705	1356	58.3	27	11.3	<1.0	<1.0	<1.0	<1.0

[a]*General medium of* Lactobacillus.
Data from Portilla et al. (2009)

concentrations higher than 5 g L^{-1}, represses the bioconversion of xylose by some lactobacilli strains (Garde et al., 2002); although the fermentation of xylose and arabinose at the same time as hexoses is not an obstacle during the fermentation of these hemicellulosic sugars by *L. pentosus* (Portilla et al., 2009).

6 Vegetable oil residues

The oil extraction process generates a great amount of waste, for which disposal is becoming a growing problem due to the high chemical oxygen demand of these residues. Moreover, the wastewater from oil extraction contains toxic substances for some microorganisms like polyphenols, but also contains valuable substances like fatty acids, nitrogen, or sugars that can be employed as nutrients to produce more ecological biosurfactants (Mercade et al., 1993; Mercade & Manresa 1994; Sarubbo et al., 1999, 2007). Therefore, Sarubbo et al. (2007) demonstrated that *Candida lipolytica* is able to synthesize a biosurfactant in a cultivation medium supplemented with canola oil and glucose as a carbon source. In this medium, *C. lipolytica* produces a protein-lipid-polysaccharide that reduces the surface tension of the media to 30 mN m^{-1} and with emulsifying properties comparable to commercial emulsifiers (Sarubbo et al., 2007). In addition, this yeast can produce bioemulsifiers using babassu oil as substrate (Sarubbo et al., 1999), whereas Pekin et al. (2005) reported the ability of *Candida Bombicola* to produce sophorolipids (extracellular) when grown on Turkish corn oil and honey. On the other hand, Benincasa and Accorsini (2008) found that *Pseudomonas aeruginosa* LBI produces surface-active rhamnolipids when it is cultivated on waste from the sunflower-oil process under different conditions. These authors reported that the biosurfactants produced by *P. aeruginosa* reduce the superficial and interfacial tensions between fluids, offering advantages over their chemical counterparts, especially because of their ecological acceptability.

Alternatively, Cha et al. (2008) isolated a new bacterial strain from activated sludge, identified as *Pseudomonas aeruginosa* EMS1; this strain produces a biosurfactant when grown on acidified soybean oil as the sole carbon source. The production of this biosurfactant occurs during the exponential phase. Velraeds et al. (1996), working with *Lactobacillus* strains, also observed that maximum biosurfactant production occurs during the exponential phase. Moreover, several strains of *Pseudomonas* also are able to grow on olive oil mill effluent and accumulate rhamnolipids (Mercade et al., 1993; Mercade & Manresa 1994), whereas *Serratia* sp or *Rhodococcus* sp are not able to produce biosurfactants growing in this medium.

7 Other agricultural residues

Cheese whey, obtained after casein precipitation in cheese making, presents one of the major disposal problems of the dairy industry. The biological oxygen demand of whey is very high (40,000 mg Kg^{-1} or more), producing a high environmental impact when it is disposed of. In this way, Rodrigues et al. (2006b) showed that whey is an economical alternative to synthetic media for biosurfactant production by *Lactococcus lactis 53* and *Streptococcus thermophilus A*. The utilization of whey as a carbon source for biosurfactant production has a double advantage: it reduces the cost of biosurfactants, and it lessens the environmental impact of this kind of residue (Dubey & Juwarkar, 2001; Rodrigues et al., 2006b).

Table 4.5 Physicochemical Composition of Cassava Wastewater

Components (unit)	Concentration	Components (Units)
Total carbohydrates	36.2	$g\ L^{-1}$
Reducing sugars	13.9	$g\ L^{-1}$
Total nitrogen	1.92	$g\ L^{-1}$
Phosphorus	246	$mg\ L^{-1}$
Potassium	894	$mg\ L^{-1}$
Magnesium	212	$mg\ L^{-1}$
Sulfur	288	$mg\ L^{-1}$
Aluminum	150	$mg\ L^{-1}$
Boron	148	$mg\ L^{-1}$
Copper	4.0	$mg\ L^{-1}$
Iron	0.3	$mg\ L^{-1}$
Manganese	4.2	$mg\ L^{-1}$
Zinc	1.6	$mg\ L^{-1}$
Cadmium	1.4	$mg\ L^{-1}$
Nickel	<0.01	$mg\ L^{-1}$
Lead	<0.01	$mg\ L^{-1}$
Ammonia [NH_4^+]	131	$mg\ L^{-1}$
Nitrate [NO_3^-]	16.2	$mg\ L^{-1}$

Data from Cavalante et al. (2008)

Another interesting residue, as carbon source, used to obtain biosurfactants is cassava wastewater. Cassava wastewater presents a good level of carbohydrates, as well as a high level of important micronutrients for microbial development, since it does not require supplementation (Cavalcante et al., 2008). Table 4.5 shows the physicochemical composition of cassava wastewater (Cavalcante et al., 2008). For instance, *Bacillus subtilis* LB5a strain is able to grow and to produce surfactant on cassava waste, reducing the surface tension of the medium to 26.6 mN m^{-1} and giving a crude surfactant concentration of 3.0 g L^{-1} after 48 h (Nitschke & Pastore, 2006). Moreover, this biosurfactant is capable of forming stable emulsions with various hydrocarbons. A preliminary chemical characterization revealed that the biosurfactant produced by *Bacillus subtilis* grown in cassava waste has a lipopeptide composition.

Consequently, based on the data reported previously, it can be concluded that use of low-cost by-products as a carbon source generates more ecological biosurfactants in comparison with chemical surfactants, with a high commercial value.

SYMBOLS AND TERMINOLOGY

DNA deoxyribonucleic acid
MELs mannosylerythritol lipids
mN m^{-1} mili Newton / metro (units employed to measure the surface tension)

STLs succinoyl trehalose lipids
Mn manganese
K potassium
Na sodium
Ca calcium
Mg magnesium
Al aluminum
Ni nickel
Cu copper
Zn zinc
Fe iron

Glossary

Activated sludge active biological material produced after the treatment of sewage and industrial wastewaters

Babassu oil a clear light yellow vegetable oil extracted from the seeds of the babassu palm (*Attalea speciosa*), which grows in the Amazon region of South America. The oil extracted by babassu palm has properties similar to coconut oil.

Barley bran husks lignocellulosic residue generated from the selection of barley before the malting process. It can be hydrolyzed by chemical and/or enzymatic stages to obtain sugar solutions.

Bioemulsifiers substances produced by living cells that stabilize an emulsion by increasing their kinetic stability. One class of emulsifiers is known as surface-active substances or surfactants.

Biosurfactants surface-active substances synthesized by living cells. They have the properties of reducing surface tension, stabilizing emulsions, and promoting foaming and are generally nontoxic and biodegradable.

Cassava waste also called yucca, or manioc; a woody shrub of the *Euphorbiaceae* (spurge family) native to South America that is extensively cultivated as an annual crop in tropical and subtropical regions for its edible starchy tuberous root, a major source of carbohydrates

Cellulose an organic compound with the formula $(C_6H_{10}O_5)_n$, a polysaccharide consisting of a linear chain of several hundred to over 10,000 $\beta(1\rightarrow4)$ linked D-glucose units

Corn cobs lignocellulosic residue generated after removing the cereal crop. It can be hydrolyzed by chemical and/or enzymatic stages to obtain sugar solutions.

Critical micelle concentration the concentration of surfactants above which micelles are spontaneously formed

Flavolipid biosurfactant produced by *Flavobacterium* sp, composed by citric acid and two cadaverine molecules, which is quite different from the polar moieties found in any of the currently reported classes of biosurfactants

Grape marc hydrolyzates sugar solution obtained after the treatment of grape marc with acids or enzymes

Grape marc lignocellulosic residue obtained during wine-making technology after pressing the grapes

Greenhouse effect the heating of the surface of a planet or moon due to the presence of an atmosphere containing gases that absorb and emit infrared radiation

Hemicellulosic sugars matrix of polysaccharides present in almost all plant cell walls along with cellulose. While cellulose is crystalline, strong, and resistant to hydrolysis, hemicellulose has a random, amorphous structure with little strength. It is easily hydrolyzed by dilute acid or base as well as by enzymes. Xylose is always the sugar monomer present in the largest amount.

Hydrolysis reaction during which one or more water molecules are split into hydrogen and hydroxide ions

Lactic acid chemical additive with several applications in the food industry as an acidulant or preservative

Lignin a complex chemical compound most commonly derived from wood and an integral part of the secondary cell walls of plants. Lignin is a cross-linked racemic macromolecule with hydrophobic and aromatic properties.

Lignocellulosic residues agricultural wastes made up of cellulose, lignin, and hemicelluloses
Lipopeptides a molecule consisting of a lipid connected to a peptide
Lipoproteins a biochemical assembly that contains both proteins and lipids
Liposan polymeric biosurfactant produced by *Candida lipilytica*, composed of carbohydrates and proteins
Mannosylerythritol lipids (MELS) glycolipid biosurfactants abundantly produced by *Basidiomycetous* yeasts such as strains of *Pseudozyma* or fungi such as *Ustilago scitaminea*
Photo-oxidation the oxidation of a polymer surface in the presence of oxygen or ozone that is facilitated by radiant energy such as UV or artificial light
Polycyclic aromatic hydrocarbon chemical compounds that consist of fused aromatic rings and do not contain heteroatoms or carry substituent. They are of concern because some compounds have been identified as carcinogenic, mutagenic, and teratogenic.
Rhamnolipids biosurfactants from the glycolipid group produced, among other species, by *Pseudomona aeruginosa*
Sophorolipids glycolipids secreted by some yeast species and in particular *Candida (Torulopsis) bombicola*. They contain the sugar sophorose.
Succinoyl trehalose lipids (STLs) glycolipid biosurfactants produced, among other species, by *Rhodococcus* sp
Surfactin lipopeptide biosurfactant produced by *Bacillus subtilis*
Trimming vine shoots lignocellulosic residue from vineyard that can be hydrolyzed by chemical and/or enzymatic stages to obtain sugar solutions

References

Abalos, A., Pinazo, A., Infante, M.R., Casals, M., Garcia, F., Manresa, A., 2001. Physicochemical and antimicrobial properties of new rhamnolípids producted by *Pseudomonas aeruginosa AT10* from soybean oil refinery wastes. Langmuir 17, 1367–1371.

Ahimou, F., Jacques, P., Deleu, M., 2000. Surfactin and iturin A effects on *Bacillus subtillis* surface hydrophobicity. Enzyme. Microb. Technol. 27, 749–754.

Batista, S.B., Mounteer, A.H., Amorim, F.R., Totola, M.R., 2006. Isolation and characterization of biosurfactant/bioemulsifier-production bacteria from petroleum contaminated sites. Bioresour. Technol. 97, 868–875.

Benincasa, M., Accorsini, R., 2008. *Pseudomonas aeruginosa* LBI production as an integrated process using the wastes from sunflower-oil refining as a substrate. Bioresour. Technol. 99, 3843–3849.

Bodour, A.A., Guerrero-Barajas, C., Jiorle, B.V., et al., 2004. Structure and characterization of flavolipids, a novel class of biosurfactants produced by *Flavobacterium* sp. Strain MTN11. Appl. Environ. Microbiol. 70 (1), 114–120.

Bognolo, G., Biosurfactants as emulsifying agents for hydrocarbons, 1998. Colloid Surf. A: Physicochem. Eng. Aspects 152, 41–52.

Calvo, C., Silva-Castro, G.A., Uad, I., García Fandiño, C., Laguna, J., González-López, J., 2008. Efficiency of the EPS emulsifier produced by *Ochrobactrum anthropi* in different hydrocarbon bioremediation assays. J. Ind. Microbiol. Biotechnol. 35 (11), 1493–1501.

Cavalcante, F.F., Nunes, A., Pastore, G.M., 2008. Production of biosurfactant by *Bacillus subtilis* LB5a on a pilot scale using cassava wastewater as substrate. J. Ind. Microbiol. Biotechnol. 35, 1071–1078.

Cha, M., Lee, N., Kim, Minju, Kim, Mia, Lee, S., 2008. Heterologous production of *Pseudomonas aeruginosa* EMS1 biosurfactant in. Pseudomonas putida. Bioresource Technol. 99, 2192–2199.

Cirigliano, M.C., Carman, M.G., 1985. Purification and characterization of liposan, a bioemulsifier from Candida lipolytica. Appl. Environ. Microbiol. 50, 846–850.

Cooper, D.G., MacDonald, C.R., Duff, S.J.B., Kosaric, N., 1981. Enhanced production of surfactin from *B. subtillis* by continuous product removal and metal cation additions. Appl. Environ. Microbiol. 42, 408–412.

Cruz, J.M., Moldes, A.B., Bustos, G., Torrado, A., Domínguez, J.M., 1997. Integral utilization of barley husk for the production of food additives. J. Sci. Food. Agric. 87 (6), 1000–1008.

Das, K., Mukherje, R.K., 2007. Comparison of lipopeptide biosurfactants production by *Bacillus subtilis* strains in submerged and solid state fermentation systems using a cheap carbon source: some industrial applications of biosurfactans. Process Biochemistry 42, 1191–1199.

Dubey, K., Juwarkar, A., 2001. Destillery and curd whey wastes as viable alternative sources for biosurfactant production. World J. Microbiol Biotechnol. 17, 61–69.

Garde, A., Jonsson, G., Schmidt, A.S., Ahring, B.K., 2002. Lactic acid production from wheat straw hemicellulose hydrolyzate by *Lactobacillus pentosus* and Lactobacillus brevis. Bioresour. Technol. 81 (3), 217–223.

García-Ochoa, J.A., Casas, J.A., 1997a. Procedimiento para la obtención de soforolípidos a partir de glucosa y aceite utilizando células desnergetizadas de. Candida Bombicola. No PATENTE: ES 2103687 A1 ESPAÑA.

García-Ochoa, J.A., Casas, J.A., 1997b. Procedimiento para la obtención de soforosa por. Candida Bombicola. No PATENTE: ES 2103688 ESPAÑA.

Golek, P., Bednarski, W., Brzozowski, B., Dziuba, B., 2009. The obtaining and properties of biosurfactants synthesized by bacteria of the genus. Lactobacillus. Annals of Microbiol. 59, 119–126.

Kjällstrand, J., Ramnäs, O., Petersson, G., 1998. Gas chromatographic and mass spectrometric analysis of 36 lignin-related methoxyphenols from uncontrolled combustion of wood. J. Chromatogr. 824 (2), 205–210.

Mahnaz, A., Mahsa, R., Ashraful, N., Moorie, L., Manoochehr, S., 2004. Rhamnolipid production by *Pseudomonas aerugiosa MM1011* from sugar beet molasses. Asian J. Microbiol. Biotechnol. Environ. Sci. 6 (2), 203–207.

Mercade, M.E., Manresa, M.A., 1994. The use of agroindustrial by-products for biosurfactant production. J. Am. Oil. Chem. Soc. 71, 61–64.

Mercade, M.E., Manresa, M.A., Robert, M., Espuny, M.J., Andres, C., Guinea, J., 1993. Olive oil mill effluent: New substrates for biosurfactant production. Bioresource Technology 43, 1–6.

Mercier, P., Yerushalmi, L., Rouleau, D., Dochain, D., 1992. Kinetics of lactic acid fermentation on glucose and corn by Lactobacillus amylophilus. J. Chem. Technol. Biotechnol. 55, 11–121.

Moldes, A.B., Torrado, A.M., Barral, M.T., Domínguez, J.M., 2007. Evaluation of biosurfactant production from various agricultural residues by Lactobacillus pentosus. J. Agric. Food. Chem. 55 (11), 4481–4486.

Morita, T., Fukuoka, T., Imura, T., Kitamoto, D., 2009. Production of glycolipid biosurfactants by *Basidiomycetous* yeast. Biotechnol. Appl. Biochem. 1, 39–49.

Neu, T.R., Hartner, T., Poralla, K., 1990. Surface active properties of viscosin a peptidolipid antibiotic. Appl. Microbiol. Biotechnol. 32, 518–520.

Nitschke, M., Pastore, G.M., 2006. Production and properties of a surfactant obtained from Bacillus subtilis grown on cassava wastewater. Bioresource Technology 97 (2), 336–341.

Pekin, G., Vardar-Sukan, F., Kosaric, N., 2005. Production of sophorolipids from *Candida bombicola* ATCC 22214 using Turkish corn oil and honey. Engineering in Life Sciences 5 (4), 357–362.

Portilla-Rivera, O., Moldes, A.B., Torrado, A.M., Domínguez, J.M., 2007. Lactic acid and biosurfactants production from hydrolyzed distilled grape marc. Process Biochemistry 42 (6), 1010–1020.

Portilla-Rivera, O., Torrado, A., Domínguez, J.M., Moldes, A.B., 2008a. Stability and emulsifying capacity of biosurfactans obtained from lignocellulosic sources using Lactobacillus pentosus. J. Agric. Food. Chem. 56, 8074–8080.

Portilla-Rivera, O., Rivas, B., Torrado, A., Moldes, A.B., Domínguez, J.M., 2008b. Revalorization of vine trimming wastes using *Lactobacillus acidophilus* and Debaryomyces hansenii. J. Sci. Food. Agric. 88, 2298–2308.

Portilla-Rivera, O., Torrado, A.M., Moldes, A.B., Domínguez, J.M., 2009. Minerals and organic nitrogen present in grape marc hydrolyzates enhance xylose consumption by. Lactobacillus pentosus. Applied Biochemistry Biotechnology 152 (2), 262–274.

Robert, M., Mercade, M.E., Bosch, M.P., Parra, J.L., Espuny, M.J., Manresa, M.A., Guinea, J., 1989. Effect on the carbon source on biosurfactant production by *Pseudomonas aeruginosa* 44T. Biotechnol. Lett. 11, 871–874.

Rodrigues, L., Moldes, A., Teixeira, J., Oliveira, R., 2006a. Kinetic study of fermentative biosurfactant production by *Lactobacillus* strains. Biochemical Eng. J. 28 (2), 109–116.

Rodrigues, L.R., Teixeira, J.A., Oliveira, R., 2006b. Low-cost fermentative medium for biosurfactant production by probiotic bacteria. Biochemical Eng. J. 32 (3), 135–142.

Ruggeri, C., Franzetti, A., Bestetti, G., et al., 2009. Isolation and characterization of surface active compound-producing bacteria from hydrocarbon-contaminated environments. Int Biodeterior Biodegradation 63 (7), 936–942.

Sarubbo, L.A., Farias, C.B.B., Campos-Takaki, G.M., 2007. Co-utilization of canola oil and glucose on the production of a surfactant by *Candida lipolytica*. Curr. Microbiol. 54 (1), 68–73.

Sarubbo, L.A., Porto, A.L.F., Campos-Takaki, G.M., 1999. The use of babassu oil as substrate to produce bio-emulsifiers by *Candida lipolytica*. Can. J. Microbiol. 45 (5), 423–426.

Tokumoto, Y., Nomura, N., Uchivama, H., et al., 2009. Structural characterization and surface-active properties of a succinoyl trehalose lipid produced by *Rhodococcus sp.* SD-74. J. Oleo Sci. 2, 97–102.

Tsutsui, T., Hayashi, N., Maizumi, H., Huff, J., Barrett, J.C., 1997. Benzene-, catechol-, hydroquinone- and phenol-induced cell transformation, gene mutations, chromosome aberrations, aneuploidy, sister chromatid exchanges and unscheduled DNA synthesis in Syrian hamster embryo cells. Mutat. Res. 373 (1), 113–123.

Van der Vegt, W., Van der Mei, H.C., Noordmans, J., Busscher, H.J., 1991. Assessment of bacterial biosurfactant production through axisymmetric drop shape analysis by profile. Appl. Microbiol. Biotechnol. 35, 766–770.

Velraeds, M., Van der Mei, H., Reid, G., Bussche, H., 1996. Inhibition of initial adhesion of uropathogenic *Enterococcus faecalis* by biosurfactants from *Lactobacillus* isolates. Appl. Environ. Microbiol. 62, 1958–1963.

Willumsen, P.A.E., Karlson, U., 1997. Screening of bacteria, isolated from PAD-contaminated soils, for production of biosurfactant and bioemulsifiers. Biodegradation 7, 415–423.

CHAPTER

5 Use of Cloud Point Extraction with Ethoxylated Surfactants for Organic Pollution Removal

B. Haddou[*], J.P. Canselier[†], C. Gourdon[†]

[*] *U. S. T. Oran-MB, Faculté de chimie, Département de génie Chimique BP 1505, M'Nouar, Oran, Algérie,*
[†] *Université de Toulouse, INPT, UPS; Laboratoire de Génie Chimique, 4 allée Emile Monso, BP 84234, 31432 Toulouse Cedex 4, France*

CHAPTER OUTLINE

The Role of Colloidal Systems in Environmental Protection. http://dx.doi.org/10.1016/B978-0-444-63283-8.00005-3

1 Cloud point extraction (CPE): principle

Cloud point extraction (CPE) is one of the most promising, environmentally friendly, energy-saving processes for aqueous effluent purification. Among other advantages, it appears to be an answer to the current restricted use of organic solvents (volatile organic compounds, VOC), often toxic, on a large scale. In fact, in water/polyethoxylated non-ionic surfactant systems, the existence of a lower consolute point curve allows the separation of a surfactant-rich phase (coacervate) and a dilute phase, in which surfactant concentration is close to its critical micelle concentration (cmc $\approx$ 10^{-5} M) (Degiorgio 1985, Falbe 1986, Myers 1988). The origin of phase separation in CPE with increasing temperature lies in a delicate balance between repulsive and attractive surfactant–surfactant and surfactant–water interactions, associated with entropy increase. Above the cloud point (CP) of the solution, the high solubilizing power of the coacervate allows concentration of hydrophobic or amphiphilic substances, thence CPE (Scamehorn et al. 1989, Hinze et al. 1993, Quina et al. 1999). This operation, first applied to metal ion separation in the presence of a chelating agent (Watanabe et al. 1978) offers an interesting opportunity to get rid of organic pollutants (hydrocarbons, e.g., polycyclic aromatic hydrocarbons) (Li et al. 2003), VOC (Sakulwongyai et al. 2000, Kimchuwanit et al. 2000) phenols (Akita and Takeuchi 1995, de Barros Neto 1999, de Barros Neto et al. 2001), alcohols, amines (Duarte 2005, Duarte et al. 2005), dyes (Purkait et al. 2005), molecules of biological interest (Bordier 1981), herbicides (Carabias-Martinez et al. 2003), various organic compounds (Akita et al. 1995, 1996, Ferrera et al. 2004) and/or heavy metal-containing electrolytes under cationic or anionic forms, without chelating agent (Duarte 2005, Canselier et al. 2006, Duarte et al. 2008). Scaling-up of CPE has been attempted with a continuous rotating disk contactor (Trakultamupatam et al. 2004) or a variation of a mixer-settler device (De Barros Neto 1999).

Non-ionic surfactant micelles are also able to desorb insoluble organic species adsorbed on a solid (e.g., soil) through micellar solubilization of the pollutant, followed by its concentration into the coacervate (Komaromy-Hiller et al. 1995, Pastewski et al. 2006). Desorption is achieved by diffusion of the molecules from the solid surface modified by the surfactant into the micelles.

Compared with other extraction processes involving micelles (microemulsions, reverse micelles (Leung et al. 1988, Friberg et al. 1989, Hatton 1989, Canselier et al. 1995, Regalado et al. 1996, Stevens et al. 2001), coacervate extraction, another designation for CPE, can be viewed as a more environmentally benign separation process because the former two involve organic solvents and/or chelating agents.

2 CPE of soluble species

CPE is considered to be convenient and environmentally safe alternative to extraction with organic solvents. Such method offers some advantages over conventional liquid-liquid extraction, including high extraction efficiency, ease of waste disposal and the use of non-toxic and less dangerous reagents. In order to study the extraction of phenol or benzyl alcohol (representing a dissolved pollution) from their aqueous solutions, we chose to work with polyethoxylated alcohols (CiEj), easily and quickly biodegradable (Falbe, 1986).

2.1 Phenol and benzyl alcohol

2.1.1 Occurrence in effluents

In this review, we present the results of CPE applied to two pollutants that are frequently present in wastewater and industrial aqueous effluents: phenol and benzyl alcohol. The first compound is a weak acid ($pK_a = 9.82$) (Weast et al. 1982), and the second is a neutral species ($pK_a = 15.4$, i.e., higher than that of water) (Fasman 1997). Both possess a benzene ring in their chemical formula. Generally, the resistance of such molecules to biodegradability is higher than that of aliphatic ones. Phenol is manufactured (worldwide production: 8,800,000 tons in 2005) and used on a large scale in the chemical industry: For example, it is an important intermediate in the synthesis of Nylon 6, through caprolactam, aspirin, and numerous pharmaceutics, as well as synthetic aromas (thymol, eugenol, vanillin, etc.) (phenol, 2009). Its solubility in water is 82.8 g/L at 25°C (Carey 1990, March 1992), and its log P is 1.5 (Sangster 1989). Although phenol has been used as an antiseptic and an anesthetic, its toxic and ecotoxic effects led the Environmental Protection Agency (EPA) to limit its allowed concentration in surface waters to 0.3 mg/L (toxicological profile for phenol, 2008). Benzyl alcohol is largely used as a solvent for inks, paints, lacquers, and epoxy resin coatings and as a precursor to various esters (soap and perfume additives)(benzyl alcohol, 2009). Its aqueous solubility is almost half that of phenol (40 g/L at 25°C) (Philip et al. 1997), but its log P, equal to 1.05 (Sangster 1989), is also lower, which shows that the variation of log P is not always contrary to the aqueous solubility of organic compounds. Until now, no water quality standards or criteria exist in the United States or Canada, and the European Commission considers benzyl alcohol as presenting practically no health hazard at moderate doses (group ADI of 5 mg/kg bw) (European Commission 2002). However, this compound was chosen for its structure rather close to that of phenol and its different functionality.

The extraction of phenol and benzyl alcohol (representing a dissolved pollution) from their aqueous solutions was investigated. The effects of temperature, surfactant concentration, and addition of sodium chloride and sodium sulphate on the extent of solute extraction were also studied. The initial phenol concentration was 0.15 wt.%, which is 5000 times higher than the required phenol concentration in crude water. This water demands physical and/or chemical treatments in order to get the characteristics of potable water. Benzyl alcohol concentration was also chosen on this basis.

2.1.2 Choice of the non-ionic surfactants

The use of surfactants is growing rapidly with the increase in the demographic growth and the improvement of the way of life and hygiene conditions, especially in third world countries. But surfactants have a negative effect on surface water and wastewater quality. They decrease the amount of dissolved oxygen at the water–atmosphere interface, and the ecotoxicity of a number of them is not negligible: 10 mg/L is enough to inhibit bacterial flora, and 50 mg/L to inhibit phytoplankton (Angelier 2000). Surface tension lowering changes the respiratory mechanism in fish gills. The presence of surfactants also involves the death of insects living on the water surface, such as hemipters (Angelier 2000). As a consequence, the use of biodegradable surfactants became necessary to let nature play its role in the biodegradability mechanisms. Hammerton (1955) proposed that the nature of surfactant hydrophobic groups was the determinant factor for biodegradability processes, while the nature and position of hydrophilic groups had less effect. The first work done at the Laboratoire de Génie Chimique, Toulouse (De Barros Neto 1999, De Barros Neto et al. 2001) focused on the use of ethoxylated alkylphenols (alkylphenol ethoxylates, APE) in phenol extraction. However, these surfactants are no longer used for detergence applications in many countries and are going to be superseded progressively because of the toxic intermediates generated from their biodegradation process, namely, alkylphenols (Thiele 1997, Jonkers 2005, Zoller 2004, Guenther et al. 2006). More recently, ethoxylated alcohols (alcohol ethoxylates, AE), easily and quickly biodegradable, have been used as good alternatives to APE (Haddou et al. 2003, Canselier et al. 2006, Haddou et al. 2006, Haddou et al. 2007).

In general, surfactants suitable for CPE should be slightly soluble in water (to reduce surfactant loss in the dilute phase and possess a high-solubilizing power for hydrophobic compounds); have a low cloud point (to reduce heating requirement); form an easily separable phase; be readily available at low cost; and show low toxicity and easy biodegradability. Several non-ionic surfactants were investigated (Haddou et al. 2003, Canselier et al. 2006, Haddou et al. 2006, 2007, Talbi et al. 2009, Ghouas et al. 2010, Haddou et al. 2011, Ghouas et al. 2012). The effects of temperature and surfactant concentration, as well as effluent pH, were studied. In some cases, to avoid excessive heating and to reduce the energy costs of a high-scale CPE process, extraction tests were performed in the presence of salts (sodium sulphate or sodium chloride). Our work also focused on the comparison between APE and AE on target solute (soluble species and dispersed pollutants) extraction.

2.1.3 Results

2.1.3.1 Phase diagrams of binary and pseudobinary systems

Before starting a CPE test, it is necessary to plot the phase diagram of the water/surfactant/solute system, which allows locating the CPE domain (two-phase zone) exactly. In a homologous series of AE (C_iE_j), surfactant properties vary regularly with the carbon number, i, of the hydrophobic chain and the degree of ethoxylation, j. For example, the CP increases with j and decreases when i increases (Mitchell et al. 1983, Degiorgio 1985, Rupert 1992, Duarte and Canselier 2005). Thus, the critical temperature of n-$C_{12}E_4$ ($t_c \cong 4$°C) is lower than the ambient one, while that of n-$C_{12}E_6$ is located above 40°C (Mitchell et al. 1983; Fig. 5.1a). A branched chain is normally less hydrophobic than a linear one, but when n-$C_{12}E_6$ and Oxo-$C_{13}E_9$, or Oxo-$C_{10}E_3$ ($T_c = 2$°C at 1 wt.% in water) and Oxo-$C_{13}E_9$ are compared, the influence of the hydrophilic part seems prevalent. As already observed with an AE (Donbrow et al. 1976), Figures 5.1a-c show that phenol strongly lowers the cloud point temperature. Furthermore, as with APE (Donbrow et al. 1976, De Barros Neto et al. 2000, 2001), this phenomenon is especially noticeable for low surfactant concentrations. This indicates a significant interaction between phenol and the polar head group of the surfactant. Like phenol, benzyl alcohol lowers the CP of a non-ionic surfactant, namely, Oxo-$C_{13}E_9$ (Figure 5.1d). Its influence is less important, however, and does not depend on the benzyl alcohol/surfactant ratio, since the shape of the curve is preserved. This difference can be related to the weaker acidity of benzyl alcohol (less polar O-H bond). Its interaction with the polar head group of the surfactant (reducing surfactant solubility) is thus less significant.

When a small amount of sodium chloride is added (less than 0.01 M) to non-ionic surfactant solutions, the CP remains constant (Valaulikar et al. 1985). On the contrary, at concentrations higher than 0.1 M, sodium chloride lowers the CP (De Barros Neto et al. 1998). This phenomenon, illustrated with Oxo-$C_{13}E_9$ in Figure 5.1d, is due to the salting-out of the surfactant, induced by the solvated electrolyte. Salting-out favors the weakening of surfactant headwater interaction. It seems that sodium chloride lowers the CP by a continuous medium change, more than that by direct interaction with the ethylene oxide (EO) groups of the surfactant (Saito et al. 1967, Schott et al. 1983, Akita et al.

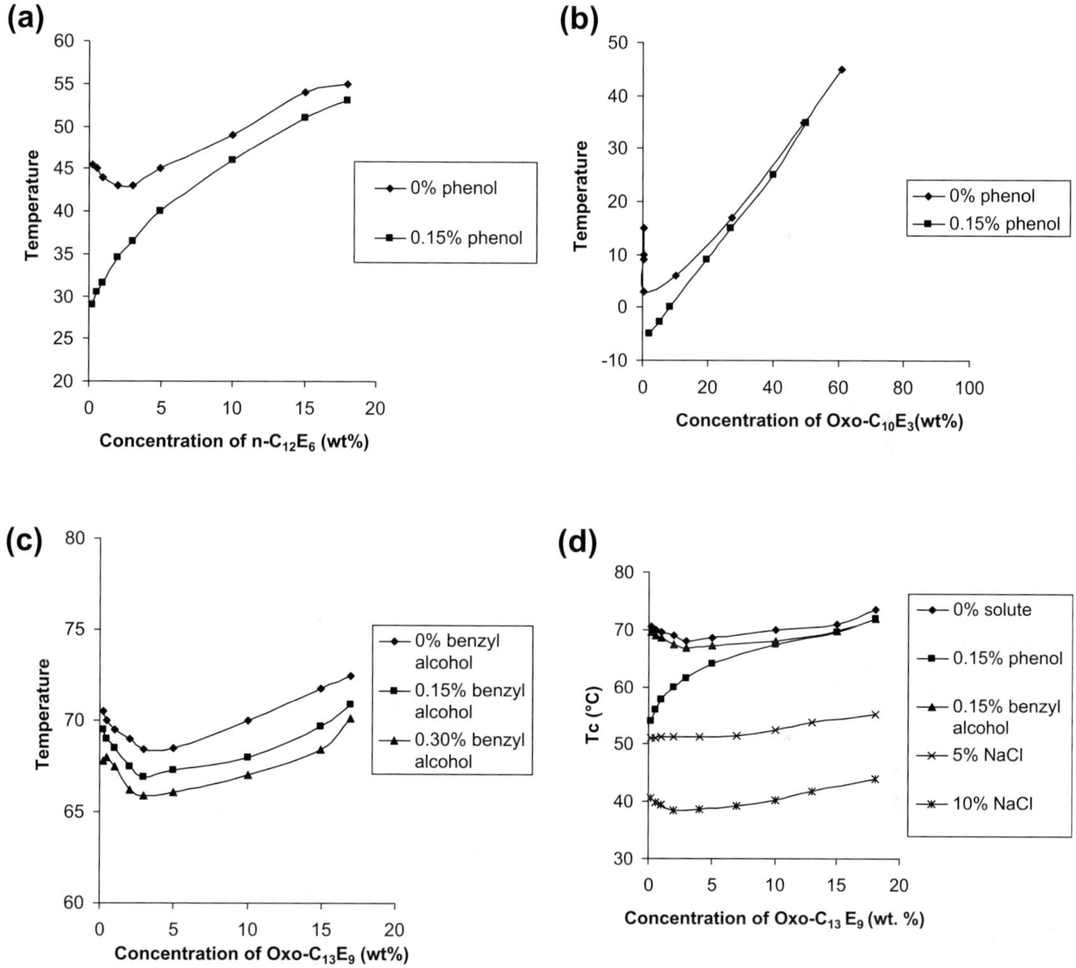

FIGURE 5.1

Effect of solutes on cloud point curves: (a) $C_{12}E_6$/phenol; (b) Oxo-$C_{10}E_3$/phenol; (c) Oxo-$C_{13}E_9$/benzyl alcohol; (d) Effect of phenol, benzyl alcohol and NaCl on the cloud point of Oxo-$C_{13}E_9$.

(from: Haddou et al., 2003, with permission).

1995, De Barros Neto et al. 1998). Figure 5.1d shows how 10 wt.% of NaCl (1.709 M) are able to lower the CP of a 1 wt.% $C_{13}E_9$ solution from 69°C to 38°C. Therefore, it is possible to adjust the CP to room temperature by simple control of surfactant and salt concentrations, thus reducing the heating energy cost for large-scale CPE. Salt addition can also have an application in the extraction of heat-sensitive products. Moreover, electrolyte does not distribute equally between the dilute and coacervate phases, and its addition increases the density of the dilute phase; this makes the separation of both phases easier in some non-ionic surfactant systems (Watanabe et al. 1978, Schott et al. 1984, Pramauro 1990).

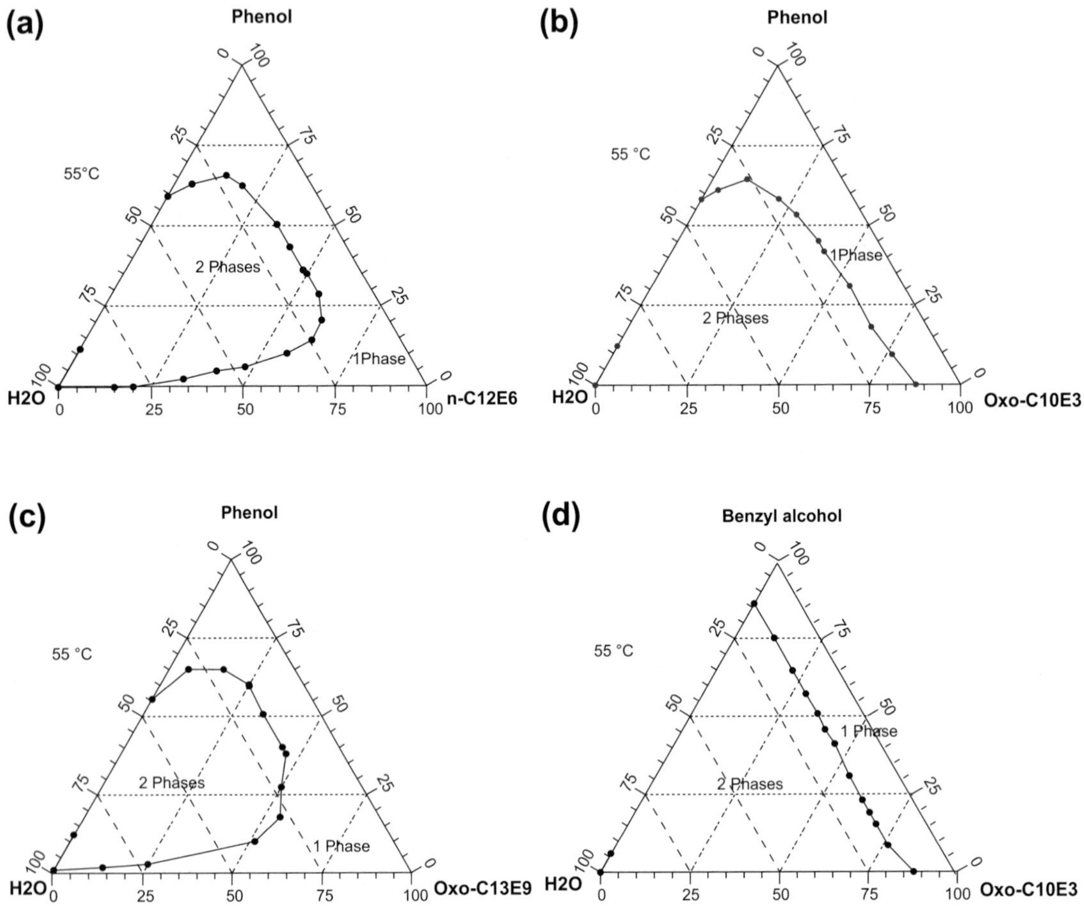

FIGURE 5.2

Isothermal phase diagrams at 55°C: (a) $H_2O/C_{12}E_6/C_6H_5OH$; (b) H_2O/Oxo-$C_{10}E_3/C_6H_5OH$; (c) H_2O/Oxo-$C_{13}E_9/C_6H_5OH$; (d) H_2O/Oxo-$C_{10}E_3/C_6H_5CH_2OH$.

(from: Haddou et al., 2003, with permission).

2.1.3.2 Isothermal diagrams: ternary systems

By analogy with solvent extraction, phenol and benzyl alcohol being the "solutes," water and surfactant are considered as the "diluent" and the "solvent," respectively. At 55°C, three of these diagrams (systems H_2O/n-$C_{12}E_6/C_6H_5OH$, H_2O/Oxo-$C_{10}E_3/C_6H_5OH$, H_2O/Oxo-$C_{10}E_3/C_6H_5CH_2OH$) belong to type II (two of the binary systems partially miscible)(Figures 5.2a,b,d) and the fourth (system H_2O/Oxo-$C_{13}E_9/C_6H_5OH$) to type I (only one of the binary systems partially miscible) (Figure 5.2c). In all cases, the second one-phase zone, near the H_2O apex, is too narrow to be visible on this scale (the aqueous solutions get cloudy as soon as the first drop of surfactant is added).

With respect to phenol, the behavior of the three surfactants studied is very close: Indeed, the mixtures remain turbid up to concentrations of about 75 % phenol, and only the shape and extent of the lower part of the two-phase domain (low-phenol concentration area) differ (Figs. 5.2a,b,c). As mentioned above, benzyl alcohol is much less water-soluble than phenol; water is also more soluble in phenol than in benzyl alcohol. The two-phase zone is thus smaller in the diagram of the $H_2O/\Phi OH/$ Oxo-$C_{10}E_3$ system than in that of the $H_2O/\Phi CH_2OH/$Oxo-$C_{10}E_3$ one (Figures 5.2b and d).

2.1.3.3 Extraction results

The extraction results of a 0.15 wt.% solution of solute (phenol, benzyl alcohol) with different surfactants, according to two variables—wt.% surfactant (X_t), and temperature (T)—are expressed by the four following "responses" (Y): percentage of extracted solute (E), residual concentrations of solute ($X_{s,w}$) and surfactant ($X_{t,w}$) in the dilute phase, and coacervate volume fraction at equilibrium (ϕ_c). For each parameter determined, by considering central composite designs (Box et al. 1987), the results were analyzed by an empirical fitting. In this method, the experimental values can be used to adjust the polynomial model constants. The models were checked. The quadratic correlation was chosen to give the slope and regression coefficient (R^2) closest to unity.

$$Y = a_0 + a_1X_t + a_2T + a_{12}X_tT + a_{11}X_t^2 + a_{22}T^2 \tag{5.1}$$

However, in what follows, one cannot allot physical significance to the portions of horizontal planes corresponding to the maximum value of the response.

2.1.3.3.1 Phenol. As an example, we will expose the results obtained with $C_{12}E_6$. The quadratic equations for the four properties (E, $X_{s,w}$, $X_{t,w}$, and ϕ_c), whose reliability was checked, are:

$$E = 75.826 + 11.714X_t - 1.651T - 0.04X_tT - 0.39X_t^2 - 0.018T^2 \tag{5.2}$$

$$X_{s,w} = \left(12.2 - 1.4X_t - 0.1271T - 0.0020X_tT - 0.0541X_t^2 - 0.0015T^2\right)/100 \tag{5.3}$$

$$X_{t,w} = \left(52.5 + 1.5X_t - 2T - 0.0229X_tT + 0.0027X_t^2 + 0.0195T^2\right)/100 \tag{5.4}$$

$$\phi_C = \left(64.9 + 16.3X_t - 2.6T - 0.0024X_tT + 0.0013X_t^2 + 0.0273T^2\right)/100 \tag{5.5}$$

Figure 5.3 represents the three-dimensional isoresponse curves of the studied properties smoothed by the quadratic model (Eqs. 5.2–5.5). Figure 5.3a shows that, as with APE (De Barros Neto et al. 2001) or ethoxylated methyl dodecanoates (Materna et al. 2002), the extent of phenol extraction, E, increases with X_t. In the present work, it reaches 95% for 10% $C_{12}E_6$. On the other hand, E decreases with a temperature rise: This is probably due to the increase of the solubility of phenol in water with temperature, which makes it not so easily extractable into the coacervate. However, Materna et al. (2001) observe that the extent of phenol extracted (E) with nonaethoxylated methyl dodecanoate ($C_{12}E_9CH_3$) goes through a maximum at about 20°C above the cloud point. With an APE, Gullickson et al. (1989) found an increase of 4-*t*-butylphenol concentration and partition ratio in the coacervate between 30 and 50°C, but no change of the fraction of solute transferred to this phase. The behavior of hexan-1-ol (concentration and partition ratio in the coacervate showing a maximum around 40°C)(Gullickson et al. 1989) could be explained by a minimum of its aqueous solubility near this temperature (Nishino et al. 1978). The percentage of residual phenol in the dilute phase, $X_{s,w}$, decreases when X_t increases, but increases slightly with T (Figure 5.3b); thus, a significant temperature rise beyond the CP has a negative effect on the extraction. Figure 5.3c shows that the residual concentration of surfactant, $X_{t,w}$, is lower at low surfactant concentration and goes through a minimum versus T. In order to increase the concentration

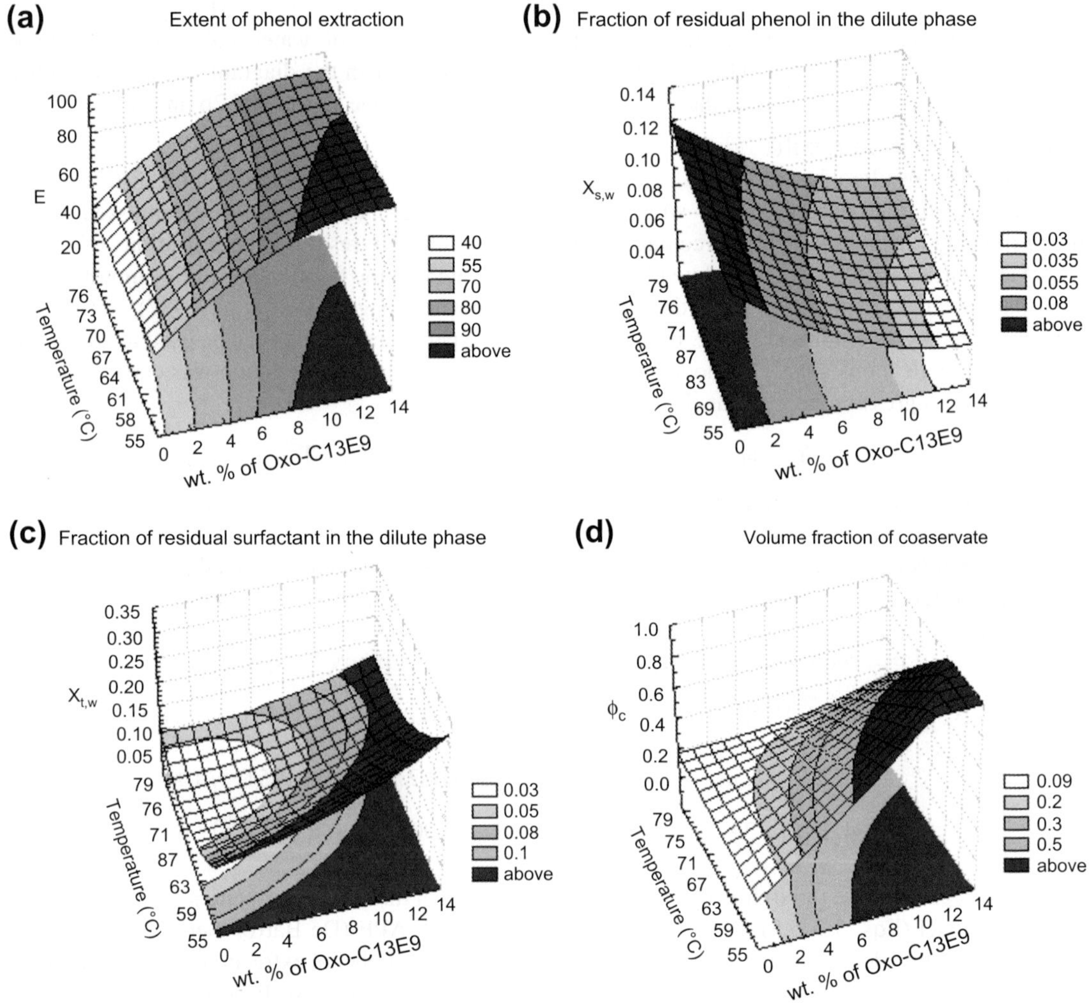

FIGURE 5.3

Extraction of phenol with $C_{12}E_6$. Three-dimensional isoresponse curves smoothed by a quadratic model (Eqs. 5.2–5.5): (a) $E(\%) = f(X_t, T)$; (b) $X_{s,w}\ (\%) = f(X_t, T)$; (c) $X_{t,w}\ (\%) = f(X_t, T)$; (d) $\phi_C = f(X_t, T)$.

(from: Haddou et al., 2003, with permission).

factor of the solute, we try to obtain a minimal volume fraction of coacervate (ϕ_c). However, logically, ϕ_c increases with X_t and decreases when temperature increases (Figures 5.1a and 5.3d).

2.1.3.3.2 Benzyl alcohol. For the properties E, $X_{s,w}$, $X_{t,w}$, and ϕ_c regarding the extraction of benzyl alcohol with Oxo-$C_{10}E_3$, the quadratic model provides the following equations:

$$E = 24.853 + 11.739X_t + 0.025T - 0.135X_tT - 0.346X_t^2 - 0.012T^2 \quad (5.6)$$

$$X_{s,w} = \left(12.1 - 1,6X_t - 0.027T + 0.0541X_tT - 0.0005X_t^2\right)/100 \quad (5.7)$$

$$X_{t,w} = \left(8.4 + 1.8X_t - 0.0716T - 0.0493TX_t + 0.0013X_t^2 + 0.0685T^2\right)/100 \quad (5.8)$$

$$\phi_C = \left(5.2 + 10.8X_t - 0.0397T - 0.1613TX_t + 0.0527X_t^2 + 0.002T^2\right)/100 \quad (5.9)$$

The behavior of the four properties versus X_t and T is generally similar to that observed for phenol extraction with $C_{12}E_6$, with some difference for $X_{s,w}$ (Figure 5.4). In fact, the percentage, E, of phenol extracted with Oxo-$C_{10}E_3$ is higher than that of benzyl alcohol under the same conditions (Haddou 2003; Figure 5.4a). Consequently, the value of $X_{s,w}$, is higher for benzyl alcohol than for phenol (Figure 5.4b). This is in agreement with the partition coefficients of both solutes between octan-1-ol and water ($\log P_{\Phi CH2OH} = 1.05$; $\log P_{\Phi OH} = 1.50$)(Sangster 1989). Besides, it has already been observed that, with compounds more efficient on cloud point lowering, higher extraction yields can be reached (Akita et al. 1996).

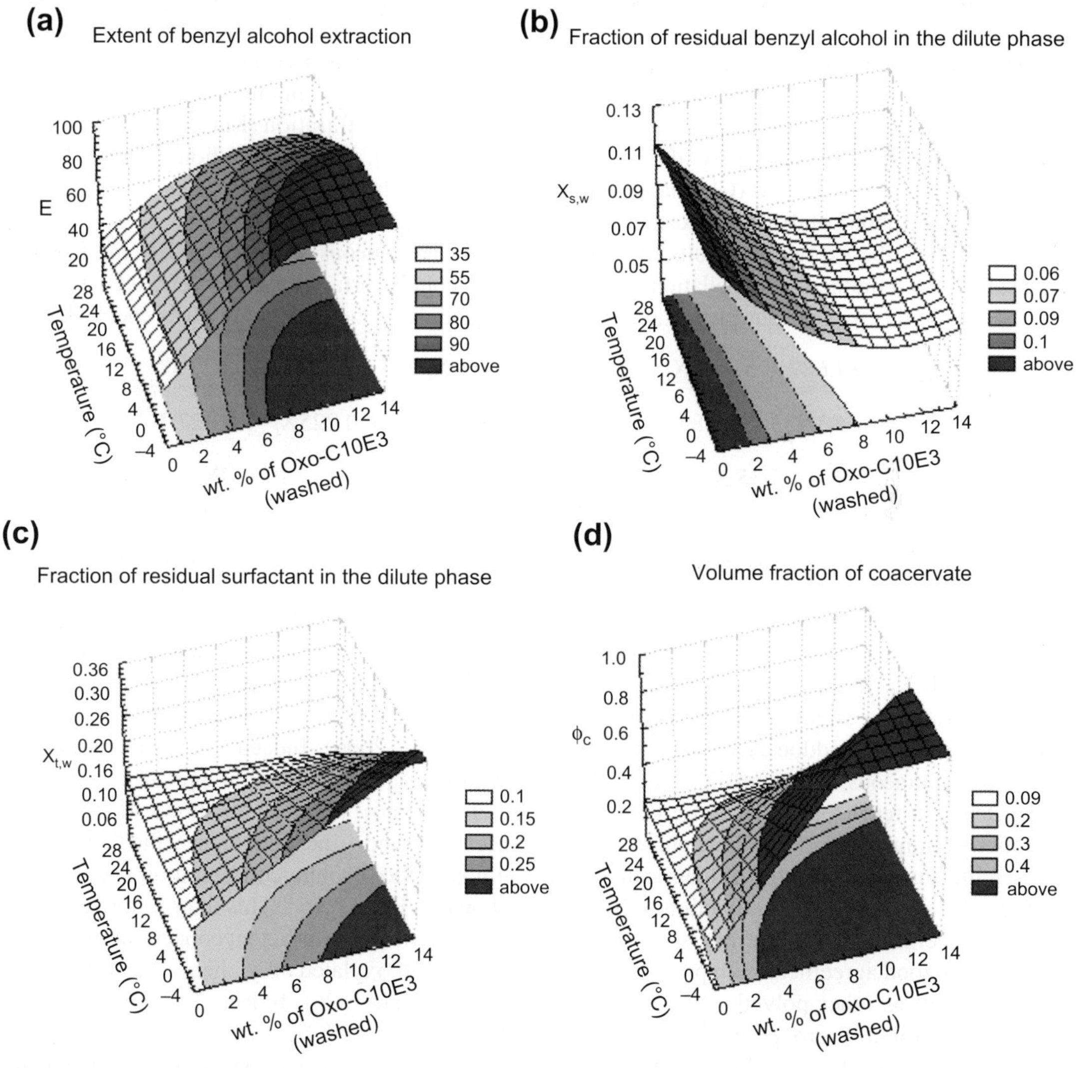

FIGURE 5.4

Extraction of benzyl alcohol with Oxo-$C_{10}E_3$. Three-dimensional isoresponse curves smoothed by a quadratic model (Eqs.6 to 9): (a) $E(\%) = f(X_t, T)$; (b) $X_{s,w}(\%) = f(X_t, T)$; (c) $X_{t,w}(\%) = f(X_t, T)$; (d) $\phi_C = f(X_t, T)$.

(from: Haddou et al., 2003, with permission).

2.2 Humic acids

Humic substances are polyelectrolytic macromolecules, having high molecular weights (Soumia et al. 2010, Riccardo et al. 2009, Katsumat et al. 2008). They are significant in aquatic systems for several reasons. They give yellow brown color to water (Jaroslaw et al. 2004) and they can form complexes with metals (Paciolla et al. 2002, Achour et al. 2011) and organic pollutants such as pesticides (McCarthy 1989). They are precursors to the formation of mutagenic halogenated compounds in water after chlorination (Polak et al. 2011). Humic acid especially represents the major advantage of the natural organic matter in soil and surface water (Uyguner et al. 2005). However, its presence in raw water can significantly affect water quality during the purification process (Jianfeng et al. 2006). It is widely agreed that trihalomethanes (THMs) can be generated, as disinfection by-products, by step chlorination of water containing humic acid (Stevene et al. 1979). Several research studies have been carried out as an alternative for the degradation of aquatic humic substances (Takeshi et al. 2006, Badis et al. 2010). Besides, different techniques for the treatment of release contaminated with humic acid have been proposed, such as biological treatment (Silvia et al. 2012), filtration (Katsoufidou et al. 2005 and 2008), adsorption (Achour et al. 2011, Polak et al. 2011), ozone oxidation (Singer 1990), coagulation and ion exchange (Ghernaout et al. 2009), electrochemistry (Chiang, et al. 2000), and (heterogeneous) photocatalysis (Al-Rasheed et al. 2003, Selcuk et al. 2004).

We will describe extraction results of humic acid from its aqueous solutions at 100 mg/L using Lutensol ON 30 (commercial appellation of Ethoxylated Alcohols surfactants equivalent to $C_{10}E_3$) The quadratic equations for the four properties (E, $X_{s,w}$, $X_{t,w}$, and ϕ_c), whose reliability was checked, are:

$$E_{(Lutensol)} = 14.687 + 5.592\ X_t + 2.607\ T - 0.033\ X_t\ T - 0.216X_t^2 - 0.027\ T^2 \quad (5.10)$$

$$X_{s,w(Lutensol)} = 83.753 - 2.485\ X_t - 2.543T - 0.051\ X_t\ T + 0.225X_t^2 + 0.027\ T^2 \quad (5.11)$$

$$X_{t,w(Luntensol)} = 0.506 + 0.035\ X_t - 0.023\ T - 1.046\ 10^{-3}X_t\ T - 1.159\ 10 - 3X_t^2 + 3.11\ 10^{-4}\ T^2 \quad (5.12)$$

$$\phi_{C(Lutensol)} = 0.2939 + 0.0009\ X_t - 0.0136\ T - 0.002\ X_tT - 0.0054\ X^2 - 0.0006\ T^2 \quad (5.13)$$

Figure 5.5 represents the three-dimensional isoresponse curves of the studied properties smoothed by the quadratic model (Eqs. 5.10–5.13). Figure 5.5a shows that the extent of humic acid extraction (E) increases with X_t. In this work, E reaches 98% for 10 % Lutensol ON30. However, a temperature increase only has a slight effect on humic acid extraction. This trend was also observed in other extraction systems (Haddou et al. 2003, 2006, Canselier et al. 2006). Indeed, the temperature rise induces simultaneous and opposite effects: It increases the concentration of solute in the micellar aggregates as a result of the decrease of ϕ_C (Haddou et al. 2006). Figure 5.5b shows that the concentration of humic acid in the dilute phase, $X_{s,w}$, decreases as X_t and T increase. Hence, the first contact between the surfactant and the effluent can allow an almost 60-fold solute concentration reduction.

The behavior of ($X_{t.w}$) as a function of X_t and T is shown in Figure 5.5c. This figure shows that the residual concentration of surfactant is low at high temperature and low surfactant concentration. These results are in good agreement with previous studies using polyethoxylated alkylphenols (Haddou et al. 20011) as well as other polyethoxylated alcohols (Haddou et al. 2006). Indeed, heating desolvates surfactant hydrophilic groups gradually and thus reduces its hydrophilic character. In general, the desolvation energy of the surfactant molecule can be associated with its energy transfer from the hydrophilic (aqueous solution) to the hydrophobic medium (micellar core). According to Figure 5.5d, the smoothed value of ϕ_c using Eq. 5.13 is low at high temperature and low surfactant concentration.

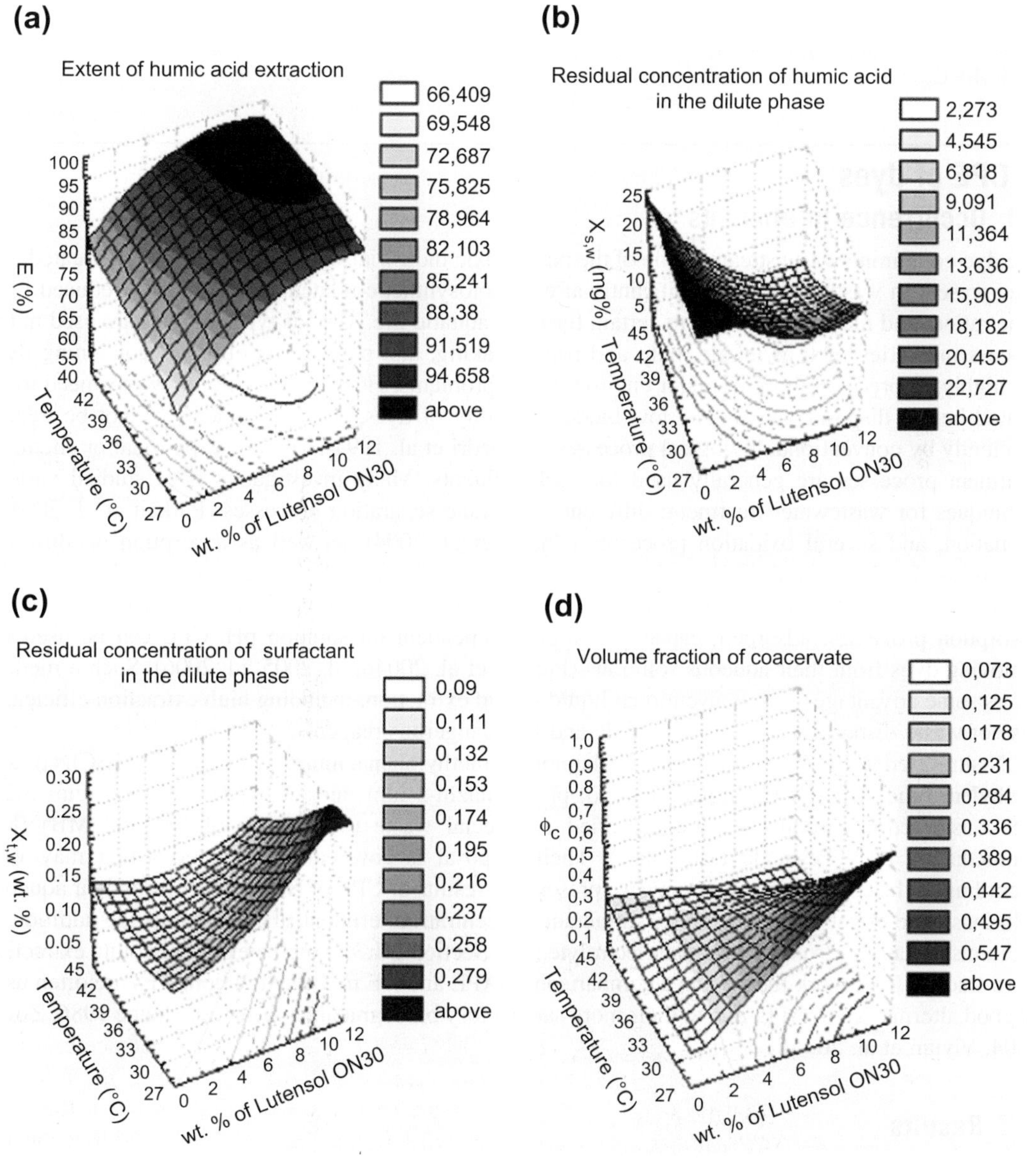

FIGURE 5.5

Extraction of humic acid with Lutensol ON 30. Three-dimensional isoresponse curves smoothed by a quadratic model (Eqs. 5.10 to 5.13): (a) E(%) = f(X_t, T); (b) $X_{s,w}$(%) = f(X_t, T); (c) $X_{t,w}$ (%) = f(X_t, T); (d) ϕ_C = f(X_t,T). *(from: Ghouas et al., 2012, with permission).*

However, high surfactant concentrations induce more surfactant loss in the dilute phase (Fig. 5.5c). Although the surfactant is biodegradable, this loss is not economical. As pointed out previously, the optimization of the process requires a compromise between the four studied parameters E, $X_{s,w}$, $X_{t,w}$, and ϕ_c (Haddou et al. 2003, Haddou et al. 2006, Canselier et al. 2006, Haddou et al. 2011). Indeed, a low surfactant concentration should be used to yield a smaller volume fraction of coacervate. On the basis of this finding, optimal values of ϕ_c (i.e., 0.1 and 0.3) were obtained using 4 wt.% Lutensol ON 30 at 40°C.

3 CPE of dyes

3.1 Occurrence in effluents

The dye-containing waste stream is one of the major toxic industrial wastes. Unless properly treated, the dyes present in wastewater can significantly affect photosynthetic-related activity, due to reduced light penetration, and may also be toxic to certain forms of aquatic life. Various types of dyes are used in the process industries such as textile, pulp and paper making, and paints. The effluents containing dyes are highly colored, resulting in major environmental problems. Hence these colored wastes need to be treated before disposal. Due to the low biodegradability of dyes, colored effluents cannot be treated efficiently by conventional biological processes (Seshadri et al. 1994). Therefore, physical or chemical treatment processes are generally used for such effluents. Many investigators have studied various techniques for wastewater treatment: different membrane separation processes (Purkait et al. 2004a), ozonation, and several oxidation processes (Zhang et al. 2004), as well as adsorption on different adsorbents (Bouberka et al. 2006, id. 2007). All these methods can be examined critically: For example, in membrane separation processes, care is needed to avoid membrane fouling, and, in the case of adsorption processes, adsorbent capacity is highly dependent on solution pH. CPE can be used for removing dyes from their aqueous solutions (Purkait et al. 2004b, id. 2005, id. 2006). Such a method offers some advantages over conventional liquid–liquid extraction, including high extraction efficiency, ease of waste disposal, and use of nontoxic and less dangerous reagents.

Neutral red (NR, 3-amino-7-(dimethylamino)-2-methylphenazinium chloride, $C_{15}H_{17}ClN_4$) and methylene blue (MB, 3,7-bis(dimethylamino)-phenothiazin-5-ium chloride, $C_{16}H_{18}ClN_3S$)(Figure 5.27) are considered hazardous and are classified as special waste (ecotoxicity of NR and MB) (MP Biomedicals, PAN Pesticide Data Base). Their disposal is covered by regulations that may vary according to the country. In this review, we present the results of CPE of NR and MB from their aqueous solutions. The effects of temperature, surfactant concentration, effluent pH, as well as the addition of sodium sulphate (a strong salting-out electrolyte,; see Section 2.1.3.1) on the extent of solute extraction were studied. This work focused on a comparison of APE and AE in dye extraction. AEs are often used as good alternatives to APE, due to their more satisfactory biodegradable properties (Falbe 1986, Zoller 2004, Vivian et al. 2004).

3.2 Results

3.2.1 Phase diagrams

The phase diagrams of the water/surfactant/dye systems cannot be plotted: The dye prevents the CP detection of such systems. Hence, it was not possible to study the solute effect on the surfactant CP curves. Nevertheless, as in the case of phenol and benzyl alcohol extraction (Haddou et al. 2003, id. 2006), it is necessary first to plot the phase diagrams of water/surfactant systems, which allow locating the CPE domain (tow-phase zone) of each surfactant (Figure 5.6), at least approximately. As explained in Section 2.1.3.1, the CP increases with the EO number and decreases when the

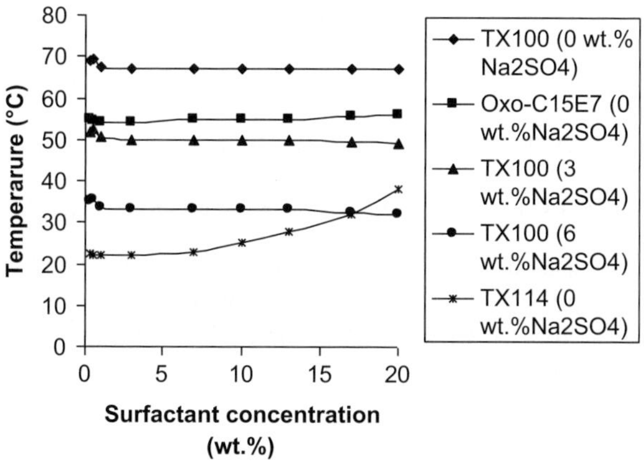

FIGURE 5.6

Effect of sodium sulphate on the cloud point temperature of Oxo-$C_{15}E_7$, TX-100 and TX-114. *(from Talbi et al., 2009, with permission).*

carbon number of the hydrophobic chain increases. Thus, the critical temperature of TX-114 (iso-octylphenol polyethyleneglycol ether EO $\approx$ 7.5) is about 24°C, while that of TX-100, a higher ethoxylate of the same family (EO $\approx$ 9), is located above 68°C (Figure 5.6). The critical temperature of Dehydol LT 7 (a commercial surfactant obtained from Oxo-alcohol ethoxylation, Oxo-$C_{15}EO_7$) is about 54°C. This surfactant does not belong to the same family. Organic solubilizates can interact with the surfactant polar head group or with its hydrophobic chain. According to their chemical nature, organic compounds can modify surfactant cloud points (Sharma et al. 1989, Abdullah et al. 1996, Sakulkongwyai et al. 2000, De Barros Neto et al. 2001, Haddou et al. 2006). In our case, phase separation in the water/surfactant/dye systems was possible above the CP of the water/surfactant systems. Indeed, at low concentration, NR and MB must not show a significant effect on the CP of the surfactants studied. Figure 5.6 also shows the CP lowering of TX-100 by sodium sulphate.

3.2.2 Extraction results

The extraction results of NR and MB from their aqueous solutions (100 mg/L and 7 mg/L, respectively) with different surfactants, according to two variables—X_t (wt.% surfactant) and temperature (T)—are expressed by three "responses" (Y): percentage of extracted solute (E), residual concentration of solute in the dilute phase ($X_{s,w}$), and coacervate volume fraction at equilibrium (ϕ_c)(De Barros Neto et al. 2001, Haddou et al. 2003, id. 2006). For each parameter determined by a central composite design (Box et al. 1987), the results are submitted to an empirical smoothing. The models were checked by plotting computed values against experimental ones. As in the case of phenol and benzyl alcohol extraction, a quadratic correlation (Eq. 5.1) was chosen to give the slope and regression coefficient (R^2) closest to unity.

3.2.2.1 Extraction of neutral red

The quadratic equations for the three properties (E, $X_{s,w}$, and ϕ_c), whose reliability was checked, are:

$$E_{(TX-114)} = 126.134 + 6.569X_t - 3.14T - 0.0004167X_tT - 0.326X_t^2 + 0.043T^2 \quad (5.14)$$

$$E_{(Oxo-C15E7)} = 172.373 + 7.762X_t - 3.301T - 0.007222X_tT - 0.345X_t^2 + 0.025T^2 \quad (5.15)$$

$$X_{s.w(TX-114)} = 24.24 - 6.483X_t + 0.252T + 0.009815X_tT + 0.284X_t^2 - 0.001806\,T^2 \quad (5.16)$$

$$X_{s.w(Oxo-C15E7)} = 1134 + 30.525X_t - 39.49T - 0.557X_tT + 0.303X_t^2 + 0.345T^2 \quad (5.17)$$

$$\phi_{c(TX-114)} = (-416 + 83\,X_t + 25T + 0.05556X_tT - 3.611\,X_t^2 - 0.4167T^2)/1000 \quad (5.18)$$

$$\phi_{c(Oxo-C15E7)} = (4073 + 156\,X_t - 116T - 1.944X_tT + 0.6944X_t^2 + 0.8333T^2)/1000 \quad (5.19)$$

Figure 5.7 represents the three-dimensional isoresponse curves of the studied properties smoothed by the quadratic model (Eqs. 5.14–5.19). Figures 5.7a and 5.7b show that the extent of NR extraction, E, increases with X_t. In this work, E reached 99% for 13 wt.% TX-114 and 98% for the same

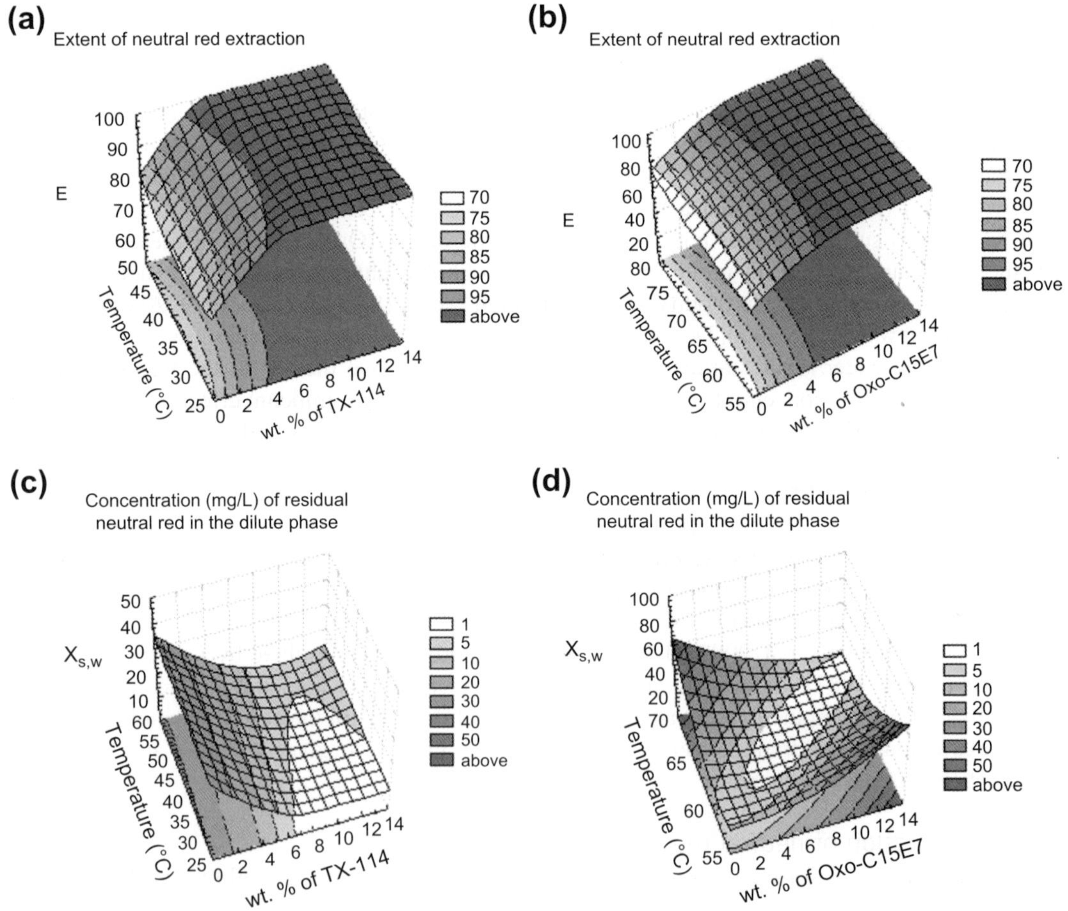

FIGURE 5.7

Three-dimensional isoresponse curves smoothed by a quadratic model (Eqs. 5.14–5.19): (a) $E_{(TX-114)} = f(X_t, T)$; (b) $E_{(Oxo-C15E7)} = f(X_t, T)$; (c) $X_{s,w\,(TX-114)} = f(X_t, T)$; (d) $X_{s,w\,(Oxo-C15E7)} = f(X_t, T)$; (e) $\phi_{C(TX-114)} = f(X_t, T)$; (f) $\phi_{C(Oxo-C15E7)} = f(X_t, T)$.

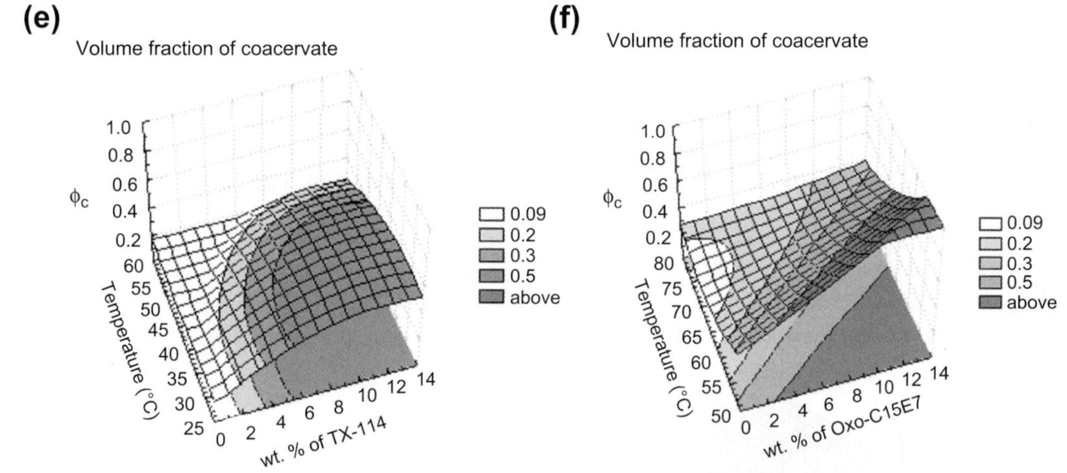

FIGURE 5.7 cont'd.

concentration of Oxo-$C_{15}E_7$. On the other hand, E decreases slightly with rising temperature, which is probably due to the increase of the solubility of the dye in water. This increase makes NR extraction into the coacervate more difficult at high temperature. In Figures 5.7c and 5.7d, it is shown that the concentration of residual NR, $X_{s,w}$, decreases as X_t increases, but slightly increases with T. Thus, beyond the CP, a significant temperature rise has an opposite effect on the extraction (Haddou et al. 2003, id. 2006). Hence, at high surfactant concentrations ($X_t = 13$ wt.%) and at temperatures slightly higher than the CP ($T \leq T_c + 5°C$), the NR concentration in the effluent can be reduced about 500-fold with TX-114 and 50-fold with Oxo-$C_{15}E_7$. In order to increase the concentration factor of solute, a minimal volume fraction of coacervate (ϕ_c) should be obtained when temperature increases. According to Figures 5.7e and 5.7f, the value of ϕ_c is low at high temperature and at low surfactant concentration. But rising temperature is unfavourable to efficient extraction extent (Figs. 5.7a and 5.7b), while high surfactant concentrations increase ϕ_c (Figs. 5.7e and f). So, the optimization of the process requires a compromise between the three studied parameters E, $X_{s,w}$, and ϕ_c.

3.2.2.2 Extraction of methylene blue

For properties E, $X_{s,w}$, and ϕ_c regarding MB extraction with TX-114 and Oxo-$C_{15}E_7$, the quadratic model provides the following equations:

$$E_{(TX-114)} = -169.982 + 18.405X_t - 7.604T - 0.229X_tT - 0.284X_t^2 + 0.104T^2 \quad (5.20)$$

$$E_{(Oxo-C15E7)} = -53.174 + 16.134X_t + 2.742T - 0{:}18X_tT - 0.08X_t^2 - 0.024T^2 \quad (5.21)$$

$$X_{s.w(TX-114)} = -4.481 - 0.956X_t + 0.519T + 0.011X_tT + 0.018X_t^2 - 0.00714T^2 \quad (5.22)$$

$$X_{s.w(Oxo-C15E7)} = (-7851 - 653X_t + 501T + 5.556X_tT + 14X_t^2 - 4.722T^2)/1000 \quad (5.23)$$

$$\phi_{c(TX-114)} = (-101 + 98X_t - 4.75T - 1.25X_tT + 0.1389X_t^2 + 0.05T^2)/1000 \quad (5.24)$$

$$\phi_{c(Oxo-C15E7)} = (1093 + 127X_t - 35T - 1.667X_tT + 0.1389X_t^2 + 0.2778T^2)/1000 \quad (5.25)$$

The behavior of the three properties versus X_t and T is similar to that observed for NR extraction (Figure 5.8). In fact, even though the initial concentration of NR in effluent is higher than that of MB, and in spite of the lower solubility of the latter in water, the extracted percentage of MB with both surfactants is lower than that of NR (Figures 5.7 and 5.8), especially when Oxo-$C_{15}E_7$ is used (Figures 5.7b and 5.8b). Consequently, the values of $X_{s,w}$ for MB are lower than those for NR. MB concentration in the synthetic effluent is reduced about 10-fold with TX-114 and halved using Oxo-$C_{15}E_7$.

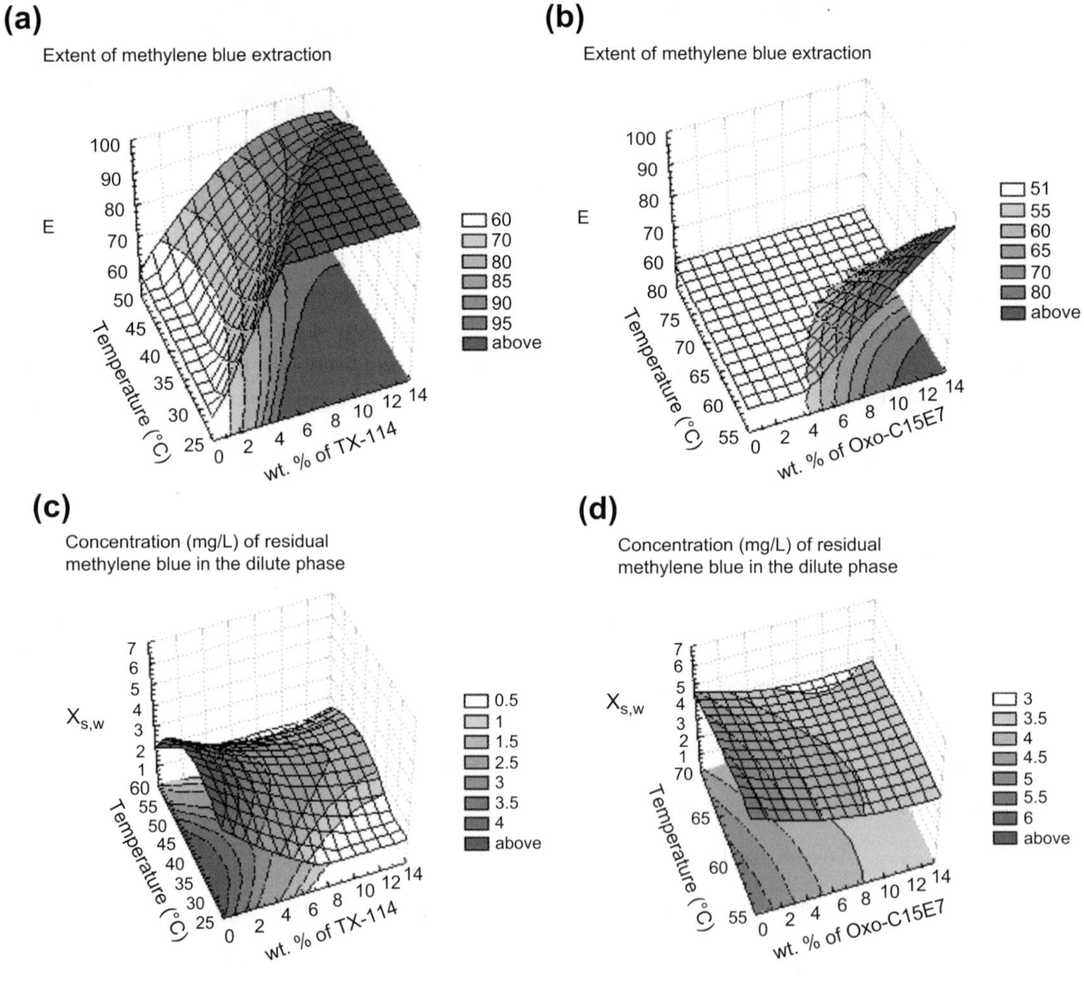

FIGURE 5.8

Three-dimensional isoresponse curves smoothed by a quadratic model (Eqs. 5.20–5.25): (a) $E_{(TX-114)} = f(X_t, T)$; (b) $E_{(Oxo-C15E7)} = f(X_t, T)$; (c) $X_{s,w\ (TX-114)} = f(X_t, T)$; (d) $X_{s,w\ (Oxo-C15E7)} = f(X_t, T)$; (e) $\phi_{C(TX-114)} = f(X_t, T)$; (f) $\phi_{C(Oxo-C15E7)} = f(X_t, T)$.

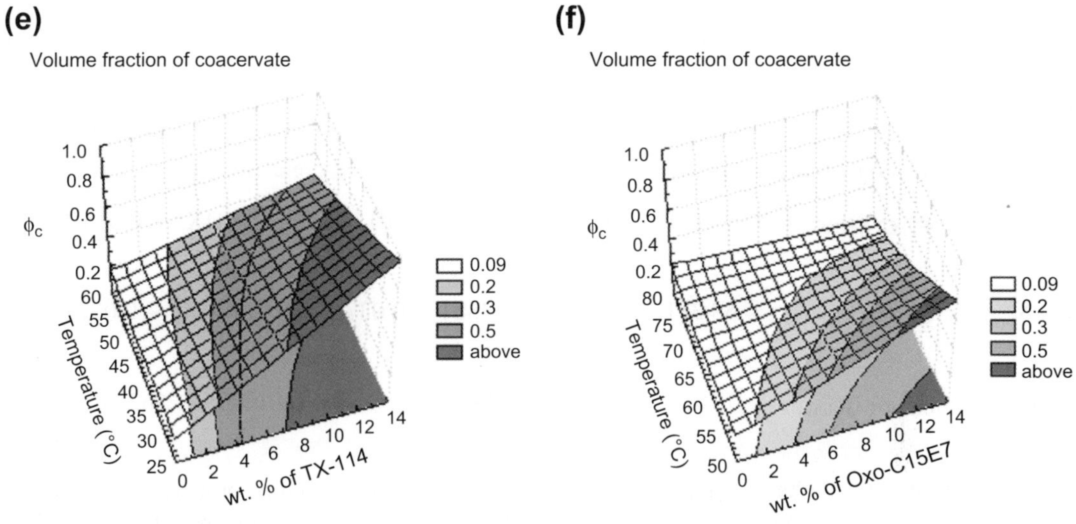

FIGURE 5.8 cont'd.

3.2.2.3 Extraction orange G using cationic-non-ionic mixed micelles

Surfactants used in practical applications virtually always consist of surfactant mixtures. For example, they are used in detergent formulations to increase washing activity and as emulsifiers to stabilize emulsions. Interactions between different surfactants can lead to synergism or to antagonism, depending on the kinds of surfactants (Halide et al. 2005). Figure 5.9 shows the synergistic effect of Oxo-$C_{10}E_3$/CTAB system toward CPE of Orange G (1-Phenylazo-2-naphthol-6,8-disulfonic acid disodium salt or 7-Hydroxy-8-phenylazo-1,3-naphthalenedisulfonic acid disodium salt). Figure 5.9

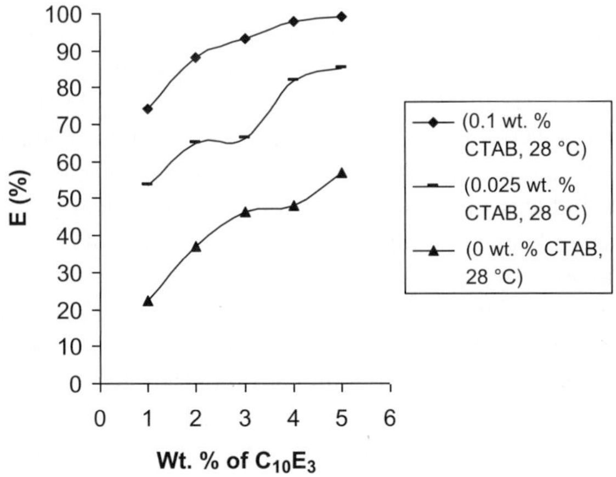

FIGURE 5.9

Effect of CTAB on the extraction extent in the systems: H2O/Oxo-$C_{10}E_3$/Orange G.

shows that the extraction extent (E) is highly improved with 0.025 wt.% CTAB and even more with 0.1 wt.%, compared with that obtained using a neutral micellar system (no CTAB). Indeed, having a charge opposite to that of Orange G, which bears two sulphonate groups and a phenolic –OH, CTAB seems to be Orange G-substantive (Mateja et al., 2008).

The three "responses," E, $X_{s,w}$, and ϕ_C, were investigated for the system Orange G/Oxo-$C_{10}E_3$/0.1 wt % CTAB. The quadratic equations, whose reliability was checked, are as follows:

$$E_{(Oxo-C10E3/CTAB,OG)} = 82.334 + 22.46X_t - 1.278T + 7.5\ 10^{-3}X_t\,T - 2.804X_t^2 + 0.015\ T^2 \quad (5.26)$$

$$X_{s.w(Oxo-C10E3/CTAB,OG)} = 88.331 - 112.3X_t + 6.388T - 0.037X_tT + 14.044X_t^2 - 0.077T^2 \quad (5.27)$$

$$\phi_{C(Oxo-C10E3/CTAB,OG)} = -0.182 + 0.32X_t + 0.011T - 0.005X_tT - 0.01\ X_t^2 - 1.562\ 10^{-4}T^2 \quad (5.28)$$

Figure 5.10 represents the three-dimensional isoresponse curves of the studied properties fitted to the quadratic model (Eqs. 5.26–5.28). The behavior of the three properties with respect to X_t and T is generally similar to that observed for neutral micellar systems (Figures 5.7 and 5.8). However, the difficulty of Orange G extraction has been overcome: The extraction extent of Orange G with this mixed micellar system is higher than that reached using non-ionic surfactants.

4 CPE of dispersed pollutants

4.1 Cutting oils

4.1.1 General

The environmental pollution problem is still relevant as many industrial activities continue to generate various pollutants, including dissolved and dispersed organic substances that could create significant harmful effects. Faced with more restrictive regulations, companies must treat their effluents before disposal (Bliefert et al. 2001). Hydrocarbons, for example, are among the potential environmental pollutants: Most municipal sewers limit their discharge to 100 ppm. Unless properly treated, such compounds spread on water surface and form oil films, preventing gas exchange between air and water, which, as a result, induces organism suffocation due to a lack of oxygen. Carbon dioxide produced by breathing accumulates in living cells and decreases the pH of the medium in which they live. Oil dispersion is favored by interfacial tension lowering by chemical dispersants forming emulsions or microemulsions, which increase the interfacial area (Bhardwaj et al. 1993, Er-Raioui et al. 2009, Kartsev et al. 2009). Moreover, because of its great capacity of penetration in the ground, oily wastewater, without further decontamination before disposal, constitutes a very serious threat to groundwater. In general, remedies for contamination by oily wastewater comprise mechanical, physical-chemical, and/or biological treatments (Hosny 1992, Gourlay-Francé et al. 2008, Tran et al. 2009). During oil slicks, emulsions are formed as a result of the presence of natural surfactants in crude oil (e.g., asphaltenes). Phase separation needs the addition of a demulsifier in order to reach an optimum formulation corresponding to the minimum emulsion stability (Rondon et al. 2006).

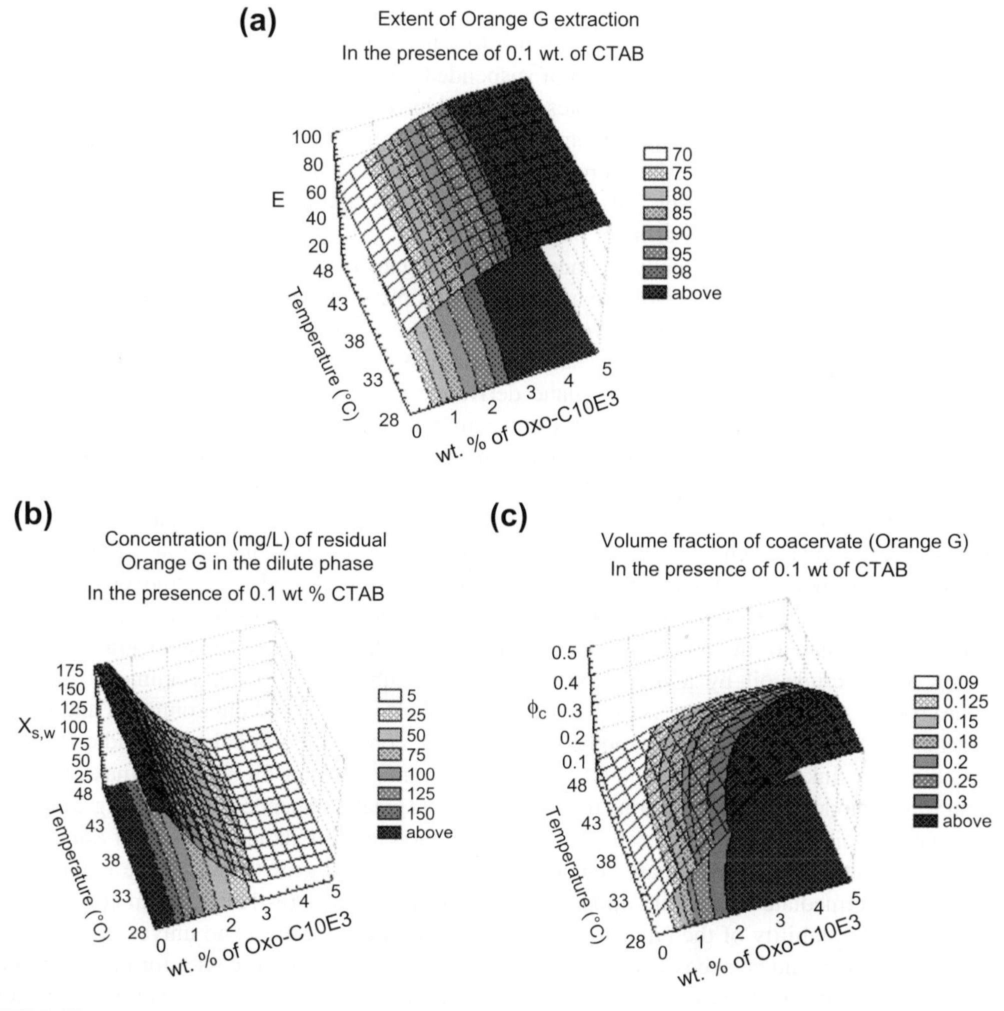

FIGURE 5.10

Three-dimensional isoresponse curves smoothed by a quadratic model (Eqs. 5.26 to 5.28): (a) $E_{(Oxo\text{-}C10E3/CTAB,\ OG)} = f(X_t, T)$; (b) $X_{s,w\ (Oxo\text{-}C10E3/CTAB,\ OG)} = f(X_t, T)$; (c); $\phi_{C(Oxo\text{-}C10E3/CTAB,\ OG)} = f(X_t, T)$.

Among the most stable emulsions, cutting fluids are potential pollutants and dangerous products for the environment. They are oil-in-water (O/W) emulsions employed in machining processes such as metal turning, milling, and drilling, in particular when the metal removal operations are conducted at high speeds and low pressures. Regulation of heat generation and lubrication of the contact point are achieved by delivering these so-called cutting oils.

A cutting fluid concentrate usually contains a mineral oil, a surfactant mixture, and in some cases water and various additives included to meet the specifications of commercial concentrates, such as resistance to bacterial growth and low-corrosion capacity (Zhu et al. 1995, Bataller et al. 2004, Al Sabbagh et al. 2006). During their use, cutting fluids lose their properties and effectiveness because of their thermal degradation and the production of suspended metal particles. The oils therefore have to be replaced periodically, and the organic wastes generated have to be taken away. A significant amount of these products has caused serious contamination of natural ecosystems, due to the presence of chlorate, boron, and heavy metals. Therefore, it is necessary to treat these effluents before disposal, but the recent developments of cutting oil formulation (e.g., synthetic or semisynthetic emulsions) somewhat complicate the issue of effluent purification.

Mineral oil molecules are eventually degraded by microorganisms, but the rate of abiotic elimination, mainly determined by the surface/volume ratio of oil droplets in water, cannot be neglected. For instance, surfactants have been used in two original methods of recovering crude oil, oil processing products, or oil residues spilled on a water surface. The first one consists of making an anti-adhesive polysiloxane matrix from a hydrolyzable silane derivative reacting with a polyoxyethylenic surfactant, trapping the hydrocarbon into the matrix and removing the silicone/hydrocarbon solid residues from the water surface by conventional mechanical means (e.g., netting)(Pelletier et al. 1994). In the second one, a ferrofluid is prepared by adsorbing sodium oleate on magnetite (Fe_3O_4), then by dispersing the particles into crude oil, kerosene, or cyclohexane and adding a non-ionic surfactant to the dispersion. By using the ferrofluid, which is an effective dispersant, and an oil slick drawing apparatus equipped with permanent magnets, the oil can be selectively removed and recovered without adversely affecting the marine ecosystem (Kim et al. 1998). It also makes sense that the rate of bioremediation increases with the dispersion state of the oil. Thus, a nonylphenol ethoxylate enhances the biodegradation of 2% (w/v) Bow River crude oil by a mixed bacterial culture substantially (Van Hamme et al. 1999). A psychrotrophic *Acinetobacter sp.* is able to grow on n-dodecane in cold environments and produce biosurfactants (Margesin et al. 1999). Another biosurfactant, consisting of a polysaccharidic chain with hydrophobic ester-linked fatty acid (C_{10}-C_{14}) substitutions, may disrupt oil slicks by making a stable emulsion whose biodegradation by marine microorganisms is greatly enhanced by the strong adhesion of the latter onto the oil surface (Crescenzi et al. 2002). Oxidation rates of alkanes by cells of *Rhodococcus sp. 094* in the stationary growth phase (but not in the exponential growth phase) are stimulated by surfactants of intermediate hydrophile-lipophile balance (HLB) values (8–12)(Bruheim et al. 1997).

Because of the stability of the emulsions used as cutting fluids, there is no universal solution for their treatment, and it can be necessary to combine one or two treatment processes for highly effective purification. Conventional treatment methods such as evaporation, membrane, or chemical separation have major disadvantages since they generate a concentrated stream that is more harmful than the original waste (Chang et al. 2001, Portela et al. 2001, Ji et al. 2004, Ayotamuno et al. 2007, Robinson et al. 2008). Conventional biological wastewater treatments are often used for the elimination of organic pollution, but, in the case of cutting oil emulsions, usually corresponding to a petroleum cut, the high complexity of such hydrocarbon mixtures and its aromatic or naphtenic nature make the organic phase not easily biodegradable. Moreover, the antibacterial agents, frequently used in commercial cutting-oil formulations, increase the difficulty of the biological treatment (Janknecht et al. 2004, Perez et al. 2006). Micro- and ultrafiltration were largely investigated, and the application of different ultrafiltration membranes was inspected (Deepak et al. 1994, Chang et al. 2001, Faibish et al. 2001, Janknecht et al. 2004, Jezabel et al. 2007). However, in membrane separation processes, care is needed to avoid membrane fouling, which may be reversible or irreversible (Faibish et al. 2001). Since conventional treatment methods (evaporation, phase separation, filtration) are often inefficient or

Table 5.1 Emulsifiable Concentrate Composition of Tasfalout 22B Cutting Oil (Talbi et al. 2009)

Component	W t. (%)
Oil (spindle)	78
Sodium sulfonates	5–9.9
Borate alkenylamide	5–9.9
Alkyl amide	5–9.9
Hexylene glycol	1–4.9
Diethyl glycol monobutyl ether	1–4.9
Alkanolamine	2.3
Dodecanol	0.5–4.5
N-N dimethylene bis-morpholine	2.4

environmentally unacceptable, the development and application of new methods are necessary. CPE was tried in our laboratories to treat and purify effluents containing cutting-oil emulsions. For the above-mentioned reasons (Section 2.1.2), we chose to work with AE (C_iE_j). In special conditions, cutting oil emulsions can be broken using non-ionic surfactant and heat. Hence, demulsification and CPE occur simultaneously. Thus, dispersed pollution (mineral oils) and dissolved pollution (surfactant, co-surfactant, and cutting oil additives) can be concentrated and eliminated simultaneously into the coacervate phase (Haddou et al. 2003). On the basis of these findings, this review presents the results of CPE of three mineral cutting oil wastes—HMP (TotalFinaElf, France), HMI (Motul, France) and Tasfalout 22B (NAFTEC, Algeria, [the approximate composition of the emulsifiable concentrate of which is given in Table 5.1]). The effects of temperature, surfactant concentration, effluent pH, as well as the addition of sodium sulphate on oil extraction, were investigated.

4.1.2 Extraction results

Two commercial AEs were used throughout this work: Oxo-$C_{15}E_7$ (Lutensol AO7, BASF), whose cloud point at 1 wt.% in water is 47°C, and Oxo-$C_{10}E_3$ (SEPPIC, France, Tc = 2°C at 1wt.% in water) after washing with water. When 2 wt.% cutting-oil emulsifiable concentrate were added to water, an O/W emulsion was obtained. Hence, the phase diagrams of the water/surfactant/cutting oil concentrate system cannot be plotted. The milky aspect of this emulsion also prevents CP detection of such a system. Therefore, before starting CPE tests of cutting oil wastes, it was necessary to plot the phase diagrams of the water/surfactant systems as a function of temperature. The amount of organic pollutants was measured as the chemical oxygen demand (COD) value. The total residual COD of the dilute phase after CPE, COD_R, includes that due to the surfactant remaining in this phase (concentration close to its critical micelle concentration (CMC)), COD_T, and that due to soluble pollution in water and not extracted by the surfactant, COD_S.

Hence, COD_R is:

$$COD_R = COD_T + COD_S \tag{5.29}$$

The extraction results by the two non-ionic surfactants, Oxo-$C_{10}E_3$ and Oxo-$C_{15}E_7$, according to two variables—surfactant wt.% (X_t), and temperature (T)—are expressed by the three following "responses" (Y): total residual chemical oxygen demand (COD_R), residual COD_T (due to surfactant), and coacervate volume fraction at equilibrium (ϕ_c). After applying a central composite design (Box et al. 1987), the results were analyzed by an empirical fitting (Section 2.1.3.3).

4.1.2.1 Total residual chemical oxygen demand (COD_R)

The initial COD (COD_I) of HMP and HMI effluents were about 55 g O_2/L, whereas that of the Tasfalout 22B effluent was about 35 g O_2/L. The quadratic model for COD_R versus X_t and T gives:

$$COD_{R(HMP,OxoC15E7)} = 14090 + 88.454X_t - 394.889T - 1.72X_tT + 5.846X_t^2 + 2.975\,T^2 \quad (5.30)$$

$$COD_{R(HMI,OxoC15E7)} = -27470 - 1969X_t + 813.295T + 2.647X_tT + 72.957X_t^2 - 4.056\,T^2 \quad (5.31)$$

$$COD_{R(T22B,OxoC15E7)} = 216.247 - 72.698X_t + 18.334T + 0.687X_tT + 7.85X_t^2 + 0.073T^2 \quad (5.32)$$

$$COD_{R(T22B,OxoC10E3)} = 5183 + 519.169X_t - 207.84T - 14.674X_tT + 8.692X_t^2 + 3.71T^2 \quad (5.33)$$

Figure 5.11 represents the corresponding three-dimensional isoresponse curves smoothed by the quadratic models (Eqs. 5.30–5.33). At constant X_t, COD_R increases with temperature (Figures

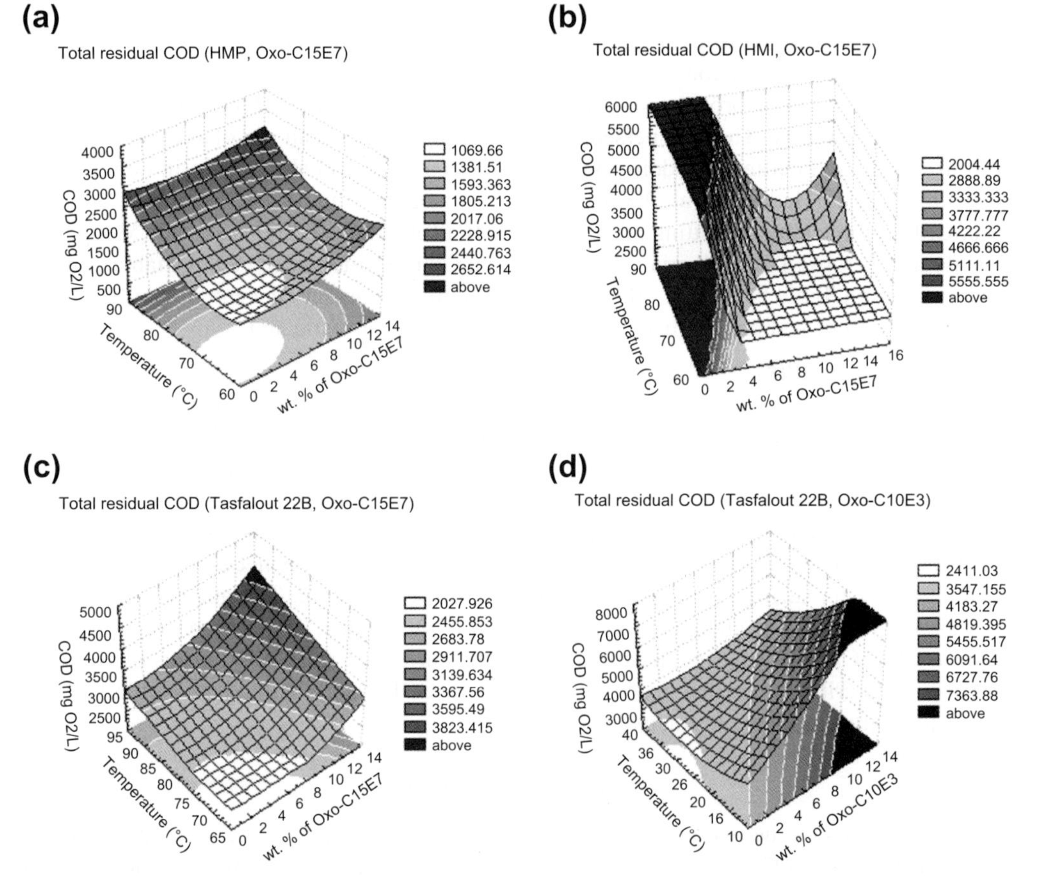

FIGURE 5.11

Three-dimensional isoresponse curves smoothed by a quadratic model (Eqs. 5.30–5.33): (a) $COD_{R(HMP, OxoC15E7)} = f(X_t, T)$; (b) $COD_{R(HMI,OxoC15E7)} = f(X_t, T)$; (c) $COD_{R(T22B,OxoC15E7)} = f(X_t, T)$; (d) $COD_{R(T22B, OxoC10E3)} = f(X_t, T)$.

(from: Talbi et al., 2009, with permission).

Table 5.2 Some Experimental Results of Cutting Oil Extraction (COD_R, COD_T, ϕ_C and COD_R/COD_I)

	(X_T, T)	COD_R(mg O_2/L)	COD_T(mg O_2/L)	ϕ_C	COD_I / $COD_R^{(a)}$
HMP, Oxo-$C_{15}E_7$	(4, 60)	1130	150	0.29	49
	(4, 65)	980	110	0.25	57
	(13, 70)	1530	354	0.80	36
	(4, 60)	690	234	0.27	79
	(4, 65)	2290	210	0.22	24
	(13, 80)	3140	259	0.25	48
	(4, 65)	1690	75.5	0.41	21
HMI, Oxo-$C_{15}E_7$	(4, 70)	1810	108	0.32	19
T22B, Oxo-$C_{15}E_7$	(13, 70)	3190	171	0.61	11
T22B, Oxo-$C_{10}E_3$	(4, 10)	5170	892	0.10	6.8
	(4, 15)	4310	751	0.13	8.2
	(13, 20)	6840	854	0.20	5.1

[a]: *COD_I = 55.1 (g O_2/L) for HMI, 54.5 for HMP and 35.2 for Tasfalout 22B.*

5.11a–c; Table 5.2). That is probably due to the increase of the water solubility of certain additives of the emulsion. Thus, a significant temperature rise beyond the cloud point has a negative effect on the extraction with Oxo-$C_{15}E_7$ (Figures 5.11a–c), whereas lukewarm temperatures allow maximal reduction of effluent COD using Oxo-$C_{15}E_7$ (Figure 5.11d). The total residual COD_R values obtained with HMI emulsion are, approximately, twice higher than those obtained in the case of HMP effluent (Figures 5.11a,b; Table 5.2). The most favorable areas for cloud point extraction are thus located in the lightest zones on the graphs in Figure 5.11. The best performance-cost compromise of the process, based on minimum surfactant consumption, led to optimal surfactant concentrations lying between 2 and 5 wt.% for HMP and Tasfalout 22B emulsions, whereas the optimal surfactant concentrations seem to be between 5 and 8 wt.% for the HMI emulsion. In these zones, the COD_R values are about 1100 mg O_2/L for the HMP emulsion, and 2000 mg O_2/L for HMI and Tasfalout 22B emulsions. The CPE of such soluble and dispersed pollutants allows about a 78-fold COD reduction (COD_I/COD_R) in optimal conditions (Table 5.2). Comparing Figures 5.11c. and 5.11d, one can notice that COD_R values obtained with Oxo-$C_{15}E_7$ are lower than with Oxo-$C_{10}E_3$ (Table 5.2). This latter surfactant is less effective than Oxo-$C_{15}E_7$ for Tasfalout 22B emulsion extraction. However, extraction using Oxo-$C_{10}E_3$ can be achieved at room temperature with less energy cost.

4.1.2.2 Chemical oxygen demand due to the surfactant residual concentration (COD_T)

The quadratic equations for COD_T, whose reliability was checked, are:

$$COD_{T(HMP,OxoC15E7)} = 726 + 73.392X_t - 18.298T - 0.869X_tT + 0.958X_t^2 + 0.118T^2 \quad (5.34)$$

$$COD_{T(HMI,OxoC15E7)} = 912.413 + 26.308X_t - 16.973T - 0.085X_tT - 0.0515X_t^2 + 0.074T^2 \quad (5.35)$$

$$COD_{T(T22B,OxoC15E7)} = 1113 + 26.667X_t - 38.797T - 0.636X_tT + 1.026X_t^2 + 0.362T^2 \quad (5.36)$$

$$COD_{T(T22B,OxoC10E3)} = 1036 + 49.97X_t - 33.96T - 0.866X_tT - 0.432X_t^2 + 0.29T^2 \quad (5.37)$$

Figure 5.12 represents the three-dimensional isoresponse curves of the studied property (COD_T), for different couples of cutting oil emulsion-surfactant, smoothed by the quadratic model (Eqs. 5.34–5.37.

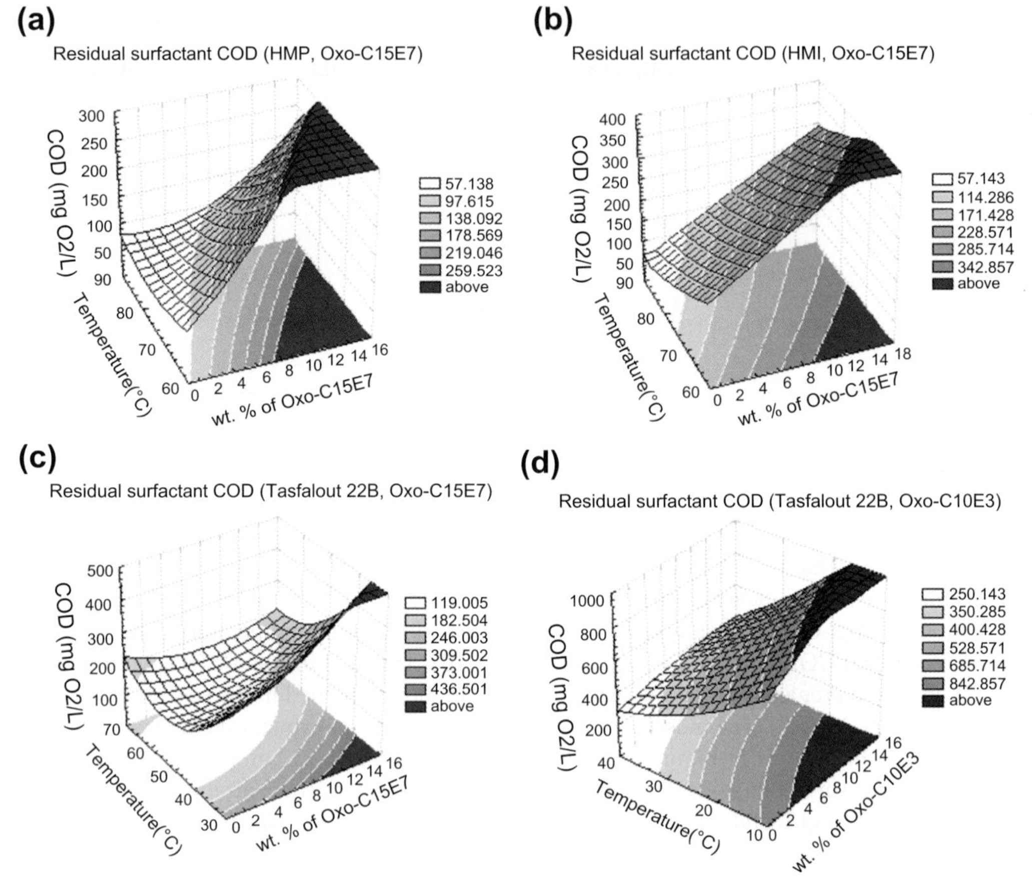

FIGURE 5.12

Three-dimensional isoresponse curves smoothed by a quadratic model (Eqs. 5.34–5.37): (a) $COD_{T(HMP,OxoC15E7)} = f(X_t, T)$; (b) $COD_{T(HMI,OxoC15E7)} = f(X_t, T)$; (c) $COD_{T(T22B,OxoC15E7)} = f(X_t, T)$; (d) $COD_{T(T22B,OxoC10E3)} = f(X_t, T)$.

(from: Talbi et al., 2009, with permission).

Figures 5.9a–d show that the COD_T is lower at low surfactant concentration, X_t, and goes through a minimum at high temperature. The behavior of COD_T versus X_t and T is rather similar for all cutting oil emulsion–surfactant couples. However, COD_T due to Oxo-$C_{10}E_3$ is higher than that obtained with Oxo-$C_{15}E_7$ (Figure 5.11; Table 5.2). Thus, the first surfactant is more soluble in water than the second. Because of satisfactory surfactant biodegradability, COD_T can be almost annihilated naturally.

4.1.2.3 Coacervate volume fraction at equilibrium (ϕ_C)

Regarding the different cutting oil emulsion-surfactant couples, the quadratic model provides the following equations:

$$\phi_{C(HMP,OxoC15E7)} = 0.306 + 0.24X_t - 0.01T - 2.783\ 10^{-3}X_tT + 1.306\ 10^{-3}\ X_t^2 + 0.085\ 10^{-3}T^2 \tag{5.38}$$

$$\phi_{C(HMI,OxoC15E7)} = 1.535 - 0.045\ X_t - 0.026\ T + 3.238\ 10^{-4} - X_tT + 1.875\ 10^{-3-}X_t^2 + 0.11\ 10^{-3-}T^2 \quad (5.39)$$

$$\phi_{C(T22B,OxoC15E7)} = 1.81 + 0.059X_t - 0.029\ T - 0.01\ 10^{-3-}X_tT - 1.062\ 10^{-4-}X_t^2 + 0.072\ 10^3T^2 \quad (5.40)$$

$$\phi_{C(T22B,OxoC10E3)} = -0.023 + 0.019X_t + 4.667\ 10^{-3}T - 0.111\ 10^{-3}X_tT - 1.389\ 10^{-4}X_t{}^{2-}8.889\ 10^{-5}T^2 \quad (5.41)$$

The behavior of ϕ_c according to X_t and T is generally similar for all the cutting oil emulsion-surfactant couples (Figure 5.13). According to Figures 5.13a–d, the value of ϕ_c is low at high

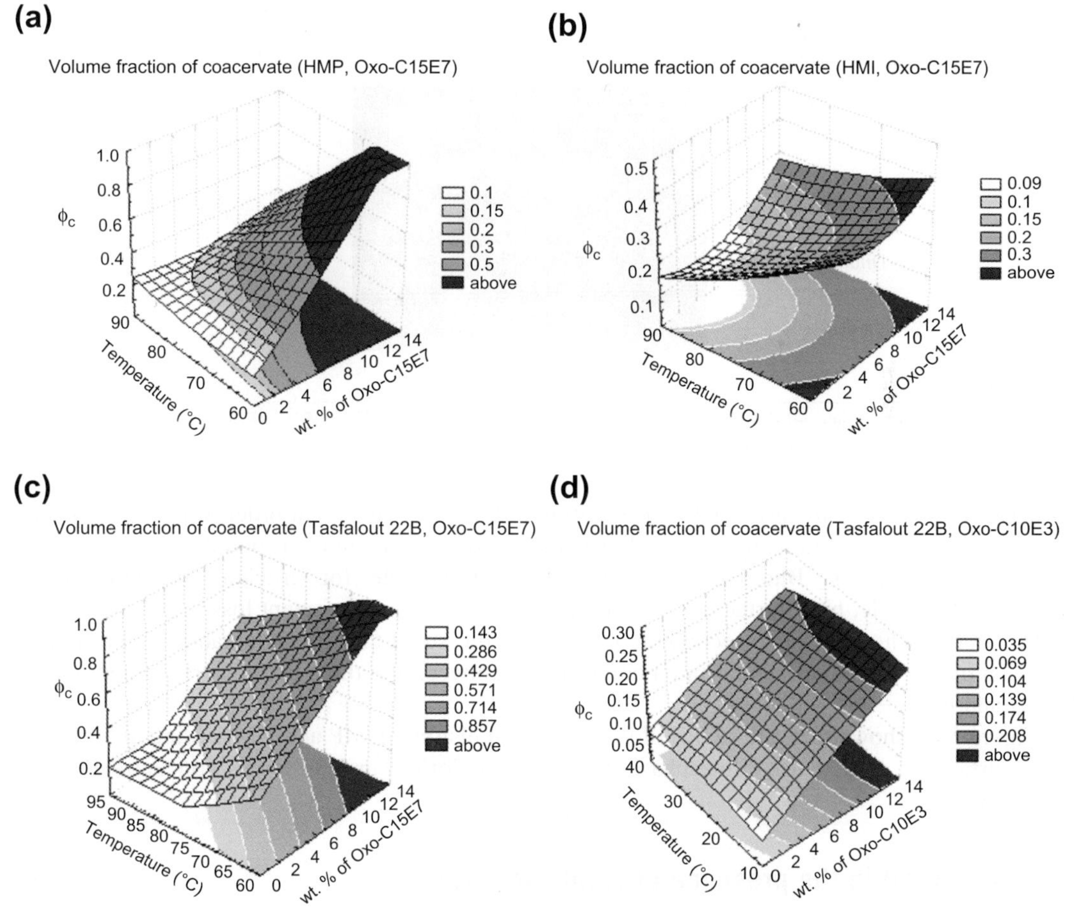

FIGURE 5.13

Three-dimensional isoresponse curves smoothed by a quadratic model (Eqs.5. 38–5.41): (a) $\phi_{c(HMP,\ OxoC15E7)} = f(X_t, T)$; (b) $\phi_{c(HMI,OxoC15E7)} = f(X_t, T)$; (c) $\phi_{c(T22B,OxoC15E7)}\ f(X_t, T)$; (d) $\phi_{c(T22B,OxoC10E3)} = f(X_t, T)$.

(from: Talbi et al., 2009, with permission).

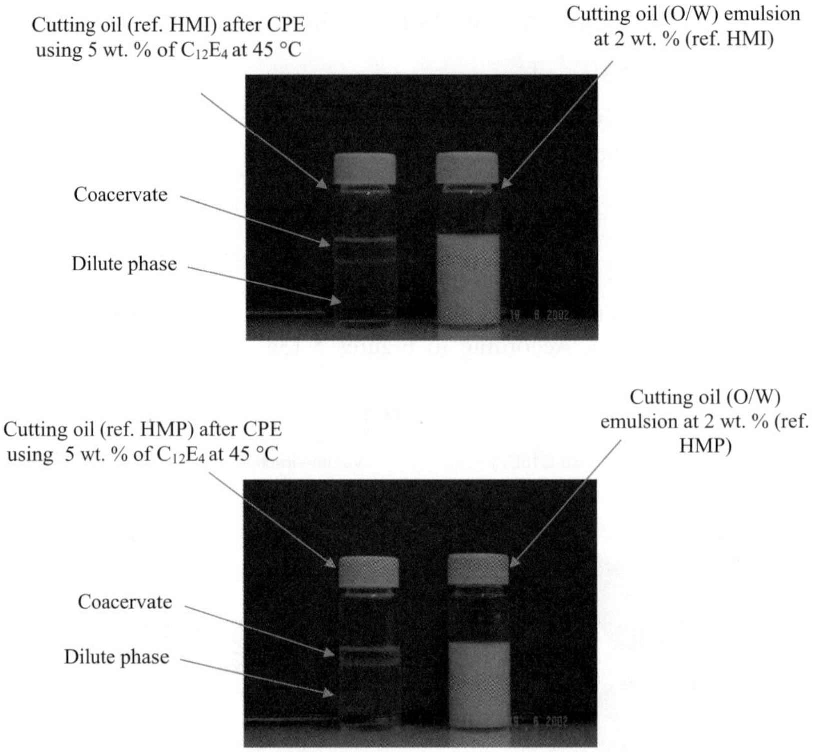

FIGURE 5.14

Cutting oil emulsions (HMI, HMP) before and after CPE using 5 wt.% $C_{12}E_4$ at 45°C. (For color version of this figure, the reader is referred to the online version of this book.)

temperature and at small surfactant concentration (Table 5.2). Hence, in order to increase the concentration factor of solute, a minimal volume fraction of coacervate (ϕ_c) should be obtained when temperature increases. But high temperature is unfavorable for efficient extraction extent (Figure 5.11), while, high surfactant concentrations induce more surfactant loss in the dilute phase (Figure 5.12). This loss is not economic even when the surfactant is biodegradable. So the optimization of the process needs to find a compromise between the three parameters studied: COD_R, COD_T and ϕ_c.

Figure 5.14 shows that, after CPE of cutting oil emulsions (HMP and HMI) using another nonionic surfactant, $C_{12}E_4$, the dilute phase obtained is colorless and the oil is concentrated in the coacervate.

4.2 Cutting oil in the presence of solid (Granite)

4.2.1 Description of the system

As another example of a dispersed pollution, this section is devoted to mud polluted by cutting oil (“mineral oil” and “synthetic oil”), used for tool lubrication in granite transformation (sawing and

Table 5.3 Crude Effluent (Oily Granite Mud) Characteristics

Sample	Oil Type	Granite Type	pH of the Liquid Phase	COD of the Liquid Phase (mgO_2/L)	Mud Density (g/cm^3)	Liquid Phase Appearance
a	Mineral ref. HMP	Sidobre granite	6.80	9540	1.0166	Milky
b	Mineral ref. HMI	Sidobre granite from abroad	6.85	8120	1.0127	Milky
c	Synthetic ref. HMS	Sidobre granite from abroad	7.60	16140	1.0112	Colorless

buffing) in Sidobre (Regional Natural Park of the Haut Languedoc, Tarn, France). Mud is made up of fine particles dispersed in water. After coarse decantation, one can distinguish visually:

a liquid "phase" (milky emulsion)
a solid phase, called the cheek, which is a mixture of particles with a size varying from a few μm to about 1 mm.

CPE tests were performed with three samples of crude effluent (oily granite mud). The sample characteristics are presented in Table 5.3.

4.2.2 Extraction results

4.2.2.1 Chemical oxygen demand (COD)

After CPE, the total residual chemical oxygen demand of the dilute phase, COD_R, includes that due to the surfactant remaining in this phase, COD_T, and that due to soluble pollution in water and not extractable by the surfactant, COD_S (Eq. 5.29).

Figures 5.15a and b show that CPE of mineral oil (HMP, HMI) with 1 wt.% $C_{10}E_3$ lowers the initial COD value considerably. This lowering is mainly due to COD_S reduction. In the case of synthetic oil (HMS), COD_S decrease is less important (Figure 5.15c). Indeed, such oil contains more water-soluble species than mineral oil and therefore is hardly extractable by the surfactant. The surfactant residual concentration in the dilute phase increases with the surfactant percentage used for CPE, which induces a COD_R increase in the dilute phase: At high surfactant concentration (8 wt.%), it can exceed the initial sample COD (Figure 5.15b). Therefore, optimal surfactant concentration is about 1 wt.%: Beyond this concentration, the initial pollution (oil) is replaced by the surfactant. One can also notice in Figures 5.15a–c that the DCO_T is significantly high: Excluding COD_T (biodegradable COD), the remaining COD_R is very low; with 1 wt.% surfactant, it is about 500 mg O_2/L in the case of a mineral oil effluent.

4.2.2.2 Solid phase and coacervate mass fraction

Here we are dealing with an initial state consisting of three phases: a solid pseudophase since granite is a rock and not a unique phase in the crystallographic sense, and two liquid (organic and aqueous) phases. After CPE, the final state is constituted of three phases: the solid phase and two aqueous phases if all the oil is solubilized in the coacervate, but possibly of four phases if oil is in excess. Whether or

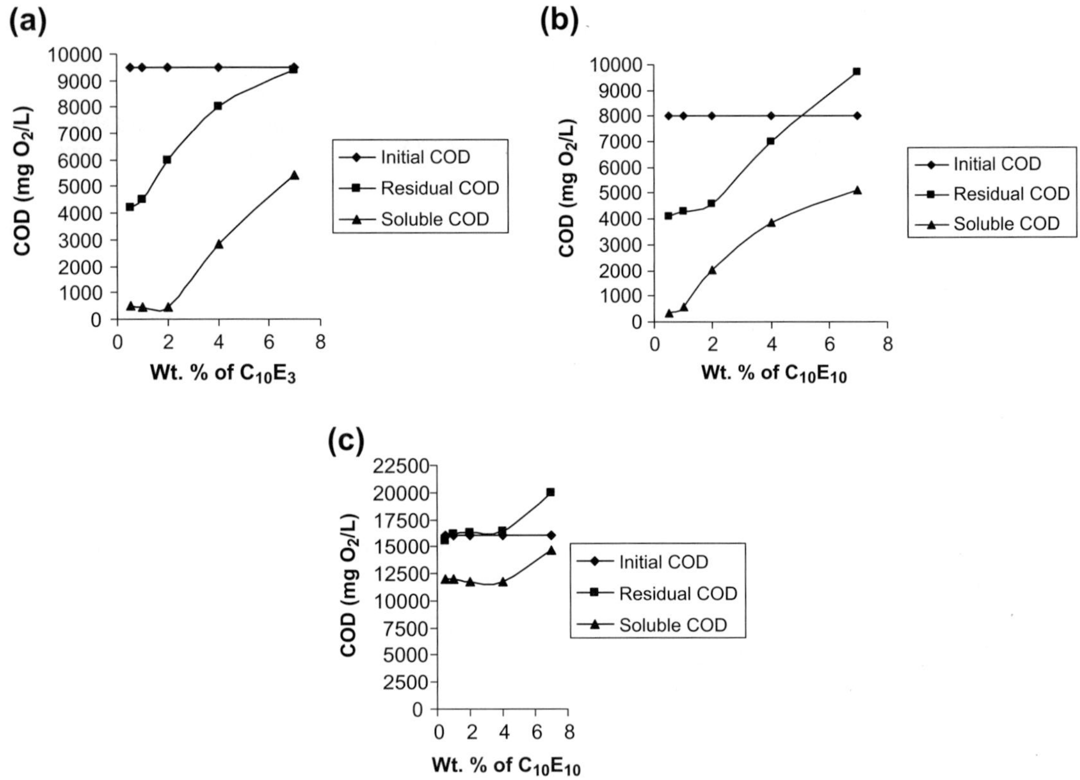

FIGURE 5.15

Initial, residual (DCO_R) and soluble (DCO_S) chemical oxygen demands of the dilute phase after CPE using $C_{10}E_3$ at 20 °C of: (a) sample "e" (containing HMP mineral oil); (b) sample "f" (containing HMI mineral oil); (c) sample "d" (containing HMS synthetic oil).

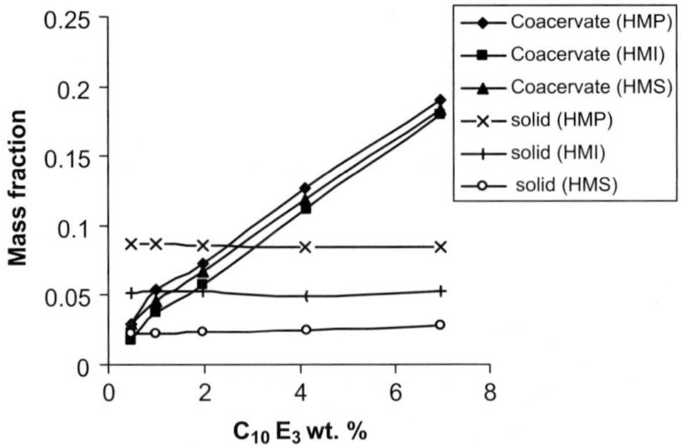

FIGURE 5.16

Mass fractions of coacervate and solid phase after CPE using $C_{10}E_3$ at 20°C of: sample "e" (containing HMP mineral oil); sample "f" (containing HMI mineral oil); and sample "d" (containing HMS synthetic oil).

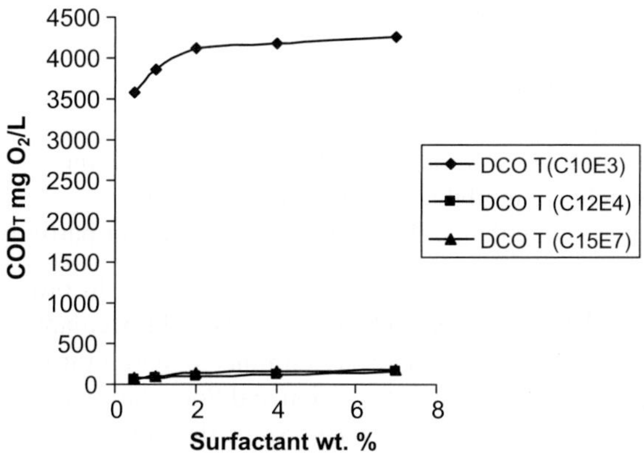

FIGURE 5.17

Residual chemical oxygen demand of the dilute phase after CPE of sample “e” (containing HMP mineral oil) using $C_{10}E_3$ at 20°C, $C_{12}E_4$ at 50°C and $C_{15}E_7$ at 60°C.

not the oil is completely solubilized, demulsification and CPE occur simultaneously. In order to increase the concentration factor of the solute, a minimal coacervate volume (or mass) fraction is the target. According to Figure 5.16, the coacervate mass fraction increases with the initial surfactant concentration. However, at a constant surfactant concentration, the coacervate mass fraction is nearly the same for the three samples studied.

4.2.2.3 Chemical oxygen demand due to the surfactant (COD_T)

Figure 5.17 shows that the COD due to Oxo-$C_{10}E_3$ (COD_T) is higher than that due to $C_{12}E_4$ or Oxo-$C_{15}E_7$. Such results can be explained by the fact that Oxo-$C_{10}E_3$ is more soluble in water than the other two surfactants: Therefore, its residual concentration in the dilute phase after the CPE process is significantly higher. Furthermore, this commercial surfactant obtained from Oxo-alcohol ethoxylation (chosen for its low cloud point) needs to be freed from the residual polyols by washing it with water at 90°C. Indeed, washed $C_{10}E_3$ allows obtaining surfactant COD_T around 500 mg O_2/L (Figure 5.12d). Thus, the COD_T values obtained using $C_{12}E_4$ and Oxo-$C_{15}E_7$ are around 100 mg O_2/L (Figure 5.17). Hence, the surfactant choice is a very important parameter in the scaling-up of the CPE process.

4.3 Silk-screen printing inks

4.3.1 General

In silk-screen printing, patterns are drawn on silk-screens that are contacted with high quantities of inks and a complex mixture of specific products. After use, the saturated settings are washed in an ultrasonic bath in order to remove—in one stage—inks and UV-protective agents. The perfectly clean silk-screens are then rinsed and dried to be reused. The ultrasonic washing process generates a large quantity of a complex effluent containing

- inks, which can decant and make mud
- a screen protector agent against UV, which can decant and make mud
- screen-drying products, which can decant and make mud

- fixer agents, eliminated by mechanical washing (by grids)
- specific aspersion solvents, which accumulate in the bath
- surfactants, which accumulate in the bath
- oxidizing agent: $NaIO_4$, reduced to $NaIO_3$
- sulfuric acid, which accumulates in the bath, inducing pH reduction
- an anti-foam agent, which accumulates in the bath
- water treatment additives (coagulants, flocculants), which decant partially in mud

For economic reasons, the ultrasonic bath needs to be regenerated in order to increase its lifetime. Hence, water coming out of the bath is treated by coagulants and flocculants in a settler, then passes in a evaporator where organic volatile compounds are eliminated, and returns to the bath to be reused in the silk-screen washing process. However, coagulation and flocculation treatment remains insufficient to eliminate inks and the other products from water. Since the liqueur (effluent) composition is very complex and changes frequently, the pH control in the bath near the colloid isoelectric pH is difficult. The replacement of this fragile technique (coagulation, flocculation) should facilitate ultrasonic bath recycling.

Promising results were obtained with a CPE test on the liqueur coming out of the bath.

4.3.2 Extraction results

In the ultrasonic bath, sodium periodate ($NaIO_4$), used as an emulsion remover, is reduced to iodate ($NaIO_3$). This reaction is faster at higher temperature. However, in the bath or during the drying process, working below 40°C is recommended. To respect this temperature constraint, Oxo-$C_{10}E_3$ was used to achieve CPE tests. In fact, the cloud point of a 1 wt.% aqueous solution of this surfactant is about 2°C (Figure 5.1b). However, the liqueur, whose composition is very complex, contains chemical

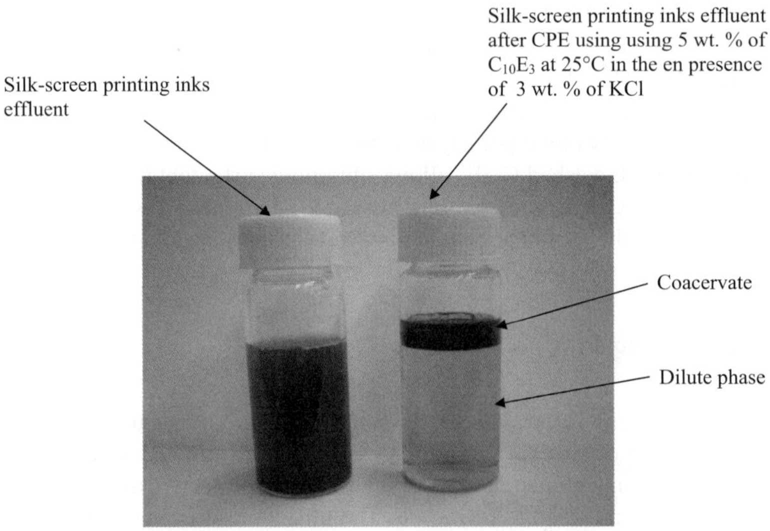

FIGURE 5.18

Silk-screen printing ink effluent before and after CPE using 5 wt.% $C_{10}E_3$ at 25°C in the presence of 3 wt.% KCl. (For color version of this figure, the reader is referred to the online version of this book.)

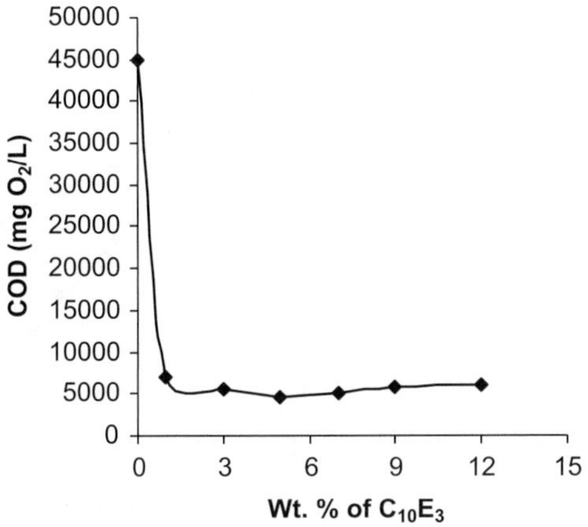

FIGURE 5.19

Residual chemical oxygen demand of the dilute phase after CPE of silk-screen printing ink effluent using various concentrations of $C_{10}E_3$ at 25°C in the en presence of 3 wt.% KCl.

species which increase the cloud point of the nonionic surfactant. Therefore, the addition of salt (KCl) to the liqueur/Oxo-$C_{10}E_3$ system allows CPE of silk-screen printing inks below 40 °C (Figure 5.18). In the conditions shown in Figure 5.18, a minimal volume fraction of coacervate (ϕ_c) is obtained (about 20%v/v), which allows increasing the solute concentration factor. After CPE using Oxo-$C_{10}E_3$, the dilute phase obtained is nearly colorless and the inks are concentrated in the coacervate (Figure 5.18).

Figure 5.19 shows that COD decreases from 45 g/L to 5g/l after CPE. The optimal surfactant concentration is about 4 wt.%; above this concentration, the COD of the dilute phase increases, due to an increase of the residual surfactant concentration in the dilute phase.

5 Salting-out effect on CPE

It is well known that, when dissolved in water, most electrolytes decrease the solubility of an organic compound. This phenomenon (reduced solubility) is called the salting-out effect, the reverse being the salting-in effect. Compounds that increase aqueous solubility are called hydrotropes or chaotropes (Da Silva et al. 1999). The CP decrease is generally due to dehydration of the EO chain by the salt (Pandya et al. 1993). Anions with a low lyotropic number, high hydration, or strong water-structure-making tendency decrease the CP. For example, SO_4^{2-} is twice more effective than Cl^- (Saito et al. 1967, Schott et al. 1983, id. 1984, Pandya et al. 1993, Sakulwongyai et al. 2000): Note that in Figure 5.20, in order to have phase separation at 38°C with Oxo-$C_{13}E_9$, it is necessary to double the mass concentration of NaCl (10 wt.% = 1.709 M) relative to that of Na_2SO_4 (5 wt.% = 0.352 M).

The coacervate volume fraction evolution as a function of NaCl concentration in the H_2O/Oxo-$C_{13}E_9$/C_6H_5OH system is shown in Figure 5.21. Sodium chloride addition induces coacervate volume fraction reduction due to the cloud point lowering of the surfactant solution. At constant surfactant and phenol concentrations, the presence of NaCl lowers T_c (Figure 5.1d) and reduces ϕc, thus increasing the values of (T-T_C). Therefore, in the presence of NaCl, small coacervate volumes

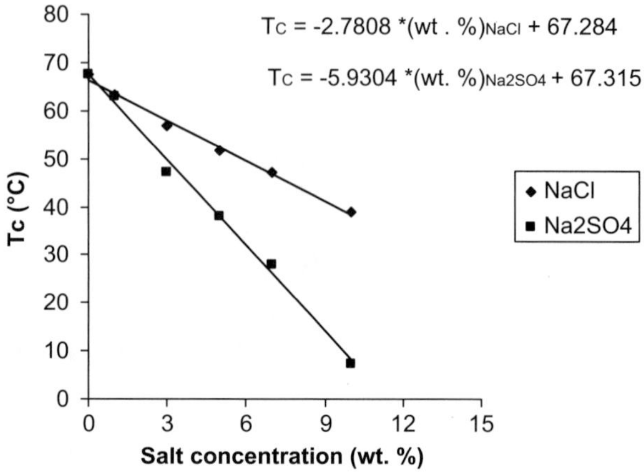

FIGURE 5.20

Effect of electrolytes on the cloud point temperature of a 3 wt.% Oxo-$C_{13}E_9$ solution.

(from: Haddou et al., 2006, with permission).

with high surfactant concentrations are obtained. The dotted part of the curve in Figure 5.21 represents a slow-phase separation rate, due to the almost equal densities of the coacervate and dilute phase.

Frankewich and Hinze (1994) studied the effect of NaCl ionic strength on the extraction extent of phenol and 4-chlorophenol using C_8E_9 in the presence and absence of 0.1 M (0.55 wt.%) NaCl, and showed that this salt had no effect on their extraction extents (15.2% against 16.8 % for phenol and 79.9 % against 79.0 % for 4-chlorophenol). These results are in good agreement with previous ones (Fendler 1982, Horvath et al. 1992). But, according to Saito and Shinoda (1967), the addition of 0.17 M (1 wt.%) NaCl to non-ionic surfactant solutions increases their hydrocarbon solubilization capacity by lowering the cmc, therefore increasing the micelle number in this concentration region. In our work, the influence of NaCl concentrations higher than 0.17 M on the extraction extent of phenol was determined (Figure 5.22). The extraction percentage of phenol increases up to 96%, with 25% NaCl using Oxo-$C_{13}E_9$ at 50°C.

It may be observed from Figure 5.23 that the extraction of MB increases significantly when an electrolyte is added to the water/surfactant/MB system. As mentioned earlier, the salting-out effect

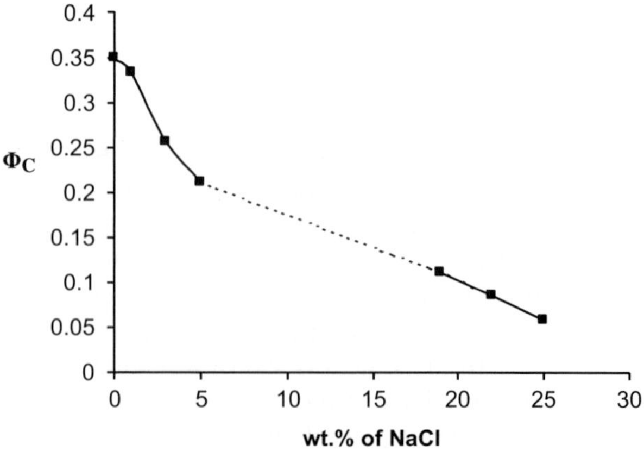

FIGURE 5.21

Effect of sodium chloride on the coacervate volume fraction (ϕ_c) in the H_2O/Oxo-$C_{13}E_9$/phenol system (T = 50°C; 0.15 wt.% phenol; 4 wt.% Oxo-$C_{13}E_9$); along the dotted line part of the graph, phase separation is very slow.

(from: Haddou et al., 2006, with permission).

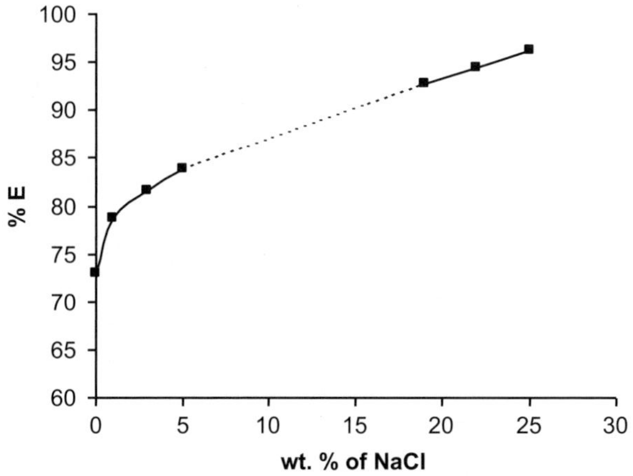

FIGURE 5.22

Effect of sodium chloride on the extraction extent in H_2O/Oxo-$C_{13}E_9$/phenol system ($T = 50°C$; 0.15 wt.% phenol; 4 wt.% Oxo-$C_{13}E_9$); along the dotted line part of the graph, phase separation is very slow.

(from: Haddou et al., 2006, with permission).

resulting from the solvated electrolyte depends on hydrogen bonding between water molecules and surfactant polar head groups as well as with the dye hydrophilic part. Hence, in the presence of electrolyte, dye and surfactant molecules are less soluble in water. Therefore, at 3 wt.% sodium sulphate solution, the surfactant and MB solubilized in micelles are pushed out into the coacervate phase. As a result, small coacervate volumes with high surfactant and dye concentrations are obtained in these conditions (Haddou et al. 2007).

Figure 5.24 shows the decrease of COD_R as a function of Na_2SO_4 concentration in a cutting oil emulsion/surfactant/aqueous solution system. For example, at 15 wt.% Na_2SO_4 in the HMP/Oxo-$C_{15}E_7$ system, the oil, the solutes (cutting oil additives), solubilized in micelles, and the surfactant are pushed out into the coacervate (salting-out effect). Therefore the dilute phase becomes less concentrated in organic molecules and the COD_R value falls to about 0.5 g O_2/L. This concentration is almost 110 times less than that of the initial effluent (55 g O_2/L).

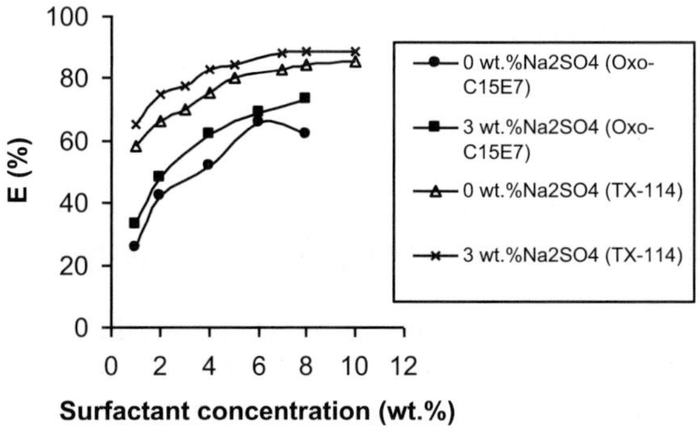

FIGURE 5.23

Effect of sodium sulphate on the extraction extent in the systems: H_2O/TX-114/methylene blue and H_2O/Oxo-$C_{15}E_7$/methylene blue ($T = 35$ °C for TX-114; $T = 55$ °C for Oxo-$C_{15}E_7$; 7 mg/L methylene blue).

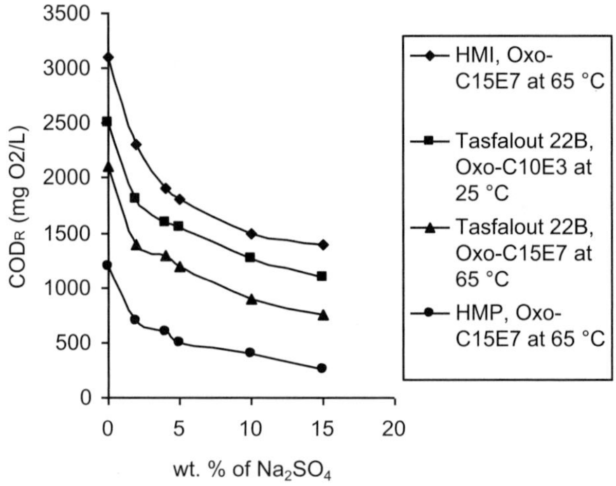

FIGURE 5.24

Effect of sodium sulphate on the total residual chemical oxygen demand (COD_R) in the dilute phase after extraction.

(from: Talbi et al., 2009, with permission).

6 Scaling-up and process

In order to check the feasibility of CPE with large flow-rate waste streams, it is necessary, first, to be able to operate the CPE process in a continuous, multistage extraction device (e.g., column, mixer-settler), and second, to develop methods for surfactant recovery (coacervate regeneration), since surfactants represent the major cost of a CPE process (Kungsanant et al., 2009).

As regards the first issue, phase separation is the limiting step of the process, although, in spite of the very low-density difference between the two phases, the interface appears almost immediately after mixing. More often, both phases remain cloudy during one or several hours. To overcome this difficulty, electrolytes could be added to make the dilute phase heavier, but the presence of salts modifies phase equilibrium versus temperature (see Section 5). Another means of accelerating phase separation is the performance of ultrasonic-assisted CPE: A pilot-scale process of three polycyclic aromatic hydrocarbons (PAHs), anthracene, phenanthrene, and pyrene from polluted water was operated continuously in an extraction column with a volume up to 500 mL (Yao and Yang, 2008). In addition, Ingram et al. (2012) implemented the CPE in a continuous stirred extraction column, thus demonstrating that surfactant/water systems represent a suitable alternative to conventional solvents and can effectively be processed in continuous extraction columns.

7 Coacervate regeneration

It therefore appears highly desirable to recycle the surfactant contained in the coacervate to reuse it. Let us consider the cases of acidic or alkaline solutes, on the one hand, and those of neutral pollutants, on the other hand.

The solute–micelle interactions are strongly influenced by solute ionization. Therefore, the metal chelate distribution between the dilute phase and coacervate varies with the pH of the non-ionic surfactant solutions (Watanabe et al. 1987). The deprotonation of a weak acid or the protonation

Table 5.4 Conditions of Coacervate Regeneration

Concentration of $Ca(OH)_2$ in the Coacervate (g/L)	pH of the Coacervate	Percentage of Phenol Release from the Coacervate (%)	Ca C_2O_4 Concentration in the Dilute Phase (ppm)
0	6.56	10.52	6.5
0.12	9.85	24.97	6.6
0.24	10.07	29.34	6.7
0.36	11.03	38.37	6.7
0.65	11.52	42.49	6.4
0.75	12.02	45.48	6.8
1.25	12.30	45.87	6.8
1.53	12.40	46.22	6.7

of a weak base generates ions, which are more soluble in water than neutral species. Consequently, only a small amount of ionized solute can be extracted. Thus, pH is the key-parameter for surfactant regeneration. After a first extraction of phenol at 10°C, the pH of the coacervate was increased above phenol pK_a using $Ca(OH)_2$ to lead to complete dissociation of the solute (Table 5.4). The ions generated are more easily dissolved in water than solubilized in the micelles (Figure 5.25). The previous coacervate is then separated into two new phases at 25°C. The solute can be concentrated in a small quantity of aqueous phase, in order to yield a new coacervate phase containing most of the surfactant. In order to reuse the surfactant, it is necessary to decrease its pH and to precipitate the base. Therefore, it is better to choose an acid forming an insoluble salt with the base cation (Ca^{2+}), for example, oxalic acid, $H_2C_2O_4$ (Table 5.4). Hence, 46% of phenol extracted at 10°C can be released from the coacervate to a new dilute phase (at 25°C and pH = 12.4), this pH being the highest reached using $Ca(OH)_2$ with a solubility limit of 1.53 g/L (Gutschick 1995). Under these conditions, the dissociated phenol is predominant. After filtration, the concentration of CaC_2O_4, calculated from the initial mass of $H_2C_2O_4$ and from the mass balance of the acid–base reaction (Table 5.4), is close

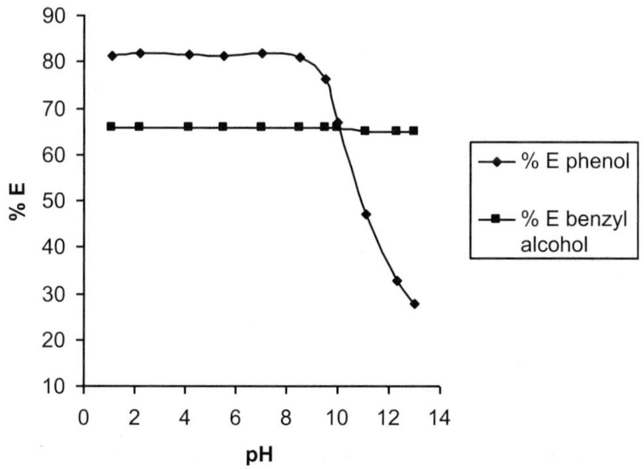

FIGURE 5.25

Effect of pH on phenol and benzyl alcohol extraction percentage, E, in the $H_2O/C_{10}E_3$/solute systems (5 wt.% Oxo-$C_{10}E_3$/solute: 0.15 wt.%) at 10°C.

(from: Haddou et al., 2006, with permission).

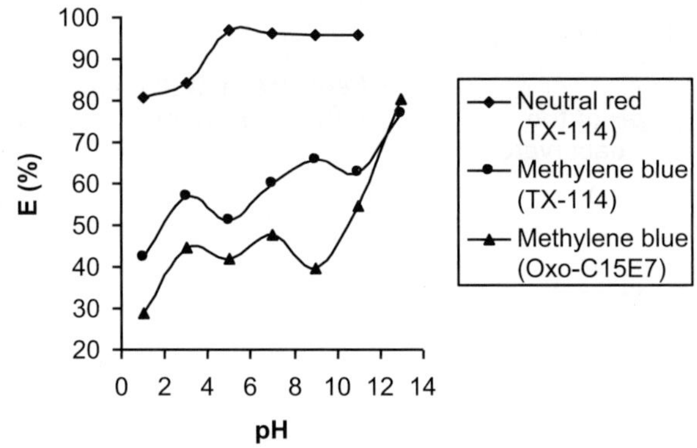

FIGURE 5.26

Effect of pH on the extraction percentage, E, in the systems: H_2O/TX-114/Neutral red, H_2O/TX-114/methylene blue, H_2O/Oxo-$C_{15}E_7$/methylene blue (4 wt. % surfactant; 70 mg/L neutral red; 7 mg/L methylene blue; T = 35 °C for TX-114; T = 55°C for Oxo-$C_{15}E_7$).

to its aqueous solubility at 25°C (6.8 ppm)(Gutschick 1995). In the case of a neutral solute such as benzyl alcohol, the change of pH does not affect its dissociation, if any. Unlike phenol extraction, benzyl alcohol extraction extent remains constant at all pH values (Figure 5.25). In this case, surfactant regeneration is not possible with a pH change.

The effects of the pH of the solution on the extent of NR and MB extractions are shown in Figure 5.26 for 70 mg/L and 7 mg/L of feed dye, respectively, using 4 wt.% surfactant (TX-114 at 35°C or Oxo-$C_{15}E_7$ at 55°C). Extraction rates of neutral red and methylene blue are lower in acidic pH and increase with pH. The pK_a values of the present dyes are found to be around 6.8 for NR and 8.2 for MB.

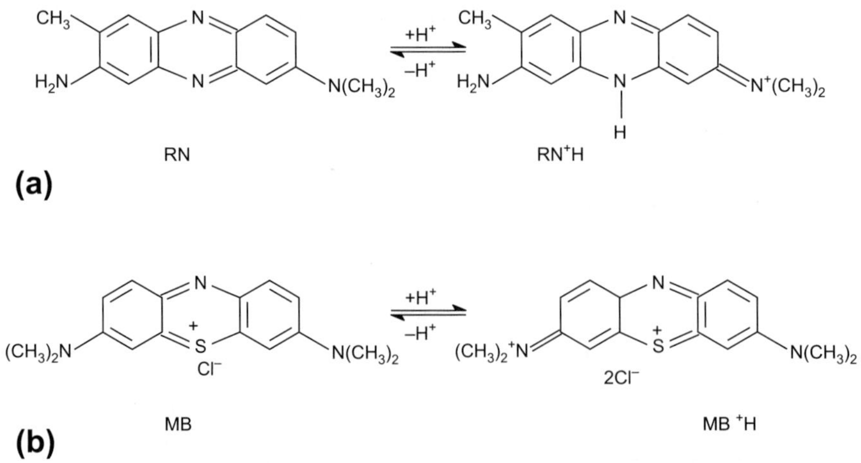

FIGURE 5.27

Structures in alkaline and in acidic media of (a) Neutral Red (NR) (b) Methylene Blue (MB).

At pH lower than their pK_a values, the studied dyes are protonated, as shown in Figure 5.27, and their ionic character increases, leading to less solubilization of both dyes in the hydrophobic micelles. At higher pH (above pK_a values) the dyes are deprotonated, and their behavior looks like that of a hydrophobic molecule, so that they are easily solubilized in the micelles. In this way, dye solubilization is more important at alkaline pH, leading to an increase in dye extraction. The effect of solution pH on MB extraction extent is more significant, and the dissociated form of MB appears to be more hydrophilic than the NR cation.

With neutral, volatile compounds, for example, VOC, a non-ionic surfactant, could be recovered effectively from contaminated coacervates by co-current vacuum stripping in a column packed with glass Raschig rings; that is, over 87% of all studied VOCs (benzene, toluene, ethylbenzene, 1,2 dichloroethane, trichloroethylene, and tetrachloroethylene) were removed from a coacervate solution within a single-stage operation. The effects of surfactant concentration,temperature and solute type were studied (Kungsanant et al., 2009).

8 Conclusion

Two-aqueous-phase extraction in the presence of polyethoxylated nonionic surfactants (CPE) is able to eliminate soluble and dispersed pollutants from effluents efficiently. This method was carried out on H_2O/C_iE_j/solute, and $H_2O/C_i\Phi E_j$/solute systems simply by contacting dilute aqueous solutions of soluble species (phenol, benzyl alcohol, dyes), O/W emulsions ,or more complex mixtures (cutting oils, silk-screen printing inks) with pure or dissolved surfactant. The presence of phenol or benzyl alcohol lowers the cloud point of non-ionic surfactants, the influence of phenol being much more marked at low surfactant concentration. A first contact between surfactant and effluent solutions can give extraction percentages, E, close to 95% for phenol and 90% for benzyl alcohol, provided that the surfactant concentration is high enough. Benzyl alcohol, though less water-soluble than phenol, is less easy to extract. Thus, the aqueous solubility of the solute is not the only parameter controlling solubilization and extraction efficiency.

In dye extraction, a first contact between surfactant and neutral red solutions yields extraction percentages, E, around 99% with TX-114 and 98% with Oxo-$C_{15}E_7$., provided that the surfactant concentration is higher than 10 wt.%. Although the solubility of methylene blue in water is lower than that of neutral red, the methylene blue is more difficult to extract than the neutral red. In all cases, the extraction extents of both dyes with TX-114 are higher than with Oxo-$C_{15}E_7$, but this latter surfactant is safer, regarding biodegradability.

On the other hand, a significant temperature rise beyond the cloud point has a negative effect on E. However, in order to increase the concentration factor of a solute, a minimal volume fraction of coacervate (ϕ_c) should be obtained at higher temperatures. NaCl and Na_2SO_4 reduce the coacervate volume fraction because of the cloud point lowering of the surfactant solution; they also increase the solute extraction percentage.

Surfactant recycling after CPE depends mainly on the extracted species. In the case of a weak acid such as phenol, the coacervate regeneration is possible by a pH change, but in the case of a neutral substance, as benzyl alcohol, surfactant recycling becomes difficult. Both studied dyes (neutral red and methylene blue) are less extractable in acid solutions, but the effect of pH is more significant on methylene blue extraction extent. Indeed, pH can be a key-parameter for surfactant regeneration in dye extraction. Surfactant recycling seems to be mainly possible for methylene blue extraction (extraction in alkaline effluent solution and regeneration in acid media).

A concentrated micellar solution (coacervate) behaves like a classical organic solvent. Two-aqueous-phase extraction is also able to eliminate dispersed pollutants (e.g., cutting oils, silk-screen printing inks) from effluents efficiently. After extraction, the residual COD in the dilute phase, mainly due to surfactant and oil additives, is close to that tolerated in industrial effluents (2 g/L).

Since water-based cutting fluid mixtures contain 90 to 98% water, the originality of the proposed environmentally benign method, which can be employed, for example, as a previous step to direct flame incineration or catalytic incineration, will permit the separation of the pollutants from water, thus avoiding the destruction of the whole waste. On the other hand, due to the presence of sufficient oil in the coacervate, the incineration fuel cost may be substantially reduced. CPE can also be used before an evaporation–distillation process, where evaporation must be accelerated by heating. Unless waste heat is readily available from some in-plant source, the cost of evaporation of the total effluent stream is prohibitive. Evaporation of water from concentrated coacervate is easier after CPE when the volumes have been considerably reduced. Cost can also be reduced using natural solar evaporation. Therefore, CPE appears competitive with other processes of purification of effluents containing soluble or dispersed organic pollutants (physicochemical distillation, ultrafiltration)(Lata et al. 1978, Ralph et al. 1978, Quemeneur et al. 1980).

SYMBOLS AND TERMINOLOGY

ADI Acceptable daily intake
AE alcohol ethoxylate(s)
APE alkylphenol ethoxylate(s)
bw body weight
C_iE_j abbreviated formula of an alcohol ethoxylate (i carbon atoms in the hydrophobic chain, j ethylene oxide groups)
cmc critical micelle concentration
COD chemical oxygen demand
COD_I initial chemical oxygen demand
COD_R total residual chemical oxygen demand
COD_S chemical oxygen demand due to soluble pollution in water
COD_T chemical oxygen demand due to the surfactant
CP cloud point (temperature)
CPE cloud point extraction
E percentage of extracted solute
EO ethylene oxide (number)
EPA environmental Protection Agency
HLB hydrophile-lipophile balance
HMI mineral cutting oil (Motul, France)
HMP mineral cutting oil (TotalFinaElf, France)
HMS synthetic cutting oil (Motul, France)
log P logarithm of the partition coefficient of solute between octan-1-ol and water
MB methylene blue

NR neutral red
O/W oil in water
$\mathbf{pK_a}$ $\log_{10} K_a$ (acidity constant)
$\mathbf{R^2}$ regression coefficient
$\mathbf{t_c}$ critical cloud point temperature
T temperature (°C)
$\mathbf{T_c}$ cloud point temperature
Tasfalout 22B mineral cutting oil from NAFTEC, Algeria
UV ultraviolet light
VOC volatile organic compounds
wt. weight
$\mathbf{X_{s,w}}$ residual concentration of solute in the dilute phase
$\mathbf{X_{t,w}}$ residual concentration of surfactant in the dilute phase
$\mathbf{X_t}$ surfactant wt. %
Y response
$\boldsymbol{\phi_c}$ coacervate volume fraction at equilibrium

Glossary

Amphiphilic qualifies a molecule containing both polar and nonpolar parts.
Biosurfactant surfactant produced by microorganisms.
Catalytic incineration waste disposal method in which a catalyst is used to increase the rate of the combustion reactions.
Central composite design experimental design, useful in response surface methodology, for building a quadratic model for the response variable(s) without the need of a complete three-level factorial design.
Chaotrope (or chaotropic agent) substance that disrupts the three-dimensional structure of a solvent (e.g., water) or a macromolecule (e.g., protein, DNA,...). It is a structure-breaking agent.
Chelating agent polydentate ligand (usually a neutral or electrically charged organic compound) able to coordinate to a metal atom or ion, thus forming a chelate (complex).
Chemical oxygen demand (COD) analytical test measuring the amount of organic compounds in water. A known excess amount of the oxidant (usually potassium dichromate) is added to the sample being analyzed. After a refluxing digestion step, the initial concentration of organic substances in the sample is calculated from the determination of the residual amount of oxidant in the sample.
Cloud point temperature above (or below) which a surfactant solution becomes turbid and eventually separates into two aqueous phases, the dilute phase and the coacervate.
Coacervate surfactant-rich phase appearing above (or below) the cloud point of a surfactant solution.
Critical micelle concentration surfactant concentration above which micelle formation becomes appreciable.
Cutting oil (fluid) O/W emulsion used in metal turning, milling, drilling. Allows regulation of heat generation and ensures lubrication between tool and metal part.
Direct flame incineration the simplest method of incineration, applied when the waste materials readily burn when mixed with air. The incinerator comprises a burner, a mixing zone, and a combustion chamber.
Hemipter (*hemiptera*) insects belonging to the hemimetabolic (undergoing incomplete metamorphosis) order (e.g., true bugs).
Hydrophile-lipophile balance empirical scale defining the nature of emulsifying agents according to their hydrophilic/lipophilic character.
Hydrophilic "water-loving"; for example, qualifies the polar head of a surfactant.
Hydrophobic "water-hating"; for example, qualifies the hydrocarbon tail of a surfactant.

Hydrotrope substance able to solubilize hydrophobic compounds in aqueous solutions without forming micelles, contrary to surfactants, but possibly able to self-aggregate in a stepwise process.

Lyotrope cations and anions are ranked along the so-called lyotropic (or Hofmeister) series according to the decreasing reduction of protein solubility and, for the first terms, their decreasing salting-out power.

Micelle organized surfactant aggregate (association colloid) grouping tens to hundreds of surfactant molecules; direct micelles expose their polar head toward water or a water-like solvent; reverse micelles expose their hydrocarbon tail toward an organic solvent.

Microemulsion homogeneous (transparent or translucent),isotropic, thermodynamically stable mixture containing both rather high oil and water ratios, at least a surfactant and, possibly, a co-surfactant (or co-solvent) and electrolytes.

Physicochemical distillation (reactive distillation) process whereby the chemical reactor is also the still. Separation of the product from the reaction mixture does not need a further step, which saves energy (for heating) and materials.

Phytoplankton the autotrophic component of the plankton community, therefore composed of organisms producing complex organic molecules from inorganic molecules using photosynthesis (e.g., diatoms).

Salting-in increase of the solubililty of a solute in the presence of electrolyte(s).

Salting-out decrease of the solubililty of a solute in the presence of electrolyte(s).

Silk-screen printing ink special ink used on a screen devoted to the printing of posters or other flat color designs by means of a stencil method operated through a piece of silk or other fine cloth (all parts of the design not to be printed are stopped out by an impermeable film).

Solubilization incorporation of a slightly soluble solute into surfactant micelles resulting in its (apparent) solubility increase.

Ultrafiltration filtration technique using membranes of adequate pore size to separate macromolecules or association colloids (from a few nm to several hundred nm) from their medium.

References

Abdullah, A.M., Nasr, E.A., 1996. Effect of oilfield chemicals on the cloud point of non-ionic surfactants. Colloids & Surfaces 125, 5–18.

Achour, T., Chafia, A.R.T., Didier, H., Safia, D., 2011. Evaluation of adsorption capacities of humic acids extracted from Algerian soil on polyaniline for application to remove pollutants such as Cd(II), Zn(II) and Ni(II) and characterization with cavity microelectrode. J. Environ. Sci. 23, 1095–1103.

Akita, S., Takeuchi, H., 1995. Cloud point extraction of organic compounds from. Aqueous solution with nonionic surfactants. Sep. Sci. Technol. 30, 833–846.

Akita, S., Takeuchi, H., 1996. Equilibrium distribution of aromatic compounds between aqueous solution and coacervate of nonionic surfactant. Sep. Sci. Technol. 31, 401–412.

Al Sabbagh, A.M., Maysour, N.A., Nasser, N.M., 2006. Some cutting oil formulations based on local prepared emulsifiers. Part I: Preparation of some emulsifiers based on local raw materials to stabilize cutting oil emulsions. J. Disp. Sci. Technol. 27, 239–250.

Angelier, E., 2000. Ecologie des eaux courante. Technique & Documentation, Paris.

AL-Rasheed, R., Cardin, D.J., 2003. Photocatalytic degradation of humic acid in saline waters. Part 2. Effects of various photocatalytic materials. Appl. Catal. A. 246, 39–48.

Ayotamuno, M.J., Okparanma, R.N., Ogaji, S.O.T., Probert, S.D., 2007. Chromium removal from flocculation effluent of liquid-phase oil-based drill-cuttings using powdered activated carbon. Appl. Energy 84, 1002–1011.

Badis, A., Ferradji, F.Z., Boucherit, A., Fodil, D., Boutoumi, H., 2010. Removal of natural humic acids by decolorizing actinomycetes isolated from different soils (Algeria) for application in water purification. Desalination 259, 216–222.

Bataller, H., Lamaallam, S., Lachaise, J., Graciaa, A., Dicharry, C., 2004. Cutting fluid emulsions produced by dilution of a cutting fluid concentrate containing a cationic/nonionic surfactant mixture. J. Mater. Processing Technol. 152, 215–220.

Benzyl alcohol, 2009. http://en.wikipedia.org/wiki/Benzyl_alcohol.

Bhardwaj, A., Hartland, S., 1993. Applications of surfactants in Petroleum Industry. J. Disp. Sci. Technol. 14 (1), 87–116.

Bliefert, C., Perraud, R., 2001. Chimie de l'environnement. De Boeck, Belgium.

Bordier, C., 1981. Phase separation of integral membrane proteins in Triton X-114 solutions. J. Biol. Chem. 256, 1604–1607.

Bouberka, Z., Kacha, S., Kameche, M., Elmaleh, S., Derriche, Z., 2005. Sorption study of an acid dye from an aqueous solutions using modified clays. J. Hazard. Mater. 119 (1-3), 117–124.

Bouberka, Z., Khenifi, A., Benderdouche, N., Derriche, Z., 2006. Removal of Supranol Yellow 4GL by adsorption onto Cr-intercalated montmorillonite. J. Hazard. Mater. 133 (1-3), 154–161.

Box, G., Draper, N., 1987. Empirical model building and response surfaces. John Wiley & Sons, New York.

Bruheim, P., Bredholt, H., Eimhjellen, K., 1997. Bacterial degradation of emulsified crude oil and the effect of various surfactants. Can J. Microbiol. 43 (1), 17–22.

Canselier, J.P., Mans, C., Llorens, J., 1995. Los tensioactivos en las operaciones unitarias de separación, (I) Operaciones que involucran una fase gaseosa. Ingenieria Química. November, pp. 183-188; id., Operaciones que no involucran fases gaseosas, December, pp. 149–155.

Canselier, J.P., Gourdon, C., Nogueira Duarte, L.J., De Barros Neto, E.L., Haddou, B., Gumila, C., 2006. Procédé d'extraction sans solvant des polluants organiques et métalliques. French Patent, 06/03632, WO-2007–122158.

Carabias-Martinez, R., Rodriguez-Ganzal, J., Dominguez-Alvarez, C., Garcia-Mendez, J.P., 2003. Prediction of the behavior of organic pollutants using cloud point extraction. J. Chromatogr. A. 1005, 23–34.

Carey, F.A., 1990. Advanced organic chemistry, third ed. Plenum, New York.

Chang, I.S., Chung, C.M., Han, S.H., 2001. Treatement of oily wastewater by ultrafiltration and ozone. Desalination 133, 225–232.

Chiang, L.C., Chang, J.E., Wen, T.C.N., 2000. Destruction of refractory humic acid by electromechanical oxidation process. Water Sci. Technol. 42, 225–232.

Crescenzi, F., Buffagni, M., D'Angeli, E., Porcelli, F., 2002. Oil and hydrocarbon spills III. Water Studies Series 11, 245–251.

Da Silva, R.C., Spitzer, M., da Silva, L.H.M., Loh, W., 1999. Investigations on the mechanism of aqueous solubility increase caused by some hydrotropes. Thermochim. Acta. 328, 161–167.

De Barros Neto, E.L., 1999. Extraction liquide-liquide à deux phases aqueuses: Systèmes Phénol/Tensioactif polyéthoxylés. Thesis, INP Toulouse.

De Barros Neto, E.L., Canselier, J.P., 1998. Cloud point behaviour and phase separation in liquid coacervate extraction. Journ. Com. Esp. Deterg. 28, 433–444.

De Barros Neto, E.L., Canselier, J.P., 2000. Diagramme de phases de systèmes eau/tensioactif polyéthoxylé/phénol. XXVIes JEEP, Marseilles. Proceedings, 32–36.

De Barros Neto, E.L., Canselier, J.P., Gourdon, C., 2001. Organic solvent-free extraction of phenol through liquid-coacervate systems. Solvent Extraction for the 21st Century, SCI, London 1, 171–176.

Deepak, D., Baargavar, F., 1994. Biodegradation kinetics of metal cutting oils. Chem. Eng. J. 56, 91–96.

Degiorgio, V., 1985. Nonionic micelles. In: Corti, M., Degiorgio, V. (Eds.), Physics of amphiphiles: Micelles, vesicles and microemulsions. North-Holland, Amsterdam, pp. 303–335.

Donbrow, M., Azaz, E., 1976. Solubilization of phenolic compounds in non-ionic surface-active agents. J. Colloid Interface Sci. 57, 20–27.

Duarte, L.J.N., 2005. Extraction à deux phases aqueuses à l'aide d'alcools polyéthoxylés en vue de l'élimination de polluants organiques et d'ions métalliques. Thesis, INP Toulouse.

Duarte, L.J.N., Canselier, J.P., 2005. Cloud point, critical micelle concentration and HLB relationships for alcohol ethoxylates. PharmaChem. 4 (3), 36–39.

Duarte, L.J.N., Canselier, J.P., 2008. Two-aqueous phase extraction for the removal of organic pollutants ant metal ions. CHEMPOR 2008–10th International Chemical and Biological Engineering Conference. Proceedings on disk, Braga (Portugal).

Er-Raioui, H., Bouzid, S., Marhraoui, M., Saliot, A., 2009. Hydrocarbon pollution of the Mediterranean coastline of Morocco. Ocean & Coastal Management 52 (2), 124–129.

European Commission: Health & Consumer Protection Directorate-General, September 2002. Opinion of the Scientific Committee on Food on Benzyl Alcohol. http://ec.europa.eu/food/fs/sc/scf/out138_en.pdf.

Faibish, R.S., Cohen, Y., 2001. Fouling and rejection behaviours of ceramic and polymer- modified ceramic membranes for ultrafiltration of oil in water emulsions and microemulsions. Colloids and Surface, A: Physicochem. Eng. Aspects 191, 27–40.

Falbe, J., 1986. Surfactant in consumer products: Theory, technology and application. Springer-Verlag, Berlin.

Fasman, G.A., 1997. Handbook of biochemistry and molecular biology. CRC Press, Cleveland (OH).

Fendler, J.H., 1982. Membrane mimetic chemistry. Wiley Interscience, New York.

Ferrera, Z.S., Sanz, C.P., Santana, C.M., Rodriguez, J.J.S., 2004. The use of micellar systems in the extraction and pre-concentration of organic pollutants in environmental samples. Trends Anal. Chem. 23 (7), 479–489.

Frankewich, R.P., Hinze, W.L., 1994. Evaluation of the factors affecting nonionic surfactant mediated phase separations. Anal. Chem. 66 (7), 944–954.

Friberg, S.E., Neogi, P., 1989. Microemulsion-based separations. In: Scamehorn, J.F., Harwell, J.H. (Eds.), Surfactant-based separation processes: Surfactant science series, vol. 33, Marcel Dekker, New York, pp. 119–138.

Ghernaout, D., Ghernaout, B., Saiba, A., Boucherit, A., Kellil, A., 2009. Removal of humic acids by continuous electromagnetic treatment followed by electrocoagulation in batch using aluminium electrodes. Desalination 239 (1-3), 295–308.

Ghouas, H., Haddou, B., Bouabdesselam, H., Bouberka, Z., Derriche, Z., 2010. Elimination of fuel spills from effluent using cloud point extraction methods. J. Hazard. Mater. 180, 188–196.

Ghouas, H., Haddou, B., Kameche, M., Derriche, Z., Gourdon, C., 2012. Extraction of humic acid by coacervate: Investigation of direct and back processes. Journal of Hazardous Materials, Elsevier Sciences, 205–206: 171–178.

Gourlay-Francé, C., Lorgeoux, C., Tusseau-Vuillemin, M.H., 2008. Polycyclic aromatic hydrocarbon sampling in wastewaters using semi-permeable membrane devices: Accuracy of time-weighted average concentration estimations of truly dissolved compounds. Chemosphere 73 (8), 1194–1200.

Guenther, K., Kleist, E., Thiele, B., 2006. Estrogen-active nonylphenols from an isomer-specific viewpoint: a systematic numbering system and future trends. Anal. Bioanal. Chem. 384, 542–546.

Gullickson, N.D., Scamehorn, J.F., Harwell, J.H., 1989. Liquid coacervate extraction. In: Scamehorn, J.F., Harwell, J.H. (Eds.), Surfactant-based separation processes, Surfactant science series, vol. 33, Marcel Dekker, New York, pp. 139–154.

Gutschick, K.A., 1995. In: Kroschwitz, J.I. (Ed.), Lime and limestone. Kirk-Othmer encyclopedia of chemical technology, fourth ed., vol. 15. John Wiley & Sons, New York, pp. 319–359.

Haddou, B., 2003. Purification d'effluent par extraction à deux phases aqueuses. Thesis INP Toulouse.

Haddou, B., Canselier, J.P., Gourdon, C., 2003. Purification of effluents by two-aqueous phase extraction. Trans. I. Chem. E. 81 (A), 1184–1192.

Haddou, B., Canselier, J.P., Gourdon, C., 2006. Cloud point extraction of phenol and benzyl alcohol from aqueous stream. Sep. Purif. Technol. 50 (1), 114–121.

Haddou, B., Taibi, A., Bouberka, Z., Bouabdesselam, H., Derriche, Z., 2007. Separation of neutral red and methylene blue from wastewater using two-aqueous phase extraction. Sep. Sci. Technol. 42 (12), 2677–2691.

Haddou, B., Guitri, N., Debbab, A., Gourdon, C., Derriche, Z., 2011. Cloud point extraction of Orange II and Orange G using neutral and mixed micelles: Comparative approach using experimental design. Separation Sci. Technol. 46 (5), 734–743.

Halide, A.C., Igdem, K., 2005. Spectrophotometric studies of anionic dye–cationic surfactant interactions in mixture of cationic and nonionic surfactants. Spectrochimica. Acta. 61 (A), 961–966.

Hammerton, C., 1955. Observation on the decay of synthetic anionic detergent in natural waters. J. Appl. Chem. 5, 517–521.

Hatton, T.A., 1989. Reversed micellar extraction of proteins. In: Scamehorn, J.F., Harwell, J.H. (Eds.), Surfactant-based separation processes: Surfactant science series, vol. 33, Marcel Dekker, New York, pp. 57–89.

Hinze, W.L., Pramauro, E., 1993. A critical review of surfactant mediated phase separation (Cloud point extraction): Theory and application. Crit. Rev. Anal. Chem. 24 (2), 133–177.

Horvath, W.J., Huie, C.W., 1992. Salting-out surfactant extraction of porphyrins and metalloporphyrins from aqueous non-ionic surfactant solutions. Talanta 39, 487–492.

Hosny, A.Y., 1992. Separation of oil from oil/water emulsions using an electroflottation cell with insoluble electrodes. Filtration and Separation 29 (5), 419–423.

Ingram, T., Storm, S., Glembin, P., Bendt, S., Huber, D., Mehling, T., Smirnova, I., 2012. Aqueous surfactant two-phase systems for the continuous countercurrent cloud point extraction. Chemie Ingenieur Technik 84 (6), 840–848.

Janknecht, P., Lopez, A.D., Mendes, A.M., 2004. Removal of industrial cutting oil from oil emulsions by polymeric Ultra- and microfiltration membranes. Environ. Sci. Technol. 38, 4878–4883.

Jaroslaw, W., Didier, R., Joanna, S., Korneliusz, M., Sixto, M., Jean, V., Didier, G., Surmacz, J., 2004. Solar photocatalytic degradation of humic acids as a model of organic compounds of landfill leachate in pilot-plant experiments: Influence of inorganic salts. App. Cat. B: Environ. 53, 127–137.

Jezabel, S.O., Juan, R.P., Nebot, E., Martinez de la Ossa, E., 2007. Hydrothermal oxidation: Application to the treatment of different cutting fluid wastes. J. Hazard. Mater. 144, 639–644.

Jianfeng, F., Zhao, Y., Wu, Q., 2006. Optimising photoelectrocatalytic oxidation of fulvic acid using response surface methodology. J. Hazard. Mater. 144, 499–505.

Ji, G.D., Yangb, Y.S., Zhouc, Q., Sunc, T., Nia, J.R., 2004. Phytodegradation of extra heavy oil-based drill cuttings using mature reed wetland: Aan in situ pilot study. Environ. Int. 30, 509–517.

Jonkers, N., Laane, R.W.P.M., De Voogt, P., 2005. Sources and fate of nonylphenol ethoxylates and their metabolites in the Dutch coastal zone of the North Sea. Mar. Chem. 96, 115–135.

Kartsev, V.N., Shtykov, S.N., Bogomolova, I.V., Ryzhov, I.P., 2009. Thermodynamic stability of microemulsion based on sodium dodecylsulfate. J. Mol. Liquids 145 (3), 173–176.

Katsoufidou, K., Yiantsios, S.G., Karabelas, A.J., 2005. A study of ultrafiltration membrane fouling by humic acids and flux recovery by backwashing: Experiments and modeling. J. Memb. Sci. 266, 40–50.

Katsoufidou, K., Yiantsios, S.G., Karabelas, A.J., 2008. An experimental study of UF membrane ouling by humic acids and sodium alginate solutions: the effect of backwashing on flux recovery. Desalination 220, 214–227.

Katsumata, H., Sada, M., Kaneco, S., Suzuki, T., Ohta, K., Yobiko, Y., 2008. Humic acid degradation in aqueous solution by the photo-Fenton process. Chem. Eng. J. 137, 255–230.

Kim, T.O., Shin, H., Suh, G.J.H., Lee, W.G., Lee, Y.K., 1998. Ferrofluid to combat oil slicks. British Patent, GB 2319023.

Kimchuwanit, W., Osuwan, S., Scamehorn, J.F., Harwell, J.H., Haller, K.J., 2000. Use of a micellar-rich phase coacervate to extract trichlorethylene from water. Sep. Sci. Technol. 35 (13), 1991–2002.

Komaromy-Hiller, G.B., von Wandruszka, R., 1995. Decontamination of oil-polluted soil by cloud point extraction. Talanta 42 (I), 83–88.

Lata, P., Bahadur, P., 1978. Flocculation of oil-in-water emulsion by detergents. Colloid Polym. Sci. 63, 65–67.

Kungsanant, S., Kitiyanan, B., Rirksomboon, T., Osuwan, S., John F. Scamehorn, J.F., 2009. Recovery of nonionic surfactant from VOC-contaminated coacervate phase solutions by co-current vacuum stripping: Effect of surfactant concentration, temperature, and solute type. Sep. Purif. Technol. 66, 510–516.

Leung, R., Hou, M.J., Shah, D.O., 1988. Microemulsion: formation, structure, properties, and novel applications. In: Wasan, T.D., Ginn, M.E., Shah, D.O. (Eds.), Surfactants in chemical/process engineering: Surfactant science series, vol. 28, Marcel Dekker, New York, pp. 315–367.

Li, J.L., Chen, B.H., 2003. Equilibrium partition of polycyclic aromatic hydrocarbons in a cloud-point extraction process. J. Colloid Interface Sci. 263, 625–632.

March, J., 1992. Advanced organic chemistry, fourth ed. John Wiley & Sons, New York.

Margesin, R., Chinner, S., 1999. Biological decontamination of oil spills in cold environments. J. Chem. Technol. Biotechnol. 74, 381–389.

McCarthy, J.F., 1989. Bioavailability and toxicity of metals and hydrophobic organic contaminants in aquatic humic substances: Influence on fate and treatment of pollutants. In: Suffet, I.H., McCarthy, P. (Eds.), Am. Chem. Soc.. Washington, DC, pp. 263–279.

Materna, K., Milosz, I., Miesiac, I., Cote, G., Szymanowski, J., 2001. Removal of phenols from aqueous streams by the cloud point extraction technique with oxyethylated methyl dodecanoates as surfactants. Environ. Sci. Technol. 35, 2341–2346.

Materna, K., Szymanowski, J., 2002. Separation of phenols from aqueous micellar solutions by cloud point extraction. J. Colloid Interface Sci. 255, 195–201.

Mateja, K., Barbara, S., 2008. The influence of nonionic surfactant structure on the thermodynamics of anionic dye-cationic surfactant interactions in ternary mixtures. Dyes and Pigments 79, 59–68.

Mitchell, D.J., Tiddy, G.J.T., Waring, L., Bostock, T., McDonald, M.P., 1983. Phase behavior of polyoxyethylene surfactants with water—mesophase structures and partial miscibility (cloud points). J. Chem. Soc. Faraday Trans. I. 79, 975–1000.

MP Biomedicals, Safety data file; Product code: 16911. www.mpbio.com/includes/msds/ansi/fr/16911-fr-ansi.pdf.

Myers, D., 1988. Surfactant science and technology. VCH, Weinheim.

Nishino, N., Nakamura, M., 1978. Minima in the solubilities of normal alkane derivatives with a polar group in water. Bull. Chem. Soc. Jap. 51 (6), 1617–1620.

Paciolla, M.D., Kolla, S., Jansen, S.A., 2002. The reduction of dissolved iron species by humic acids and subsequent production of reactive oxygen species. Ad. Environ. Res. 7, 169–178.

PAN Pesticide Database, CASNumber 61-73-4: www.pesticideinfo.org/List_AquireAll.jsp

Pandya, K., Lad, K., Bahadur, P., 1993. Effect of additives on the clouding behaviour of an ethylene oxide-propylene oxide block copolymer in aqueous solution. Pure Appl. Chem. A30 (1), 1–18.

Pastewski, S., Hallmann, E., Medrzycka, K., 2006. Physicochemical aspects of the application of surfactants and biosurfactants in soil remediation. Environ. Eng. Sci. 23, 579–588.

Pelletier, E., Siron, R., 1994. Silicone based polymers as spill treatment agent. World Patent, WO 9420421.

Perez, M., Rodriguez-caro, R., Romero, D., Sales, L.I., 2006. Anaerobic haemophilic of cutting oil wastewater: Effect of co-substrate. Biochem. Eng. J. 26, 250–257.

Phenol, 2009. http://fr.wikipedia.org/wiki/Phénol_(molécule). http://en.wikipedia.org/wiki/Phenol.

Philip, H.H., William, M.M., 1997. Handbook of physical properties of organic chemicals. Lewis Publishers, New York.

Polak, J., Bartoszek, M., Żadło, M., Kos, A., Sułkowski, W.W., 2011. A comparative study of the adsorption of humic acids, fulvic acid and phenol onto Bacillus subtilis and activated sludge, Chemosphere 84, 1548–1555.

Portela, J.R., Lopez, J., Nebot, E., 2001. Martinez de la Ossa E. Elimination of cutting oil wastes by promoted hydrothermal oxidation. J. Hazard. Mater. 88, 96–106.

Pramauro, E., 1990. Concentration and removal of chloroaromatic pollutants using micelle-mediated methods. Ann. Chim. (Ital.) 80', 101–109.

Purkait, M.K., Dasgupta, S., De, S., 2004a. Removal of dye from wastes using micellar-enhanced-ultrafiltration and recovery of surfactant. Sep. Purif. Technol. 37 (1), 81–92.

Purkait, M.K., Vijay, S.S., Dasgupta, S., De, S., 2004b. Separation of congo red by surfactant mediated cloud point extraction. Dyes and Pigment 63 (2), 151–159.

Purkait, M.K., Banerjee, S., Mewara, S., Dasgupta, S., De, S., 2005. Cloud point extraction of toxic Eosin dye using Triton X-100 as nonionic surfactant. Water Res. 39, 3885–3890.

Purkait, M.K., Dasgupta, S., De, S., 2006. Performance of TX-100 and TX-114 for the separation of chrysoidine dye using cloud point extraction. J. Hazard. Mater. 137, 827–835.

Quemeneur, F., Schlumpf, J.P., 1980. Traitement des huiles soluble par ultrafiltration. Entropie 23, 22–29.

Quina, F.H., Hinze, W.L., 1999. Surfactant mediated cloud point extraction: An environmentally bening alternative separation approach. Ind. Eng.Chem. Res. 38, 4150–4168.

Ralph, C., Robert, L.P., 1978. Breaking emulsion in navy bilge collection and treatment Systems. Environ. Sci. Technol. 12 (5), 584–590.

Regalado, C., Asenjo, J.A., Pyle, D.L., 1996. Studies on the purification of peroxidase from Horseradish roots using reverse micelles. Enzyme Microb. Technol. 18, 332–339.

Riccardo, S., Alessandro, P., 2009. Molecular characteristics of humic acids extracted from compost at increasing maturity stages. Soil Biol. Biochem. 41, 1164–1172.

Robinson, J.P., Snape, C.E., Kingman, S.W., Shang, H., 2008. Thermal desorption and pyrolysis of oil contaminated drill cuttings by microwave heating. J. Anal. Appl. Pyrolysis 81 (1), 27–32.

Rondon, M., Bouriat, P., Lachaise, J., Salager, J.L., 2006. Physicochemical phenomenology of demulsifier action. Energy & Fuels 20 (4), 1600–1604.

Rupert, L.A.M., 1992. A thermodynamic model of clouding in water/alcohol ethoxylate mixtures, J. Colloid Interface Sci. 153, 92–105.

Saito, H., Shinoda, K., 1967. The solubilization of hydrocarbons in aqueous solution of non-ionic surfactants. J. Colloid Interface Sci. 24, 10–15.

Sakulwongyai, S., Trakultamupatam, P., Scamehorn, J.F., Osuwan, S., Christian, S.D., 2000. Use of surfactant coacervate phase to extract chlorinated aliphatic compounds from water: extraction of chlorinated ethanes and quantitative comparison to solubilization in micelles. Langmuir. 16, 8226–8230.

Sangster, J., 1989. Octanol-water partition coefficients of simple organic compounds. J. Phys. Chem. Ref. Data 18 (3), 1111–1229.

Scamehorn, J.F., Harwell, J.H., 1989. Surfactant-based separation processes, Surfactant science series, vol. 33, Marcel Dekker, New York.

Schott, H., Royce, A.E., 1983. Effect of inorganic additive on solutions of non-ionic surfactant VI: Further cloud point relations. J. Pharm. Sci. 73, 793–799.

Schott, H., Alan, E.R., Hans, K., 1984. Effect of inorganic additives on solutions of nonionic surfactants. J. Colloid Interface Sci. 98 (1), 196–201.

Seshadri, S., Bishop, P.L., Agha, A.M., 1994. Anaerobic/aerobic treatment of selected azo dyes in water. Waste Manag. 15, 127–137.

Sharma, K.D., Sudha, G., Suri, S.K., Randhawa, H.S., 1989. Studies on mixed surfactant systems: Effect of some anionic surfactants on the cloud point of poly(nona-)oxyethylated nonylphenol. J. Am. Oil Chem. Soc. 66 (7), 1015–1017.

Selcuk, H., Sene, J.J., Sarikaya, H.Z., Bekbolet, M., Anderson, M.A., 2004. An innovative photocatalytic technology in the treatment of river water containing humic substances. Water Sci. Technol. 49, 153–158.

Singer, P.C., 1990. Assessing ozonation research needs in water treatment. J. Am. Water Works Assoc. 82, 78–88.

Silvia, S., Gabriella, P. Fabrizio, 2011. Perspective on the use of humic acids from biomass as natural surfactants for industrial applications. Biotechn. Adv. 29 (6), 913–922.

Soumia, A., Abdelmajid, J., Abdelilah, M., Mohamed, G., Peter, W., Mohamed, H., 2010. Structural study of humic acids during composting of activated sludge-green waste: Elemental analysis, FTIR and ^{13}C NMR. J. Hazard. Mater. 177, 524–529.

Stevene, A.A., Slocum, C.J., 1976. Chlorination of organics in drinking water. J. Am. Water Works Assoc. 68, 615–623.

Stevens, G.W., Perera, J.M., Grieser, F., 2001. Interfacial aspects of metal-ion extraction in liquid-liquid systems. Rev. Chem. Eng. 17 (2), 87–110.

Takeshi, M., Muneaki, T., Kazuhiko, Y., 2006. Adsorbability and photocatalytic degradability of humic substances in water on Ti-modified silica. J. Col. Int. Sci. 297, 678–686.

Talbi, Z., Haddou, B., Bouberka, Z., Derriche, Z., 2009. Simultaneous elimination of pollutants dissolved and dispersed from cutting oil wastes using two-phase aqueous extraction methods. J. Hazard. Mater. 163 (2-3), 748–755.

Thiele, B., Gunther, K., Schwuger, M.J., 1997. Alkylphenol ethoxylates: trace analysis and environmental behavior. Chem Rev. 97, 3247–3272.

Toxicological Profile for Phenol, September 2008. United States Public Health Service. Agency for Toxic Substances and Disease Registry. www.atsdr.cdc.gov/toxprofiles/tp115-c1.pdf (updated December 15, 2008).

Trakultamupatam, P., Scamehorn, J.F., Osuwan, S., 2004. Scaling up cloud point extraction of aromatic contaminants from wastewater in a continuous rotating disk contactor. II. Effect of operating temperature and added electrolyte. Sep. Sci. Technol. 39 (3), 501–516.

Tran, L.H., Drogui, P., Mercier, G., Blais, J.F., 2009. Coupling extraction–flotation with surfactant and electrochemical degradation for the treatment of PAH contaminated hazardous wastes. J. Hazard. Mater. 170 (2-3), 1218–1226.

Valaulikar, B.S., Manohar, C., 1985. The mechanism of clouding in Triton X-100: The effect of additives. J. Colloid Interface Sci. 108 (2), 403–406.

Van Hamme, J.D., Ward, O.P., 1999. Influence of chemical surfactants on the biodegradation of crude oil by a mixed bacterial culture. Can J. Microbiol. 45 (2), 130–137.

Vivian, R.D., Stephanie, L.C., Kristin, S., Denis, H.L., Niels, C., Watanabe, H., Kamidate, T., Kawamorita, S., Haraguchi, K., Miyajima, M., 1987. Distribution of nickel (II), Cadmium (II) and Copper (II) Chelates of (2-2-pyridylazo) –5-methylphenol in two phases separated from micellar solution of nonionic surfactant. Anal. Sci. 3, 433–436.

Uyguner, C.S., Bekbolet, M., 2005. Evaluation of humic acid photocatalytic degradation by UV–vis and fluorescence spectroscopy. Cat Today 101, 267–274.

Vivian, R.D., Stephanie, L.C., Kristin, S., Denis, H.L., Niels, C.B.B., 2004. Evaluating the toxicity of Triton X-100 to protozoan, fish, and mammalian cells using fluorescent dyes as indicators of cell viability. Ecotoxicol. Environ. Safety 57, 375–382.

Watanabe, H., Tanaka, H., 1978. A nonionic surfactant as a new solvent for liquid-liquid extraction of zinc (II) with 1-(2-Pyridylazo)-2-naphtol. Talanta 25, 585–589.

Weast, R.C., Astle, M.J., 1982. Handbook of chemistry and physics. CRC Press, Boca Raton, FL.

Yao, B., Yang, L., 2008. Pilot-scale ultrasonic assisted cloud point extraction of polycyclic aromatic hydrocarbons from polluted water. Sep. Sci. Technol. 43 (6), 1564–1580.

Zhang, F., Yediler, A., Liang, X., Ketttrup, A., 2004. Effect of dye additives on the ozonation process and oxidation by-products: A comparative study using hydrolyzed C. I. Reactive red 120. Dyes and Pigment 60, 1–7.

Zhu, S.C., Yang, C.Z., Aurelle, Y., Canselier, J.P., Cotteret, J., 1995. Cutting oils: Disposal procedures and search for less-polluting formulations. Journ. Com. Esp. Deterg. 26, 191–208.

Zoller, U., 2004. The survival and distribution of alkylphenol ethoxylate surfactants in surface water and ground water. In: Zoller, U. (Ed.), The case of Israel—is there an environmental problem? Handbook of Detergents, part A. Marcel Dekker, New York.

CHAPTER 6

Adsorption of Synthetic Surfactants from Aqueous Solutions on Natural Adsorbents

Alisa Sineva
Faculty of Chemistry, Lomonosov Moscow State University, Moscow

CHAPTER OUTLINE

The Role of Colloidal Systems in Environmental Protection. http://dx.doi.org/10.1016/B978-0-444-63283-8.00006-5

1 Introduction

Today we face the problem of creating new approaches to protect the environment based on mobilizing nature for its purification. The ability of water to regenerate is high, but several standing and current waters have already lost the renewal ability owing to a high concentration of pollutants. The problems of purifying water in reservoirs polluted by toxic chemical substances, as well as purifying freshwater, are extremely relevant in the 21st century. In this chapter, the objects of our study are synthetic surfactants (detergents), water, and natural adsorbents.

Colloidal systems can often be composed of either two phases (liquid–solid, liquid–gas, solid–gas), or three phases (liquid–solid–gas). Peculiarities in the interactions in these systems are defined by properties of the interfaces. Hence, at small distances from the surface of the finely dispersed solid phase, water properties differ from usual values. In colloidal systems, this phase is characterized by large specific surfaces. Anomalous properties of "bounded" water in finely dispersed systems may be explained utilizing the difference in intermolecular bonding near solid surfaces.

Colloid chemistry involves knowledge of certain boundary chemical disciplines to develop the methods of managing the properties of disperse systems and surface phenomena on them. In particular, disperse systems are widespread in nature; thus, colloid chemistry offers many possibilities for solving practical tasks of current importance, including protection of the environment. At the same time, all purification methods applied so far only decrease pollution, but real self-purification can be only achieved in the biosphere. For the modern ecological situation complicated by social economical problems, the optimization of technogenesis supposes optimal use of the environment when, on the one hand, industrial effect is achieved and, on the other hand, the environment is not polluted.

2 Problems of change and pollution of water resources

2.1 Water as the important element of the biosphere and the necessary factor of alive organisms

Water is an important component of the biosphere and one of the essential factors of existence of living organisms. Most of the water, namely, 95%, exists as a world ocean, which occupies nearly 70% of the world surface and contains 1300×10^6 km^3 of water. The surface water (lakes, rivers) contains only 0.182×10^6 km^3, and living organisms contain 0.001×10^6 km^3 of water. Accordingly, water on average consists of 80% of body mass of multicellular organisms. Only 3% of the world's water resources are freshwater. The Aral Sea is the fourth largest resource of water in the world. Today more than two-thirds of its lake water volume has disappeared because of irrigation water needs over the last decades.

Lake Baikal is the largest lake in Russia and one of the largest lakes in the world. It contains roughly 20% of the world's surface fresh high-quality water. The surface area of this lake, at 31,500 km^2, is in ninth place in the world. Baikal Lake contains 23,600 km^3 of water, or 20% of surface unfrozen freshwater, and it can provide 1.5 billion people with drinking water. Nevertheless, the water of highest quality lies more than 400–500 m below the lake surface (Annual 2000). Water as a natural medium is important in all biochemical and industrial processes. Russian scientist

Vladimir Vernadsky wrote, "Water stays separately in the history of our planet. There is no natural object that could be compared with water relative to the greatest geological processes." The demand for water is high; but the need to dump water that contains undesirable admixtures is also high (Parfit 1993). Water of bad quality affects human health and of course can only be used in very limited amounts.

The main hygienic requirements for drinking water quality is absence of pathogenic microorganisms and of toxic and additional substances; color; and flavor. Traditional technological water treatments includes chlorine disinfection or disinfection with chlorine substances (chlorine dioxide, bleach powder, sodium hypochlorite). In the first water treatment stage, the microorganisms are separated during the processes of coagulation, clarification, and filtration. The negative effect of chlorinated top water on human health was determined over 20 years ago. It may by caused by chlorine organic products with mutagenic and carcinogenic activity. Other effective methods such as ozone treatment and ultraviolet radiation are proposed for the substitution of chlorine disinfection. These methods allow separating surface-active substances, phenols, oil products, and some inorganic substances (salts of manganese, iron, ammonia-nitrogen, and sulphides) from water. In practice, the chlorination stage by minimal chlorine concentration might be introduced in water treatment. (Drinking 1996).

2.2 Earth's need for water

Water is the most consumed natural resource on Earth. In 1970, total water consumption was nearly 1000 m^3 per one inhabitant per year (and as a point of comparison, fuel consumption is 100 times less than water consumption);the same parameter is in the range of (2.5–7.5) $\times$ 1000 m^3 per year. Water's unique physical and chemical properties (as heat transfer medium, solvent, and transport agent) determine its importance many growing branches of industry. Power engineering, which uses enormous quantities of water for the cooling systems, occupies one of the first places in this rating, followed by mineral resources and the manufacturing industry (metallurgy and chemical industry). Agriculture's very high water consumption can be explained by the fact that water is consumed not only on irrigated lands but also on arid lands. Production of crops requires six times more water than all other water consumption needs. Moreover, water for cattle breeding needs to be taken into account for drinking, sanitary, and hygienic purposes.

Household use of water has increased greatly in recent times. In 1970 in Russia, the average person in urban areas consumed about 150 liters of water per day, and in rural areas daily individual use consumption did not exceed 54 liters. However, with the worldwide increase in the urban population, as a whole on our planet and especially in developed countries, improvement of these standards is expected. Indeed, in 1995, water consumption in Moscow was 450 liters per person per day; after metering was set up in homes and apartments, this figure declined to 225 liters per person per day (2009 data). In Europe, water consumption is around 120–160 liters per day.

The contradictions between global water needs and the capabilities to meet those needs can be summarized as follows: Although Earth has a lot of water, freshwater accounts for only 2–3% of the hydrosphere. In turn, use of most of the freshwater is difficult. One of the limiting factors is its inaccessibility. An even more significant obstacle is the slow renewability of these resources. The decisions regarding modern water management are possible only through an integrated approach. Water management should include both organic methods for preventing water pollution (sewage treatment, including in the fields of irrigation, reducing harmful emissions by introducing a closed-cycle production and legal measures for water protection) and methods of active removal of those contaminants that flow in the aquatic environment with the existing technology and cannot be completely eliminated. These methods include promotion processes and mechanisms of natural self-purification of water bodies, development of biological control of pollution, and changes in the range of industrial products by increasing the proportion of less toxic and sustainable products as well as efficiently distributing production and human settlements, taking into account characteristics of water objects, etc.) (Ryabchikov 1976).

2.3 Water pollution (natural processes and anthropogenic influences)

Water pollution is an environmental problem in both developing and industrial countries. Nearly half of the world's rivers are polluted because 95% of wastewater quantity, according to UNESCO, are in ground and surface waters without purification.

Both natural processes and anthropogenic influences affect water reservoirs. Understanding the technogenic evolution of water reservoirs is key to the large-scale planning of reservoirs and the solution of regional pollution problems. One example is Lake Baikal (Sineva 2002), which illustrates the chemical aspects of dumping (putting wastes into the lake) as well as their sources, both natural and artificial (technogenic). The natural sources originate in complex biogeochemical cycles of migration of substance and energy (i.e., biogenic migration of atoms, according to Vernadsky). Both are known to impact water quality and, therefore, the ecosystem of the lake via water, atmospheric emissions, and biological depositions. The technogenic sources in the Baikal region are listed in Table 6.1. Monitoring of mineral substances and organic matter in Lake Baikal, following the state of the lake from 1982 to 1998, is presented in the same table. The main pollutants are organic substances, including oil products, pesticides, surfactants, lignins, heavy metals, and biogens, which interact with water and enter to the water depth and bottom sediment; see Table 6.1 (Annual 2000).

Colloidal transformations of pollutants under dumping are as follows:

1. Formation of raw and fine dispersed suspensions and sols, including solid particles, soils, silts, ions of heavy metals, and organic substances (cellulose and lignin).
2. Coagulation processes in suspensions, for example, clay suspensions, under the action of increasing mineralization of water due to multicharged ions (Ca^{2+}, Mg^{2+}, Al^{3+}, Fe^{3+}, SO_4^{2-}, PO_4^{3-}).
3. Selective adsorption of ions of heavy metals on organic substances by the iron oxides and phosphorus during the formation of bottom sediments.
4. Adsorption of organic substances of both natural (proteins and lipids) and technogenic character (aliphatic and aromatic alcohols and acids present in liquors, lignins modified under pulping, detergents, etc.) on the water–air and water–solid boundaries; decrease in surface (and interphase) tension resulting in liophilization and stabilization of disperse systems). In the presence of multicharge ions, adsorption is accompanied by ion exchange.
5. Formation of different migration forms of oil hydrocarbons in water (thin films, emulsions, and associated forms with the suspensions and gas bulbs, as well as adsorbed by bottom sediments, and also in the form of oil aggregates.

Table 6.1 Dynamics of Total Pollutants Generated by Baikal Pulp and Paper Mill, Jsc and Discharged with Wastewater and Drain Water in 1998–1999

	Discharged, Tons		
Pollutants	**1998**	**1999**	**Diff., Tons**
BOD total	424.8	331.4	−93.4
Solid particles	74.3	104.5	+30.2
Petroleum hydrocarbons	1.2	1.5	+0.3
Nitrogen ammonia	14.4	1.9	−12.5
Lignin	145.9	185.9	+40
Chloroform	0.97	1.96	+0.99
Formaldehyde	0.59	0.35	−0.24
Sulfate soap	40.9	67.0	+26.1
Methanol	10.9	12.4	+1.5
Nitrates	11.9	13.4	+1.5
Sulfur organic substances	0.55	0.48	−0.07
Mercury	0.03	0	−0.03
SSSA	0.59	1.47	+0.88
Sulfates	4215	4808	+593
Turpentine	2.95	4.2	+1.25
Chlorine organic substances	25.8	32.8	+7
Phosphorus total	0	0	0
Chlorides	1596	2767	+1171
Aluminum	2.37	2.6	+0.23
Phenols	0.78	0.26	−0.52

6. Foaming on the waves as a result of the formation of thin film of organic substances (lipids and proteins from natural and technogenic wastes as well as surfactants); carbon dioxide formation under oxidation of organic substances, and breathing of water animals probably take part in this process.
7. Spontaneous formation of thermodynamically stable micelle systems on the base of surfactants from civil wastes and oils.

We analyzed the probable influence of colloid-chemical transformation of pollutants on the properties and quality of water, and their well-known toxicological action was not considered. The following changes in the properties of Baikal water may be possible:

1. Turbidity, because of the presence of insoluble and colloid substances of organic and inorganic origin as well as the excess of coagulant (aluminium- and iron-ions) when sewage is treated through the coagulation method.
2. Color of water in the formation of suspensions of silt, sulfides, ions of iron (III), and organic compounds of lignin and humic compounds.

3. Decrease in surface tension at dumps where the sewage contains surfactants and detergents.
4. Increase in water filtration through clay suspensions in the volume of water bodies when the coagulation structure is formed in suspension.
5. Increase in the toxicological danger of water due to accumulation (solubilization) of insoluble in water oil products, hydrocarbons and the like in surfactant micelles are insoluble in water.

In summary, dumping results in essential change in the hydrochemical and hydrobiological parameters of water ecosystems. The ecosystem either adapts to changed conditions or decays leading in appearance of dead zones. In general, the scale of pollution probably determines the situation.

Research results reveal a connection between the colloidal transformations of pollutants and the change in water quality of Baikal Lake (Sineva 2002).

Long-term data showed that, despite a significant influx of pollution agents to Baikal, the stability of its lake water quality would be proposed in future. The pelagic waters of Baikal are good for living organisms, but its low-level mineralization leads to conclusions that its long-term use as drinking water may adversely affect human health (Annual 2000).

2.4 Water and sustainable development

According to UNESCO data (World Water Development Report), nearly 2 billion people in 48 countries will be impacted by water deficiencies in the middle of this century. Conservation of water resources is one of the most important global challenges. In accordance with instructions from the European Union Water Framework Directive (2001 and 2008), sustained development programs may guarantee stable natural recourses management throughout the world for future generations.

In 2008, the EU–Russia Cooperation Programme on Harmonization of Environmental Standards II was created. Warren (2008) has described this program in depth. For example:

> *An important principle of the EU Water Framework Directive is that, irrespective of whether or not a water body meets its water quality objective it should not be allowed to deteriorate. This means that in principle no increase in the polluting load discharged to a water body should be allowed. Any new discharge should be offset by reductions in existing discharges.*
>
> *Conservative pollutants—those that do not degrade and are not assimilated or attenuated in the river—will be present a long way downstream and discharges downstream may add to their concentration. Upstream discharge permits need therefore to take account of the impact of the discharge downstream.*
>
> *For works in areas sensitive to eutrophication towns of population greater than 10,000 must also incorporate nutrient (nitrogen and phosphorus) removal.*
>
> *The EU intention is to eliminate altogether the priority hazardous substances and to reduce the environmental concentrations of the other priority substances as far as possible*
>
> *The use of phosphate-free detergents has helped to reduce phosphorus levels in sewage effluent but a phosphorus removal stage is needed in the municipal waste water treatment plants located in areas sensitive to eutrophication (e.g., the Baltic and Black Sea basins).*
>
> *The EU Water Framework Directive was adopted in 2000. River basin plans are due in 2009 and to be implemented between 2009 and 2015 so that objectives are achieved by 2015*

3 Surfactants as a factor of surface- and groundwater pollution

3.1 Colloidal and biotic properties of synthetic surfactants

Detergents are among the most widespread pollutants of water bodies. It can be said that the role of detergents in surface- and groundwater pollution is comparable to that of oil pollution of the world's oceans. The total amount of detergents (SSAA in Table 6.1) discharged with wastewater and drain water into Lake Baikal in 1999 was almost equal to the amount of petroleum hydrocarbons, namely, 1.47 tons and 1.5 tons, accordingly. Detergents have taken a special place among chemical products produced owing to their toxic effects on aquatic systems as well as using purification systems for wastewater treatment. Being very toxic and resistant to biodegradation, detergents have become one of the most widespread pollutants. The concentration of surface active substances (SAS) in wastewater has increased to 5–15 mg/L in developed nations. Moreover, separating detergents from wastewater is an inefficient process, so that 50% to 60% of the initial substances remain in natural waters. Practically, the concentration of these substances in river water is 0.3–1.0 mg/L, and sometimes as high as 1.5 mg/L. Nevertheless, the maximum permissible concentration (MPC) of detergents in water must be less than 0.3 mg/L. Detergents are very persistent compounds, with 20% to 50 % of their initial concentration detected in water after six months. The effect of detergents on the aquatic ecosystems varies greatly.

1. Change for the worse of organoleptic properties of water, namely, color, odor, taste, and foaming ability; when the concentration range of detergents is 1–3 mg/L the water acquires an unpleasant taste and odor.
2. An adverse effect on the oxygen balance; when the concentration of detergent is 1 mg/L in lowland rivers, the level of aeration may decrease to 60%.
3. Change of the chemical processes in water systems.
4. Poisoning effects on aquatic organisms, depression of biological activity (the lethal concentration of detergents is 3–5 mg/L for fish and nearly 1 mg/L for plankton). When the concentration of anionic detergents in water is 120 mg/L algae growth slows down. The synergistic effect of detergents and other pollutants must be taken into account here as well.
5. Aesthetic value of water objects declines, thereby limiting their recreational use.
6. Difficulty of navigation, especially, in the case of small boats.

The use of detergents causes by-effects that are related to their production. In this case, the effectiveness of detergents increases in the presence of substances such as phosphates, silicates, sodium carbonate, and carboxymethyl cellulose. These substances, in turn, affect the water quality, causing eutrophication of the water bodies. The presence of such substances in water complicates the purification of water. For example, the concentration of phosphates more than 1 mg/L reduces coagulation when in the presence of aluminium. Use of biodegradable detergents that decompose during purification up to 80–85% may solve the problem of water pollution by synthetic surfactants (Ryabchikov 1976).

Continued pollution of natural waters is connected with the specific colloidal properties of synthetic surfactants, including surface activity, ability to form foams and emulsions as well as micelle formation, solubilization, and stabilization of other pollutants in water.

Most of the surfactants used in households and in industry end up in the sewers. The biodegradation rate in sewage plants determines the volume of surfactant that reaches the environment. The rate of biodegradation, in combination with the degree of aquatic toxicity, determines the environmental impact. The Organization for Economic Cooperation and Development (OECD) has issued guidelines and directives regarding aquatic toxicity, biodegradability, and bioaccumulation. Aquatic toxicity may be measured in fish, daphnia, and algae. Toxicity is given as LC_{50} (for fish) or EC_{50} (for daphnia or algae), where LC and EC stand, respectively, for lethal and effective concentrations. Values below 1 mg/L after 96-h testing on fish and algae and 48 h on daphnia are considered toxic. Environmentally benign surfactants should preferably be above 10 mg/L.

Biodegradation is a process carried out through bacteria. By a series of enzymatic reactions, a surfactant molecule is ultimately converted into carbon dioxide, water, and oxides of the other elements. If a product does not undergo natural biodegradation, then it is stable and persists in the environment. For surfactants the rate of biodegradation varies from 1 to 2 h for fatty acids, via 1–2 days for linear alkyl benzenesulfonates and two months for branched alkyl benzenesulfonates.

Two criteria are important with respect to testing for biodegradation: primary degradation and ultimate degradation. Primary degradation of surface-active agents relates to the loss of surfactant properties. More important from an ecological point of view is the test for ultimate biodegradation, the popular modified Sturm test (*OECD* Test 301 B for ultimate biodegradation).

Bioaccumulation can be measured directly in fish but is more often calculated from a model experiment. A surfactant is considered to be biodegradable if it satisfies the following criteria:

1. At least 60% conversion into CO_2 in 28 days.
2. Conversion from 10% to 60% CO_2 within 10 days.

Tests for aquatic toxicity, ready biodegradation, and bioaccumulation do not give complete details on the environmental impact of a surfactant. In order to obtain the full picture, ecological tests should be carried out on each individual end product (Holmberg et al. 2002).

3.2 Production and application of synthetic detergents in the 21st century

Due to their excellent performance, anionic surfactants in the late 20th century held first place in worldwide surfactant production–consumption. It amounted to about 12 million metric tons (mmt) per year: 8.3 mmt of soap, 2.0 mmt of linear (LAS*)*, branched alkyl benzenesulfonates *(*BABS*)* (1.77 mmt of LAS in 1995), 0.37 mmt of alkyl sulphates, and 0.25 mmt of alkyl ether sulfates. Anionic surfactants dominate the market, accounting for 65% of synthetic surfactant production, with LAS topping the "big six" surfactant list. The excellent cleaning performance of anionic synthetic surfactants in households, their greater volume of application, low manufacturing cost, rapid and high biodegradability, and low aquatic toxicity are the main reasons why their prevalence should continue in the future (Piasecki 2006).

Non-ionic surfactants are produced in large quantities. They hold second place in worldwide surfactant consumption, amounting to over eight million tons a year in the 1990s when their share of the total use of surfactants was about 35%. Of the non-ionic surfactants produced, the polyoxyethylene alcohols hold third place after soap and LAS on the "big six" surfactant list. Their production exceeded 800,000 tons in the 1990s. Polyoxyethylene alkyl phenols appeared on the market in the 1940s with a high-production volume (300,000 tons in 1990).They are widely used primarily because of their excellent properties: good detergency at low temperatures of washing and in phosphate-free detergents; low foaming abilities; and resistance to hard water in the presence of dilute acids and bases. The main use of these synthetic surfactants is in household and institutional detergents, the care product sector, and in many industrial and agriculture areas.

As to the biodegradation properties of non-ionic surfactants, data are needed to estimate the maximal permissible surfactant concentration (PEC). The majority of non-ionic surfactants are more chemically degradable than biodegradable surfactants. Chemically degradable surfactants are generally toxic to water organisms, and their impact on a given aquatic organism is concentration-dependent. Nevertheless, chemically degradable surfactants may serve as surfactant-based media in organic synthesis in some separation processes, in the biochemical and biomedical field, and so on (Burczyk 2006).

In 2007 Russia consumed the following raw materials:

1. Anionic surfactants: nearly 60,000 tons of LAS, including nearly 5000 tons of secondary alkyl sulfonates; nearly 5000 tons of alkyl sulfonates; nearly 3000 tons of alfa-olefin sulfonates; alkyl sulfates, and so on.
2. Non-ionic surfactants: nearly 20,000 tons of alkyl phenol ethoxylates (neonols); nearly 5000 tons of ethoxylated alcohols.
3. Zwitterionic surfactants: nearly 2000 tons of betaines and others.
4. Cationic surfactants: nearly 1000 tons of quaternary ammonium-based derivatives.
5. Sodium tripolyphosphate: nearly 120,000 tons.

Currently, it is most important to control the biodegradability of anionic and non-ionic surfactants, as well as limiting the phosphate content (Divakova 2007).

3.3 Water pollution and its consequences for the aquatic ecosystem

Some surfactants and surfactant-containing detergents were produced to affect the filtration rate of mussels. With regard to the ecological value of this effect, it is important to elaborate on whether detergents act on aquaculture living things, among which the mollusks are of primary importance. Studies in this area focus on the effect of surfactants and detergents on the filtration activity of mollusks grown in mariculture conditions, particularly the oysters *Crassostrea gigas Thunberg*. The average annual production of marine mollusks, grown in aquaculture, is mush higher than that of the mariculture of marine fish and crustaceans.

As surfactants, sodium dodecyl sulfate (SDS), tetradecyltrimethyl ammonium bromide (TDTMAB), and some synthetic detergents were selected for our study. SDS is a constituent of many commercial detergents and gets into water bodies as a contaminant. SDS was found to inhibit filtration activity in *Crassostrea gigas,* as a result of which within 12 to 20 min the control and experimental concentrations of suspensions in water differed more than two times. The cationic surfactant—quaternary ammonium-containing compound TDTMAB—likewise suppressed the filtration in *C. gigas*. As a result of the action of this xenobiotic, the control and experimental concentrations of the suspensions differed, in 5 min, more than two times, whereas in 20 min they differed more than seven times. The revealed inhibiting effects became apparent at the surfactant concentrations $\geq$ 0.5 mg/l observed in water ecosystems. Since the water filtration rate by invertebrates is quite notable and renders a conditioning effect on the water medium, the decrease in filtration drastically affects the ecosystem. The derived results on the filtration activity of mollusks can notably influence the parameters, characteristics, and components of water ecosystems containing macrozooplankton, phytoplankton, bacterioplankton, suspended solids, dissolved inorganic nitrogen, transparency (Secchi disk), and soluble reactive phosphorus (Ostroumov 2001a).

Similar results were derived in the study of the effect of detergents on the filtration activity of the limnetic mollusk (pipi), *Unio tumidus Philipsson* (1788), collected from the Moscow River's gritty bottom, in the upstream water (upper the town of Zvenigorod). It was shown that detergents (50 mg/L) decelerated water filtration by limnetic mollusks, that is, inhibited their filtration activity and decreased the withdrawal of suspension particles from water.

Typical components of detergents are synthetic surfactants and phosphorus compounds. In a number of widely applied preparations ("Bio-C," "Crystal," "Kashtan," etc.), tripolyphosphate content is 3–40 mass %. Since phosphorus compounds stimulate phytoplankton growth, the effect of detergent-phosphates at certain concentrations on phytoplankton organisms is higher than the surfactant's potential inhibiting effect. Hence, detergents carry two types of detriment to ecosystems: the danger of the stimulating effect of phosphates on phytoplankton and the danger of the inhibiting effect on filtrating living things (filterers) stimulating phytoplankton propagation. Summation of the two effects can increase the danger of the ecosystem disorder.

Therefore, the work has disclosed increased danger for the ecosystem under study upon synecological summation of the effects of anthropogenic factors both on the populations of plankton organisms and on filterers. For this reason, it is necessary to analyze the overall situation using the synecological approach: the study of interacting populations of plankton organisms, their consumers–filterers, and detergents affecting them (Ostroumov 2001b).

Although the ecological impacts of surfactants have not yet been studied extensively, we think that surfactants may interact actively with components of water ecosystems at concentrations of 0.01–1%. Surfactants appeared to be membrane tropic pollutants that disturb lipid–lipid, lipid–protein, and protein–protein interactions in the cells and subcellular structures. In this way, surfactants may have significant impacts on aquatic natural systems.

4 Self-purification and quality of water

4.1 The mechanism of formation of water quality: physical, chemical, and biotic processes

Water regeneration ability due to evaporation and filtering through the soil and also through microbiological purification is relatively high. However, because of the high concentrations of harmful substances, numerous standing and running waters have already lost their ability to regenerate. Modern tendencies in the artificial wastewater treatment are connected with the application of the principles of natural purification in the engineering process. True purification exists only in the biosphere, and humans need to learn to use its methods.

Self-purification of water in reservoirs involves three steps: physical, chemical, and biological. The third step of self-purification is run by heterotrophic organisms. However, other groups of organisms cannot be excluded. In natural biological processes, a large group of aerobic microorganisms oxidize organic matter to simple mineral components with the use of enzyme systems. A bacterium plays an important role in self-purification from the organic biologically degradable compounds; water-plants, mushrooms, simple and high plants also take part in this process. However, as yet there is no clear understanding of the possibilities and conditions of maximum action of hydrobionts (micro- and macrofites as well as mushrooms) on the decomposition of xenobionts. Some toxicants may simply alternate among the water mass, aquatic organisms. and the ground; their real detoxification may not occur.

Recently, the concept of aquatic ecosystems was formulated as an analogy of bioreactors contributing to water purification with the participation of major representatives of hydrobionts. Since the volume of Lake Baikal's ecosystems is huge, scaling has resulted and the role of biofiltering processes leading to self-purification has increased. The important role of the biodiversity of organisms in water ecosystem was stressed in Ostroumov (2000).

In later work, he showed that the water filtering by invertebrates (for example, marine mollusks) is notable and produces a conditioning effect on the water medium (Ostroumov 2001a). Similar results were revealed in the study of filtering activity of the limnetic mollusks (pipi) *Unio tumidus Philipsson* (1788), collected from the Moscow River (Ostroumov 2001b).

4.2 Natural processes of auto-purification of water in an ecosystem and main principles of technology of water purification

These processes are listed in Table 6.2, which is based on Ostroumov (2004). The author's task was as follows:

1. To compare some technologies (stages) of water purification processes in ecosystems.
2. To apply this comparison for evaluation of some natural processes.

The question of the economic effect of the water treatment process is discussed below. Here we compare some technologies (stages) of water treatment and natural processes of self-purification in water ecosystems and to make an economic assessment of the value equivalent of some natural processes of the self-purification of water. First, the analogy between the processes of self-purification of water in natural ecosystems and technological methods of water treatment for water consumption is applied. Such a comparison is shown in Table 6.2. In (Ostroumov 2004), the question of the economic evaluation of water filtration invertebrate filter feeders, hydrobiologic and ecological processes, bactericidal nature was analyzed. For this purpose, the data shown in Table 6.3 were used because the natural water disinfection process can be compared to a few lines in the table. For the economic evaluation of natural processes as chlorinated or dechlorinated, one can take 30% of the average cost of the procedure in line 2 of Table 6.3 (i.e., 30% of 7 is equal to 2.1 dollars per cubic meter). Similarly, natural processes can be compared to rows in line 3 of Table 6.3 (ozonation) and in line 4 of the table (UV treatment), whereas 30% of the average cost of ozonation and UV treatment of water mass ecosystems totaled, respectively, \$7.6575 + 1.065 per 1 cubic meter. Aquatic organisms also, reduce the amount of aquatic bacteria in water filtration, direct water. This process is similar to the technological process of liquid filtration using a filter with a pore size smaller than the size of bacteria. You should consider the process of direct water filtration valuation, which is USD 3.10 per 1 cubic meter, because we have the same 30% of average value of all the ways bactericidal treatment (see line 5 of Table 6.3). Thus, the minimum score for the natural processes of disinfecting water is USD 13.92 per 1 cubic meter. In terms of rubles, at 29 rubles for 1 USD, this corresponds to \$404 per 1000 liters. Note that we estimate the equivalent of only one type of other natural processes similar to other water treatment processes. So, the resulting estimates of bactericidal natural processes were close to this average value.

Table 6.2 Comparison Between Technological and Natural Prossesses of Water Purification

Process	Technologies for the Water Purification	Self-purification of Natural Water Ecosystems
1. Sedimentation of suspended matter	Sedimentation of suspended particals in storage reservoirs, with slowly passing the water. Coarse particles sedimentate to the bottom as aggregates with bacteria, the bacteria decrease in number by 75%, the number of bacterium coli may decrease in 50 times	Gravimetric sedimentation of seston particles
2. Acceleration of suspended matter and bacteria	Coagulation. Acceleration of the sedimentation by coagulant agents (aluminum sulfate or ferric sulfate). The forming flocs sedimentate to the bottom and carry away suspended particles and bacteria. The number of bacteria may decrease till 90%.	The formation of sea and estuaine snow. Some aquatic organisms form mukus, which may remove fine particles, and, probably, some of the bacteria
3. Water filtration, remove of suspended particles and some bacteria	Water filtration through the sand filtre. The filter created from sand, gravel and crushed stones. The film on the sand surface accumulate small particles and up to 99% bacteria	Active water filtration by filterfeeding aquatic organisms
4. Bactericidal treatment	Chlorination. Active chlorine is oxidizing agent and effective against almost all bacteria	Natural UV radiation; Some aquatic organisms may produce the bacteria.
	Ozonation	Reaktive oxygen species in natural water bodies and streams
	UV radiation treatment	Natural UV radiation
5. Removal of inorganic nitrogen	Biological denitrification in continuous-flow bioreactors	Natural population of bacteria; sorption by plants, and aquatic organisms

Table 6.3 The Cost of Various Methods of Water Treatment

Type of Processing	Type of Processing Capacity for One Thousand Cubic Meters Per Day				Average cost, USD per m^3
	4.0	20.0	40.0	400	
Chlorination	8.2	5.0	4.2	3.5	5.2
dechlorination	11.5	7.0	5.0	4.5	7
ozonation	42.5	23.2	20.7	15.7	25.5
UV radiation treatment	4.5	3.5	3.2	3.0	3.6
average for all types of water treatment					10.3

5 Natural adsorbents

5.1 Natural organic and inorganic adsorbents

Fine-grained minerals and rocks that are widely distributed in nature may be effective sorbents for water (soils) purification from toxic substances. Natural inorganic adsorbents possess adsorption ability without special technological preparation. The adsorption ability of these adsorbents can be caused by physical adsorption in the developed porous or layered structure of adsorbents by ion-exchange processes, or by formation of hydrogen bonding, etc. Often some factors simultaneously interact. Natural adsorbents differ greater variety owing to various conditions of formation of rocks and on the properties do not concede synthetic ones. Cheapness and wide prevalence of natural adsorbents do them, in many cases, suitable for purification of industrial sewage.

Natural organic adsorbents (peat, brown coals, lignin) adsorb surfactants fairly effectively. However, due to strong swelling in water and, the difficulty of recycling, these materials have not found application in sewage treatment processes. The amphiphilic molecules of detergents may absorb both on hydrophobic and on hydrophilic inorganic natural adsorbents, such as clay and various mineral oxides. So, for example, use of glauconite and loess as adsorbents allows clearing the sewage polluted with surfactants in various regions where there are deposits of the mentioned adsorbents. For adsorption clearing from anionic surfactants, the most effective adsorbents are those on which physical adsorption is accompanied with the acid–base interaction of functional groups on their surface with anionic surfactants. A similar situation can take place with the application of dolomite (Klimenko 1974).

Dispersed systems on the base of different natural material existing in ultra and micro-heterogeneous state are characterized by an enormously developed surface that allows their use as adsorbents. High adsorption, ion exchange, and filtering properties, as well as low cost and large occurrence of natural disperse materials, make useful in water purification systems (Mashkova 1998).

The dispersed form of carbon in lithosphere greatly exceeds all others. The most widespread adsorbent for water treatment is activated carbon. This adsorbent has a high adsorption capacity for removing organic compounds from aqueous mediums, and it is the most widely used form in water treatment. Nevertheless, looking for new low-cost adsorbents activated interest in the carbon alternative (Gupta 2006). Natural disperse materials, being micro-heterogeneous, are used in water treatment systems.

Precambrian shungite ores containing noncrystalline carbon are widespread on Karelia territory. Shungite ores are classified into five types by their carbon content, which may be up to 99 mass %, the most interesting being types I and III shungite. Between Karelian shungite ores, the Shunga deposits are classified as type I. They contain almost pure glass-like carbon (96–99 mass %) with a few impurities of the other elements (Table 6.4). This type of ore isrepresented as one with relatively small veins and inclusions. They are not commercial but are widely used in investigations as a standard of shungite carbon. The type III shungite ores with carbon content of nearly 30 mass % are the most widespread (Table 6.4, based on Rozhkova 2004). The Zazhogino deposit containing these ores is commercial and is currently under development. Shungite carbon deposits of Karelia are effective catalysts, adsorbents and filters in water treatment as well as inorganic binders and building materials.

The high microporosity and developed surface jointly with high-strength properties define the wide opportunities offered by application of shungite carbon. Since its maximum achievable dispersion (the average particle size) is determined by the size of the structural element (globule), which remains constant for shungite of a given deposit, the specific surface area and the amount and distribution of functional groups on the surface can be varied by selecting appropriate conditions for processing. The thermal treatment in inert atmosphere does not lead to change in shungite structural parameters and does not affect the specific surface area until 1000°C. At the same time, the thermal treatment in the oxidative or reductive atmosphere is effective. The nontraditional modification methods of carbon materials, such as low-temperature oxidation with ozone and glow discharged plasma treatment, are of special interest.

Table 6.4 Chemical Composition and Other Properties of Shungite Samples

Properties		SH-1 (Shunga Deposit)	SH-3 (Zazhogino Deposit)
Carbon, mass %		96…98	28…32
Minerals, mass %		$SiO_2 < 3.0$ $Al_2O_3 < 0.4$	SiO_2: 60 Al_2O_3: 3.5 $Fe_2O_3 + FeO$: 2 TiO_2: 0.2 $MgO + CaO$: 1.0 $K_2O + Na_2O$: 1.0
Microelements	> 0.1 %	S Cl Na	Al Fe Mg K Ti Ca Na
	~ 0.1 %	Si Al As Cu	Mn
	< 0.1 %	Mg K Fe Cr V Ni Ag Ti Co	–
Rare earths		Sc Pd Y Zr La Pr	–
Density, g/cm^3		1.90	2.23
Porosity, %	Open	4.7	–
	Closed	8.1	–
Pore size, nm		4…6	2…10
Specific surface area (BET), m^2/g		2…5	36

5.2 Activation of shapes of shungite

The type I shungite from the Leningrad Oblast deposit that was mechanically activated by means of a MV–0.005 toroid mill (Russia) was selected for the present work. We have assumed that samples of that shungite before activation could be regarded as a type I shungite with a BET (Brunauer - Emmett - Teller) surface area of 2–5 m^2/g, while the same activated shungite had a larger area of 13.1 m^2/g (Sineva 2007).

In our more recent work, we selected the type III shungite (from the Zazhogino deposit) produced by Geotech VIMS with BET surface area of 4.2 m^2/g. The same ore was selected after preliminary activation using the Aronov vibration mill with the parameters of activation as follows: 2 drams; oscillation frequency 50 Hz; amplitude fluctuations 5 mm; dispersion time 5 minutes (1 minute intermittently in 1 minute). In addition, in the case of mechanical-chemical treatment, all mechanical activation parameters were fixed, but the treatment was held in the presence of *n*-hexane additive. The data show that mechanically and mechanically-chemically activated (in the presence of *n*-hexane additive) type III shungite have BET surface area of 40.2 and 31.0 m^2/g, accordingly (Table 6.5) (Sineva 2007).

The influence of mechanical-chemical treatment on coal is described in the literature (see Khrenkova 1993). With the example of coal powders of different content, the influence of time of the mechanical treatment has been shown. Differential particle-size distribution curves have one maximum that points to the presence of one of the most probable fractions in the poly-disperse system. Upon longer treatment, the maximum on the curve shifts toward the smaller radii of the coal particles. And upon further treatment, the enlargement of coal particles may be observed, and the maximum of the differential curves shifts toward bigger radii of the particles. Such changes are confirmed by data on the modification of the specific surface area on pounding in a vibrating mill during mechanical treatment. The outer specific surface area of the dispersity specimens increases with the enlargement of the mechanical treatment time, reaches maximum value (the dispersion process prevails), and then decreases (the aggregation process prevails). The destruction of coal particles and their aggregation occur on different durations of pounding.

Upon dispersion of coals, their activity toward different reagents increases. This increase can be caused by the growth of both the outer and inner surface—as a result of new pores forming or disclosing earlier inaccessible pores—and due to the changing physicochemical structure of the newly formed surfaces.

During mechanical treatment, the initial porous structure of the coals is transformed: Macropores are completely destroyed, while and meso- and micropores are only partially destroyed. The accessibility of

Table 6.5 Physicochemical Properties of Composite Material Shungite of the Types I and III

Adsorbent	Surfactant	S, m^2/g		$\Delta G^0{}_a$, kJ/mol		$\Delta G^0{}_{ma}$, kJ/mol
Shungite I	SDS	6.5		−34.2		−22.7
	CTAB	3.1		−34.3		−30.0
Activated $S_{BET} = 13.1$ m^2/g	TX-100	2.4		−49.6		−19.2
Shungite III		*n*	*a*	*n*	*a*	
	SDS	–	31	–	−24.6	−22.5
Non-activated (*n*) $S_{BET} = 4.2$ m^2/g	CTAB	5.2	41	−30.9	−33.9	−28.0
Activated (*a*) $S_{BET} = 40.2$ m^2/g	TX-100	3.4				−19.5

micro- and mesopores improves; that is why the measurable volume of these pores increases. Along with the transformation of the original porous structure, new secondary pores appear, which are newly formed microcracks. As a consequence, the specific surface area also increases in comparison to the initial one.

When keeping the dispersive coals, their micropore volume and the specific surface area decrease. The closing of the newly formed cracks, when the load is removed and pore accessibility diminishes, can be one reason for this phenomenon. Such changes probably occur because of shungite dispersion on mechanical shungite activation.

6 Laws of adsorption of the surfactants from aqueous solution at air–water and water–adsorbent interfaces

Physicochemical and biological methods are applied for the purification of sewage from surfactants (detergents). Physicochemical purification includes ion exchange; adsorption on activated charcoals; and inert materials or natural adsorbents; foam separation; coagulation. The adsorption of molecules and ions is known to play an essential role in many aspects of environment protection.

The adsorption ability of low-molecular micelle-forming surfactants — ionic (sodium dodecylsulfate (SDS), and cetyltrimethylammonium bromide (CTAB)) and non-ionic (Triton X-100, shortly TX-100) — on powders of nonactivated, mechanically activated and mechanically-chemically activated type III shungite in comparison to the activated shungite of type I were investigated. Shungite of type I as adsorbent was studied earlier (Sineva 2007). As the initial solutions, the 0.01 M solutions of ionic surfactants and 0.00005 M solution of TX-100 were taken. CTAB and SDS were purchased from Serva Feinbiochemica (Germany), while TX-100 was used as received. The deionized water purified with Millipore Plant (France) with conductivity no more than 18.2 MΩ/cm was used for the preparation of all solutions.

In our work, two types of shungite were chosen as natural adsorbents because there are no data reported on adsorption of surfactants on shungite except that in our work (Sineva 2007). The most widespread Karelia shungite are shungite of type III containing about 30 mass % of carbon and over 60 mass % of SiO_2 (Table 6.4). The Zazhogino deposit of these rocks is a commercial one. Activated and nonactivated shungite were chosen for the investigation, and, in addition, both types shungite were activated mechanically and mechanically-chemically activated.

The values of the critical micelle concentration (CMC) for the studied surfactant solutions were determined from surface tension measurements. Among other methods, for ionic surfactants the conductance-measuring method was used, while for TX-100 the spectroscopic method was chosen for estimation of CMC as well as for measure of adsorption on type III shungite, with the aim of Agilent 8453 UV-visible spectroscopy systems at room temperature. Their CMC values were (in mol/L): 7.9×10^{-3} (SDS), 2.61×10^{-3} (CTAB), and 3.15×10^{-4} (TX-100). The average CMC are in good agreement with the results mentioned previously.

Our earlier measurements (Sineva 2007) of the dynamic surface tension of both water and aqueous solutions of surfactants was determined for adsorption times up to 10 to 150 sec. The absence of dynamic surface tension in CTAB, SDS, and TX-100 solutions and the absence of minimum on the graphs of the equilibrium surface tension in the area close to the CMC (Sineva 2007) allows us to infer a conclusion about "surface-chemical" purity in the studied surfactants (Lunkenheimer 2006).

6.1 Adsorption isotherms and adsorption model

In the cases of nonactivated and activated type III shungite, isotherms of adsorption of the studied surfactants at both the air/solution and solution/solid interfaces were obtained from the surface tension isotherms. The kinetics of adsorption of surfactants at times of supervision from 2 h to 27 days was investigated.

The surface tension versus concentration isotherms of micelle-forming surfactants had the appearance shown in Figure 6.1 for the surfactant studied (TX-100). The adsorption isotherm was obtained using the Gibbs equation, which, in the case of ionic surfactants, may be written as

$$\Gamma = -\frac{1}{2}\frac{c}{RT}\frac{d\sigma}{dc}.$$

where Γ is the surfactant surface excess at the air–water interface, σ is surface tension, c is bulk volume concentration, R is gas constant, and T is absolute temperature. In the case of TX-100, the factor of 1/2 is absent in the Gibbs equation (Rusanov 1992). This procedure also allowed obtaining the adsorption isotherms of all the surfactants at the water–air interface, as shown in Figure 6.2 for CTAB, as well as at the water/adsorbent interface (Figures 6.3–6.7). The adsorption isotherms obtained for all the surfactants adsorbed onto solid surfaces of shungite of the types I and III were classified as Langmuir type (L class). Among them, classical profiles with plateau adsorption were obtained on activated mechanically solids, in the case of the duration of the adsorption process, several days (or one week); thus, practically, one week is sufficient time to reach adsorption equilibrium at the solid/water interface

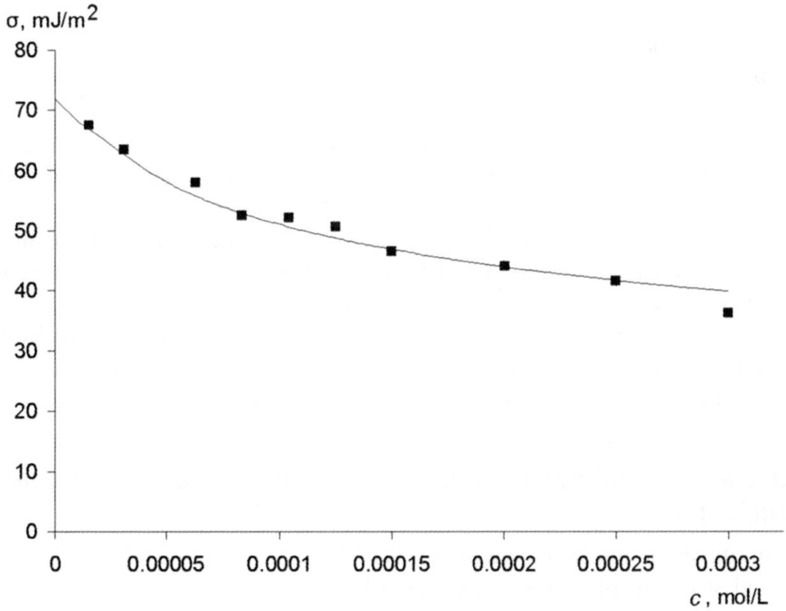

FIGURE 6.1

Surface tension (σ) vs. concentration (c) of Trtion X-100 at 22.5 °C.

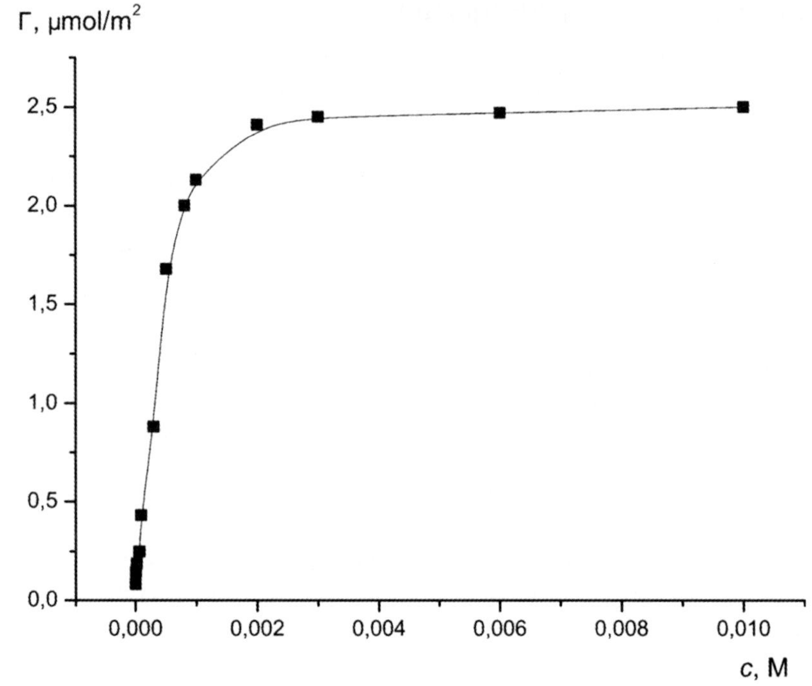

FIGURE 6.2

Adsorption isotherm of CTAB at the water–air interface.

(Figure 6.5). In other cases, the profiles of adsorption were obtained with a second plateau. Finally, a maximum adsorption was measured for CTAB on shungite of type I (activated), for prolonged periods of time (days or weeks) (Fig. 7).

The experimental data obtained for adsorption were fitted to the Langmuir equation and to the Gibbs–Langmuir model (Bae 1999), and the following adsorption parameters were estimated for both the air/water and the solid/water interfaces: the adsorption maxima Γ_{max} and Γ^*_{max}, the adsorption activities k and k^*, and the area occupied by a surfactant molecule in the close-packed layer (or its cross-sectional area) at the air/water interface, s_1, and the specific surface area of adsorbent, S. Calculation of the adsorption parameters is described in detail in Sineva, 2007.

6.2 Interparticle (intermolecular) interactions in the system adsorbate/adsorbent/water

In our work, we establish a connection between measured parameters (a surface tension and adsorption of surfactants) and structure of atoms, molecules of an adsorbent, and adsorbate: the forces and character of interparticle interactions. The research theme is important for determining the mechanism of interaction of molecules of surfactants on the water–solid interface. The analyses of the received data for interpretation of adsorption of surfactants at a molecular level within the limits of the certain

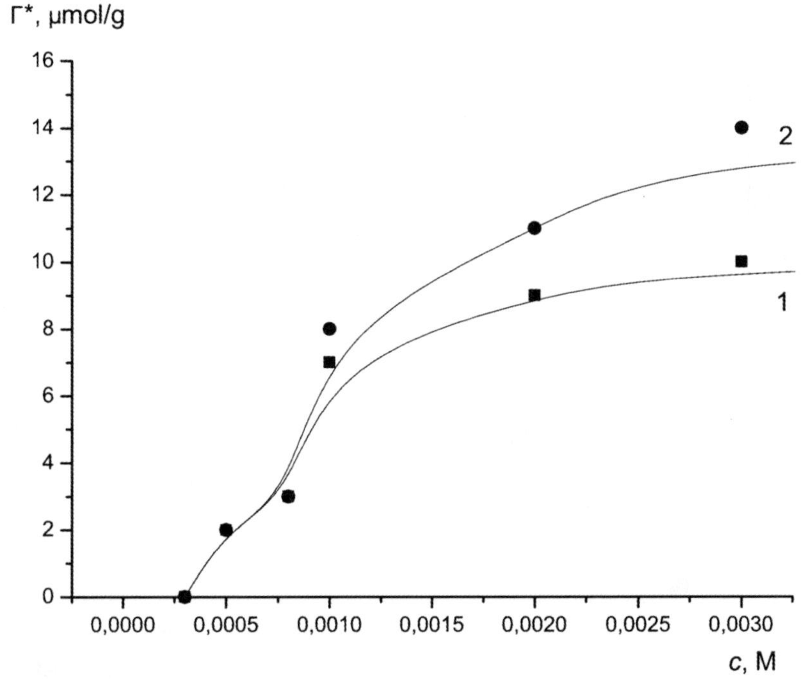

FIGURE 6.3

Adsorption isotherms of CTAB at the water–non-activated type III shungite interface after 2 h (1) and 7 days (2).

theoretical model is necessary, namely, the Gibbs–Langmuir model, which describes isotherms of adsorption for the water–air and water–adsorbent interfaces.

The adsorption process is formed through the action of surface forces. Of all the forces, there are unique universal forces operating between all molecules in any system, dispersion forces of an attraction (van der Waals forces); there are alternative interactions that are hydrogen bonding. The result of these interactions is physical adsorption, which was observed in types I and III shungite for the studied surfactants with their various factors of adsorption activity.

In forming chemical bonds between an adsorbent surface and the adsorbed molecules, adsorption is referred to as chemisorption. Data obtained in work for CTAB on the type III shungite denote that a hydrophilic mineral component of shungite (silicon oxide) adsorption is caused by electrostatic forces, whereas adsorption on carbon nanoparticles of shungite is attributed to dispersion forces (Sineva 2008).

Chemisorption is generally limited by the formation of a monolayer, and the specific surface at chemisorption is defined by the area occupied by the adsorption centers on a surface. At physical adsorption, adsorbed film can be bilayer and multilayer, and the specific surface of an adsorbent depends on the area of a molecule of adsorbate with its certain orientation onto surfaces.

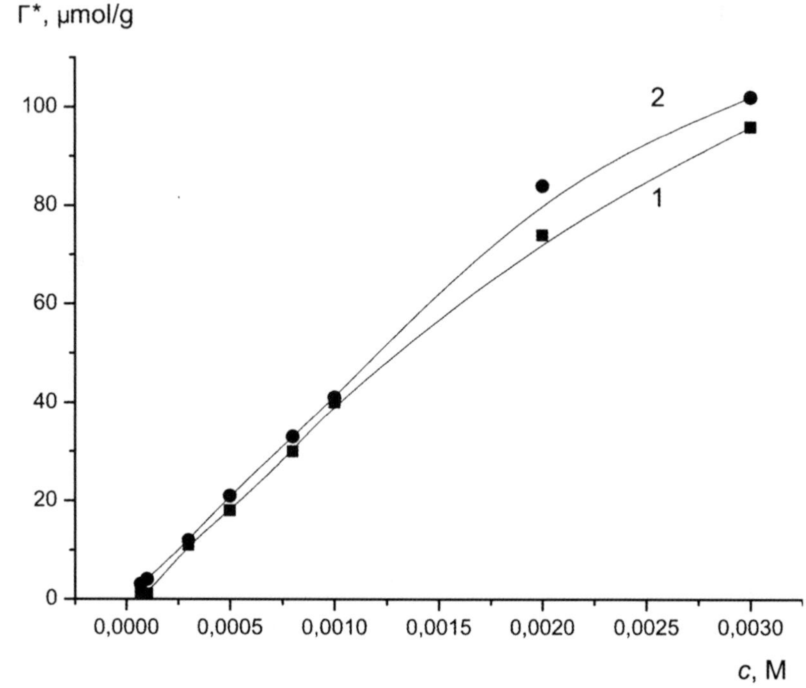

FIGURE 6.4

Adsorption isotherms of CTAB at the water–activated type III shungite interface after 2 h (1) and 7 days (2).

6.3 Surface area determination of the adsorbents

One of the important constants of adsorbed surfactants is the area occupied by a molecule or parking area s_1, on the surface of an adsorbent. This characteristic is necessary for determining the specific surface area S and for clarifying the interaction mechanism between adsorbed molecules and a solid. Depending on the adsorption mechanism, the determination of the area s_1 accounting for each adsorption center differs. In the case of chemisorption, s_1 is determined by the area occupied by the adsorption centers on the surface. In the case of physical adsorption, it mainly depends on the area of the adsorbed molecules oriented toward the surface in a certain way.

The orientation of the molecules in the adsorbed layer generally depends on the chemical nature of surfactants, their concentration in solution, the interactions between the adsorbent and adsorbate, and the lateral interactions between the molecules of adsorbed surfactants.

If the interaction energy between solvent and adsorbate is higher than the interaction energy between adsorbate and adsorbent, the S-shaped adsorption isotherm may be obtained. Such isotherm shape takes place when the molecules adsorbed on the surface have vertical orientation. The adsorbed molecules are lying flat on the surface, and the isotherm is defined as L-shaped (i.e., Langmurian isotherm) (Nechaev 1989).

As for S obtained in the case of surfactant adsorption from solutions, it has been shown that this value for the activated type I shungite sample weakly depends on the type of micelle-forming

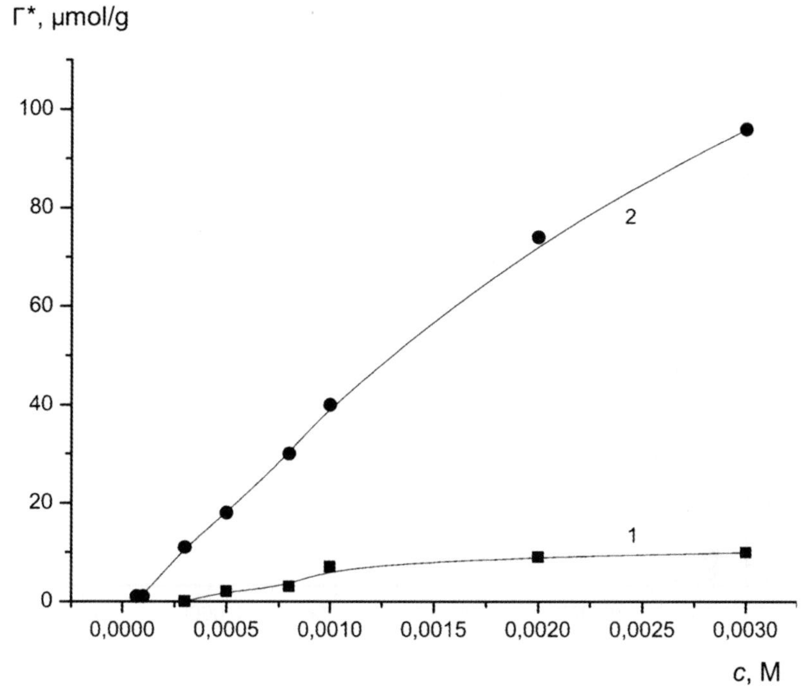

FIGURE 6.5

Adsorption isotherms of CTAB at the water–non activated (1) and the water–activated (2) type III shungite after 7 days.

surfactant and the size of its molecules. The *S* obtained appeared to be in agreement with those reported in Rozhkova (2004) where values of the BET surface areas were found to be between 2 and $5 m^2/g$ for type I shungite. In our work, *S* values for mechanically activated type I shungite samples determined by the adsorption method were found to be lower than the BET surface area of the same sample ($13.1 m^2/g$) (Sineva 2007). However, the methods of the specific surface area determination based on adsorption from solutions have methodical errors, too. As a whole, the successful application of the adsorption method from solutions is mainly suitable for estimating relative specific surface areas of similar materials.

6.4 Dependence of adsorption from type of activation of adsorbent and duration of adsorption process

As shown in Figure 6.6, the differences of values of adsorption maximum of CTAB on the activated type III shungite were found in experiment (1.05×10^{-4} mol/g) in comparison with inactivated shungite (1.32×10^{-5} mol/g). It is connected with an increase in a specific surface area (approximately in 10 times) as a result of mechanical activation. The results will be coordinated with data according to a specific surface of samples on BET (Table 6.5), namely, 40.2 and 4.2 m^2/g. As is known,

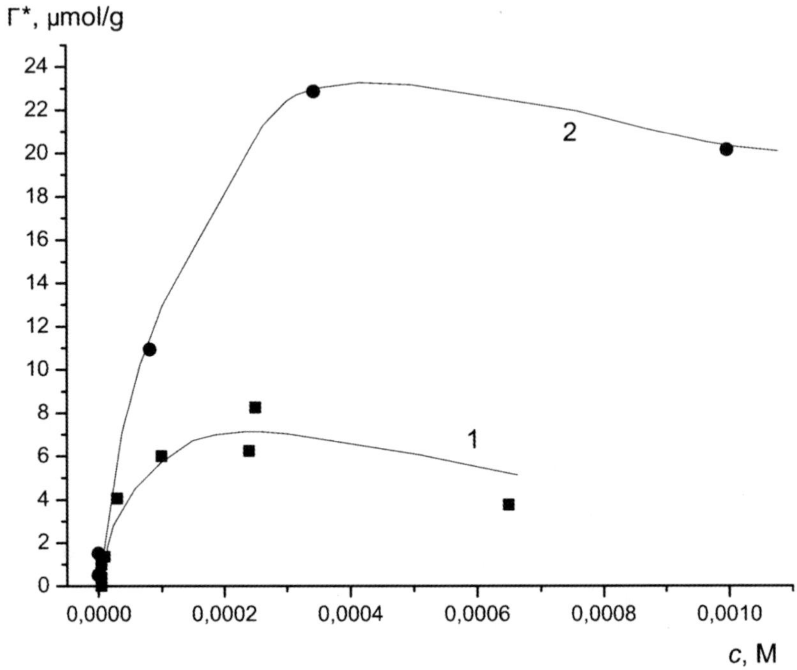

FIGURE 6.6

Adsorption isotherms of CTAB at the water–type I shungite interface after 2 h (1) and 27 days (2).

cationic surfactants adsorb strongly to most surfaces, and their main uses are related to in-situ surface modification. The reason for it is that the majority of different surfaces, metals, minerals, plastics, and membranes are negatively charged. In The adsorption of cationic CTAB on the hydrophilic mineral component (SiO_2) of the type III shungite is, probably attributed to electrical forces, while adsorption at the hydrophobic carbon part of surface is attributed to dispersion forces. As a result, on type III shungite for a given surfactant, it is observed simultaneously as physical adsorption and chemisorption.

With regard to the *S* of activated type III shungite obtained in the case of CTAB, it has been shown that these values are in good agreement with the BET surface area values (Table 6.5). Adsorption of this surfactant on the activated type I shungite, as well as its specific surface area (Sineva 2007), were much lower (approximately 15 times slower) than those found for the activated type III shungite after 2 h of adsorption processes (Figure 6.5, curve 1 and Figure 6.7, curve 1). The peculiarity of the type III shungite structure is known to be the existence of a penetrating network between two phases—carbon and silica—which are the main components of shungite rock and have a great surface area.

Ionic surfactants usually do not adsorb onto solid surfaces of like charges. As for anionic SDS adsorption on the type III shungite, it was investigated at the other conditions of activation of shungite. The study of SDS adsorption on the samples of the type III shungite was carried out under mechanical-chemical activation. In this case, adsorption occurred under conditions of mechanical-chemical treatment during which all mechanical activation parameters were fixed, but

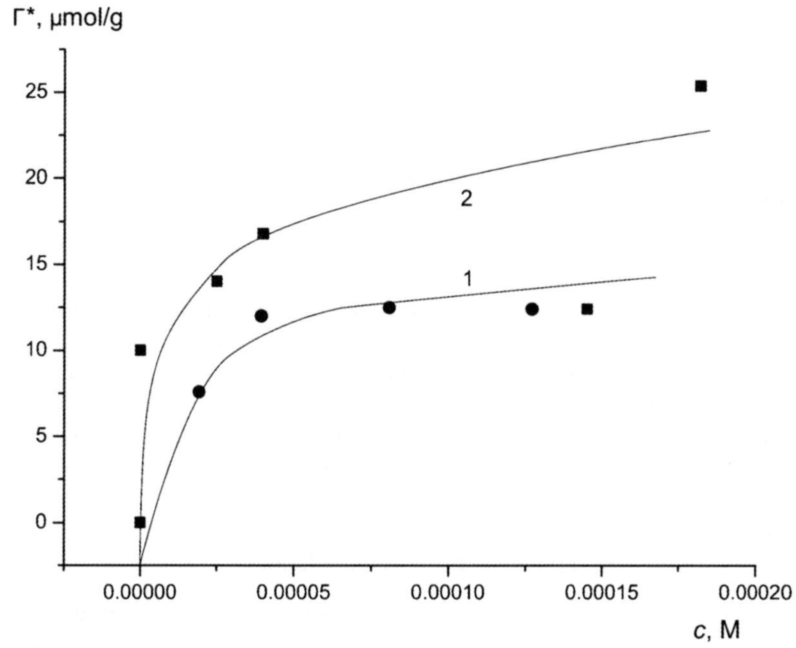

FIGURE 6.7

Adsorption isotherms of TX-100 at the water–non-activated type III shungite after 3 days (1) and 12 days (2).

it was held in the presence of *n*-hexane additive (40 mL). An increase in the specific surface area caused by the thin hydrocarbon film preventing the aggregation of shungite nanoparticles was expected. However, the experiment revealed the negligible adsorption of SDS from water solutions during 16 days, probably due to the blocking of active centers on both the mineral and carbon components of shungite by hexane film. After the removal of hexane, the SDS adsorption was measured in the regular way. The specific surface area of the shungite III powder obtained 16 days later was found to be 31 m^2/g, which is lower than the BET area of activated powder (Table 6.5) (Sineva 2008).

As outlined in Treiner (2006) and Partyka (1986), non-ionic surfactants such as TX-100 adsorb readily on silica, but their adsorption on hydrophilic surfaces is generally lower than that of ionic surfactants because of the weak bonding between the hydrophilic sites and the surfactant molecules.

The results obtained for TX-100 adsorption on the inactivated type III shungite are not comparable in terms of the duration (3 and 12 days, respectively) (Figure 6.7) of the adsorption process with data on adsorption of the same surfactant on type I shungite. The values of maximum adsorption of all the surfactants on the activated type I shungite found earlier, as well as its specific surface area (13.1 m^2/g), were several times lower than those found for the activated type III shungite (40.2 m^2/g). As a whole, it has been shown that CTAB *S* values of activated types I and III shungite are in good agreement with the BET surface area values, while SDS and TX-100 *S* values are lower than the BET specific surface area. This fact may be explained by the dependence of *S* values on both the type and conditions of adsorbent activation as well as duration of the process of adsorption from solution (Sineva 2008).

7 Thermodynamic evaluation of the adsorbent efficiency for water purification

As has been shown, the relative energetic characterization of adsorption and the accompanying micellization of the surfactants in water is important for the selection of adsorbents that are effective in purifying sewage from surfactants. First, we consider the theoretical base of the processes of adsorption at the air/water and water/solid interfaces (e.g., adsorbent).

There are two types of adsorption from solutions on solids: molecular adsorption (surfactant molecules are adsorbed) and chemisorption. Molecular adsorption is typical for non-ionic surfactant solutions, while chemisorption is typical for ionic surfactants solutions. The possibility of molecular adsorption of a substance can be defined on the basis of the Rehbinder *rule of polarities equalization.* The substance C can be adsorbed on the interface of A and B if the adsorption layer decreases the polarity difference at the interface of these phases (Sineva 2007).

For the adsorption of organic surfactants from water solutions, several cases are possible.

1. On hydrophobic solids (coal, paraffin, etc.), hydrocarbon chains that are pushed out from the water are adsorbed, the polar groups of the molecules staying in water. In this case, adsorption has a physical (reversible) nature.
2. On hydrophilic solids (e.g., ionic crystals), the polar groups of the surfactant molecules are adsorbed, with hydrocarbon chains appearing in water. When a compact monolayer is formed, essentially a new water /hydrocarbon interface with the surface tension until 50 mJ/m^2 appears. The Gibbs energy, which evolved fron adsorption of polar groups, should compensate this new interface surface tension, which is possible when a chemical bond of the polar groups with substrate forms (i.e., on chemical fixation on a solid). Consequently, the chemisorption of surfactants occurs in these systems.
3. If adsorption Γ is expressed through the surface concentration c, volume concentration *c*, and the adsorption layer thickness *δ*, then the Gibbs equation can be written as:

$$-\frac{d\sigma}{dc} = RT\delta\frac{c_s}{c}$$

As a measure of the surface activity, Rehbinder proposed that the *G* value to be used, which is equal to $G = \lim_{c\to 0}(d\sigma/dc)$.

For surfactants, the ratio c_s/c is almost unlimited and can be large. Consequently, the surface activity that is defined, in this case, by the equation $G = RT\delta c_s/c$ can be large. At the same time,

$$\frac{c_s}{c} = \frac{k\Gamma_{\max}}{\delta} = K_a$$

where *k* is adsorption activity. Thus, Henry law can be applied for the equilibrium of the adsorption layer and the solution volume (at low volume and surface concentrations). The Henry law describes the distribution of a substance between two phases with a constant partition coefficient K_a. That is why the regions of linear dependences $\sigma(c)$ and $\Gamma(c)$ are often called the Henry region (Shchukin 2001). Then

$$\Delta G^0_{ads} = RT\delta \frac{k\Gamma_{max}}{\delta} = RTk\Gamma_{max}$$

$$\Delta G^0_{ads} = -RT \ln K_a = -RT \ln \frac{k\Gamma_{max}}{\delta} = -RT \ln \frac{\Gamma_{max}}{\alpha\delta}$$

where $\alpha = 1/k$ and k is adsorption activity of surfactant onto the solid surface.

Two approaches are used most often for the modern interpretation of micelle formation. Micelle formation is described as the phase separation or, more often, from the idea of chemical equilibrium, where the mass law is applicable. From the position of chemical equilibrium, a micelle is taken as an aggregate from j-ions of surfactant and $j - z_m$ combined counter-ions where z_m is micelle charge and j is aggregation number.

The standard condition for different ions is chosen in such way that the mole fraction is 1, and the solution is close to infinitely diluted (i.e., all activity coefficients are equal to 1).

The thermodynamic parameters calculated with the use of mass law equations are usually marked by *mf* indices. ΔG^o_{mf} in the absence of electrolyte additives is specified by the equation:

$$\Delta G^o_{mf} = \left(2 - \frac{z}{j}\right) RT \ln CMC.$$

If $z = 0$, that is, j counter-ions are strictly bound with the micelle imparting it the zero charge, the last equation turns to

$$\Delta G^o_{mf} = 2RT \ln CMC.$$

If $z = j$, that is, counter-ions are bound with the micelle, then

$$\Delta G^o_{mf} = RT \ln CMC$$

where n_1 is aggregation number of surfactants, n_2 is aggregation number of counter-ions, and n is the total number of aggregations of all the ions and $b_2 = 1$. For a surfactant that is 1–1-electrolyte when $n = n_1 + n_2 (1 + \beta)$ and $b_2 = 1$, the standard concentration $c_{1S} = 1$ mol/L,

$$\Delta G^0_{mf} = 2RT(1 + \beta)\ln CMC$$

where β is the degree of binding of counter-ions for ionic surfactants, which is equal to 0.87 (Rusanov 1992).

The standard Gibbs energy change of CTAB adsorption on shungite of type I obtained was −34.34 kJ/mol, indicating that this process was more favorable than surfactant micellization (−30 kJ/mol). The values of $\Delta G^0{}_{ads}$ of SDS and TX-100 on shungite were −34.23 and −49.6 kJ/mol, respectively, while $\Delta G^0{}_{mf}$ for these surfactants were found to be equal to −22.7 and −19.9 kJ/mol, respectively. As for two Gibbs energies, in the case of surfactant adsorption as mechanically activated shungite of type III, the standard Gibbs energy obtained for CTAB and SDS adsorption were found to be equal to −33.9 and −24.6 kJ/mol, accordingly, while the standard Gibbs energy of micelle formation, in the case of the same surfactants was −28.0 and −22.5 kJ/mol, respectively. Thus, comparison of two Gibbs energies shows that adsorption on shungite of types I and III of each surfactant studied is thermodynamically more favorable than their micelle formation (Sineva et al. 2007).

8 Ecological and resource-keeping aspects of surfactants and detergents

The known compositions of synthetic detergents (SD) consist mainly of the mixtures of anion active surfactant (*anionic*) and non-ionic surfactant (*non-ionic*), constituting 10–40% of the total composition of detergents. Alkyl benzenesulfonates, alkyl sulfates, and alkyl sulfonates produced in the form of sodium, potassium, and triethanolamine salts, as well as disodium salts of monoalkyl sulfosuccinic acid, have the most important value among the anionic salts. Non-ionic in the composition of SD are ethoxylated alkyl phenols, ethoxylated fat acids, and alkylolamids (Sineva 1998).

As to sulfate surfactants, they are rapidly hydrolyzed by acids in an autocatalytic process. The other types of ionic surfactants are stable unless extreme conditions develop. Today, the single most important type of non-ionic is fatty alcohol ethoxylates, which are used in liquid and powder detergents as well as in a variety of industrial applications. Fatty alcohol ethoxylates are stable in the pH range 3–11. They undergo slow oxidation in air, but some oxidation products are more irritating to the skin than the intact surfactant (Holmberg 2002).

Besides, the surfactant, the SD, contains different organic and inorganic additives (extenders) such as sodium tripolyphosphates, phosphorus organic complexons, sodium silicates, sulfates and carbonates, carboxymethylcellulose, optical bleacher, antistatic, and perfume.

An important part of the ecological monitoring system is the control of surface waters. In connection with this, the task is to estimate and predict the state of water reservoirs and flaws as the natural surroundings. Following this task, the ecological and resource-keeping aspects of production and application of synthetic detergents were considered in our earlier work (Sineva 1998).

In this connection, we reminder about soaps again. Soap is still the largest single type of surfactant. It is produced by saponification of natural oils and fats: soap-salt of a fat acid and alkaline metal. The most effective soap is soap that comes from coconut oil containing lauric acid. This soap is one of the standard substances used in estimating washing action. It is ecologically safest, and it possesses good washing, foaming, and emulsifier action, but as is true of all other soaps it is not effective in the presence of rigid salts—calcium, magnesium, barium, and iron. Therefore, only the soft water of soap is it possible to use for household washing functions. Under the right conditions, soaps are excellent surfactants. Cosmetic soaps are a special kind of washing-up liquid containing not only an odorant for a pleasant smell but also various oils or disinfectants.

We have suggested the following program regarding industrial surfactants:

1. Direct substitution of most biologically soft surfactants, that is, soaps (fat acid salts and soap-like surfactant (alkyl sulfates, ethoxylates, and their sulfates), by other surfactants and
2. The search for ecologically safe and economically effective ways of purifying daily sewage from remaining fat acid salts, suggested as the main components of detergents (Sineva 1998).

We adhere to the method of purification of wastewaters from fat acid salts as well as the regeneration of fat acid salts (as well as the regeneration of fat acid salts) by means of consecutive reactions of soap decomposition in the sewage with the separation of fat acids rendered in (Chemistry 1992):

$$2RCOONa + H_2SO_4 \rightarrow 2RCOOH + Na_2SO_4$$

the etherification of isolated fat acids by glycerol:

$$3RCOOH + CH_2OH–CHOH–CH_2OH \rightarrow CH_2OCOR–CHOCOR–CH_2OCOR + 3H_2O$$

and trapping the products of reaction (2) with the following saponification:

$$CH_2OCOR–CHOCOR–CH_2OCOR + 3NaOH \rightarrow CH_2OH–CHOH–CH_2OH + 3RCOONa$$

To conduct the above reactions in the dilute solutions of soap and soaplike surfactant with concentrations that are from ten to a hundred times higher than the high allowable concentration of anionic surfactant (0.2–0.8 mg/L), the respective experiments with the determination of product yield have been planned in (Chemistry 1992). The product yield has been determined for the given reactions with the use of different qualitative methods of analysis. In particular, nonsaponified fat has been measured refractometrically while the formed or remained surfactant has been measured by the means of tensiometry, conductometry and with other methods.

On the basis of preliminary experiments, the creation of close local cycles applying the following scheme of "washing–purification of sewage from the surfactant–regeneration of surfactants" is suggested for everyday chemical functions (bath-house, laundry) (Sineva 1998).

Since 1990_s the trend has been from fatty alcohols as well as methyl ester sulfonates toward natural-based surfactants (Piasecki 2006).

As for Russia in the 21st century, annual sales include the following: nearly 700,000 tons of detergents and fabric care aids, nearly 220,000 tons of cleaning agents, nearly 59,000 tons of auto cosmetics, and nearly 800,000,000 tons of domestic chemicals in conventional containers. The following regulations have been issued in Europe:

1. Regulation on dangerous substances
2. Regulation on dangerous chemical production
3. Regulation on powder detergents

All of these regulations are designed to regulate the production, classification, packing, and marking of products. For example, the first instruction concerns dangerous substances and of course is an essential one. According to this instruction, all substances are divided into two groups: known and new substances. Substances are called known substances if they have been on the European market since the 1980s. The new instruction set for new substances is now being prepared. The second instruction is closely related to the first one, and it states that chemical production shall be taken to be dangerous if only one dangerous substance enters into the composition. More restrictive instructions for powder detergents came into effect in 2005 and introduced stricter controls for the process of biodegradation of washing powders.

9 Conclusions

The water purification of reservoirs polluted by toxic chemical substances as well as of freshwater poses an enormous problem for the 21st century. This chapter reported on the maximum adsorption of surfactants, namely, sodium dodecyl sulphate, cetyltrimethylammonium bromide, and Triton X-100 on natural adsorbents—nonactivated, mechanically and mechanically and chemically activated shungite of types I and III.

It was found that the maximum adsorption of surfactants depends on the type and conditions of shungite activation as well as on the duration of the adsorption process (i.e., the kinetic process).

A comparison of the standard Gibbs energy of adsorption (ΔG^{0}_{ads}) for equilibrium adsorption of cationic surfactant on the shungite during 6 to 7 days of adsorption, with the value of the standard Gibbs energy of micellization (ΔG^{0}_{mf}) of surfactants in water, suggests a thermodynamically favorable adsorption process. Therefore, type III shungite, both nonactivated and activated, can be recommended as an effective adsorbent for water treatment and for purification of drinking water.

Traditional adsorption isotherms, combined with the thermodynamic analysis of the standard Gibbs energy of adsorption and accompanying micelle formation of the surfactants in water, seem to be useful for an understanding of the laws of water treatment and the rational selection of adsorbents based on the interface phenomena.

Also included in this chapter are data on the industrial perspective and on the risk assessment aspects as an example, comparing some technologies of water treatment for water consumption as well as the processes of self-purification of water in natural ecosystems.

Acknowledgments

The author is grateful to E. V. Lazareva for providing language help and to F. A. Kulikov-Kostyshko who was very helpful in preparing the manuscript.

References

Annual report of governmental commission on Baikal lake (2000). Moscow.

Bae, S., Haage, K., Wantke, K., Motschmann, H., 1999. On the factor in Gibbs equation for ionic surfactants. J. Phys. Chem. B vol. 103, 1045.

Burczyk, B., 2006. In: Somasundaran, P. (Ed.), Encyclopedia of surface and colloid science, vol. 1. Taylor & Francis, Boca Raton, FL, p. 917.

Chemistry of fats, 1992. Kolos, Moscow (in Russian).

Divakova, N.A., 2007. Safety of detergents and other household products. SÖFW, No 1, 56 (in Russian).

Drinking water, 1996. Information collection No. 2 VINITI RAN, Moscow, p. 110. (in Russian).

Holmberg, K., Jönsson, B., Kronberg, B., Lindman, B., 2002. Surfactants and polymers in aqueous solutions. John Wiley & Sons, New York.

Gupta, V.K., Ali, I., 2006. In: Somasundaran, P. (Ed.), Encyclopedia of surface and colloid science, vol. 1. Taylor & Francis, Boca Raton, FL, p. 129.

Koganovsky, A.M., Klimenko, N.A., 1974. Phisicochemical methods of purification of waste water from surfactants. Naukova dumka, Kiev, 1974 (in Russian).

Khrenkova, T.M., 1993. Mechanochemical activation of coals. Nedra, Moscow (in Russian).

Lunkenheimer, K., 2006. In: Somasundaran, P. (Ed.), Encyclopedia of surface and colloid science, vol. 8. Taylor & Francis, Boca Raton, FL, p. 58.

Mashkova, L.P., Sineva, A.V., 1998. Adsorption of toxic organic and inorganic ions on natural minerals. *Ecology of cities*: Conference proceedings. Rhodes 1998, 247.

Nechaev, E.A., 1989. Chemisorption of organic substances on oxides and metals. Vysshaya shkola, Kharkov (in Russian).

Ostroumov, S.A., 2000. Water ecosystem: Big size diversified bioreactor with the function of water self-purification. Doklady Biology vol. 374, 427.

Ostroumov, S.A., 2001a. Effect of amphiphilic chemicals on filter-feeding marine organisms. Doklady Biol. vol. 378 (No. 2), 248.

Ostroumov, S.A., 2001b. Responses of *Unio tumidus* to mixed chemical preparations and the hazard of synecological summation of anthropogenic effects. Doklady Biol. vol. 380 (No. 5), 492.

Ostroumov, S.A., 2004. Facts and concepts of the ecology. In: Aquatic ecosystems and organisms 5. Max Press, Moscow (in Russian).

Parfit, M., 1993. Troubled waters run deep. National Geographic vol. 184 (No. 5A), 78 (in Russian).

Partyka, S., Lindheimer, M., Zaini, S., Keh, E., Brun, B., 1986. Langmuir vol. 2, 101.

Piasecki, A., 2006. Biodegradable and chemically degradable anionic surfactants. In: Somasundaran, P. (Ed.), Encyclopedia of surface and colloid science, vol. 1. Taylor & Francis, Boca Raton, FL, p. 894.

Rozhkova, N.N., Emel'yanova., G.I., Gorlenko, L.E., Lunin, V.V., 2004. Shungite carbon and its modification. Ros. Khim. Zh. vol. 48, 107 (in Russian).

Rusanov, 1992. Micelle formation in aqueous solutions of surfactants. Khimiya, St. Petersburg (in Russian).

Ryabchikov, A.M., 1976. Relevant problems of changes of nature abroad. Izd. MGU, Moscow, 1976. (in Russian).

Shchukin, E.D., Pertsov, A.V., Amelina, E.A., Zelenev, A.S., 2001. Colloid and surface chemistry. In: Mobius, D., Miller, R. (Eds.). Elsevier Science, Amsterdam.

Sineva, A.V., 1998. Ecological and resource keeping aspects of productions and application of surfactants and detergents. In: Ecology of cities. Conference proceedings, Rhodes, pp. 231–233.

Sineva, A.V., 2002. To preserve Lake Baikal is our task. In: Managing protected areas in a changing world. Proceedings of 4th International Conf. on Science and Management of Protected Areas, Moscow, 1387–1393.

Sineva, A.V., Parfenova, A.M., Fedorova, A.A., 2007. Colloids and surfaces A: Physicochem. Eng. Aspects. vol. 306, 68.

Sineva, A.V., Parfenova, A.M., Kolbanev, I.V., 2008. Natural minerals in water purification from ionic and non-ionic surfactants. Abstracts of 3rd Internationcal Conference on Colloid Chemistry and Physicochemical Mechanics, Moscow.

Treiner, V., 2006. In: Somasundaran, P. (Ed.), Encyclopedia of surface and colloid science, vol. 1. Taylor & Francis, Boca Raton, FL, p. 732.

Warren, S., 2008. EU–Russia Cooperation Programme.

CHAPTER

7 Co-Contaminant-Aided Removal of Organics from Produced Water Using Micellar-Enhanced Ultrafiltration

Ali Deriszadeh, Maen M. Husein, Thomas G. Harding

Department of Chemical and Petroleum Engineering, University of Calgary, Calgary, Alberta, Canada T2N 1N4

CHAPTER OUTLINE

1 Introduction

Industrial contaminated waters may contain organic and/or inorganic contaminants with types and concentrations depending on the source. These contaminants have adverse effects on industrial operations; including scaling and corrosion in pipelines (Quagraine et al. 2005; Cline 1998; Simms et al.

The Role of Colloidal Systems in Environmental Protection. http://dx.doi.org/10.1016/B978-0-444-63283-8.00007-7

1992). Furthermore, drinking water sources and aquatic life may be negatively affected if such water is discharged without proper treatment (Veil et al. 2004; Cline 1998). In Northern Alberta (Canada), for example, the expansion of oil sands processing has resulted in an exponential increase in the magnitude of tailings ponds; artificial ponds collecting the process affected water from oil sands operations known as produced water (Chalaturnyk et al. 2002). Produced water is contaminated with different soluble and suspended organics as well as dissolved heavy metals and suspended clays (Hum et al. 2006). From an environmental perspective, there is a requirement to reclaim the produced water and thereby decrease the footprint of these contaminants on native water bodies. From an industrial standpoint and since most of the thermal heavy oil recovery operations are water or steam assisted, alternative water resources are necessary for the sustainability of the oil sands operations. Therefore, reliable and cost-effective water treatment technologies that enable produced water recycling are of crucial importance.

Many techniques are currently employed in the oil and gas industry in order to decrease the contaminant level in produced water and enable water recycling (Shpiner et al. 2007; Bridle 2005; Heins and Peterson 2005; Jan and Reed 1989; Thomas et al. 1987). Depending on the required water quality, the contaminated water may undergo single or multiple treatment steps such as coagulation, flocculation, sedimentation, filtration, precipitation, ion exchange, distillation, reverse osmosis (RO) filtration, activated carbon adsorption, hot/warm lime softening, and chemical oxidation. These technologies are often classified under primary, secondary and tertiary treatments depending on the treatment objectives. Primary treatment consists of removing total suspended solids (TSS) and free oil, while secondary treatment involves further removal of TSS and partial reduction of total dissolved solids (TDS). Tertiary treatment removes dispersed oil, and residual TDS and dissolved organics, respectively (Schroeder 1977).

Membrane technology has been used progressively in the last few decades as an effective separation tool due to its compact size, selectivity, and relatively low energy requirements (Baker 2004). Membrane filtration is a physical separation process associated with rejecting molecules of particular size and allowing the passage of smaller-sized molecules. Figure 7.1 shows the classification of membrane technologies according to membrane size. Membrane filtration can follow a cross flow or a dead-end scheme (Marcotte et al. 1998; Zaidi et al. 1992). Figure 7.2 illustrates the two operation modes and the manner in which a solute deposits on the surface of the membrane. It is evident from Figure 7.2 that in dead-end filtration, where the feed and permeate are in line, the formation and growth of a cake are more pronounced compared to cross-flow filtration. The shearing effect from the feed stream in cross-flow filtration limits the accumulation of materials on the membrane surface and hence, leads to a steady flux and resistance. The most efficient cross-flow filtration zone can be identified from

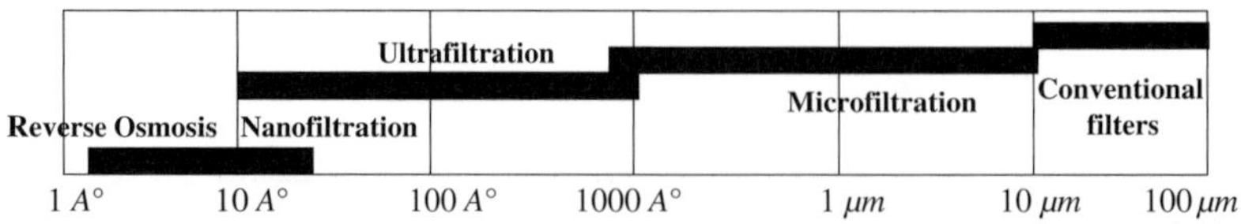

FIGURE 7.1

Classification of the membranes according to pore size (1 $A°$ is 10^{-10} m and 1 μm is 10^{-6} m) (modified from Baker, 2004).

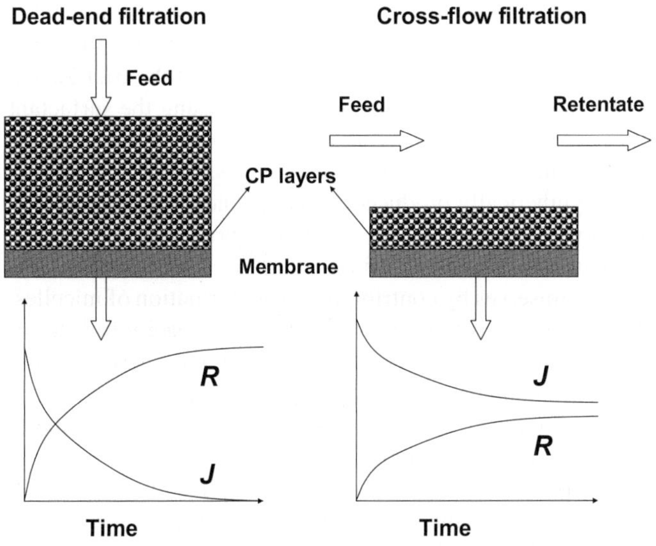

FIGURE 7.2

Dead-end and cross-flow filtration operation modes (J: Flux and R: resistance).

a plot of permeate flux versus transmembrane pressure (TMP). The permeate flux changes linearly up to a limiting TMP, above which a plateau begins to occur (Bacchin et al. 2002a,b). At low TMP the permeate flux is low, and accumulation of materials in the concentration polarization (CP) layer is small (Deriszadeh et al. 2008). As the TMP increases, on the other hand, the flux increases leading to increased concentration polarization and formation of a gel layer on the surface of the membrane (Bacchin et al. 2002a). Operating above the limiting TMP results in an increase in the thickness of the accumulated materials and, consequently, constant permeate flux, though at a higher pumping cost (Van den Berg and Smolders 1990). This phenomenon is a significant and universal drawback of membrane filtration.

Micellar enhanced ultrafiltration (MEUF) is a low-energy physicochemical membrane-based separation technique that adds a chemical aspect to the physical separation of a membrane process (Deriszadeh et al. 2009, 2008; Purkait et al. 2006, 2005, 2004; Lee et al. 2005; Yang et al. 2005; Juang et al. 2003; Jadhav et al. 2001; Fillipi et al. 1999; Markels et al. 1995; Dunn et al. 1989, 1985). MEUF couples the dissolution of organic contaminants within the micelles, or the binding of inorganic contaminants to the head group of oppositely charged surfactants, with the size-dependent physical separation (Dunn et al. 1989). The main objective of involving the chemical step is to increase the size of the contaminants to be removed. Therefore, formation of micelles and application of the proper membrane are key factors for an effective MEUF process. On one hand, formation of micelles requires a certain minimum concentration of surfactant to ensure aggregation of surfactant monomers into micelles. However, increasing the surfactant concentration may cause surfactant precipitation as well as CP layer formation and fouling of the membrane (Deriszadeh et al. 2008). Fouling reduces permeate flux and can shorten, or even terminate, the lifetime of a membrane. In addition, increasing surfactant concentration in the MEUF feed increases the possibility of surfactant leakage to the permeate side, which results in back contamination (Deriszadeh et al. 2008). Hence, best performance of MEUF hinges on using an optimum surfactant concentration that allows the highest possible dissolution of organics, while maintaining the highest permeate flux and lowest back contamination. Traditionally, this major drawback has been addressed by using mixed micellar solutions, that is, adding non-ionic surfactant to an ionic one in order to reduce the CMC of the ionic surfactant (Li et al. 2006; Bielska and Szymanowski 2004; Tung et al. 2002). Lowering the CMC is equivalent to decreasing the surfactant dosage, while maintaining the required effectiveness. Addition of an extra chemical to the contaminated water, however, may pose its own environmental and economic challenges.

In Part I of our study we showed, through synthetically produced water samples, that amphiphilic organic co-contaminants play a similar role to that of a non-ionic surfactant and form "mixed micelles" leading to an improvement in the performance of MEUF (Deriszadeh et al. 2008; 2009). This improvement entailed contaminants removing themselves by contributing to the formation of micelles, meanwhile lowering the CMC of the surfactant and increasing its solubilization capacity for the organics. Consequently, lower surfactant dosage could be employed, and, in turn, higher permeate flux was obtained coupled with lower fouling and back contamination. The improved MEUF performance observed in Part I was shown to apply to field-produced water samples from operations in Alberta (Canada: Field A sample) and California (USA: Field B sample). Although these field tests help demonstrate the industrial relevance of the proposed process, the industrial applicability of MEUF hinges on the cost of surfactant needed, recyclability of the surfactant, and its cost, as well as the environmental impact of the chemicals added. Ideally, both lower cost and environmental impact can be attained if naturally occurring surfactants in produced water can be activated. Chemicals will still need to be added, but those chemicals would mainly impact the pH and the ionic strength of the produced water. The risk associated with adding a surfactant not native to the produced water is twofold: environmental and economic. The economic risk is associated with the cost of the process and its feasibility, whereas the environmental risk relates to back contamination and whether the newly added chemical will promote side reactions that would produce hazardous by-products.

The analysis in this work also includes mathematical models assuming constant and variable concentration polarization (CP) layer resistance developed on the basis of Darcy's law and the resistance in-series model. These models were shown to apply to different synthetic and field-produced water samples and for both polymeric and ceramic membranes.

2 Experimental

2.1 Materials

Cetylpyridinium chloride, CPC (96-100 wt.% pure, Fisher Scientific, ON) was used as the cationic surfactant for all the MEUF runs. *P*-xylene (99 wt.% pure, Fisher Scientific, ON) was used as a model contaminant in the synthetic produced water study. Other organic contaminants used to form synthetically produced water were pentanoic acid $C_5H_{10}O_2$ (99 wt.% pure, Fisher Scientific, ON), heptanoic acid $C_7H_{14}O_2$ (99 wt.% pure, Fisher Scientific, ON), octanoic acid, $C_8H_{16}O_2$ (99 wt.% pure, Fisher Scientific, AB), and decanoic acid $C_{10}H_{20}O_2$ (99 wt.% pure, Fisher Scientific, AB). A mixture of naphthenic acids that represent higher molecular weights and that better represent contaminants in produced water was known as NAs practical grade (Fisher Scientific, ON). According to the manufacturer, NAs was a mixture of naphthenic acids with carbon number ranging from 11 to 18. Methanol (99% pure, VWR, ON) was used to break the micelles and release *p*-xylene for the high-performance liquid chromatography (HPLC) measurements.

Table 7.1 Specification for PAN Polymeric and TiO_2 Ceramic Membranes

Polymeric Membranes		Ceramic Membranes	
Length (cm)	34.7	Length (cm)	25
Diameter of the membrane module (cm)	4.5	Membrane OD (cm)	1.0
Membrane effective surface area (m^2)	0.2	Membrane effective surface area (m^2)	0.013
Number of hollow fibers per module	~300	Number of channels	7
Inner diameter of the hollow fibers (cm)	0.08		

The field water samples consisted of two produced waters from oil production sites belonging to Alberta (Canada) and California (USA). These samples were subject to Part II of the investigation.

2.2 Membranes

Polyacrylonitrile (PAN) hollow fiber membranes (Pall Corporation, ON) with three molecular weight cutoffs (MWCO) of 6, 13, and 50 kDa were used as representatives of hollow fiber polymeric membranes. PAN is hydrophilic and was used because hydrophilic materials allow for the easy passage of water through the membrane pores (Qtaishat et al. 2009; Kosaraju and Sirkar 2007). Titanium oxide (TiO_2) ceramic membranes (Sterlitech, WA) having different MWCO of 50 kDa with seven channels were also tested. Ceramic membranes have the advantage of high thermal resistance and are better suited to produce water treatment (Benito et al. 2007; Mahesh Kumar et al. 2007; Velterop and Keizer 1991). Table 7.1 shows specifications for the membranes involved in this work.

2.3 Experimental methods

2.3.1 Cross-flow mode

Figure 7.3 is a schematic representation of the cross-flow MEUF experimental setup. The feed stream was pumped from the feed tank through a pulsation dampener to the membrane module using a peristaltic pump (Masterflex L/S, Cole-Parmer Instrument Co., QC). During passage of the feed inside the hollow fibers of the polymeric membranes, or inside the channels of the ceramic membranes, a permeate stream passes through the membrane pores by virtue of the pressure difference between the

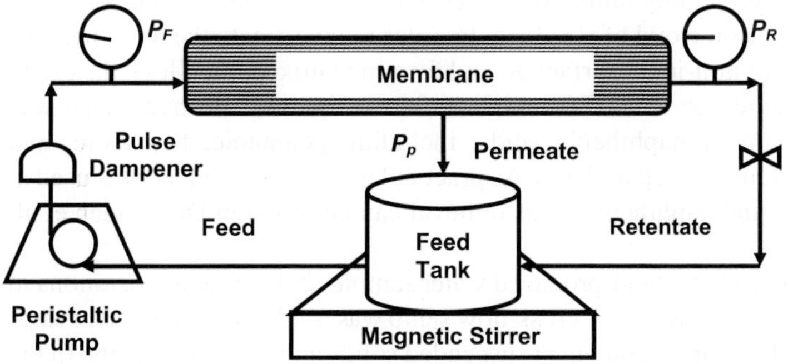

FIGURE 7.3

Micellar-enhanced ultrafiltration (MEUF) experimental setup.

inner and the outer sides of the hollow fibers/channels. The transmembrane pressure (TMP) is the driving force for the permeate flux and can be calculated from Eq. E1, since the permeate side is open to atmosphere.

$$TMP = \frac{P_F + P_R}{2} \tag{E1}$$

The flow rates of the permeate and the retentate were determined by measuring the time required to collect certain volumes of these streams in a graduated cylinder. The recovery ratio, *r*, was defined as the ratio of permeate to feed volumetric flow rates according to Eq. E2 (Seader and Henley 2006). The recovery ratio varies between 0 and 1 and indicates the fraction of the inlet feed that can be collected as treated water. In the experimental apparatus, the permeate and the retentate streams were recycled back to the feed tank in order to ensure a steady-state operation. A magnetic stirrer was also employed in the feed tank to maintain homogeneous feed. The percent rejection (%*R*) of the contaminants and the surfactant was calculated using Eq. E3 (Dunn et al. 1985). The concentrations of the contaminants and the surfactant in the feed, permeate and retentate streams, were measured as outlined below.

$$r = \frac{\dot{V}_P}{\dot{V}_F} \tag{E2}$$

$$\%R = \left(1 - \frac{C_{i,p}}{C_{i,f}}\right) \times 100 \tag{E3}$$

Each experiment lasted about one hour, with a sampling time of approximately 15 minutes. Samples of feed, permeate, and retentate were collected at a time interval to satisfy steady-state conditions at each data point. Three independent runs were performed for all experiments, and the 95% confidence intervals are shown in the figures and tables.

Experimental procedure

The initial experiments focused on treating synthetically produced water samples prepared by adding organic contaminants to DI-water. Contaminant types and concentrations were typical of field-produced water and involved aromatic compounds and naphthenic acids (Chalaturnyk et al. 2002; Rogers et al. 2002; Carey et al. 1992; Giesbrecht et al. 1992; Janks and Cadena 1992). Removal of *p*-xylene, a model aromatic contaminant, as a sole contaminant was studied first using polymeric PAN hollow fiber membranes. CPC was chosen as a surfactant since it has a low CMC and a high-dissolution capacity for organic compounds. Moreover, CPC is a cationic surfactant, so it better interacts with the negative electron cloud of *p*-xylene. In order to investigate the hypothesis of naphthenic acid mimicking the role of non-ionic surfactant and forming "mixed micelles," the effect of adding octanoic acid to the synthetic *p*-xylene/CPC sample was investigated. This hypothesis was further scrutinized using a spectrum of naphthenic acids, including pentanoic, heptanoic, and octanoic acids as well as a separate sample prepared by NAs practical grade. No *p*-xylene was used at this stage. More details on *p*-xylene and naphthenic acids removal can be found in Deriszadeh et al. (2008; 2009).

Part II of the investigation considered two field-produced water samples belonging to operations in Alberta (Canada) and California (United States). The cross-flow setup was used, but pretreatment steps were deemed necessary since the field samples contained suspended solids and free oil droplets. In the

MEUF part of the treatment, the pretreated field streams were mixed with 10 mM CPC and then passed through polymeric and ceramic membranes. More details on these experiments can be found in Deriszadeh et al. (2010).

The regeneration of the membrane, following each UF or MEUF run, was carried out by flushing the membrane with a CPC micellar solution to remove any possible deposited organic solutes followed by flushing with DI-water. The DI-water flux after regeneration was compared with the original DI-water flux, and a decision was made to terminate regeneration accordingly.

2.3.2 Batch experiments

The batch runs were performed using the same setup shown in Figure 7.3; however, the permeate stream was collected in a separate vessel, and the initial feed was concentrated by returning the retentate stream to the feed tank. These runs involved a feed solution containing only CPC in DI-water. For these experiments, TMP was kept constant, and the permeate flux and $\%R$ were recorded as functions of the concentration factor. The concentration factor for the batch experiments was calculated following Eq. E4 (Seader and Henley 2006).

$$CF = \frac{V_{F,0}}{V_{F,t}} \tag{E4}$$

The steady fluxes from the two separate batch experiments at constant transmembrane pressures were modeled to calculate the compressibility index, n, for the concentration polarization (CP) layer. The resistance of the CP layer, R_c, for each constant TMP batch experiment can be calculated from Eq. E5 using the steady flux, J, viscosity, μ, and the intrinsic membrane resistance, R_m. The R_c values are then used to determine the corresponding α_0 using Eq. E6. The mass of the deposited materials on the surface of the membrane, m, in the CP layer was found from the mass balance, knowing the initial and final volumes and the corresponding CPC concentration in the feed tank. The effective surface areas of the membranes, A, were provided by the manufacturers and these are shown in Table 7.1. The compressibility index, n, was found from the slope of the log-log plot of α_0 versus the corresponding TMP as given from Eq. E7. Two steady flux data points obtained from two separate batch experiments, at constant TMP, were used to establish the log α_0 versus log TMP plot (Juang et al. 2008; Piron et al. 1995).

$$R_c = \frac{TMP}{J\mu} - R_m \tag{E5}$$

$$R_c = \frac{m}{A}\alpha_0 \tag{E6}$$

$$\alpha_0 = \alpha(TMP)^n \tag{E7}$$

2.4 Analyses

The concentration of the surfactant and the different contaminants in the feed, permeate and retentate, were measured using a total organic carbon analyzer, gas chromatography, and high-performance liquid chromatography. The total organic carbon analyzer (TOC-Ve, Mandel, ON) was calibrated and used to measure the overall concentration of organics in the feed, permeate and retentate streams.

Three replicates were prepared ,and the 95% confidence interval was determined for the TOC measurements. The concentration of the organic contaminants in field water samples was found by subtracting the corresponding CPC concentration provided by the GC from the TOC measurements. A gas chromatograph, GC (HP 6890, Agilent Technologies, ON), was used to measure the feed, permeate and retentate concentrations of CPC, pentanoic acid, heptanoic acid, and NAs practical grade. Two different GC columns were used; one to detect CPC and another for naphthenic acids. More details on GC calibration and columns can be found in Deriszadeh et al. (2008; 2009).

High-performance liquid chromatography (1100 series, Agilent Technologies, ON) having Zorbax Eclipse XDB-C18 column (Agilent Technologies, ON) was used to determine the concentrations of *p*-xylene, octanoic acid, and decanoic acid in the feed, permeate and retentate. No *p*-xylene peaks could be detected by HPLC in the presence of CPC micellar solution. It is believed that *p*-xylene molecules were encapsulated within the micelles and passed through the HPLC column without being adsorbed in the column. Therefore, methanol was added in a 1:1 volume ratio to break the micellar system and release the *p*-xylene. More details on the HPLC operation and calibration can be found elsewhere (Deriszadeh et al. 2008).

A Krüss tensiometer (Krüss K12, Krüss Instruments, NC) was used to measure the critical micelle concentration (CMC) of the surfactant solution and the mixtures of the surfactant and the organic contaminants. The surfactant solution was added to a certain volume of DI-water as an incremental volume based on logarithmic dosing. The plot of surface tension (ordinate) as a function of the logarithm of surfactant concentration (abscissa) was constructed to determine the CMC (Deriszadeh et al. 2009).

Dynamic light scattering, DLS (Zetasizer, Nano series, Malvern Instruments, PA), which is also known as Photon Correlation Spectroscopy (PCS), was used to measure the size of the CPC micelles before and after dissolution of the organic contaminants. Due to the continuous motion of the micelles, the speckle pattern changed resulting in a fluctuation in the intensity in the dark and bright areas. Moreover, an important feature of the Brownian motion during DLS measurements is that smaller particles travel more quickly than larger particles. The famous Stokes-Einstein equation defines the relationship between the size of the particles and their speed due to Brownian motion (Castillo et al. 1994; Russel 1981).

Rotational rheometer/viscometer (HAAKE RotoVisco 1 Typo 003-5363, Thermo Scientific, MA) was used to measure the viscosities of the different water samples. These viscosities were used as input to the model. Generally, aqueous surfactants with spherical micelles show Newtonian behavior; thus, viscosity can be measured by taking a ratio of shear stress to the shear rate.

3 Results and discussion

3.1 Part I: Synthetic samples

3.1.1 Removal of P*-xylene*

P-xylene as a sole contaminant

The preliminary experiments were performed on synthetically produced water having *p*-xylene as model organic contaminant. *P*-xylene belongs to the aromatic hydrocarbons, which have acute toxicity and are a component of oil and grease in produced water (Simms et al., 1992). The PAN hollow fiber membranes with different MWCO were used.

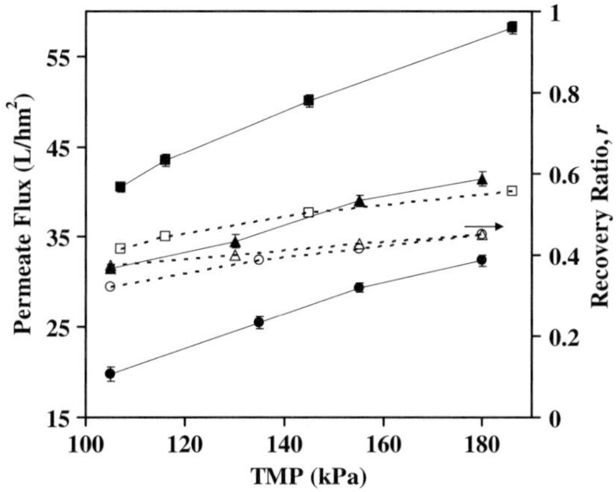

FIGURE 7.4

Permeate flux (solid symbols) and recovery ratio (open symbols) as a function of TMP for feed of (●;○) 6 kDa, (■;□)13 kDa and (▲;Δ) 50 kDa PAN membranes {[*p*-xylene] = 90 ppm, [CPC] = 24mM, 2 L feed stock stirred at 300 rpm}.

Screening experiments performed with feed containing 90 ppm of *p*-xylene just about completely dissolved in 24 mM CPC solution, showed that the 13 kDa PAN membrane gave the highest permeate flux. Figure 7.4 is a plot of the permeate flux and recovery ratio as functions of TMP for 6, 13, and 50 kDa polymeric membranes. The low permeate flux of the 6 kDa membrane is attributed to a very small pore size, while the low permeate flux of the 50 kDa one is attributed to internal pore plugging (Markels et al., 1995). Figure 7.5 is a schematic representation of internal pore plugging and back contamination arising from employing membranes with large pore sizes. In order to support this hypothesis, the size of the 90 ppm *p*-xylene in 24 mM CPC micelles was measured using DLS and compared to the membrane pore size. Figure 7.6 shows an increase in the average size of CPC micelles from an initial value of about 1 nm to 5 nm upon dissolution of *p*-xylene in the micelles. As per the manufacturer, the 6 and 13 kDa MWCOs have mean pore sizes of 0.6 and 1.3 nm, and hence, were sufficiently small to block passage of the 5 nm CPC micelles containing the *p*-xylene. On the other hand, the average pore size of the 50 kDa membrane, 5 nm, is comparable to the average micellar size. Furthermore, the back contamination shown schematically in Figure 7.5 was confirmed by the rejection data of CPC and *p*-xylene as functions of TMP shown in Table 7.2.

In summary, the screening experiments led to the following conclusions. There exists an optimum membrane pore size, depending on the type and size of the solutes in produced water, above which

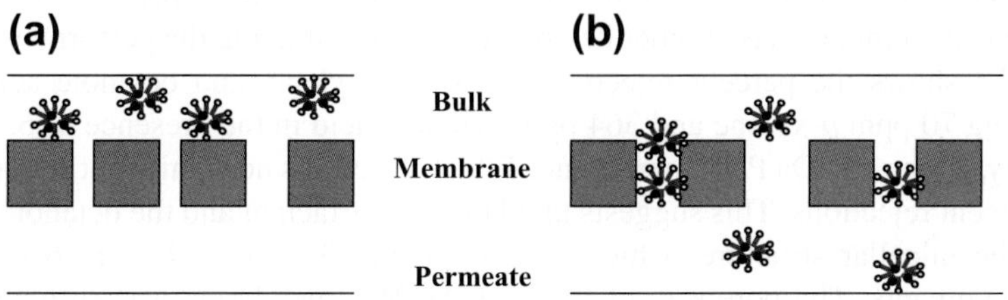

FIGURE 7.5

Schematic representation of (a) efficient rejection of the solutes with appropriate pore size membrane, and (b) the phenomenon of internal pore plugging. (For color version of this figure, the reader is referred to the online version of this book.)

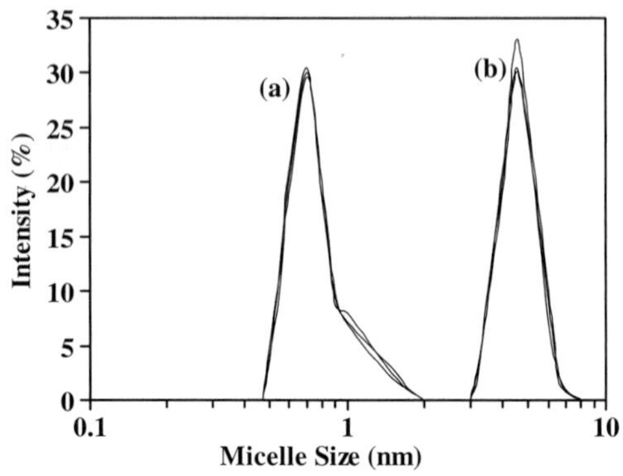

FIGURE 7.6

DLS measurements of micelle size of (a) 24 mM CPC solution and (b) 24 mM CPC solution containing 90 ppm *p*-xylene (3 replicates are shown for each measurement).

pore plugging and flux decline occur. The increase in the size of the micelles in the presence of organic contaminants allows for using larger pore size membranes, which, in turn, provides higher permeate flux. On the other hand, a membrane with efficient rejection and the highest possible pore size is most desirable.

Removal of *P*-xylene in the presence of an amphiphilic co-contaminant

When 2.5 mM (364 ppm) octanoic acid was added to the synthetically produced water sample, it was possible to dissolve up to 70 ppm *p*-xylene in 12 mM CPC solution. The higher dissolution capacity of CPC, when mixed with octanoic acid, is believed to be due to expansion in the size of the CPC micelles leading to a larger hydrocarbon core (Rosen, 2004). Moreover, the critical micelle concentration (CMC) of CPC surfactant decreased in the presence of octanoic acid, as shown in Figure 7.7. Higher contaminant dissolution capacity and lower CMC of CPC created by the octanoic acid co-contaminant resulted in lowering CPC feed concentration in order to maximize permeate flux. Figure 7.8 shows that, indeed, decreasing the CPC concentration from 24 to 6 mM improved the permeate flux and recovery ratio significantly over the whole range of operating TMP, which is believed to be due to lower CP layer resistance (Deriszadeh et al. 2008). Nevertheless, the percent rejection of CPC surfactant and organic contaminants is another key parameter in evaluating the performance of a filtration process. Table 7.3 shows the percent rejection of CPC, *p*-xylene, and octanoic acid for synthetic samples containing 70 ppm *p*-xylene and 364 ppm octanoic acid in the presence of 6, 12, and 24 mM CPC, respectively, by the 13 kDa PAN membrane. Table 7.3 reveals no significant difference in CPC or octanoic acid percent rejections. This suggests that both the surfactant and the octanoic acid integrated favorably into the micellar structure, which got larger and formed at lower concentration in the presence of contaminants. The pore size of the 13 kDa PAN membrane was sufficient to reject the resultant micelles. Naphthenic acids are amphiphilic compounds and are capable of stabilizing colloidal structures at the interface of hydrophilic/hydrophobic media (Acevedo et al. 1999; Marquez et al. 1995). It is anticipated that they would be accommodated in the CPC micelles, shield the repulsive forces between the positively charged head groups of the surfactant, and, consequently,

Table 7.2 Percent Rejection and Average Permeate Concentration of CPC and *p*-xylene using 6, 13, and 50 kDa Polymeric Membranes for MEUF Experiments on 90 ppm *p*-xylene in 24 mM CPC

		6 kDa Polymeric Membrane		13 kDa Polymeric Membrane		50 kDa Polymeric Membrane	
TMP (kPa)		CPC	*p*-xylene	CPC	*p*-xylene	CPC	*p*-xylene
105	$\%R_i$	99.8 ± 0.11	96.4 ± 0.51	99.7 ± 0.08	95.0 ± 0.15	92.9 ± 0.37	91.9 ± 0.56
	$[i]$ [a]	0.14 ± 0.01	7.9 ± 0.09	0.19 ± 0.07	8.8 ± 0.15	4.50 ± 1.28	14.5 ± 0.08
135	$\%R_i$	99.7 ± 0.09	95.4 ± 0.40	99.5 ± 0.67	94.4 ± 0.06	91.5 ± 0.16	88.8 ± 1.01
	$[i]$ [a]	0.16 ± 0.02	8.6 ± 0.06	0.24 ± 0.10	9.6 ± 0.08	5.07 ± 1.22	19.1 ± 0.12
165	$\%R_i$	99.7 ± 0.08	95.1 ± 0.42	99.5 ± 0.32	93.7 ± 0.07	80.7 ± 0.62	85.9 ± 0.91
	$[i]$ [a]	0.12 ± 0.02	8.7 ± 0.06	0.23 ± 0.06	9.4 ± 0.10	10.80 ± 2.28	22.5 ± 0.10
190	$\%R_i$	99.7 ± 0.10	94.4 ± 0.35	99.3 ± 0.09	92.6 ± 0.10	74.1 ± 0.67	82.5 ± 0.62
	$[i]$ [a]	0.12 ± 0.02	9.1 ± 0.06	0.28 ± 0.12	9.8 ± 0.07	13.80 ± 2.20	27.2 ± 0.06

[a] *[i] is the species average concentration in the permeate stream. CPC [=] mM, p-xylene [=] ppm.*

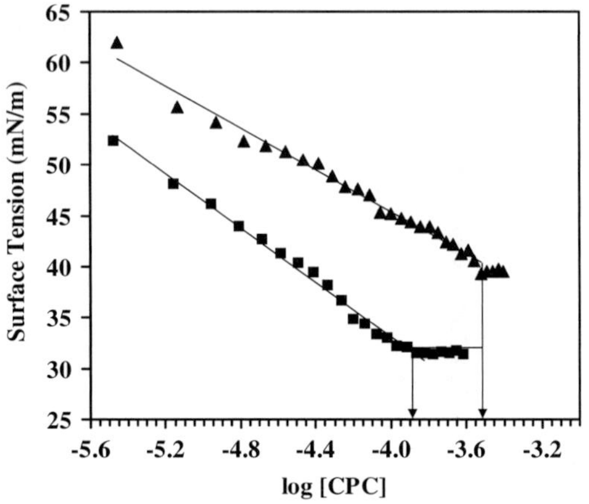

FIGURE 7.7

Surface tension as a function of CPC concentration for (▲) CPC, CMC = 0.32 mM and (■) mixed CPC and octanoic acid, CMC= 0.14 mM.

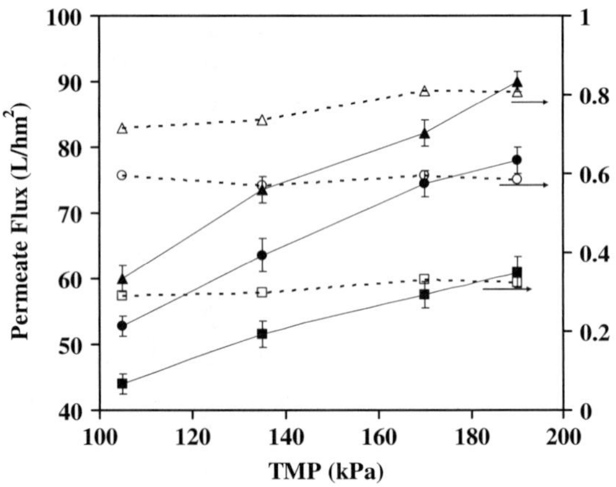

FIGURE 7.8

Permeate flux (solid symbols) and recovery ratio (open symbols) as a function of TMP for feeds containing 70 ppm *p*-xylene in (■;□) 24 mM CPC, (●;○) 12 mM CPC and 364 ppm octanoic acid, and (▲;Δ) 6 mM CPC and 364 ppm octanoic acid {13 kDa membrane; 2 L feed stock stirred at 300 rpm}.

reduce the CMC of the surfactant. Similar observations on reduction of CMC and increase in aggregation number upon the addition of an electrolyte were reported by Rosen (2004), Nagarajan and Ruckenstein (1991), and Stasiuk and Schramm (1996). However, the *p*-xylene percent rejection decreased from 97% for the feed containing 24 mM CPC to about 90% when CPC concentration was decreased to 6 mM. It seems likely that at 6 mM CPC, no sufficient space was available to accommodate all *p*-xylene molecules and some remained undissolved in the micelles. The *p*-xylene percent rejection at 12 mM CPC feed was approximately equal to that of the 24 mM CPC, suggesting that 12 mM CPC accommodated most of *p*-xylene and octanoic acid at their respective concentrations.

The conclusion drawn from considering the flux and the rejection results as functions of TMP is that the optimum feed chemistry corresponded to 12 mM CPC, 364 ppm octanoic acid and 70 ppm

Table 7.3 Percent Rejection and Average Permeate Concentration of CPC, *p*-xylene, and Octanoic Acid using 13 kDa Polymeric Membrane for MEUF Experiments {*p*-xylene = 70 ppm, CPC = 24, 12, and 6 mM, Octanoic Acid ($C_8H_{16}O_2$) = 364 ppm}

		70 ppm *p*-xylene in 24 mM CPC		70 ppm *p*-xylene in 12 mM CPC and 364 ppm Octanoic Acid			70 ppm *p*-xylene in 6 mM CPC and 364 ppm Octanoic Acid		
TMP (kPa)		CPC	*p*-xylene	CPC	*p*-xylene	$C_8H_{16}O_2$	CPC	*p*-xylene	$C_8H_{16}O_2$
105	$\%R_i$	99.6 ± 0.02	97.1 ± 0.06	99.3 ± 0.07	94.1 ± 0.10	99.3 ± 0.14	99.1 ± 0.06	90.3 ± 0.19	98.8 ± 0.19
	$[i]$ [a]	0.20 ± 0.05	4.9 ± 0.10	0.27 ± 0.11	7.0 ± 0.12	2.15 ± 0.51	0.16 ± 0.04	9.5 ± 0.11	4.1 ± 0.41
135	$\%R_i$	99.5 ± 0.06	96.6 ± 0.07	99.5 ± 0.08	94.2 ± 0.03	99.6 ± 0.13	99.2 ± 0.04	89.2 ± 0.13	99.1 ± 0.1
	$[i]$ [a]	0.22 ± 0.07	5.1 ± 0.06	0.22 ± 0.09	7.2 ± 0.08	1.15 ± 0.14	0.13 ± 0.06	10.3 ± 0.15	2.8 ± 0.16
165	$\%R_i$	99.5 ± 0.02	95.9 ± 0.09	99.7 ± 0.06	93.5 ± 0.02	99.7 ± 0.12	99.4 ± 0.21	88.1 ± 0.06	99.5 ± 0.15
	$[i]$ [a]	0.22 ± 0.05	5.6 ± 0.09	0.16 ± 0.04	7.6 ± 0.07	1.04 ± 0.03	0.08 ± 0.06	10.5 ± 0.05	1.7 ± 0.49
190	$\%R_i$	99.4 ± 0.07	95.1 ± 0.08	99.6 ± 0.12	93.4 ± 0.17	99.6 ± 0.14	99.1 ± 0.19	87.1 ± 0.04	99.4 ± 0.20
	$[i]$ [a]	0.24 ± 0.10	6.8 ± 0.07	0.19 ± 0.05	7.9 ± 0.24	1.02 ± 0.04	0.09 ± 0.06	11.3 ± 0.06	2.1 ± 0.11

[a] *[i] is the species average concentration in the permeate stream. CPC [=] mM, p-xylene and $C_8H_{16}O_2$ [=] ppm.*

p-xylene. This optimum composition entailed formation of larger micelles at lower CMC and consequently led to 20% higher permeate flux compared with the feed containing 24 mM CPC, while maintaining a similar percent rejection. The CPC and octanoic acid percent rejections, on the other hand, were independent of *p*-xylene feed concentration.

3.1.2 Removal of naphthenic acids

Effect on CMC and micellar size

In order to provide a more detailed study of the role of amphiphilic contaminants in the performance of MEUF, synthetically produced water having a spectrum of naphthenic acids, typically existing in produced water, were run in a MEUF setup. The first step was to confirm that naphthenic acids, regardless of their carbon number and molecular weight, show the same effect of reducing CMC and increasing micellar size of the CPC surfactant as did the octanoic acid. In order to test this hypothesis, several stock solutions of 6 mM CPC containing 0.32 mM of individual acyclic naphthenic acids, 0.32 mM of a mixture of acyclic naphthenic acids ($C_5H_{10}O_2$, $C_7H_{14}O_2$, $C_8H_{16}O_2$ and $C_{10}H_{20}O_2$, 0.08 mM each), and 0.32 mM NAs practical grade were prepared for CMC measurements.

As Table 7.4 shows, all naphthenic acids decreased the CMC of CPC, and the ability to reduce the CMC increased with increasing the carbon number of the naphthenic acid. Moreover, it seems that the effect of coexistence of naphthenic acids on the CMC reduction is cumulative. Table 7.4 also quotes the concentration of naphthenic acids in the aqueous solution that corresponds to CMC. This concentration decreased as the carbon number increased, leading to the conclusion that longer chain naphthenic acids are more effective in saturating the air–water interface, forcing more of the surfactant molecules into the bulk solution. Naphthenic acids may also reduce the CMC of CPC by participating at the micellar interface, thus screening the repulsion forces among the surfactant head groups (Stasiuk and Schramm 1996). In addition to decreasing the CMC of CPC, the DLS measurements of Figures 7.9 and 7.10 show an increase in the micelle size when naphthenic acids are added. The DLS measurements used solutions of 6 mM CPC containing individual or mixtures of naphthenic acids at 0.32 mM concentrations. Also, the DLS trend in Figure 7.9 shows that the size of the CPC micelles increased as the carbon number of the naphthenic acid increased. The increase in the CPC micellar size further supports the hypothesis of naphthenic acids participating at the micellar interface, thus screening the repulsion forces among the surfactant head groups. This leads to increasing the aggregation number of

Table 7.4 Critical Micelle Concentrations (CMC) of CPC Solutions Containing Different Naphthenic Acids

Contents	$[C_nH_{2n+Z}O_2]$ at CMC (mM)	CMC (mM)
CPC (control sample)		0.32 ± 0.01
CPC and $C_5H_{10}O_2$	0.0079	0.14 ± 0.02
CPC and $C_7H_{14}O_2$	0.0071	0.13 ± 0.01
CPC and $C_8H_{16}O_2$	0.0064	0.12 ± 0.01
CPC and $C_{10}H_{20}O_2$	0.0012	0.016 ± 0.006
CPC in a mixture of $C_5H_{10}O_2$, $C_7H_{14}O_2$, $C_8H_{16}O_2$ and $C_{10}H_{20}O_2$	0.0017	0.019 ± 0.006
CPC in NAs practical grade	0.0008	0.018 ± 0.005

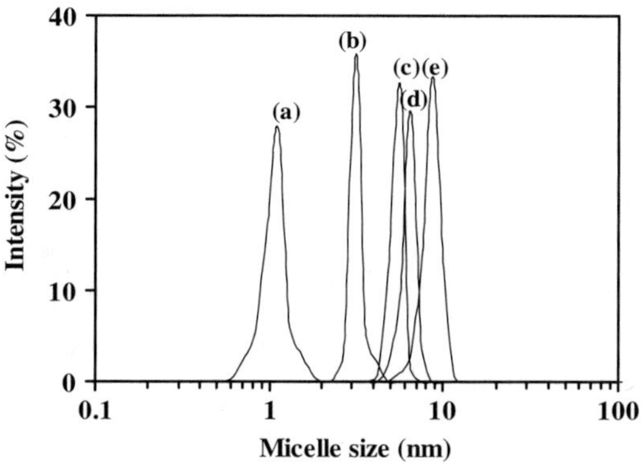

FIGURE 7.9

DLS measurements of micelle size of (a) 6 mM CPC, and a mixture of 6 mM CPC and (b) 0.32 mM $C_5H_{10}O_2$, (c) 0.32 mM $C_7H_{14}O_2$, (d) 0.32 mM $C_8H_{16}O_2$, and (e) 0.32 mM $C_{10}H_{20}O_2$.

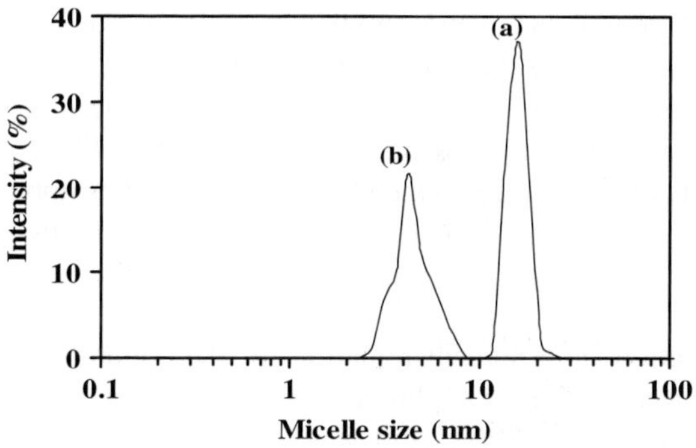

FIGURE 7.10

DLS measurements of micelle size of (a) feed A: 300 ppm mixture of acyclic naphthenic acids: ($C_5H_{10}O_2$, $C_7H_{14}O_2$, $C_8H_{16}O_2$ and $C_{10}H_{20}O_2$) 75 ppm each in 10 mM CPC, and (b) feed B: 300 ppm of NAs practical grade in 10 mM CPC.

the CPC micelles by bonding more of the CPC monomers in the micellar structure. Figure 7.10 shows the increase in the CPC micelle size upon dissolving mixtures of naphthenic acids. A decrease in CMC and an increase in micellar size promote solubilization of otherwise less soluble organics at lower CPC concentrations.

Effect on the removal

Experiments were carried out for the removal of mixtures of acyclic and cyclic naphthenic acids from synthetically produced water. Two different synthetically produced water samples were prepared: One contained 300 ppm mixture of equal concentration of 75 ppm of $C_5H_{10}O_2$, $C_7H_{14}O_2$, $C_8H_{16}O_2$ and $C_{10}H_{20}O_2$ in 10 mM CPC solution (Feed A), and the other contained 300 ppm NAs practical grade, which according to the manufacturer contains cyclic and acyclic naphthenic acids, in 10 mM CPC solution (Feed B). These concentrations were typical of produced water samples (Rogers et al. 2002). The concentration of naphthenic acids in produced water may significantly vary from field to field, and the concentrations used here were representative (Rogers et al. 2002).

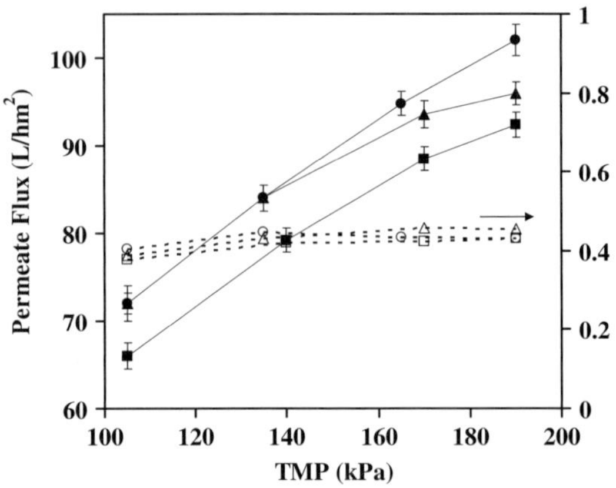

FIGURE 7.11

Permeate flux (solid symbols) and recovery ratio (open symbols) as a function of TMP for (■;□) feed A: 300 ppm mixture of acyclic naphthenic acids: ($C_5H_{10}O_2$, $C_7H_{14}O_2$, $C_8H_{16}O_2$ and $C_{10}H_{20}O_2$) 75 ppm each in 10 mM CPC, (▲;Δ) feed B: 300 ppm NAs practical grade in 10 mM CPC, and (●;O)10 mM CPC (Control sample).

Figure 7.11 shows the permeate flux and recovery ratio as functions of TMP for Feeds A and B and a control sample of 10 mM CPC. The permeate flux for Feeds A and B tend to show a fairly linear behavior. Nonetheless, deviation from the straight line is observed for TMPs higher than approximately 170 kPa. The permeate flux for Feed A, which contained acyclic naphthenic acids, was smaller than that of NAs practical grade. The DLS measurements above show that CPC micelles are larger for Feed A than Feed B. Dissolution of straight chain naphthenic acids in Feed A most likely changes the shape of the CPC micelles from spherical to elongated-cylindrical, which can cause extra resistance to the flow and lowers the permeate flux. Similar effects of the change in the shape of the micelles upon dissolution of long-chain solubilizates was reported by Rosen (2004). The recovery ratio for the two feeds was essentially constant, suggesting that the recovery ratio was mostly affected by CPC concentration.

Table 7.5 shows the average permeate concentrations and percent rejection for the components used in the two synthetic feed samples. Permeate concentration of naphthenic acids decreased as the carbon number increased. The permeate concentration of pentanoic acid was the highest, whereas those of decanoic acid and NAs practical grade were the lowest. Nonetheless, there was not a big difference in the CPC permeate concentration for Feeds A and B. An increasing trend in percent rejection of acyclic naphthenic acids can be related to the increase in the molecular weight, in accordance with the decrease in the solubility trend of these species in water (Yepez 2007). Consequently, the percent removal of pentanoic acid was the lowest. NAs practical grade, which was a mixture of higher-molecular-weight naphthenic acids (carbon number > 11) with higher hydrophobicity, tended to dissolve in the micelles to a higher extent. Percent rejection as high as 99.8% was observed for NAs practical grade. The percent rejection of CPC and NAs were independent of TMP probably because the membrane pores of the 13 kDa membrane were small enough to be effective in filtering out these contaminants when considering their micellar size.

3.2 Part II: Field samples

3.2.1 Effect on CMC and micellar size

The existence of suspended solids and free oil droplets in the field samples mandated an ultrafiltration preliminary treatment step using a 50 kDa PAN membrane followed by a 50 kDa ceramic membrane.

Table 7.5 Percent Rejection and Average Permeate Concentration of CPC, $C_5H_{10}O_2$, $C_7H_{14}O_2$, $C_8H_{16}O_2$, $C_{10}H_{20}O_2$ and NAs Practical Grade for 13 kDa Polymeric Membrane and Two Synthetic Produced Water Samples; feed A: 300 ppm Mixture of Acyclic Naphthenic Acids ($C_5H_{10}O_2$, $C_7H_{14}O_2$, $C_8H_{16}O_2$ and $C_{10}H_{20}O_2$) 75 ppm Each in 10 mM CPC, and Feed B: 300 ppm NAs Practical Grade in 10 mM CPC

		Feed A: 300 ppm (75 ppm each $C_5H_{10}O_2$, $C_7H_{14}O_2$, $C_8H_{16}O_2$ and $C_{10}H_{20}O_2$) in 10 mM CPC					Feed B: 300 ppm NAs practical grade in 10 mM CPC	
TMP (kPa)		CPC	$C_5H_{10}O_2$	$C_7H_{14}O_2$	$C_8H_{16}O_2$	$C_{10}H_{20}O_2$	CPC	NAs
105	$\%R_i$	97.2 ± 0.06	81.5 ± 2.53	94.5 ± 0.31	98.2 ± 0.15	99.6 ± 0.11	97.1 ± 0.01	99.8 ± 0.02
	$[i]$ [a]	0.28 ± 0.02	13.83 ± 1.91	4.14 ± 0.22	1.37 ± 0.12	0.27 ± 0.08	0.29 ± 0.04	0.84 ± 0.01
135	$\%R_i$	97.5 ± 0.02	85.7 ± 1.18	94.1 ± 0.39	92.1 ± 0.61	99.2 ± 0.10	97.2 ± 0.02	99.7 ± 0.01
	$[i]$ [a]	0.25 ± 0.01	10.74 ± 0.88	4.49 ± 0.29	5.98 ± 0.46	0.59 ± 0.07	0.28 ± 0.05	0.85 ± 0.01
165	$\%R_i$	97.7 ± 0.02	86.7 ± 0.21	91.4 ± 0.36	87.3 ± 0.56	98.1 ± 0.05	97.1 ± 0.01	99.7 ± 0.02
	$[i]$ [a]	0.23 ± 0.01	9.95 ± 0.16	6.42 ± 0.27	9.53 ± 0.42	1.49 ± 0.04	0.28 ± 0.03	0.84 ± 0.01
190	$\%R_i$	97.7 ± 0.02	87.8 ± 0.32	89.8 ± 0.47	88.8 ± 0.18	97.6 ± 0.05	97.2 ± 0.03	99.7 ± 0.01
	$[i]$ [a]	0.23 ± 0.01	9.11 ± 0.24	7.69 ± 0.35	8.42 ± 0.14	1.81 ± 0.04	0.27 ± 0.07	0.81 ± 0.01

[a] *[i] is the species average concentration in the permeate stream. CPC [=] mM, naphthenic acid [=] ppm.*

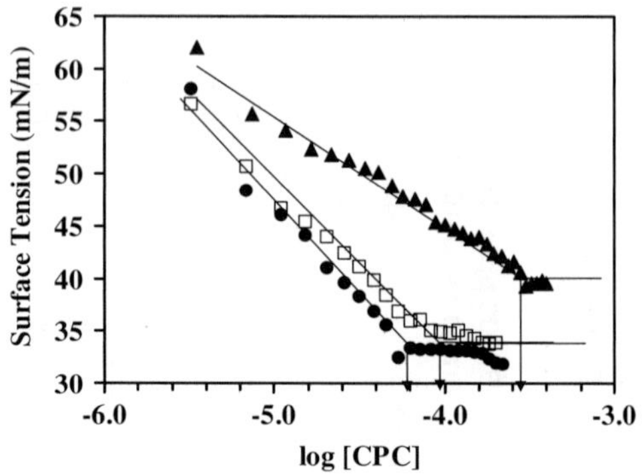

FIGURE 7.12

Surface tension as a function of CPC concentration for (▲) CPC, CMC = 0.32 mM, (●) mixed CPC and Canadian field sample, CMC = 0.07 mM, and, (□) mixed CPC and the US field sample, CMC = 0.08 mM.

This step was performed in a batchwise mode and at a constant TMP of about 150 kPa. Minor reduction in total organic carbon, from 103 to 80 ppm for Field A sample and from 78 to 66 ppm for Field B sample, resulted. The samples looked clear after filtration. GC analyses of the pretreated Field A and B samples showed peaks at approximate retention times observed for naphthenic acids having a carbon number between 7 and 10. Surface tension measurements in Figure 7.12 show that the critical micelle concentration of CPC was reduced from 0.32 mM in DI-water to 0.07 mM in Field A solution and to 0.08 mM in Field B solution. Moreover, DLS measurements in Figure 7.13 show that the average size of CPC micelles increased from 1 nm in DI-water to 6 nm in Field A solution and to 7 nm in Field B solution.

3.2.2 Removal runs

Feeds consisting of preliminary treated Fields A and B were mixed with CPC to a final CPC concentration of 10 mM and introduced into the MEUF setup. Figures 7.14(a) and (b) show the permeate

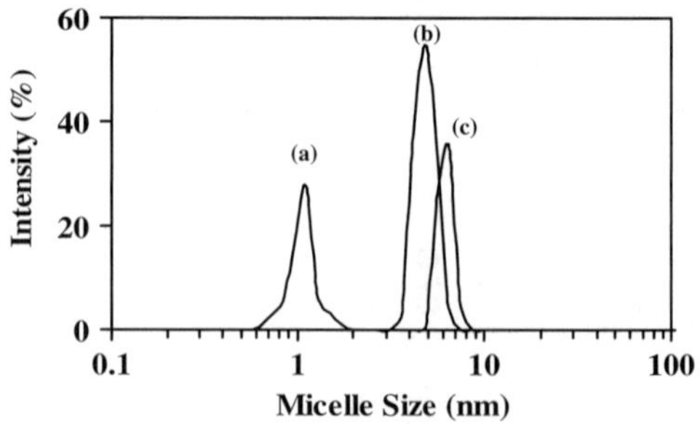

FIGURE 7.13

DLS measurements of micelle size of (a) 10 mM CPC, and a mixture of 10 mM CPC and permeate of step 2 for (b) Field A sample, and (c) Field B sample.

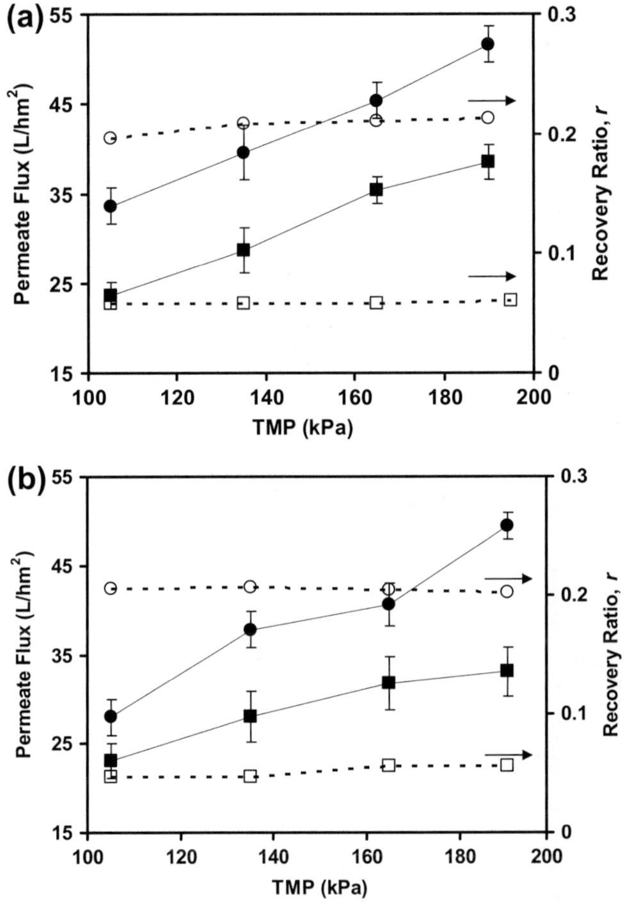

FIGURE 7.14

Permeate flux (solid symbols) and recovery ratio (open symbols) as functions of TMP for Step 3 treatment using (●;○) 13 kDa polymeric, (■;□) 50 kDa ceramic on (a) Field A sample, and (b) Field B sample {[CPC] = 10 mM}.

flux and recovery ratio as functions of TMP for Field A and Field B samples, respectively, with the 13 kDa PAN and the 50 kDa ceramic membranes. Actually, these two membranes have been shown to have similar average pore sizes (Deriszadeh et al. 2010). The higher permeate flux from the polymeric PAN can be related to the relatively higher hydrophilicity of PAN in a hollow fiber structure module as well as larger effective surface area of the 13 kDa PAN membrane. Consequently, the polymeric membrane gave higher recovery ratios. On the other hand, higher percent rejections for the CPC and organic contaminants were achieved by the 50 kDa ceramic membrane, as indicated in Table 7.6. Regardless of the membrane type, MEUF treatment gave remarkably higher removal of the organics when compared to the ultrafiltration pretreatment.

It is evident from Table 7.6 that higher percent rejection of CPC, and subsequently lower back contamination, was observed for the runs involving the field samples—that is, in the presence of contaminants, as opposed to 10 mM CPC in DI-water feed. In addition, the 50 kDa ceramic membrane was more effective in rejecting the organic contaminants at high TMPs, probably due to the

Table 7.6 Percent Rejection and Average Permeate Concentration of CPC and Organic Contaminants Following MEUF Treatment using 13 kDa Polymeric and 50 kDa Ceramic Membranes for the Control Sample of 10 mM CPC and the UF Permeates of Field A and B Samples Mixed in 10 mM CPC

		13 kDa Polymeric 10 mM CPC (Control Sample)	50 kDa Ceramic 10 mMCPC (Control Sample)	13 kDa Polymeric 10 mM CPC Mixed Field A Sample		50 kDa Ceramic 10 mM CPC Mixed in Field A Sample		13 kDa Polymeric 10 mM CPC Mixed in Field B Sample		50 kDa Ceramic 10 mM CPC Mixed in Field B Sample	
TMP (kPa)		**CPC**	**CPC**	**CPC**	**Organics**	**CPC**	**Organics**	**CPC**	**Organics**	**CPC**	**Organics**
105	$\%R_i$	96.90 ± 0.05	96.60 ± 0.06	98.10 ± 0.07	96.10 ± 0.05	97.70 ± 0.05	93.8 ± 0.6	98.20 ± 0.05	96.2 ± 0.1	99.70 ± 0.02	97.7 ± 0.1
	$[i]^a$	0.310 ± 0.004	0.34 ± 0.02	0.190 ± 0.002	2.7 ± 0.3	0.23 ± 0.03	4.3 ± 0.8	0.18 ± 0.06	2.7 ± 0.2	0.020 ± 0.006	1.8 ± 0.5
135	$\%R_i$	96.80 ± 0.01	96.40 ± 0.02	97.9 ± 0.1	94.20 ± 0.05	97.40 ± 0.04	84.1 ± 0.7	98.00 ± 0.04	91.3 ± 0.1	99.40 ± 0.03	98.1 ± 0.1
	$[i]^a$	0.320 ± 0.005	0.36 ± 0.02	0.200 ± 0.007	4.1 ± 0.1	0.26 ± 0.07	11.1 ± 0.6	0.25 ± 0.03	5.8 ± 0.4	0.06 ± 0.01	1.4 ± 0.4
165	$\%R_i$	96.70 ± 0.01	96.30 ± 0.02	97.60 ± 0.08	88.90 ± 0.04	96.10 ± 0.04	85.4 ± 0.2	97.50 ± 0.06	94.90 ± 0.05	98.60 ± 0.05	96.20 ± 0.05
	$[i]^a$	0.320 ± 0.001	0.37 ± 0.01	0.230 ± 0.005	8.4 ± 0.2	0.30 ± 0.02	10.2 ± 0.6	0.29 ± 0.05	3.5 ± 0.2	0.14 ± 0.05	2.9 ± 0.4
190	$\%R_i$	96.80 ± 0.01	95.80 ± 0.02	97.30 ± 0.05	80.90 ± 0.09	96.60 ± 0.02	79.5 ± 0.3	97.4 ± 0.8	92.30 ± 0.05	97.90 ± 0.04	95.80 ± 0.05
	$[i]^a$	0.310 ± 0.004	0.42 ± 0.03	0.260 ± 0.006	13.4 ± 0.8	0.34 ± 0.02	14.4 ± 0.5	0.31 ± 0.02	5.4 ± 0.1	0.21 ± 0.02	4.4 ± 0.7

[a] *[i] is the species average concentration in the permeate stream. CPC [=] mM, organic contaminants [=] ppm.*

hydrophobic nature of membrane material. In general, industry prefers the ceramic membranes because of their high durability, high thermal resistance, and relatively easier cleaning protocols.

4 Modeling

Darcy's law is commonly used to describe fluid flow in porous media (Gray and Miller 2004; Fetter 1999; Whitaker 1986). A typical membrane filtration, with the exception of reverse osmosis, encompasses the passage of fluids through porous materials. Hence, Darcy's law has been widely used to model flux as a function of TMP for micro and ultrafiltration processes (Chen and Kim 2006; Chen et al. 2005; Kim and Yuan 2005; Bacchin et al. 2002a,b; Sethi and Wiesner 1997; Field et al. 1995; Van der Berg and Smolders 1990). A modification to Darcy's law known as the "resistance in-series," E8, better describes the flux trend as a function of TMP (Juang et al. 2008; Rai et al. 2006; Purkait et al. 2004; Tansel et al. 2000; Cheng et al. 1997). The term $\sum R$ stands for the sum of all the resistances obstructing the flow to the permeate side, including membrane (R_m), internal pore plugging (R_{pp}), concentration polarization (R_c), gel layer (R_g), and the resistance caused when fouling occurs on the surface of the membrane (R_f) (Field et al. 1995; Davis and Leighton 1987).

$$J = \frac{(TMP)}{\mu \sum R} \tag{E8}$$

The intrinsic membrane resistance (R_m) is a property of the membrane and varies with membrane type, material, and void fraction. The membrane resistance is proportional to the reciprocal of membrane permeability. The values of R_m are typically found from the slope of the flux versus TMP for the DI-water using Eq. E8 when R_m is the only resistance (Juang et al. 2008). The R_m values for the 13 kDa polymeric PAN and 50 kDa ceramic TiO_2 membranes were found to be 2.77×10^{12} (1/m) and 2.36×10^{12} (1/m); respectively. The viscosity term in Equation E8 was set at 0.001 Pa.sec for DI-water.

Resistance due to concentration polarization (CP) originates from the retention of solute material near the surface of the membrane. Although the cross-flow operation mode creates shear that limits the growth of the CP or gel layer, no complete elimination of the CP layer is possible. The existence of the CP layer leads to a continuous decline in the flux as the TMP increases. Bacchin et al. (2002b) referred to the point where the slope starts declining as the critical flux.

The reversible CP layer can form an irreversible fouled layer with the corresponding resistance, R_f, upon increasing TMP (Van den Berg and Smolders 1990). Since the operating conditions were monitored and fouling was avoided in the current runs, R_f was not taken into account in the model. Generally, the regions where the R_f becomes effective and can cause a significant drop in the flux, and consequently shorten the lifetime of the membrane, is called the limiting flux region. This region is characterized by a constant flux that is independent of the TMP.

Internal pore plugging occurs when the particle size of the solutes is smaller than the membrane pore diameter. This phenomenon causes bridging and complete blockage of the membrane pores and diminishes the performance and lifetime of the membrane (Markels et al. 1995). The corresponding resistance, R_{pp}, can be completely eliminated by selecting a membrane with pores smaller than the solute size and made out of materials that minimize the solute–membrane interactions. Because the appropriate pore size membrane was applied for each of the produced water samples used in this study, the resistance due to internal pore plugging was assumed to be negligible.

Table 7.7 Viscosities of the Different Synthetic and Field MEUF Feed Solutions

MEUF Feed Content	Viscosity (Pa.s)
20 mM CPC	0.0024445 ± 0.00015
10 mM CPC	0.0016728 ± 0.00015
Feed A {300 ppm (75 ppm each) $C_jH_{2j}O_2$; j = 5, 7, 8, and 10 in 10 mM CPC}	0.0016576 ± 0.00022
Feed B {300 ppm NAs practical grade in 10 mM CPC}	0.0016754 ± 0.00016
Field A sample following pretreatment in 10 mM CPC	0.0035521 ± 0.00014
Field B sample following pretreatment in 10 mM CPC	0.0036158 ± 0.00014

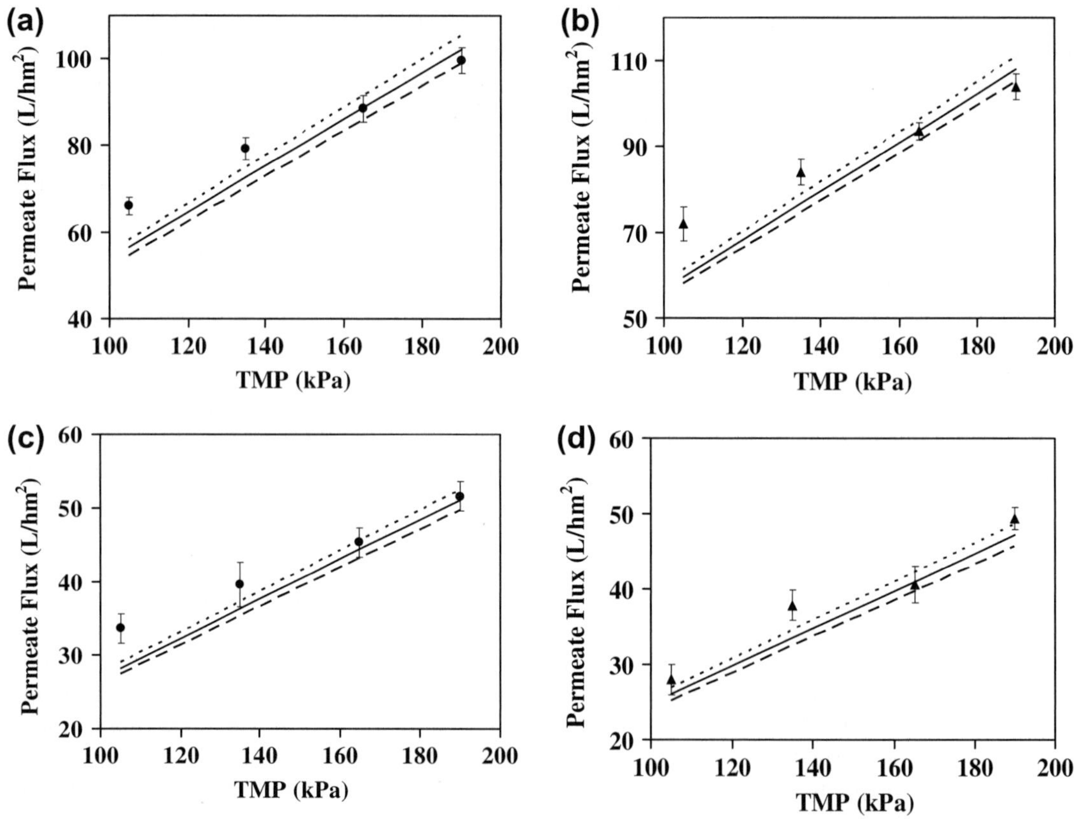

FIGURE 7.15

Permeate flux versus TMP for 13 kDa PAN membrane and feed solutions of (a) Feed A: 300 ppm (75 ppm of each $C_5H_{10}O_2$, $C_7H_{14}O_2$, $C_8H_{16}O_2$, and $C_{10}H_{20}O_2$ in 10 mM CPC, (b) Feed B: 300 ppm NAs practical grade in 10 mM CPC, (c) Field A sample, and (d) Field B sample{Lines model:(▬▬) linear model, (▬ ▬) model: (R_c) +10%, (▪ ▪ ▪) model: (R_c) -10%; Points: experiments}.

Following the discussion above, Eq. E8 can be revised to produce Eq. E9. The resistance of the membrane and that of the CP layer constitute the total resistance obstructing the flow to the permeate side.

$$J = \frac{(TMP)}{\mu(R_m + R_c)} \tag{E9}$$

In order to investigate the validity of E9, a model fit was attempted involving R_c as the only fitted parameter for every feed stream, with R_m obtained as stated above and μ independently measured for the different feeds. Results are shown in Table 7.7. At low TMPs, when the effect of R_c is small, viscosity had a major impact on the flux decline when compared with DI-water. The model fits using

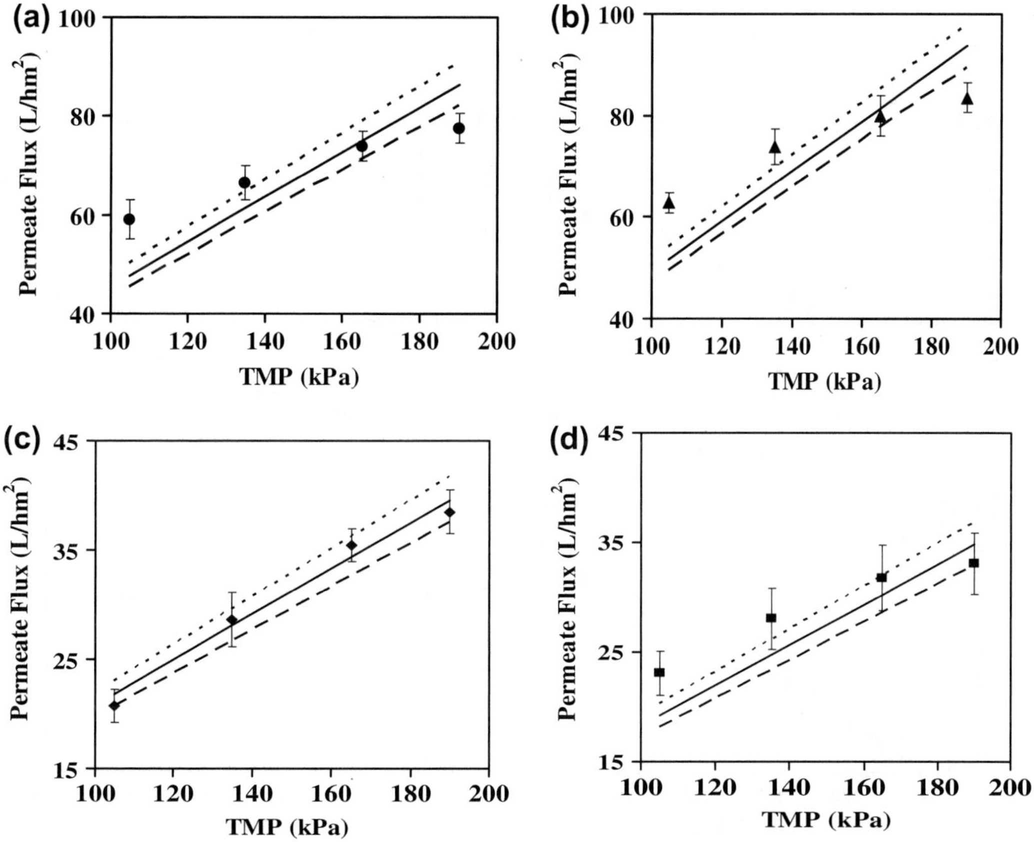

FIGURE 7.16

Permeate flux versus TMP for 50 kDa TiO_2 membrane and feed solutions of (a) Feed A: 300 ppm (75 ppm of each $C_5H_{10}O_2$, $C_7H_{14}O_2$, $C_8H_{16}O_2$, and $C_{10}H_{20}O_2$ in 10 mM CPC, (b) Feed B: 300 ppm NAs practical grade in 10 mM CPC, (c) Field A sample, and (d) Field B sample{Lines model:(——) non-linear model, (— —) model: (R_c) +10%, (- - -) model: (R_c) -10%; Points: experiments}.

Eq. E9 for synthetic Feeds A and B, and the real Field A and B samples are presented in Figures 7.15 and 7.16 for the 13 kDa polymeric and 50 kDa ceramic membranes, respectively. The model, however, did not seem to capture the experimental trends very well. This is probably because the model assumes a constant R_c for a given feed, while it is known that R_c is a function of TMP. Therefore, a second model was developed, with R_c taken as a function of TMP. Boerlage et al. (2003), Carrere et al. (2001), Piron et al. (1995), and Nakanishi et al. (1987) suggested that R_c changes with TMP, according to $\beta(TMP)^n$. Equation E9 was therefore modified to Eq. E10 below.

$$J = \frac{(TMP)}{\mu(R_m + \beta(TMP)^n)} \tag{E10}$$

The compressibility index, n, was independently estimated from the batch experiments and varies between 0 and 1 (Nakanishi et al. 1987). Steady flux data, which were reached after about 2 hours of

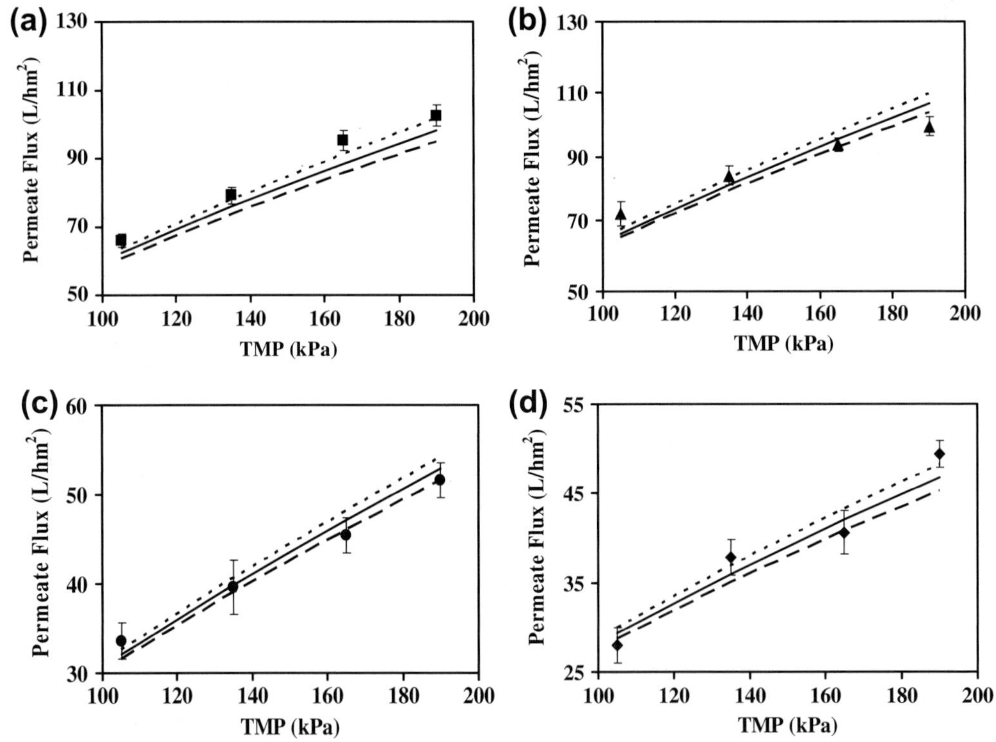

FIGURE 7.17

Permeate flux versus TMP for 13 kDa PAN membrane and feed solutions of (a) Feed A: 300 ppm (75 ppm of each $C_5H_{10}O_2$, $C_7H_{14}O_2$, $C_8H_{16}O_2$, and $C_{10}H_{20}O_2$ in 10 mM CPC solution, (b) Feed B: 300 ppm NAs practical grade in 10 mM CPC solution, (c) Field A sample, and (d) Field B sample {Lines model:(——) non-linear model, (— —) model: (β) +10%, (- - -) model: (β) -10%; Points: experiments}.

the batch experimental runs, were used to estimate n to be 0.8. Consequently, β was the only fitted parameter in Eq. E10.

The model fits, using Eq. E10, for synthetic Feeds A and B, and the real Field A and B samples are presented in Figures 7.17 and 7.18 for the 13 kDa polymeric and 50 kDa ceramic membranes, respectively. As shown in these figures, Eq. E10 better represents the experimental data points at the tested TMPs. The comparison of the flux behavior for corresponding feeds in Figures 7.15 through 7.18 confirm the ability of Eq. E10 to capture the linear and leveling response of the flux versus TMP compared to corresponding behaviors using Eq. E9.

Sensitivity analysis was carried out by investigating the effect of a 10% increase or decrease of the fitted parameter, β. The results were shown by dotted lines in Figures 7.15 through 7.18. The sensitivity analysis showed that β cannot be chosen randomly and that a 10% variation in the fitted parameter can lead to the model over- or underpredicting the experimental points. In addition, the model seems to be more sensitive to the fitted parameter at high TMPs. This again confirms that the CP

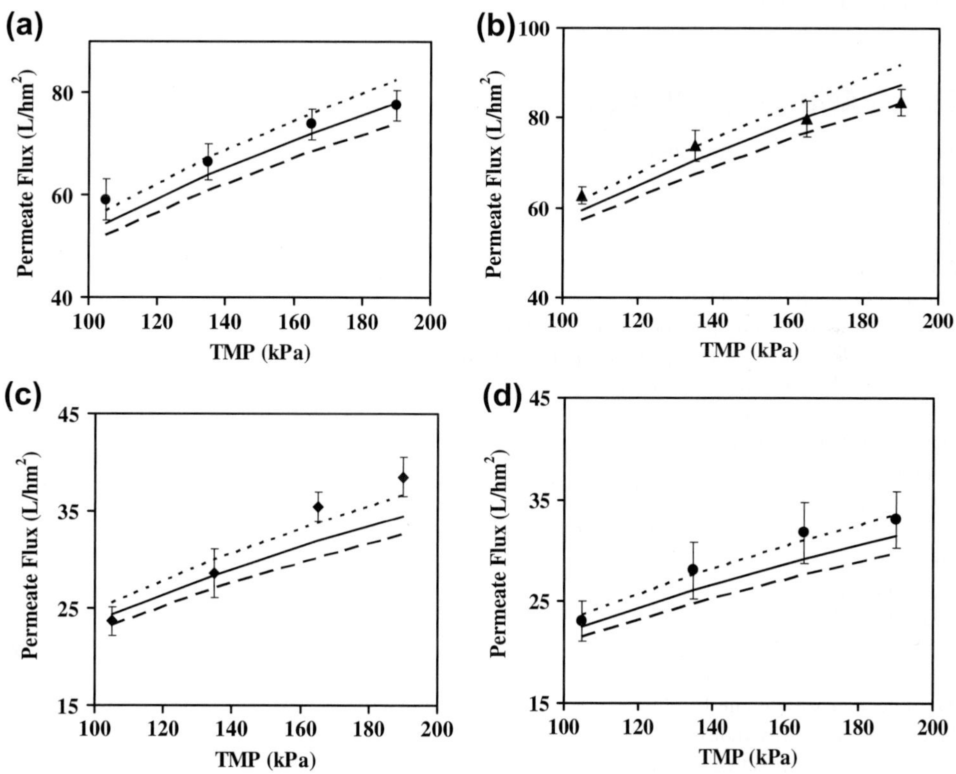

FIGURE 7.18

Permeate flux versus TMP for 50 kDa TiO_2 membrane and feed solutions of (a) Feed A: 300 ppm (75 ppm of each $C_5H_{10}O_2$, $C_7H_{14}O_2$, $C_8H_{16}O_2$, and $C_{10}H_{20}O_2$ in 10 mM CPC solution, (b) Feed B: 300 ppm NAs practical grade in 10 mM CPC solution, (c) Field A sample, and (d) Field B sample {Lines model:(——) non-linear model, (— —) model: (β) +10%, (- - -) model: (β) -10%; Points: experiments}.

layer had a more pronounced impact on the permeate flux. It also supports the conclusion that viscosity had a major role in reducing permeate flux at low TMPs.

5 Conclusions

This chapter describes an experimental study on the role of organic co-contaminants, especially the amphiphilic ones, in improving the performance of micellar-enhanced ultrafiltration. This improvement is twofold. First, organic co-contaminants contributed to increasing the size of the surfactant micelles, which enabled the use of a larger pore size membrane. This, in turn, led to higher treated water flux. Second, amphiphilic co-contaminants not only increased the size of the surfactant micelles, but also participated in the formation of "mixed micelles." The "mixed micelles" possessed lower critical micellar concentration and higher solubilization capacity of other organic contaminants. Consequently, lower surfactant concentration could be applied, giving rise to higher permeate flux, while maintaining high rejection.

SYMBOLS AND TERMINOLOGY

CF concentration factor, dimensionless
$C_5H_{10}O_2$ pentanoic acid
$C_7H_{14}O_2$ heptanoic acid
$C_8H_{16}O_2$ octanoic acid
$C_{10}H_{20}O_2$ decanoic acid
$C_nH_{2n}O_2$ acyclic naphthenic acids
$C_{i,F}$ feed concentration of component i, ppm, or mM
$C_{i,p}$ permeate concentration of component i, ppm, or mM
CMC critical micelle concentration, mM
CP concentration polarization
CPC cetylpyridinium chloride
DLS dynamic light scattering
GC gas chromatography
HPLC high performance liquid chromatography
j carbon number
J permeate flux, $Lh^{-1}m^{-2}$
J_w DI-water flux, $Lh^{-1}m^{-2}$
m mass of the deposited solutes, kg
MEUF micellar-enhanced ultrafiltration
MWCO molecular weight cutoff
n CP layer compressibility index
NAs naphthenic acids (practical grade)
PAN polyacrylonitrile
P_F gauge pressure of the feed stream, kPa
P_P gauge pressure of the permeate stream, kPa
P_R gauge pressure of the retentate stream, kPa
***p*-xylene** para xylene
r recovery ratio, dimensionless

R_c CP layer resistance, m^{-1}
R_f fouled layer resistance, m^{-1}
R_g gel layer resistance, m^{-1}
R_m membrane resistance, m^{-1}
R_{pp} pore plugging resistance, m^{-1}
TSS total suspended solids
TDS total dissolved solids
TiO_2 titanium oxide
TMP transmembrane pressure, kPa
TOC total organic carbon
UF ultrafiltration
$V_{F,0}$ feed volume at time zero, ml
$V_{F,t}$ feed volume at time t, ml
$\dot{V}_F$ feed volumetric flow rate, $mlmin^{-1}$
$\dot{V}_P$ permeate volumetric flow rate, $mlmin^{-1}$
α CP layer resistance coefficient, $mkg^{-1}Pa^{-n}$
α_0 specific resistance of the CP layer, mkg^{-1}
β fitted parameter, $m^{-1}Pa^{-0.8}$
μ solution viscosity, Pa.s
μ_w DI-water viscosity, Pa.s

Acknowledgments

The authors would like to thank the Natural Sciences and Engineering Research Council of Canada (NSERC) for their financial support.

References

Acevedo, S., Escobar, G., Ranaudo, M.A., Khazen, J., Borges, B., Pereira, J.C., Mendez, B., 1999. Isolation and characterization of low and high molecular weight acidic compounds Cerro Negro extra heavy crude oil. Role of these acids in the interfacial properties of the crude oil emulsions. Energy Fuels 13 (2), 333–335.

Bacchin, P., Meireles, M., Aimar, P., 2002b. Modelling of filtration: from the polarised layer to deposit formation and compaction. Desalination 145, 139–146.

Bacchin, P., Si-Hassen, D., Starov, V., Clifton, J., Aimar, P., 2002a. A unifying model for concentration polarization, gel-layer formation and particle deposition in cross-flow membrane filtration of colloidal suspensions. Chem. Eng. Sci. 57, 77–91.

Baker, R.W., 2004. Membrane technology and applications, second ed. John Wiley & Sons, New York.

Benito, J.M., Sánchez, M.J., Pena, P., Rodríguez, M.A., 2007. Development of a new high porosity ceramic membrane for the treatment of bilge water. Desalination 214, 91–101.

Bielska, M., Szymanowski, J., 2004. Micellar-enhanced ultrafiltration of nitrobenzene and 4-nitrophenol. J. Membr. Sci. 243, 273–281.

Boerlage, S.F.E., Kennedy, M.D., Aniye, M.P., Abogrean, E., Tarawneh, Z.S., Schippers, J.C., 2003. The MFI-UF as a water quality test and monitor. J. Membr. Sci. 211, 271–289.

Bridle, M., 2005. Treatment of SAGD-produced water without lime softening. International Thermal Operations and Heavy Oil Symposium Proceedings, A149–A154.

Carey, J., Zaidi, A., Ribo, J., 1992. Specific toxic organics in produced waters from in-situ heavy oil recovery operations in Western Canada. Produced Water: Technological/Environmental Issues and Solutions, 133–150.

Carrère, H., Blaszkow, F., Roux de Balmann, H., 2001. Modelling the clarification of lactic acid fermentation broths by cross-flow microfiltration. J. Membr. Sci. 186, 219–230.

Castillo, R., Garza, C., Ramos, S., 1994. Brownian motion at the molecular level in liquid solutions of C_{60}. J. Phys. Chem. 98, 4188–4190.

Chalaturnyk, R.J., Scott, J.D., Özüm, B., 2002. Management of oil sands tailings. Pet. Sci. Technol. 20, 1025–1046.

Chen, J.C., Kim, A.S., 2006. Monte Carlo simulation of colloidal membrane filtration: Principal issues for modeling. Adv. Colloid Interface Sci. 119, 35–53.

Chen, J.C., Elimelech, M., Kim, A.S., 2005. Monte Carlo simulation of colloidal membrane filtration: Model development with application to characterization of colloid phase transition. J. Membr. Sci. 255, 291–305.

Cheng, T.W., Yeh, H.M., Gau, C.T., 1997. Resistance analyses for ultrafiltration in tubular membrane module. Sep. Sci. Technol. 32, 2623–2640.

Cline, J.T., 1998. Treatment and discharge of produced water for deep offshore disposal. Presented at the API Produced Water Management Technical Forum and Exhibition, Lafayette, LA. November 17–18, (1998).

Davis, R.H., Leighton, D.T., 1987. Shear-induced transport of a particle layer along a porous wall. Chem. Eng. Sci. 42, 275–281.

Deriszadeh, A., Harding, T.G., Husein, M.M., 2008. Role of naphthenic acid contaminants in the removal of p-xylene from synthetic produced water by MEUF. Process Saf. Environ. Prot. 86, 244–251.

Deriszadeh, A., Harding, T.G., Husein, M.M., 2009. Improved MEUF removal of naphthenic acids from produced water. J. Membr. Sci. 326, 161–167.

Deriszadeh, A., Husein, M.M., Harding, T.G., 2010. Oil sands produced water treatment by micellar enhanced ultrafiltration. Environ. Sci. Technol. 44, 1767–1772.

Dunn, R.O., Scamehorn, J.F., Christian, S.D., 1985. Use of micellar-enhanced ultrafiltration to remove dissolved organics from aqueous streams. Sep. Sci. Technol. 20, 257–284.

Dunn, R.O., Scamehorn, J.F., Christian, S.D., 1989. Simultaneous removal of dissolved organics and divalent metal ions from wastewater using micellar-enhanced ultrafiltration. Colloids Surf. 35, 49–56.

Fetter, C.W., Fetter Jr, C.W., 1999. Contaminant hydrogeology, second ed. Prentice Hall, New York.

Field, R.W., Wu, D., Howell, J.A., Gupta, B.B., 1995. Critical flux concept for micro filtration fouling. J. Membr. Sci. 100, 259–272.

Fillipi, B.R., Brant, L.W., Scamehorn, J.F., Christian, S.D., 1999. Use of micellar enhanced ultrafiltration at low surfactant concentrations and with anionic-nonionic surfactant mixtures. J. Colloid Interface Sci. 213, 68–80.

Giesbrecht, G.D., Chang, F., Nelson, R., 1992. Water reuse testing at Shell's Peace River complex. Produced water: Technological/Environmental issues and Ssolutions, 561–568.

Gray, W.G., Miller, C.T., 2004. Examination of Darcy's law for flow in porous media with variable porosity. Environ. Sci. Technol. 38, 5895–5901.

Heins, W., Peterson, D., 2005. Use of evaporation for heavy oil produced water treatment. J. Can. Pet. Technol. 44, 26–30.

Hum, F., Tsang, P., Harding, T.G., Kantzas, A., 2006. Review of produced water recycle and neneficial reuse. Paper No. 19 of the Alberta Energy Futures Project, Institute of Sustainable Energy Environment and Economy. University of Calgary.

Jadhav, S.R., Verma, N., Sharma, A., Bhattacharya, P.K., 2001. Flux and retention analysis during micellar-enhanced ultrafiltration for the removal of phenol and aniline. Sep. Purif. Technol. 24, 541–557.

Jan, R.J., Reed, T.G., 1989. New caustic process for softening produced water for steam generation. Prod. Oper. Eng. 1989, 519–525.

Janks, J.S., Cadena, F., 1992. Investigations into the use of modified zeolites for removing benzene, toluene and xylene from saline produed water. Produced Water: Technological/Environmental Issues and Solutions, 473–487.

Juang, R.-S., Chen, H.-L., Chen, Y.-S., 2008. Resistance-in-series analysis in cross-flow ultrafiltration of fermentation broths *Bacillus subtilis* culture. J. Membr. Sci. 323, 193–200.

Juang, R.S., Xu, Y.Y., Chen, C.L., 2003. Separation and removal of metal ions from dilute solutions using micellar-enhanced ultrafiltration. J. Membr. Sci. 218, 257–267.

Kim, A.S., Yuan, R., 2005. A new model for calculating specific resistance of aggregated colloidal cake layers in membrane filtration processes. J. Membr. Sci. 249, 89–101.

Kosaraju, P.B., Sirkar, K.K., 2007. Novel solvent-resistant hydrophilic hollow fiber membranes for efficient membrane solvent back extraction. J. Membr. Sci. 288, 41–50.

Lee, J., Yang, J.S., Kim, H.J., Baek, K., Yang, J.W., 2005. Simultaneous removal of organic and inorganic contaminants by micellar enhanced ultrafiltration with mixed surfactants. Desalination 184, 395–407.

Li, C.-W., Liu, C.-K., Yen, W.-S., 2006. Micellar-enhanced ultrafiltration(MEUF) with mixed surfactants for removing Cu(II) ions. Chemosphere 63, 353–358.

Mahesh Kumar, S., Madhu, G.M., Roy, S., 2007. Fouling behavior, regeneration options and on-line control of biomass-based power plant effluents using microporous ceramic membranes. Sep. Purif. Technol. 57, 25–36.

Marcotte, M., Grabowski, S., LeBlanc, D.I.M., Darley, K., 1998. Separation technologies R&D for food industry, A Canadian Survey. Can. Agri-Food Res. Counc.

Markels, J.H., Lynn, S., Radke, C.J., 1995. Cross-flow ultrafiltration of micellar surfactant solutions. AIChE Journal 41, 2058–2066.

Marquez, N., Anton, R., Graciaa, A., Lachasie, J., Salager, J.L., 1995. Partitioning of ethoxylated alkylphenol surfactants in microemulsion oil-water systems. Colloid. Surf. A: Physicochem. Eng. Aspect. 100, 225–231.

Nagarajan, R., Ruckenstein, E., 1991. Theory of surfactant self-assembly: A predictive molecular thermodynamic approach. Langmuir 7, 2934–2969.

Nakanishi, K., Tadokoro, T., Matsuno, R., 1987. On the specific resistance of cakes of microorganisms. Chem. Eng. Comm. 62, 187–201.

Piron, E., René, F., Latrille, E., 1995. A cross-flow microfiltration model based on integration of the mass transport equation. J. Membr. Sci. 108, 57–70.

Purkait, M.K., DasGupta, S., De, S., 2004. Removal of dye from wastewater using micellar-enhanced ultrafiltration and recovery of surfactant. Sep. Purif. Technol. 37, 81–92.

Purkait, M.K., DasGupta, S., De, S., 2005. Separation of aromatic alcohols using micellar-enhanced ultrafiltration and recovery of surfactant. J. Membr. Sci. 250, 47–59.

Purkait, M.K., DasGupta, S., De, S., 2006. Micellar-enhanced ultrafiltration of eosin dye using hexadecyl pyridinium chloride. J. Hazard. Mater. B 136, 972–977.

Qtaishat, M., Khayet, M., Matsuura, T., 2009. Guidelines for preparation of higher flux hydrophobic/hydrophilic composite membranes for membrane distillation. J. Membr. Sci. 328, 193–200.

Quagraine, E.K., Peterson, H.G., Headley, J.V., 2005. In-situ bioremediation of naphthenic acids contaminated tailing pond waters in the Athabasca oilsands region-demonstration field studies and plausible options: A review. J. Environ. Sci. Health. 40, 685–722.

Rai, P., Rai, C., Majumdar, C.G., DasGupta, S., De, S., 2006. Resistance in series model for ultrafiltration of mosambi (*Citrus sinensis* (*L.*) Osbeck) juice in a stirred continuous mode. J. Membr. Sci. 283, 116–122.

Rogers, V.V., Liber, K., MacKinnon, M.D., 2002. Isolation and characterization of naphthenic acids from Athabasca oil sands tailing ponds water. Chemosphere 48, 519–527.

Rosen, M.J., 2004. Surfactants and interfacial phenomena, third ed. John Wiley & Sons, New York.

Russel, W.B., 1981. Brownian motion of small particles suspended in liquids. Ann. Rev. Fluid Mech. 13, 425–455.

Schroeder, E.D., 1977. Water and wastewater treatment, first ed. McGraw-Hill, New York. Series in Water Resources and Environmental Engineering.

Seader, J.D., Henley, E.J., 2006. Separation process principles, second ed. John Wiley & Sons, New York.

Sethi, S., Wiesner, M.R., 1997. Modeling of transient permeate flux in cross-flow membrane filtration incorporating multiple particle transport mechanisms. J. Membr. Sci. 136, 191–205.

Shpiner, R., Vathi, S., Stuckey, D.C., 2007. Treatment of oil field produced water by waste stabilization ponds. Water Sci. Technol. 55, 265–271.

Simms, K., Zaidi, A., Bhargara, O., 1992. A protocol for determining oil and grease in produced water. Environ. Sci. Res. 46, 455–471.

Stasiuk, E.N.B., Schramm, L.L., 1996. The temperature dependence of the critical micelle concentration of foam-forming surfactants. J. Colloid Interfac. Sci. 178, 324–333.

Tansel, B., Bao, W.Y., Tansel, I.N., 2000. Characterization of fouling kinetics in ultrafiltration systems by resistances in series model. Desalination 129, 7–14.

Thomas, S.A., Yost, M.E., Cathey, S.R., 1987. Silica removal from steamflood-produced water: south Texas tar sand pilot. SPE Prod. Eng. 2, 131–136.

Tung, C.-C., Yang, Y.-M., Chang, C.-H., Maa, J.-R., 2002. Removal of copper ions and dissolved phenol from water using micellar-enhanced ultrafiltration with mixed surfactants. Waste Manage. 22, 695–701.

Van Den Berg, G.B., Smolders, C.A., 1990. Flux decline in ultrafiltration processes. Desalination 77, 101–133.

Veil, J.A., Puder, M.G., Elcock, D., Redweik, R.J., 2004. A White Paper describing produced water from production of crude oil, natural gas and coal bed methane. Prepared for U.S. Department of Energy.

Velterop, F.M., Keizer, K., 1991. Development of a high temperature resistant module for ceramic membranes. Key Eng. Mater. 61-62, 391–394.

Whitaker, S., 1986. Flow in porous media II: The governing equations for immiscible, two-phase flow. J. Trans. Porous Media. 1, 105–125.

Yang, J.S., Baek, K., Yang, J.W., 2005. Crossflow ultrafiltration of surfactant solution. Desalination 184, 385–394.

Yepez, O., 2007. On the chemical reaction between carboxylic acids and iron including the special case of naphthenic acids. Fuel 86, 1162–1168.

Zaidi, A., Simms, K. Kok, Nelson, S.R., 1992. Recent advances in the application of membrane technology for the removal of oil and suspended solids from produced waters. Produced Water: Technological/Environmental Issues and Solutions, 489–501.

CHAPTER 8

Heavy Metals Uptake from Aqueous Effluents by Novel Adsorbent Derived from Tannin Extracts: Role of Tannin Source

Sánchez-Martín, J. Beltrán-Heredia, V. Encinas-Sánchez
Department of Chemical Engineering and Physical Chemistry, University of Extremadura, 06071 Badajoz, Spain

CHAPTER OUTLINE

1 Introduction

Concerns about the presence of heavy metals in aqueous effluents are well known, and they appear to be linked to a large variety of industries (Yantasee et al. 2007). Some of them are related to metal finishing, electroplating, plastics, pigments, and mining, which threaten the environment and human lives very severely (Kurniawan et al. 2006). It remains a challenge for researchers to find new and

The Role of Colloidal Systems in Environmental Protection. http://dx.doi.org/10.1016/B978-0-444-63283-8.00008-9

cheaper mechanisms for removing heavy metals from wastewater; these metals represent a harmful and noxious contamination because they tend to accumulate in the food chain.

Heavy metal removal from aqueous solution can be achieved by several conventional methods: chemical precipitation (Tünai and Kabdasly 1994), coagulation and flocculation (Charerntanyarak 1999), adsorption onto vegetal wastes (Ngah and Hanafiah 2008), and special treatments such as nanofiltration (Bruggen and Vandecasteele 2003) or reverse osmosis. All of these treatments may be effective when applied in very high metal concentrations, but when used at low levels (1–100 $mg \cdot L^{-1}$) they are usually expensive and ineffective.

Adsorption technology has recently become a real alternative to traditional wastewater treatment due to its relative simplicity and efficiency. Moreover, a large number of low-cost adsorbents from wastewater treatment can be found, which make this method strongly attractive since they are obviously cheap, widespread, and highly efficient (Wingenfelder et al. 2005).

In the context of pollution remediation, natural raw materials are an interesting potential source of low-cost adsorbents (Altenor et al. 2009; Demirbas 2008; Wang et al. 2005). One class of such materials is the *tannins*, a term that covers many families of chemical compounds. Their name derives from their traditional use in tanning animal skins for the production of leather (Convington 1997). Today they are also widely used in tannin-modified adhesive formulations (López-Suevos and Riedl 2003), as adsorbents for pollution control of industrial effluent (Vázquez et al. 2002), and as flocculants (Beltrán-Heredia and Sánchez-Martín 2008). Their natural origin is as secondary metabolites of plants (Schofield et al. 2001), occurring in the bark, fruit, leaves, and so on. Whereas *Acacia* and *Schinopsis* bark constitutes the principal source of tannins for the leather industry, the bark of other, nontropical trees such as *Quercus ilex*, *Q. suber*, *Q. robur*, *Castanea sativa,* and *Pinus pinaster* can also be tannin-rich.

From a chemical point of view, there are three kinds of tannins: hydrolysable, condensed, and combined (Haslam 1989). These products are chemically complex, and they are usually taken from a natural matrix, without undergoing a very exhaustive purification. As a result, it is difficult to define their structure with any precision, although probable approaches can be found in literature (e.g., Pizzi 2008) (Figure 8.1).

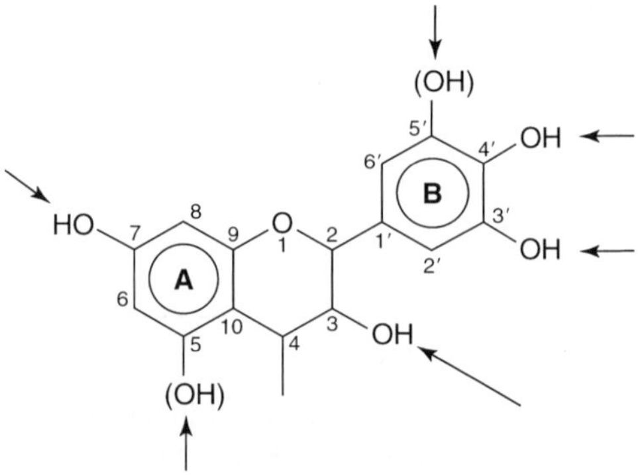

FIGURE 8.1 Probable flavonoid unit in a condensed tannin.

Tannin gelation is a chemical procedure that immobilizes tannins in an insoluble matrix (Pizzi 1994) so that their properties of interest, for example, metal chelation, are then available in an efficient adsorbent agent. In addition, the material resulting from their gelation (sometimes called *tannin rigid resin*) presents interesting properties in terms of resistance, nonflammability, and mechanical undeformability (Tondi et al. 2009a; Zhao et al. 2003). This process has been widely described in the scientific literature and in patents. Examples of basic gelation can be found in the scientific literature (Nakano et al. 2001; Kim and Nakano 2005; Tondi et al. 2009b) and in patents such as U.S. patent 5,158,711(Shirato and Kamei 1992). In addition, acid gelation has been described by other workers (Vázquez et al. 1994; 2006).

The applicability of these products is more than evident, and gelation has been reported to be effective in removing heavy metals (Nakano et al. 2001; Kim and Nakano 2005; Sengil and Özacar, 2009) as well as a wide variety of cationic substances and dangerous products (Vázquez et al. 2006; Sánchez-Martín et al. 2010). However, a full study on how to optimize the tannin gelation in order to produce a more efficient *tannin rigid resin* for heavy metal removal has not yet been produced. That is the scope of the present work.

To this end, we have selected four tannin sources (*Acacia mearnsii de Wild*, called *Weibull black*; *Cupressus sempervivens*, called *Cypress*; *Schinopsis balansae*, called *Quebracho*; and *Pinus pinaster*, called *Pine*). These extracts were gelified with two doses of formaldehyde, and they were tested with each metal. The Analysis of Variance (ANOVA) tests identified the best formaldehyde dosage, the best tannin sources, and the most removable metals. Finally, we have performed a theoretical modelization under the Langmuir hypothesis for the most relevant systems.

2 Materials and methods

2.1 Tannin extracts and the gelation process

Acacia mearnsii de Wild (Weibull black) and *Schinopsis balansae* (Quebracho) tannin extracts were kindly supplied by TANAC Inc. (Brazil).

Tannins from cypress and pine were lab-extracted according to the normalized method described elsewhere (Sánchez-Martín et al. 2011a,b). First, 100 g of pine bark or cypress husks were milled in a cutting mill (RETSCH, SM 2000 model) ,and they were put in 600 mL of tap water. Then, 5 g of NaOH (PANREAC) were added, and the mixture was stirred in a magnetic stirrer at 90°C for 1 h. Solids were separated by filtration, and a liquid fraction was dried in an oven (65°C) overnight. The result was considered the tannin extract.

Tanningels were prepared according to the basis of Nakano et al. (2001): 5 g of tannin extract were dissolved in 32 mL of NaOH (PANREAC) 0.125 mol L^{-1} and 30 mL of distilled water at 80°C. When the mixture was homogeneous, a certain amount of formaldehyde was added. This depends on the running experiment. The reaction was kept at the same temperature for 8 hours until polymerization was considered completed. Then, the apparent gummy product was led to complete evaporation of the remaining water and dried in an oven (again at 65°C) overnight. After drying, tannin rigid gels were crushed and sieved to produce 38–53 μm diameter particles. They were washed successively with distilled water and HNO_3 0.01 mol L^{-1} (PANREAC) to remove unreacted sodium hydroxide. Finally, the adsorbent was again oven dried. This preparation method and that described by Yurtsever and Sengil (2010) differ mainly concerning the amount of formaldehyde. In our case, the scope is to optimize this variable.

2.2 Model compounds and metal analysis

Metal salts were supplied by Merck. Although each one presents different degrees of hydration, the final concentration of metal in the bulk solution is calculated only on the basis of the heavy metal ion. The specific chemical compounds were: $Cd(NO_3)_2.4H_2O$, $CoCl_2.6H_2O$, $Cr(NO_3)_3.9H_2O$, $Cu(NO_3)_2.2.5H_2O$, $HgCl_2$, $MnSO_4.H_2O$, $Ni(NO_3)_2.6H_2O$, $Pb(NO_3)_2$, $UO_2SO_4.4.5H_2O$, and $Zn(NO_3)_2.6H_2O$.

Up to four different analytical methods were used to measure the metal concentration. For Cd^{2+}, Mn^{2+}, Co^{2+}, Ni^{2+}, Zn^{2+}, and Ni^{2+}, the classical analytical process developed by Qi and Zhu (1985) was used: 1 mL of sample was put into a 25-mL flask, and 5 mL of borate buffer (sodium borate and sodium carbonate in equal quantities, pH 10) were added. Subsequently, 2 mL of 4-(2-pyridyl-azo)-resorcinol (PAR) 5 10^{-4} M were added, and the mixture was diluted to the mark with distilled water. Then, the solution was mixed, and absorbance at 505 nm in a 1-cm optical glass cell was measured after 20 min.

Pb^{2+} content was determined by modifying this method according to Pollard et al. (1959). PAR was used in a five times higher concentration than in the previous case, and absorbance was measured at 530 nm after 30 min.

UO_2^{2+} was analytically determined according to Pollard et al. (1959): 5 mL of sample were mixed with 10 mL of pH-8 buffered solution and 2 mL of PAR $2.5 \cdot 10^{-3}$ M. This mixture was heated for 10 min at 115°C in perfectly sealed vials. Once the temperature of the solution decreased, it was put in a 25 mL-flask and distilled water was added up to the mark. Immediately, absorbance was determined by visible spectroscopy at 530 nm in a 1-cm optical glass cell.

Cr^{3+} was determined according to Yotsuyanagi et al. (1973). This step was carried out as follows: 1 mL of sample was put into a vial, and 1 mL of pH-5 buffered solution and 3.5 mL of PAR $2.5 \cdot 10^{-3}$ M and 10 mL of distilled water were added. The sealed vials were then heated for 2 h at 115°C. Absorbance was determined in the same way as indicated for the case of UO_2^{2+}.

Finally, Hg^{2+} was determined according to Manchón-Vizuete et al. (2005): 5 mL of sample were introduced in a 25-mL flask and 10 mL of HCl 6 M were added. Distilled water was added up to the mark, and absorbance was determined at 231 nm in a 1-cm quartz cell.

Spectrophotometric determinations were carried out with an HEλIOS UV/VIS spectrophotometer.

2.3 Equilibrium studies

In order to test the ability of each *tanningel* in the removal of heavy metals, a standard protocol of adsorption was developed. Samples of aqueous solutions with the corresponding pollutants concentration (25 mg L^{-1}) were put into a 100 mL-flask. A fixed amount of *tanningel* was added to each flask (50 mg) with different volumes of contaminated aqueous solutions. A period of 7 days was established for guaranteed equilibrium.

2.4 Kinetic studies

In order to confirm the equilibrium period of these adsorbents, kinetic approaches must be carried out. To this end, 100 mL of solutions with initial concentrations of 50 $mg \cdot L^{-1}$ of UO_2^{2+} , Pb^{2+}, and Hg^{2+} were treated with 50 mg of PFC, CFC, WFC, and QFC. The presence of heavy metals in the bulk solution was measured at different intervals, and the decay was modeled according to three kinetic

Table 8.1 Gelation Experiments

Tannin Extract	Formaldehyde ($mmol \cdot g^{-1}$)	Does it Gelify?	Symbol
Cypress	1	No	
Cypress	3.68	Yes	CFC
Quebracho	1	Yes	QFD
Quebracho	3.68	Yes	QFC
Pine	1	Yes	PFD
Pine	3.68	Yes	PFC
Weibull	1	Yes	WFD
Weibull	3.68	Yes	WFC

models: Lagergren, Ho, and Elovich. Similar trials were performed with other metals (results not shown).

Both equilibrium and kinetic studies were carried out under strict thermal control and magnetic stirring.

2.5 Mathematical and statistical processes

Design of experiments was carried out by using SPSS 14.0 for Windows (SPSS).

The basis of the statistical method is found in the interaction of variables. Briefly, it can be summarized as follows:

1. Up to eight different combinations regarding the two variables (tannin extract and formaldehyde concentration) were attempted. Only seven of them were feasible combinations, since the gelation of cypress extract with diluted formaldehyde was not possible (Table 8.1).
2. Each feasible combination was synthesized twice and subsequently tested on each heavy metal solution. The resulting target variable was the percentage removal of metal.
3. The results were evaluated under subsequent ANOVA tests: one for the reproducibility of the synthesis (indistinguishable observation) and another one for evaluating the effective influence of formaldehyde concentration, tannin source, and metal type. This drove us to identify the best adsorbents and the most removable metals. All of these variables were statistically significant; hence the conclusions are consistent.

3 Results and discussion

3.1 Kinetics of heavy metal removal

The efficiency of the adsorption process must be measured according to a parameter called adsorption capacity (q), defined by Eq. 8.1:

$$q = \frac{(C_0 - C_1)V}{W} \tag{8.1}$$

where

C_0 is the initial contaminant concentration (mg L^{-1}),
C_1 is the equilibrium contaminant concentration in bulk solution (mg L^{-1}),
V is the volume of solution (L), and
W is the adsorbent mass (g).

Adsorption capacity tends to increase up to a maximum q level, which is reached after the equilibrium period. The determination of this period is performed according to kinetic studies and can be easily modeled by three kinetic hypotheses: the Lagergren model (pseudo-first order) (Lagergren 1898), the Ho model (pseudo-second order) (Ho and McKay 1998), and the Elovich model (Aroua et al. 2008).

The first experimental model (Lagergren) is given by Eq. 8.2:

$$q = q_e - q_e.e^{-K_l.t} \tag{8.2}$$

where

q_e is the amount of adsorbate retained per adsorbent mass in equilibrium ($mg \cdot g^{-1}$),
K_l is the first-order Lagergren constant (min^{-1}), and
t is the experimental time (min).

Ho's model is given by Eq. 8.3:

$$q = \frac{t}{\frac{1}{h} + \frac{t}{q_e}} \tag{8.3}$$

where h is Ho's constant ($mg \cdot g^{-1} \cdot min^{-1}$).

Finally, Elovich's model is presented by Eq. 8.4:

$$q = \frac{1}{\beta_e} \ln(\alpha_e \cdot \beta_e) + \frac{1}{\beta_e} \ln t \tag{8.4}$$

where

α_e is the initial adsorption rate ($mg \cdot g^{-1} \cdot min^{-1}$) and
β_e is the desorption constant, ($g \cdot mg^{-1}$).

Table 8.2 shows the analytical results of these adjustments for four *tanningels* and three metals, since results were similar in the rest of the systems. As can be appreciated, the large majority of the involved systems presented high correlation factors r^2, in many cases above 0.95. The three models fit reasonably well to each pair of metal-*tanningel*. Attending to the standard deviation factor (data not shown), the best model for removal of Hg^{2+} is the hypothesis of Elovich, whereas for the other two metals (UO_2^{2+} and Pb^{2+}) Ho's model is the best one.

3.2 General heavy metal removal test

After confirming that the equilibrium period was enough with 1 week of stirring, several considerations may be done in order to guarantee the statistical significance. The first consideration has to do with the reproducibility of the trials.

Table 8.2 Analytical Adjustments for Kinetic Models

	Hg^{2+}	UO_2^{2+}	Pb^{2+}	Hg^{2+}	UO_2^{2+}	Pb^{2+}
			Lagergren model			
		PFc			**QFc**	
q_e (mg/g)	302	119	208	244	76	109
K_L (min^{-1}) $\cdot 10^3$	2.34	4.67	24.79	0.63	0.87	16.58
r^2	0.873	0.908	0.984	0.902	0.894	0.997
		WFc			**CFc**	
q_e	226	90	84	164	135	176
K_L(min^{-1}) $\cdot 10^3$	0.41	8.85	8.48	14.65	18.19	47.61
r^2	0.944	0.974	0.964	0.797	0.964	0.944
			Ho model			
		PFc			**QFc**	
q_e	330	135	228	269	83	112
h (mg/g·min)	0.97	0.77	8.17	0.26	0.10	4.06
r^2	0.946	0.966	0.996	0.949	0.939	0.988
		WFc			**CFc**	
q_e	262	94	88	181	148	190
h (mg/g·min)	0.13	1.42	1.25	3.48	3.65	14.89
r^2	0.967	0.993	0.997	0.887	0.982	0.992
			Elovich model			
		PFc			**QFc**	
α_e (mg/g·min)	4.07	2.12	43.36	1.05	0.41	192.66
β_e (g/mg)	0.020	0.041	0.030	0.023	0.075	0.084
r^2	0.991	0.991	0.873	0.964	0.95	0.762
		WFc			**CFc**	
α_e (mg/g·min)	0.62	292.72	177.10	64.68	20.23	535.80
β_e (g/mg)	0.024	0.132	0.135	0.047	0.470	0.050
r^2	0.944	0.807	0.829	0.912	0.869	0.861

The ANOVA test can be carried out considering just two variables in order to analytically explain the target variable *Percentage removal*: One may be the specific category of each *tanningel,* and the other one can be the particular trial (numbered by 1 or 2). Figure 8.2 presents the graphical expression of this ANOVA, where each point represents the estimated margin means of the removal percentage for the ten metals. As can be seen, the two lines are subsequently crossed, which is due to the absence of significance of this variable. This is analytically confirmed by the *p*-value of this parameter (0.99). Consequently, reproducibility exigencies are adequately satisfied.

3.2.1 Formaldehyde content in the gelation process

There is a similar consideration regarding the formaldehyde content in the gelation process. As can be observed from Table 8.1, two different doses were applied: One corresponds to 1 mmol g^{-1}, and

FIGURE 8.2 Indistinguishability of the adsorption trials.

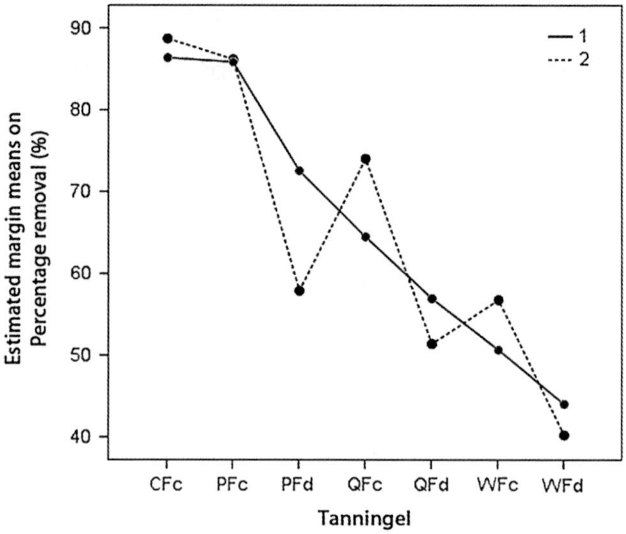

Table 8.3 Significant Variables in the Anova Test for Three Parameters (statistical summary)

Variable	F-value	*p*-value
Tannin	26.868	0
Heavy metal	25.011	0
Formaldehyde	30.092	0
Metal x Tannin Interaction	2.25	0.04
r^2 (%)	87.1	

the other corresponds to 3.68 mmol g^{-1}. These levels were selected according to previous works (Sánchez-Martín et al. 2011b). Again, the ANOVA test gives us a definitive conclusion when it sets the *p*-value of this parameter equal to 0 (Table 8.3). That means there is a clear and significant influence of this variable, and, according to Tukey's homogeneous groups, the level that drives us to the best adsorbent is 3.68 mmol g^{-1}, that is, the upper level.

3.2.2 Tannin extract

A preliminary approach to the influence of the tannin extract in the adsorption equilibrium *tanningel*-heavy metal is to perform a classical multiple comparison taking into account only the specific metal and each tannin extract type. This approach is shown in Figure 8.3.

In this figure, columns present the increasing metal removal regardless of the rest of the variables, such as formaldehyde concentration. It is an unclear representation because it contains a huge number of percentage removal data. Therefore, a more specific analysis must be carried out.

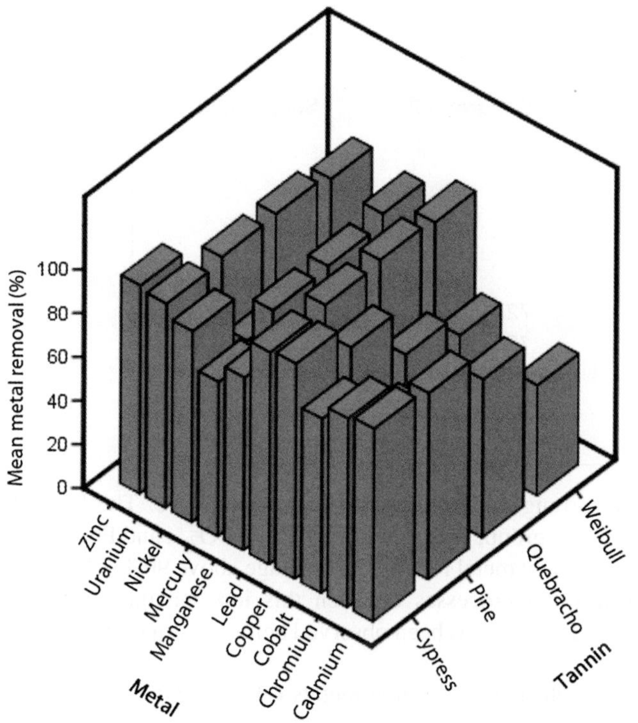

FIGURE 8.3 General comparison tannin extract and heavy metal removal.

(For color version of this figure, the reader is referred to the online version of this book.)

The ANOVA test is done involving the whole system. That is, our phenomenon (heavy metal adsorption from aqueous effluents by *tanningels*) may be explained according to the three variables: tannin extract, heavy metal, and formaldehyde concentration. That will lead us to the traditional Tukey's homogeneous groups; those will help us to identify the best adsorbents and the most removable metals. Table 8.3 shows the main statistic data for this ANOVA test.

According to this statistical study, one can approach the influence of the specific tannin extract in the final response. Four separate and well-differenced groups are presented according to Tukey's Honestly Significant Difference (HSD) test, one for each tannin extract. This test shows us that the most effective tannin is Cypress, followed by Pine and Quebracho. Finally, the least effective tannin extract is the one derived from black wattle, named Weibull (*Acacia mearnsii de wild*). The means removal percentages that stated these differences were 87.5, 75.6, 61.7, and 47.9, respectively. These differences have been observed previously in other works (Sánchez-Martín et al. 2011b), but have never referred to such a variety of heavy metals.

3.2.3 Heavy metal ion

On the other hand, Table 8.4 presents the HSD Tukey's homogeneous groups according to metal removal. This shows us how the chemical species of metal influences the final interaction between metal and *tanningel*.

Table 8.4 Tukey's Homogeneous Groups According to Metal Type in Heavy Metal Removal (percentage values)

Metal	Subgroup 1	Subgroup 2	Subgroup 3
Cobalt	39.4		
Manganese	41.4		
Zinc	50.1		
Nickel	50.2		
Chromium	50.3		
Cadmium		72.1	
Copper		72.5	
Mercury		84.2	84.2
Uranium			94.6
Lead			99.5

As can be observed, three subgroups are arranged according to heavy metal type. The highest affinity to *tanningels* is presented by the last metal group, which involves Hg^{2+}, UO_2^{2+}, and Pb^{2+}. It is remarkable that this last group presents a mean heavy metal removal percentage above 90%. This may be due to the intrinsic relationships in adsorption processes between this ion and the *tanningel* themselves, which may enhance these linkings. This fact has been observed before (Palma et al. 2003), though with a significantly lower efficiency.

According to these preliminary results, we selected these three metals (Hg^{2+}, UO_2^{2+}, and Pb^{2+}), and we performed an equilibrium study on the four *tanningels*: PFC, WFC, QFC, and CFC.

3.3 Isotherms and theoretical modelization

In order to evaluate not only the efficacy of the treatment but also the efficiency, the percentage removal must be complemented by other objective target. Traditionally, adsorption capacity (q) was defined for this scope according to the Eq. 8.1 in Section 3.1.

The physical system heavy metal-adsorbent may be ruled by adsorption-like relationships. Taking into account the need to characterize how these adsorbents work on metal removal, a model must be presented in order to fully characterize the adsorption process. To this end, adsorption capacity q should be a reference measure of the efficiency of the removal of the specific contaminant in relationship with the amount of adsorbent. This is extensively used in such processes and must link q and the remnant adsorbate concentration.

The main adsorption model is the one Langmuir presented in the early 20th century (Langmuir 1916). Because it is theoretically deduced, it continues to be feasible and appropriate. It assumes that the molecules striking the surface have a given probability of adsorbing. Similarly, molecules already adsorbed have a given probability of desorbing. At equilibrium, equal numbers of molecules desorb and adsorb at any time. The probabilities are related to the strength of the interaction between the adsorbent surface and the adsorbate. That is the physical meaning of Eq. 8.5:

$$q = q_{\max}\frac{C_1}{b + C_1} \tag{8.5}$$

where

q_{max} is the maximum q capacity (mg of metal·[mg of adsorbent]$^{-1}$) and
b is the Langmuir adsorption constant (mg of metal·L^{-1}).

This expression is usually applied in its linear form (Mohan et al. 2005), which gives us the goodness of the model, but it is not recommended for a precise estimation of parameters. Reliable values of q_{max} and b are only obtained through nonlinear least-squares data adjustment, since there is no need of homocedasticity, homogeneity of variance, or normal distribution of errors (Kumar et al. 2008).

Accordingly, these nonlinear adjustments were carried out for up to 12 adsorption systems that include the three best adsorbents plus the one derived from *Acacia mearnsii* (Weibull black) and the three heavy metals that present more affinity by the adsorbents; they are UO_2^{2+}, Pb^{2+}, and Hg^{2+}. The experimental data were obtained in similar stirring conditions, at 20°C and after an equilibrium period of 1 week.

As an example, Figure 8.4 presents the isotherm for PFC- UO_2^{2+}. This is a representative curve, very similar to the rest of them. At first glance, it presents the typical shape of the L-class in the specific classification (Giles et al. 1974): a concave initial region followed by a saturation zone in the last part of the curve. This is usual in those systems where the adsorption becomes more difficult as the pollutant concentration increases.

Table 8.5 presents the statistical summary for every adsorption system. As can be checked, the values of q_{max} confirm the hierarchy of adsorbents. The least efficient is the one derived from *Acacia mearnsii* (labeled WFC), which presents a mean q_{max} level of 159 mg·g^{-1}, whereas PFC and CFC present 226 and 229, respectively.

Graphically, the adequacy of each adjustment can be appreciated under a linear plot. This is represented in Figure 8.5, where predicted q (according to Langmuir's expression and the corresponding parameter data in Table 8.5) is plot versus experimental q values. The nearer to the angle bisection the points are placed, the more accurate the adjustment is. As can be appreciated, the large

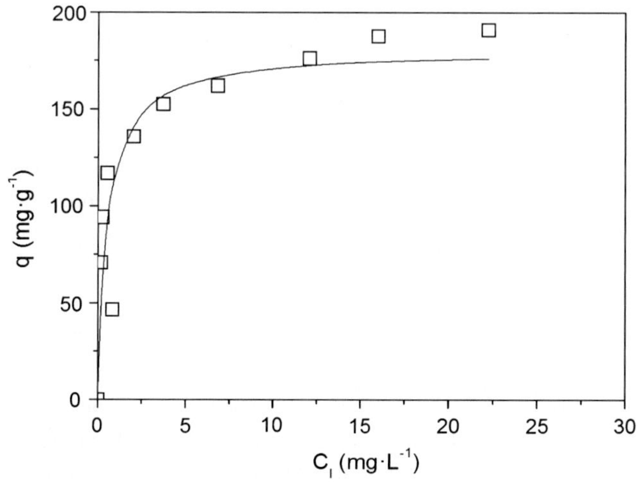

FIGURE 8.4 Isotherm for PFC- UO_2^{2+} adsorption system.

Table 8.5 Nonlinear Langmuir Adjustment Data for Each Adsorption System (units in text)

Metal	CFC	PFC	QFC	WFC
Pb^{2+}	$q_{max} = 287.9$ $b = 1.15$ $r^2 = 0.87$	$q_{max} = 188.5$ $b = 0.41$ $r^2 = 0.89$	$q_{max} = 125.3$ $b = 0.35$ $r^2 = 0.97$	$q_{max} = 98.8$ $b = 0.67$ $r^2 = 0.98$
UO_2^{2+}	$q_{max} = 225.7$ $b = 2.44$ $r^2 = 0.81$	$q_{max} = 179.4$ $b = 0.49$ $r^2 = 0.82$	$q_{max} = 154.9$ $b = 3.37$ $r^2 = 0.97$	$q_{max} = 155.2$ $b = 7.34$ $r^2 = 0.94$
Hg^{2+}	$q_{max} = 175.4$ $b = 7.78$ $r^2 = 0.95$	$q_{max} = 311.5$ $b = 6.10$ $r^2 = 0.81$	$q_{max} = 622.3$ $b = 61.1$ $r^2 = 0.96$	$q_{max} = 223.2$ $b = 14.1$ $r^2 = 0.98$

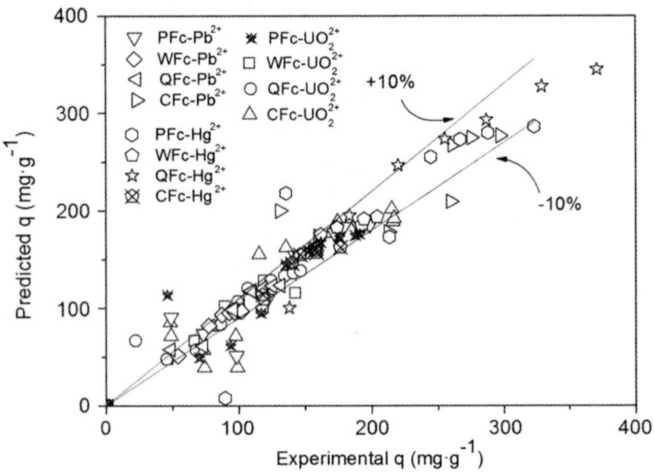

FIGURE 8.5 Predicted versus experimental *q* values.

majority of experiments are placed inside an error margin of ±10%, which is commonly accepted as a significant accuracy level.

If compared with other adsorbent materials, *tanningels* present a very competitive adsorption capacity, since the values obtained in the current work are inside the range of those presented by Babel and Kurniawan (2003), including a wide variety of waste materials, or by Bhattacharyya and Gutpta (2008) regarding the use of natural and modified clays.

4 Conclusions

Tanningels have been confirmed as promising adsorbents for removal of several heavy metal ions from aqueous effluents. Statistical approaches have pointed out that a high content of formaldehyde is needed for the most efficient *tanningel,* and cypress was found to be the best tannin source, followed by pine, quebracho, and acacia. On the other hand, the best *tanningels* performed a very significant metal uptake with lead, uranium, and mercury. Every adsorption system responded to Langmuir's model, with a very high correlation factor and significantly high adsorption capacities.

Glossary

Acacia mearnsii de Wild a fast-growing leguminous tree native to Australia. Common names for it include Black Wattle, *Acácia-negra* (Portuguese), Australian Acacia, *Australische Akazie* (German), *Swartwattel* (Afrikaans), and *Uwatela* (Zulu). It is considered the first tannin feedstock due to the high content of polyphenols in its bark.

Adsorbent whatever material that is used in adsorption process; that is, the accumulation of atoms or molecules on the surface of a material. This process creates a film of the adsorbate (the molecules or atoms being accumulated) on the adsorbent's surface.

Freundlich model an adsorption isotherm, which is a curve relating the concentration of a solute on the surface of an adsorbent, to the concentration of the solute in the liquid with which it is in contact. This model has an experimental basis and does not include a saturation zone.

Gelification a physicochemical modification that implies immobilizing the target substance by making it become gel. Usually, this is achieved by polymerizing with a gelifying agent, such as, in the case of tannins, formaldehyde.

Langmuir model an adsorption isotherm that describes quantitatively the buildup of a layer of molecules on an adsorbent surface as a function of the concentration of the adsorbed material in the liquid in which it is in contact. Its shape is a gradual positive curve that flattens to a constant value. It often represents an initial surface adsorption, followed by a condensation effect resulting from extremely strong solute–solute interaction.

Schinopsis balansae a hardwood tree that forms forests in the subtropical Gran Chaco ecoregion of northeastern Argentina, and Paraguay. It is also found in the wild Pantanal vegetation of Brazil. It is the second most important tannin feedstock, after *Acacia mearnsii* de Wild.

Tannins astringent, bitter plant polyphenols that either bind and precipitate or shrink proteins. They are present in the large majority of plant barks, fruits, or pods, and they are used mainly in the tanning industry.

Acknowledgments

This investigation was supported by the COMISIÓN INTERMINISTERIAL DE CIENCIA Y TECNOLOGÍA (CICYT) CTQ 2010-14823/PPQ project.

References

Altenor, S., Carene, B., Emmanuel, E., Lambert, J., Ehrhardt, J.J., Gaspard, S., 2009. Adsorption studies of methylene blue and phenol onto *vetiver* roots activated carbon prepared by chemical activation. J Hazard Mater. 165 (1–3), 1029–1039.

Aroua, M.K., Leong, S.P.P., Teo, L.Y., Yin, C.Y., Daud, W.M.A.W., 2008. Real-time determination of adsorption of Lead (II) onto palm shell-based activated carbon using ion selective electrode. Biores Technol. 99 (13), 5786–5792.

Babel, S., Kurniawan, T.A., 2003. Low-cost adsorbents for heavy metals uptake from contaminated water: A review. J Hazard Mater. 97, 219–243.

Bahattacharyya, K.G., Gupta, S.S., 2008. Adsorption of a few heavy metals on natural and modified kaolinite and montmorillonite: A review. Advan Coll Interfac Sci. 140, 114–131.

Beltrán-Heredia, J., Sánchez-Martín, J., 2008. Removing heavy metals from polluted surface water with a tannin-based flocculant agent. J Hazard Mater. 165 (1–3), 1215–1218.

Bruggen, B., Vandecasteele, C., 2003. Removal of pollutants from surface water and groundwater by nanofiltration: Overview of possible applications in the drinking water industry. Environ Poll. 122 (3), 435–445.

Charerntanyarak, L., 1999. Heavy metals removal by chemical coagulation and precipitation. Water Sci Technol. 39 (10–11), 135–138.

Convington, A.D., 1997. Modern tannin chemistry. Chem Soc Rev. 26 (2), 111–126.

Demirbas, A., 2008. Heavy metal adsorption onto agro-based waste materials: A review. J Hazard Mater. 157 (2–3), 220–229.

Giles, C.H., D'Silva, I.A., Easton, I.A., 1974. A general treatment and classification of the solute adsorption isotherm. Part II. Experimental interpretation. J Colloid Interface Sci. 47 (3), 766–778.

Haslam, E., 1989. Plant polyphenols—vegetables and tannins revisited. Cambridge University Press, Cambridge.

Ho, Y.S., McKay, G., 1998. The kinetics of sorption of basic dyes from aqueous solutions by *sphagnum* moss peat. Can J Chem Eng. 76 (4), 822–827.

Kim, Y.H., Nakano, Y., 2005. Adsorption mechanism of palladium by redox within condensed-tannin gel. Water Res. 39 (7), 1324–1330.

Kumar, K.V., Porkodi, K., Rocha, F., 2008. Isotherms and thermodynamics by linear and non-linear regression analysis for the sorption of methylene blue onto activated carbon: Comparison of various error functions. J Hazard Mater. 151 (2-3), 794–804.

Kurniawan, T.A., Chan, G.Y.S., Lo, W.-H., Babel, S., 2006. Physicochemical treatment techniques for wastewater laden with heavy metals. Chem Eng J. 118 (1–2), 83–98.

Lagergren, S., 1898. Zur theorie der sogenannten adsorption geloster stoffe. Kungliga Svenska Vetenskapsakademiens. 24 (4), 1–39.

Langmuir, I., 1916. The constitution and fundamental properties of solids and liquids. Part I. Solids. J Am Chem Soc. 38 (11), 2221–2295.

López-Suevos, F., Riedl, B., 2003. Effects of *Pinus pinaster* bark extracts content on the cure properties of tannin-adhesives and on bonding of exterior grade MDF. J Adhes Sci Technol. 17 (11), 1507–1522.

Manchón-Vizuete, Macías-García, A., Nadal-Gisbert, A., Fernández-González, C., Gómez-Serramo, V., 2005. Adsorption of mercury by carbonaceous adsorbents prepared from rubber of tyre wastes. J Hazard Mater. 119 (1–3), 231–238.

Mohan, D., Singh, K.P., Ghosh, D., 2005. Removal of α-picoline, β-picoline and γ-picoline from synthetic wastewater using low cost activated carbons derived from coconut shell fibers. Environ Sci Technol. 39 (13), 5076–5086.

Nakano, Y., Takeshita, K., Tsutsumi, T., 2001. Adsorption mechanism of hexavalent chromium by redox within condensed-tannin gel. Water Res. 35 (2), 496–500.

Ngah, W.S.W., Hanafiah, M.A.K.F.W.S., 2008. Adsorption of copper on rubber (*Hevea brasiliensis*) leaf powder: Kinetic, equilibrium and thermodynamic studies. Biochem Eng J. 39 (3), 521–530.

Palma, G., Freer, J., Baeza, J., 2003. Removal of metal ions by modified *Pinus radiata* bark and tannins from water solutions. Water Res. 37 (20), 4974–4980.

Pizzi, A., 1994. Advanced wood adhesives technology. Marcel Dekker, New York.

Pizzi, A., 2008. Monomers, polymers and composites from renewable sources. In: Belgacem, M., Galdini, A. (Eds.), Tannins: major sources, properties and applications. Elsevier, Amsterdam.

Pollard, F.H., Hanson, P., Geary, W.J., 1959. 4-(2-pyridylazo)-resorcinol as a possible analytical reagent for the colorimetric estimation of cobalt, lead and uranium. Anal Chim Acta. 20, 26–31.

Qi, W.B., Zhu, L.Z., 1985. Spectrophotometric determination of trace amounts of cadmium and zinc in waste water with 4-(2-pyridylazo)-resorcinol and mixed ionic and non-ionic surfactants. Talanta. 32 (10), 1013–1015.

Sánchez-Martín, J., Beltrán-Heredia, J., Carmona-Murillo, C., 2011a. Adsorbents from *Schinopsis balansae*: Optimisation of significant variables. Ind Crop Prods. 33 (2), 409–417.

Sánchez-Martín, J., Beltrán-Heredia, J., Gibello-Pérez, P., 2011b. Adsorbent biopolymers from tannin extracts for water treatment. Chem Eng J. 168 (3), 1241–1247.

Sánchez-Martín, J., González-Velasco, M., Beltrán-Heredia, J., Gragera-Carvajal, J., 2010. Salguero-Fernández-J. Novel tannin-based adsorbents in removing cationic dye (methylene blue) from aqueous solution. Kinetics and equilibrium studies. J Hazard Mater. 174 (1–3), 9–16.

Schofield, P., Mbugua, D.M., Pell, A.N., 2001. Analysis of condensed tannins: a review. Anim Feed Sci Technol. 91 (1), 21–40.

Sengil, I.A., Özacar, M., 2009. Competitive biosorption of Pb^{2+}, Cu^{2+} and Zn^{2+} ions from aqueous solutions onto valonia tannin resin. J Hazard Mater. 166 (2-3), 1488–1494.

Shirato, W., Kamei, Y., 1992. Insoluble tannin preparation process, waste treatment process and adsorption process using tannin. U.S. Patent 5,158,711.

Tondi, G., Oo, C.W., Pizzi, A., Thevenon, M.F., 2009b. Metal absorption of tannin-based rigid foams. Ind Crops Prods. 29 (2–3), 336–340.

Tondi, G., Zhao, W., Pizzi, A., Du, G., Fierro, V., Celzard, A., 2009a. Tannin-based rigid foams: A survey of chemical and physical properties. Biores Technol. 100 (21), 5162–5169.

Tünai, O., Kabdasly, N.I., 1994. Hydroxide precipitation of complexed metals. Water Res. 98 (10), 2117–2124.

Vázquez, G., Alonso, R., Freire, S., González-Álvarez, J., Antorrena, G., 2006. Uptake of phenol from aqueous solutions by adsorption in a *Pinus pinaster* bark packed bed. J Hazard Mater. 133 (1–3), 61–67.

Vázquez, G., Antorrena, G., González, J., Doval, M.D., 1994. Adsorption of heavy metal ions by chemically modified *Pinus pinaster* bark. Biores Technol. 48 (3), 251–255.

Vázquez, G., González-Álvarez, J., Freire, S., López-Moreno, M., Antorrena, G., 2002. Removal of cadmium and mercury ions from aqueous solutions by sorption on treated *Pinus pinaster* bark: Kinetics and isotherms. Biores Technol. 82 (3), 247–251.

Wang, L.K., Hung, Y.T., Shammas, N.K., 2005. Physicochemical treatment processes. Humana Press, Totowa, NJ.

Wingenfelder, U., Hansen, C., Furrer, G., Schulin, R., 2005. Removal of heavy metals from mine waters by natural zeolites. Environ Sci Technol. 39 (12), 4606–4613.

Yantasee, W., Warner, C.L., Sangvanich, T., Addleman, R.S., Carter, T.G., Wiacek, R.J., Fryxell, G.E., Timchalk, C., Warner, M.G., 2007. Removal of heavy metals from aqueous systems with thiol functionalized superparamagnetic nanoparticles. Environ Sci Technol. 41 (14), 5114–5119.

Yotsuyanagi, T., Takeda, Y., Yamashita, R., Aomura, K., 1973. The extraction-spectrophotometric determination of chromium (III) with 4-(2-pyridylazo)-resorcinol. Anal Chim Acta. 67 (2), 297–306.

Yurtsever, M., Sengil, I.A., 2010. Biosorption of Pb(II) ions by modified quebracho tannin resin. J Hazard Mater. 174 (1–3), 58–64.

Zhao, W., Fierro, V., Pizzi, A., Du, G., Celzard, A., 2003. Effect of composition and processing parameters on the characteristics of tannin-based rigid foams. Part II: Physical properties. Mater Chem Phys. 123 (1), 210–217.

CHAPTER

9

The Contribution of the Coagulation Process in Controlling Microbial Risk and Disinfection By-products Formation in Drinking Water

Luigi Rizzo

Department of Civil Engineering, University of Salerno, 84084 Fisciano (SA), Italy

CHAPTER OUTLINE

The Role of Colloidal Systems in Environmental Protection. http://dx.doi.org/10.1016/B978-0-444-63283-8.00009-0

1 Introduction

Natural water resources used for potable purposes are characterized by different levels of purity depending on the type of water source. Typically, groundwater is a higher quality water source compared to surface water; accordingly, no specific treatment is required other than disinfection to guarantee a disinfectant residual inside the distribution network to protect water from microbial regrowth. In contrast, surface water may be polluted by several organic and inorganic contaminants, suspended particles, and microorganisms, making a specific water treatment necessary (Figure 9.1).

Low turbidity (typically < 10 NTU) surface water may be effectively treated by the direct filtration process, where coagulants can be added directly in the pipeline (Figure 9.1A). When turbidity is high (> 10 NTU), specific units for coagulation and sedimentation steps can significantly improve turbidity removal, and an oxidation step may be useful to protect the filtration bed from algae and bacterial colonization (Figure 9.1B). When metals (at concentrations higher than the respective limits set by regulations) and micropollutants occur in the raw water, preoxidation and adsorption units can also be used to further improve the quality of the water (Figure 9.1C).

The coagulation process has been historically used in drinking water treatments to remove colloids from surface water in order to decrease turbidity. Moreover, the discovery of harmful disinfection by products (DBPs) in chlorinated waters as well as the role of natural organic matter (NOM) in the formation of DBPs (Rook, 1974; Bellar et al., 1974) increased the importance of the coagulation process, which is now typically optimized to improve NOM removal. In the present review, the disinfection of drinking water is first discussed to address specific issues such as bacterial regrowth and DBPs. Then the coagulation process is shortly introduced by addressing its mechanisms and the use of conventional coagulants in drinking water treatments. Afterward, the importance of an effective

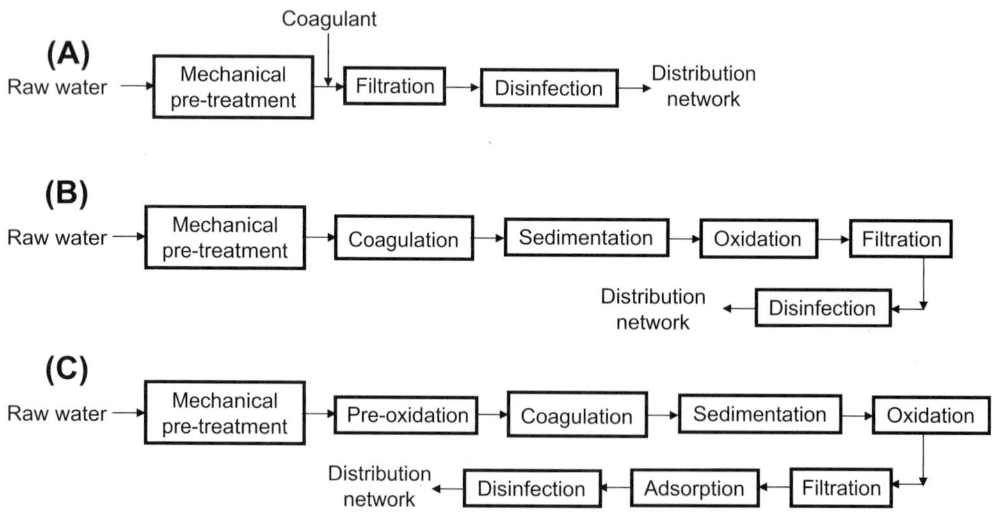

FIGURE 9.1

Surface water treatment processes for low-turbidity surface water (direct filtration) (A), high-turbidity surface water (B), metal oxidation, and micropollutants removal (C).

removal of turbidity and NOM is stressed. Finally, the main advances in the coagulation process are discussed, with particular regard to new, prepolymerized, and natural organic coagulants.

2 Disinfection of drinking water

Disinfection of drinking water is considered one of the major public health successes of the 20th century due to the dramatic reduction observed in the mortality from waterborne diseases. Even so, the threat of waterborne diseases has not been entirely eliminated. Within the past 50 years several waterborne disease outbreaks were documented; since the 1960s, *Escherichia coli* and *Giardia lamblia* were identified as microorganisms responsible for waterborne outbreaks. In 1975, a waterborne disease outbreak at Carter Lake National Park in Oregon was attributed to *Escherichia coli* (Craun, 1981). The biggest outbreak took place in Milwaukee in 1993 (when 403,000 people fell ill); it was associated with both water quality deterioration and decreased efficiency of coagulation/filtration processes (Kramer et al., 1996; MacKenzie et al., 1994).

2.1 Disinfection and bacterial regrowth

The disinfection process can accomplish different tasks according to raw water characteristics. Some disinfection processes (typically ozonation and UV radiation) can only act as *primary disinfection* systems because they can effectively inactivate several microorganisms, but they cannot guarantee a residual disinfectant inside the distribution network that is useful in protecting water from microbial regrowth. So, when ozonation and UV radiation are used as primary disinfectants in the treatment of surface water as well as low-quality groundwater, a subsequent *final disinfection* step (implementing chlorine, chlorine dioxide, or chloramines) should be used to send water to distribution network. In contrast, high-quality groundwater does not need any specific treatment, and, typically, final disinfection is used just to control the risk of microbial regrowth inside the distribution network.

The occurrence of microorganisms in the drinking water distribution network can be due to (i) infiltrations in the pipe system and (ii) disinfection treatment inefficiency. The inefficiency of the disinfection process can increase the number of bacteria in the distribution network (breakthrough). In particular, bacteria only affected by the disinfection process can restore their vital functions as soon as environmental conditions in the drinking water system are suitable. The factors affecting bacterial regrowth in the distribution system were found to be related to disinfectant residual, temperature, and nutrient level (LeChevallier et al., 1996; Power et al., 1997; Volk and LeChevallier, 1999; Bresner et al., 2002). Several works have documented the presence of bacteria also when chlorine residual up to 2.5 mg/L was detected (McFeters et al., 1986; LeChevallier et al., 1987; Lund and Ormerod, 1995; Rizzo et al., 2004a). The residual free chlorine concentration, temperature, and occurrence of total coliform bacteria were monitored in the distribution network of a municipality in the province of Salerno (Italy) by taking water samples from 25 uniformly and spatially distributed locations (Rizzo et al., 2004a). The free residual chlorine concentrations were found to vary in the 0.01–0.25 mg Cl_2/l range; the occurrence of total coliforms was found to significantly increase (40%) during the summer season because of increased temperatures (11°C), although free residual chlorine concentrations were at the same level or even higher than in the winter season, according to the results achieved by Pintar and Slawson (2003). Moreover, batch tests were carried out on the same tap water (groundwater source) to evaluate both (1) the regrowth of coliform bacteria affected by free chlorine residual concentrations using more selective (m-Endo) and less selective (m-T7) media and (2) the effect of the addition of carbon source on the regrowth of coliform bacteria. The investigators showed that a chlorine dose of up to 0.035 mg/L harmed coliform bacteria and at least 4 h incubation was needed for regrowth. However, when the dose of chlorine was increased to 0.07 mg/L, the coliform bacteria were not able to regrow and in addition, bacteria regrowth increased by increasing the C:N ratio from 1:40 to 11:40 (Rizzo et al., 2004a).

Groundwater, however, is typically expected to affect bacteria regrowth less than surface water. Accordingly, Niquette et al. (2001) found out that the drinking water distribution system in Brussels, fed by groundwater, is less susceptible to promoting bacterial regrowth; also, the regrowth was notably decreased in treated surface water when biodegradable DOC levels were below 0.25 mg C/l. Moreover, Boualam et al. (2002) found that coliform bacteria lost their culturability in diluted river water (0.5–1.5 mg/L of DOC) and algal bloom suspensions (1.3–2.5 mg/L of DOC); at lower initial DOC values, the culturability was lost more rapidly.

The material used to make the pipes that are part of the distribution system has also been found to affect micro-organism type and occurrence. Schwartz et al. (1998) demonstrated that the type of pathogen depended on the material used in the distribution system. LeChevallier et al. (1996) reported a positive relationship between the occurrence of coliform and iron pipes subjected to corrosion. Zacheus et al. (2000) demonstrated that the accumulation of biofilm on different surface materials was quite similar among polyvinyl chloride, polyethylene, and stainless steel pipes. On the other hand, soft deposits were recorded as the key site of microbial growth, mainly heterotrophic bacteria, actinomycetes, and fungi, in drinking water distribution networks (Zacheus et al., 2001). Finally, laboratory experiments confirmed that chloramine could decay rapidly in the presence of new copper pipe, providing a possible explanation for the rapid disinfectant loss and microbial regrowth (Nguyen et al., 2012).

2.2 Disinfection by products

Unfortunately, the addition of chemical disinfectants results in the formation of DBPs dangerous to human health. The formation of DBPs sparks remarkable concern when surface water is used for potable purposes. In fact, these water sources typically include some undesirable/dangerous substances (e.g., NOM, metals such as iron and manganese, bacteria, and viruses) that make a specific treatment by physical, chemical, and physical-chemical processes essential. The removal of metals, odors, and microorganisms is typically achieved by oxidation/disinfection agents such as chlorine (chlorine gas and hypochlorites), chlorine dioxide, chloramines, and ozone, which can react in different ways with substances that naturally occur in the water source to form DBPs (Table 9.1).

The major categories of DBPs formed during disinfection with chlorine are trihalomethanes (THMs) and haloacetic acids (HAAs), whose formation results from the reaction among free residual chlorine, bromide, and NOM. These DBPs together with chlorite and bromate are included among the regulated DBPs (Roccaro and Vagliasindi, 2007). Unfortunately, the levels of many emerging DBPs are more genotoxic than some of the regulated DBPs. Richardson et al. (2007) identified three categories of DBPs of particular interest. Among these, Category 1 contains eight DBPs with some or all of the toxicologic characteristics of human carcinogens: four regulated in the United States (bromodichloromethane, dichloroacetic acid, dibromoacetic acid, and bromate) and four unregulated DBPs (formaldehyde, acetaldehyde, MX, and N-nitrosodimethylamine (NDMA)).

Table 9.1 Main Disinfection by-Products and their Carcinogenic Effects

Disinfectant	Main DBPs		Cancer Descriptor[a]
Chlorine (gas and hypochlorites)	Trihalomethanes	Chloroform	L/N
		Bromodichloromethane	L
		Dibromochloromethane	S
		Bromoform	L
	Haloacetic acids	Monochloroacetic acid	I
		Dichloroacetic acid	L
		Trichloroacetic acid	S
		Monobromoacetic acid	-
Chlorine dioxide		Chlorite	D
Ozone		Bromate	B2
Chloramines		N-Nitrosodimethylamine	B2[b]

[a]*According to the USEPA (2006a): Carcinogenic to humans (H); Likely to be carcinogenic to humans (L); Likely to be carcinogenic above a specified dose but not likely to be carcinogenic below that dose because a key event in tumor formation does not occur below that dose (L/N); Suggestive evidence of carcinogenic potential (S); Inadequate information to assess carcinogenic potential (I); Not likely to be carcinogenic to humans (N); B2 indicates sufficient evidence in animals and inadequate or no evidence in humans; (D) Not classifiable as to human carcinogenicity.*
[b]*According to USEPA (2006b).*

The coagulation process is widely used in drinking water treatments. Since it can remove NOM before oxidation/disinfection with chlorine, the concentration of chlorination by-products can be decreased accordingly.

3 The coagulation process

Natural waters include a lot of impurities varying in size, by about six orders of magnitude, from a few Angstrom (dissolved substances) to a few hundred microns (suspended solids). Settling suspended solids can be removed by gravitational force in a sedimentation tank, but many contaminants are too small to be effectively removed in this way. The coagulation process, which promotes the aggregation of smaller particles into larger aggregates, allows an effective separation from the water by a subsequent sedimentation and/or filtration unit. The process by which the particles in suspension are modified in order to increase their tendency to aggregation is known as destabilization.

3.1 Coagulation mechanisms

Destabilization of colloidal suspension can take place by means of different coagulants in different ways. The main destabilization mechanisms are the following: double-layer compression, surface-charge neutralization, enmeshment, and interparticle bridging (O'Melia, 1972).

Colloidal particles carry an electrical (primary) charge that can be either positive or negative depending on the pH and ionic content of the suspension; in the typical pH range values of natural water and wastewater, the average charge is negative. Since the primary charge on the particles must be counterbalanced, an equivalent excess of ions of opposite charge accumulates close to the interfacial region; this layer, together with the primary charge, forms an electrical double layer. The mechanism of double-layer compression takes place if enough ions (e.g, an electrolyte such as NaCl), opposite in sign to the net charge of the particle surface, are added to the suspension. The diffuse layer is therefore compressed, and the energy required to a small particle approaching to one other decreases.

The destabilization mechanism by surface-charge neutralization occurs by reducing the net surface charge of the particles in a colloidal suspension. This process can take place using coagulants such as metal salts. These compounds form hydrolysis products characterized by a charge opposite in sign to the net surface charge of the particles; therefore adsorption can occur.

The enmeshment mechanism can occur when coagulants such as metal salts (e.g., alum, ferric chloride) or metal oxide or hydroxide (e.g., CaO, $Ca(OH)_2$) are added in sufficiently high concentration to result in a fast precipitation of a metal hydroxide or metal carbonate and colloids can be enmeshed in these precipitates as they settle down.

Destabilization by bridging typically occurs when high-molecular-weight polymer entraps more particles, forming an interparticle bridge. Two conditions are necessary in order for this mechanism to take place: (1) the polymer must be long enough to extend beyond the electrical double layer and (2) the particle to be entrapped must have sites available for adsorption.

3.2 Conventional coagulants

The most widely used coagulants are aluminium and iron salts. In aqueous solutions, the small, highly positive ions contained in these salts form strong bonds with oxygen atoms (hydrolysis). If a sufficiently high amount of metal ions is present in the solution, the hydrolysis process continues, producing complex polynuclear species that can form microcrystals and metal hydroxide precipitate able to adsorb particles. The solubility of the metal hydroxide precipitates is one of the most important

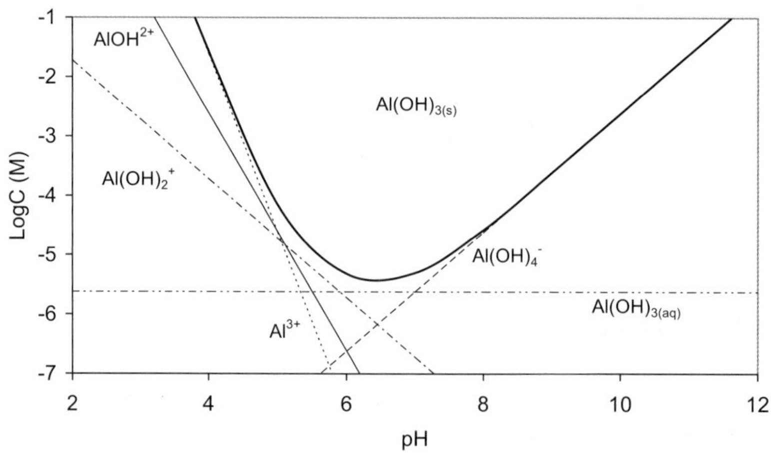

FIGURE 9.2

Aluminium solubility diagram plotted using stability constants reported by Benjamin (2002).

factors that must be considered in improving the efficiency of the coagulation process as well as in controlling the metal residual in finished water. For instance, when aluminium salt is dissolved in water, it results in the formation of well-known hydrolysis products (Al^{3+}, $AlOH^{2+}$, $Al(OH)_2^+$, $Al(OH)_4^-$, $Al(OH)_{3(s)}$) whose presence strongly depends on the pH level. At a low pH, the aluminium salt-based solution is characterized by positively charged aluminium soluble species (Al^{3+}, $AlOH^{2+}$, $Al(OH)_2^+$), whereas at a high pH a negatively charged, a soluble product ($Al(OH)_4^-$) is produced (Figure 9.2).

The major drawback in the coagulation of natural waters is the inability to control the nature of the coagulant species formed (Jiang, 2001). In the pH range of natural waters (6–8), extremely rapid and uncontrolled hydrolysis, and frequently, very rapid precipitation occur as metal salts are dissolved in water. In order to improve the effectiveness of metal salts, these can be partially hydrolyzed prior to the addition to raw water in order to form polymeric metal species (prepolymerized coagulants). Prepolymerized coagulants can work efficiently in a wide range of pHs and changes in water temperature and raw water characteristics.

4 NOM and turbidity removal by coagulation

The coagulation process has been used primarily in drinking water treatments to remove turbidity. Since the discovery of DBPs such as THMs in chlorinated waters (Rook, 1974; Bellar et al., 1974), and consequently the role of NOM in the formation of such by-products, the removal of NOM by coagulation has gained growing attention. The NOM fraction responsible for DBP formation is a mixture of humic and fulvic acids, which are typically characterized by means of total organic carbon (TOC), dissolved organic carbon (DOC), and UV absorbance at 254 nm (UV_{254}) measurements. Moreover, in the effort to better characterize the NOM fraction responsible for the formation of chlorination by-products, new aggregate parameters were developed, such as specific ultraviolet absorbance (SUVA; the absorbance at 254 nm per unit of DOC in mg/L) (Edzwald et al., 1985) and the differential absorbance (Korshin et al., 2002).

4.1 Enhanced coagulation

In the United States, in order to protect the consumer of drinking water from DBP risk, the U.S. Environmental Protection Agency (USEPA,1998) promulgated Stage 1 of the Disinfectants and Disinfection By-Products Rule establishing both maximum contaminant levels for four DBPs categories (TTHM, HAAs, bromate, and chlorite) and TOC removal requirements for water systems that use conventional filtration treatments. Moreover, water treatment plants that cannot meet the criteria delineated in this rule are allowed to establish alternative TOC removal requirements implementing the Jar test procedure known as enhanced coagulation (USEPA, 1999). This term refers to the process of improving the removal of DBP precursors in a conventional water treatment plant. The USEPA has developed a TOC-based performance standard for enhanced coagulation (and enhanced precipitative softening) using a two-step system. Step 1 requires removal of a specific percentage of influent TOC to demonstrate compliance, based on the TOC and alkalinity of the source water. If an enhanced coagulation plant does not achieve the specified TOC removal, it must proceed to Step 2, which requires Jar testing to set an alternative TOC percent removal for determining compliance.

4.2 NOM versus turbidity removal

Although the formation of DBPs focuses on NOM removal, turbidity is also an important parameter because turbidity spikes may result in the breakthrough of pathogens. The role of turbidity as an indicator of a pathogen's occurrence is related to (1) the electrostatic proprieties of the surface of some microorganisms (Drozd and Schwartzbrod, 1996; Brush et al., 1998; Butkus et al., 2003) and (2) the possibility that they can be removed by coagulants according to the same mechanisms by which turbidity is removed during the coagulation process (Bustamante et al., 2001).

The Long Term 1 Enhanced Surface Water Treatment Rule of the USEPA (2002) set very stringent limits on effluent turbidity in both conventional and direct filtration plants. According to that regulation, the turbidity values of 95% of samples taken each month must be less than or equal to 0.3 NTU and never more than 1 NTU to improve control of microbial pathogens, including specifically the protozoan *Cryptosporidium*. Table 9.2 summarizes some data available in the scientific literature dealing with NOM and turbidity removal during the coagulation of drinking water.

According to the above-discussed issues, the coagulation process should be optimized for the purposes of: (1) maximizing NOM removal (in order to reduce DBP formation and bacterial regrowth in drinking water systems), (2) maximizing turbidity removal (in order to control the risk of pathogen breakthrough), (3) minimizing residual coagulant, and (4) minimizing sludge production. The optimization procedure involves types of coagulant, coagulation pH, and coagulant dosage. The pH of coagulation is dependent on raw water quality and coagulant type.

5 Advances in the coagulation process

The advances in the coagulation process are mainly related to the need (1) to improve the coagulants' efficiency and (2) to find new, natural, environmental-friendly coagulants. The following subsections review prepolymerized, natural organic and nonconventional (potassium ferrate, magnetic anion exchange resin [MIEX®, registered trademark of Orica, Australia], and titanium salts) coagulants.

5.1 Prepolymerized coagulants

One of the main drawbacks of the metal salt coagulation-based process is its inability to control the nature of the coagulant species that rapidly forms after the coagulant is added to the solution. In order to solve this problem, several prepolymerized coagulants have been developed and investigated.

Polyaluminium chloride (PACl) is probably the most well known and most frequently used of the prepolymerized coagulants. PACl was found to be more effective than alum and ferric chloride in removing TOC and total THM formation potential (THMFP) from high alkalinity surface water (Rizzo et al., 2004b). Moreover, PACl (60 mg/L) decreased toxicity to *D. magna* better than alum at the same concentration in coagulated surface water with high amount of alkalinity (150 mg $CaCO_3$/l), bromide (274 μg/l), and chloride (98 mg/L) (Selcuk et al., 2007). The effect of alkalinity/pH and PACl speciation was investigated according to turbidity and NOM removal by Yan et al. (2008). Three kinds of river waters in China with low, moderate, and high alkalinity were selected. At neutral and basic pH values, PACls with higher basicity (ratio of OH^-/Al) were found to be more efficient for turbidity and NOM removal. At slightly acidic pH, PACls with lower basicity were found to be more efficient. Optimal NOM removal was achieved at pH 5.5–6.5 for all PACls investigated. In order to improve

Table 9.2 NOM and Turbidity Removal During Coagulation of Drinking Water

Water Source	Coagulant and Doses (mg/L)	NOM Removal (%)	Turbidity Removal (%)	Final Turbidity (NTU)	Comments	References
Surface water: 4.81 mg TOC/l, 102 μg TTHM/l	Alum: 44–152	29–54 (TOC) 39–64 (UV_{254})	-	0.73–0.99	The best NOM and turbidity removal were achieved under optimized coagulation. 73% TTHM removal was also detected.	Miltner et al. 1994
Surface water: pH 6.8, 2.83 mg TOC/l (2.62 mg DOC/l), UV_{254} 0.008 1/cm, alkalinity 21 mg/L, turbidity 9.3 NTU	PACl: 20 Ferric chloride: 20	29 (DOC) 46 (UV_{254}) 45 (DOC) 96 (UV_{254})	90 88	0.9 1.1	The NOM and turbidity removal data refer to optimized coagulation conditions.	Volk et al. 2000
Surface water: pH 7.6, 7.91 mg TOC/l (5.59 mg DOC/l), UV_{254} 0.210 1/cm, alkalinity 41 mg/L, turbidity 39 NTU	Alum: 30 Ferric chloride: 20	44 (DOC) 69 (UV_{254}) 58 (DOC) 85 (UV_{254})	97 98	1.0 0.7	The NOM and turbidity removal data refer to optimized coagulation conditions.	Volk et al. 2000
Surface water: pH 7.8, 2.69 mg TOC/l (2.75 mg DOC/L), UV_{254} 0.060 1/cm, alkalinity 95 mg/L, turbidity 1.5 NTU	Ferric chloride: 20 PACl: 20	18 (DOC) 35 (UV_{254}) 5 (DOC) 8 (UV_{254})	62 56	0.6 0.6	The NOM and turbidity removal data refer to optimized coagulation conditions.	Volk et al. 2000
Surface water: pH 8.0, 2.70 mg TOC/l (2.63 mg DOC/l), UV_{254} 0.065 1/cm, alkalinity 133 mg/L, turbidity 2.0 NTU	Ferric chloride: 35 Alum: 35	32 (DOC) 62 (UV_{254}) 4 (DOC) 43 (UV_{254})	81 58	0.4 0.8	The NOM and turbidity removal data refer to optimized coagulation conditions.	Volk et al. 2000
Surface water: pH 8.3, 2.75 mg TOC/l (2.50 mg DOC/l), UV_{254} 0.071 1/cm, alkalinity 214 mg/L, turbidity 21 NTU	Alum: 40 Ferric chloride: 30	10 (DOC) 55 (UV_{254}) 28 (DOC) 66 (UV_{254})	95 98	1.1 0.4	The NOM and turbidity removal data refer to optimized coagulation conditions.	Volk et al. 2000

Continued

Table 9.2 NOM and Turbidity Removal During Coagulation of Drinking Water—cont'd

Water Source	Coagulant and Doses (mg/L)	NOM Removal (%)	Turbidity Removal (%)	Final Turbidity (NTU)	Comments	References
Surface water: pH 7.49±0.14; 2.81±0.19 mg TOC/l; alkalinity 190.3±2.79 mg/L; turbidity 10.66±6.74 NTU	Alum: 30-80 Ferric chloride: 30–80	16–46 (TOC) 22–46 (TOC)	50–93 37–95	0.8–6.0 0.6–5.0	The best TOC and turbidity removals were achieved at 80 mg/L of coagulant dose.	Rizzo et al. 2005
Surface water: pH 8.6, UV_{254} 0.126 1/cm, alkalinity 200 mg/L, turbidity 4.9 NTU	Aluminium chloride: 0.1 mM (as Al)	16.5 (UV_{254})	75	1.2	The coagulation process was optimized according to UV_{254} removal.	Yan et al. 2009
Surface water: pH 8.2, UV_{254} 0.069 1/cm, alkalinity 115 mg/L, turbidity 8.8 NTU	Aluminium chloride: 0.12 mM (as Al)	27 (UV_{254})	91	0.8	The coagulation process was optimized according to UV_{254} removal.	Yan et al. 2009

NOM removal, the coagulation process might be optimized according to basicity, speciation, and dosage of coagulant as well as alkalinity of raw water.

Polyaluminium iron sulphate and polyferric sulphate (PFS) were compared to conventional coagulants such as ferric sulphate and aluminium sulphate, for the coagulation of algal-type model waters and a lowland surface water containing algae and NOM (Jiang and Graham 1998). The performance of prepolymerized coagulants was found to be significantly higher compared to conventional coagulants. Polyaluminium ferric chloride (PAFC) removed turbidity in kaolin suspension better than PACl but worse than PFS for coagulant doses less than 0.14 mmol/L (Gao et al. 2002). PAFC decreased turbidity more than PACl and PFS for coagulant doses higher than 0.14 mmol/L because PFS restabilized the kaolin suspension for relatively higher coagulant doses. Polyaluminium chloride sulphate was seen to effectively remove NOM and turbidity at a fixed bacicity ([OH]/[Al]) value of 2.0 and SO_4^{2-}/Al^{3+} ratio of 0.0664 (Gao and Yue, 2005).

5.2 Natural organic coagulants

The use of aluminium-based coagulants, particularly in drinking water treatments, has raised great concern about the release of aluminium into the effluent: Aluminium may be linked with Alzheimer's disease as well as carcinogenic properties (Miller et al., 1984; Mallevialle et al., 1984; McLachlan, 1995). Moreover, the impact of ferric salts and synthetic polymers on human health has not yet fully been investigated. In this context, the use of natural organic coagulants (NOCs) has gained increasing interest because they do not release any dangerous chemicals into the treated water.

NOCs can be classified according to the natural source from which they are extracted: plants seeds (e.g, *Moringa oleifera*, *Prosopis juliflora*, *Strychnos potatorum*, *Mais*), plants (e.g., *Cactus latifaria*, *Cacataceous opuntia*), and animals (e.g., chitosan is produced from chitin, which is extracted from the outer skeleton of insects, crabs, shrimps, and lobsters). Because of the high costs of conventional coagulants, NOCs have been proposed for water treatment in developing countries in order to decrease turbidity (Jahn, 1988). Among these sources, *Moringa oleifera* has been the most studied plant for water treatment (Muyibi and Evison, 1995; Ndabigengesere et al., 1995; Okuda et al., 2001; Sharma P. et al., 2006). *Moringa oleifera* is a tropical plant belonging to the family *Moringaceae*, which contains an active coagulating compound in its seeds. *Moringa oleifera*, commonly known as the horse-radish tree or drumstick tree, is native to India butcan now be found throughout the tropics. The effectiveness of *Moringa oleifera* as a coagulant is due to the presence of water-soluble cationic proteins in its seeds. These proteins are densely charged cationic dimmers (the protein complex is made up of two subunits with a positive charge) with a molecular weight of about 6.5–13 kDa and isolelectric pH value of 10 and 11 (Gassenschmidt et al., 1995; Ndabigengesere et al., 1995). The main mechanisms of coagulation seem to be adsorption and neutralization of the charge (Ndabigengesere et al., 1995). *Moringa oleifera* seeds do not significantly affect pH, conductivity, and alkalinity, but they increase orthophosphate, nitrate, and organic matter; moreover, the volume of sludge produced is considerably less compared to alum (Ndabigengesere and Narasiah, 1998). The main drawback in the use of NOCs is the release of organic matter, which acts as a precursor of THMs in chlorinated waters. In order to reduce the release of organic matter, *Moringa oleifera* seeds should be used only after the active proteins (Ndabigengesere and Narasiah, 1998) have been adequately purified. Okuda et al. (2001) showed that coagulant extracted from *Moringa oleifera* seeds by salt solution (NaCl) did not increase DOC after coagulation.

The encouraging results achieved in water treatment by *Moringa oleifera* have increased interest in new plants with coagulant properties. Diaz et al. (1999) tested two indigenous Venezuela plants, *Cactus latifaria* and *Prosopis juliflora,* on synthetic waters. Both coagulants showed comparable removal of turbidity, with final turbidity values close to the required standard of 5 NTU starting from high (100–200 NTU) and low (30–40 NTU) initial turbidities. Zhang et al. (2005) used a cactus plant (*Cactaceous opuntia*) originating from torrid areas and the subtropics as a coagulant. The optimum dosage according to turbidity removal (94%) was found to be about 50 mg/L. Moreover, in contrast to previous studies on *Moringa oleifera* (Ndabigengesere and Narasiah, 1998; Okuda et al., 2001), this coagulant was able to remove organic matter. In addition, the ability of seed extracts of several species of chestnut and acorn to act as natural coagulants was investigated using asynthetic turbid water (Šćiban et al., 2009). Active components were extracted from ground seeds of Horse chestnut and acorns of some species of the family *Fagaceae*. The seed extracts from European chestnut and common oak acorn were found to be the most efficient: 80% and 70% turbidity removals were achieved in low and medium initial turbidities water, respectively, at the lowest coagulant dose (0.5 ml/L).

Vegetable tannins are polyphenolic products of plant origin that occur in almost every part of the higher plants. Their effectiveness in water treatments depends mainly on their chemical structure and degree of tannin modification. Özacar and Sengil (2003) comparatively investigated tannin extracted from vanolia with a synthetic anionic polyelectrolyte (AN913) as a coagulant aid in drinking water treatments. Both alum + tannin and alum + AN913 showed significant improvement of turbidity removal from synthetic water compared to alum only. In particular, tannin was more effective than AN913 as a coagulant aid, and it significantly reduced the required doses of the alum. Moreover, tannin and alum together significantly improved sludge filterability compared to alum only.

Among NOCs, chitosan, a linear cationic polymer of high molecular weight obtained by deacetylation of chitin, has also garnered increasing interest. Chitin is a cellulose biopolymer widely distributed in nature, especially in marine invertebrates, insects, fungi, and yeasts. Coagulation experiments on synthetic and real water samples were carried out using chitosan, PACl, and a mixed coagulant (chitosan and PACl at various ratios) (Pan et al., 1999). The experiments performed on raw water at pH 7 showed that chitosan and the mixed coagulant (1:1 ratio) have a similar degree of turbidity removal, which was far better than PACl alone. One of the limiting factors in using chitosan as a coagulant in water treatment is the release of organic matter; so the choice of the solvent to be used to prepare a chitosan solution will affect the amount of organic matter that will be released during the coagulation process.

Batch tests with synthetic turbidity water samples were performed to evaluate the coagulation efficiency of chitosan prepared with different concentrations of acetic acid (HA) (which is the most common solvent for chitosan) and hydrochloric acid (HCl) (Huang et al., 2000). Although the chitosan/HA coagulant was as efficient as the chitosan/HCl coagulant, the TOC after coagulation with chitosan/HA significantly increased. Chitosan efficiency was also tested in a coagulation-direct filtration pilot plant in order to identify optimum conditions for the treatment of low turbidity, soft, humic waters (Eikebrokk, 1999). The optimum removal of NOM (around 50%) using chitosan was obtained for a 7.5 mg/L coagulant dose at pH 5.5. Subsequent chlorination steps may affect the by-products' formation and final toxicity. Accordingly, in a recent study, chitosan was compared to conventional coagulants (aluminium sulphate and ferric chloride) in terms of turbidity and NOM removal, as well as acute toxicity to *Daphnia magna* in coagulated and coagulated/chlorinated surface water (Rizzo et al., 2008). All coagulants decreased toxicity to *D. magna* from 100% to 0%

immobilization under actual conditions. When waters were spiked with humic acids, however, final toxicity was significantly affected according to coagulant type and dose. Moreover, the chlorination step after coagulation also increased toxicity, depending on the coagulant (chitosan > ferric chloride > aluminium sulphate).

In a subsequent study, the efficiency of chitosan, alum, and ferric chloride used as a pretreatment to granular media filtration was investigated according to the removal of particles, turbidity, *Cryptosporidium parvum oocysts,* and *oocyst*-sized polystyrene microspheres (Brown and Emelko, 2009). Under optimized operating conditions, all coagulants generally resulted in similar turbidity and particle reductions and median reductions of *C. parvum oocysts* (~4-log) and *oocyst*-sized microspheres (~3–4-log). At all doses investigated, chitosan produced high turbidity and particle removal by filtration (filter effluent turbidities consistently < 0.1 NTU and <10 total particles ≥ 2 μm) that were comparable to those achieved when in-line filtration was preceded by alum and $FeCl_3$ coagulation during stable (optimized) operations.

5.3 Other coagulants

Because of the high interest in the coagulation process, several new coagulants have been proposed or investigated, including potassium ferrate, Magnetic Anion Exchange Resin (MIEX®), and titanium salts.

5.3.1 Potassium ferrate

Potassium ferrate (VI) (K_2FeO_4) was shown to be really effective as a disinfectant and coagulant. Under acidic conditions, its redox potential (2.2 V) is grater than ozone (2.0 V), which is the strongest oxidant among all the oxidants/disinfectants typically used in water and wastewater treatments. Moreover, ferrate(VI) ions are reduced to Fe(III) ions or ferric hydroxide, showing coagulant properties. As a coagulant, ferrate(VI) removed 10% to 20% more UV_{254} and DOC from water than ferric sulphate using the same doses in the natural pH range (6 and 8), as well as reduced THMFP to less than 100 μg/l at low doses (2 mg as Fe/l) (Jiang et al., 2006).

Ferrate (VI) can also effectively contribute to algae removal. Ma and Liu (2002) investigated the effect of potassium ferrate pretreatment in algae removal by alum coagulation and reported that ferrate can significantly increase algae removal efficiency (around 72% at 5 mg K_2FeO_4/l and 60 mg alum/l) even after just a short period of preoxidation time (1 min). Yuan and coworkers (2002) investigated the use of ferrate in the removal of mycrocystin (the most commonly occurring toxin produced by cyanobacteria) from raw water. At a 40 mg FeO_4^{2-}/l dosage, most of the high-molecular-weight organic compounds were precipitated and TOC removal was about 50%, but for lower-molecular-weight organic compounds, a higher coagulant dosage was required and a lower TOC removal was obtained. Potassium ferrate (VI) was found to be more efficient than sodium hypochlorite in the inactivation of *Escherichia coli* spiked in tap water (Jiang et al., 2007). In order to achieve the same *Escherichia coli* killing efficiency, lower dose and contact time were required, the performance was less affected by solution pHs, and the disinfection rate of the ferrate(VI) was faster than that of sodium hypochlorite. Potassium ferrate was also found to be effective in the oxidation of sulfamethoxazole, a worldwide-applied antibacterial drug (Sharma et al., 2006). Its removal was found to fit a first-order kinetic with a half-life of 2 min at pH 7 using more Fe(VI) concentration (10 μM) than the sulfamethoxazole in water.

The potential of ferrate to oxidize NDMA precursors during water treatment was also investigated (Lee et al., 2008). In this work, the NDMA formation potentials (NDMA-FP) of the model NDMA precursors and natural waters were measured with and without preoxidation by Fe(VI) and, for most of the NDMA precursors with the exception of dimethylamine, a significant reduction of the NDMA-FP (495%) was observed after complete transformation of the NDMA precursor. Moreover, preoxidation of several natural waters with a high dose of Fe(VI) (0.38 mM) led to removal of the NDMA-FP in the 46–84% range.

5.3.2 Magnetic anion exchange resin (MIEX®)

Magnetic ion-exchange resin is a strong base anion resin with ammonia functional groups, consisting of 180 mm beads of a macroporous, polyacrylic structure (Singer and Bilyk, 2002). Raw water from nine different water utilities across the United States was tested in terms of THM and HAA precursor removals by MIEX® coagulation (Singer and Bilyk, 2002). The water samples varied widely with pH (6.1–8.2), alkalinity (4–155 mg/L as $CaCO_3$), TOC (2.6–26.4 mg/L), SUVA (1.4–4.5 l/(mg•m)), turbidity (1.9–55 NTU), and bromide (13.7–236 μg/l). The use of MIEX® significantly decreased the alum dose and increased THM and HAA precursor removals. However, subsequent experiments showed that MIEX® coagulation and MIEX® followed by alum coagulation can yield similar results; this finding suggests that coagulation of MIEX®-treated water does not provide additional removal of organic carbon (Boyer and Singer, 2005).

The impact of MIEX® resin on ozone demand and bromate formation was also investigated (Johnson and Singer, 2004). The results indicated that pretreatment of the water with the MIEX® resin significantly reduces DOC, UV absorbance, color, and, to some extent, bromide. Moreover, MIEX® pretreatment of the water prior to ozonation showed that ozone demand and formation of bromate during subsequent ozonation can be significantly decreased.

Drikas et al. (2009) assessed the impact of MIEX® pretreatment, followed by either coagulation or microfiltration, on the effectiveness of pilot granular-activated carbon (GAC) filters for the removal of the taste and odor compounds, 2-methylisoborneol and geosmin, from a surface drinking water source over a two-year period. Reduction of empty bed contact time in all but one GAC filter resulted in a breakthrough of spiked 2-methylisoborneol and geosmin, with initial results inconclusive regarding the impact of MIEX® pretreatment.

5.3.3 Titanium salts

Although water coagulation by titanium salts was first proposed by Upton and Buswell in 1937, only recently has there been renewed interest in their use for water treatment (Shon et al. 2007; Okour et al., 2009; Zhao et al., 2011; Wu et al., 2011). In particular, Wu et al. (2011) evaluated the efficiency of titanium salt coagulation in a water treatment plant at pilot scale (160 m^3/d), using Yangtze River water (turbidity: 40–49 NTU; alkalinity: 80–89 mg/L as $CaCO_3$; pH: 7.0–7.8; UV_{254} absorbance: 0.072–0.074 1/cm; temperature: around 23°C) as a water source. When titanium sulfate was dosed at 16–32 mg $Ti(SO_4)_2$/L (0.07–0.13 mmol/L) with pH adjusted to 5.4–5.6, the pilot plant showed an overall UV_{254} removal of 54–57%, compared to 41% removal achieved by PACl (22 mg/L initial dose).

Additional interest in the use of titanium salts as coagulant is related to the possibility of reducing the amount of sludge to be disposed off by recovering titanium to produce a valuable by-product, TiO_2, which can be used as a catalyst in a photocatalytic process for water treatment (Zhao et al., 2011).

Zhao et al. (2011) comparatively investigated the coagulation behavior of titanium tetrachloride ($TiCl_4$) and PACl in terms of turbidity and NOM removal from surface water. The results showed DOC removal as high as 78.4% for $TiCl_4$ and 41.0% for PACl. Sludge produced from $TiCl_4$ flocculation was calcined at 600°C to produce TiO_2, and the photocatalytic activity in removing reactive brilliant red after 2 h irradiation time was found to be as high as 91.4%, compared to 92.9% of commercially available TiO_2 P-25 (Zhao et al., 2011).

In spite of the good efficiency of titanium salts in water treatment compared to conventional/alternative coagulants, two issues should be addressed in evaluating their suitability as a water treatment: (1) the cost of the coagulant and, most important, (2) the effect of titanium release on water toxicity after treatment.

6 Conclusions

The role of the coagulation process in drinking water treatments is fundamental because an incorrect operation can increase the formation of DBPs if a subsequent oxidation/disinfection step is implemented. The process should be optimized according to coagulant type, dosage, and pH, taking into account raw water characteristics, particularly alkalinity and TOC. Prepolymerized coagulants have been proved effective in removing NOM and turbidity, showing higher efficiencies than conventional coagulants. However, the release of metals during the coagulation process has increased interest in environmental friendly alternatives such as NOCs. Because of the high costs of conventional coagulants, NOCs can be a sustainable alternative in developing countries, taking into account that the plants from which active ingredients are extracted can be easily found, grown, and prepared. Unlike NOCs extracted from tropical plants and seeds, chitosan is expensive (up to 10 times higher) compared to conventional coagulants, so its use on a large scale is currently problematical. Anyway, if NOCs are used in drinking water treatments, particular care should be taken in preparing the coagulant in order to minimize the release of organic matter that can increase the formation of DBPs.

Abbreviations

DBPs disinfection by-products
DOC dissolved organic carbon
GAC granular activated carbon
HA acetic acid
HAAs haloacetic acids
HCl hydrochloric acid
NDMA N-nitrosodimethylamine
NDMA-FP N-nitrosodimethylamine formation potential
NOCs natural organic coagulants
NOM natural organic matter
PACl polyaluminium chloride
PAFC polyaluminium ferric chloride
PFS polyferric sulphate
SUVA specific UV absorbance
THMFP trihalomethane formation potential

THMs trihalomethanes
TOC total organic carbon
UV_{254} UV absorbance at 254 nm wavelength

Glossary

Bacterial regrowth the recovery of disinfected-injured bacterial cells that have entered the distribution system from the water source or water treatment plant
Biofilm a layer of microorganisms in the aquatic environment held together in a polymeric matrix typically found attached to a support (e.g., inside wall of water pipes)
Breakthrough the increase in bacterial numbers in the distribution system resulting from viable bacteria passing through the disinfection barrier
Coliform bacteria family of bacteria used as indicators of microbial water contamination
Disinfection by-products compounds resulting from the reaction among substances naturally occurring in waters (e.g., humic acids, bromide, iron, manganese) and chemicals used for disinfection process (e.g., chlorine, chlorine dioxide, ozone, chloramines)
Granular activated carbon (GAC) filter a filter typically used in water treatment to remove by adsorption a wide range of dissolved contaminants as well as colloids
Natural organic coagulants coagulants extracted from plants seeds, plants, and animals
Prepolymerized coagulants metal-based coagulants specifically designed to control the nature of coagulant species rapidly formed after coagulant addition to the solution

References

Bellar, T.A., Lichtenberg, J.J., Kroner, R.C., 1974. The occurrence of organohalides in chlorinated drinking water. J. Am. Water Works Assn 66 (12), 703–706.

Benjamin, M.M., 2002. Water chemistry. McGraw-Hill, New York.

Boualam, M., Mathieu, L., Fass, S., Cavard, J., Gatel, D., 2002. Relationship between coliform culturability and organic matter in low nutiritive waters. Water Res. 36, 2618–2626.

Boyer, T.H., Singer, P.C., 2005. Bench-scale testing of a magnetic ion exchange resin for removal of disinfection by-product precursors. Water Res. 39, 1265–1276.

Bresner, M., Gauthier, V., Servais, P., Camper, A., 2002. Explaining the occurrence of coliforms in distribution systems. J. Am. Water Works Assn 94 (8), 95–109.

Brown, T.J., Emelko, M.B., 2009. Chitosan and metal salt coagulant impacts on *Cryptosporidium* and microsphere removal by filtration. Water Res. 43, 331–338.

Brush, C.F., Walter, M.F., Anguish, L.J., Ghiorse, W.C., 1998. Influence of pretratment and experimental conditions on electrophoretic mobility and hydrophobicity of *Cryptosporidium parvum oocysts*. Appl. Environ. Microbiol. 64 (11), 4439–4445.

Bustamante, H.A., Shanker, S.R., Pashley, R.M., Karaman, M.E., 2001. Interaction between *cryptosporidium oocysts* and water treatment coagulants. Water Res. 35, 3179–3189.

Butkus, M.A., Bays, J.T., Labare, M.P., 2003. Influence of surface characteristics on the stability of. Cryptosporidium parvum oocysts. Appl. Environ. Microbiol. 69 (7), 3819–3825.

Craun, G.F., 1981. Outbreaks of waterborne disease in the U.S. J. Am. Water Works Assn 73 (7), 360–369.

Diaz, A., Rincon, N., Escorihuela, A., Fernandez, N., Chacin, E., Forster, C.F., 1999. A preliminary evaluation of turbidity removal by natural coagulants indigenous to Venezuela. Process Biochem. 35, 391–395.

Drikas, M., Dixon, M., Morran, J., 2009. Removal of MIB and geosmin using granular activated carbon with and without MIEX pre-treatment. Water Res. 43, 5151–5159.

Drozd, C., Scwartzbrod, J., 1996. Hydrophobic and electrostatic cell surface properties of. Cryptosporidium parvum. Appl. Environ. Microbiol. 62 (4), 1227–1232.

Edzwald, J.K., Becker, W.C., Wattier, K.L., 1985. Surrogate parameters for monitoring organic matter and THM precursors. J. Am. Water Works Assn 77 (4), 122–132.

Eikebrokk, B., 1999. Coagulation-direct filtration of soft, low alkalinity humic waters. Water Sci. Technol. 40 (9), 55–62.

Gao, B., Yue, Q., Miao, J., 2002. Evaluation of polyaluminium ferric chloride (PAFC) as a composite coagulant for water and wastewater treatment. Water Sci. Technol. 47 (1), 127–132.

Gao, B., Yue, Q., 2005. Effect of SO_4^{2-}/Al^{3+} ratio and OH^-/Al^{3+} value on the characterization of coagulant polyaluminium chloride sulphate (PACS) and its coagulation performance in water treatment. Chemosphere 61, 579–584.

Gassenschmidt, U., Jany, K.D., Tauscher, B., Niebergall, H., 1995. Isolation and characterization of a flocculating protein from *Moringa oleifera* Lam. Biochim. Biophys. Acta. 1243, 477–481.

Huang, C., Chen, S., Pan, J.R., 2000. Optimal condition for modification of chitosan: a biopolymer for coagulation of colloidal particles. Technical note. Water Res. 34, 1057–1062.

Jahn, S.A.A., 1988. Using *Moringa* seeds as coagulants in developing countries. J. Am. Water Works Assn 80, 43–50.

Johnson, C.J., Singer, P.C., 2004. Impact of a magnetic ion exchange resin on ozone demand and bromate formation during drinking water treatment. Water Res. 38, 3738–3750.

Jiang, J.-Q., 2001. Development of coagulation theory and pre-polymerized coagulants for water treatment. Sep. Purif. Methods 30 (1), 127–141.

Jiang, J.-Q., Graham, N.J.D., 1998. Preliminary evaluation of the performance of new pre-polymerised inorganic coagulants for lowland surface water treatment. Water Sci. Technol. 37 (2), 121–128.

Jiang, J.-Q., Wang, S., Panagoulopoulos, A., 2006. The exploration of potassium ferrate(VI) as a disinfectant/coagulant in water and wastewater treatment. Chemosphere 63, 212–219.

Jiang, J.-Q., Wang, S., Panagoulopoulos, A., 2007. The role of potassium ferrate(VI) in the inactivation of *Escherichia coli* and in the reduction of COD for water remediation. Desalination 210, 266–273.

Korshin, G.V., Wu, W.W., Benjamin, M.M., Hemingway, O., 2002. Correlations between differential absorbance and the formation of individual DBPs. Water Res. 36, 3273–3282.

Kramer, M.H., Herwaldt, B.L., Craun, G.F., Calderon, R.L., Juranek, D.D., 1996. Waterborne disease: 1993 and 1994. J. Am. Water Works Assn 88 (3), 66–80.

LeChevallier, M.W., Babcock, T.M., Lee, R.G., 1987. Examination and characterization of distribution systems biofilms. Appl. Environ. Microbiol. 53 (12), 2714–2724.

LeChevallier, M.W., Welch, N.J., Smith, D.B., 1996. Full-scale studies of factors related to coliform regrowth in drinking water. Appl. and Environ. Microbiol. 62 (7), 2201–2211.

Lee, C., Lee, Y., Schmidt, C., Yoon, J., Von Gunten, U., 2008. Oxidation of suspected N-nitrosodimethylamine (NDMA) precursors by ferrate (VI): Kinetics and effect on the NDMA formation potential of natural waters. Water Res. 42, 433–441.

Lund, V., Ormerod, K., 1995. The influence of disinfection processes on biofilm formation in water distribution systems. Water Res. 29, 1013–1021.

Ma, J., Liu, W., 2002. Effectiveness and mechanism of potassium ferrate(VI) preoxidation for algae removal by coagulation. Water Res. 36, 871–878.

MacKenzie, W.R., Neil, Hoxie, J., Proctor, M.E., Gradus, M.S., Blair, K.A., Peterson, D.E., Kazmierczak, J.J., Addiss, D.G., Fox, K.R., Rose, J.B., Davis, J.P., 1994. A massive outbreak in Milwaukee of Cryptosporidium infection transmitted through the public water supply. N. Engl. J. Med. 331, 161–167.

Mallevialle, J., Brichet, A., Fiessinger, F., 1984. How safe are organic polymers in water treatment. J. Am. Water Works Assn. 76 (6), 431–436.

McFeters, G.A., Kippin, J.S., LeChevallier, M.W., 1986. Injured coliforms in drinking water. Appl. Environ. Microbiol. 51 (1), 1–5.

McLachlan, D.R.C., 1995. Aluminium and the risk for Alzheimer's disease. Environmetrics 6, 233–275.

Miller, R.G., Kopfler, F.C., Kelty, K.C., Strober, J.A., Ulmer, N.S., 1984. The occurrence of aluminium in drinking water. J. Am. Water Works Assn. 76 (1), 84–91.

Miltner, R.J., Nolan, S.A., Dryfuse, M.J., Summers, R.S., 1994. Evaluation of enhanced coagulation for DBP control. In: Proceedings, American Society of Chemical Engineers, National Conference on Environmental Engineering. Boulder, CO.

Muyibi, S.A., Evison, L.M., 1995. Optimizing physical parameters affecting coagulation of turbid water with *Moringa oleifera* seeds. Water Res. 29, 2689–2695.

Ndabigengesere, A., Narasiah, K.S., 1998. Quality of water by coagulation using *Moringa Oleifera* seeds. Water Res. 32, 781–791.

Ndabigengesere, A., Narasiah, K.S., Talbot, B.G., 1995. Active agents and mechanism of coagulation of turbid waters using. Moringa Oleifera. Water Res. 29, 703–710.

Nguyen, C., Elfland, C., Edwards, M., 2012. Impact of advanced water conservation features and new copper pipe on rapid chloramine decay and microbial regrowth. Water Res. 46, 611–621.

Niquette, P., Servais, P., Savoir, R., 2001. Bacterial dynamics in the drinking water distribution system of Brussels. Water Res. 35, 675–682.

Okuda, T., Baes, A.U., Nishijima, W., Okada, M., 2001. Isolation and characterization of coagulant extracted from *Moringa Oleifera* seed by salt solution. Water Res. 35, 405–410.

Okour, Y., Shon, H.K., Saliby, I.E., 2009. Characterization of titanium tetrachloride and titanium sulfate flocculation in wastewater treatment. Water Sci. Technol. 59 (12), 2463–2473.

O'Melia, C.R., 1972. Coagulation and flocculation. In: Weber Jr., W.J. (Ed.), Physicochemical processes for water quality control. John Wiley & Sons, New York.

Özacar, M., Sengil, A., 2003. Evaluation of tannin biopolymer as a coagulant aid for coagulation of colloidal particles. Colloids Surf. A. 229, 85–96.

Pan, J.R., Huang, C., Chen, S., Chung, Y.-C., 1999. Evaluation of a modified chitosan biopolymer for coagulation of colloidal particles. Colloids Surf. A. 147, 359–364.

Pintar, K.D.M., Slawson, R.M., 2003. Effect of temperature and disinfection strategies on ammonia-oxidizing bacteria in a bench-scale drinking water distribution system. Water Res. 37, 1805–1817.

Power, K.N., Schneider, R.P., Marshall, K.C., 1997. The effect of growth conditions on survival and recovery of *Klebsiella Oxytoca* after exposure to chlorine. Water Res. 31, 135–139.

Richardson, S.D., Plewa, M.J., Wagner, E.D., Schoeny, R., DeMarini, D.M., 2007. Occurrence, genotoxicity, and carcinogenicity of regulated and emerging disinfection by-products in drinking water: A review and roadmap for research. Mutation Res./Rev. in Mutation Res. 636, 178–242.

Rizzo, L., Belgiorno, V., Meric, S., 2004b. Organic THMs precursors removal from surface water with low TOC and high alkalinity by enhanced coagulation. Water Sci. Technol.: Water Supply 4 (5–6), 103–111.

Rizzo, L., Belgiorno, V., Napoli, R.M.A., 2004a. Regrowth evaluation of coliform bacteria injured by low chlorine doses using selective and nonselective media. J. Environ. Sci. Health. Part A 39 (8), 2081–2092.

Rizzo, L., Belgiorno, V., Casale, R., 2005. Simultaneous compliance of TOC and turbidity related to pathogen breakthrough and THMs control by enhanced coagulation. Global Nest: the Int. J. 7 (1), 77–85.

Rizzo, L., Di Gennaro, A., Gallo, M., Belgiorno, V., 2008. Coagulation/chlorination of surface water: a comparison between chitosan and metal salts. Sep. Purif. Technol. 62, 79–85.

Roccaro, P., Vagliasindi, F.G.A., 2007. Regulations. In: Nikolaou, A., Rizzo, L., Selcuk, H. (Eds.), Control of Disinfection by-products in drinking water systems. Nova Science Publisher, New York.

Rook, J.J., 1974. Formation of haloforms during chlorination of natural waters. Treat. Water Treat. Exam 23, 234–243.

Schwartz, T., Hoffmann, S., Obst, U., 1998. Formation and bacterial composition of young, natural biofilms obtained from public bank-filtered drinking water systems. Water Res. 32, 2787–2797.

Šćiban, M., Klašnja, M., Antov, M., Škrbić, B., 2009. Removal of water turbidity by natural coagulants obtained from chestnut and acorn. Bioresour. Technol. 100, 6639–6643.

Selcuk, H., Rizzo, L., Nikolaou, N., Meric, S., Bekbolet, M., Belgiorno, V., 2007. DBPs formation and toxicity monitoring in different origin water treated by ozone-alum/PACl coagulation. Desalination 210, 31–43.

Sharma, P., Kumari, P., Srivastava, M.M., Srivastava, S., 2006. Removal of cadmium from aqueous system by shelled *Moringa oleifera* Lam. seed powder. Bioresour. Technol. 97, 299–305.

Sharma, V.K., Mishra, S.K., Ray, A.K., 2006. Kinetic assessment of the potassium ferrate(VI) oxidation of antibacterial drug sulfamethoxazole. Chemosphere 62, 128–134.

Shon, H.K., Vigneswaran, S., Kim, I.S., Cho, J., Kim, G.J., Kim, J.B., Kim, J.H., 2007. Preparation of titanium dioxide (TiO2) from sludge produced by titanium tetrachloride (TiCl4) flocculation of wastewater. Environ. Sci. Technol. 41, 1372–1377.

Singer, P.C., Bilyk, K., 2002. Enhanced coagulation using a magnetic ion exchange resin. Water Res. 36, 4009–4022.

Upton, W.V., Buswell, A.M., 1937. Titanium salts in water purification. Ind. Eng. Chem. 29, 870–871.

USEPA., December 16, 1998. (1998). National Primary Drinking Water Regulations: Disinfectants and disinfection byproducts: Final Rule, Washington, DC. Federal Register 63 (241), 69390–69476.

USEPA., 1999. Enhanced coagulation and enhanced precipitative softening guidance manual. Office of Water. EPA 815-R, 99–012.

USEPA, 2002. Long Term 1 Enhanced Surface Water Treatment Rule (LT1ESWTR). Federal Register 67 (9). January 14, 2002/ Rules and Regulations, Cincinnati.

USEPA, 2006a. 2006 Edition of the Drinking Water Standards and Health Advisories. Office of Water, U.S. Environmental Protection Agency, Washington, DC, EPA 822-R-06-013.

USEPA, 2006b. EPA Integrated Risk Information System (IRIS). N-Nitrosodimethylamine. www.epa.gov/iris/subst/0045.htm.

Volk, C., Bell, K., Ibrahim, E., Verges, D., Amy, G., LeChevallier, M., 2000. Impact of enhanced and optimized coagulation on removal of organic matter and its biodegradable fraction in drinking water. Water Res. 34, 3247–3257.

Volk, C.J., LeChevallier, M.W., 1999. Impacts of the reduction of nutrient levels on bacterial water quality in distribution systems. Appl. Environ. Microbiol. 65 (11), 4957–4966.

Wu, Y.-F., Liu, W., Gao, N.-Y., Tao, T., 2011. A study of titanium sulfate flocculation for water treatment. Water Res. 45, 3704–3711.

Yan, M., Wang, D., Yu, J., Ni, J., Edwards, M., Qu, J., 2008. Enhanced coagulation with polyaluminium chlorides: Role of pH/alkalinity and speciation. Chemosphere 71, 1665–1673.

Yan, M., Wang, D., Ni, J., Qu, J., Ni, W., Van Leeuwen, J., 2009. Natural organic matter (NOM) removal in a typical North-China water plant by enhanced coagulation: Targets and techniques. Sep. Purif. Technol. 68, 320–327.

Yuan, B.-L., Qu, J.-H., Fu, M.-L., 2002. Removal of cyanobacterial microcystin-LR by ferrate oxidation-coagulation. Toxicon 40, 1129–1134.

Zacheus, O.M., Iivanainen, E.K., Nissinen, T.K., Lehtola, M.J., Martikainen, P.J., 2000. Bacterial biofilm formation on polyvinyl chloride, polyethylene and stainless steel exposed to ozonated water. Water Res. 34, 63–70.

Zacheus, O.M., Lehtola, M.J., Korhonen, L.K., Martikainen, P.J., 2001. Soft deposits, the key site for microbial growth in drinking water distribution networks. Water Res. 35, 1757–1765.

Zhang, J., Zhang, F., Luo, Y., Yang, H., 2005. A preliminary study on cactus as coagulant in water treatment. Process Biochem. 41, 730–733.

Zhao, Y.X., Gao, B.Y., Cao, B.C., Yang, Z.L., Yue, Q.Y., Shon, H.K., Kim, J.-H., 2011. Comparison of coagulation behavior and floc characteristics of titanium tetrachloride (TiCl4) and polyaluminum chloride (PACl) with surface water treatment. Chem. Eng. J. 166, 544–550.

CHAPTER

Coagulation–Flocculation Method for the Treatment of Pulp and Paper Mill Wastewater

10

Tjoon Tow Teng, Sook San Wong, Ling Wei Low

School of Industrial Technology, Universiti Sains Malaysia, 11800 Penang, Malaysia

CHAPTER OUTLINE

1 Introduction

With the rapid increase in world population and urban industrial development, the world demand for paper and paper board was estimated to be over 420 million tons by 2010, with an average growth of about 2.8% yearly (Mbendi Information Services 2009).

The pulp and paper industry is a highly water-intensive industry with a freshwater consumption as high as 60 m^3 per ton of paper produced (Thompson et al. 2001). It produces equally large amounts of wastewater whose characteristics depend on the type of manufacturing process used. The pulp and paper mill effluents are highly loaded with corrosive and toxic materials that, if untreated or poorly treated, will greatly affect aquatic life. They contain high organic materials that cause high biochemical oxygen demand (BOD) and chemical oxygen demand (COD), high adsorbable organic

The Role of Colloidal Systems in Environmental Protection. http://dx.doi.org/10.1016/B978-0-444-63283-8.00010-7

Table 10.1 Characteristics of a Pulp and Paper Mill Wastewater

Parameters	Value*
pH	7.3–8.3
Suspended solids (mg l^{-1})	5240
Turbidity (NTU)	4770
SCOD[a] (mg l^{-1})	318
TCOD[b] (mg l^{-1})	3087

*Average values over 20 samples.
[a]Soluble chemical oxygen demand.
[b]Total chemical oxygen demand.

halides (AOX), suspended solids (colloids), metal ions, tannins, lignin, and derivatives, fatty acids, and so on (Ali and Sreekrishnan 2001; Lacorte et al. 2003). Nomylphenolic compounds were found in the textile and pulp and paper effluents (Berryman et al. 2004). A recent study showed that pulp and paper mill effluents contain some toxic and neuroactive substances that could affect fish reproduction (Rana et al. 2004; Basu et al. 2009). Table 10.1 shows the chemical characteristics of a pulp and paper mill's wastewater (Wong 2006). The parameters are determined according to the *Standard Methods for the Examination of water and wastewater* (APHA 2000).

In view of the highly water-intensive nature of pulp and paper mills, measures to reduce operating costs by means of wastewater reclamation have received increasing attention in many parts of the world where water resources are scarce (Yang and Abbasspour 2007; Li et al. 2009; Kujawski 2009). An effective wastewater treatment technology for the pulp and paper mill effluents is needed to reduce freshwater consumption and to meet the more stringent requirements of the environmental regulatory body on the quality of the effluent to be discharged into receiving waters.

Basically, three treatment processes have been widely used for pulp and paper mill effluents: the physicochemical, biological, and combined processes (Pokhrel and Viraraghavan 2004). These processes have their respective advantages and limitations. Thermal treatment has also been used in which elevated temperatures help decompose some contaminants. Metallic species are decomposed to an elemental form such as ash or pure gases; organic compounds are decomposed to carbon dioxide, water, halides, or halogen gases (Belhateche 1995).

Physicochemical processes such as screening, flotation, sedimentation, adsorption, membrane filtration, coagulation-flocculation, ion exchange, ozonation, and advanced oxidation remove suspended solids, colloids, colors, and toxic materials, and reduce COD of the wastewaters. Biological processes such as aerobic and anaerobic treatment processes reduce BOD and COD of the wastewaters. Combined processes are hybrid systems that make use of the advantages of two or more processes.

Pulp and paper mills are facing a common problem of underdesigned wastewater plants to cope with ever growing production. The high-capacity plant and new alternative treatment systems such as the membrane filtration process and ion exchange methods always involve high cost. Chemically enhanced primary treatment (CEPT) is a simple and efficient technology (Harleman and Murcott 1999; Mahvi et al. 2005; Xu et al. 2005) for initial stage treatment of industrial wastewaters in removing suspended solids, organic loads, and reducing COD. A schematic flow diagram of chemically

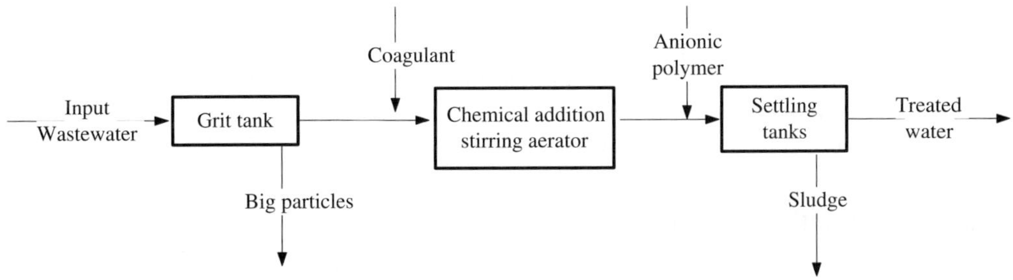

FIGURE 10.1

Schematic diagram of chemically enhanced primary treatment (CEPT).

enhanced primary treatment is shown in Figure 10.1. Development of highly efficient coagulation-flocculation process becomes an important alternative to tackle the problem.

2 The coagulation-flocculation process for wastewater treatment

The words "coagulation" and "flocculation" originated from the Latin *coagular* and *floccus,* respectively. *Coagular* means to put two or more entities together, whereas *floccus* means tufts of wool. The flocculation process involves condensation into fluffy masses (Huang and Pan 2002). Coagulation is the process of neutralizing electric charge on colloids and keeping them in suspension. Flocculation is the process of bringing together the microscopic particles to form larger agglomerations through binding action of flocculants (Ebeling et al. 2003).

The destabilization of microscopic colloidal particles by inorganic salts (coagulation) and/or by polymeric flocculants (flocculation) is a phenomenon of scientific interest and also has industrial importance. The system undergoes an irreversible process controlled by hydrodynamic and physicochemical conditions. The movements of colloidal particles in a liquid are subjected to Brownian motion, fluid flow, and external forces (Huang and Pan 2002). The electrostatic repulsive force between colloidal particles in a stable suspension prevents colliding particles from aggregation. The aggregation of the colloidal particles is induced by the introduction of salt or polymeric flocculant into the liquid. This salt or flocculant causes the colloid to become unstable, which leads to the formation of bonds between colliding particles resulting in the formation of aggregates.

The effectiveness of the coagulation–flocculation process depends on the following factors: type of coagulant and its dosage, solution pH, temperature and ion strength, mixing time and speed, and concentration and nature of the organic compounds in the wastewater (Muralidhara 1986; Randtke 1988; Taylor et al. 2002).

Coagulation is a process that is highly pH dependent. According to Randtke (1988), the influence of solution pH on chemical coagulation is a balance of two competitive forces: (1) between H^+ and metal hydrolysis products for organic ligands, and (2) between OH^- and organic anions for metal hydrolysis products. At very low pH, protons overcome the metal hydrolysis products for organic ligands, and poor removal occurs. At high pH, OH^- competes with organic compounds for metal adsorption sites, and precipitation of metal hydroxides prevails. Further increase in alkalinity causes the coagulating species to become less positively charged, thus diminishing their attraction to the anionic organic compounds. The growth rate of aggregates is reduced at high pH, leading to a very low or negligible coagulation rate (Stephenson and Duff 1996).

Water temperature can affect fluid and particle motion to some extent. A drop in temperature causes: (1) rate reduction of chemical reaction of added coagulants, (2) viscosity increase and hence slower particle-settling velocity, and (3) reduction in colloid activity (Casey 1997).

2.1 Coagulants and flocculants

Many materials have been used as coagulants. Among the commonly used are inorganic salts such as alum (aluminium sulfate), ferric chloride, ferric sulfate, lime (Metcalf and Eddy 2004), polyaluminium chloride (PACl), and magnesium chloride (Tan et al. 2000). Conventional chemical flocculants are synthetic polymers such as polyacrylates, and polyacrylamides (Nasser and James 2007; Aguilar et al. 2005). Polymeric flocculants are long, flexible chains of high molecular mass, usually in the range of 1×10^6 to 25×10^6. Some natural and bioflocculants produced from microorganisms are found to be excellent flocculants or coagulant aids (Deng et al. 2005; Gong et al. 2008; Ghosh et al. 2009). Suitable coagulant and dosage at appropriate pH not only improves the water characteristics but also decreases the settling time and cost involved.

3 Experimental studies

The experimental study consists of two parts: a bench scale study and a pilot scale study. The bench scale study uses Jar test equipment to investigate the best combination of coagulant and flocculant, and the optimal conditions that produce the best performance in terms of highest removal efficiency. The pilot scale study applies information obtained from the bench scale study to a pilot flow system that will be useful for industrial scale-up.

3.1 Bench scale study

A bench scale experiment for pulp and paper mill wastewater treatment by the coagulation–-flocculation process was carried out using Jar test equipment, as shown in Figure 10.2. It consists of six-spindle multiple stirrer units with stainless steel paddles.

A coagulant of certain dosage was introduced into 500 ml of wastewater of preadjusted pH. It was stirred for 2 min at 200 rpm. A flocculant of a certain dosage was then added to the mixture with a slow mixing of 15 min at 40 rpm. The flocs formed was allowed to settle for 30 min. After the flocs settled, the supernatant was withdrawn for total suspended solids (TSS), turbidity, and COD determinations. The remaining portion of the treated wastewater sample was used for sludge volume index (SVI) determination.

The turbidity of a sample in NTU was measured by a turbidity meter. The TSS concentration was determined by filtering a well-mixed sample through a glass fiber filter, and the residue retained on the filter was weighed after drying it in an oven at 102°C for 60 min. The standard method (5220 D. closed reflux, calorimetric method) was used to determine the COD value of a sample. The SVI was determined using a 1000 ml Incoff cone.

The performance of the coagulation–flocculation process for pulp and paper mill wastewater treatment was determined through percentage turbidity reduction and percentage COD reduction.

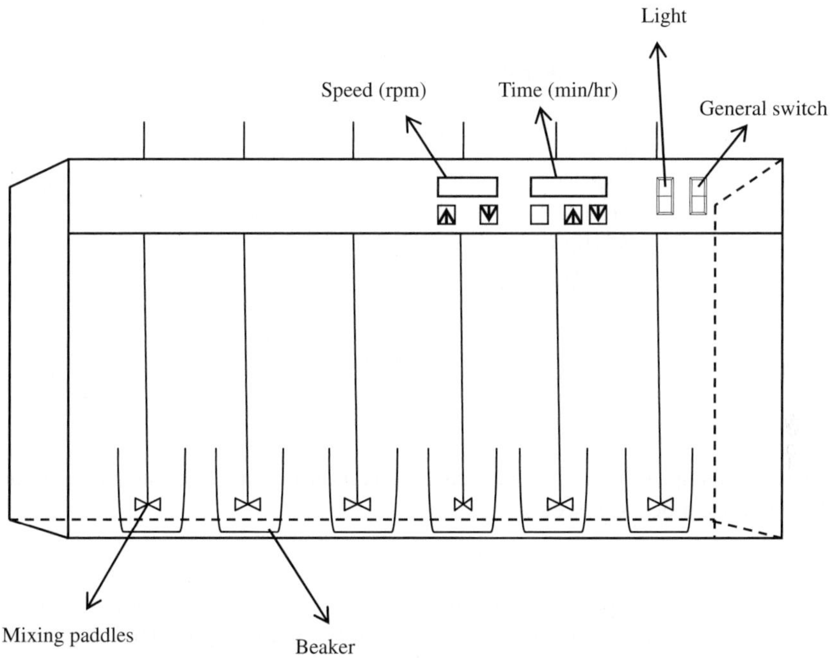

FIGURE 10.2

Schematic diagram of jar test equipment.

Three systems were studied for the treatment of pulp and paper mill wastewater: single coagulant (alum and polyaluminum chloride), single flocculant (nine types of cationic and anionic polyacrylamides), and coagulant + flocculant.

Alum (aluminum sulfate) and polyaluminum chloride (PACl) were used as coagulants, with dosages ranging from 0 to 2000 mg l^{-1}. PACl is a prehydrolyzed inorganic coagulant prepared by reacting NaOH or Na_2CO_3 with concentrated $AlCl_3$ solution at mild temperature (Huang and Pan 2002).

Five commercial cationic polyacrylamides (C-PAM) and four anionic polyacrylamides (A-PAM) in a wide range of molecular weights and charge density were used as flocculants, with dosages ranging from 0 to 15 mg l^{-1}. Table 10.2 lists the polyacrylamides used.

For the system of coagulant + flocculant, the best C-PAM and A-PAM, with the highest efficiency of turbidity reduction and COD reduction, coupled with both coagulants, alum and PACl, were studied.

3.2 Response surface methodology

Response surface methodology (RSM) is a statistical design technique for experimentation used to obtain optimal situations given specified ranges of factors affecting the process. Optimization of the coagulation–flocculation process for pulp and paper mill wastewater treatment can be performed using RSM with an experimental design such as Central Composite Face-Centered Design (CCFD). For a process with multiple variables, RSM performs optimization through systematic variation of all variables to yield an experimental design with a minimum number of experiments (Montgomery 2004).

Table 10.2 The Properties of Polyacrylamides Used

Polyacrylamide	Molecular Weight	Charge	Charge Density
Organopol 5415	Very high	Cationic	Low
Organopol 5020	Low	Cationic	Medium
Organopol 5470	High	Cationic	High
Organopol 5450	High	Cationic	Medium
Chemfloc 1515C	Medium	Cationic	Medium
Organopol 5540	Very high	Anionic	Low
Chemfloc 430A	High	Anionic	High
AN 913	High	Anionic	Low
AN 913SH	Very high	Anionic	Low

RSM produces relationships of one or more measured responses y as functions of input factors x. The response surface can then be expressed as

$$y = f(x_1, x_2, x_3, \ldots..x_k) \tag{10.1}$$

The CCFD used for treating pulp and paper mill wastewater by the coagulation–flocculation process involved two different factors: coagulant dosage and solution pH. The responses y in the coagulation–flocculation process for pulp and paper mill wastewater treatment are total suspended solids, SVI, and water recovery. The quadratic equation for predicting the optimal conditions can be expressed as (Mason et al. 2003):

$$y = \beta 0 + \sum_{i=1}^{k} \beta ixi + \sum_{i=1}^{k} \beta iix_i^2 + \sum_{i\leq j}^{k} \sum_{j}^{k} \beta ijxixj + \varepsilon \tag{10.2}$$

where i and j are linear and quadratic coefficients, respectively, β is the regression coefficient, k is the number of input factors, and ε is the random error.

3.3 Pilot scale study

A pilot scale unit such as that shown in Figure 10.3 is used after the bench scale study has been performed. It comprises a chemical treatment unit and a filtration unit. The unit is capable of handling 400 l of pulp and paper mill wastewater. A plate-and-frame filter press is installed to separate the treated water from the sludge, at a mass flow rate of 15 l/dm^2.min. The pilot scale experiment was applied to the best system obtained from the bench scale study.

4 Results and discussion

4.1 The single-coagulant system

When a single coagulant is used, the coagulant dosage and solution pH play an important role in determining the coagulation efficiency. When an inorganic coagulant is introduced into the

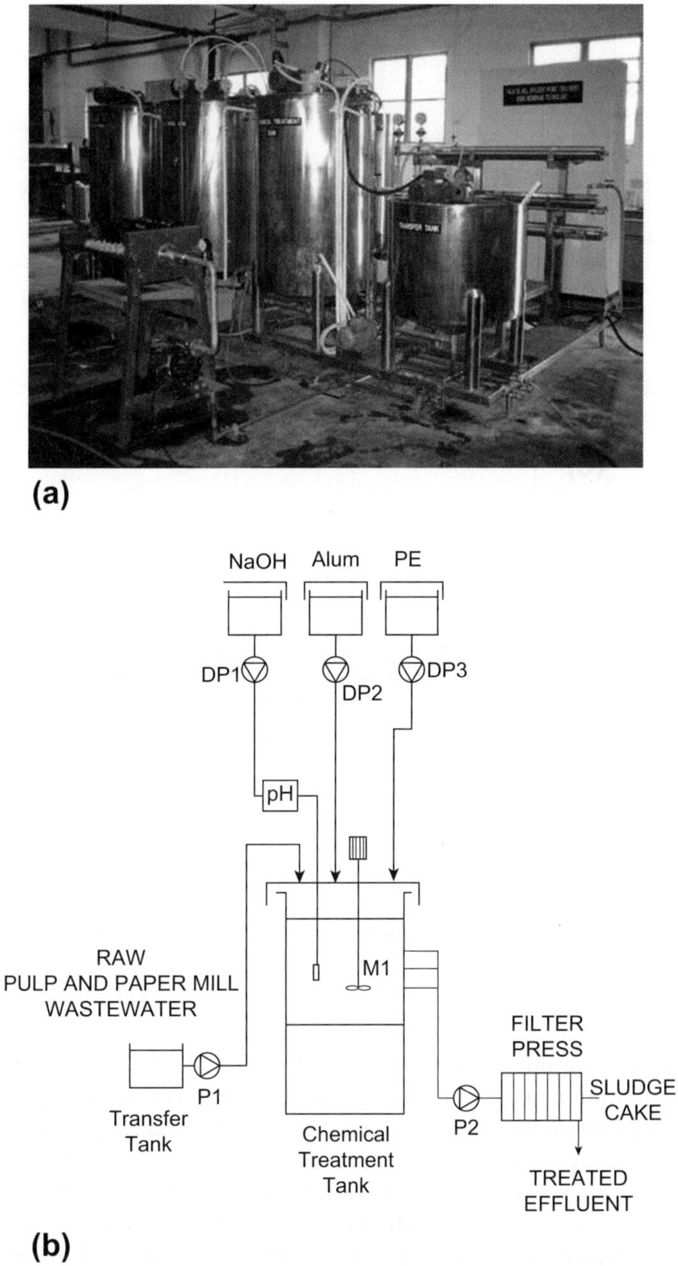

FIGURE 10.3

Pilot scale apparatus: (a) the pilot plant, (b) schematic diagram.
(P1: Pump 1; P2: Pump 2; PE: Polymer Tank; DP1: NaOH Dosing Pump; DP2: Alum Dosing Pump; DP3: PE Dosing Pump; M1: Treatment Tank Mixer) (Wong 2006). (For color version of this figure, the reader is referred to the online version of this book.)

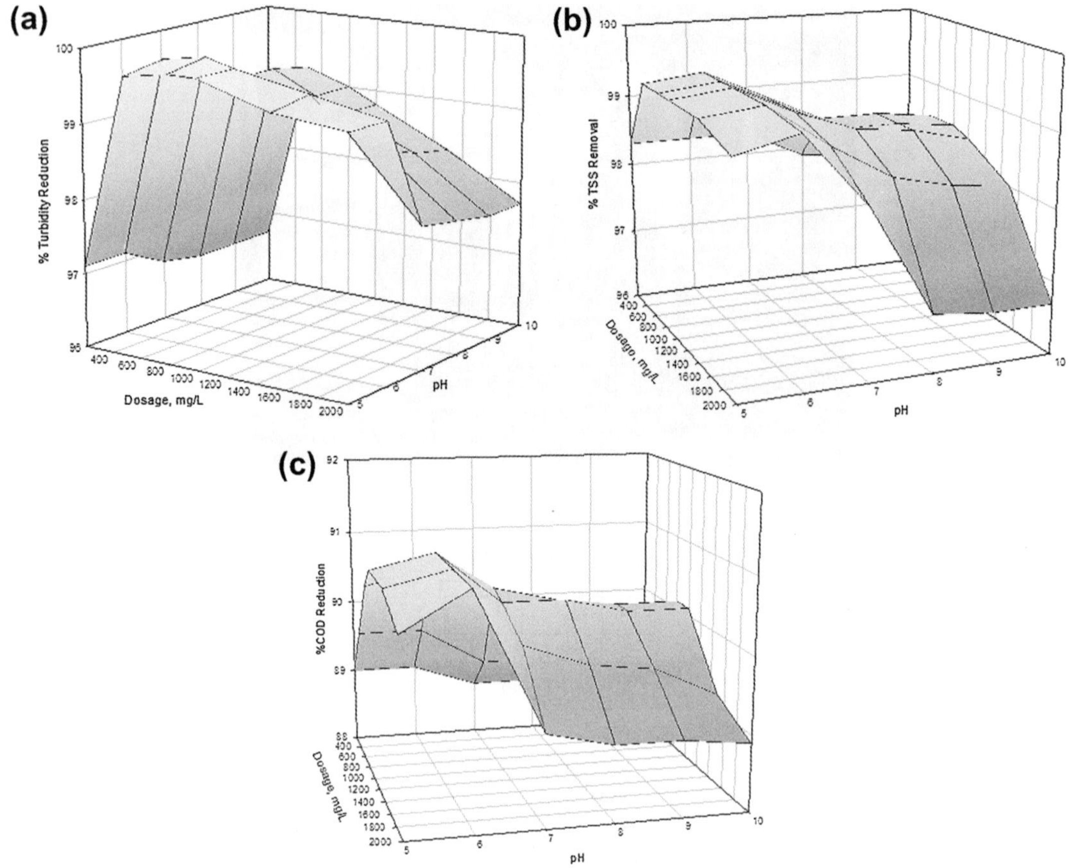

FIGURE 10.4

Effects of alum dosage and pH on (a) turbidity reduction, (b) TSS removal, and (c) COD reduction.

(From Ahmad et al. 2008. With permission.)

wastewater, metal hydroxide precipitates from which take-away OH^- ions in the solution and the solution pH drop. An optimum pH at which the coagulant performs best should be determined. Alum was used as the coagulant in treating pulp and paper mill wastewater, and the effects of pH adjustment and alum dosage on turbidity, TSS, and COD are presented in Figure 10.4(a–c), respectively. Generally, alum works best in slightly acidic conditions. Figure 10.4 shows that turbidity reduction, TSS removal, and COD reduction efficiencies increase with alum dosage until about 1000 mg l^{-1}, beyond which the efficiencies begin to decrease. The optimum pH occurs at about 6.0. The corresponding turbidity reduction is 99.8%, TSS removal is 99.4%, and COD reduction is 91%.

At low dosages, alum may be used for double-layer compression and charge neutralization of colloidal particles. At very high dosages, the overdosing of alum not only neutralizes the negative

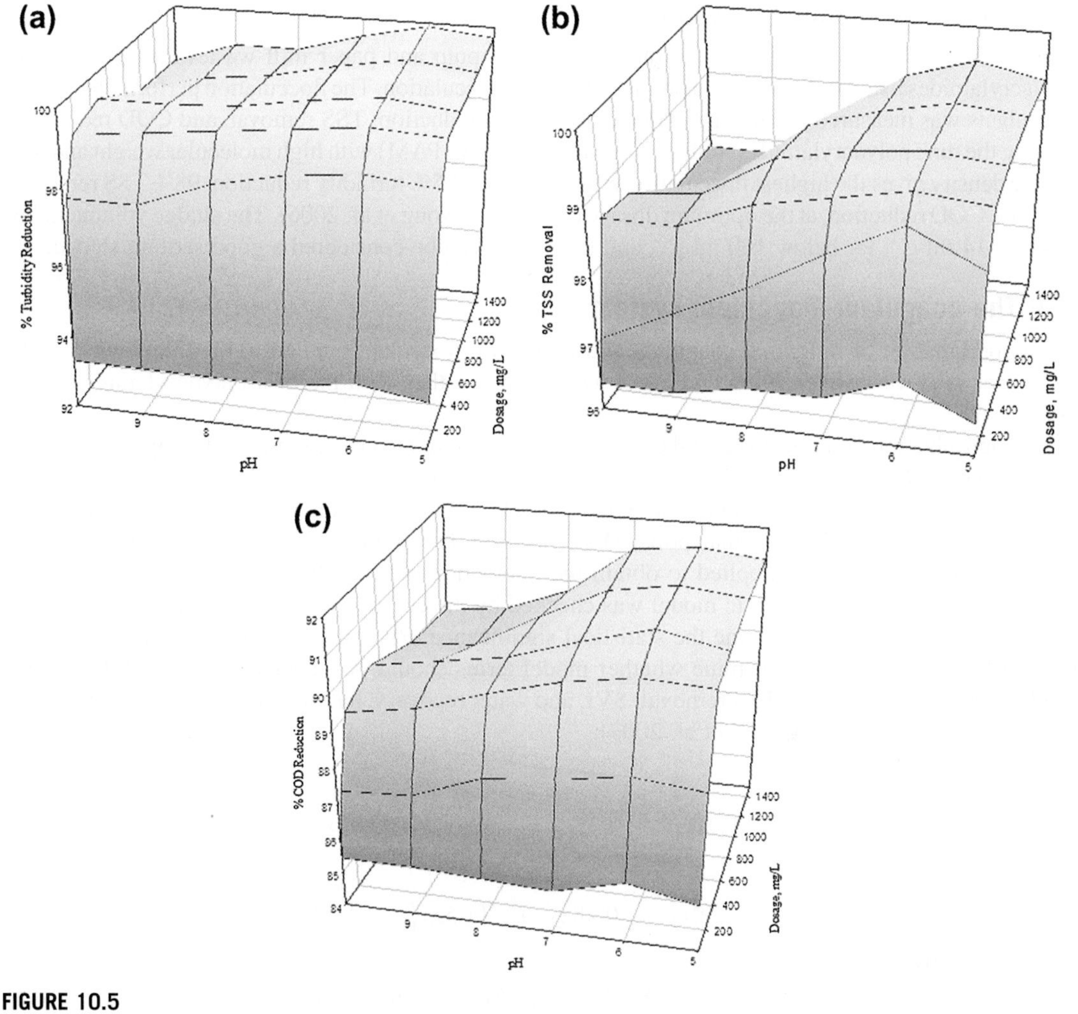

FIGURE 10.5

Effects of PACl dosage and pH on (a) turbidity reduction, (b) TSS removal, and (c) COD reduction.

(From Ahmad et al. 2008. With permission.)

charges on the particles, but adds a net positive charge, creating the same repulsive tendencies among particles as in the initial conditions, and restabilization occurs.

When PACl is used as the coagulant, the optimum pH and coagulant dosage is 6.0 and 500 mg l^{-1}, respectively. The corresponding turbidity reduction, TSS removal, and COD reduction are 99.9%, 99.5%, and 91.3%, respectively (see Figure 10.5).

The optimum pH of both coagulants is 6.0. At low pH, the aggregates are small in size. The main dissolved Al species from PACl available for coagulation are monomeric Al species such as Al^{+3} and $Al(OH)^{+2}$. Some dimeric Al species such as $Al_2(OH)_2^{+4}$ may also be present. Colloidal destabilization occurs mainly through charge neutralization (Huang and Pan 2002).

4.2 The single-flocculant system

Experiments have been conducted for the treatment of pulp and paper mill wastewater by the nine polyacrylamides listed in Table10.2 in polymer-induced flocculation. The flocculation performance of the flocculants was measured through percentage of turbidity reduction, TSS removal, and COD reduction. Among the nine polyacrylamides studied, Organopol 5415 (C-PAM) with high molecular weight and low charge density gives the highest flocculation efficiency, with 95% turbidity reduction, 98% TSS removal, and 93% COD reduction, at the optimum dosage of 5 mg l^{-1} (Wong et al. 2006). The sludge volume index (SVI) is 14 ml g^{-1}, far below 100 ml g^{-1}, and the sludge can be considered a good settling sludge.

4.3 The coagulant–flocculant system

For a combination of inorganic coagulant and polymeric flocculant applied to the treatment of pulp and paper mill wastewater, alum and PACl were used together with the best polyacrylamide, Organopol 5415.

An optimization process via central composite face-centered RSM experimental design was carried out for coagulant dosage and pH in order to achieve the highest TSS removal, lowest SVI, and highest water recovery using alum coupled with Organopol 5415. For regression, the Design Expert Software (version 6.0, Stat-Ease, Inc., Minneapolis, MN) was used. For graphical analysis of the data, analysis of variances (ANOVA) was applied to obtain the interaction between the responses and the process variables. The fit of the quadratic model was checked by the coefficient of determination R^2, and the Fisher F-test was used to examine the statistical significance. *P*-value (probability) with a 95% confidence level was used to determine whether model terms should be selected or rejected. Quadratic models for the responses for TSS removal, SVI, and water recovery, based on $R^2 > 0.99$ obtained from ANOVA, are as follows (Ahmad et al. 2007):

- TSS removal (y_1):

$$y_1 = 99.23 + 1.07x_1 - 0.35x_2 - 2.26x_1^2 - 0.81x_2^2 - 0.025x_1x_2 \quad (10.3)$$

- *SVI* (y_2):

$$y_2 = 38.41 - 11.60x_1 + 0.48x_2 + 11.55x_1^2 + 1.40x_2^2 - 0.35x_1x_2 \quad (10.4)$$

- Water recovery (y_3):

$$y3 = 82.23 - 2.05x_2 - 3.95x_2^2 \quad (10.5)$$

where x_1 and x_2 are coagulant dosage and pH, respectively.

Statistical parameters obtained from analysis of variances are listed in Table 10.3. For all three responses—TSS removal, SVI, and water recovery—the values of Prob > F < 0.0001, indicating that the model equations represent well the response surfaces in the range of study. All values of R^2 are close to 1.0. An R^2 value must be in good agreement with the adjusted R^2 for a significant model (Nordin et al. 2004). The coefficient of variance (CV), the ratio of standard error of estimate to the mean of the response, is a measure of the model's reproducibility. Generally, if CV < 10.0, the model can be considered reasonably reproducible. Table 10.3 shows that the CV values of the three responses are well below 1.0.

The response surface plots for the three responses TSS removal, SVI, and water recovery are presented in Figure 10.6. The response surface for TSS removal (Figure 10.6a) shows a clear peak,

Table 10.3 ANOVA for the Quadratic Models

Variables	TSS Removal	SVI	Water Recovery
Significant terms	x_1, x_2, x_1x_2, x_1^2, x_2^2	x_1, x_2, x_1x_2, x_1^2, x_2^2	x_2, x_2^2
R^2	0.9986	0.9998	0.9949
Adjusted R^2	0.9976	0.9996	0.9939
Prob>F	< 0.0001	< 0.0001	<0.0001
Adequate precision	83.5020	190.5510	63.3890
Standard deviation, S.D.	0.0790	0.2000	0.2000
Coefficient of variance, CV	0.0800	0.4400	0.2400
PRESS	0.1500	2.0400	0.6500
Probability for lack of fit 0.7202	0.1206	0.2406	

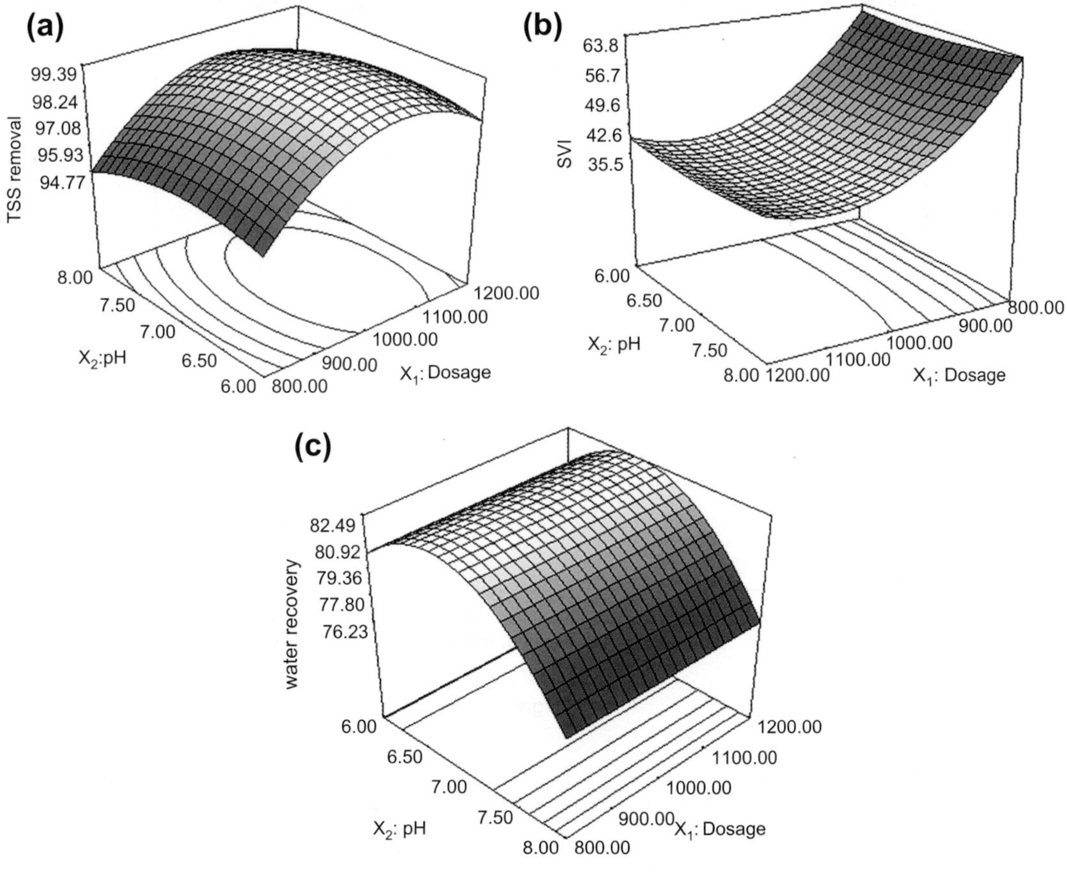

FIGURE 10.6

DESIGN-EXPERT plots. Response surface plot for (a) TSS removal, (b) SVI, and (c) water recovery.

(From Ahmad et al. 2007. With permission.)

which indicates the optimum condition for maximum TSS removal within the design boundary. The maximum percentage TSS removal of 99% corresponds to a coagulant dosage of about 1045 mg l^{-1} at a pH of around 6.75. The lowest SVI of 37 ml g^{-1} (Figure 10.6b), and the maximum water recovery of about 82.5% (Figure 10.6c) also occur under the same optimal conditions.

In order to study the effect of the flocculant dosage on the performance of the coagulation–-flocculation process, alum at a dosage of 500 mg l^{-1} and pH 6.0 together with Organopol 5415 with varying dosages were used. The treatment performance is shown in Figure 10.7 (Ahmad et al. 2008). The result of the treatment using PACl at a dosage of 200 mg l^{-1} at pH 6.0, together with Organopol 5415 as flocculant, is given in Figure 10.8. Both Figures 10.7 and 10.8 show that the removal and reduction efficiencies of turbidity, TSS, and COD do not change significantly with flocculant dosage. Thus Organopol 5415 of dosage 1.0 mg l^{-1} was used in the following experiments to determine the optimal dosages of coagulant. Coagulant dosage was increased from 50 to 2000 mg l^{-1},with a fixed flocculant dosage of 1.0 mg l^{-1} and a pH of 6.0. The results for turbidity reduction, TSS removal, and COD reduction are shown in Figure 10.9. Figure 10.9a shows that the turbidity reduction efficiency for

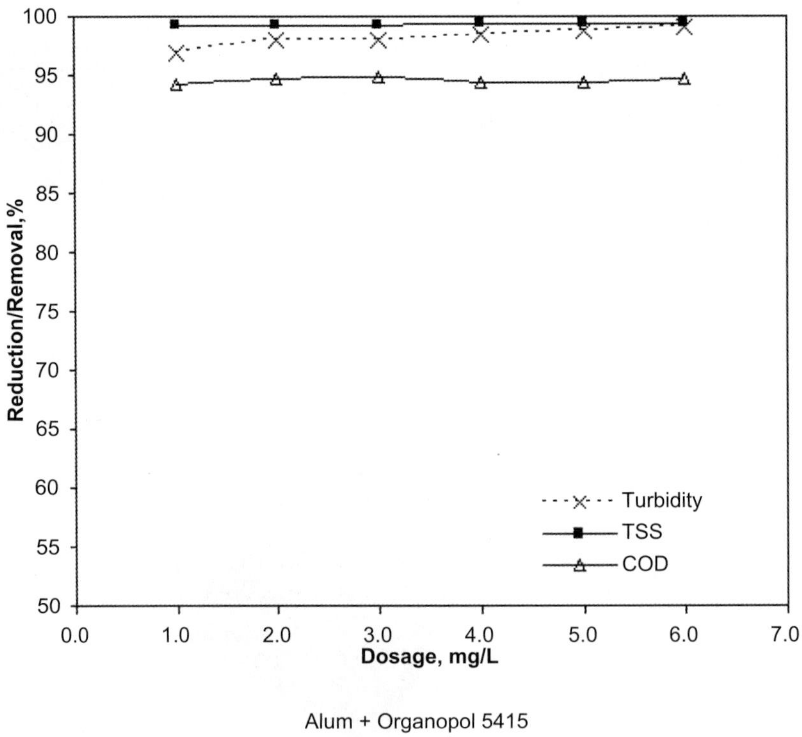

FIGURE 10.7

Effect of flocculant dosage on turbidity reduction, TSS removal and COD reduction in alum/Organopol 5415 system. Alum dosage = 500 mg l^{-1} and pH 6.0.

(From Ahmad et al. 2008. With permission.)

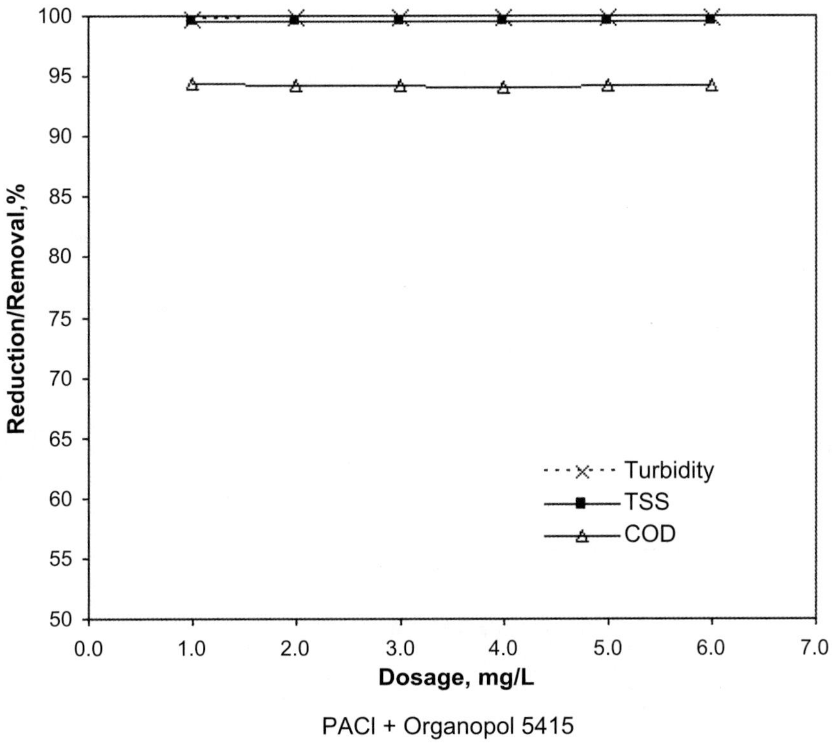

FIGURE 10.8

Effect of flocculant dosage on turbidity reduction, TSS removal and COD reduction in PACl/Organopol 5415 system. PACl dosage = 200 mg l^{-1} and pH 6.0.

(From Ahmad et al. 2008. With permission.)

alum at a dosage of 50 mg l^{-1} is improved from 80 to 96% when flocculant is added. Thus a smaller amount of alum is needed to achieve an acceptable reduction in turbidity. PACl also shows similar improvement in performance; at a dosage of 50 mg l^{-1}, the turbidity reduction efficiency increases from 95 to 99% when flocculant is added. The effects of adding flocculant on TSS removal and COD reduction are shown in Figs 10.9b and c, respectively. Trends similar to that of turbidity reduction apply to both TSS removal and COD reduction.

Both alum + Organopol 5415 and PACl + Organopol 5415 systems have the same optimal coagulant dosage of 1000 mg l^{-1}.

Flocculants are capable of bridging smaller particles to form larger, stronger, and higher density flocs with better settling rate. Settling time in the Jar test is measured by recording time for the flocs formed to reach half the solution height. The results are shown in Figure 10.10. Adding flocculant has a significant effect on the settling time when alum and PACl are used as coagulants. The bridging mechanism of the high-molecular-weight polyacrylamide may cause the flocculant to coil the flocs to become more compact, stronger, and larger (Negro et al. 2006). When the floc size becomes larger, it

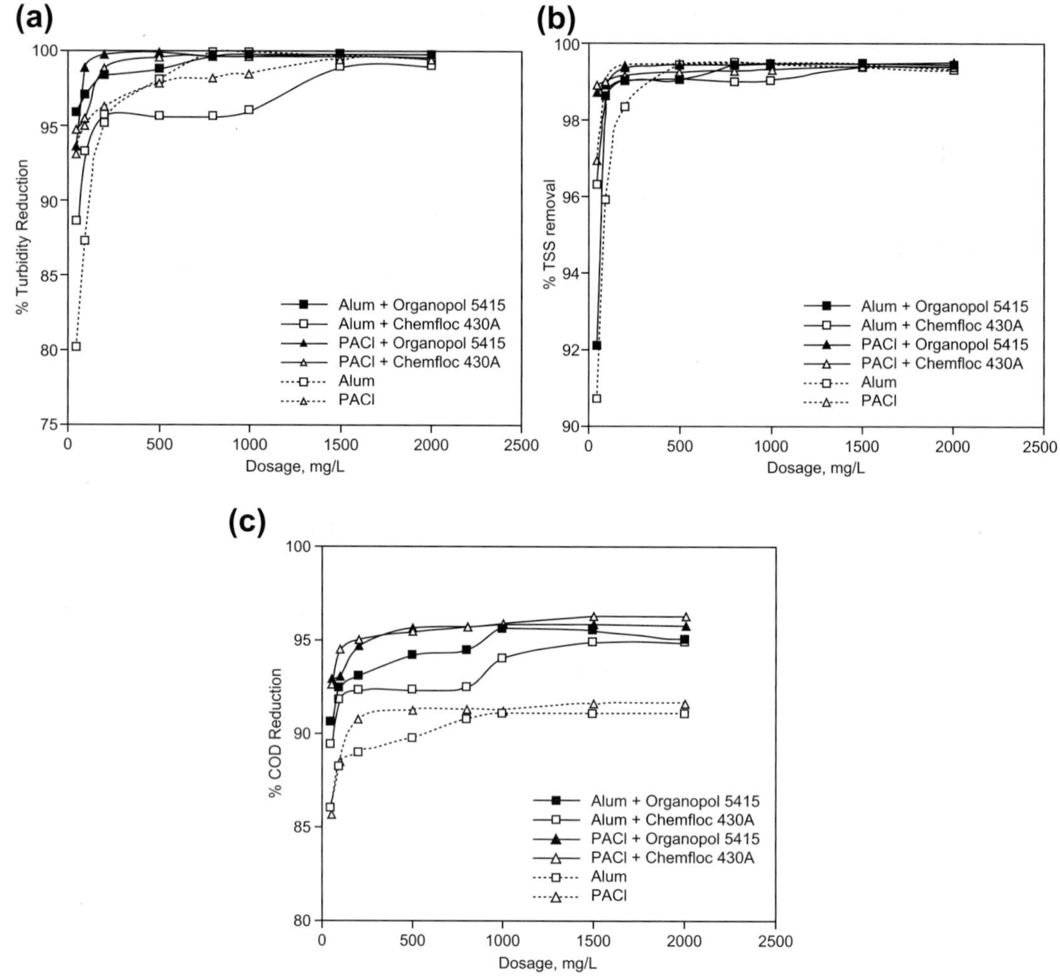

FIGURE 10.9

Effects of flocculant addition on (a) turbidity reduction, (b) TSS removal, and (c) COD reduction.

favors the flocs settling speed and it shortens the settling time of the flocs. The flocs generated by alum settle faster than that by PACl.

The sludge generated in the coagulation–flocculation process derives from the amount of organic matter and total colloids removed and the compounds formed from coagulating action. The amount of sludge produced is measured through the sludge volume index (SVI). The SVI of the sludge produced from the optimal conditions is shown in Figure 10.11. When a single coagulant is used, PACl generates slightly less sludge than alum. However, when cationic Organopol 5415 is used together with alum, the SVI drops from 63 mg l^{-1} to about 38 mg l^{-1}.

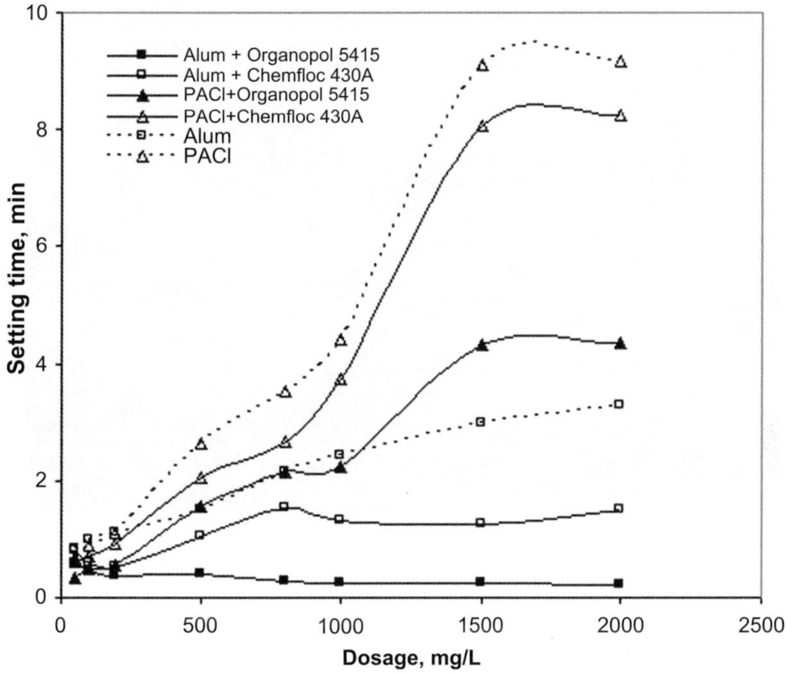

FIGURE 10.10

Effect of flocculant addition on settling time.

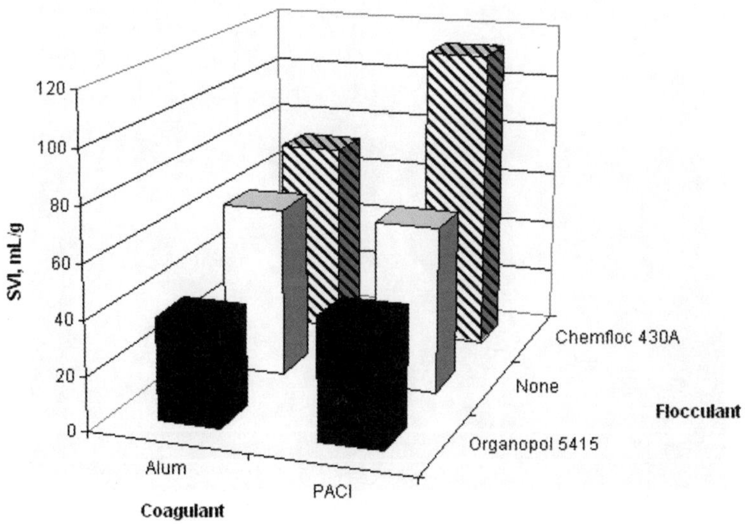

FIGURE 10.11

Effect of flocculant addition on SVI.

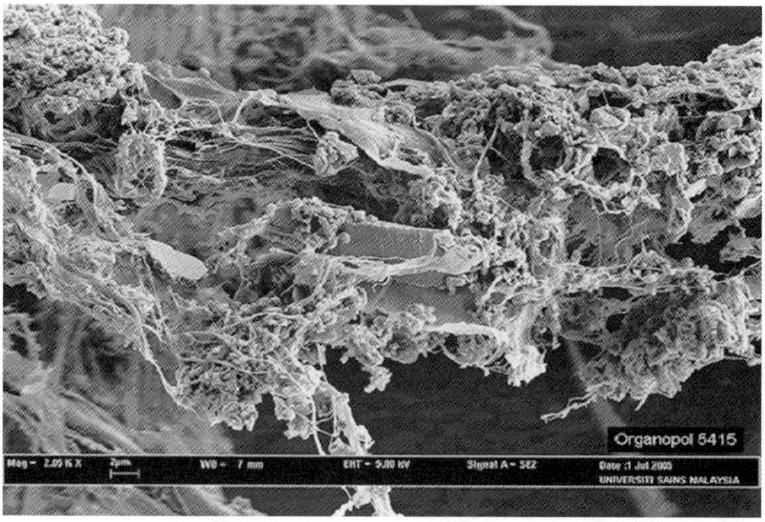

FIGURE 10.12

SEM of the flocs produced by Organopol 5415 at the dosage of 5 mg l^{-1} (Magnification 2005×).

FIGURE 10.13

SEM of the flocs produced by Alum + Organopol 5415 at 1000 mg l^{-1} of alum 1 mg l^{-1} of Organopol 5415 and pH 6.0 (Magnification 2000×).

Scanning electron microscopy (SEM) was used to study the characteristics of the flocs formed in the process. The SEM of the flocs produced by using Organopol 5415 alone at a dosage of 5 mg l^{-1} is shown in Figure 10.12, while that produced by the alum + Organopol 5415 system at 1000 mg l^{-1} of alum and 1 mg l^{-1} of Organopol 5415 at pH 6.0 is shown in Figure 10.13. Figure 10.12 shows the closely packed pulp and paper wastewater particles with the long-chain polyacrylamide as marked by the arrow. Pulp and paper wastewater particles are negatively charged. Cationic Organopol 5415 has effective charge-neutralizing ions, and diffusion occurs slowly into the fiber pores. Figure 10.13 shows that flocs formed by alum coupled with Organopol 5415 are different from that produced by flocculant alone. The flocs are more compact and smaller than those with flocculant alone. The presence of polymer chain is difficult to observe from the SEM due to the low dosage used.

4.4 Pilot plant study

A pilot scale study was conducted for the alum + Organopol 5415 system under the optimal conditions obtained from the bench scale study. The performance of the coagulation–flocculation process including filtration for solid–liquid separation was evaluated based on turbidity reduction, TSS removal, and COD reduction. The results are shown in Figure 10.14. The turbidity reduction, TSS removal, COD reduction, and water recovery exceed 98%, 99%, 95%, and 96%, respectively. The final values of turbidity, TSS, and COD of the filtrate are 3.50 NTU, 21.78 mg l^{-1}, and 165.48 mg l^{-1}, respectively.

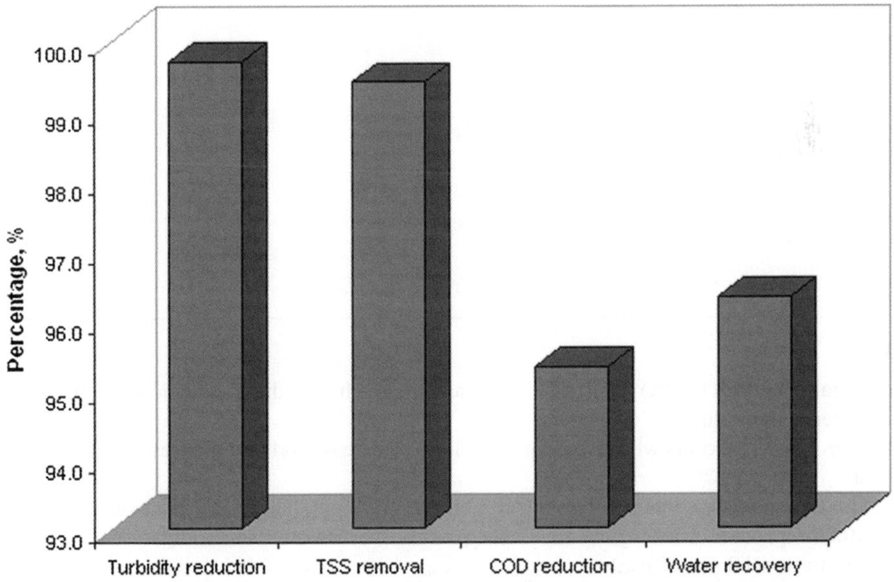

FIGURE 10.14

Percentage reduction of turbidity, removal of TSS, reduction of COD and water recovery obtained from the filter press process in pilot plant. (For color version of this figure, the reader is referred to the online version of this book.)

5 Conclusions

The coagulation–flocculation process is effective in treating pulp and paper mill wastewater with respect to turbidity reduction, TSS removal, COD reduction, and water recovery. When coagulant is used alone, the optimal dosages for alum and PACl are 1000 mg l^{-1} and 500 mg l^{-1}, respectively. The optimal pH is 6.0 for both coagulants. The turbidity reduction and TSS removal for both alum and PACl are > 99%. The COD reductions for alum and PACl are 90.7% and 91.3%, respectively.

When a single flocculant is used, a cationic Organopol 5415 among eight other polyacrylamides gives the best performance, achieving 95% turbidity reduction, 98% TSS removal, and 93% COD reduction. The sludge volume index is 14 ml g^{-1}, and the optimal dosage is 5 mg l^{-1}.

For the coagulation–flocculation system, alum and PACl are used together with Organopol 5415. The inclusion of Organopol 5415 enhances alum and PACl performances with respect to the reduction/removal efficiencies of turbidity, TSS, and COD. Alum + Organopol 5415 produces flocs that settle in less than 1 minute. The SVI drops from 63 ml g^{-1} to 38 ml g^{-1} when alum acts together with Organopol 5415. It is concluded that the alum + Organopol 5415 system is the best among all systems studied. It achieves high efficiency in terms of turbidity reduction, TSS removal, and COD reduction, with low SVI and short settling time.

A pilot scale study using alum together with Organopol 5415 for pulp and paper mill wastewater achieves 99.7% turbidity reduction, 99.4% TSS removal, 95.3% COD reduction, and 96.3% water recovery. The flocs produced are small, strong, and compact.

SYMBOLS

k number of input factors
R correlation coefficient
β regression coefficient
ε random error
x input factor
y measured response

Glossary

Adsorbable organic halides (AOX) a measure of the amount of chorine that is chemically bound to the soluble organic matter in the effluent

Adsorption a process that occurs when a gaseous or a liquid solute (adsorbate) adheres on the surface of a solid (adsorbent)

Advanced oxidation a process that involves the combination of more than one oxidant such as UV/ozone

Aerobic treatment process a biological treatment process that involves microorganisms requiring oxygen environment for breakdown of waste

Alum (aluminium sulfate) a chemical used as an inorganic coagulant in the water/wastewater treatment process

Anaerobic treatment process a biological treatment process that involves microorganisms that do not require oxygen environment for breakdown of waste

Analysis of variance (ANOVA) a statistical technique used to analyze the variation in the experimental data in order to determine the extent of variation attributed to each factor

Biochemical oxygen femand (BOD) a measure of the amount of oxygen needed by microorganisms in water bodies to decompose the biodegradable organic matters
Brownian motion random movement of tiny particles in a gas or liquid
Charge neutralization a process in which the net electrical charge of particles in an aqueous solution has been neutralized by the action of an equal number of opposite charges of another material
Chemically enhanced primary treatment (CEPT) a treatment process where low-dose chemicals, metal salts, or polymers (polyelectrolytes) are added to primary sedimentation basins
Chemical oxygen femand (COD) the amount of oxygen needed to oxidize the organic matters in water/wastewater
Coagulation a water/wastewater treatment process of destabilizing colloidal particles with the use of coagulants to facilitate particle growth during flocculation
Flocculation a water/wastewater treatment process in which particle collisions are stimulated in order to form large particles
Flotation a process in which suspended matters are raised to the surface of a liquid to facilitate its removal
Hybrid system a combined process that makes use of the advantages of two or more processes
Industrial scale-up a process to convert a process/equipment from a laboratory scale or pilot plant to a full-scale industrial unit
Ion-exchange a process of exchanging one ion for another ion in a charged, solid substrate
Jar test a laboratory bench scale procedure to provide useful information for a water/wastewater treatment plant's coagulation/flocculation. Parameters involved include solution pH, chemical dose, stirring speed, and settling time.
Ozonation an oxidation treatment process that uses ozone (O_3); an unstable gas that can destroy bacteria and viruses for water/wastewater treatment
Polyacrylamide (PAM) a polymeric material that can be used as a flocculant in a water/wastewater treatment unit
Polyaluminium chloride (PACl) a coagulant used in a water/wastewater treatment
Membrane filtration a separation process that utilizes membranes as the medium
Nonylphenolic compounds these compounds are degraded from alkylphenol polyethoxylates (APEOs), non-ionic surfactants used largely in industries such as textile, pulp and paper. They are toxic to invertebrate, fish, mammals and algae.
Response surface methodology (RSM) a statistical experimental design used for optimizing the factors affecting the process
Sedimentation a partial separation of suspended solid particles from a liquid through settling by an external force, usually gravity
Sludge volume index (SVI) a measure for the settling characteristics of sludge
Soluble chemical oxygen demand (SCOD) the amount of oxygen needed to oxidize the organic matters in water/wastewater without the suspended solids
Total chemical oxygen demand (TCOD) the amount of oxygen needed to oxidize the organic matters in water/wastewater including the suspended solids
Total suspended solid (TSS) a measure of the amount of suspended solid particles in water/wastewater
Turbidity a cloudiness of a liquid caused by tiny solids, particles, and other pollutants. It is usually measured in terms of the Nephelometric Turbidity Unit (NTU)

References

Aguilar, M.I., Saez, J., Llorens, M., et al., 2005. Improvement of coagulation-flocculation process using anionic polyacrylamide as coagulant aid. Chemosphere 58, 47–56.

Ahmad, A.L., Wong, S.S., Teng, T.T., Zuhairi, A., 2007. Optimization of coagulation-flocculation process for pulp and paper mill effluent by response surface methodological analysis. J. Hazard. Mater 145, 162–168.

Ahmad, A.L., Wong, S.S., Teng, T.T., Zuhairi, A., 2008. Improvement of alum and PACl coagulation by polyacrylamides (PAMs) for the treatment of pulp and paper mill wastewater. Chem. Eng. J 137, 510–517.

Ali, M., Sreekrishnan, T.R., 2001. Aquatic toxicity from pulp and paper mill effluents: a review. Adv. Environ. Res. 5, 175–196.

APHA, 2000. Standard methods for the examination of water and wastewater, twentieth ed. American Public Health Association, Washington, DC.

Basu, N., Ta, C.A., Waye, A., et al., 2009. Pulp and paper mill effluents contain neuroactive substances that potentially disrupt neuroendocrine control in fish reproduction. Environ. Sci. Technol. 43, 1635–1641.

Belhateche, D.H., 1995. Choose appropriate wastewater treatment technologies. Chem. Eng. Prog., 32–51. August.

Berryman, D., Houde, F., DeBlois, C., Shea, M.O', 2004. Nonylphenolic compounds in drinking and surface waters downstream of treated textile and pulp and paper effluents: A survey and preliminary assessment of their potential effects on public health and aquatic life. Chemosphere 56, 247–255.

Casey, T.J., 1997. Unit treatment processes in water and wastewater engineering. John Wiley & Sons, Chichester, England.

Deng, S.G., Yu, G., Ting, Y.P., 2005. Production of bioflocculant by Aspergillus parasiticus and its application in dye removal. Colloids and Surfaces B: Biointerfaces 44, 179–186.

Ebeling, J.M., Sibrell, P.L., Ogden, S.R., Summerfelt, S.T., 2003. Evaluation of chemical coagulation-flocculation aids for the removal of suspended solids and phosphorus from intensive recirculating aquaculture effluent discharge. Aquacul. Eng 29, 23–42.

Ghosh, M., Ganguli, A., Pathak, S., 2009. Application of a novel biopolymer for removal of *Salmonella* from poultry wastewater. Environ. Technol. 30, 337–344.

Gong, W.-X., Wang, S.-G., Sun, X.-F., Liu, X.-W., Yue, Q.-Y., Gao, B.-Y., 2008. Bioflocculant production by culture of *Serratia ficaria* and its application in wastewater treatment. Bioresource Technol. 99, 4668–4674.

Harleman, D.R.F., Murcott, S.E., 1999. The role of physical-chemical wastewater treatment in the mega-cities of the developing world. Water Sci. Technol 40, 75–80.

Huang, H.C., Pan, J.R., 2002. Coagulation approach to water treatment. Encyclopedia of surface and colloid science. Marcel Dekker, New York.

Kujawski, D., 2009. Trends in oil refinery wastewater reclamation. Poll. Eng. 41, 22–25.

Lacorte, S., Latorre, A., Barcelo, D., Rigol, A., Malmqvist, A., Welander, T., 2003. Organic compounds in paper-mill process waters and effluents. Trends Anal. Chem. 22, 725–737.

Li, Fangyue, Wichmann, K., Otterpohl, R., 2009. Review of the technological approaches for grey water treatment and reuses. Sci. Total Environ. 407, 3439–3449.

Mahvi, A.H., Vaezi, F., Balodor, A., Basseri, S., 2005. Pilot-plant testing of a CEPT system application for the largest industrial complex in Iran. J. Applied Sci. 5, 177–181.

Mason, R.L., Gunst, R.F., Hess, J.L., 2003. Statistical design and analysis of experiments with applications to engineering and science, second ed. John Wiley & Sons, New York.

Mbendi Information Services (Pty) Ltd, 2009. An Mbendi profile: world: pulp and paper industry — world pulp and paper. http://www.mbendi.co.za/indy/pulp/p2005.htm (accessed 25.07.09.).

Metcalf, L., Eddy, H.P., 2004. Wastewater engineering treatment and reuse. McGraw-Hill, New York.

Montgomery, D.C., 2004. Design and analysis of experiments. John Wiley & Sons, New York.

Muralidhara, H.S., 1986. Advances in solid-liquid separation. Battelle Press, Columbus, Ohio.

Nasser, M.S., James, A.E., 2007. Effect of polyacrylamide polymers on floc size and rheological behaviour of kaolinite suspensions. Colloids and Surfaces A: Physicochemical and Eng. Aspects 301, 311–322.

Negro, C., Sanchez, L.M., Fuente, E., Blanco, A., Tijero, J., 2006. Polyacrylamide induced flocculation of a cement suspension. Chem. Eng. Sci 61, 2522–2532.

Nordin, M.Y., Venkatesh, V.C., Sharif, S., Elting, S., Abdullah, A., 2004. Application of response surface methodology in describing the performance of coated carbide tools when turning AISA 104 steel. J. Mater. Process. Technol. 145, 46–58.

Pokhrel, D., Viraraghavan, T., 2004. Treatment of pulp and paper mill wastewater—a review. Sci. Total Environ. 333, 37–58.

Rana, T., Gupta, D., Kumar, S., et al., 2004. Toxic effects of pulp and paper-mill effluents on male reproductive organs and some systemic parameters in rats. Environ. Toxicol. Pharm. 18, 1–7.

Randtke, S.J., 1988. Organic contaminant removal by coagulation and related process combinations. J. Am. Water Works Assoc. 80, 40–56.

Stephenson, R.J., Duff, S.J.B., 1996. Goagulation and precipitation of a mechanical pulping effluent. I. Removal of carbon, colour and turbidity. Water Res. 30, 781–792.

Tan, B.H., Teng, T.T., Mohd Omar, A.K., 2000. Removal of dyes and industrial dye wastes by magnesium chloride. Water Res. 34, 597–601.

Taylor, M.L., Morris, G.E., Self, P.G., Smart, R.St.C., 2002. Kinetics of adsorption of high molecular weight anionic polyacrylamide onto kaolinite: The flocculation process. J. Colloid Interface Sci. 250, 28–36.

Thompson, G., Swain, J., Kay, M., Forster, C.F., 2001. The treatment of pulp and paper mill effluent: A review. Bioresource Technol. 77, 275–286.

Wong, S.S., 2006. Pre-treatment of pulp and paper mill wastewater by coagulation-flocculation method. M.Sc. thesis, Universiti Sains Malaysia, Malaysia.

Wong, S.S., Teng, T.T., Ahmad, A.L., Zuhairi, A., Najafpour, G., 2006. Treatment of pulp and paper mill wastewater by polyacrylamide (PAM) in polymer induced flocculation. J. Hazard. Mater. B135, 378–388.

Xu, G.R., Zhang, W.T., Li, G.B., 2005. Adsorbent obtained from CEPT sludge in wastewater chemically enhanced treatment. Water Res. 39, 5175–5185.

Yang, H., Abbaspour, K.C., 2007. Analysis of wastewater reuse potential in Beijing. Desalination 212, 238–250.

CHAPTER

Arsenic Removal from Water by the Coagulation Process

11

Shaoxian Song, Marisol Gallegos-Garcia

Instituto de Metalurgia, Universidad Autónoma de San Luis Potosí, Av. Sierra Leona 550, San Luis Potosí, C. P. 78210, México

CHAPTER OUTLINE

The Role of Colloidal Systems in Environmental Protection. http://dx.doi.org/10.1016/B978-0-444-63283-8.00011-9

1 Arsenic and its removal from water

Arsenic is the 20th most abundant element in the Earth's crust and is widely distributed throughout nature as a result of weathering, dissolution, fire, volcanic activity, and anthropogenic input.

It finds application in the manufacture of herbicides and pesticides. Other industries using arsenic include wood and hide preservatives, lead shot manufacture, phosphate detergents builder, and presoaks used in many fertilizers. One of the present usages of arsenic as a therapeutic agent is in cancer treatment. Recently randomized clinical trials in the United States led the Food and Drug Administration (FDA) to approve arsenic trioxide for treatment against leukemia (Peters et al., 1996).

Arsenic in nature comes from soil erosion and minerals leaching. Soil erosion and agricultural runoff are large contributors to the arsenic concentration in sediments. High arsenic levels (8.6–13.2 mg/g) have been reported to be associated with sediments, and a potential exists that it may be released in hazardous amounts to the overlying waters (Viraraghavan et al., 1999). Industrial effluents are a major source of arsenic in the environment. Arsenic and arsenical compounds are found in effluents from the metallurgical industry (production of arsenic-containing wastes during smelting and mining operations), glassware and ceramic industries, dry and pesticide manufacturing industries, petroleum refining, rare earth industry, and other organic and inorganic chemical industries. The environmental impact of the use of arsenical compounds, at least locally, will remain for some years.

1.1 Arsenic in water

Arsenic rarely occurs in a free state; it is largely found in combination with sulfur, oxygen, and iron (Driehaus et al., 1994; Jain and Ali, 2000; Sato et al., 2002). Water is one the major means of transport of arsenic in the environment. Arsenic occurs naturally in water in many parts of the world. Natural processes, including soil erosion, mineral leaching, and weathering, are responsible for introducing arsenic into surface waters (Welch et al., 1988). Arsenic-enriched geothermal waters can also significantly increase the amounts of arsenic in surface waters (Seong et al., 2004).

Arsenic contamination of surface and subsurface waters is a worldwide problem, with reported studies in a large number of countries including Italy (Frau and Ardau, 2003), New Zealand (Roddick-Lanzilotta et al., 2002), Bengal, Chile, Taiwan, (Zhang et al., 2003), Vietnam, and the United States (Tempel et al., 2000). In arsenic-enriched environments, a major concern is the potential for mobilization and transport of this toxic element to groundwater and drinking water supplies. In Bangladesh more than 100 million people have been reported to be at risk due to the consumption of arsenic from groundwater with high arsenic concentration, around 0.5–5 mg/l arsenic (Jiang, 2001; Meng et al., 2001). In the United States, over 350,000 people may be drinking water containing more than 50 μg/l of arsenic, and over 2.5 million people could be supplied with water having arsenic levels over 25 μg/l. A survey of 114 wells in the Tainan region of the southwest coast of Taiwan showed that arsenic concentration in water ranges from 0.6 to 2.0 mg/l (Viraraghavan et al., 1999). A survey of groundwater supplies in Halifax, Canada, revealed consistent maximum contaminant level (MCL) violation in various communities. In at least 10% of the samples analyzed, the arsenic contamination was in excess of 500 μg/l, while 23% of the sample analyzed was found to contain arsenic higher than 250 μg/l (Meranger et al., 1984).

In 1958, an international guideline established an allowable level of 200 μg/l for arsenic in drinking water. However, recognition of the harmful health risk of arsenic exposure has resulted in the standard being reduced to 50 μg/l and more recently to 10 μg/l. In most developing nations, drinking water guidelines are still in the range of 40–50 μg/l, mainly because of the inadequate testing facilities for lower concentrations commonly available.

Arsenic is usually found in natural water in the forms of the soluble arsenic species As(III) (arsenite) and As (V) (arsenate). The distribution of arsenic species in natural water is mainly dependent on redox potential and pH (Ioannis and Zouboulis, 2004). Under oxidizing conditions such as those prevailing in surface waters, the predominant species is pentavalent arsenic, which is present chiefly in oxyanionic forms ($HAsO_4^{-2}$ and $H_2AsO_4^-$) with $pk_a = 2.19$ and $pk_b = 6.94$, respectively, throughout the pH range typically encountered in water treatment. The trivalent arsenic is a thermodynamically stable form ($As(OH)_3$, $As(OH)_4^-$, AsO_2OH^{-2}, and AsO_3^{-3}), under anoxic conditions (Jiang, 2001). The pentavalent arsenic species are predominant and stable in oxygen-rich aerobic environments, whereas the trivalent arsenite species are predominant in moderately reducing anaerobic environments such as groundwater (Viraraghavan et al., 1999). The pH value determines the predominant species present in water and is therefore expected to be a very influential parameter during arsenic removal treatments. In natural watercourses, organic forms of arsenic can also be found, such as monomethylarsonate and dimethylarsinate. Its release in natural waters also depends on redox conditions, especially the oxidation of sulphites and the reduction of iron phases.

Although thermodynamics can accurately predict changes in a given nonequilibrium condition, it gives no insight into the rate at which those conditions will occur. In general, As(III) and As(V) acid–base reactions can be assumed to occur instantaneously, whereas time periods for the changes between oxidation states are indeterminate in the natural water (Jiang, 2001). For instance, the conversion of As(III) to As(V) in oxygenated water is thermodynamically favored, yet the rate of the transformation may take days, weeks, or even months, depending on the specific conditions.

Both organic and inorganic forms of arsenic can be detected in natural water systems. Methylated or organic arsenic occurs at concentrations less than 1 μg/l and is not of major significance in drinking water treatment (Edwards, 1994).

1.2 Toxicity of arsenic in water

The presence of trace arsenic levels in public drinking water supplies has not previously caused alarm. Arsenic is both toxic and carcinogenic. Inorganic forms of arsenic dissolved in drinking water are the most significant forms of natural exposure (Jiang, 2001). Organic forms of arsenic that may be present in food such as shellfish are much less toxic to humans and include such forms as monomethyl arsenic acid, dimethyl arsenic acid, and arseno-sugars.

Many studies have indicated that arsenic ingestion may result in internal malignancies, including cancers of the kidney, bladder, liver, lung, and other organs. The Public Health Service Agency for Toxic Substances and Disease Registry in the United States has compiled a comprehensive review of studies related to arsenic and cancer. With respect to exposure through drinking water, it has stated that "An association between environmental exposure to arsenic through drinking water and skin cancer has been observed and confirmed; two cases of bladder cancer also were described, with latent periods of eight to 20 years. The latent periods for two cases of skin cancer related to arsenic in drinking-water were 20 and 23 years, and the concentrations or uptake of arsenic were reported to be 1.2 and 1 mg per day, respectively, with an estimated total ingested dose of about 8 g in one study. Epidemiological studies in areas with different frequencies of black-foot disease and where drinking-water contained 0.35–1.14 mg/l arsenic revealed elevated risks for cancers of the bladder, kidney, skin, lung, liver and colon in both men and women" (O'Connor, 2002).

Arsenic also has been associated with a broad range of noncancerous health effects. These include cardiovascular, neurological, pulmonary, immunological, endocrine, reproductive, and developmental effects. Vascular system effects have also been observed, including peripheral vascular disease, which in its most severe form results in gangrene or blackfoot disease. Other potential effects include neurologic impairment (Lomaquahu and Smith, 1998). The primary route of exposure to arsenic for humans is ingestion. Exposure via inhalation is considered minimal, although in some regions elevated levels of airborne arsenic occur periodically (Hering and Chiu, 1998). To lower the cancer risk to one out of ten thousands, which is considered to be an acceptable risk, the concentration of arsenic in drinking water needs to be reduced to at least 20 μg/l.

1.3 Arsenic removal processes

In some arsenic-affected areas, substitution of the drinking water source with a safe and easily available source may not be possible during part or all of the year, or may be very expensive. Arsenic removal from water may be a more appropriate water supply option.

A variety of treatment processes have been developed for removing arsenic from water. The commonly used technologies include coagulation, adsorption, ion exchange, membrane separation, biological treatment, and solar distillation. (The coagulation process will be described in detail later in this chapter.) The adsorption process for arsenic removal is realized through arsenic species on special solid adsorbents, such as activated carbon (Dambies et al., 2002), activated alumina (Jiang, 2001), and iron powders. It is a mass transfer process in which a substance is transferred from the liquid phase to solid surfaces and becomes bound by chemical or physical forces (Zhang, et al., 2003). In the ion-exchange process, arsenic is eliminated from water through the reversible interchange of ions between solid and water phases, where there is no permanent change in the structure of the solid (Korngold et al., 2001). The membrane separation uses semipermeable membranes that are selectively permeable to water and certain solutes to separate arsenic-borne particles from water. There are many different membrane alternatives, including microfiltration, reverse osmosis, electrodialysis, ultrafiltration, and nanofiltration (Feenstra et al., 2007). The biological treatment transforms, stabilizes, and/or removes arsenic by means of microorganisms (Johnston and Heijnen, 2001).

2 The coagulation process for arsenic removal

In colloid chemistry, coagulation or electrolytic coagulation is defined as the aggregation of colloidal or fine particles in a medium through the addition of electrolytic ions. Through this mechanism the electrolytic ions compress the electrical double layer of the particle surfaces and thus reduces the repulsion of the electrical double layer between the particles in the medium, resulting in the lowering of the potential energy barrier between the particles and thus the proximity of the particles. It has been widely applied in water treatment, solid–liquid separation, and so on.

2.1 The coagulation process

Coagulation or coagulation/precipitation is a common and effective process for eliminating As(V) from contaminated water. It is schematically represented in Figure 11.1. This process includes three or four steps, namely, electrolytic coagulation, adsorption, sedimentation, or filtration.

Electrolytic Coagulation: The coagulation step in the process is similar to that applied to water treatment (Huang and Pan, 2002). Usually, there are colloidal solid particles suspended in arsenic-contaminated water. The colloidal particles in the water aggregate into coagulates in the presence of electrolytic ions (Al^{3+} or Fe^{3+}) that are termed coagulants.

Adsorption: In the coagulates, the Al^{3+} or Fe^{3+} ions appear in the Stern layers of the colloidal particle surfaces. The arsenate ion (As(V)) in the water would form ferric arsenate ($FeAsO_4$) or aluminum arsenate ($AlAsO_4$) with the Al^{3+} or Fe^{3+} ions and thus would adsorb on the coagulates through chemical adsorption. This step is also termed precipitation or co-precipitation (Johnston and Heijnen, 2001).

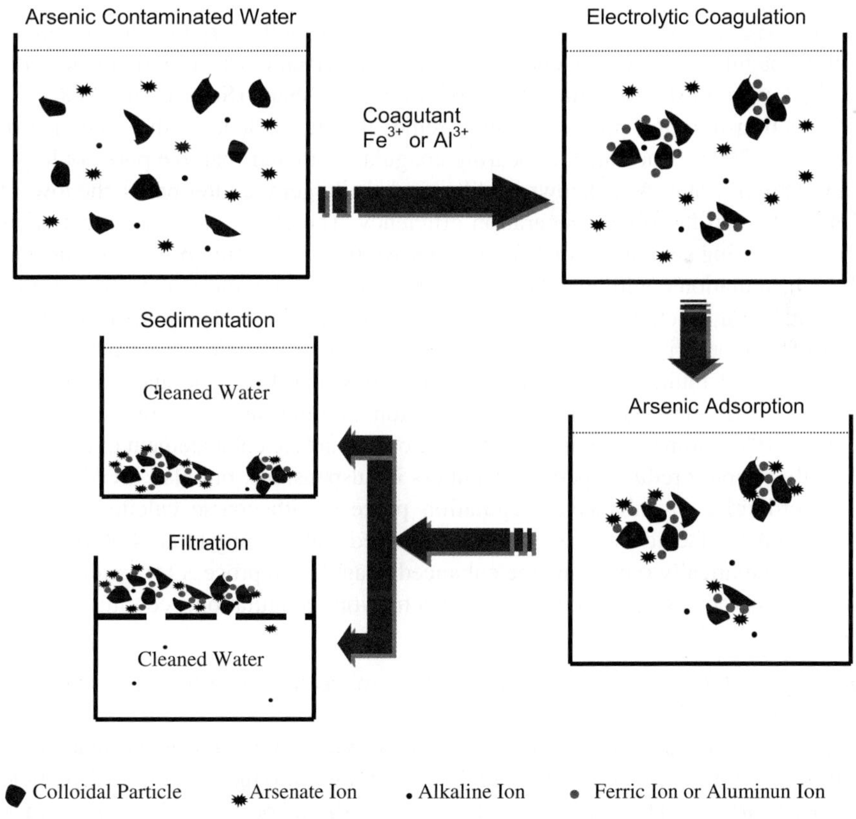

FIGURE 11.1

Schematic representation of coagulation process for arsenic removal from water. (For color version of this figure, the reader is referred to the online version of this book.)

Sedimentation: The coagulates settle down in this step to be separated from cleaned water, while the As(V) is removed from water together with the arsenic-bearing coagulates.
Filtration: In the most cases, arsenic-bearing coagulates settle down very slowly because of the fine size. Therefore, sedimentation is not an effective way to separate the coagulates from water. It is needed to apply filtration to separate the coagulates from water and thus to eliminate arsenic from water (Feenstra et al., 2007). The filter media (porous materials) used in this step could be clothes, porous rubbers, porous ceramics and membranes, and so on.

In the early 1970s, the coagulation process was found to be efficient for removing As(V) from water (Gulledge and O'Connor, 1973). Since then, numerous research studies have been performed to improve the process and to understand the mechanisms by which the As-contaminated water can be cleaned. Today it is common process to eliminate arsenic from water by using ferric salts or alum as the coagulant.

2.2 The enhanced coagulation process

Arsenic-bearing coagulates from electrolytic coagulation are usually very fine. It was reported that the arsenic-bearing coagulates from the coagulation of high-arsenic water with ferric sulfate as the coagulant at pH 5 were of 0.5 to 4 mm in size with irregular shape (Song et al., 2006).

In the coagulation-filtration process, arsenic removal is dependent on the pore size of the filter medium (Han et al., 2003), since arsenic-bearing coagulates smaller than the pore could pass through the filter and remain in water. As is known, in filtration, the larger the filter pores, the lower the capital and operation costs, and the higher separation efficiency. Therefore, there is a great significance in enlarging arsenic-bearing coagulates in order to improve the coagulation process for arsenic removal.

Electrolytic coagulation could be enhanced by adjusting pH and electrolyte concentration and by optimizing coagulation kinetics (Hiemenz and Rajagopalan, 1997). The enhancement can also be realized by adding appropriate coarse particles, in which fine particles in suspensions coat coarse particles in the form of multilayers. Such coagulates consist of both fine and coarse particles. This phenomenon was early observed in mineral flotation systems and was termed "slime coating" (Fuerstenau et al., 1979; Somasundaran, 1980). The coarse particle enhancement of coagulation of fine particles not only strongly reduces particle numbers in suspensions, but also greatly increases coagulate size. Accordingly, an enhanced coagulation process with coarse calcite particles has been developed to eliminate arsenic from arsenic-contaminated water (Song et al., 2006).

Figure 11.2 schematically represents the enhanced coagulation process for removing arsenic from water. This process consists of four steps: electrolytic coagulation, coating, adsorption, and sedimentation.

Electrolytic Coagulation: This step is exactly the same as the common coagulation process mentioned above.
Coating: The small coagulates generated in the electrolytic coagulation step coat a coarse mineral on the surfaces, forming much larger coagulates that are a mixture of colloidal particles and coarse mineral particles. The minerals used in this step should be charged positively in a wide pH range. In other words, the PZC (point of zero charge) of the mineral should be at a high pH, such as calcite. Therefore, the coating could be realized through electrostatic attraction between the small negatively charged coagulates and the coarse positively charged mineral. The particle size

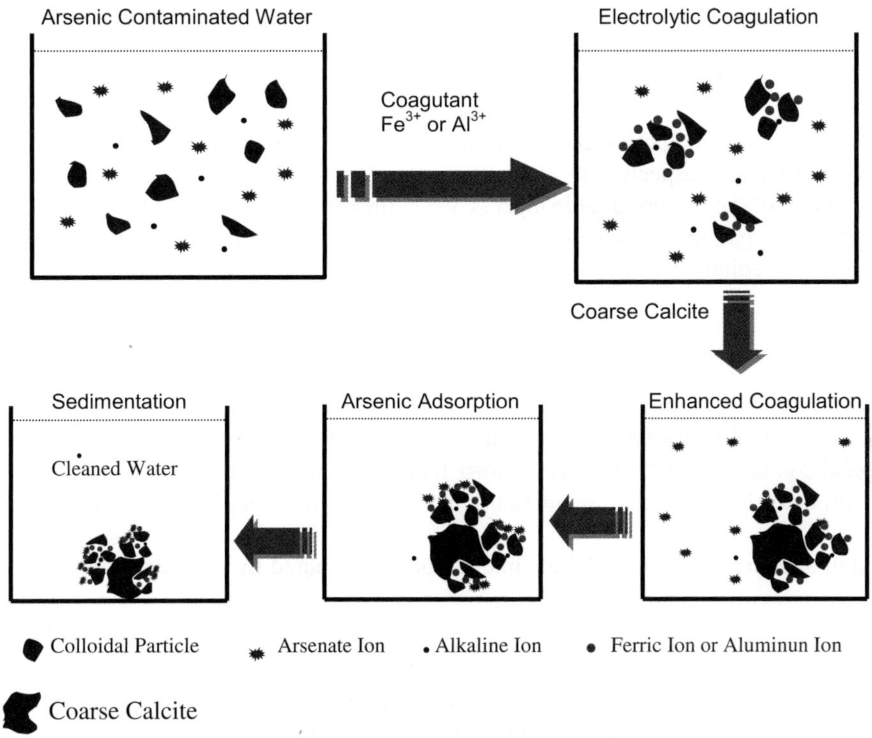

FIGURE 11.2

Schematic representation of enhanced coagulation process for arsenic removal from water. (For color version of this figure, the reader is referred to the online version of this book.)

of the mineral could be in the 38–150 μm range. Clearly, the coarser the mineral particles, the larger the mixed coagulates.

Adsorption: In the same way, As(V) adsorbs on the coagulates through chemical adsorption. The As(V) is concentrated on the coarse mineral particles together with the small coagulates.

Sedimentation: Since the arsenic-bearing coagulates settle down together with the coated coarse particles, they would have a very high velocity of sedimentation. Accordingly, a low-cost water/solid separation method, sedimentation, would be effective in removing the arsenic-bearing coagulates from contaminated water, producing a cleaned water.

3 Mechanisms of coagulation

3.1 Particle interaction and colloid stability

The aggregative stability of colloidal suspensions is due to the existence of a potential energy barrier between particles preventing the proximity of the particles, as explained by the DLVO (Derjaguin-Landau-Verwey-Overbeek) theory (Derjaguin and Landau, 1941; Verwey and Overbeek, 1948).

According to this theory, the net potential energy of interaction between particles in a medium is expressed as follows:

$$V_T = V_R + V_A \tag{11.1}$$

where V_T, V_R, and V_A are the potential energies of interaction of net, electric double layer, and van der Waals between particles, respectively.

The interaction of the electric double layer is due to the overlapping of the electric double layers on the two approaching particles in a medium. The potential energy of the electric double layer between two particles in a medium is expressed by (Gregory, 1975):

$$V_R = 32\pi\varepsilon_0\varepsilon\, a\left(\frac{kT}{ev}\right)^2 \tanh\left(\frac{ev\psi_1}{4kT}\right)\tanh\left(\frac{ev\psi_2}{4kT}\right)\exp(-\kappa h) \tag{11.2}$$

where ε_0 is vacuum dielectric permittivity; ε is relative dielectric permittivity of the double layer; k is Boltzmann constant; T is absolute temperature; e is elementary charge; v is ionic valence of electrolyte; ψ_1 and ψ_2 are the OHP (Outer Helmholtz Plane) potentials or zeta potentials of particle 1 and particle 2, respectively; κ is Debye reciprocal length; and h is the shortest distance between the two particles.

The potential energy of van der Waals interaction is computed by (Gregory, 1981):

$$V_A = -\frac{aA_{132}}{12}\left(\frac{1}{h} - \frac{5.32}{\lambda}\ln\left(1 + \frac{\lambda}{5.32h}\right)\right) \tag{11.3}$$

where λ is the characteristic wavelength of the atoms and A_{132} is the Hamaker constant of particle 1 and particle 2 in medium 3, which may be obtained by the following expression:

$$A_{132} = \left(\sqrt{A_{11}} - \sqrt{A_{33}}\right)\left(\sqrt{A_{22}} - \sqrt{A_{33}}\right) \tag{11.4}$$

where A_{11}, A_{22}, and A_{33} are the Hamaker constants of particle 1, particle 2, and medium 3 in vacuum.

Based on Eqns. 11.1, 11.2, and 11.3, the net potential energy of interaction between particles in a medium could be calculated as the function of the distance. The result could be illustrated in the graph of potential energy, as shown in Figure 11.3. The potential energy barrier is the criterion to evaluate the stability of colloidal suspensions. It is accepted that in a colloidal system in which particle motion is totally governed by Brownian movement, no aggregation takes place if a potential energy barrier is bigger than 15kT. In contrast, a strong aggregation occurs if it is less than 5kT because Brownian motion endows the particles with sufficient kinetic energy to overcome the potential energy barrier. Such a phenomenon is termed *rapid coagulation*, in which a collision between two particles leads to permanent contact between the colliding particles. When repulsive potential energy barriers are 5~15kT, a weak aggregation that is called slow coagulation takes place, in which the collision rate between particles is low because of the hindrance of the energy barrier.

3.2 Mechanisms of coagulation

The mechanisms of coagulation for arsenic removal are the same as those for water treatment, which might be due to electric double-layer compression, adsorption-charge neutralization, and/or enmeshment.

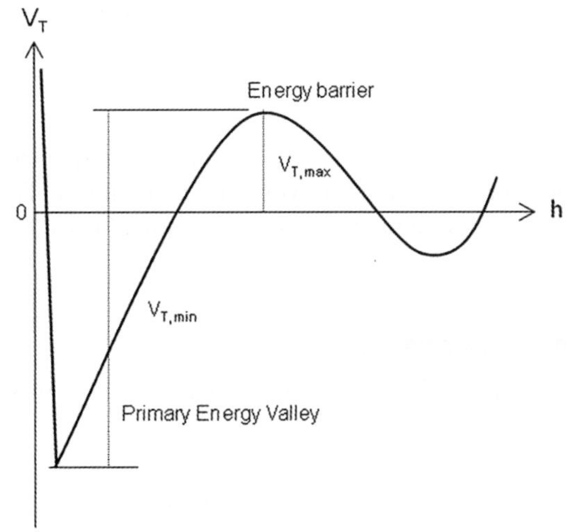

FIGURE 11.3

Potential energy curve of interaction between colloidal particles.

3.2.1 Electrical double-layer compression

The presence of electrolytic ions in a colloidal dispersion would compress the electrical double layer of the colloidal particle surfaces. This compression would reduce the absolute value of the zeta potential and the thickness of the electrical double layer. In other words, the absolute value of ψ would decrease and the value of κ would increase, leading to the reduction of the potential energy of electrical double-layer repulsion between particles on the basis of Eq. 11.2. On the other hand, the potential energy of van der Waals does not change with the presence of electrolytic ions. According to the DLVO theory (Eq. 11.1), the reduction of electrical double-layer repulsion would lower the potential energy barrier and thus the coagulation of particles. Furthermore, the distance between particles in which electrical double-layer repulsion acts would also decrease, leading the potential energy barrier to locate at a short distance. This observation indicates a stronger coagulation.

Compression of the electrical double layer is related to the valence of the electrolytic ions as well as the dosage. The higher the valence, the less the dosage of the electrolyte required for colloidal coagulation. The Schulz-Hardy rule gives the relationship between the electrolytic ion valence and the dosage in an electrolytic coagulation.

3.2.2 Adsorption-charge neutralization

Ferric and aluminum ions in aqueous solutions would hydrolyze to form the species of hydrolysis. These species could adsorb on colloidal particles surfaces through the attractive forces of coulombic interaction, dipole interactions, hydrogen bonding, and van der Waals interaction. This adsorption is termed the Stern-layer adsorption (Legaly, 2006). In general, colloidal particles in aqueous solutions are negatively charged. When the positively charged hydrolysis species adsorb on negatively charged colloidal particles, charge neutralization occurs, and the absolute value of the zeta potential of the particles decreases. As already stated, reduction of the zeta potential would lead to a decrease in the potential energy barrier between the particles and thus the electrolytic coagulation of the colloidal particles.

3.2.3 Enmeshment

If a large amount of Fe^{3+} or Al^{3+} is added to a colloidal particle suspension, the ions hydrolyze and polymerize to form insoluble iron or aluminum hydroxide precipitates. The hydroxide particles that are thermodynamically stable at room temperature sweep colloidal particles from water to form large coagulates.

3.3 Adsorption of arsenic on coagulates

As(V) in water has the dissociation reaction as follows (Raven et al., 1998):

$$H_3AsO_4 \leftrightarrow H^+ + H_2AsO_4^- \tag{11.5}$$

$$H_2AsO_4^- \leftrightarrow H^+ + HAsO_4^{2-} \tag{11.6}$$

$$H_2AsO_4^{2-} \leftrightarrow H^+ + AsO_4^{3-} \tag{11.7}$$

In the pH of water higher than 7 (which is common for underground water), $HAsO_4^{2-}$ and AsO_4^{3-} are the dominant species in water.

In other hands, the ions of Fe^{3+} or Al^{3+} adsorb on colloidal particles in the form of the hydrolysis species of the ions, for instance, $Fe(OH)^{2+}$ and $Fe(OH)_2^+$. The hydrolysis species would have a chemical reaction with $HAsO_4^{2-}$ or AsO_4^{3-} to form precipitates. This reaction would result in the chemical adsorption of As(V) on the colloidal particles, which could be represented as follows:

$$\text{Particle}\big|Fe(OH)^{2+} + HAsO_4^{3-} \rightarrow \text{Particle}|FeAsO_4 + H_2O \tag{11.8}$$

If Fe^{3+} or Al^{3+} has hydrolyzed and polymerized to form insoluble iron or aluminum hydroxide precipitates, the As(V) species directly adsorb on the precipitate solid through chemical adsorption. This adsorption can be represented as follows:

$$Fe(OH)_3\big|_{(solid)} + HAsO_4^{3-} \rightarrow Fe(OH)_3|FeAsO_4 + H_2O \tag{11.9}$$

3.4 Coarse particles enhancing coagulation

The enhancement of electrolytic coagulation by adding specialized coarse particles has been demonstrated by numerous experimental observations (Leja, 1982). Raju et al. (1991) studied the effect of coarse hydrophobic quartz on the hydrophobic agglomeration of quartz fines (−5 μm) induced by dodecylamine. In the absence of coarse quartz, the concentration of −5 μm quartz in the suspension only decreased to 330 mg/l from 400 mg/l after 15 minutes stirring at 500 rev/min. With the addition of −38+20 μm quartz, the concentration of −5 μm quartz in the suspension sharply declined to 50 mg/l from 400 mg/l at the same reagent scheme and mechanical conditioning. In addition, it has been found that the size fraction of −38+20 μm enhanced more strongly the hydrophobic agglomeration than other size fractions.

The mechanisms of the positive effect of coarse particles in the coagulation of colloidal particles might be attributed to the increase of aggregation rate. In a turbulent flow, collision rate between coarse particle and fine particle is very different from that between fine particles themselves. The calculation showed that the former might be 1000- to 10,000-fold greater than the latter. As the collision rate closely correlates with the aggregation rate in an aggregation system, the presence of coarse particles results in a considerable increase of the aggregation rate, leading to a much strong coagulation.

4 Important factors in the coagulation process

4.1 Coagulant

The main coagulants used for arsenic removal from water are ferric chloride, ferric sulfate, and alum (aluminum sulfate), although many kinds of coagulants are used for water treatment.

Ferric chloride and ferric sulfate are the most frequently used coagulants. In aqueous acidic solutions, ferric ion is easy to hydrolyze to form mononuclear $Fe(OH)^{2+}$ and $Fe(OH)_2^+$, and polynuclear $Fe_2(OH)_2^{4+}$ and $Fe_3(OH)_4^{5+}$. The polynuclear species are formed by bridging mononuclear entities through alkyl linkage, and the amorphous $Fe(OH)_3$ precipitates from the polymerization and nucleation of the species. As already stated, the mononuclear and polynuclear species and the $Fe(OH)_3$ precipitate are the key entities for removal of arsenic from water. In the range of pH > 6, the coagulation of colloidal particles is mainly due to the enmeshment mechanism. At the lower pH values, a combination of charge neutralization by soluble positively charged ferric hydroxospecies and enmeshment by $Fe(OH)_3$ precipitate could occur, depending on the iron dosage and pH value (Johnson and Amirtharajah, 1983).

Recently, polymeric ferric sulfate, polymeric alum, and poly-alumino-iron sulfate have been attracting more interest as the coagulant for water treatment, which are more effective and less sensitive to dosage change.

4.2 Coagulation pH

In the coagulation process for arsenic removal from water, the pH value in water greatly affects the efficiency of arsenic removal and the consumption of coagulant. Figure 11.4 illustrates arsenic removal from high-arsenic water as a function of pH by coagulation with 100 mg/l ferric sulfate and microfiltration with the membrane (Song et al., 2006). As can be seen, in the acidic range, very high arsenic removal (about 99%) was achieved, while in the alkaline range, arsenic removal declined sharply with the increase of pH. This observation might be due to the formation of $Fe(OH)_3$ precipitate in the alkaline range, which negatively affects the adsorption of As(V).

4.3 Coarse calcite particles

Figure 11.5 illustrates arsenic removal from high-arsenic water through coagulation with ferric ions as the coagulant and filtration with filter papers (2.5 μm aperture) as a function of coarse calcite addition at pH 6 and 100 mg/l ferric sulfate addition sulfate addition (Song et al., 2006). As can be observed, without coarse calcite, arsenic removal was only 85%. Such a low percentage should be attributed to small arsenic-borne coagulates passing through the filter papers. As the increase of coarse calcite

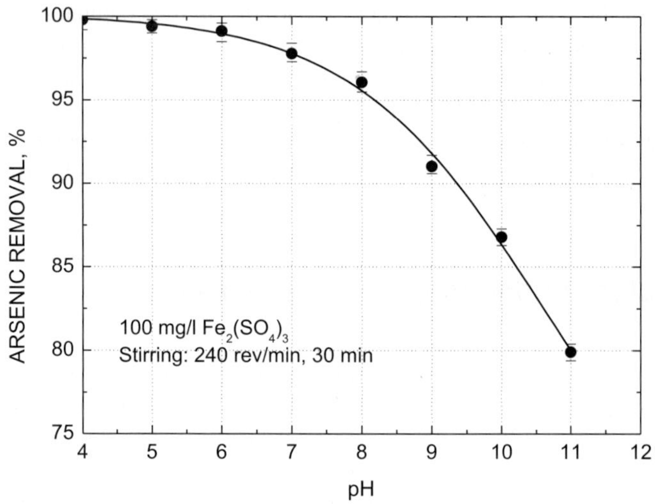

FIGURE 11.4

Arsenic removal from contaminated water with coagulation process as the function of pH in the presence of ferric ion (Song et al., 2006).

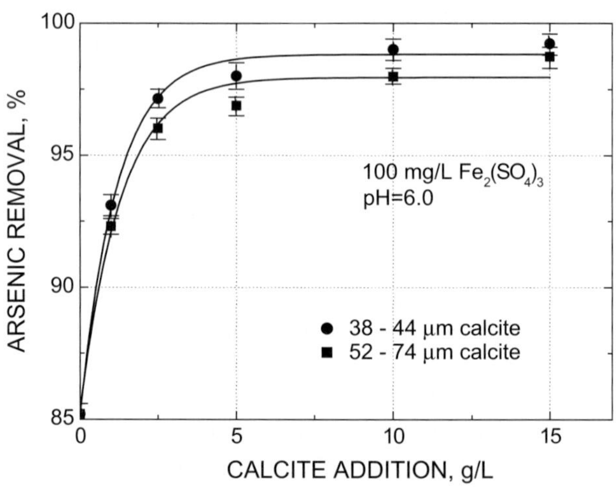

FIGURE 11.5

Arsenic removal from high-arsenic water by enhanced coagulation process as the function of calcite addition.

addition, arsenic removal increased until it reached about 99%. This increase was very sharp in the small calcite addition range but was mild after 2.5 g/l calcite addition. Obviously, the coagulation could be greatly enhanced by adding coarse calcite particles, and thus arsenic can be effectively removed from the water by filtration in a large pore filter medium. Accordingly, this process was termed enhanced coagulation (Song et al., 2006). Also, Figure 11.5 shows that 38–44 μm calcite enhanced arsenic removal more strongly than 52–74 μm calcite at the same amount of addition. The mechanism might be due to the larger surface area on smaller particles, leading to more coatings of arsenic-borne coagulates on calcite surfaces.

At pH 5, the water produced by the enhanced coagulation and filtration with filter papers contained 13.2 μg/l arsenic, which was almost the same as that produced by coagulation with the same amount of ferric sulfate and filtration with the membrane. It indicates that coarse calcite in the enhanced coagulation acts solely to increase the size of arsenic-borne coagulates.

5 Application of the coagulation process

5.1 Arsenic removal from drinking water

The coagulation–filtration process has traditionally been used to remove solid particles from drinking water supplies. Early it was found that arsenic ions in water could be simultaneously removed in the process. For example, in one coagulation–filtration plant, the arsenic concentration in water could be reduced from 7.5 μg/l to 3.5 μg/l (EPA, 2000). Since most of the coagulation–filtration water treatment plants use ferric ions or aluminum ions as the coagulants, this process has been widely applied to remove arsenic from tap water.

In general, the drinking water with the arsenic-contaminated problem is from groundwater. For instance, the well water in the Sonargaon area of Bangladesh has an arsenic concentration higher than 50 μg/l (Wickramasinghe et al., 2004). The coagulation–filtration process has been tested to eliminate arsenic from underground drinking water in many laboratories and field studies around the world. This process consists of a mixer and a microfiltration module. The coagulant (ferric ions or aluminum ions) with an appropriate dosage is added into the mixer to form arsenic-bearing coagulates, followed by the microfiltration to separate the coagulates from water. The arsenic is concentrated in the solid waste, leading the cleaned water to be assayed in a lower arsenic concentration.

Wickramasinghe et al. (2004) reported the experimental results on arsenic removal from groundwater by the coagulation-–filtration process. Ferric chloride or ferric sulfate was used as the coagulant at pH 6.2–6.8, and a 0.22 μm pore size membrane was used as the filter medium in the vacuum filtration. For a groundwater with the arsenic concentration of 68 μg/l, the process removed about 88% arsenic from the groundwater, leaving the arsenic concentration of about 6 μg/l in the cleaned water.

5.2 Arsenic removal from high-arsenic water in the mine drainage system

In Mexico, hundreds years of mining activities for gold, silver, and metallic sulfide minerals left a considerable quantity of tailings, some of which have a high arsenic content. Owing to rain, wind, and other natural effects, the arsenic dispersed into the nearest watercourses in the form of arsenate ion (As (V)) and arsenite ion (As (III)), leading to a high arsenic concentration in the watercourses around the mines. For example, in the city of Matehuala in the state of San Luis Potosí, water from mine drainage systems contained up to 7000 μg/l As, which is much higher than the maximum contaminant level (MCL) for arsenic in drinking water in Mexico (20 μg/l). Accordingly, the high-arsenic water from mine drainage systems would be a great environmental concern through the contamination of shallow groundwater and surface water around the areas. The water has to be treated to remove arsenic effectively, in order to minimize the environmental effects.

The coagulation–filtration process is the most promising process for removing arsenic from high-arsenic water because of its low cost and high efficiency. We tested the enhanced coagulation (with ferric sulfate as the coagulant and coarse calcite) and filtration (with filter paper as the filtration medium) process to eliminate arsenic from high-arsenic water in a Mexican mine drainage system (Song et al., 2006). The As-contaminated water contained 5071 μg/l As. The treatment included the addition of 100 mg ferric sulfate per liter of water and 5 g coarse calcite (38–74 μm) per liter of water, while a mild vibration was applied at the water pH of 6 to 7 and filtration with filter paper (2.5 μm aperture) was used as the medium to separate the solid from water. The experimental results have shown that over 99% arsenic could be removed from the high-arsenic water through the enhanced coagulation–filtration process. The residue arsenic concentration in the cleaned water was about 13 μg/l.

The coagulation-filtration process is generally acknowledged to be the best option for removing arsenic from high-arsenic-contaminated water. However, it is not effective for the deep elimination of arsenic from water that will produce cleaner water with very low arsenic concentration. For instance, if the requirement is to reduce the arsenic concentration in water from 30 μg/l to 5 μg/l, other processes, such as adsorption and ion exchange, would be better than the coagulation–filtration process with respect to efficiency and economy.

SYMBOLS

a radius of the particle (m)
A_{11} Hamaker constants of the particle 1 in vacuum (J)
A_{22} Hamaker constants of the particle 2 in vacuum (J)
A_{33} Hamaker constants of the medium 3 in vacuum (J)
A_{132} Hamaker constants of the particle 1 and particle 2 in medium 3 (J)
e elementary charge
h the shortest distance between the two particles (m)
k Boltzmann constant, 1.38×10^{-23} J/K
T absolute temperature (K)
v ionic valence of electrolyte
V_A potential energy of van der Waals interaction between particles (J)
V_R potential energy of electric double-layer interaction between particles (J)
V_T net potential energy of interaction between particles (J)
ε_0 vacuum dielectric permittivity
ε relative dielectric permittivity of double layer
ψ_1 OHP (Outer Helmholtz Plane) potentials or zeta potentials of particle 1 (V)
ψ_2 OHP (Outer Helmholtz Plane) potentials or zeta potentials of particle 2 (V)
κ Debye reciprocal length (m^{-1})
λ characteristic wavelength of the atoms (m)

Glossary

Adsorption the accumulation of molecules or ions on an interface
Arsenic the chemical element that has the symbol As with atomic number 33
As(III) the soluble arsenic species of arsenic, which exists in the form of arsenite

As(V) the soluble arsenic species of arsenic, which exists in the form of arsenate
Calcite a carbonate mineral with the chemical formula of $CaCO_3$
Coagulant the chemical reagents that arise from the formation of the coagulation of colloidal particles in a dispersion
Coagulate the cluster of colloidal particles in a medium
Coagulation a process of contact and adhesion whereby colloidal particles in a dispersion form larger-size clusters
Colloidal particle solid or liquid particles less than 1 micrometer in size
DLVO theory the theory that explains the stability of a colloidal dispersion, which was presented by two Russian scientists (Derjaguin and Landau) and two Netherlands scientists (Verwey and Overbeek)
MCL maximum contaminant level of a chemical element or component in water or air, which is established by a government or an organization
Mine drainage system water-draining system affected by mining activities, which is contaminated by the leaching of minerals in water

Acknowledgment

Financial support for this work came from the Consejo Nacional de Ciencia y Tecnología (CONACyT) of Mexico under grants No. 123487 and No. 155148 and is gratefully acknowledged.

References

Dambies, L., Vincent, T., Guibal, E., 2002. Treatment of arsenic-containing solutions using chitosan derivaties: uptake mechanism and sorption performance. Water Res. 36 (15), 3699–3710.

Derjaguin, B.V., Landau, L., 1941. Theory of molecular interaction. Acta Physicochem. 14, 633–668.

Driehaus, W., Srith, R., Jekel, M., 1994. Oxidation of arsenate(III) with manganese oxides in water treatment. Water Res. 29 (1), 279–305.

Edwards, M., 1994. Chemistry of arsenic removal during coagulation ad Fe-Mn oxidation. J. Am. Water Works Assoc. 79 (10), 81–84.

EPA., 2000. Arsenic removal from drinking water by coagulation/filtration and lime softening plants. Prepared by Battelle under contract 68-C7-0008 for EPA ORD.

Feenstra, L., van Erkel, J., Vasak, L., 2007. Arsenic in groundwater: Overview and evaluation of removal methods. Utrecht: Int. Groundwater Res. Assess. Centre, 2–23.

Frau, F., Ardau, C., 2003. Geochemical controls on arsenic distribution in the *baccu locci* stream catchment (Sardinia, Italy) affected by past mining. Appl. Geochem. 18 (9), 1373–1386.

Fuerstenau, D.W., Chander, S., Abouzeid, A.M., 1979. The recovery of fine particles by physical separation methods. In: Somasundaran, P., Arbiter, N. (Eds.), Beneficiation of mineral fines—Problems and research needs. AIME, New York, pp. 3–59.

Gregory, J., 1975. Interaction of unequal double layers at constant charge. J. Colloid Interface Sci. 51, 44–51.

Gregory, J., 1981. Approximate expressions for retarded van der Waals interaction. J. Colloid Interface Sci. 83, 138–145.

Gulledge, J.H., O'Connor, J.T., 1973. Removal of arsenic (V) from water by adsorption on aluminum and ferric hydroxides. J. AWWA 65 (8), 548–552.

Han, B., Zimbron, J., Runnells, T.R., Shen, Z., Wickramasinghe, S.R., 2003. New arsenic standard spurs search for cost-effective removal techniques. J. AWWA 95 (10), 109–118.

Hering, J.G., Chiu, V.Q., 1998. The chemistry of arsenic: Treatment and implications of arsenic speciation and occurrence. AWWA Inorganic contaminants workshop, San Antonio, TX, pp. 23–24.

Hiemenz, P.C., Rajagopalan, R., 1997. Principles of colloid and surface chemistry, third ed. Marcel Dekker, New York.

Huang, C., Pan, J.R., 2002. Coagulation approach to water treatment. In: Hubbard, A.T. (Ed.), Encyclopedia of surface and colloid science, vol. 1. Marcel Dekker, New York, pp. 1049–1064.

Ioannis, A.K., Zouboulis, A.I., 2004. Application of biological processes for the removal of arsenic from groundwaters. Water Res. 38 (1), 17–26.

Jain, C.K., Ali, I., 2000. Arsenic: occurrence, toxicity and speciation techniques. Water Res. 34 (17), 4304–4312.

Jiang, J.Q., 2001. Removing arsenic from groundwater for the developing world—a review. Water Sci. Technol. 44 (6), 89–98.

Johnson, P.N., Amirtharajah, A., 1983. Ferric chloride and alum as single and dual coagulants. J. AWWA 75, 232–239.

Johnston, R., Heijnen, H., 2001. Safe water technology for arsenic removal. In: Ahmed, M.F., et al. (Eds.), Technologies for arsenic removal from drinking water. Bangladesh. Bangladesh University of Engineering and Technology, Dhaka, pp. 7–12.

Korngold, E., Belayev, N., Aronov, L., 2001. Removal of arsenic from drinking water by anion exchange. Desalination 141 (1), 81–84.

Lagaly, G., 2006. Colloid clay science. In: Bergaya, F., Theng, B.K.G., Lagaly, G. (Eds.), Handbook of clay science. Elsevier, Amsterdam.

Leja, J., 1982. Surface chemistry of froth flotation. Plenum Press, New York.

Lomaquahu, E.S., Smith, A.H., 1998. Feasibility of new epidemiology studies on arsenic exposures at low levels. AWWA Inorganic Contaminants Workshop, San Antonio, TX, pp. 23–24.

Meng, X., Korfiatis, P.G., Christodoulatos, C., Bang, S., 2001. Treatment of arsenic in Bangladesh well water using a household co-precipitation and filtration system. Water Res. 35 (12), 2805–2810.

Meranger, J.C., Subramanian, K.S., McCurdy, R.F., 1984. Arsenic in Nova Scotian groundwater. Sci. Total Environ. 39 (1–2), 49–55.

O'Connor, J.T., 2002. Arsenic in drinking water, Part 2. Human exposure and health effects. Water Eng. Manag. 149 (3), 35–37.

Peters, G.R., McCurdy, R.F., Hindmarsh, T.J., 1996. Environmental aspects of arsenic toxicity. Crit. Rev. Environ. Sci. Technol. 33 (6), 457–493.

Raju, G.B., Subrahmanyam, T.V., Sun, Z., Forsling, W., 1991. Shear-flocculation of quartz. Int. J. Mineral Process. 32, 283–294.

Raven, K.P., Jain, A., Loeppert, R.H., 1998. Arsenite and arsenate adsorption on ferrihydrite: Kinetics, equilibrium and adsorption envelopes. Envir. Sci.Technol. 32, 344–349.

Roddick-Lanzilotta, A.J., McQuillan, A.J., Craw, D., 2002. Infrared spectroscopic characterisation of arsenate(V) ion adsorption from mine waters, macraes mine, New Zealand. Appl. Geochem 17 (64), 445–454.

Sato, Y., Kang, M., Kamei, T., Magara, Y., 2002. Performance of nanofiltration for arsenic removal. Water Res. 36 (13), 3371–3377.

Seong, K.N., Ujan, Z., Le Clech, P., 2004. Arsenic removal technologies for drinking water treatment. Rev. Environ. Sci. Bio/ Technol. 3 (1), 43–43.

Somasundaran, P., 1980. Role of surface chemistry of fine sulphides in their flotation. In: Jones, M.J. (Ed.), Complex sulphide ores. Institute of Mining and Metallurgy, London, p. 126.

Song, S., Lopez-Valdivieso, A., Hernandez-Campos, D.J., Peng, C., Monroy-Fernandez, M.G., Razo-Soto, I., 2006. Arsenic removal from a mine drainage system by enhanced coagulation with ferric ions and coarse calcite. Water Res. 40 (2), 364–372.

Tempel, R.N., Shevenell, L.A., Lechler, P., Price, J., 2000. Geochemical modeling approach to predicting arsenic concentrations in a Mine Pit Lake. Appl. Geochem. 15 (4), 475–492.

Verwey, E.J., Overbeek, J.G., 1948. Theory of the stability of lyophobic colloid. Elsevier, Amsterdam.

Viraraghavan, T., Subramanian, K.S., Aruldoss, J.A., 1999. Arsenic in drinking water-problems and solutions. Water Sci. Tech. 40 (2), 69–76.

Welch, A.H., Lico, M.S., Hughes, J.L., 1988. Arsenic in ground water of the western united states. Ground Water 26 (3), 333–347.

Wickramasinghe, S.R., Han, B., Zimbron, J., Shen, Z., Karim, M.N., 2004. Arsenic removal by coagulation and filtration: Comparison of groundwaters from the United States and Bangladesh. Desalination 169, 231–244.

Zhang, Y., Yang, M., Huang, X., 2003. Arsenic(V) removal with a Ce(IV)-droped iron oxide adsorbent. Chemosphere 51, 945–952.

CHAPTER

12

Optimization and Modeling of Decolorization and COD Reduction of Reactive Dye Solutions by Coagulation Processes with SIWW's Coagulant

Abdelkader Anouzla, Younes Abrouki, Salah Souabi, Mohamed Safi, Hicham Rhbal

Laboratoire de Génie de l'Eau et de l'Environnement, Université Hassan II, Faculté des sciences et techniques de Mohammedia, Maroc

CHAPTER OUTLINE

1 Introduction

Synthetic dyes are widely used in the textile and dyestuff industries such as textile dyeing, paper printing, food, cosmetics, pharmaceutical, and color photography (Padmavathy et al., 2003). Nearly 10^4 different dyes and pigments are used industrially, and 0.7 million tons of synthetic dyes are

The Role of Colloidal Systems in Environmental Protection. http://dx.doi.org/10.1016/B978-0-444-63283-8.00012-0

produced each year throughout the world (Zhang et al., 2005). Among the available dyes, about 50% of the industrial dyes produced in the world are azo dyes (Kulkarni et al., 1985).

Reactive group azo dyes are used mostly in textile dyeing because of their superior fastness to the applied fabric, high photolytic stability, and resistance to microbial degradation. However, reactive dyes are particularly problematic because the dyes exhibit low levels of fixation with the fiber and about 10–20% of the total dye used in the dyeing process remains in the spent dye bath with accessory chemicals (Gouvea et al., 2000).

Treatment of the textile dye containing effluent is difficult and ineffective using conventional biological processes (Peralta-Zamora et al., 1999) because many synthetic dyes are very stable to light and temperature and resistant to microbial attack. Several physicochemical methods have been used, including adsorption techniques (Barka et al., 2009), photodegradation (Daneshvar et al., 2003), chemical oxidation (Arslan et al., 2000), membrane processes (Magara et al., 1998), and coagulation–flocculation (Harrelkas et al., 2009) for color removal from effluent.

Coagulation–flocculation followed by sedimentation and filtration is used worldwide in the wastewater treatment process before discharging the treated water to the river. Many coagulants are widely used in the conventional wastewater treatment processes (Boisvert et al., 1997). These coagulants can be inorganic coagulants (e.g., aluminium sulphate, ferric chloride, and polyaluminium chloride), synthetic organic polymers (e.g., polyacrylamide derivatives), or naturally occurring flocculants (e.g., microbial flocculants). These coagulants and flocculants are used for different purposes depending on their chemical characteristics (Okuda et al., 1999).

The present work tests the ability of steel industrial wastewater (SIWW) (Anouzla et al., 2009) $FeCl_3$ rich as a coagulant to remove the reactive dye solution. The experiments were planned by using statistical design methods (Myers and Montgomery, 1995). The dye concentration, coagulant dosage, and initial pH were chosen as process parameters. The regression models obtained were used in a constrained optimization to find optimum process conditions for maximum decolorization, and COD reduction efficiency of Reactive Orange 16.

2 Materials and methods

2.1 Materials

The color index international (C.I.) Reactive Orange 16 was purchased from Sigma-Aldrich Co with a molecular structure as given in Figure 12.1. The characteristics of reactive dye orange 3R are summarized in Table 12.1.

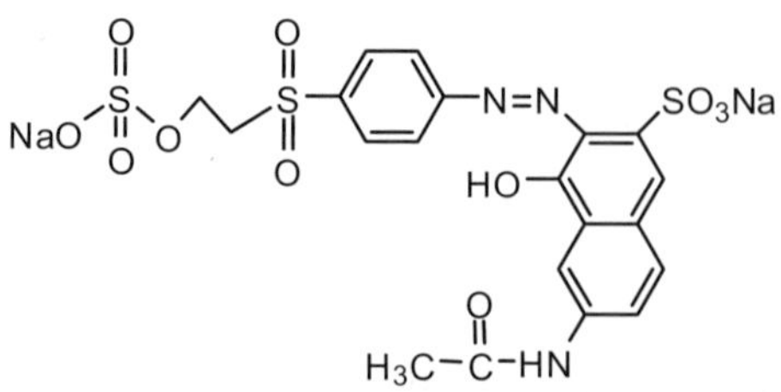

FIGURE 12.1

Molecular structure of Reactive Orange 16.

Table 12.1 The Characteristics of Reactive Orange 16

Parameter	Value
Color index	Reactive Orange 16
Chromophore Molecular weight	Monoazo
(g/mol)	617.54
λ_{max} (nm)	494
Melting point Mmp (°C)	> 300
Chemical formula	$C_{20}H_{17}N_3Na_2O_{11}S_3$

Table 12.2 The Characteristics of SIWW

Parameter	Value
$FeCl_3$ (mg/L)	1787
Conductivity (mS/cm)	26200
pH value	< 1

The SIWW was taken from Magreb Steel (Morocco society) and was used as a novel coagulant in this study. The characteristics of SIWW are given in Table 12.2. The decolorization and COD reduction of dye solutions were examined through coagulation processes by the Jar test experiment.

2.2 Analytical procedure

The standard stock solution of 1000 mg/L, including Reactive Orange 16 dye, was prepared in distilled water. The standard working solutions were prepared by further dilution of standard stock solutions. The solutions were prepared for calibration curves by further dilution of standard working solutions. The initial pH of standard working solutions was adjusted by the addition of NaOH or H_2SO_4 to a desired value in the range of 7–11. Samples of dye solutions were measured using a UV/Visible spectrophotometer (Model 7800 UV/VIS) at the beginning and at the end of the experiment. COD removal was detected by using standard methods at the beginning and at the end of the experiment (Clesceri et al., 1998). The percentages of decolorization and COD reduction were calculated as follows. Eq. 12.1:

$$\text{Decolourization\%}/\text{COD reduction\%} = (1 - C/C_0) \times 100 \tag{12.1}$$

where C_0 is initial dye concentration/COD and C is final dye concentration/COD after treatment by coagulation with SIWW.

2.3 Experimental design

A central composite rotatable design for k independent variables was employed to design the experiments (Istadi and Amin, 2005) in which the variance of the predicted response, $\hat{Y}$, at some points of independent variables, X, is only a function of the distance from the point to the design center.

The design of the experiment is intended to reduce the number of experiments and to arrange the experiments with various combinations of independent variables. In the rotatable design, the standard error, which depends on the coordinates of the point on the response surface at which Y is evaluated and on the coefficients β, is the same for all points that are the same distance from the central point.

These designs consist of a 2^k factorial (Clarke and Kempson, 1997) augmented by $2*k$ axial points $(\pm\alpha, 0, 0)$, $(0, \pm\alpha, 0)$, $(0, 0, \pm\alpha)$ and 2 center points $(0, 0, 0)$. The value of α for rotatability depends on the number of points in the factorial portion of the design, which is given in Eq. 12.2:

$$\alpha = (N_F)^{1/4} \tag{12.2}$$

where N_F is the number of points in the cube portion of the design ($N_F = 2^k$, k is the number of factors). Since there are three factors, the N_F number is equal to 2^3 (= 8) points, while α is equal to $(8)^{1/4}$ (= 1.682) according to Eq. 12.2.

In this study, the responses were decolorization efficiency (Y_{DEC}) and COD reduction (Y_{COD}) of dye solutions. Each response was used to develop an empirical model (Ravikumar et al., 2005) that correlated the response to the coagulation processes activated variables using a second-degree polynomial equation as given by Eq. 12.3:

$$\widehat{Y} = \beta_0 + \beta_1 X_1 + \beta_2 X_2 + \beta_3 X_3 + \beta_{12} X_1 X_2 + \beta_{13} X_1 X_3 + \beta_{23} X_2 X_3 + \beta_{11} X_1{}^2 + \beta_{22} X_2{}^2 + \beta_{33} X_3{}^2 \tag{12.3}$$

where β_o is the constant coefficient, β_i the linear coefficients, β_{ij} the interaction coefficients, and β_{ii} the quadratic coefficients.

STATGRAPHICS-Plus Version 4 was used for regression analysis of the experimental data to fit the equations developed and also to evaluate the statistical significance of the equations.

3 Results and discussion

3.1 Development of the regression model equation

Preliminary experiments were carried out to screen the appropriate parameters and to determine the experimental domain. From these experiments, the effects of coagulant dosage (X_1), initial pH of reactive dye solutions (X_2), and dye concentration (X_3) were investigated with regard to two responses: the decolorization efficiency and COD reduction. The parameter levels and coded values are given in Table 12.3. The experimental design matrix and the corresponding experimental parameters and

Table 12.3 Study Field and Coded Factors

Natural Variables (x_j)	Unit	Coded Variables X_1, X_2, X_3*				
		$-\alpha$	-1	0	1	$+\alpha$
x_1 = Coagulant dosage	mL	0.73	1	1.4	1.8	2.07
x_2 = Initial pH	–	7.32	8	9	10	10.68
x_3 = Dye concentration	mg/L	65.91	100	150	200	234.09

*The coded values $X_j = \pm 1$ are obtained by the equation: $X_j = (x_j - \bar{x}_j)/\Delta$.

Table 12.4 Experimental Design and Results for Dye Removal

Order		Coded Units of Variable			Responses*	
Logical	Actual	X_1	X_2	X_3	Y_{DEC} exp.	Y_{COD} exp.
1	10	1	1	1	95	85
2	12	1	1	−1	98	89
3	16	1	−1	1	92	79
4	6	1	−1	−1	94	82
5	2	−1	1	1	47	46
6	5	−1	1	−1	66	58
7	9	−1	−1	1	67	61
8	4	−1	−1	−1	85	76
9	7	1.68179	0	0	97	82
10	15	−1.68179	0	0	47	42
11	13	0	1.68179	0	89	75
12	8	0	−1.68179	0	98	83
13	1	0	0	1.68179	69	54
14	11	0	0	−1.68179	93	78
15	14	0	0	0	97	84
16	3	0	0	0	97	85

Decolorization efficiency (Y_{DEC}) and COD reduction (Y_{COD}) of Reactive Orange 16.

response value are shown in Table 12.4. STATGRAPHICS-Plus computer software was used to model and optimize the experimental results. At the end of the coagulation processes with SIWW, the final empirical models in terms of coded factors, after excluding the insignificant terms for decolourization efficiency ($\hat{Y}_{DEC}$) and COD reduction ($\hat{Y}_{COD}$) of Reactive Orange 16, are shown in Eq. 12.4 and Eq. 12.5, respectively.

$$\hat{Y}_{DEC} = 97.07 + 14.51X_1 - 3.45X_2 - 6.03X_3 - 9.02{X_1}^2 + 5.75X_1X_2 + 4X_1X_3 - 5.84{X_3}^2 \quad (12.4)$$

$$\hat{Y}_{COD} = 84.12 + 11.81X_1 - 2.45X_2 - 5.445X_3 - 7.04{X_1}^2 + 5.75X_1X_2 - 5.62{X_3}^2 \quad (12.5)$$

A positive sign in front of the terms indicates a synergistic effect, whereas a negative sign indicates an antagonistic effect. The quality of the model developed was evaluated based on the correlation coefficient value. The R value for Eq. 12.4 and Eq. 12.5 was 0.995 and 0.971, respectively. Both R values obtained were relatively high, suggesting that there was a good agreement between the experimental and the predicted values from the models. The R^2 value was 0.989 for Eq. 12.4 and 0.942 for Eq. 12.5. Thus, 98.9 % and 94.2 % of the total variation in the decolorization efficiency and COD reduction of reactive azo dye, respectively, were attributed to the experimental variables studied. The statistical results showed that the above models were adequate to predict the decolorization efficiency and COD reduction of reactive azo dye by coagulation processes with SIWW within the range of variables studied.

3.2 Decolorization efficiency of reactive orange 16

The adequacy of the models was further justified through analysis of variance (ANOVA). The ANOVA for the model for decolorization efficiency of Reactive Orange 16 by coagulation processes with SIWW is listed in Table 12.5. From this ANOVA, the Model F-value of 155.64 implied that the model was significant. Values of *Prob.* $> F$ less than 0.05 indicated that the model terms were significant. In this case, coagulant dosage (X_1), initial pH of reactive dye solutions (X_2), dye concentration (X_3), and the interaction terms (X_1^2, X_3^2, X_1X_2, and X_1X_3) were significant model terms, whereas the interaction terms (X_2^2 and X_2X_3) were insignificant to the response.

A geometrical representation for the decolorization efficiency of reactive azo dye by coagulation processes with SIWW, which is dependent on the coagulant dosage and initial pH of reactive dye solutions, for constant values of the dye concentration ($X_3 = 0$), is indicated in Figure 12.2 As Figure 12.2 shows, when the coagulant dosage increases and the initial pH of reactive dye solutions decreases together, or when the coagulant dosage increases and the initial pH of reactive dye solutions remains unchanged, then the decolorization efficiency of reactive azo dye increases up to 100%.

3.3 COD reduction of reactive orange 16

For COD reduction of Reactive Orange 16 by coagulation processes with SIWW, the ANOVA for this response is listed in Table 12.6. From this ANOVA, the Model F-value of 27.95 implied that the model was significant as well. In this case, coagulant dosage (X_1), initial pH of reactive dye solutions (X_2), dye concentration (X_3), and the interaction terms (X_1^2, X_3^2 and X_1X_2) were significant model terms, whereas the interaction terms (X_1X_3, X_2^2, and X_2X_3) were insignificant to the response.

Figure 12.3 shows the three-dimensional response surfaces that were constructed to show the effects of the coagulant dosage and the initial pH of reactive dye solutions on the COD reduction of

Table 12.5 Analysis of Variance for Decolorization Efficiency of Reactive Orange 16

Source of Variation	Sum of Squares	Degree of Freedom	Mean Square	F_{exp}	Significance Test
Regression	4835.7268	9	537.302978	155.640744	***
β_1	2873.25	1	2873.25	832.29	***
β_2	162.689	1	162.689	47.13	***
β_3	496.722	1	496.722	143.89	***
β_{11}	753.312	1	753.312	218.21	***
β_{12}	264.5	1	264.5	76.62	***
β_{13}	128	1	128	37.08	***
β_{22}	18.5768	1	18.5768	5.38	NS
β_{23}	0.5	1	0.5	0.14	NS
β_{33}	315.471	1	315.471	91.38	***
Residue	20.7132	6	3.4522	–	–

***Prob. $\leq$ 0.01, NS: Not significant.

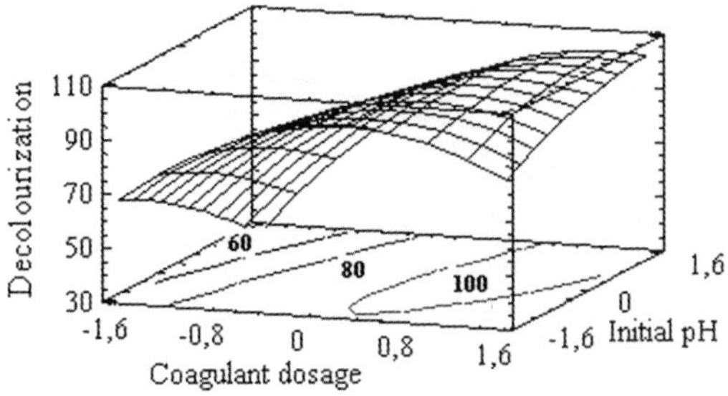

FIGURE 12.2

Response surface plot showing the effect of coagulant dosage and initial pH of reactive dye solutions on Reactive Orange 16 decolorization at fixed dye concentration (150 mg/mL).

Table 12.6 Analysis of Variance for COD Reduction of Reactive Orange 16

Source of Variation	Sum of Squares	Degree of Freedom	Mean Square	F_{exp}	Significance Test
Regression	3336.3468	9	370.7052	27.9449903	***
β_1	1904.43	1	1904.43	143.56	***
β_2	81.951	1	81.951	6.18	*
β_3	404.914	1	404.914	30.52	***
β_{11}	458.766	1	458.766	34.58	***
β_{12}	264.5	1	264.5	19.94	***
β_{13}	50	1	50	3.77	NS
β_{22}	9.76499	1	9.76499	0.74	NS
β_{23}	0.5	1	0.5	0.04	NS
β_{33}	292.902	1	292.902	22.08	***
Residue	79.5932	6	13.2655	—	—

***Prob. $\leq$ 0.01.
*Prob. $\leq$ 0.05, NS: Not significant.

reactive azo dye by coagulation processes with SIWW. The dye concentration was fixed at zero level. From this geometrical representation, when the coagulant dosage increases and the initial pH of reactive dye solutions remains unchanged, then the COD reduction of reactive azo dye increases up to 93%.

3.4 Process optimization

The main objective of this research was to determine the experimental conditions required to remove dye color and COD. Then, using the above methodology for experimental design, the ranges of the parameters required to obtain optimum conditions were determined.

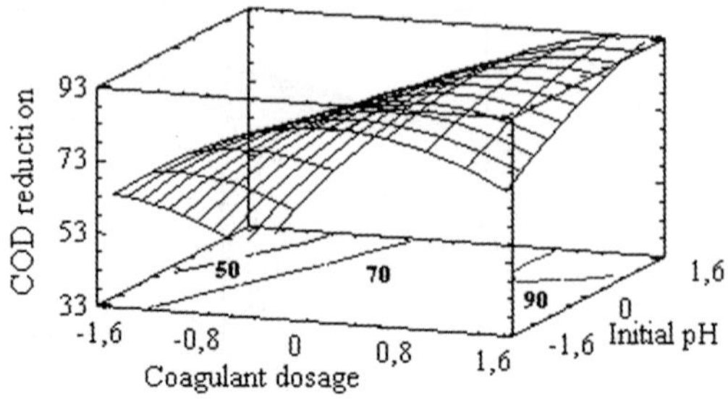

FIGURE 12.3

Response surface plot showing the effect of coagulant dosage and initial pH of reactive dye solutions on Reactive Orange 16 COD reduction at fixed dye concentration (150 mg/mL).

In this optimization study, decolorization efficiency and COD reduction of the Reactive Orange 16 solutions were chosen as the objective function. Furthermore, optimum conditions are often calculated in the presence of some constraints that ensure that they will be more realistic. If the model used in the optimization study is an empirical one, high and low levels of the process parameters in the experimental design are inevitably considered as explicit constraints, in order to avoid extrapolation. Thus, the optimization problem (Sayan, 2006) defines both responses as follows.

- Maximize;

$$\widehat{Y}_{\mathrm{DEC}}/\widehat{Y}_{\mathrm{COD}} \tag{12.6}$$

- Constraints on the parameters X;

$$-\alpha_{\mathrm{i}} < X_{\mathrm{i}} < \alpha_{\mathrm{i}} \qquad i = 1, 2, 3 \tag{12.7}$$

The investigation of Eqns. 12.4 and 12.5 showed that, if $X_1 = 1$, $X_2 = 0.5$ and $X_3 = 0$; the value prediction of decolorization efficiency and COD reduction are 100% and 91%, respectively. The experimental checking in this point, that is, under conditions such as coagulant dosage = 1.8 mL, initial pH of reactive dye solutions = 9.5, and dye concentration = 150 mg/L with $Y_{\mathrm{DEC}} = 98\%$ and $Y_{\mathrm{COD}} = 90\%$, confirms this result.

4 Conclusion

The present study clearly demonstrated that SIWW is a very effective coagulant for dye removal. The response surface methodology provides the best operating conditions for maximizing dye removal. The graphical response surface and contour plot were used to locate the optimum point. A 2^3 full factorial central composite rotatable design was successfully employed for experimental design and analysis of results.

SYMBOLS AND TERMINOLOGY

SIWW steel industrial wastewater
ANOVA analysis of variance
***F*-value** Fischer's value
Prob. probability
R^2 square of the correlation coefficient
COD chemical oxygen demand (mg/L)
$\hat{Y}$ predicted response
X_1 pH of reactive orange 16 solutions
X_2 coagulant dosage
X_3 dye concentration for reactive orange 16 solutions
Y_{DEC} decolorization efficiency of reactive orange 16 solutions (%)
Y_{COD} COD reduction of reactive orange 16 solutions (%)

Greek Letters

β_0 constant coefficient
β_i linear coefficients
β_{ij} cross-product coefficients
β_{ii} quadratic coefficients
λ wavelength (nm)

Glossary

Analysis of variance (ANOVA) a collection of statistical models, and their associated procedures, in which the observed variance is partitioned into components due to different explanatory variables

Central composite design (CCD) an experimental design, useful in response surface methodology, for building a second-order (quadratic) model for the response variable without needing to use a complete three-level factorial experiment

Chemical oxygen demand (COD) the quantity of a specified oxidant that reacts with a sample under controlled conditions

Coagulant a substance that aids or produces coagulation

Coagulation the destabilization of colloidal particles brought about by the addition of a chemical reagent called as a coagulant

Coagulation–flocculation a physical-chemical technique commonly used for water and wastewater treatment

Experimental design the design of all information-gathering exercises where variation is present, whether or not they are under the full control of the experimenter

Flocculant substance added to a suspension to enhance aggregation of the suspended particles

Flocculation the agglomeration of destabilized particles into microfloc and afterward into bulky floccules that can be settled called floc

Jar test test intended to simulate the coagulation/flocculation process in a water treatment plant

Response surface methodology (RSM) a technique for designing experiments to help researchers build models, evaluate the effects of several factors, and achieve optimum conditions for desirable responses, in addition to reducing the number of experiments

Steel industrial wastewater (SIWW) a wastewater $FeCl_3$ rich used as a coagulant

References

Anouzla, A., Abrouki, Y., Souabi, S., Safi, S., Rhbal, H., 2009. Colour and COD removal of disperse dye solution by a novel coagulant: Application of statistical design for the optimization and regression analysis. J. Hazard. Mater. 166 (2), 1302–1306.

Arslan, I., Balcioglu, I.A., Bahnemann, D.W., 2000. Advanced chemical oxidation of reactive dyes in simulated dyehouse effluents by ferrioxalate-Fenton/UVA and TiO_2/UV-A processes. Dyes Pigm. 47 (3), 207–218.

Barka, N., Assabbane, A., Nounah, A., Laanab, L., Ichou, Y.A., 2009. Removal of textile dyes from aqueous solutions by natural phosphate as a new adsorbent. Desalination 235 (1), 264–275.

Boisvert, J., To, T., Berrak, A., Julicocur, 1997. Phosphate adsorption in flocculation processes of aluminium sulphate and poly-aluminium-silicatesulphate. Water Res. 31 (8), 937–1946.

Clarke, G.M., Kempson, R.E., 1997. Introduction to the design and analysis of experiments. Arnold, London.

Clesceri, A., Greenberg, L., Eaton, A., 1998. Standard methods for the examination of water and wastewater. American Public Health Association, Washington, DC.

Daneshvar, N., Salari, D., Khataee, A.R., 2003. Photocatalitic degradation of azo dye acid red 14 in water: Investigation of the effect of operational parameters. J. Photochem. Photobio. Chem. 157 (1), 111–116.

Gouvea, C.K., Wypych, F., Moraes, S.G., Duran, N., Nagata, N., Peralta-Zamora, P., 2000. Semiconductor-assisted photocatalytic degradation of reactive dyes in aqueous solution. Chemosphere 40 (4), 433–440.

Harrelkas, F., Azizi, A., Yaacoubi, A., Benhammou, A., Pons, M.N., 2009. Treatment of textile dye effluents using coagulation-flocculation coupled with membrane processes or adsorption on powdered activated carbon. Desalination 235 (1), 330–339.

Istadi, M., Amin, N.S., 2005. A hybrid numerical approach for multi-responses optimization of process parameters and catalyst compositions in CO_2 OCM process over CaO-MnO/CeO_2 catalyst. Chem. Eng. J. 106 (3), 213–227.

Kulkarni, S., Blackwell, C., Blackard, A., Stackhocese, C., Alexander, M., 1985. Textile dyes and dyeing equipment, classification, properties and environmental aspects. Research Triangle Park, NC.

Magara, Y., Kunikane, S., Itoh, M., 1998. Advanced membrane technology for application to water treatment. Water Sci. Technol. 37 (10), 91–99.

Myers, R.H., Montgomery, D.C., 1995. Response surface methodology: Process and product optimization using designed experiments. John Wiley & Sons, New York.

Okuda, T., Baes, A.U., Nishijima, W., Okada, M., 1999. Improvement of extraction method of coagulation active components from *Moringa oleifera* seed. Water Res. 33 (15), 3373–3378.

Padmavathy, S., Sandhya, S., Swaminathan, K., Subrahmanyam, Y.V., Chakrabarti, T., Kaul, S.N., 2003. Aerobic decolorization of reactive azo dyes in presence of various cosubstrates. Chem. Biochem. Eng. 17 (2), 147–151.

Peralta-Zamora, P., Kunz, A., Moraes, S.G., Pelegrini, R., Moleiro, P.D., Reyes, J., 1999. Degradation of reactive dyes: A comparative study of ozonation, enzymatic and photochemical processes. Chemosphere 38 (4), 835–852.

Ravikumar, K., Pakshirajan, K., Swaminathan, T., Balu, K., 2005. Optimization of batch process parameters using response surface methodology for dye removal by a novel adsorbent. Chem. Eng. J. 105 (3), 131–138.

Sayan, E., 2006. Optimization and modeling of decolorization and COD reduction of reactive dye solutions by ultrasound-assisted adsorption. Chem. Eng. J. 119 (2), 175–181.

Zhang, H., Duan, L., Zhang, Y., Wu, F., 2005. The use of ultrasound to enhance the decolorization of the C.I. Acid Orange 7 by zero-valent iron. Dyes Pigm. 65 (1), 39–43.

CHAPTER

13

Modified and Nonmodified TiO_2 Nanoparticles for Environmental Applications

Vassiliki Belessi[*], **Dimitris Petridis**[†]

[*] *Department of Graphic Arts, Technological Educational Institution of Athens, Agiou Spyridonos Street, 12210 Egaleo, Athens, Greece,* [†] *Institute of Materials Science, NCSR "Demokritos," 15310 Aghia Paraskevi Attikis, Athens, Greece*

CHAPTER OUTLINE

1 Introduction

Environmental pollution represents one of the biggest risks to the survival of humankind. So, every attempt that seeks to fight environmental pollution must find supporters throughout the world. In many countries, environmental pollution has caused a shortage of suitable water for drinking and agricultural activities, creating problems to both public health and the economy. For these reasons, the elimination of toxic chemicals from wastewater and drinking water is one of the most important aims of pollution control. In addition, since only 1% of the freshwater and approximately 0.007% of all water on Earth are available for consumption, then the need for cleaning and reusing wastewater is an obvious strategy to pursue (Lefort 1996). In this direction, various chemical, physical, and biological treatment methods have been developed in order to monitor, remediate, reduce, or even restrict pollution. These methods

The Role of Colloidal Systems in Environmental Protection. http://dx.doi.org/10.1016/B978-0-444-63283-8.00013-2

include adsorption, nanofiltration, precipitation, coagulation-flocculation, reverse osmosis, oxidation with ozone, chlorine or hydroxen peroxide, and use of anion exchange membranes and bacterial cells (Robinson et al. 2001, Cooper 1993, Slokar and Majcen Le Marechal 1998, Vandevivere et al. 1998). However, the ineffectiveness of conventional methods to destroy some pollutants has led to the development of other efficient wastewater treatment processes such as advanced oxidation processes (AOPs) (Hoffmann et al. 1995, Linsebigler et al. 1995, Ollis et al. 1991). Among them, adsorption and photocatalysis play a significant role in water remediation (Dabrowski 2001, Qu 2008, Hoffmann et al. 1995). Adsorption is considered a very promising method because it is rather simple and easy to apply, does not produce toxic substsances, and the capital cost needed for its application is relatively low. However, if the amount of water that needs to be purified is large, then the application of adsorption is limited due to the high cost of some adsorbents (Figueiredo et al. 1998, Crini 2006). Semiconductor-mediated photocatalysis is a highly recognized advanced oxidation process that is widely used in environmental applications (Hoffmann et al. 1995). In particular, this promising technology acquires a significant momentum with regard to exploiting the energy of solar light (S. Malato et al. 2002).

In both applications (adsorption and photocatalysis), TiO_2 is the most promising photocatalyst because of the strong decomposing power of organic pollutants in water. Because TiO_2 absorbs only a small part ($\sim$5%) of solar energy, the need arises to develop novel modified titania that will extend its wavelength response into the visible region. Moreover, as TiO_2 is generally a powder with very small particle size, its use requires an additional and expensive separation step involving the removal of the photocatalyst from the treated water. A way to overcome this problem is to exploit magnetically separable photocatalysts that enable the easy recovery of the adsorbent/photocatalyst by an external magnetic field (Beydoun et al. 2000, Belessi et al., 2009a). Undoubtedly, the as-prepared magnetic photocatalysts should be stable and not suffer from photodissolution. Furthermore, the ability of these derivatives to be reused several times without significant decrease of their photocatalytic efficiency renders them promising candidates for practical use in wastewater treatment technologies. Another approach is the immobilization of TiO_2 as bound particles on solid supports or on solid films (Arabatzis et al. 2003). However, attachment of TiO_2 on films does not solve the problem because the exposed surface area for photocatalytic degradation is much lower than this on slurry surfaces.

It is clear that there is plenty of room for development of novel materials based on titania that could resolve these technical limitations. This review paper focuses on the use of modified and nonmodified titania that has found application in wastewater treatment technologies involving adsorption and photocatalysis in aqueous solution.

2 Wastewater treatment technologies

2.1 Adsorption

In this section we describe research works based on the adsorption of metals (Gao et al. 2004, Kim et al. 2003, Xie and Gao 2009), organic acids (Pettibone et al. 2008), and dyes (Bourikas et al. 2005, Belessi et al. 2009b) on the surface of TiO_2.

2.1.1 Adsorption of heavy metals onto TiO_2 surfaces

Heavy metals are among the most serious pollutants in natural and treated water. It has been recognized that TiO_2 is an effective adsorbent for removing heavy metal cations from aqueous solutions. For example, the effective removal of toxic Cu^{2+} from aqueous solutions by adsorption on anatase-type TiO_2 particles was described by Kim et al. (2003). The adsorbed amount of Cu^{2+} was found to increase rapidly with increasing pH from 2 to 5, and at higher values it remains constant. The adsorption of Cu^{2+} depended on the concentration of hydroxyl groups on the TiO_2 surface and thus on the medium pH. These findings are in good agreement with the adsorption data of Barakat (2005) who studied the adsorption of Cu^{2+} and CN^- with and without their corresponding complexes at TiO_2-solution interfaces (Barakat 2005, Barakat et al. 2004). The extended adsorption of Cu^{2+} at low pH values was attributed to an inner-sphere type of complex formed between Cu^{2+} and TiO_2 surface. The adsorption curve shifted to higher pH values with increasing total Cu^{2+} concentration. However, the probable reason for the efficient removal of Cu^{2+} at pH>8 was not the adsorption but the precipitation of $Cu(OH)_2$. Furthermore, the copper adsorption in the presence of CN^- was studied. In this case, it was found that the formation of Cu^{2+}-hydroxide species was prohibited even at the alkaline pH region due to the negative role of CN^- on the adsorption of Cu^{2+}.

Gao et al. (2004) studied the adsorption of Cd^{2+} on anatase samples of various sizes over a broad range of pH levels. It was found that the adsorption mechanism of Cd^{2+} by anatase nanoparticles was different from the corresponding mechanism on larger particles. Specifically, the surface adsorption of Cd^{2+} decreased by decreasing TiO_2 particle size. The authors suggested that this behavior could be attributed to intraparticle electrostatic interactions or caused by Ti-site disorder for nanoparticles with size lower than 20 nm. However, we should note here that these results are not in agreement with other studies (Zhang et al. 1999, Wang et al. 2005, Pettibone et al. 2008).

The adsorption of toxic Cr^{6+} onto TiO_2 surfaces and the effect of adsorption for the removal of Cr^{6+} using TiO_2 photocatalysis have been studied by many scientists (Yang and Lee 2006, Uyguner and Bekbolet 2004). The results for the Cr^{6+} adsorption by TiO_2 in the presence of humic acid indicate that the adsorption capacity of TiO_2 is greatly influenced by the presence of humic acid (Uyguner and Bekbolet 2004). As the presence of Cr^{6+} inhibited the adsorption of humic acid on TiO_2 this led to a decreased adsorption capacity of humic acid on TiO_2 with increasing Cr^{6+} concentration.

The effect of crystalline structures of TiO_2 on the adsorption of Pb^{2+} and Cd^{2+} on such nanoparticles was studied by Xie et al. (Xie and Gao 2009). It was found that both chemisorption (for pH values lower than the point of zero charge) and physisorption (for pH values higher than the point of zero charge) were taking place during the adsorption of cations on the titania surface. The chemisorption capacity was higher in the anatase phase. It was suggested that parameters such as the higher surface energy, surface roughness, and unsaturated oxygen bonds of anatase determined this behavior.

The adsorption of molybdenum on anatase particles was first studied by Saripalli et al. (2002). It was shown that the adsorption was maximized at low pH and surface loadings. In accordance with other researchers, the authors suggested that a constant capacitance model could successfully describe the adsorption of MoO_4^{-2} on the oxide surface.

TiO_2 has also been studied extensively as an adsorbent for arsenic removal (Bang et al. 2005, Dutta et al. 2004, 2005, Pena et al. 2006). Specifically, two commercially available TiO_2 samples (Degusssa P25 and Hombicat UV100) were studied for the adsorption of As^{5+} and As^{3+} anions on their surface by Dutta et al. (2004). It was found that the adsorption of both arsenate (As^{5+}) and arsenite (As^{3+}) anions was higher on the surface of Hombicat UV100 particles, probably due to the higher surface area of Hombicat UV100 (334 m^2/g) compared to Degussa P25 (55 m^2/g). It was also suggested that the surface charge of TiO_2, the form of arsenic species, and pH were the determining factors for the adsorption of arsenic species on the surface of TiO_2. Specifically, it was found that the adsorption capacity for As^{5+} was higher at low pH, while that for As^{3+} was higher at high pH (Dutta et al. 2004). Similar observations were made by other groups (Dutta et al. 2005, Yang et al. 1999, Bissen et al. 2001), and this trend was explained in terms of electrostatic considerations. However, Bang et al. (Bang et al. 2005) have noted that the observed strong adsorption of arsenic species on TiO_2 at pH higher than the point of zero charge does not take place via electrostatic interactions but rather is due to surface complexation. In accordance with these data, Pena et al. (Pena et al. 2006) examined the adsorption mechanism of arsenic species by nanosized TiO_2 using a combination of analytical techniques and proposed the formation of the negatively charged inner-sphere surface complexes $Ti_2AsO_4^-$ and $Ti_2AsO_3^-$ for As^{5+} and As^{3+}, respectively.

Furthermore, the adsorption capacity of arsenic species was found to be affected by the origin of water when granular TiO_2 was used (Bang et al. 2005). This type of titanium oxide was reported to be efficient in eliminating arsenic in groundwater. By examining groundwater from the United States, Band et al. found that the adsorption of As^{5+} ions (41.1 mg/g) on TiO_2 was bigger compared to As^{3+} (32.4 mg/g). On the other hand, experiments on Bangladesh's groundwater showed that both arsenic species were adsorbed on TiO_2 with comparable adsorption capacities ($\sim$40 mg/g). Similarly, Pena et al. proposed that nanocrystaline TiO_2 in the anatase form is a good adsorbent for As removal. Factors such as the high surface area (329 m^2/g) of titania and the high affinity of surface hydroxyl groups are claimed to be the main parameters for this behavior (Pena et al. 2006).

The combination of adsorption and photocatalysis was proposed by Dutta et al. (2004, 2005) as an effective method for treating and removing arsenic from contaminated water. Specifically, it was suggested that the photocatalytic oxidation of the more toxic As^{3+} to As^{5+} under basic conditions and the subsequent adsorption of As^{5+} on TiO_2 surface under acidic conditions could drastically reduce the arsenic content in water.

The role of adsorbed arsenic species on the TiO_2 surface in the oxidation of As^{3+} to As^{5+} during TiO_2 photocatalysis was assessed by Xu et al. (2005). The analysis of the reaction solutions and catalyst extracts showed that the main portion of arsenic was powerfully adsorbed by TiO_2 surfaces. Also, it was confirmed that the photocatalytic oxidation of As^{3+} takes place rapidly on the TiO_2 surfaces. Furthermore, they reported that hydroxyl radicals were important in this process.

TiO_2 has also found application as an appropriate material for the adsorption of the lanthanide Eu^{3+} (Tan et al. 2009) where the pH plays a key role. At low pH values, the fulvic acid favors the adsorption of Eu^{3+} on TiO_2, while the ionic strength does not affect this process. According to the authors, these experimental findings suggest the formation of an inner-sphere surface complexation of Eu^{3+} at the titania surface.

Olsson et al. (2003) investigated the effect of various parameters such as pH and plutonium concentration for the adsorption of Pu^{6+} by TiO_2. In agreement with the study of Tan et al. (2009), they concluded that the mechanism of adsorption does not depend on ionic strength. So, the authors assigned these findings to the formation of an inner-sphere complex.

2.1.2 Adsorption of organic acids onto TiO_2 surfaces

Zhang et al. (1999) examined the adsorption of organic acids onto nanosized anatase titania (6–16 nm), and based on thermodynamic arguments they found that the smaller the particle size, the higher the adsorption capacity. However, this trend contradicts the findings from the adsorption of heavy metals on anatase particles as described previously (Gao et al. 2004). Moreover, Pettibone et al. (2008), who studied the effect of nanoparticle size of TiO_2 on the adsorption of oxalic and adipic organic acids, argue that there is no general rule as to who predicts that the smaller the TiO_2 particles the higher their adsorption capacity.

2.1.3 Adsorption of dyes onto TiO_2 surfaces

The decolorization of dye effluents has received increasing attention due to the potential toxicity of dyes and their visibility in water (Vandevivere et al. 1998). Although the use of TiO_2 for removal of various dyes from aqueous media has been studied extensively (Konstantinou and Albanis, 2004, Han et al. 2009), only a few works have investigated the mechanism of dye adsorption by TiO_2 surfaces (Bourikas et al. 2005, Belessi et al. 2009). Most papers describe the correlation between dye adsorption on the photocatalyst and the rate of its degradation (Beyers et al. 2009, Aguedach et al. 2005).

The adsorption of Acid Orange 7 (AO7) by TiO_2 surfaces (Degussa P25) in aqueous TiO_2 suspensions was studied by Bourikas et al. (2005). They concluded that the adsorption of the azo-dye on TiO_2 can be nicely described by a model that takes into account the heterogeneity on the TiO_2 surfaces. Also, they claimed that the surface of titania was covered by singly ($TiOH^{1/2}$), doubly ($Ti_2O^{-2/3}$), and triply (Ti_3O^0) coordinated groups. The sulphonic group of the azo-dye was found to be responsible for the adsorption of dye on the TiO_2 surfaces. This process results in the formation of a bidentate inner-sphere surface complex. The role of pH in the adsorption process was important, and the optimum pH of sorption was approximately 2.

In accordance with these data, Belessi et al. reported that the adsorbed dye amount of the commercial azo-dye Reactive Red 195 (RR195) on mesoporous TiO_2 nanoparticles reached its maximum value at pH 3 (Belessi et al. 2009b). The authors, in agreement with other studies, attributed the higher uptake values found at lower pH values to electrostatic attractions between the negatively charged dye anion and the positively charged surface of TiO_2 ($Ti^{IV}–OH_2^+$) (Aguedach 2005, Cicek 2007, Wang and Ku 2007). In this process, the potential of the interface plays an important role on the dye adsorption (Bourikas et al. 2005). It was also found that the nanosized TiO_2 had a high surface area (155 m^2/g) and was an efficient adsorbent. Specifically, the value of the dye loading per unit surface area of titania (5.7×10^{-22} mmol/nm^2) was much higher than that of various porous carbons or of other TiO_2 systems (Aguedach 2005, Asouhidou et al. 2009, Bourikas et al. 2005). Finally, the highest percentage of dye removal (100%) was observed for adsorbent concentration, initial dye concentrations, and contact time values of 2.0 g/l, 10–75 mg/L, and 15 min, respectively. Additional measurements revealed a reduction of 81.4% for chemical oxygen demand (COD), verifying that the dye was efficiently removed due to an adsorption mechanism.

2.1.4 Adsorption on a modified titania surface

Hybrids of TiO_2 with organic or inorganic materials are expected to be promising solids because of their better performance compared to pure TiO_2 in the adsorption and photocatalysis processes. For example, the unique adsorption ability of chitosan toward various metal cations has been exploited to prepare a hybrid film of chitosan/TiO_2. This sample exhibited high sorption capacity for the removal of Pd^{2+} from aqueous solutions (Tao et al. 2009). Also, attempts to anchor TiO_2 on porous supports including silica, clay, zeolite, and activated carbon, have been very successful (Belessi et al. 2007, Machado et al. 2006, Bhattacharyya et al. 2004, Xu and Langford 1995, 1997). The supports are anticipated to enhance catalytic activity through properties ranging from surface acidity to reactive intermediate stabilization.

Furthermore, a high dispersion of the catalytically active phase on the supports is required. In order to meet this requirement, the supports should mainly have a well-developed inner structure and preferentially a high surface area (Dabrowski 2001). It is clear that catalysis and adsorption are interdependent and constitute the most significant area of surface science.

Thus, a catalyst should be a good adsorbent, and a porous solid should act as adsorbent, catalyst, and catalyst support. Thus, characteristic parameters of materials such as the chemical character, size and surface areas, porous structure, mechanical properties, and thermal stabilities are determinant in adsorption and catalysis (Dabrowski 2001).

TiO_2 due to its polar nature cannot efficiently absorb nonpolar organic molecules. This problem could be avoided by loading TiO_2 on a high surface area support, in which preconcentration of organic pollutants can take place. The adsorbed substances are then transferred to TiO_2 by diffusion where they are decomposed together with their resulting intermediates, at rates that depend on their adsorption strength. Thus, by supporting TiO_2 on adsorbents, higher specific surface areas with more effective adsorption sites are provided than unsupported TiO_2. Furthermore, the adsorbents are also acting as reaction sites for the out-diffusion of oxidizing species (Fujishima et al. 2008).

For example, TiO_2-mounted on activated carbon was studied for the removal of phenol from aqueous solution under UV irradiation (Tryba et al. 2003). It was found that the adsorption of phenol on the activated carbon as well as its photocatalytic degradation by TiO_2 is crucial for its removal. The surface area of the catalysts significantly affects the adsorption that was found to take place even during the photocatalysis. During this process, the mineralization was not complete and by-products were formed.

Mesoporous silica (MCM-41), microporous (β-zeolite), and pillared montmorillonite with high surface areas were used as adsorbents for TiO_2 loading by Bhattacharyya et al. (2004). The sol-gel method was used for preparation of TiO_2, and the loading amount of TiO_2 ranged between 10% and 80%. The supported photocatalysts were tested and found to be efficient for the photocatalytic degradation of azo-dye Orange II. All of them were more effective than naked-TiO_2 prepared by the sol-gel method. Degussa P25 was found to be a better photocatalyst than TiO_2-pillared montmorillonite. The other two supported catalysts (TiO_2-MCM-41 and TiO_2-β-zeolite) were more active compared with the commercial photocatalyst. The adsorption of Orange II was probably stronger on TiO_2-pillared montmorillonite and, thus, hindered the mobility of the adsorbed dye to the photoactive TiO_2 sites (Yoneyama and Torimoto, 2000). Finally, it is clear that the kind of substrate is an important factor for the photodegradation of the dye.

A study of TiO_2 supported on four different substrates (silica, alumina, zeolite A, and zeolite ZSM5) was reported by Xu and Langford (Xu and Langford, 1995). It was observed that for the photo-oxidation of 4-chlorophenol, the ZSM5 was the most active substrate even at low titania loading (< 2 wt%). The high efficiency at low Ti loading suggested that light collection was favored by the adsorbed molecules having only small losses on the supports, and/or the light adsorption was efficient across the surface in the thin film of titanium oxide. The authors also found a correlation between the catalytic activity of the TiO_2-zeolite A sample and the zeolite crystallinity. Specifically, they ascribed any observed high reactivity of such a supported photocatalyst to the high crystallinity of the zeolite. Xu and Langford, based on their experimental data, claimed that the adsorption capacity and structure of zeolite were determinant parameters for the enhanced photocatalytic activity of the supported catalysts. A few years later, the same research group reported a more detailed study of the photocatalytic degradation of acetophenone in an aqueous medium in the presence of TiO_2 supported on microporous zeolites of type X and Y and on mesoporous molecular sieves of the MCM41 type. There was evidence that the crystallinity of the zeolite was critical for the photocatalytic efficiency of the system (Xu and Langford, 1997).

The remarkable properties of carbon nanotubes (CNTs) opened the possibility of using them in adsorption and heterogeneous catalysis (Lee and Sigmunt 2003, Planeix et al. 1994, Gao et al. 2009). Yu et al. (2005) examined how mixtures of TiO_2 (Degussa P25) with oxidized carbon CNTs affect the adsorption and activity of TiO_2 for the photocatalytic degradation of three azo-dyes. The oxidized CNTs were used after direct mixing with Degussa P25 in different weight ratios aiming to increase the adsorption and the photocatalytic activity of TiO_2 compared with activated carbon. It was found that the mixtures of CNTs and TiO_2 had superior adsorption capacity for the tested dye, followed by the corresponding mixtures of activated carbon and TiO_2. Lower adsorption capacities were observed for TiO_2 and CNTs. The authors concluded that the positive effect of CNTs on the adsorption capacity of TiO_2 was better than that of activated carbon due to strong interactions between their surfaces. Furthermore, the mixtures of TiO_2 and CNTs were catalytically more active than TiO_2 and the corresponding mixtures of TiO_2 with activated carbon. It was noted that activated carbon due to its different structure compared with CNTs could not cause an increment of hydroxyl radicals with P25. On the other hand, CNTs could interact through adsorption or binding with P25. Porous CNTs also played an important role in picking up photoexcited electrons generated in TiO_2 improving, thus producing the charge separation between electrons and holes because of free movement of electrons through the walls of nanotubes.

Ilisz et al. explored the possibility of combining adsorption and photocatalysis as a potential water treatment process (Ilisz et al. 2002). An organo-clay (derived from hexadecylpyridiniumchloride and montmorillonite) and TiO_2 (Degussa P25) were used as adsorbent and photocatalyst, respectively, for the decomposition of 2-chlorophenol. Initially, the pollutant was adsorbed on the organo-clay, while consequently a direct photolytic or heterogeneous photocatalytic process was used for its elimination. According to the results of this study, such a combination of adsorption and photocatalysis was satisfying. Furthermore, the repetitive use of the adsorbent showed that it could be reused successfully for several cycles.

Another aspect of the design of a titanium oxide catalyst is its surface modification to make it more hydrophobic (Burns et al. 1999, Inumaru et al. 2004, Esumi et al. 1997, 1998, Yuan et al. 2000). Esumi et al. (1997) found that a commercial titania modified with an anionic surfactant (sodium dodecyl sulphate, SDS) favored the adsorption of the heavy metal cation Cu^{2+} in the acidic range of pH (3-5) and the removal of non-ionic compounds (2-naphtol). Similarly, it was found that when titania (Degussa P25) modified with dodecyl sulphate chains (at low pH values) the tested organic compounds 1,2-dichlorobenzene and phenanthrene were efficiently captured on its surface (Yuan et al. 2000). The authors proposed that SDS-coated TiO_2 could be used for the photocatalytic degradation of organic compounds, exploiting its good adsorptive performance. However, it is known that cationic, anionic, and non-ionic surfactants with long hydrocarbon chains are easily cleaved photochemically and that their complete mineralization is possible. Accordingly, the stability of surfactants plays an important role in the degradation of hydrophobic organic compounds. For this reason, the same research group in another work (Yuan et al. 2001) used as surfactants the potassium perfluorooctylsulfonate and SDS for titania surface modification. It was found that the fluorocarbon-based surfactant, contrary to SDS, was resistant under UV irradiation in the presence of TiO_2 and also that it promoted both the adsorption and photodecomposition of the

1,2-dichlorobenzene. The surfactant SDS caused enhanced adsorption of the hydrophobic organic compounds but prevented their degradation on the surface of the photocatalyst because it was competed with them.

Inumaru et al. (2004) described how the combined action of selective adsorption with photocatalysis can lead to an efficient degradation of dilute contaminants in the presence of other extremely concentrated compounds. Specifically, they focused on the preparation of titanium oxide (Degussa P25) surface-modified with alkyl groups. The commercial catalyst was found to be less active than the as-prepared n-octyl-grafted TiO_2 (C8-TiO_2) for the photocatalytic degradation of 4-nonylphenol in the presence of concentrated phenol. The authors attributed this behavior to the selective adsorption of various pollutants on the tested photocatalyst. Specifically, the molecules of 4-nonylphenol, because of their strong hydrophobicity, entered into the hydrophobic space between the n-octyl groups surrounding the catalyst surface. On the other hand, TiO_2 was prevented by molecules such as phenol to come near its surface due to their lower hydrophobicity. It is obvious that the surface modified TiO_2 was surrounded mainly by the molecules of 4-nonylphenol facilitating their degradation. However, the decomposition of alkyl groups or their transformation into other species during photocatalysis is not a preferable reaction.

2.2 Photocatalysis

Photocatalysis is a very promising technology belonging to advanced oxidation processes (AOPs). The method is of particular interest for the destruction of toxic and hazardous pollutants at low concentrations as an alternative route to other well-established oxidation technologies, such as ozonolysis or oxidation with H_2O_2 under UV irradiation and so on (Hoffmann et al. 1995, Ollis et al. 1991, Vidal et al. 1999). Numerous studies have focused on the photocatalytic degradation of large categories of organic compounds involving, for example, pesticides (Konstantinou and Albanis 2003) and dyes (Konstantinou and Albanis 2004).

Because the detailed mechanism of photocatalysis at semiconductor phases has been previously reviewed by many authors (Herrmann 1999, Mills and Le Hunte 1997, Kamat 1993, Hoffmann et al. 1995, Linsebigler et al. 1995), it will be discussed here very briefly.

Among various semiconductors (e.g., TiO_2, ZnO, Fe_2O_3, CdS, ZnS), TiO_2 is the most extensively studied since it is widely available, it is physically and chemically stable with high oxidizing power, and it possesses electronic and optical properties while it is inexpensive and nontoxic. Its bandgap energy (Eg $>$ 3.2 eV) necessitates UV-excitation ($\lambda < 388$ nm) to induce charge separation within the particle, therefore restricting the possible exploitation of sunlight or visible light in photocatalytic reactions (Goswami 1997). To increase the activity of TiO_2 and to extend its wavelength range response to the visible region, different preparation methods have been developed for modification of the TiO_2 nanoparticles that are summarized in the following paragraphs. Water treatment, air purification, self-cleaning, and self-sterilizing surfaces are some selected examples of photoactive TiO_2 applications (Fujishima et al., 2008).

The illumination of TiO_2 particles with light energy equal to or greater than its bandgap energy (e.g., 3.2 eV) promotes an electron from the semiconductor valence band to the conduction band (e_{cb}^-), leaving an electron vacancy or positively charged hole (h_{vb}^+) in the valence band. The fate of generated charges may vary. Some of them will recombine, while others will migrate on titanium oxide surface participating in oxidation-reduction reactions with adsorbed species (Linsebigler et al. 1995).

Trapping charge carriers is necessary during photocatalysis in order to reduce recombination and enhance the possibility of interfacial charge transfer (Hoffmann et al. 1995, Linsebigler et al. 1995).

The positively charged holes (h_{vb}^{+}) may oxidize organic molecules or react with surface-bound H_2O or OH^- to produce hydroxyl radicals. The e_{cb}^{-} could reduce the organic substrate or be picked up by oxygen adsorbed on the Ti^{3+}-surface or be dissolved in water to generate superoxide radical anions ($O_2^{\bullet -}$) as indicated by the following equations (13.1–134).

$$TiO_2 + h\nu \rightarrow e_{cb}^{-} + h_{vb}^{+} \quad (13.1)$$

$$(O_2)_{ads} + e_{cb}^{-} \rightarrow (O_2^{\bullet -})_{ads} \quad (13.2)$$

$$Ti^{4+} - OH + h_{vb}^{+} \rightarrow Ti^{4+} - OH^{\bullet} \quad (13.3)$$

$$\text{compound} + h_{vb}^{+} \rightarrow \text{compound}_{\text{oxidized}} \quad (13.4)$$

Photoinduced holes and electrons have been detected in TiO_2 surfaces using the electron paramagnetic resonance (EPR) technique (Hurum et al. 2003, Micic et al. 1993a,b, Berger et. al. 2005, Nakaoka et al. 1997, Howe and Gratzel 1985). It has been suggested that trapped holes are the main species in oxidative photocatalytic processes and have been detected as O^- species (Berger et al. 2005, Tamaki et al. 2007). Trapped electrons have been detected as Ti^{3+} species, but it has been shown that the main part of photoexcited electrons remains in the conduction band (Berger et al. 2005, Howe and Gratzel 1985). Also it has been suggested the existence of various types of species in the following forms $Ti - O^{\bullet} - Ti - OH$ (subsurface oxygen radicals connected to surface hydroxyls) (Hurum et al. 2003, Howe and Gratzel 1985, Fujishima et al. 2008), $Ti - O - Ti - O^{\bullet}$ (surface oxygen anion radicals intermediate covalently bound to titanium) (Micic et al. 1993a,b, Fujishima et al. 2008), $O^{\bullet -}$ (lattice radical) (Hurum et al. 2003, Berger et al. 2005, Howe and Gratzel 1985, Fujishima et al. 2008). The organic substrate reacts with the produced radicals, and then based on the initial conditions other species are formed which can be further oxidized or reduced (Pelizzetti and Minero 1999, Turchi and Ollis 1989, Childs and Ollis 1980).

3 Nanostructured TiO_2 and modified TiO_2 for environmental applications

3.1 TiO_2

Anatase, brookite, and rutile are the three main crystal structures of titanium dioxide. Each of these structures presents different physical properties and therefore, different applications. Anatase and rutile are easily prepared compared with brookite and have been studied extensively as photocatalysts. However, pure rutile with particle size less than 14 nm is difficult to obtain in contrast to anatase (Zhang and Banfield 1998, 2000). This explains why there are few reports on the preparation of pure rutile nano titania (Wu et al. 1999, Zhang et al. 2003, Cheng et al. 1995a, Kumar et al. 1993, Li et al. 2002a).

Factors affecting the overall catalytic performance of TiO_2 include preparation method, calcination temperature, particle size and particle shape, specific surface area, ratio between the anatase and rutile crystal phases, light intensity and the nonuniformity of irradiation field, degree of particle agglomeration, solution pH, and possibly others.

Li et al. describes the formation mechanism of nanosized rutile, anatase, and a mixture of anatase and brookite titania (Li et al. 2004). The synthesis of rutile particles of size less than 14 nm was realized by following a simple but effective preparation method. An aqueous $TiOCl_2$ solution was used as a precursor, and it was found that the effect of pH was determinant for the formation of the various crystal phases. It was observed that the formation of nanosized anatase with size less than 11 nm was promoted in a weak acidic environment in regard to rutile. Furthermore, the SO_4^{2-} ions inhibited the formation of both rutile and brookite.

In many instances, the anatase has been considered to be less active or even inactive compared with the rutile phase (Ohno et al. 2001, Liu et al. 2009). It has been pointed out that this may be due to the surface affinity of rutile against organic substrates, the lower surface adsorbed water, the lower surface hydroxylation, and the higher rates of the electron–hole pair recombination on the surface of the rutile phase (Fujishima et al. 2008, Liu et al. 2009, Riegel and Bolton 1995).

However, Song et al. (2006) found that a nanosized rutile powder with a high specific surface area ($\sim$200 m^2/g) was a more effective catalyst than anatase ($\sim$200 m^2/g) with similarly shaped particles and the reference material Degussa P25. This photocatalyst was studied for the degradation of 4-chlorophenol and Cu- and Pd-EDTA solutions under UV irradiation. The authors proposed that this finding could be due to the preparation method of the samples (precipitation of titanium oxide using $TiOCl_2$ as a precursor), which may cause fewer defects on rutile surface.

As far as the particle size is concerned, it is known that if the semiconductor size decreases, the band gap increases, resulting in larger redox potentials (Hoffmann et al. 1995). So, the use of nanosized TiO_2 particles promotes high photoefficiencies for systems in which the rate-limiting step is charge transfer (Hoffmann et al. 1992a,b). Enhanced photoefficiency is also expected for a semiconductor with small particle size due to the smaller diffusion distance of the electrons and holes toward the surface (Hagfeldt and Gratzel, 1995). The smaller size of titanium oxide crystallites corresponds to larger porosity and higher surface area, which may possess more suitable active sites for adsorption compared to a sample of smaller surface area and as consequence may enhance photoactivity (Belessi et al., 2007). However, contradictory results have also been reported for the contribution of size-quantized semiconductors to the increased or decreased photoefficiency (Hoffmann et al. 1992a,b, 1995, Anpo et al. 1987, Nosaka et al. 1990), while Wang et al. (1997) demonstrated that there exists an optimal particle size of about 10 nm in nanocrystalline TiO_2 systems for maximum photocatalytic efficiency.

Finally, the shape of TiO_2 nanoparticles (nanotubes, nanowires, nanorods, etc.) is expected to play a role in its photoactivity. For example, Manna et al. (2002) has noted that a greater density of active sites and a greater interfacial charge carrier transfer rate is observed in nanorods compared to nanospheres.

Many reports have been issued regarding the preparation of active TiO_2 for environmental applications (Li et al. 2004). However, Degussa P25, which is a mixture of $\sim$ 80% anatase and $\sim$ 20% rutile with high specific surface area ($\sim$ 50 m^2/ g), has been considered the most active formulation of TiO_2, although the reasons for the high activity of P25 for the oxidation of many organic compounds are not clear. Bickley et al. (1991) attributed the high photocatalytic activity of the P25 powder to the interaction between different phases of rutile and anatase. According to their model, the photogenerated electrons move from anatase to a lower electron trapping site in rutile. It was proposed that this electron transfer reduces the recombination of anatase and leads to more efficient electron-hole separation as well as higher photoactivity. However, the data supporting this hypothesis are controversial (Hurum et al. 2003, 2006). Hurum et al. (2003) proposed that the mechanism according to which a reduced recombination rate of anatase takes place is the following: The electrons are transferred from rutile to anatase lattice trapping sites of lower energy and then to surface trapping sites, while at the same time the holes are moving toward the surface.

The combined action of anatase and rutile phases was further studied by Ohno et al. (2003), who examined commercial titanium oxide samples containing only either anatase or rutile or mixtures of anatase and rutile in various ratios. The samples were studied for the photocatalytic oxidation of naphthalene in a mixed solution of acetonitrile and water. It was found that the samples that contain mixtures of anatase and rutile were the most active for the studied reaction, which means that a synergistic effect may take place. The mixtures resulted either from a physical mixing of the two phases (anatase and rutile) or after calcination of an anatase sample. Furthermore, the authors, based on their experimental results, concluded that electrons are transferred from rutile particles to anatase in accordance with Hurum et al. (2003). Similarly, Hurum et al. (2003), using the EPR technique, attributed the high catalytic activity of Degussa P25 to a large extent to the synergistic activation of the rutile phase by anatase. In a more recent work where Degussa P25 was studied by EPR, these same authors observed that an electron movement at higher temperatures could be easier through distorted interfacial sites (reactive hot spots) (Hurum et al. 2006).

Sun et al. (2003, Sun and Smirniotis 2003) proposed a mechanism in which band bending occurred at the interfaces between anatase and rutile in Degussa P25 through a Fermi level lineup, changing the relative positions of the valence and conduction bands. According to this mechanism, the band bending at the interface between anatase and rutile phases is such that it allows the holes formed in anatase to move into rutile while the movement of electrons from anatase to rutile is blocked. It has also been found that the undesirable electron tunneling processes are prevented due to the lower mobility of electrons in rutile compared to anatase (Sun et al. 2003). Furthermore, the synergistic effect has been found to be affected by the particle shape and the relative positions of the Fermi levels in isolate anatase and rutile particles (Sun and Smirniotis 2003).

Regardless of the use of TiO_2 P25, many investigations have clearly demonstrated the importance of choosing the optimum degradation parameters for obtaining high degradation rates essential for any practical application of photocatalytic oxidation processes (light intensity, pH, etc.). Additionally, many different preparation methods have been described in order to obtain high-performance TiO_2 nanoparticles, including sol-gel (Keshmiri et al. 2004), hydrothermal methods (Ma et al. 2005), precipitation (Song et al. 2006), and others. The sol-gel method is widely used for the preparation of metal oxides. Specifically, for the preparation of TiO_2, titanium alkoxides or titanium chlorides are used as precursors. Then, a homogeneous aqueous solution is prepared using the appropriate starting material while hydrolysis and condensation for the gel formation are following. Recently, this method has been used for the synthesis of homogeneous organic-inorganic hybrid materials and of mesoporous TiO_2 using surfactant or triblock or diblock copolymers as templates (Sung et al. 2006, Bosc et al. 2003, Lee et al. 2004, Hwang et al. 2001, Yun et al., 2001, Yang et al. 1998). However, sol-gel preparations (Colomer and Jurado 2000, Sugimoto et al. 2003a, b, Zhang and Gao 2003, Hay and Raval 2001, Vioux 1997, Niederberger et al. 2002) and microemulsions (Papoutsi et al. 1994, Stathatos et al. 1997) usually provide amorphous titania. Thermal treatment is therefore necessary for crystallization and determines both the final morphology and the crystal structure of titania. By controlling experimental parameters such as aging, annealing temperature, time of calcination, cooling, and washing, nanostructures with different morphologies can be obtained.

For example, nanoporous poly(ethylene oxide)-TiO_2 particles were prepared by the sol-gel method and calcined at different temperatures (Sung et al. 2006). It was found that hybrid particles were crystallized in an anatase structure (spherical) when calcined at 600°C. The particles consisted of anatase and rutile when heated in the temperature region 800–900°C. The formation of a core(anatase)/shell(rutile) structure was observed after heating at 800°C. The core/shell particles were very active for the decomposition of Rhodamine-B (more active than commercial photocatalysts). This effect was attributed to the conduct between the anatase and rutile phase, which favors the move of electron and holes in the interface between the two phases, and to the enhanced surface area of the nanoporous core.

Yu et al. (2003) used phosphoric acid and the triblock copolymer poly(ethyleneglycol)-block-poly(propylene glycol)-block-poly(ethylene glycol) as a template in order to prepare phosphated mesoporous anatase TiO_2 with high surface area. It was reported that the mesoporous framework was stabilized due to the incorporation of phosphorus from phosphoric acid. Both mesoporous TiO_2 and phosphated mesoporous TiO_2 synthesized following the same preparation method were more efficient catalysts than Degussa P25 for the photocatalytic oxidation of n-pentane in air. When phosphated mesoporous TiO_2 was calcined, it exhibited higher catalytic activity than that of calcined mesoporous TiO_2. The authors proposed three reasons to explain this behavior of the calcined phosphated mesoporous TiO_2—specifically, its extended bandgap energy, the larger surface area, and the existence of Ti ions in a tetrahedral coordination.

Liu et al. (2005) prepared mesoporous titania with anatase stucture using a sol-gel method. In this preparation method, the 1-butyl-3-methylimidazolium tetrafluoroborate solvent was used as template agent and was found to play an important role in the formation of anatase. The prepared catalysts exhibited significantly enhanced photoactivity for the decomposition of Rhodamine B dye but not higher than the corresponding activity of Degussa P25.

There are many different approaches to modification of TiO_2 in order to attain enhanced photocatalytic activities and activation of TiO_2 in the energy range of visible light. These studies include, for example, modification with transition metal ions (Dvoranova et al. 2002), noble metals (Sclafani and Herrmann 1998, Sakthivel et al. 2002, Dobosz and Sobczynski 2003, Arana et al. 2001, Di Paola et al. 2002, Arabatzis et al. 2003, Subramanian et al. 2001, Hufschmidt et al. 2002) lanthanides metals, alkaline and alkaline earth metals, nonmetals (Asahi et al. 2001, Irie et al. 2003, Kobayakawa et al. 2005, Ohno et al. 2004), addition of inner support (Matos et al. 1999), dye photosensitization on the TiO_2 surface (Vinodgopal et al. 1996, Zhao et al. 1998, Fujishima et al. 2000, Wu et al. 1998), and interaction between the particles of different semiconductor (Lo et al. 2004, Kang et al. 1999). Here we will describe two important classes of TiO_2 derivatives: the magnetically modified TiO_2 and TiO_2-clay nanocomposites.

3.2 Magnetically modified TiO_2

TiO_2 is the most commonly used weapon for treatment of pollutants in water. Because of its high specific surface area and dispersion, TiO_2 slurry has been proved to be more effective than immobilized TiO_2. The major disadvantage of the use of TiO_2 slurries comes from the costly separation process of the photocatalyst from the water. A solution to this problem is the development of photocatalysts with magnetic properties that can be easily recovered by applying a magnetic field (Beydoun et al. 2000, 2001, Beydoun and Amal 2002, Gao et al. 2003, Watson et al. 2002, Beydoun et al. 2002, Chen et al. 2001, Watson et al. 2005, Chung et al. 2004, Shchukin et al. 2002, Xu et al. 2007, Belessi et al. 2009a).

The magnetic photocatalytic systems contain mainly magnetite (Fe_3O_4), maghemite (γ-Fe_2O_3) (Beydoun et al. 2000, 2001, Belessi et al. 2009a, Qiu et al. 2007, Kurinobu et al. 2007, Song and Gao 2007, Chen and Zhao 1999), spinel-like ferrites ($CoFe_2O_4$, $NiFe_2O_4$, $Zn_{0.35}Ni_{0.65}Fe_2O_4$) (Chung et al. 2004, Shchukin et al. 2002, Xu et al. 2007), or hexaferrites ($BaFe_{12}O_{19}$) (Fu et al. 2006, Lee S-W et al. 2004) as magnetic cores.

The magnetic iron oxide-TiO_2 surfaces exhibit lower photocatalytic activity compared to TiO_2. Specifically, it has been found that the TiO_2/Fe_3O_4 or TiO_2/γ-Fe_2O_3 composites show lower photoactivity than bare TiO_2, when there is direct contact between TiO_2 and the magnetic nanoparticles. The reason is that the Fe_3O_4 or γ-Fe_2O_3 function as recombination centers for electrons and holes, especially in the case of magnetite (Beydoun et al. 2000). The mechanism involves the injection of charges from TiO_2 into γ-Fe_2O_3 and is known as induced photodissolution (Beydoun et al. 2000, Gao et al. 2003, Chen et al. 2001). At this point, we should note that the relative positions of the conduction and valence bands of the two semiconductors are important for the particular photoreactions to take place (Beydoun et al. 2000). The conduction band of γ-Fe_2O_3 is lower compared to the conduction band of TiO_2 while the valence band of γ-Fe_2O_3 is higher than the corresponding band of TiO_2 (Chen et al. 2001). As a result, the photogenerated electrons of TiO_2 can move into the conduction band of maghemite while the holes are moving into the valence band of the iron oxide where they act as recombination centres (Beydoun et al. 2000). This is also valid for the magnetite. As Büchler et al. (1998) have explained, the electron-hole recombination is facilitated by the small band gap of magnetite and the fast electron hopping between the Fe^{3+} and Fe^{2+} lattice sites.

First, Beydoun et al. (2000) showed that the combination of the two semiconductor materials (such as TiO_2/Fe_3O_4 or TiO_2/γ-Fe_2O_3) gives rise to the photodissolution of the iron oxide phase. The magnetic photocatalysts were synthesized by coating titanium dioxide particles with colloidal magnetite and magnetite nanoparticles. The coating process that involved the hydrolysis of titanium butoxide in the presence of magnetite particles took place via a sol-gel method, causing the deposition of a titanium dioxide layer onto the surface of the magnetite particles (Beydoun et al. 2000) (Table 13.1). The photocatalytic activity of the bare TiO_2 was higher than that of the prepared coated particles. This behavior was interpreted by a heterojunction between titanium dioxide (shell) and the magnetic particles (core), where the relative positions of conduction and valence bands are such that an enhanced electron-hole recombination takes place. These electronic interactions influenced both the activity and stability of the coated particles. Also, it was found that the heat treatment, which is necessary to crystallize the amorphous TiO_2 coating, lowered the specific surface area and the amount of hydroxyl groups of TiO_2 and decreased the photocatalytic activity of the magnetic photocatalysts. Thus, the higher heat treatment offers better stability but weaker magnetic properties (Beydoun et al. 2000, Beydoun and Amal 2002, Watson et al. 2002).

Later, Gao et al. (2003) investigated the system TiO_2(shell)/γ-Fe_2O_3(core) for the possible effect of titania content, phase composition, sintering temperature, and reaction conditions on the photocatalytic degradation of acridine dye solution. The magnetically separable photocatalysts (Table 13.1) were prepared by a dispersion of Fe_3O_4 in the TiO_2 sol. Then, the catalyst TiO_2/Fe_3O_4 sintered at high temperatures. It was found that this method was successful in attaching the TiO_2 shell on the Fe_2O_3 magnetic core, while at the same time the strong light absorption of γ-Fe_2O_3 was reduced. It was also found that the light absorption by γ-Fe_2O_3 nanoparticles caused lower photocatalytic activity when the amount of γ-Fe_2O_3 exceeded that of TiO_2 in the composite. The authors pointed out the unfavorable role of the phase Fe_2TiO_5, which was formed due to heat treatment, in the photoactivity of the catalysts.

Table 13.1 Magnetic Photocatalytic Systems

Type of Catalyst	Compounds	Precursors and Modification Methods	Reaction Conditions	% Magnetic Core	S (m^2/g)	Saturation Magnetization, M_s (emu/g)	k_{app} (min^{-1})	Reference
TiO_2/Fe_3O_4	Sucrose	$Fe(OH)_2$ and KNO_3 as precursors for colloidal (10-40 nm) Fe_3O_4. $FeCl_3$, $FeCl_2$, NH_4OH as precursors for nanosized (4-10 nm) Fe_3O_4. (A) Colloidal Fe_3O_4, ethanol, Titanium-n-butoxide and H_2O as precursors for TiO_2/Fe_3O_4. Heat treatment at 90 °C for 1.5 h and at 60 °C for 2 days. (B) Nanosized Fe_3O_4, Titanium-n-butoxide as precursors for TiO_2/Fe_3O_4. Heat treatment at 20 °C for 5 h and at 60°C for 2 days. Calcination at various temperatures, e.g., (i)450°C (550°C) or for 3h, (ii)550°C for 5 h.	Under near-UV illumination.	(Bi) 19 (Bii) 19	(Bi) 60 (Bii) 22	N/A N/A	N/A	Beydoun et al. 2000[1]
TiO_2/γ-Fe_2O_3	Acridine dye	Fe_3O_4 nanoparticles dispersed in 5 ml of tetra-n-butyl titanate and 25 ml ethanol mixed with 25 ml ethanol, 1 ml H_2O and 0.02 ml HCl. Calcined at 500°C for 6h. Three samples were prepared with different amounts of titania. TFa: TiO_2/γ-Fe_2O_3 = 3/7 TFb: TiO_2/γ-Fe_2O_3 = 1/1 TFc: TiO_2/γ-Fe_2O_3 = 7/3	Under either UV or visible light illumination. Concentration of Acridine dye: 2.74 10^{-5} M, pH 6.4 Catalyst dosage:1 g/L TiO_2/γ-Fe_2O_3.	N/A	N/A	N/A	k_{TFa}= 5.47×10^{-3} k_{TFb}= 2.61×10^{-2} k_{TFc}= 2.65×10^{-2}	Gao et al. 2003[2]

TiO_2/AC/γ-Fe_2O_3 (TMAC)	Phenol	Titanium-n-butoxide, isopropanol, HNO_3 as precursors for TiO_2 sol. $FeSO_4$, $FeCl_3$ and NH_4OH as precursors for γ-Fe_2O_3. 3g commercial AC (activated carbon), 200 ml γ-Fe_2O_3 (3g/L) as precursors for MAC (magnetic activated carbon). 3g MAC and titania sol as precursors for TMAC (titania coated magnetic activated carbon) with mass ratio TiO_2/AC/γ-Fe_2O_3 12:5:1. Drying at 75°C.	Under UV illumination. Concentration of phenol: 100 mg/L. Catalyst dosage:1.5 g/L.	N/A	386.42	2.21 (γ-Fe_2O_3/AC 1:5) and 1.22 (γ-Fe_2O_3/AC 1:7)	N/A	Ao et al. 2008b[3]
TiO_2/AC/Fe_3O_4 (TMAC)	Phenol	Titanium-n-butoxide, isopropanol, HNO_3 as precursors for TiO_2 sol. $FeSO_4$, $FeCl_3$ and NH_4OH as precursors for Fe_3O_4. 3g AC, 200 ml Fe_3O_4 as precursors for MAC (magnetic activated carbon). 3g MAC and titania sol as precursors for TMAC with mass ratio TiO_2/AC/Fe_3O_4 12:5:1. Drying at 75°C.	Under either UV illumination. Concentration of phenol 100 mg/L. Catalyst dosage:1.5 g/L.	N/A	373.56	6.45	k_{app} = 0.0062	Ao et al. 2008c[4]
TiO_2/AC/Fe_3O_4 (TMAC)	Reactive red X-RB	Titanium-n-butoxide, isopropanol, HNO_3 as precursors for TiO_2 sol. $FeSO_4$, $FeCl_3$ and NH_4OH as precursors for Fe_3O_4. 3g AC, 200 ml Fe_3O_4 as precursors for MAC (magnetic activated carbon). 3g MAC and titania sol as precursors for TMAC with mass ratio TiO_2/AC/Fe_3O_4 12:5:1. Drying at 75°C.	Under either UV or visible light illumination. Concentration of dye: 50 mg/L. Catalyst dosage:1 g/L.	N/A	373.56	6.45	k_{appUV} = 0.0288 $k_{app,visible}$ = 0.0225	Ao et al. 2009[5]

Continued

Table 13.1 Magnetic Photocatalytic Systems—cont'd

Type of Catalyst	Compounds	Precursors and Modification Methods	Reaction Conditions	% Magnetic Core	S (m^2/g)	Saturation Magnetization, M_s (emu/g)	k_{app} (min^{-1})	Reference
TiO_2/ $NiFe_2O_4$	(i) p-totyl sulphide and (ii) Methyl Orange	Combination of reverse micelle and hydrolysis. NH_4OH, H_2O, diiso-octylsulphoccinate, iso-octane as precursors of microemulsion system I. $FeCl_2$ $4H_2O$, $NiCl_2$ $6H_2O$, H_2O, diiso-octylsulphoccinate-iso-octane as precursors of microemulsion system II. $TiO_2/NiFe_2O_4$ was obtained with acidic hydrolysis of $TiCl_4$. Heat treatment at 90°C for 1h. Drying at 30°C for 1h.	(i) Under UV illumination (300-350 nm). Catalyst dosage: 4.43 g/L. (ii) Under UV illumination (300-350 nm). Catalyst dosage: 0.8 g/L.	10-20	N/A	3.8	N/A	Rana et al. 2005[6]
TiO_2/ $BaFe_{12}O_{19}$	Procion Red MX-5B	200 mg $BaFe_{12}O_{19}$ and 150 ml 1% PEI (polyethyleneimine) as precursors for PEI-$BaFe_{12}O_{19}$. Appropriate amount of PEI-$BaFe_{12}O_{19}$, 150 ml H_2O and 2 ml of bis-ammonium lactato dihydroxide (TALH) (pH 5.5) as precursors for $TiO_2/BaFe_{12}O_{19}$. Heat treatment at 95°C and drying at 60°C for 2 days. Calcination at 500°C for 1h.	Under UV illumination. Concentration of Procion Red MX-5B: 10 ppm. pH 6 Catalyst dosage: 0.6 g/L	92.8 (heat treated)	17.9	N/A	N/A	Lee S-W. et al. 2004[7]
TiO_2/ SiO_2/γ-Fe_2O_3	Fluoresein, Orange II and Red acid G	$FeSO_4$, $FeCl_3$ as precursors of γ-Fe_2O_3. Heating at 450 °C for 30 min. 2g of γ-Fe_2O_3 (2g/L) and silicate solution (3%) at pH 10 as precursors for SiO_2/γ-	Under either UV or visible irradiation Catalyst dosage: 2 g/L	~ 50 %	N/A	N/A	$k_{app\ Orange\ II}$ = 0.013	Chen et al. 2001[8]

		Fe_2O_3. Heating at 450°C for 30 min. 2g of SiO_2/γ-Fe_2O_3 and 2g of P25 as precursors for $TiO_2/SiO_2/\gamma$-Fe_2O_3. Calcination at 500°C for 30 min.						
γ-Fe_2O_3/ PSS^-/ PDD^+/ TiO_2	Propachlor	$Fe(NH_4)_2(SO_4)_26H_2O$, KOH, $(NH_4)_2S_2O_8$ as precursors for γ-Fe_2O_3. γ-Fe_2O_3 at pH 4.5 and poly(sodium 4-styrene sulfonate) PSS^-Na^+ as precursors for γ-Fe_2O_3/PSS^-Na^+. γ-Fe_2O_3 at pH 10 and poly-(diallyldimethylammonium) chloride PDD^+Cl^- as precursors for $[TiO_2/PDD]^+Cl^-$. Appropriate amounts of Fe_2O_3/PSS^-Na^+ and $[TiO_2/PDD]^+Cl^-$ were mixed for the preparation of γ-$Fe_2O_3/PSS^-/PDD^+/TiO_2$.	Under UV-Vis irradiation Concentration of propachlor: 5 mg/L Catalyst dosage: 0.2 g/L	(i) 2.28 (ii) 7.75 (iii) 12.87 (iv) 19.65 (v) 29.49	N/A	11.4 (for 19.69% γ-Fe_2O_3)	(i) k_{app}= 0.0495 (ii) k_{app}= 0.0538 (iii) k_{app}= 0.0749 (iv) k_{app}= 0.0938 (v) k_{app}= 0.0711	Belessi et al. 2009a[9]
TiO_2/ SiO_2/ $NiFe_2O_4$	Methylene Blue	For ultrasonic spray pyrolysis. First step: $FeCl_3$, $Ni(C_2H_3O_2)_24H_2O$ as precursors for $NiFe_2O_4$. Calcination at 800 °C-1000°C. Second step: Tetraethyl orthosilicate, ethanol as precursors for SiO_2 and titanium isopropoxide as precursor for TiO_2. Calcination at 700°C.	Under UV irradiation Concentration of Methylene Blue: 10 mg/L Catalyst dosage 0.26 g/L	(i) 49 (ii) 63 (iii) 95 (iv) 100	N/A	(i) 14.68 (ii) 23.16 (iii) 32.28 (iv) 34.32	N/A	Chung et al. 2004[10]

Continued

Table 13.1 Magnetic Photocatalytic Systems—cont'd

Type of Catalyst	Compounds	Precursors and Modification Methods	Reaction Conditions	% Magnetic Core	S (m^2/g)	Saturation Magnetization, M_s (emu/g)	k_{app} (min^{-1})	Reference
$TiO_{2-x}N_x$/ SiO_2/ $NiFe_2O_4$	Methyl Orange	$Fe(NO_3)_3$ 2.5 M, $Ni(NO_3)_2$ 1.25 M, NaOH, $FeCl_2$ 1 M as precursors for $NiFe_2O_4$. Boiling and refluxing for 2 h. $NiFe_2O_4$, $(NaPO_3)_6$, $Na_2O{\cdot}3SiO_2$ as precursors for $SiO_2/NiFe_2O_4$. Heating at 90°C, pH 10. 1g $TiO_{2-x}N_x$, 0.1 g SiO_2/ $NiFe_2O_4$ as precursors for $TiO_{2-x}N_x/SiO_2/NiFe_2O_4$. Calcination at 400°C for 30 min.	Under either UV or visible irradiation Concentration of Methyl Orange: 10 mg/L Catalyst dosage: 1.1 g/L	N/A	96.11	0.5	N/A	Xu et al. 2007[11]
TiO_2/ SiO_2/ $CoFe_2O_4$	Methyl Orange	$Co(NO_3)_2{\cdot}6H_2O$, $FeSO_4{\cdot}7H_2O$, $Fe_2(SO_4)_3{\cdot}7H_2O$, NaOH as precursors for $CoFe_2O_4$. Aging at 40°C for 4h. Heating at 900°C for 2 h. 0.5g $CoFe_2O_4$, 230 ml isopropanol, 6ml tetraethyl orthosilicate, NH_4OH as precursors for SiO_2/ $CoFe_2O_4$. Heating at 115°C for 24h. 0.5g $SiO_2/CoFe_2O_4$, 5 ml titanium bis-ammonium lactato dihydroxide, 3.6g urea as precursors for TiO_2/ $SiO_2/CoFe_2O_4$. Heating at 65°C for 4h, 97°C for 24 h, 70°C for 2 days (vacuum). A part of the sample calcined at 300°C for 1h.	Under either UV irradiation Concentration of Methyl Orange: 6 mg/L Catalyst dosage: N/A	N/A	154	8.8	For the first stage k_{app} = 0.00417 For the second stage k_{app} = 0.00324	Zhang et al. 2009[12]

TiO_2/Si/ magnetic black sand	Phenol	0.3g black sand, hexadecyl trimethyl ammonium chloride, tetraethyl orthosilicate and NH4OH as precursors for Si/black sand. Calcination at 500°C for 2 h. *(A)* 1.5 ml titanium tetra-isopropoxide, 40 ml isopropanol and 0.2g Si/ Black sand as precursors for TiO_2/Si/ Black sand through an impregnation procedure. Calcination at various temperatures for 3 h. *(B)* 10 ml isopropanol, H_2O and 1 ml titanium tetra-isopropoxide as precursors for the preparation of TiO_2 colloidal solution. 0.2g Si/ Black sand, TiO_2 colloidal solution as precursors for the preparation of TiO_2/Si/ Black sand through the direct deposition method. Heating at 90°C for 2h and drying at 80°C.	Under visible irradiation Concentration of phenol: 20 mg/L Catalyst dosage: 2.7 g/L	Bulk iron content (A) 18.6% (B) 9.1%	(A) 80 (B) 123.7	N/A	N/A	Luo et al. 2009[13]

[1]"*Reprinted from* The Journal of Physical Chemistry *B, 104, Beydoun D., R. Amal, G. K.-C. Low et al. Novel photocatalyst: Titania-coated magnetite. Activity and photodissolution., 4387-4396, Copyright (2000), with permission from American Chemical Society.*

[2]"*Reprinted from* Materials Chemistry and Physics*, 80, Gao Y., B. Chen, H. Li et al. Preparation and characterization of a magnetically separated photocatalyst and its catalytic properties, 348–355, Copyright (2003), with permission from Elsevier.*

[3]"*Reprinted from* Separation and Purification Technology*, 61, Ao Y., J. Xu, D. Fu et al. A novel magnetically separable composite photocatalyst: Titania-coated magnetic activated carbon., 436-441, Copyright (2008), with permission from Elsevier.*

[4]"*Reprinted from* Journal of Hazardous Materials*, 160, Ao Y., J. Xu, X. Shen, et al. Magnetically separable composite photocatalyst with enhanced photocatalytic activity, 295-300, Copyright (2008), with permission from Elsevier.*

Continued

Table 13.1 Magnetic Photocatalytic Systems—cont'd

Type of Catalyst	Compounds	Precursors and Modification Methods	Reaction Conditions	% Magnetic Core	S (m^2/g)	Saturation Magnetization, M_s (emu/g)	k_{app} (min^{-1})	Reference

[5]"Reprinted from Journal of Alloys and Compounds, *471, Ao Y., J. Xu, D. Fu, et al., Photocatalytic degradation of X-3B by titania-coated magnetic activated carbon under UV and visible irradiation, 33-38, Copyright (2009), with permission from Elsevier.*
[6]"Reprinted from Materials Science and Engineering*: B, 119, Rana S., R.S. Srivastava, M.M. Sorensson, et al. Synthesis and characterization of nanoparticles with magnetic core and photocatalytic shell: Anatase TiO_2-$NiFe_2O_4$, 144-151, Copyright (2005), with permission from Elsevier.*
[7]*"Reprinted from Chemistry of Materials, 16, Lee S-W, J. Drwiega, C-Y. Wu, et al. Anatase TiO_2 nanoparticle coating on barium ferrite using titanium bis-ammonium lactato dihydroxide, and its use as a magnetic photocatalyst, 1160-1164, Copyright (2004), with permission from American Chemical Society.*
[8]"Reprinted from Chemosphere, *44, Chen F., Y. Xie, J. Zhao et al. Photocatalytic degradation of dyes on a magnetically separated photocatalyst under visible and UV irradiation, 1159–1168, Copyright (2001), with permission from Elsevier.*
[9]"Reprinted from Applied Catalysis B: Environmental, *87, Belessi V., D. Lambropoulou, I. Konstantinou, et.al., Structure, photocatalytic performance of magnetically separable titania photocatalysts for the degradation of propachlor, 181–189, Copyright (2009), with permission from Elsevier.*
[10]"Reprinted from Materials Chemistry and Physics, *86, Chung Y.S., S.B. Park, D.-W. Kang, Magnetically separable titania-coated nickel ferrite photocatalyst, 375–381, Copyright (2004), with permission from Elsevier.*
[11]"Reprinted from Applied Catalysis B: Environmental, *71, Xu S., W. Shangguan, J. Yuan et.al., Preparations and photocatalytic properties of magnetically separable nitrogen-doped TiO_2 supported on nickel ferrite, 177–184, Copyright (2007), with permission from Elsevier.*
[12]"Reprinted from Materials Research Bulletin, *44, Zhang H., R. Hou, Z-L. Lu, et.al., A novel magnetic nanocomposite involving anatase titania coating on silica-modified cobalt ferrite via lower temperature hydrolysis of a water-soluble titania precursor, 2000-2008, Copyright (2009), with permission from Elsevier.*
[13]"Reprinted from Applied Catalysis B: Environmental, *87, Luo M., D. Bowden, P. Brimblecombe, Preparation of black sand-based magnetic photocatalysts for photocatalytic oxidation of aqueous phenol, 1-8, Copyright (2009), with permission from Elsevier.*

Recently, He et al. (2008) developed Fe_3O_4(core)/TiO_2(shell) nanoparticles with magnetic performance and high visible light photocatalytic activity, using a homogeneous precipitation method. The photocatalytic action was tested for a malignant tumor therapy, and the authors suggested that it could be a promising candidate in this field. Specifically, cervical carcinoma Hela cells were taken as a model to check the killing efficiency of Fe_3O_4/TiO_2 core-shell nanoparticles. According to the experimental results, the magnetic photocatalytic system had higher absorption in the visible region and higher photoactivity than TiO_2 nanoparticles.

From a wide range of magnetic core materials, $NiFe_2O_4$ has been selected by many scientists as a core material because of its high magnetization and stability (Rana et al. 2005, Rawat et al. 2007). In order to prepare anatase TiO_2-coated $NiFe_2O_4$ nanoparticles (10–20 nm), Rana et al. (2005) used a preparation procedure whereby a hydrolysis reaction followed the reverse micelle technique (Table 13.1). The important properties of the uncoated nanocrystalline nickel ferrites, such as superparamagnetism and nonsaturation of magnetic moments at high field, were maintained from the as-prepared nanoparticles. However, the experimental results showed reduced magnetization for the $TiO_2/NiFe_2O_4$ nanoparticles compared with the uncoated $NiFe_2O_4$ nanoparticles. The samples were tested for the photocatalytic degradation of methyl orange and the oxidation of p-tolylsulphide. Also, the magnetic system TiO_2/$NiFe_2O_4$ showed photocatalytic activity that could be used for antimicrobial activity.

The hard magnetic composite $BaFe_{12}O_{19}$ (core)-anatase TiO_2 nanoparticles (shell) was developed by Lee et al. (2004). The photocatalytic activity of $TiO_2/BaFe_{12}O_{19}$ was studied for the decomposition of Procion Red MX-5B under UV illumination. Explicitly, TiO_2 nanoparticles were deposited on barium ferrite captured with polyethyleneimine (PEI) by hydrolysis and condensation of titanium bis-ammonium lactato dihydroxide (TALH) (Table 13.1). Coverage of the $BaFe_{12}O_{19}$ by the TiO_2 nanoparticles was complete. The authors noted the role of the electrostatic attractions for the successful deposition of the doubly negatively charged TALH precursor onto the positively charged polyethyleneimine-adsorbed barium ferrite particles during the coating process. In this case, protonated PEI acts as a surface-modified agent of $BaFe_{12}O_{19}$ nanoparticles that enwrapped TiO_2 closely around the core. The photocatalytic activity of the samples followed the order: Degussa P25 $>$ unheated $TiO_2/BaFe_{12}O_{19}$ $>$ heated at 500 °C $TiO_2/BaFe_{12}O_{19}$. This trend was expected because of the preservation of the surface hydroxyl groups on TiO_2 and also because of the higher specific surface area of the uncalcined composite compared to the calcined one. The fact that the weight percentage of TiO_2 in the prepared composites was only 7.2% constitutes a disadvantage of the method.

Another interesting class of magnetic photocatalysts is the magnetic activation carbons (AC) (Ao et al. 2008a,b,c, 2009). Undoubtedly, magnetic-AC composites are expected to have interesting properties that become promising when the composites are used as a support for the deposition of titania to prepare TiO_2 magnetic activated carbon (TMAC). Actually, Ao et al. (2008a,b) prepared TiO_2-γ-Fe_2O_3-AC by depositing anatase onto the surface of magnetic-activated carbon (Table 13.1). The photocatalytic activity of the as-prepared photocatalysts was studied for the degradation of phenol in aqueous solutions. It was found that photocatalysts with different γ-Fe_2O_3/AC mass ratios (1:3, 1:5, 1:7, and 1:9) exhibited different activities. The most active photocatalyst was that with γ-Fe_2O_3/AC mass ratio 1:7. However, all these samples were more catalytically active than Degussa P25. As far as the repetitive use of the photocatalysts is concerned, it was found that the degradation of phenol was high (above 80 %) even when they were used six times. Furthermore, study of the magnetic properties of the prepared photocatalysts with mass ratio 1:5 and 1:7 revealed their superparamagnetic behavior. So, it is expected that their separation from a suspended system will be an easy task.

The same preparation method was used to prepare titania-coated Fe_3O_4-AC with homogeneous distribution of anatase titania nanoparticles (Table 13.1). These samples were used for the degradation of reactive brilliant red X-RB (Ao et al. 2009) and phenol (Ao et al. 2008c) under UV (or visible) irradiation. The photocatalytic activity of the superparamagnetic TMAC was higher than Degussa P25 (Ao et al. 2008c, 2009). Moreover, the photocatalytic activity of the recycled photocatalyst was very good after six cycles (Ao et al. 2009). The authors supported the idea that no photodissolution phenomenon was taking place. The enhanced photoactivity of TMAC was attributed to a synergetic effect between titania and AC. As AC is an excellent adsorbent, it concentrates the substrates near to titania, causing a more rapid degradation upon irradiation. Simultaneously, the photoinduced electrons migrate toward the magnetic particles, reducing the recombination of the species. The above studies would be more convincing if structural analysis and magnetic measurements of the recycled photocatalysts had been performed.

In general, the widespread acceptance of magnetically modified titania is restricted owing to the photodissolution. However, related studies have shown that use of a SiO_2 layer between the Fe_3O_4 or γ-Fe_2O_3 core and the TiO_2 shell, that is, TiO_2/SiO_2/magnetic oxide, preserves the photocatalytic activity of the catalyst by preventing the injection of charges from TiO_2 particles to magnetic particles (Beydoun et al. 2002, Chen et al. 2001). In other words, the SiO_2 layer inhibits the photodissolution of the catalyst. In line with these concepts, aqueous dispersions of the photocatalytic system $TiO_2/SiO_2/\gamma$-Fe_2O_3 were evaluated for the photodegradation of three kinds of dyes (Fluoresein, Orange II, and Red acid G), under both UV and visible irradiation (Chen et al. 2001). The preparation method (Table 13.1) resulted in agglomerated γ-Fe_2O_3 units in the $TiO_2/SiO_2/\gamma$-Fe_2O_3 system, while the thin SiO_2 membrane was imperfect; thus the TiO_2 particles (shell) could directly contact the γ-Fe_2O_3 agglomerates (core) at some sites. It was found that the composite system $TiO_2/SiO_2/\gamma$-Fe_2O_3 was more active than TiO_2/γ-Fe_2O_3 but less active than Degussa-P25 for the photocatalytic degradation of both Fluoresein and Orange II, under UV and visible light irradiation. In any case, it is obvious that the SiO_2 membrane inhibits the migration of charges from TiO_2 to γ-Fe_2O_3. Concerning the repeatability of the catalysts, it was found that use of the recycled catalysts was feasible after four cycles. The authors did not provide information regarding the magnetic properties of the composite sample, but they reported that the separation of catalysts by the magnetic field was effective.

Another work reported the synthesis of the composite $TiO_2/SiO_2/Fe_3O_4$ (Song and Gao 2007). According to this preparation method, hydrolysis and condensation polymerization of the starting materials took place by using Fe_3O_4 spheres ($\sim$ 427 nm), which were prepared by the polyol solvothermal method (Deng et al. 2005). Finally, a hydrothermal treatment for the amorphous $TiO_2/SiO_2/Fe_3O_4$ was adopted in order to ensure the crystallinity of TiO_2 particles. Unfortunately, the binding of the anatase TiO_2 on the SiO_2/Fe_3O_4 ($\sim$ 443 nm) surface was imperfect; the percentage of SiO_2 in the $TiO_2/SiO_2/Fe_3O_4$ sample was 8%wt and its thickness layer was 8 nm. The authors therefore did not exclude the notion that few electronic interactions occur at the heterojunctions. The photocatalytic activities of the magnetic photocatalysts were sufficient for the degradation of methylene blue in aqueous solution. However, the photoactivity of bare TiO_2 (anatase) was higher than that of the magnetic photocatalyst. The authors, in accordance with other studies, ascribed this behavior to the absorption of the UV light by the magnetite core and to the imperfect SiO_2 membrane that allows contact between TiO_2 and γ-Fe_2O_3 particles. As far as the magnetic saturation is concerned, the composite had the lowest value compared to SiO_2/Fe_3O_4 and Fe_3O_4 particles.

The photocatalytic system $TiO_2/SiO_2/Fe_3O_4$ was also studied by Kurinobu et al. (2007). For the coating of Fe_3O_4 particles with the layers of SiO_2 (55 nm) and TiO_2, the sol-gel method was adopted and the coated particles were heated at 500°C. Methylene blue, two basic dyes, and two phenols were used as pollutants in order to estimate the efficiency of the system. The degradation of these substances was ascribed to the strong adsorption of the pollutants on the magnetic photocatalyst and to the photolytic degradation.

Similarly, $TiO_2/SiO_2/Fe_3O_4$ magnetic nanoparticles were prepared by Chen et al. (2008) and used successfully as photokilling agents. The authors showed that the immunoglobulin-bound magnetic nanoparticles could target many pathogenic bacteria and prevent their cell growth under UV irradiation.

Undoubtedly, the conclusion derived from these studies is that two factors determine the photoactivity, the stability, and the magnetic properties of the tested photocatalysts—namely, the silica layer and the heat treatment during the preparation process. The heat treatment step has important consequences for the magnetic properties of the prepared photocatalysts because it creates chemical or physical changes in the magnetic core (Beydoun et al. 2000). However, this problem is not solved using calcined magnetic cores for the preparation of photocatalysts (Beydoun et al. 2002, Shchukin et al. 2002) since the sol-gel synthesis for the preparation of titania requires a heating step for anatase growth.

Recently, magnetically separable and efficient photocatalysts were prepared, avoiding the heat treatment step during their preparation method (Table 13.1) (Belessi et al. 2009a). According to this method, negatively charged γ-Fe_2O_3 nanoparticles and positively charged nanosized TiO_2 (Degussa P25) were combined for the formation of the magnetic photocatalysts. Specifically, a solution of poly(sodium 4-styrene sulfonate) polyanion (PSS^-Na^+) was used for the encapsulation of γ-Fe_2O_3 nanoparticles, while a poly(diallyldimethylammonium) chloride solution was used as a positively charged binder to negatively charged TiO_2 particles. Finally, appropriate amounts of the negatively charged maghemite particles in the form of $[\gamma\text{-}Fe_2O_3/PSS]^-Na^+$ and of the positively charged titanium oxide dispersion in the form of $[TiO_2/PDD]^+Cl^-$ were combined for the preparation of γ-$Fe_2O_3/PSS^-/PDD^+/TiO_2$ photocatalysts. The photocatalytic activities of the composite photocatalysts with different γ-Fe_2O_3 loadings were evaluated for the degradation of the herbicide propachlor. A key component of these photocatalysts was the insulator polyelectrolyte PSS^-/PDD^+ bilayer that provides a protective membrane between maghemite and titanium dioxide, inhibiting the negative influence of γ-Fe_2O_3 on the photocatalytic process. The developed nanocomposites exhibited good catalytic activity toward organic pollutants such as chloroacetamide herbicides and were only a little less efficient than the commercial photocatalyst Degussa P25. The as-prepared magnetic photocatalysts were stable and did not suffer from photodissolution. The recycled photocatalyst was used during four repetitive cycles under UV-Vis irradiation, and no significant change in its activity was observed. It is important that the stability of the photocatalysts was certified by measurement of hysteresis loops after periods of UV-Vis illumination and from TEM images. It was found that there was no change in the hysteresis parameters derived from the magnetization loops, denoting that the system maintained its magnetic properties after UV-Vis illumination. Also, the stability of a representative sample after 3h of UV-Vis illumination was further evidenced from TEM images, in which the presence of the polyelectrolyte layers, connecting maghemite nanoparticles with anatase, were still clearly seen.

The $TiO_2/SiO_2/NiFe_2O_4$ system was developed by a relatively complicated continuous multistep spray pyrolysis process (Table 13.1). The silica layer separating titania and $NiFe_2O_4$ presented an obstacle to titania crystal growth (Chung et al. 2004). Furthermore, it was clear from TEM images that the TiO_2 particles on the magnetic core were not forming uniform layers but appeared as distinct spots. The size of the magnetic core was bigger than 1000 nm, so it is doubtful whether such large particles remain suspended. The magnetic catalyst was checked for the photodegradation of methylene blue and showed enhanced photoactivity compared to Degussa P25. Then, the authors checked the photocatalytic degradation of the tested dye using both $TiO_2/SiO_2/NiFe_2O_4$ and $TiO_2/NiFe_2O_4$ systems in order to check whether the photoactivities remained unchanged after repeated uses of the magnetic photocatalysts. Obviously, the photocatalytic activity of the $TiO_2/NiFe_2O_4$ was reduced (around 30% from the first cycle). It was found that only the photoactivity of the $TiO_2/SiO_2/NiFe_2O_4$ did not change after three catalytic cycles, thereby confirming the endurance of TiO_2 and SiO_2 layers and the role of the latter as a barrier.

Similarly, the nitrogen-doped magnetic photocatalyst $TiO_{2-x}N_x/SiO_2/NiFe_2O_4$ was studied and found to be effective for the degradation of methyl orange, under UV and visible light irradiation (Xu et al. 2007). According to the preparation method, $SiO_2/NiFe_2O_4$ particles were synthesized by a liquid phase transformation (Table 13.1). Then the nanoparticles of $TiO_{2-x}N_x$ and $SiO_2/NiFe_2O_4$ were mixed, dried, and calcined. The characterization of the composite catalyst revealed the successful attachment of $TiO_{2-x}N_x$ and $SiO_2/NiFe_2O_4$ particles, while the existence of chemical bonds between $TiO_{2-x}N_x$ and $SiO_2/NiFe_2O_4$ was not clear. In accordance with other researchers, it was proposed that the SiO_2 layer blocks the injection of charges from $TiO_{2-x}N_x$ particles to $NiFe_2O_4$. On the other hand, the photocatalytic activity of the composite was remarkable, but that of $TiO_{2-x}N_x$ was higher and the authors disputed the successful coating of silica on the $NiFe_2O_4$ particles. The recycled composites did not have any noticeable change after three cycles under UV and visible light irradiation.

Zhang et al. (2009) prepared a $TiO_2/SiO_2/CoFe_2O_4$ magnetic nanocomposite. According to this synthetic method, a layered double hydroxide precursor was used for the preparation of $CoFe_2O_4$ nanoparticles (~55 nm), while the $SiO_2/CoFe_2O_4$ particles were prepared by hydrolysis of tetraethyl orthosilicate TEOS and heat treatment at 115°C (Table 13.1). The coating of $SiO_2/CoFe_2O_4$ (~150–180 nm) by anatase nanoparticles (~12 nm) occurred by hydrolysis and condensation of titanium bis-ammonium lactato dihydroxide and formed a loose layer (~20–35 nm). It is uncertain whether a chemical linkage Si–O–Ti was created between the TiO_2 particles and the silica-coated ferrite cores. The photocatalytic activity of the magnetic photocatalyst was sufficient but lower than that of Degussa P25 and superior to that of the calcined samples toward the degradation of methyl orange.

In a similar study, the $TiO_2/SiO_2/Zn_{0.35}Ni_{0.65}Fe_2O_4$ system showed high photoactivity for the oxalic acid oxidation but lower than that of Degussa P25 (Shchukin et al. 2002). The stable magnetic core $Zn_{0.35}Ni_{0.65}Fe_2O_4$ was obtained by hydrolysis of the corresponding salts and heating of the precipitated solid. A homogeneous SiO_2 layer (25 nm) coated the magnetic core, and the coating of $SiO_2/Zn_{0.35}Ni_{0.65}Fe_2O_4$ by titania followed. The composite photocatalyst calcined at 450°C, and the anatase phase was obtained.

Recently, Luo et al. (2009) developed a different method for preparing magnetic photocatalysts that include the use of black sand as magnetic core in the composites. The surfactants hexadecyl trimethyl ammonium chloride and tetraethoxysilane were used for the coating of black sand by a SiO_2 shell. Then, for the enwraping of TiO_2 (shell) on the SiO_2/black sand (core) (Table 13.1), two different synthetic procedures were examined. The first was an impregnation, and the second was a direct deposition procedure. During this process, loosely and tightly wrapped TiO_2 particles were formed on the magnetic core. It is interesting to note that the loosely adsorbed TiO_2 particles were separated from the tightly bound TiO_2 components on the Si/black sand particles by a magnetic field. The composite prepared via impregnation exhibited higher iron percentage (18.6 wt%) than the sample prepared by direct deposition (9.1 wt%). The catalytic properties of the as-prepared photocatalysts were examined for the oxidation of aqueous phenol. Degussa P25 had higher photoactivity compared to the magnetic photocatalysts. However, it was found that when the kinetics rates were normalized per mass of TiO_2 the magnetic photocatalysts were twice as fast as Degussa P25. Finally, the reusability of the photocatalysts was checked. It was found to be adequate, though a minor deactivation of the samples was observed.

3.3 TiO_2-clay nanocomposites

The structure of titania has a great impact on its photocatalytic activity. The role of the structure was investigated in many papers. The titania is usually prepared in the form of layered and pillared titanates, microcrystalline TiO_2 pillared clays, mixed Ti/Si oxides, nanocrystals of TiO_2 dispersed in inorganic media, or on the surface of active carbons (Cheng et al., 1995b, Ding et al., 1999, Ooka et al. 1999, 2004a,b, Li et al. 2002a, Sun et al. 2002, Zhu et al. 2002, 2005, Tryba et al. 2003). Among the previously referred classes of materials, the composites derived from dispersion of TiO_2 particles on layered clays appeared to be very effective photocatalysts (Li et al. 2002a, Sun et al. 2002, Zhu et al. 2002). However, TiO_2-pillared clays exhibit poor photocatalytic activity for the degradation of large organic molecules, very likely due to the steric hindrance in the small pore entrances of TiO_2 pillared clays (Burch 1988) and the amorphous phase of TiO_2 pillars (Ding et al. 1999).

From Table 13.2, it can be seen that various types of clays such as montmorillonite (Zhu et al. 2005, Belessi et al. 2007, Li et al. 2009), laponite (Zhu et al. 2002, 2005, Li et al. 2002a), hectorite (Zhu et al. 2005, Belessi et al. 2007), saponite (Zhu et al. 2005), and vermiculite (Matchado et al. 2006) have been used. In general, clays have been regarded as inert supports. However, the experimental results from various published studies show that clays react with various acids even under moderate conditions (Corma et al. 1987, Mokaya and Jones 1995, Kaviratna and Pinnavaia 1994, Rhodes and Brown 1993), causing leaching and subsequent deposition of metal ions onto clay layers and, thus, affecting their structure and composition and creating high porous composite solids (Zhu et al. 2005, Belessi et al. 2007, Matchado et al. 2006). The acid leaching has a different impact on clays, depending on the type of clay used (Zhu et al. 2005, Belessi et al. 2007, Kaviratna and Pinnavaia 1994, Mokaya and Jones 1995, Corma et al. 1987, Theocharis et al. 1988). Clay-TiO_2 composites with different TiO_2 loadings showed size and morphology of TiO_2 that were strongly influenced by the TiO_2 loading (Belessi et al. 2007, Matchado et al. 2006). The mean crystal size of TiO_2 nanoparticles in the composite samples ranged between 1.7 and 9.0 nm and showed superior catalytic performance. Moreover, the catalytic performance for TiO_2 crystals below 10 nm remarkably enhanced the effect known as the size quantization effect (Anpo et al. 1987).

From a practical point of view, when ultrafine TiO_2 powders are used in aqueous photocatalytic processes, it is very difficult to recover them from water. Accordingly, the dispersion of titanium dioxide nanoparticles in an inorganic medium, such as layered clays, can ensure satisfactory separation as the resulting composite materials sediment much faster than raw or P25 TiO_2 nanoparticles (Zhu et al. 2002, Li et al. 2002a). Sedimentation experiments reveal that TiO_2/bentonite samples decant from the suspension much faster (10 min) compared to pure TiO_2 (12 h) (Sun et al. 2002). A brief

Table 13.2 TiO_2-clay Nanocomposites

Type of Composite Photocatalysts	Compounds	Precursors and Modification Methods	Reaction Conditions	Crystal Phase	S_{TiO2} (m^2/g)		Average Particle Size of TiO_2 (nm)	References
TiO_2- Laponite	Rhodamine-6G	4g Laponite (370 m^2/g), 200 ml H_2O and polyethylene oxide ($C_{12-14}H_{25-29}O(CH_2CH_2O)_nH$) as precursors of mixture (I). $Ti(OC_3H_7)_4$, HCl ($[Ti]/[H^+] = 1:4$) as precursors for the sol of titanium hydrate. Addition of sol to mixture (I). Heating in an autoclave at 100°C for 2 days. Calcination at 500°C for 20 h.	Under UV illumination. Concentration of dye: 4 x 10^{-5} M. Catalyst concentration: 1 g/L.	Anatase	n=5 n=7 n=9 n=12 n=30	514 627 747 525 459	3.5-9.0	Zhu et al. 2002[1]
TiO_2-Laponite TiO_2-Saponite TiO_2-Hectorite TiO_2-Montmorillonite	Phenol and dichlorophenol	4g Laponite (370 m^2/g), 200 ml H_2O and 8g polyethylene oxide ($C_{12-14}H_{25-29}O(CH_2CH_2O)_7H$) as precursors of mixture (I). $Ti(OC_3H_7)_4$, HCl ($[Ti]/[H^+] = 1:4$ and mass ratio of Ti/clay = 11mmol Ti/g) as precursors	Under UV illumination. Concentration of phenol: 2 x 10^{-5} M. Catalyst concentration: 1 g/L.	Anatase	Lap Sap. Hect. Mont.	635 452 290 225	3.9-6.9	Zhu et al. 2005[2]

		for the sol of titanium hydrate. Addition of sol to mixture (I). Heating in an autoclave at 100°C for 2 days. Calcination at 500°C for 20 h.								
TiO_2-Laponite	Sulforhodamine B	The same preparation method as above was followed.	Under UV-Vis illumination. Catalyst concentration: 1 g/L.	Anatase	343			3.9		Li et al. 2002a[3]
TiO_2-Exfoliated Vermiculite	Drimaren Red	Exfoliation of vermiculite at 1000°C for 60 sec. Composites of TiO_2-Exfoliated Vermiculite with 20, 40, and 50 wt % were prepared by impregnation of $Ti(OCH(CH_3)_2)_4$ (0.3-2.1g) in hexane on exfoliated vermiculite (2g). Drying at 80°C. Hydrolysis by HCl/ H_2O vapor. Drying at 80°C for 12 h. Calcination at 400-600°C for 1 h.	Under UV illumination Concentration of dye: 50 mg/L Catalyst concentration: 1 g/L	Anatase	Pure EV Pure EV exposed to HCl/H_2O 20% TiO_2/EV 40% 50% TiO_2/EV	17 84 120 76 89		N/A		Machado et al. 2006[4]
TiO_2-montmorillonite and TiO_2-Laponite	Dimethachlor	Precipitation of TiO_2 in various percentages per weight (15, 30, and 55 wt%) to the surfaces of two different raw clays: (i) natural clay	Under UV-Vis illumination Concentration of dimethachlor: 2 mg/L Catalyst concentration: 0.2 g/L	Anatase	TiO_2 wt% 15 30 55	Mont 156 183 183	Lap 456 443 320	Mont 3.1 4.5 4.2	Lap 1.7 2.0 2.8	Belessi et al. 2007[5]

Continued

Table 13.2 TiO_2-clay Nanocomposites—cont'd

Type of Composite Photocatalysts	Compounds	Precursors and Modification Methods	Reaction Conditions	Crystal Phase	S_{TiO2} (m^2/g)	Average Particle Size of TiO_2 (nm)	References
		(montmorillonite) and (ii) synthetic hectorite (laponite RD). The ratio of Ti/clay = 2.21 mmol Ti/g of clay (15% TiO_2), 5.36 mmol Ti/g of clay (30% TiO_2) and 15.3 mmol Ti/g of clay (55% TiO_2). $Ti(OC_3H_7)_4$, 1 M HCl with molar ratio of [Ti]/[H+] = 1:2 as precursors for sols of titanium hydrate. Peptization at 50°C. Mixing of titanium hydrate sol with a 5 wt% suspension of natural montmorillonite. Mixing of titanium hydrate sol with a 2 wt% suspension of synthetic clay (laponite RD). pH adjustment at 2–2.5. Precipitation of TiO_2. Heating at 50°C for 24h.					

		Calcination at 500°C for 1h.					
TiO_2-Montmorillonite	Methyl Orange	2g montmorillonite, 0.24g NaOH as precursors for alkaline suspension. Milling for 4h. 2g CTAB and 1 ml PEG-200 as precursors of gelatinous solution. Mixing of alkaline suspension with the gelatinous solution. TEOS was also added. Heating at 80°C for 5 days. Drying at 120°C. Calcination at 600°C for 10 h. $Ti(OC_4H_9)_4$, ethanol, HNO_3 as precursors for titanium hydrate sol. Addition of 3 ml of titanium hydrate sol to the host. Drying of the loaded catalyst at 160°C for 10 h.	Under UV illumination. Concentration of dye: 20 mg/L Catalyst concentration: 0.2 g/L	N/A	420	N/A	Li et al. 2009[6]

[1]"*Reprinted from* Chemistry of Materials, *14, Zhu H.Y., J.A. Orthman, J.-Y. Li, et al. Novel composites of TiO_2 (anatase) and silicate nanoparticles, 5037–5044, Copyright (2002), with permission from American Chemical Society.*
[2]"*Reprinted from* Applied Clay Science, *28, Zhu H.Y., J.-Y. Li, J.-C. Zhao, et al.., Photocatalysts prepared from layered clays and titanium hydrate for degradation of organic pollutants in water, 79–88, Copyright (2005), with permission from Elsevier.*
[3]"*Reprinted from* Applied Catalysis B: Environmental, *37, Li J., C. Chen, J. Zhao, et al. Photodegradation of dye pollutants on TiO_2 nanoparticles dispersed in silicate under UV–VIS irradiation, 331–338, Copyright (2002), with permission from Elsevier.*
[4]"*Reprinted from* Catalysis Communications, *7, Machado L. C. R., C. B. Torchia, R. M. Lago et al. Floating photocatalysts based on TiO_2 supported on high surface area exfoliated vermiculite for water contamination, 538-541, Copyright (2006), with permission from Elsevier.*
[5]"*Reprinted from* Applied Catalysis B: Environmental, *73, Belessi V., D. Lambropoulou, I. Konstantinou, et al. Structure and photocatalytic performance of TiO_2/clay nanocomposites for the degradation of dimethachlor, 292–299, Copyright (2007), with permission from Elsevier.*
[6]"*Reprinted from* Journal of Hazardous Materials, *165, Li F., Y. Jiang, M. Xia et al. A high-stability silica–clay composite: Synthesis, characterization and combination with TiO_2 as a novel photocatalyst for azo dye, 1219–1223, Copyright (2009), with permission from Elsevier.*

presentation of papers focused on the synthesis of TiO_2 and silicate nanocomposites with environmental applications follows.

Zhu et al. (2002) described a method to produce highly mesoporous composite nanostructures of anatase and silicate nanoparticles using Laponite. According to the preparation method (Table 13.2), appropriate amounts of Laponite, deionized water, and a non-ionic poly(ethylene oxide) (PEO) surfactant provided a homogeneous solid dispersion of silicate nanoparticles. In this mixture, an acidic sol of titanium hydrate was added, and the final product was calcined for several hours. This method allowed alteration of the size of the anatase crystals. This was achieved by tuning the acid to titanium molar ratio. Furthermore, it was found that the use of PEO surfactant with small molecular weight could enhance the porosity of the as-prepared composite (Table 13.2). The TiO_2-Laponite nanocomposites were tested for the degradation of Rhodamine 6G under UV light and found to be highly active photocatalysts compared to Degussa P25. The same author in a more recent study (Zhu et al. 2005) prepared mesoporous composite nanostructures using four different smectite clays (laponite, saponite, hectorite, and montmorillonite), which reacted with strong acidic titanium hydrate sol. The acid leaching least affected the montmorillonite, then the hectorite, and finally the saponite and Laponite. The composite solids were very active photocatalysts for the degradation of dichlorophenol and phenol from aqueous solutions.

Similarly, by reacting a titanium hydrate sol of strong acidity and a clay dispersion (Laponite) of high pH, Li et al. (2002a) prepared a composite sample. During the synthetic procedure, nanoparticles of anatase were dispersed between silicate layers without orientation. The resulting composite has a high surface area and exhibits good catalytic activity for the degradation of sulforhodamine B under UV–VIS irradiation (Li et al. 2002a) though Degussa P25 had a higher activity. At this point, however, it should be noted that the TiO_2 loading in the composite sample was 59%.

Matchado et al. (2006) used vermiculite (2:1 phyllosilicate) in order to prepare floating composite TiO_2/vermiculite photocatalysts with different loading in TiO_2 (20, 40 and 50 wt%) (Table 13.2). The two important steps in the preparation method were the impregnation of tetraisopropylorthotitanate on the vermiculate, followed by a hydrolysis step with vapors of HCl/H_2O. The as-prepared samples are porous and have the property of floating on the water surface. As a result, the best conditions concerning the illumination and oxygenation of the catalysts are ensured. The composite samples were tested for the degradation of Drimaren Red dye and exhibited increased photoreactivity. The most efficient photocatalyst was that with the higher surface area and loading in TiO_2 20 wt%. The commercial Degussa P25 was inactive under the present reaction conditions. The authors also proposed that these composites could find application in oil-spill accidents through their ability to preconcentrate the contaminant.

Recently, a series of materials with different TiO_2 loading were prepared by precipitation of TiO_2 to the surfaces of natural clay (montmorillonite) or synthetic hectorite (laponite RD). Leaching of cations from the structures of montmorillonite (Al^{3+}, Mg^{2+}) and hectorite (Mg^{2+}) was observed because of the preparation method whereby the clay suspensions react with acidic solutions (Table 13.2). In agreement with previous works, a correlation was found between the leaching of clay cations and the type of clay (Zhu et al. 2005, Kaviratna and Pinnavaia 1994, Mokaya and Jones 1995, Corma et al. 1987, Theocharis et al. 1988). Also, due to the acid leaching, the higher the percentage of TiO_2, the higher the pore volume and the specific surface area of the montmorillonite-TiO_2 samples. However, this was not true in the case of hectorite-TiO_2 samples, where the pore volume and the specific surface area increased as the content of TiO_2 decreased. The photoactivity of the TiO_2/clay composites was tested for the degradation of a chloroacetanilide herbicide (dimethachlor) in aqueous solution. Finally, it was found that all photocatalysts were efficient and had a higher removal efficiency per mass of TiO_2 compared to bare TiO_2.

Finally, Li et al. (2009) presented a preparation method (Table 13.2) for the synthesis of a TiO_2-silica clay composite whereby the loading of TiO_2 on silica-clay is approximately 5%wt. The high surface area composite photocatalyst has micropores, small and large mesopores. Their occurrence is due to usage of two different types of surfactants during the preparation method (Table 13.2). The composite catalyst was tested for the photodegradation of methyl orange. Its photoactivity was compared with that of bare TiO_2 particles, and it was found to show superior activity.

4 Conclusions

Adsorption and photocatalysis are closely related to each other, playing an important role in water remediation. Titanium oxide photocatalyst appears to be the most promising material among other semiconductors for these processes. Thus, many efforts have been made to modify TiO_2 in order to achieve improved adsorption behavior, enhanced photocatalytic activities, and activation of TiO_2 in the range of visible light. In this review, we focused on modified and nonmodified titania photocatalytic systems that have been applied in wastewater treatment technologies involving adsorption and photocatalysis. Modification of TiO_2 has been attained using many different approaches. Among these approaches, magnetically modified TiO_2 and TiO_2-clay derivatives are promising candidates for their practical use in wastewater treatment technologies. Furthermore, the magnetic photocatalysts facilitate their easy recovery from the treated water by an external magnetic field.

SYMBOLS AND TERMINOLOGY

AC activated carbon
AO7 the azo-dye Acid Orange 7
CTAB cetyltrimethyl ammonium bromide
C8-TiO_2 TiO_2 grafted with *n*-octyltriethoxysilane
Degussa P25 a commercially available TiO_2
Eg the bandgap energy of a semiconductor
E_{CB} the energy of the edge of the conduction band
$e_{CB}-$ a conduction band electron
EDTA ethylenediaminetetracetic acid
EPR electron paramagnetic resonance
EV exfoliated vermiculite
E_{VB} the energy of the edge of the valence band
Hombicat UV100 a commercially available TiO_2
h^+_{vb} an electron vacancy or positively charged hole
MAC magnetic-activated carbon
NHE normal hydrogen electrode
PE polyethylen glycol
PEI polyethyleneimine
POM polyoxometalate

PSS^-Na^+ Poly(sodium 4-styrene sulfonate) polyanion
pzc the point of zero charge of metal oxide colloid particles
RR195 the azo-dye Reactive Red 195
SDS the anionic surfactant sodium dodecyl sulphate
TALH titanium bis-ammonium lactato dihydroxide
TEOS tetraethoxysilane
TMAC TiO_2 magnetic activated carbon
UV ultraviolet radiation
UV-Vis ultraviolet-visible radiation
λ wavelength (nm)

Glossary

Activated carbon (AC) a form of carbon with high surface area and very wide range of pore sizes (micropores, mesopores, macropores). Activated carbons are used in various applications depending on their porous properties.

Advanced oxidation processes (AOPs) processes that involve the generation of highly reactive species that oxidize a wide range of organic substances.

Bandgap energy (Eg) the minimum energy of light required to produce conduction band electrons and valence band holes.

Block copolymers a kind of copolymer consisting of blocks of various polymerized monomers.

Carbon nanotubes (CNTs) allotropes of carbon with unique physicochemical properties. They are long nanotubes where the walls have graphite structure and are often capped at each end. There are mainly three types: single-walled carbon nanotubes (SWCNT), double-walled carbon nanotubes (DWCNT), and multiwalled carbon nanotubes (MWCNT) (Thostenson et al. 2001).

Chemical oxygen demand (COD) the amount of oxygen required to completely chemically oxidize any organic compound in the water.

Chitosan a nontoxic, biocompatible and biodegradable polymer derived from the partial deacetylation of chitin.

Clays layered silicates that are also described as phyllosilicates. Their structure is composed from Si-oxygen tetrahedra and Al- oxygen octahedra. They can be described by the formula $(Si_4)^{IV}(Al_4)^{VI}O_{10}(OH)_8$ where IV and VI correspond to the cation composition of the tetrahedral and octahedral layer, respectively (Molinard, 1994).

Electron paramagnetic resonance (EPR) a magnetic resonance technique for studying chemical species that have unpaired electrons.

Hectorite a smectite-type trioctahedral mineral with general formula $M_x^+(Si_8)^{IV}(Mg_{6-x}Li_x)^{VI}O_{20}(OH)_4$, where IV and VI correspond to the cation composition of the tetrahedral and octahedral layer, respectively (Molinard, 1994).

Laponite a range of synthetic silicates manufactured from pure chemicals. Laponite is closely related to the natural clay mineral hectorite $[Si_8Mg_{5.34}Li_{0.66}(Ca, Na)_{0.66}]$, a tri-octahedral sheet silicate.

Laponite RD a synthetic hectorite with an empirical formula of $[Na_{0.7}(Si_8Mg_{5.5}Li_{0.3})O_{20}(OH)_4]$.

Macropores material a material having pore diameters greater than 50 nm.

Mesoporous material a material having pore diameters between 2 nm and 50 nm.

Mesoporous silica (MCM-41) silica-walled materials that exhibit a regularly ordered two-dimensional hexagonal pore arrangement and narrow pore size distributions.

Microporous material a material having pore diameters of less than 2 nm.

Montmorillonite a smectite-type dioctahedral mineral with general formula $M_x^+(Si_8)^{IV}(Al_{4-x}Mg_x)^{VI}O_{10}(OH)_4$, where IV and VI correspond to the cation composition of the tetrahedral and octahedral layer, respectively (Molinard, 1994).

Pillared clays clays with high and permanent porosity. They are obtained by exchanging the compensating ions between the clay sheets with agents such as organic cations, metal oxide sols, and polyoxocations (Molinard, 1994)

Pollyoxometalate (POM) a large class of well-defined metal oxygen cluster anions, formed by condensation of mainly tungstates and/or molybdates with or without participation of various other elements (Baker and Glick, 1998, Kormali et al., 2007).

Saponite smectite with trioctahedral layer type and general formula $M_x^+(Si_{8-x}Al_x)^{IV}(Mg_6)^{VI}O_{20}(OH)_4$, where IV and VI correspond to the cation composition of the tetrahedral and octahedral layer, respectively (Molinard, 1994).

Semiconductors materials whose electrical conductivity is between that of a conductor and an insulator.

Smectite a 2:1 phyllosilicate (Molinard, 1994).

Superparamagnetism observed in small single-domain particles. The magnetic moments of the particles behave in a way similar to the paramagnetic system, but the total moment is several orders of magnitude larger than those of individual atoms.

Transmission electron microscopy (TEM): a microscopy technique used to evaluate the particle size and morphology of various materials.

Vermiculite 2:1 phyllosilicate (Molinard, 1994).

Zeolite: microporouse aluminosilicates of the general formula $M_{x/n}[(AlO_2)_x(SiO_2)_y]$ mH_2O (RaO C.N.R. and Gopalakrishnan, 1986).

Zeolite A an insoluble synthetic crystalline sodium aluminosilicate, $Na_{12}[(AlO_2)(SiO_2)]_{12}27H_2O$. It has been used as a builder in laundry detergents (Savitsky et al., 1981).

β-zeolite the only high-silica zeolite with a fully three-dimensional 12-ring pore system (RaO C.N.R. and Gopalakrishnan, 1986).

Zeolite X a faujasite zeolite having an Si/Al ratio between 1 and 1.5 (RaO C.N.R. and Gopalakrishnan, 1986).

Zeolite Y A faujasite zeolite having an Si/Al ratio above 1.5 (RaO C.N.R. and J. Gopalakrishnan 1986).

Zeolite ZSM5 A zeolite with the general formula $H_x[(AlO_2)_x(SiO_2)_{96-x}]$ $16H_2O$ (RaO C.N.R. and J. Gopalakrishnan 1986).

References

Aguedach, A., Brosillon, S., Morvan, J., et al., 2005. Photocatalytic degradation of azo-dyes reactive black 5 and reactive yellow 145 in water over a newly deposited titanium dioxide. Appl. Catal. B: Environ. 57, 55–62.

Anpo, M., Shima, T., Kodama, S., et al., 1987. Photocatalytic hydrogenation of CH_3CCH with H_2O on small-particle TiO_2: Size quantization effects and reaction intermediates. J. Phys. Chem. A 91, 4305–4310.

Ao, Y., Xu, J., Fu, D., et al., 2008a. A simple route for the preparation of anatase titania-coated magnetic porous carbons with enhanced photocatalytic activity. Carbon 46, 596–603.

Ao, Y., Xu, J., Fu, D., et al., 2008b. A novel magnetically separable composite photocatalyst: Titania-coated magnetic activated carbon. Sep. Purif. Technol. 61 (3), 436–441.

Ao, Y., Xu, J., Shen, X., et al., 2008c. Magnetically separable composite photocatalyst with enhanced photocatalytic activity. J. Hazard. Mater. 160, 295–300.

Ao, Y., Xu, J., Fu, D., et al., 2009. Photocatalytic degradation of X-3B by titania-coated magnetic activated carbon under UV and visible irradiation. J. Alloys Comp. 471, 33–38.

Arabatzis, I.M., Stergiopoulos, T., Bernard, M.C., Labou, D., Neophytides, S.G., Falaras, P., 2003. Silver-modified titanium dioxide thin films for efficient photodegradation of methyl orange. Appl. Catal. B: Envir 42, 187–201.

Arana, J., Gonzalez-Diaz, O., Miranda Saracho, M., et al., 2001. Photocatalytic degradation of formic acid using Fe/TiO_2 catalysts: the role of Fe^{3+}/Fe^{2+} ions in the degradation mechanism. Appl. Catal. B: Envir 32, 49–61.
Asahi, R., Morikawa, T., Ohwaki, T., et al., 2001. Visible-light photocatalysis in nitrogen-doped titanium oxides. Science 293, 269–271.
Asouhidou, D.D., Triantafyllidis, K.S., Lazaridis, N.K., et al., 2009. Sorption of reactive dyes from aqueous solutions by ordered hexagonal and disordered mesoporous carbons. Micropor. Mesopor. Mater. 117, 257–267.
Baker, L.C.W., Glick, D.C., 1998. Present general status of understanding of heteropoly electrolytes and a tracing of some major highlights in the history of their elucidation. Chem. Rev. 98, 3–50.
Bang, S., Manish, P., Lippincott, L., et al., 2005. Removal of arsenic from groundwater by granular titanium dioxide adsorbent. Chemosphere 60, 389–397.
Barakat, M.A., Chen, Y.T., Huang, C.P., et al., 2004. Removal of toxic cyanide and Cu(II) Ions from water by illuminated TiO_2 catalyst. Applied Catalysis B: Environ. 53, 13–20.
Barakat, M.A., 2005. Adsorption behavior of copper and cyanide ions at TiO_2–solution interface. J. Colloid Interface Sci. 291, 345–352.
Belessi, V., Lambropoulou, D., Konstantinou, I., et al., 2007. Structure and photocatalytic performance of TiO_2/clay nanocomposites for the degradation of dimethachlor. Appl. Catal. B: Environ. 73, 292–299.
Belessi, V., Lambropoulou, D., Konstantinou, I., et al., 2009a. Structure, photocatalytic performance of magnetically separable titania photocatalysts for the degradation of propachlor. Appl. Catal. B: Environ. 87, 181–189.
Belessi, V., Romanos, G., Boukos, N., et al., 2009b. Removal of Reactive Red 195 from aqueous solutions by adsorption on the surface of TiO_2 nanoparticles. J. Hazard. Mater. 170, 836–844.
Berger, T., Sterrer, M., Diwald, O., et al., 2005. Light-Induced charge separation in anatase TiO_2 particles. J. Phys. Chem. B 109, 6061–6068.
Beydoun, D., Amal, R., Low, G.K.-C., et al., 2000. Novel photocatalyst: Titania-coated nagnetite. Activity and photodissolution. J. Phys. Chem. B 104, 4387–4396.
Beydoun, D., Amal, R., Scott, J., et al., 2001. Studies on the mineralization and separation efficiencies of a magnetic photocatalyst. Chem. Eng. Technol. 24, 745–748.
Beydoun, D., Amal, R., 2002. Implications of heat treatment on the properties of a magnetic iron oxide–titanium dioxide photocatalyst. Mater. Sci. Eng. B 94, 71–81.
Beydoun, D., Amal, R., Low, G., et al., 2002. Occurrence and prevention of photodissolution at the phase junction of magnetite and titanium dioxide. J. Mol. Catal. A: Chem. 180, 193–200.
Beyers, E., Biermans, E., Ribbens, S., et al., 2009. Combined TiO_2/SiO_2 mesoporous photocatalysts with location and phase controllable TiO_2 nanoparticles. Appl. Catal. B: Environ. 88, 515–524.
Bhattacharyya, A., Kawi, S., Ray, M.B., 2004. Photocatalytic degradation of orange II by TiO_2 catalysts supported on adsorbents. Catalysis Today 98, 431–439.
Bickley, R.I., Gonzalez-Carreno, T., Lees, J.S., et al., 1991. A structural investigation of titanium dioxide photocatalysts. J. Solid State Chem. 92, 178–190.
Bissen, M., Vieillard-Baron, M.-M., Schindelin, A.J., et al., 2001. TiO_2-catalyzed photooxidation of arsenite to arsenate in aqueous samples. Chemosphere (44), 751–757.
Bosc, F., Ayral, A., Albouy, P.-A., et al., 2003. A simple route for low-temperature synthesis of mesoporous and nanocrystalline anatase thin films. Chem. Mater. 15, 2463–2468.
Büchler, M., Schmuki, P., Böhni, H., et al., 1998. Comparison of the semiconductive properties of sputter-deposited iron oxides with the passive film on iron. J. Electrochem. Soc. 145, 378–385.
Bourikas, K., Stylidi, M., Kondarides, D.I., et al., 2005. Adsorption of acid orange 7 on the surface of titanium dioxide. Langmuir 21, 9222–9230.
Burch, R., 1988. Pillared Clays. In: Catalysis Today, vols. 2 and 3. Elsevier, New York.
Burns, R.A., Crittenden, J.C., Hand, D.W., et al., 1999. Effect of inorganic ions in heterogeneous photocatalysis of TCE. J. Environ. Eng. 125, 77–85.

Chen, F., Zhao, J., 1999. Preparation and photocatalytic properties of a novel kind of loaded photocatalyst of TiO_2/SiO_2/γ-Fe_2O_3. Catal. Lett. 58, 245–247.
Chen, F., Xie, Y., Zhao, J., et al., 2001. Photocatalytic degradation of dyes on a magnetically separated photocatalyst under visible and UV irradiation. Chemosphere 44, 1159–1168.
Chen, W.-J., Tsai, P.-J., Chen, Y-C., 2008. Functional Fe_3O_4/TiO_2 core/shell magnetic nanoparticles as photokilling agents for pathogenic bacteria. Small 4, 485–491.
Cheng, H., Ma, J., Zhao, Z., et al., 1995a. Hydrothermal preparation of uniform nanosize rutile and anatase particles. Chem. Mater. 7, 663–671.
Cheng, S., Tsai, S.-J., Lee, Y-F., 1995b. Photocatalytic decomposition of phenol over titanium oxide of various structures. Catal. Today 26, 87–96.
Childs, L.P., Ollis, D.F., 1980. Is photocatalysis catalytic? J. Catal. 66, 383–390.
Chung, Y.S., Park, S.B., Kang, D.-W., 2004. Magnetically separable titania-coated nickel ferrite photocatalyst. Mater. Chem. Phys. 86, 375–381.
Cicek, F., Ozer, D., Ozer, A., et al., 2007. Low cost removal of reactive dyes using wheat bran. J. Hazard. Mater. 146, 408–416.
Colomer, M.T., Jurado, J.R., 2000. Structural, microstructural, and electrical yransport properties of TiO_2-RuO_2 ceramic materials obtained by polymeric dol−gel route. Chem. Mater. 12, 923–930.
Cooper, P., 1993. Removing colour from dyehouse waste waters- a critical review of technology available. J. Soc. Dyers Colour. 109, 97–100.
Corma, A., Mifsud, A., Sanz, E., 1987. Influence of the chemical composition and textural characteristics of palygorskite on the acid leaching of octahedral cations. Clay Minerals 22, 225–232.
Crini, G., 2006. Non-conventional low-cost adsorbents for dye removal: A review. Biores. Technol 97 (9), 1061–1085.
Dabrowski, A., 2001. Adsorption—from theory to practice. Adv. Colloid. Interface. Sci. 93, 135–224.
Deng, H., Li, X.L., Peng, Q., et al., 2005. Monodisperse nagnetic single-crystal ferrite microspheres. Angew. Chem. Int. Ed. 44, 2782–2785.
Di Paola, A., Garicia-Lopez, E., Ikeda, S., et al., 2002. Photocatalytic degradation of organic compounds in aqueous systems by transition metal doped polycrystalline TiO_2. Catal. Today 75, 87–93.
Ding, Z., Zhu, H.Y., Lu, G.Q., et al., 1999. Photocatalytic properties of titania pillared clays by different drying methods. J. Colloid Inter. Sci. 209, 193–199.
Dobosz, A., Sobczynski, A., 2003. The influence of silver additives on titania photoactivity in the photooxidation of phenol. Water Res. 37, 1489–1496.
Dutta, P.K., Ray, A.K., Sharma, V.K., et al., 2004. Adsorption of arsenate and arsenite on titanium dioxide suspensions. J. Colloid Inter. Sci. 278, 270–275.
Dutta, P.K., Pehkonen, S.O., Sharma, V.K., et al., 2005. Photocatalytic oxidation of arsenic (III): Evidence of hydroxyl radicals. Environ. Sci. Technol. 39, 1827–1834.
Dvoranova, D., Brezova, V., Mazur, M., et al., 2002. Investigations of metal-doped titanium dioxide photocatalysts. Appl. Catal. B: Environ. 37, 91–105.
Esumi, K., Uda, S., Goino, M., et al., 1997. Adsorption of cationic surfactants on titanium dioxide with a hydrophobic group anchor and their adsolubilization behaviors. Langmuir 13, 2803–2807.
Esumi, K., Hayashi, H., Koide, Y., et al., 1998. Adsorption of metal ion and aromatic compounds by anionic surfactant-coated particles of titanium dioxide. Coll. Surf. A 144, 201–206.
Figueiredo, S.A., Boaventura, R.A., Loureiro, J.M., 1998. Natural chitin containing materials for colour removal from textile wastewaters. In: Misaelidis, P., Macasek, F., Pinnavaia, T.J., Colella, C. (Eds.), natural microporous materials in Environmental technology, Proceedings of the NATO Science Series. Series E: Applied Sciences, Vol. 362. Kluwer Academic Publishers, The Netherlands, pp. 499–508.
Fu, W., Yang, H., Li, M., et al., 2006. Preparation and photocatalytic characteristics of core-shell structure TiO_2/$BaFe_{12}O_{19}$ nanoparticles. Mater. Lett. 60, 2723–2727.

Fujishima, A., Rao, T.N., Tryk, D.A., 2000. Titanium dioxide photocatalysis. J. Photochem. Photobiol. C Photochem. Rev. 1, 1–21.

Fujishima, A., Zhang, X., Tryk, D.A., 2008. TiO_2 photocatalysis and related surface phenomena. Surface. Sci. Rep. 63, 515–582.

Gao, Y., Chen, B., Li, H., et al., 2003. Preparation and characterization of a magnetically separated photocatalyst and its catalytic properties. Mater. Chem. Phys. 80, 348–355.

Gao, Y., Wahi, R., Kan, A.T., et al., 2004. Adsorption of cadmium on anatase nanoparticles – Effect of crystal size and pH. Langmuir 20, 9585–9593.

Gao, B., Chen, G.Z., Puma, G.L., 2009. Carbon nanotubes/titanium dioxide (CNTs/TiO_2) nanocomposites prepared by conventional and novel surfactant wrapping sol-gel methods exhibiting enhanced photocatalytic activity. Appl. Catal. B: Environ. 89, 503–509.

Goswami, D.Y., 1997. A review of engineering developments of aqueous phase solar photocatalytic detoxification and disinfection processes. J. Sol. Energ. Eng. 119, 101–107.

Han, F., Kambala, V.S.R., Srinivasan, M., et al., 2009. Tailored titanium dioxide photocatalysts for the degradation of organic dyes in wastewater treatment: A review. Appl. Catal. A 359, 25–40.

Hagfeldt, A., Gratzel, M., 1995. Light-induced redox reactions in nanocrystalline dystems. Chem. Rev. 95, 49–68.

Hay, J.N., Raval, H.M., 2001. Synthesis of organic-inorganic hybrids via the non-hydrolytic sol-gel process. Chem. Mater. 13, 3396–3403.

He, Q., Zhang, Z., Xiong, J., et al., 2008. A novel biomaterial—Fe_3O_4:TiO_2 core-shell nano particle with magnetic performance and high visible light photocatalytic activity. Opt. Mater. 31, 380–384.

Herrmann, J.M., 1999. Heterogeneous photocatalysis: Fundamentals and applications to the removal of various types of aqueous pollutants. Catal. Today 53, 115–129.

Hoffmann, A.J., Mills, G., Yee, H., et al., 1992a. Q-sized CdS: Synthesis, characterization, and efficiency of photoinitiation of polymerization of several vinylic monomers. J. Phys. Chem. 96, 5546–5552.

Hoffmann, A.J., Yee, H., Mills, G., et al., 1992b. Photoinitiated polymerization of methyl methacrylate using Q-sized ZnO colloids. J. Phys. Chem. 96, 5540–5546.

Hoffmann, M.R., Martin, S.T., Choi, W., et al., 1995. Environmental applications of semiconductor photocatalysis. Chem. Rev. 95, 69–96.

Howe, R.F., Gratzel, M., 1985. EPR observation of trapped electrons in colloidal TiO_2. J. Phys. Chem. 89, 4495–4499.

Hufschmidt, D., Bahnemann, D.W., Testa, J.J., et al., 2002. Enhancement of the photocatalytic activity of various TiO_2 materials by platinisation. J. Photochem. Photobiol. A Chem. 148, 223–231.

Hurum, D.C., Agrios, A.G., Gray, K.A., et al., 2003. Explaining the enhanced photocatalytic activity of Degussa P25 mixed-phase TiO_2 using EPR. J. Phys. Chem. B 107, 4545–4549.

Hurum, D.C., Agrios, A.G., Crist, S.E., et al., 2006. Probing reaction mechanisms in mixed phase TiO_2 by EPR. J. Electron. Spectros. Relat. Phenom. 150, 155–163.

Hwang, Y.K., Lee, K.-C., Kwon, Y-U., 2001. Nanoparticle routes to mesoporous titania thin films. Chem. Commun. 18, 1738–1739.

Ilisz, I., Dombi, A., Mogyorosi, K., et al., 2002. Removal of 2-chlorophenol from water by adsorption combined with TiO_2 photocatalysis. Appl. Catal. B: Environ. 39, 247–256.

Inumaru, K., Murashima, M., Kasahara, T., et al., 2004. Enhanced photocatalytic decomposition of 4-nonylphenol by surface-organografted TiO_2: A combination of molecular selective adsorption and photocatalysis. Appl. Catal. B: Environ. 52, 275–280.

Irie, H., Watanabe, Y., Hashimoto, K., 2003. Nitrogen-concentration dependence on photocatalytic activity of $TiO_{2-x}N_x$ powders. J. Phys. Chem. B 107, 5483–5486.

Kamat, P.V., 1993. Photochemistry on nonreactive and reactive (semiconductor) surfaces. Chem. Rev. 93 (1), 267–300.

Kang, M.G., Han, H.-E., Kim, K-J., 1999. Enhanced photodecomposition of 4-chlorophenol in aqueous solution by deposition of CdS on TiO_2. J. Photochem. Photobiol. A: Chem. 125, 119–125.

Kaviratna, H., Pinnavaia, T.J., 1994. Acid hydrolysis of octahedral Mg (super 2+) sites in 2:1 layered silicates: An assessment of edge attack and gallery access mechanisms. Clays and Clay Minerals 42 (6), 717–723.

Keshmiri, M., Mohseni, M., Troczynski, T., 2004. Development of novel TiO_2 sol–gel-derived composite and its photocatalytic activities for trichloroethylene oxidation. Appl. Catal. B: Environ. 53, 209–219.

Kim, M.-S., Hong, K.-M., Chung, J.G., 2003. Removal of Cu (II) from aqueous solutions by adsorption process with anatase-type titanium dioxide. Water Res. 37, 3524–3529.

Kobayakawa, K., Murakami, Y., Sato, Y., 2005. Visible-light active N-doped TiO_2 prepared by heating of titanium hydroxide and urea. J. Photochem. Photobiol. A 170, 177–179.

Konstantinou, I.K., Albanis, T.A., 2003. Photocatalytic transformation of pesticides in aqueous titanium dioxide suspensions using artificial and solar light: intermediates and degradation pathways. Appl. Catal. B: Environ. 42, 319–335.

Konstantinou, I.K., Albanis, T.A., 2004. TiO_2-assisted photocatalytic degradation of azo dyes in aqueous solution: kinetic and mechanistic investigations. Appl. Catal. B: Environ. 49, 1–14.

Kormali, P., Troupis, A., Triantis, T., et al., 2007. Photocatalysis by polyoxometallates and TiO_2: A comparative study. Catal. Today 124, 149–155.

Kumar, K.-N.P., Keizer, K., Burggraaf, A.J., et al., 1993. Synthesis and textural properties of unsupported and supported rutile (TiO_2) membranes. J. Mater. Chem. 3, 923–929.

Kurinobu, S., Tsurusaki, K., Natui, Y., et al., 2007. Decomposition of pollutants in wastewater using magnetic photocatalyst particles. J. Magn. Magnetic. Mater. 310, e1025–e1027.

Lee, S.-W., Sigmunt, W.M., 2003. Formation of anatase TiO_2 nanoparticles on carbon nanotubes. Chem. Commun. 6, 780–781.

Lee, S-W., Drwiega, J., Wu, C-Y., et al., 2004. Anatase TiO_2 nanoparticle coating on barium ferrite using titanium bis-ammonium lactato dihydroxide, and its use as a nagnetic photocatalyst. Chem. Mater. 16, 1160–1164.

Lee, S., Jeon, C., Park, Y., 2004. Fabrication of TiO_2 tubules by template synthesis and hydrolysis with water vapour. Chem. Mater. 16, 4292–4295.

Lefort, R., 1996. Down to the last drop. UNESCO Sources 84, 7–28.

Li, J., Chen, C., Zhao, J., et al., 2002a. Photodegradation of dye pollutants on TiO_2 nanoparticles dispersed in silicate under UV–VIS irradiation. Appl. Catal. B: Environ. 37, 331–338.

Li, Y., Fan, Y., Chen, Y., 2002b. A novel method for preparation of nanocrystalline rutile TiO_2 powders by liquid hydrolysis of $TiCl_4$. J. Mater. Chem. 12, 1387–1390.

Li, Y., Lee, N.-H., Hwang, D-S., et al., 2004. Synthesis and characterization of nano titania powder with high photoactivity for gas-phase photo-oxidation of benzene from $TiOCl_2$ aqueous solution at low temperatures. Langmuir 20, 10838–10844.

Li, F., Jiang, Y., Xia, M., et al., 2009. A high-stability silica–clay composite: Synthesis, characterization and combination with TiO_2 as a novel photocatalyst for Azo dye. J. Hazard. Mater. 165, 1219–1223.

Linsebigler, A.L., Lu, G., T. Jr., J., Yates, 1995. Photocatalysis on TiO_2 surfaces: Principles, mechanisms, and selected results. Chem. Rev. 95, 735–758.

Liu, Y., Li, J., Wang, M., et al., 2005. Preparation and properties of nanostructure anatase TiO_2 nonoliths using 1-butyl-3-methylimidazolium tetrafluoroborate room-temperature ionic liquids as template solvents. Cryst. Growth Des. 5, 1643–1649.

Liu, G., Wang, X., Chen, Z., et al., 2009. The role of crystal phase in determining photocatalytic activity of nitrogen doped TiO_2. J. Colloid Inter. Sci. 329, 331–338.

Lo, S.-C., Lina, C.-F., Wu, C.-H., et al., 2004. Capability of coupled $CdSe/TiO_2$ for photocatalytic degradation of 4-chlorophenol. J. Hazard. Mater. B 114, 183–190.

Luo, M., Bowden, D., Brimblecombe, P., 2009. Preparation of black sand-based magnetic photocatalysts for photocatalytic oxidation of aqueous phenol. Appl. Catal. B: Envir. 87, 1–8.

Ma, R., Fukuda, K., Sasaki, T., et al., 2005. Structural features of titanate nanotubes/nanobelts revealed by raman, X-ray absorption fine structure and electron diffraction characterizations. J. Phys. Chem. B 109, 6210–6214.

Machado, L.C.R., Torchia, C.B., Lago, R.M., 2006. Floating photocatalysts based on TiO_2 supported on high surface area exfoliated vermiculite for water contamination. Catal. Commun. 7, 538–541.

Malato, S., Blanco, J., Caeres, J., et al., 2002. Photocatalytic treatment of water-soluble pesticides by photo-fenton and TiO_2 using solar energy. Catal. Today 76, 209–220.

Manna, L., Scher, E.C., Li, L.-S., et al., 2002. Epitaxial growth and photochemical annealing of graded CdS/ZnS shells on colloidal CdSe nanorods. J. Am. Chem. Soc. 124, 7136–7145.

Matos, J., Laine, J., Herrmann, J.M., 1999. Association of activated carbons of different origins with titania in the photocatalytic purification of water. Carbon 37, 1870–1872.

Micic, O.I., Zhang, Y., Cromack, K.R., et al., 1993a. Photoinduced hole transfer from TiO_2 to methanol molecules in aqueous solution studied by electron paramagnetic resonance. J. Phys. Chem. 97, 13284–13288.

Micic, O.I., Zhang, Y., Cromack, K.R., et al., 1993b. Trapped holes on TiO_2 colloids studied by electron paramagnetic resonance. J. Phys. Chem. 97, 7277–7283.

Mills, A., Le Hunte, S., 1997. An overview of semiconductor photocatalysis. J. Photochem. Photobiol. A: Chem. 108, 1–35.

Molinard, A., 1994. Physicochemical and gas adsorption properties of ion exchanged alumina pillared clays. PhD Thesis, Laboratory of Inorganic Chemistry, University of Antwerpen, 1994.

Mokaya, R., Jones, W., 1995. Pillared clays and pillared acid-activated clays: A comparative-study of physical, acidic, and catalytic properties. J. Catal. 153 (1), 76–85.

Nakaoka, Y., Nosaka, Y., 1997. ESR investigation into the effects of heat treatment and crystal structure on radicals produced over irradiated TiO_2 powder. J. Photochem. Photobiol. A: Chem. 110, 299–305.

Niederberger, M., Bartl, M.H., Stucky, G.D., 2002. Benzyl alcohol and titanium tetrachloride: A versatile Reaction system for the nonaqueous and low-temperature preparation of crystalline and luminescent titania nanoparticles. Chem. Mater. 14, 4364–4370.

Nosaka, Y., Ohta, N., Miyama, H., et al., 1990. Photochemical kinetics of ultrasmall semiconductor particles in solution: effect of size on the quantum yield of electron transfer. J. Phys. Chem. 94, 3752–3755.

Ohno, T., Sarukawa, K., Matsumura, M., 2001. Photocatalytic activities of pure rutile particles isolated from TiO_2 powder by dissolving the anatase component in HF solution. J. Phys. Chem. B 105, 2417–2420.

Ohno, T., Tokieda, K., Higashida, S., et al., 2003. Synergism between rutile and anatase TiO_2 particles in photocatalytic oxidation of naphthalene. Appl. Catal. A: General 244, 383–391.

Ohno, T., Akiyoshi, M., Umebayashi, T., et al., 2004. Preparation of S-doped TiO_2 photocatalysts and their photocatalytic activities under visible light. Appl. Catal. A 265, 115–121.

Ollis, D.F., Pelizzetti, E., Serpone, N., 1991. Photocatalyzed destruction of water contaminants. Environ. Sci. Technol. 25 (9), 1522–1529.

Olsson, M., Jakobsson, A.-M., Albinsson, Y., 2003. Sorption of Pu(VI) onto TiO_2. J. Colloid Inter. Sci. 266, 269–275.

Ooka, C., Akita, S., Ohashi, Y., et al., 1999. Crystallization of hydrothermally treated TiO_2 pillars in pillared montmorillonite for improvement of the photocatalytic activity. J. Mater. Chem. 9, 2943–2952.

Ooka, C., Yoshida, H., Suzuki, K., et al., 2004a. Highly hydrophobic TiO_2 pillared clay for photocatalytic degradation of organic compounds in water. Micropor. Mesopor. Mater. 67, 143–150.

Ooka, C., Yoshida, H., Takeuchi, S., et al., 2004b. Hydrogen peroxide improving crystallinity of TiO_2 nanoparticle in layer compound. Catal. Commun. 5, 49–54.

Papoutsi, D., Lianos, P., Yianoulis, P., et al., 1994. Sol-gel derived TiO_2 microemulsion gels and coatings. Langmuir 10, 1684–1689.

Pelizzetti, E., Minero, C., 1999. Role of oxidative and reductive pathways in the photocatalytic degradation of organic compounds. Colloid Surf. A Phys. Eng. Aspects 151, 321–327.

Pena, M., Meng, X., Korfiatis, G.P., et al., 2006. Adsorption mechanism of arsenic on nanocrystalline titanium dioxide. Environ. Sci. Technol. 40, 1257–1262.

Pettibone, J.M., Cwiertny, D.M., Scherer, M., et al., 2008. Adsorption of organic acids on TiO_2 nanoparticles: Effects of pH, nanoparticle size and nanoparticle aggregation. Langmuir 24, 6659–6667.

Planeix, J.M., Coustel, N., Coq, B., et al., 1994. Application of carbon nanotubes as supports in heterogeneous catalysis. J. Am. Chem. Soc. 116, 7935–7936.

Qiu, W., Zheng, Y., Haralampides, K.A., 2007. Study on a novel POM-based magnetic photocatalyst: Photocatalytic degradation and magnetic separation. Chem. Eng. J. 125, 165–176.

Qu, J., 2008. Research progress of novel adsorption processes in water purification: A review. J. Environ. Sci. 20, 1–13.

Rana, S., Srivastava, R.S., Sorensson, M.M., et al., 2005. Synthesis and characterization of nanoparticles with magnetic core and photocatalytic shell: Anatase TiO_2-$NiFe_2O_4$. Mater. Sci. Eng. B 119, 144–151.

Rao, C.N.R., Gopalakrishnan, J., 1986. New directions in solid state chemistry. In: Cahn, R.W., Davis, E.A., Ward, I.M. (Eds.), Cambridge Solid State Science Series. Cambridge University Press, pp. 34–38.

Rawat, J., Rana, S., Srivastava, R.S., et al., 2007. Antimicrobial activity of composite nanoparticles consisting of titania photocatalytic shell and nickel ferrite magnetic core. Mat. Sci. Eng. C 27, 540–545.

Rhodes, C.N., Brown, D.R., 1993. Surface properties and porosities of silica and acid-treated montmorillonite catalyst supports: influence on activities of supported $ZnCl_2$ alkylation catalysts. J. Chem. Soc., Faraday Trans. 89 (9), 1387–1391.

Riegel, G., Bolton, J.R., 1995. Photocatalytic efficiency Variability in TiO_2 particles. J. Phys. Chem. 99 (12), 4215–4224.

Robinson, T., McMullan, G., Marchant, R., et al., 2001. Remediation of dyes in textile effluent: a critical review on current treatment technologies with a proposed alternative. Bioresour. Technol. 77, 247–255.

Sakthivel, S., Geissen, S.-U., Bahnemann, D.W., et al., 2002. Enhancement of photocatalytic activity by semiconductor heterojunctions: α-Fe_2O_3, WO_3 and CdS deposited on ZnO. J. Photochem.Photobiol. A Chem. 148, 283–293.

Saripalli, K.P., McGrail, B.P., Girvin, D.C., 2002. Adsorption of molybdenum on to anatase from dilute aqueous solutions. Appl. Geochem. 17, 649–656.

Savitsky, A.C., Wiers, B.H., Wendt, R.H., 1981. Adsorption of organic compounds from dilute aqueous solutions onto the external surface of Type A zeolite. Environ. Sci. Techol. 15 (10), 1191–1196.

Shchukin, D.G., Kulak, A.I., Sviridov, D.V., 2002. Magnetic photocatalysts of the core-shell type. Photochem. Photobiol. Sci. 1, 742–744.

Sclafani, A., Herrmann, J.-M., 1998. Influence of metallic silver and of platinum-silver bimetallic deposits on the photocatalytic activity of titania (anatase and rutile) in organic and aqueous media. J. Photochem. Photobiol. A Chem. 113, 181–188.

Slokar, Y.M., Majcen Le Marechal, A., 1998. Methods of decoloration of textile wastewaters. Dyes Pigments 37, 335–356.

Song, J.-S., Kim, H.-J., Koo, B.-K., et al., 2006. Photonic decomposition of ultrafine, rutile phased TiO_2 powder in aqueous 4-chlorophenol solutions. Smart Mater. Struct 15, S65–S73.

Song, X., Gao, L., 2007. Fabrication of bifunctional titania/silica-coated magnetic spheres and their photocatalytic activities. J. Am. Ceram. Soc. 90 (12), 4015–4019.

Stathatos, E., Lianos, P., Del Monte, F., et al., 1997. Formation of TiO_2 nanoparticles in reverse micelles and their deposition as thin films on glass substrates. Langmuir 13 (16), 4295–4300.

Subramanian, V., Wolf, E., Kamat, P., 2001. Semiconductor-metal composite nanostructures. To what extent do metal nanoparticles improve the photocatalytic activity of TiO_2 films? J. Phys. Chem. B 105, 11439–11446.
Sugimoto, T., Zhou, X., Muramatsou, A., 2003a. Synthesis of uniform anatase TiO_2 nanoparticles by gel–sol method: 3. Formation process and size control. J. Colloid Inter. Sci. 259, 43–52.
Sugimoto, T., Zhou, X., Muramatsou, A., 2003b. Synthesis of uniform anatase TiO_2 nanoparticles by gel–sol method: 4. Shape control. J. Colloid Inter. Sci. 259, 53–61.
Sun, B., Smirniotis, P.G., 2003. Interaction of anatase and rutile TiO_2 particles in aqueous photooxidation. Cat. Today 88, 49–59.
Sun, B., Vorontsov, A.V., Smirniotis, P.G., 2003. Role of platinum deposited on TiO_2 in phenol photocatalytic oxidation. Langmuir 19, 3151–3156.
Sun, Z., Chen, Y., Ke, Q., et al., 2002. Photocatalytic degradation of a cationic azo dye by TiO_2/bentonite nanocomposite. J. Photochem. Photobiol. A: Chem. 149, 169–174.
Sung, Y.-M., Lee, J.-K., Chae, W-S., 2006. Controlled crystallization of manoporous and core/shell structure titania photocatalyst particles. Cryst. Growth Des. 6, 805–808.
Tamaki, Y., Furube, A., Murai, M., et al., 2007. Dynamics of efficient electron-hole separation in TiO_2 nanoparticles revealed by femtosecond transient absorption spectroscopy under the weak-excitation condition. Phys. Chem. Chem. Phys. 9, 1453–1460.
Tan, X., Fang, M., Li, J., et al., 2009. Adsorption of Eu(III) onto TiO_2: Effect of pH, concentration, ionic strength and soil fulvic acid. J. Hazard. Mater. 168, 458–465.
Tao, Y., Ye, L., Pan, J., et al., 2009. Removal of Pb(II) from aqueous solution on chitosan/TiO_2 hybrid film. J. Hazard. Mater. 161, 718–722.
Theocharis, C.R., Jacob, K.J., Gray, A.C., 1988. Enhancement of Lewis acidity in layer aluminosilicates. Treatment with acetic acid. J. Chem. Soc., Faraday Trans. 84 (5), 1509–1516.
Thostenson, E.T., Ren, Z., Choua, T.-W., 2001. Advances in the science and technology of carbon nanotubes and their composites: a review. Compos. Sci. Technol 61, 1899–1912.
Tryba, B., Morawski, A.W., Inagaki, M., 2003. Application of TiO_2-mounted activated carbon to the removal of phenol from water. Appl. Catal. B: Env. 41, 427–433.
Turchi, C.S., Ollis, D.F., 1989. Mixed reactant photocatalysis: Intermediates and mutual rate inhibition. J. Catal. 119, 483–496.
Uyguner, C.S., Bekbolet, M., 2004. Evaluation of humic acid, chromium (VI) and TiO_2 ternary system in relation to adsorptive interactions. Appl. Catal. B: Environ. 49, 267–275.
Vandevivere, P.C., Bianchi, R., Verstraete, W., 1998. Treatment and reuse of wastewater from the textile wet-processing industry: Review of emerging technologies. J. Chem. Technol. Biotechnol. 72, 289–302.
Vidal, A., Dinya, Z., Mogyorodi Jr., F., et al., 1999. Photocatalytic degradation of thiocarbamate herbicide active ingredients in water. Appl. Catal. B: Environ. 21, 259–267.
Vinodgopal, K., Wynkoop, D.E., Kamat, P.V., 1996. Environmental photochemistry on semiconductor surfaces: Photosensitized degradation of a textile azo dye, Acid Orange 7, on TiO_2 particles using visible light. Environ. Sci. Technol. 30, 1660–1666.
Vioux, A., 1997. Non hydrolytic sol-gel routes to oxides. Chem. Mater. 9, 2292–2299.
Wang, C.-C., Zhang, Z., Ying, J.Y., 1997. Photocatalytic decomposition of halogenated organics over nanocrystalline titania. Nanostruct. Mater. 9, 583–586.
Wang, C-Y., Groenzin, H., Shultz, M.J., 2005. Comparative study of acetic acid, methanol, and water adsorbed on anatase TiO_2 probed by sum frequency generation spectroscopy. J. Am. Chem. Soc. 127, 9736–9744.
Wang, W.-Y., Ku, Y., 2007. Effect of solution pH on the adsorption and photocatalytic reaction behaviors of dyes using TiO_2 and Nafion-coated TiO_2. Colloid Surf. A: Physicochem. Eng. Aspect. 302, 261–268.

Watson, S., Beydoun, D., Amal, R., 2002. Synthesis of a novel magnetic photocatalyst by direct deposition of nanosized TiO_2 crystals onto a magnetic core. J. Photochem. Photobiol. A: Chem. 148, 303–313.

Watson, S., Scott, J., Beydoun, D., et al., 2005. Studies on the preparation of magnetic photocatalysts. J. Nanoparticle Res. 7, 691–705.

Wu, M., Long, J., Huang, A., et al., 1999. Microemulsion-mediated hydrothermal synthesis and characterization of nanosize rutile and anatase particles. Langmuir 15, 8822–8825.

Wu, T., Liu, G., Zhao, J., et al., 1998. Photoassisted degradation of dye pollutants. V. Self-photosensitized oxidative transformation of rhodamine B under visible light irradiation in aqueous TiO_2 dispersions. J. Phys. Chem. B 102, 5845–5851.

Xie, X., Gao, L., 2009. Effect of crystal structure on adsorption behaviors of nanosized TiO_2 for heavy-metal cations. Current Applied Physics 9, S185–S188.

Xu, Y., Langford, C.H., 1995. Enhanced photoactivity of a titanium (IV) oxide supported on ZSM5 and zeolite A at low coverage. J. Phys. Chem. 99, 11501–11507.

Xu, Y., Langford, C.H., 1997. Photoactivity of titanium dioxide supported on MCM41, zeolite X, and zeolite Y. J. Phys. Chem. B 101, 3115–3121.

Xu, T., Kamat, P.V., O'Shea, K.E., 2005. Mechanistic evaluation of arsenite oxidation in TiO_2 assisted photocatalysis. J. Phys. Chem. A 109, 9070–9075.

Xu, S., Shangguan, W., Yuan, J., et al., 2007. Preparations and photocatalytic properties of magnetically separable nitrogen-doped TiO_2 supported on nickel ferrite. Appl. Catal. B: Environ. 71, 177–184.

Yang, P., Zhao, D., Margolese, D.I., et al., 1998. Generalized syntheses of large-pore mesoporous metal oxides with semicrystalline frameworks. Nature 396, 152–155.

Yang, H., Lin, W-Y., Rajeshwar, K., 1999. Homogeneous and heterogeneous photocatalytic reactions involving As(III) and As(V) species in aqueous media. J. Photochem. Phtobiol. A 123, 137–143.

Yang, J.-K., Lee, S.-M., 2006. Removal of Cr(VI) and humic acid by using TiO_2 photocatalysis. Chemosphere 63, 1677–1684.

Yoneyama, H., Torimoto, T., 2000. Titanium dioxide/adsorbent hybrid photocatalysts for photodestruction of organic substances of dilute concentrations. Catal. Today 58, 133–140.

Yu, J.C., Zhang, L., Zheng, Z., et al., 2003. Synthesis and characterization of phosphated mesoporous titanium dioxide with high photocatalytic activity. Chem. Mater. 15, 2280–2286.

Yu, Y., Yu, J.C., Chan, C.-Y., et al., 2005. Enhancement of adsorption and photocatalytic activity of TiO_2 by using carbon nanotubes for the treatment of azo dye. Appl. Catal. B: Envir. 61, 1–11.

Yuan, Q., Jain, P.M., Valsaraj, K.T., 2000. Reusable adsorbents for dilute solution separation. 4: Adsorption of 1,2 dichlorobenzene and phenanthrene on a surfactant-modified semiconductor (titania) surface. Sep. Purif. Technol. 21, 9–16.

Yuan, Q., Ravikrishna, R., Valsaraj, K.T., 2001. Reusable adsorbents for dilute solution separation. 5. Photodegradation of organic compounds on surfactant-modified titania. Sep. Purif. Technol. 24, 309–318.

Yun, H.-S., Miyazawa, K., Zhou, H.S., et al., 2001. Synthesis of mesoporous thin TiO_2 films with hexagonal pore structures using triblock copolymer templates. Adv. Mater. 13, 1377–1380.

Zhang, H., Banfield, J.F., 1998. Thermodynamic analysis of phase stability of nanocrystalline titania. J. Mater. Chem. 8, 2073–2076.

Zhang, H., Banfield, J.F., 2000. Understanding polymorphic phase transformation behaviour during growth of nanocrystalline aggregates: Insights from TiO_2. J. Phys. Chem. B 104, 3481–3487.

Zhang, H.Z., Penn, R.L., Hamers, R.J., et al., 1999. Enhanced adsorption of molecules on durfaces of nanocrystalline particles. J. Phys. Chem. B 103, 4656–4662.

Zhang, Q., Gao, L., 2003. Preparation of oxide nanocrystals with tunable morphologies by the moderate hydrothermal method: Insights from rutile TiO_2. Langmuir 19, 967–971.

Zhang, H., Hou, R., Lu, Z.-L., et al., 2009. A novel magnetic nanocomposite involving anatase titania coating on silica-modified cobalt ferrite via lower temperature hydrolysis of a water-soluble titania precursor. Mater. Res. Bulletin 44 (10), 2000–2008.

Zhao, J., Wu, T., Wu, K., et al., 1998. Photoassisted degradation of dye pollutants. 3. Degradation of the cationic dye rhodamine B in aqueous anionic surfactant/TiO_2 dispersions under visible light irradiation: Evidence for the need of substrate adsorption on TiO_2 particles. Environ. Sci. Technol. 32, 2394–2400.

Zhu, H.Y., Orthman, J.A., Li, J.-Y., et al., 2002. Novel composites of TiO_2 (anatase) and silicate nanoparticles. Chem. Mater. 14, 5037–5044.

Zhu, H.Y., Li, J.-Y., Zhao, J.-C., et al., 2005. Photocatalysts prepared from layered clays and titanium hydrate for degradation of organic pollutants in water. Appl. Clay Sci. 28, 79–88.

CHAPTER

Groundwater Purification and Soil Remediation Using Carbon Colloids

14

Rashid A. Khaydarov, Renat R. Khaydarov, O. Gapurova

Institute of Nuclear Physics, Tashkent, Uzbekistan

CHAPTER OUTLINE

1 Introduction

Using metals and chemicals in process industries has generated large quantities of effluent containing high levels of toxic heavy metals. Meanwhile, mining and mineral processing operations also generate toxic liquid wastes. The presence of different organic and heavy metal contaminants in groundwater has a large environmental, public health, and economic impact. Most of the traditional technologies such as solvent extraction, activated carbon adsorption, biological degradation, and common chemical oxidation, while effective, very often are costly and time-consuming.

Rumeau et al. (Rumeau et al., 1992) were the first to introduce the method of separation by coupling ultrafiltration and complexation of metallic species with industrial water-soluble polymers. This approach implies that polymeric molecules operate as a sorbent and a carrier of metal ions. In this method metal ions are primarily bound to water-soluble polymers. The unbound ions pass through the membrane, whereas the polymers and their complexes are retained. Various weakly basic, water-soluble polymers, including chitosan, polyethylenimine (PEI), poly(diallyl dimethylammonium chloride), and sodium polystyrene sulfonate (PSS), (Juang & Chiou, 2000; Juang & Chiou, 2001; Molinari et al., 2004; Vieira et al., 2001; Steenkamp et al., 2002; Abderrahim et al., 2006) have been used for removal of metal ions from water. The main advantages of the method lie in the relatively low-energy requirements of the ultrafiltration process and high removal efficiency due to effective binding of metal ions with polymers (Juang & Shiau, 2000). On the other hand, there is complex removal of the carrier of metal ions from concentrated solution and the separation of polymeric molecules and metal ions.

Application of carbon nanoparticles and nanotubes for removal of metal ions (Cd^{2+}, Cu^{2+}, Ni^{2+}, Pb^{2+}, Zn^{2+}, etc.) from water has been described recently (Rao et al., 2007; Hudson et al., 1997). In this approach, nanoparticles or nanotubes operate as a sorbent and a carrier of metal ions. Electrolytically generated nanocarbon colloids (NCCs) have functional groups such as carbonyl, hydroxyl, and carboxyl groups forming on the surface of carbon nanoparticles (Peckett et al., 2000; Khaydarov et al., 2010; Hsu et al., 1996). The sorption capacities of these particles are high, up to 7 mmol/g (Khaydarov et al., 2010), and are purely comparable with those of cation-exchange resins. On the contrary, the sorption capacities of raw nanotubes for metal ions are very low because walls of carbon nanotubes are not reactive. But their fullerene-like tips are known to be more reactive, and the sorption capacity significantly increases after oxidation by HNO_3, NaOCl, and $KMnO_4$ solutions due to generation of –COOH, –OH, or –CO groups (Li et al., 2002; Lu & Liu, 2006; Li et al., 2003; Lu & Chiu, 2006; Lu et al., 2006; Li et al., 2006; Liang et al., 2004; Li et al., 2005; Li et al., 2003; Chen & Wang, 2006; Lu et al., 2007). The sorption capacities of carbon nanotubes are not greater than 1 mmol/g (Rao et al., 2007); that is 2 to 5 times less than those of cation-exchange resin. The main disadvantage of this method is a complexity of separation of metal ions carriers (i.e., nanoparticles with metal ions) from water.

In order to improve the separation of carriers of metal ions from treated water, the metal ions can be bound to polymeric molecules and carbon nanoparticles, forming nanocarbon-polymer composition (NCPC) in water that is able to precipitate rapidly. The idea behind the technique is to add water-soluble polymeric molecules and carbon colloids to water in such a way as to bind metal ions and simultaneously form NCPC. That leads to a significant increase in the size of NCPC species, with follow-up formation of precipitates. This sediment can be easily removed from water by filtration or centrifugation, with a follow-up procedure of extraction of the metals.

2 Materials and equipment

Copper sulphate pentahydrate ($CuSO_4 \cdot 5H_2O$), cobalt nitrate hexahydrate ($Co(NO_3)_2 \cdot 6H_2O$), nickel nitrate hexahydrate ($Ni(NO_3)_2 \cdot 6H_2O$), cadmium nitrate tetrahydrate ($Cd(NO_3)_2 \cdot 4H_2O$), zinc nitrate hexahydrate ($Zn(NO_3)_2 \cdot 6H_2O$), poly(ethylenimine) (molar weight of 10,000, and 200,000–350,000) from Sigma Aldrich, UK were used in the as-received condition.

NCC was prepared by the electrochemical method described earlier (Khaydarov et al., 2010). The process was based on the use of an inexpensive two-electrode device in which an anode and a cathode are made from high-density isotropic graphites OEG4 (Russia) (65 mm × 30 mm × 15 mm) to be transformed into carbon colloidal particles. The anode and the cathode were immersed in a plastic electrolytic cell (120 mm × 140 mm × 105 mm) filled with distilled or deionized water as the electrolyte. The distance between the electrodes was varied from 10 to 120 mm in the current density range of 0.1–3 mA/cm^2 at a constant voltage of 60 V. The electrolytic cell was installed on the magnetic stirrer in order to provide water flowing between the electrodes. That allows saturating the electrolyte with carbon colloids and discharging the gas generated on the surface of electrodes due to electrolysis of water. The process of device operation involved two repeatable consecutive steps: (1) the electrolysis during 10 min, and (2) the electrolyte stirring during 60 s. The process was executed automatically, using twin timer ST-T (Korea).

The NCC preparation process was executed in two stages: anode activation and carbon nanoparticles generation. At the first stage the electrolyte has low conductivity, the value of the electric current density is small—about 0.1–0.2 mA/cm^2—and the oxidation reaction is slow. The duration of this stage is about 50 h and depends on the quality (density) of graphite. At this stage a voltage between electrodes is high, about 60–100 V. As the reaction proceeds, the conductivity of the electrolyte is abruptly increased, the current density increases up to 10 mA/cm^2 and higher, and the oxidization reaction sets in. As a result, the carbon electrode is finely split ,with follow-up covering by the carboxyl group.

In the second stage, the electric current density between electrodes is about 3–4 mA/cm^2. The NCC is stable for at least 150 days. If the current density values are greater than 8–10 mA/cm^2, the rate of oxygen diffusion through the electrode is so high that the pressure inside the electrode causes its disintegration, the synthesized NCC is not stable, and the precipitation of NCC is observed in 2–3 weeks.

The size and shape of nanoparticles were determined with transmission electron microscopy (TEM) (LEO-912-OMEGA, Carl Zeiss, Germany). The size values were averaged over more than 200 nanoparticles from different TEM micrographs of the same sample. Conductivity and pH of solutions were measured with WTW bench Multiparameter MultiLab 540.

The concentration of PEI in solutions was determined by neutron activation analysis. Water samples were irradiated in the neutron generator of the Institute of Nuclear Physics (Tashkent, Uzbekistan). An NaJ(Tl) 63 × 63 mm detector and a 6144-channel multichannel analyzer were used for recording gamma-ray quanta. The area under γ-peak of radionuclide ^{13}N (half-life $T_{1/2} = 9.96$ minutes, energy of the γ-peak $E\gamma = 0.511$ MeV) was measured to determine concentrations of nitrogen. Based on the obtained values, the concentration of PEI in water was calculated. Radionuclides used as the label of chemical elements during the study of water purification process are given in Table 14.1. The radionuclides were prepared by irradiating salts of chemical elements in the nuclear reactor of the Institute of Nuclear Physics (Tashkent, Uzbekistan). A Ge(Li) detector with a resolution of about 1.9 keV at 1.33 MeV and a 4096-channel analyzer were used for detection of gamma-ray quanta. Areas under γ-peaks of radionuclides were measured to calculate the amount of ions.

The bonding capacity Q, mmol/g, was calculated as follows:

$$Q = \frac{(A_0 - A_e) * B}{(A_0 - A_B) * W} \tag{14.1}$$

Table 14.1 Radionuclides Used as Labels ($T_{1/2}$ — Half-Life of the Radionuclides, E_γ — Energy of the γ-Peak)

Elements	Radionuclides	$T_{1/2}$	E_γ, MeV
Cr(VI)	^{51}Cr	27.73 days	0.320
Ni(II)	^{65}Ni	2.5 h	1.480
Cu(II)	^{64}Cu	12.7 h	0.511
Zn(II)	^{65}Zn	244.1 days	1.115
Cd(II)	^{115}Cd	53.5 h	0.336
Hg(II)	^{203}Hg	46.6 days	0.279

where B is amount of metal ions, mmol; W is a weight of absorber, g; A_0 is a count rate of the initial solution; A_e is a count rate of the solution at equilibrium in condition of saturation of the absorber; and A_B is a background count rate. The distribution coefficient K_d and the removal ratio P were calculated by Eqs. 14.2 and 14.3:

$$K_d = \frac{A_0 - A_e}{A_e - A_B} * \frac{V}{W} \tag{14.2}$$

$$P = \frac{A_0 - A_e}{A_0 - A_B} * 100\% \tag{14.3}$$

where V is a total volume of the solution, ml.

3 Results and discussion

The process of removing metal ions from water comprises the following stages: (i) capture of metal ions by PEI and NCC and simultaneous formation of NCPC; (ii) sedimentation of NCPC; and (iii) removal of sediments containing NCPC with metal ions by filtration.

3.1 Formation of NCPC

Two methods of NCPC formation have been studied: (1) the NCPC formation by adding PEI solution with the concentration from 60 mg/l to 320 mg/l into the NCC solution in the concentration range 2–100 mg/l; and (2) the NCPC formation by adding a solution of NCC with the concentration from 100 mg/l to 10,000 mg/l into the PEI solution in the concentration range 50–320 mg/l.

The typical TEM image for NCC obtained is shown in Figure 14.1.

Equilibrium reaction for water solution of NCC can be written as:

$$\mathrm{Carb - kCOOH} \leftrightarrow \mathrm{Carb - kCOO^-} + \mathrm{k\,H^+} \tag{14.4}$$

where $0 \leq k \leq k_m$ (k_m is the number of carboxyl groups on the surface of carbon nanoparticle).

NCC with carboxyl groups on its surface interacts with divalent metal ions. Figure 14.2a demonstrates schematically a carbon nanoparticle with attached metal ion:

$$\mathrm{Carb - 2COO^- - Me^{2+}} \tag{14.5}$$

In aqueous solution, PEI combines with a proton according to the following equilibrium reaction:

$$\mathrm{PEI} + \mathrm{n\,H_2O} \leftrightarrow \mathrm{PEI\,H_n^{n+}} + \mathrm{n\,OH^-} \tag{14.6}$$

where $0 \leq n \leq n_m$ (n_m is the number of monomers -CH_2-CH_2-NH- contained in a single polymeric chain) and depends on pH of the solution.

The mechanism of PEI-Me^{2+} interaction can be described by the following equilibrium reaction:

$$\mathrm{PEI} + \mathrm{a\,Me^{2+}} \leftrightarrow \mathrm{PEI\,Me_a^{2a+}} \tag{14.7}$$

where $0 \leq a \leq a_m$, $a_m = n_m/4$. The idealized structure of polymeric complex PEI-metal ions is given in Figure 14.2b.

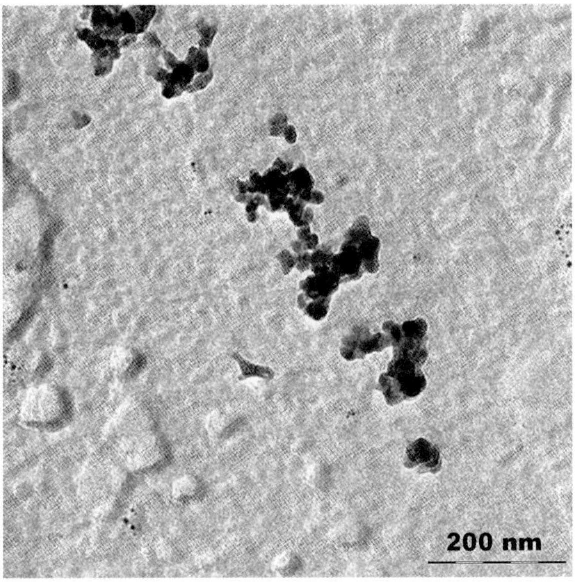

FIGURE 14.1

Typical TEM image of carbon colloids (the scale is 200 nm).

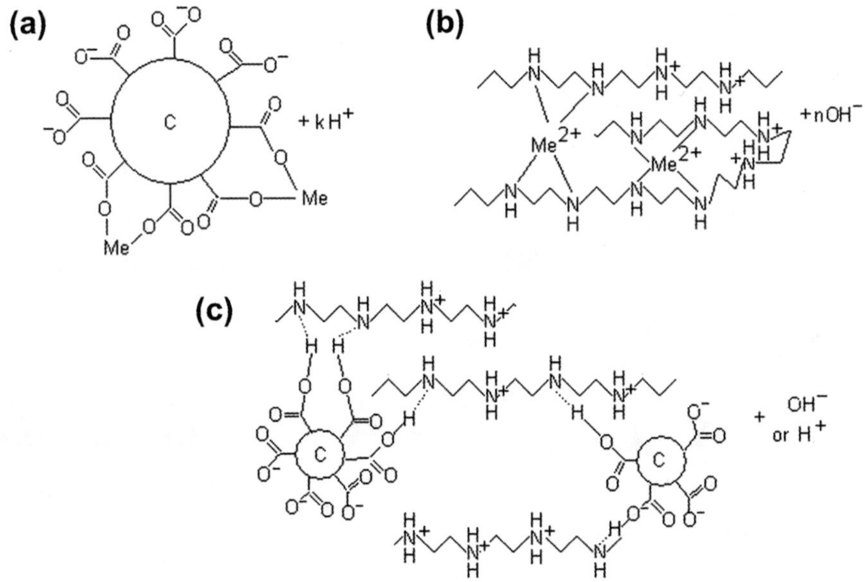

FIGURE 14.2

The structures of (a) c arbon nanoparticle with attached metal ions, (b) polymeric complex PEI-metal, (c) NCPC formation.

The reaction of NCC with PEI in water solution leads to formation of NCPC (Figure 14.2c). The conductivity and pH values of mixed solutions were measured to study the process of NCPC formation. Figure 14.3a and 14.3c demonstrate the dependence of conductivity S and pH of mixed solution with initial concentration of PEI of 100 mg/l against volume V of NCC solution with concentration of 58 mg/l, which is added to PEI solution. Dependences of conductivity S and pH of mixed solution with initial concentration of NCC of 80 mg/l against volume V of adding PEI solution with concentration of 4 mg/l are given in Figure 14.4a and 14.4c. The second-order derivatives of these curves were calculated and are presented in Figures 14.3b, 14.3d, and 14.4b and 14.4d.

When NCC is added to PEI solution, the conductivity S and pH of the solution decrease (cf. Figure 14.3, interval AB). In the interval BC, not all of the carboxyl groups can react with PEI because of the influence of charges of neighboring chains of PEI; therefore the excess of the number of carboxyl groups is required. The coagulation process begins with progressive acceleration from point C and reaches the maximal rate in point D, while the conductivity S does not change within the interval CD. After point D when all NCPC precipitates, conductivity S of the solution increases due to increasing the concentration of H^+ - ions in the solution.

When PEI solution is added to NCC, conductivity S of solution decreases and pH increases (Figure14.4, interval AB). The interval BC corresponds to the event when the excess of the number of

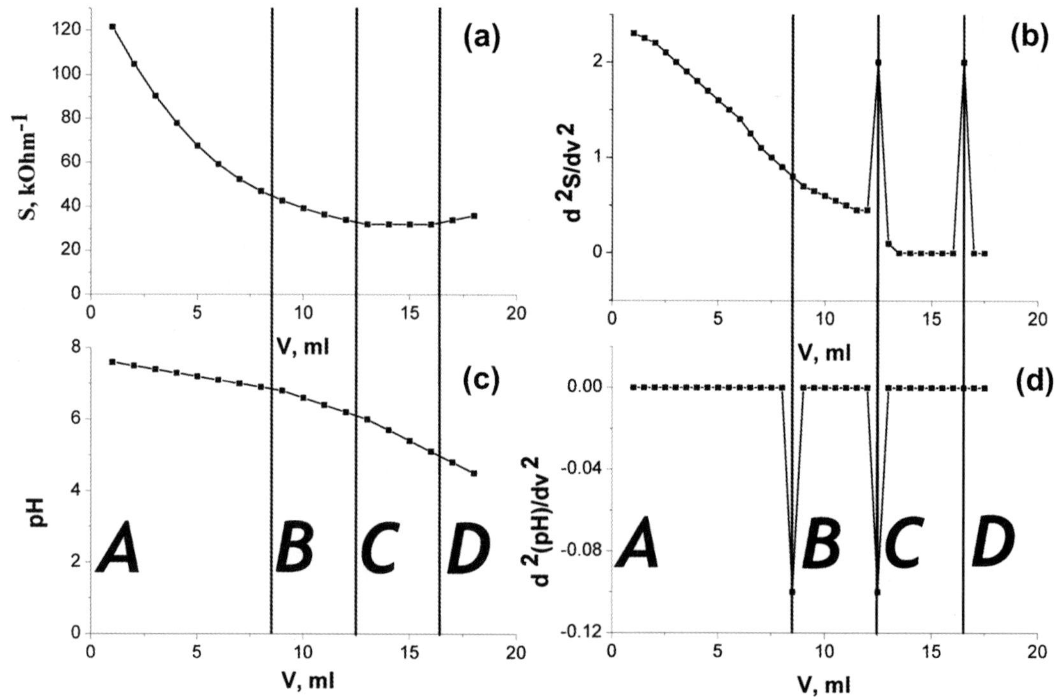

FIGURE 14.3

Dependences of conductivity S and pH of mixed solution against volume V of NCC solution which is added to PEI solution (part a and c) and second-order derivatives of these curves (part b and d).

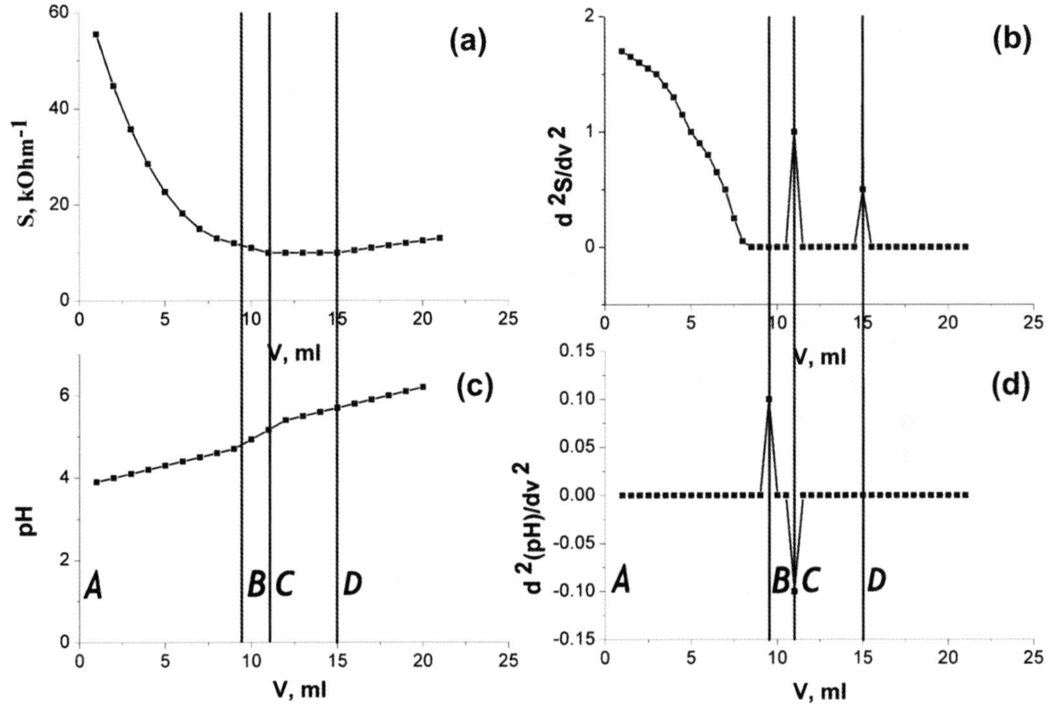

FIGURE 14.4

Dependences of conductivity S and pH of mixed solution against volume V of PEI solution which is added to NCC solution (part a and c) and second-order derivatives of these curves (part b and d).

carboxyl groups is required to react with PEI. The coagulation process begins with progressive acceleration from point C and reaches the maximal rate in point D; the conductivity S does not change in the interval CD. After point D when all NCPC precipitated, the conductivity S of solution increases due to increasing the concentration of OH^-–ions in the solution.

Experiments have shown that at point D of Figures 14.3 and 14.4 when NCPC precipitates, the ratio $(C_{pei}/C_{ncc})_D$ of concentrations of PEI C_{pei} and NCC C_{ncc} does not depend on the concentration of PEI in the interval from $2{\cdot}10^{-4}$ to 0.1%. But the specific surface of NCC increases with decreasing their diameter, and the ratio $(C_{pei}/C_{ncc})_D$ depends on the mean size of NCC (Figure 14.5). These results show that use of NCC with small sizes is more effective. For example, for NCC with size of 20 nm, the ratio $(C_{pei}/C_{ncc})_D = 0.42 \pm 0.07$ and $(C_{pei}/C_{ncc})_D = 6.7 \pm 1.1$ for NCC with the mean size of 5 nm.

Dependence of the rate of NCPC coagulation process against the ratio of concentrations of PEI and NCC C_{pei}/C_{ncc} was investigated. The results obtained for NCC with the mean size of 20 nm are given in Figure 14. 6 and show that the coagulation time within the interval CD is very short and dramatically increases with decreasing NCC concentration. This feature can be effectively used in methods of removing metal ions from water and remediation of contaminated soil.

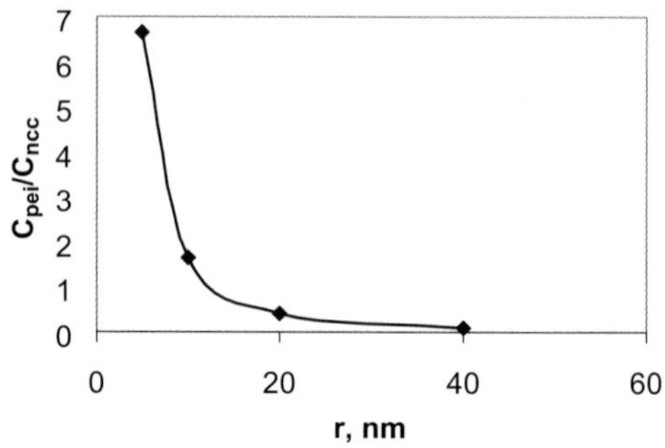

FIGURE 14.5

Dependence of the ratio $(C_{pei}/C_{ncc})D$ against the mean size of NCC. (For color version of this figure, the reader is referred to the online version of this book.)

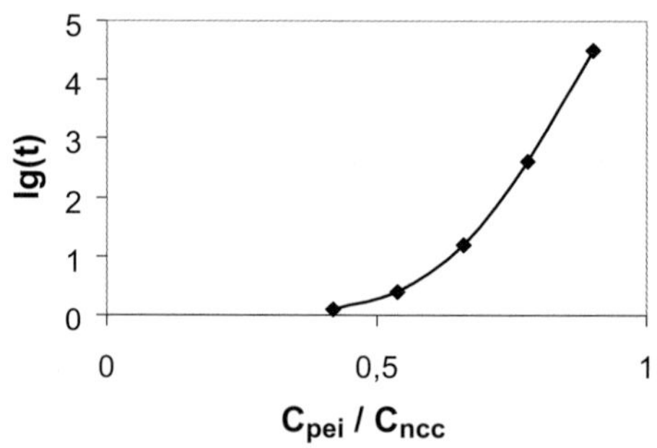

FIGURE 14.6

The rate of NCPC coagulation process against the ratio of concentrations of PEI and NCC C_{pei}/C_{ncc}. (For color version of this figure, the reader is referred to the online version of this book.)

3.2 Removing metal ions by NCPC

As mentioned earlier, the process of metal ions removal from water solutions involves an interaction of metal ions with PEI and NCC. Simultaneously, the positively charged complex Me–PEI combines with carboxyl groups of NCC with the NCPC formation as follows: (i) by hydrogen bond between carboxyl group of NCC and amine group of PEI which did not form coordination bonds with Me^{2+}, and (ii) by electrostatic interaction between negatively charged carboxyl ion of NCC and positively charged Me-PEI complex. After the interval of time depending on the ratio C_{pei}/C_{ncc}, the NCPC coagulates and can be easily removed by filtration or centrifugation.

The ability of the method to remove metal ions Zn^{2+}, Cd^{2+}, Cu^{2+}, Hg^{2+}, Ni^{2+}, CrO_4^{2-} was investigated. In these tests, the mean size of NCC was 20 nm, the ratio C_{pei}/C_{ncc} was 0.75, and PEI and metal ions concentrations were 150 mg/l and 1 mg/l, respectively. Figure 14.7 and Figure 14.8 represent the dependence of the removal ratio P against pH of solutions containing metal ions with concentration of 10 mg/l. In order to determine the bonding capacity Q of PEI and the distribution

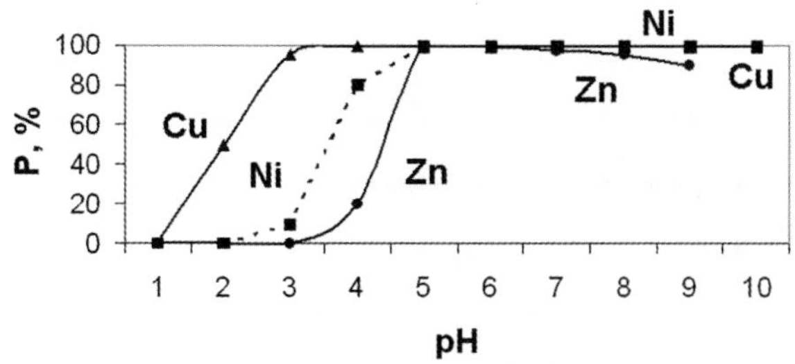

FIGURE 14.7

Dependence of the removal ratio P for Cu(II), Zn(II) and Ni(II) against pH of solutions.

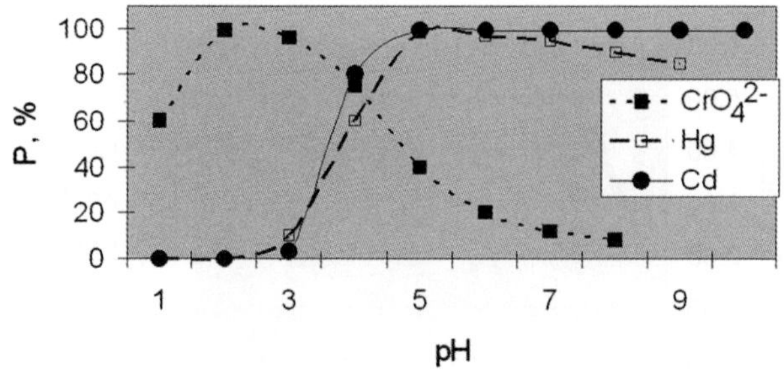

FIGURE 14.8

Dependence of the removal ratio P for Cd(II), Hg(II) and CrO42- against pH of solutions.

Table 14.2 Bonding Capacity Q and Distribution Coefficients K_d (pH = 6, V = 50 ml, W = 7.5 mg, Contact Time is 5 Minutes)

Parameters	Cr(VI)	Ni(II)	Cu(II)	Zn(II)	Cd(II)	Hg(II)
Q, mmol/g	2.0 (pH = 6) 4.0 (pH = 2)	5.5	5.7	5.2	5.0	4.4
K_d	13 (pH = 6) 1600 (pH = 2)	1100	1200	990	760	35

coefficients K_d, appropriate tests were carried out at pH = 6 (V = 50 ml, W = 7.5 mg) and the contact time of 5 minutes (see Table 14.2).

Time courses of metal ions adsorption were conducted using solutions of 10 mg/l concentration (Figures 14 9 and 14.10). The amount of metals adsorption increased rapidly during beginning of 30–40 seconds (about 90% removal). Subsequently, the adsorption rate rises gradually and reaches equilibrium (about 100% removal) after 50–70 sec. The short time required to reach equilibrium implies that NCPCs have very high adsorption efficiency and a great potential in divalent metals adsorbent application.

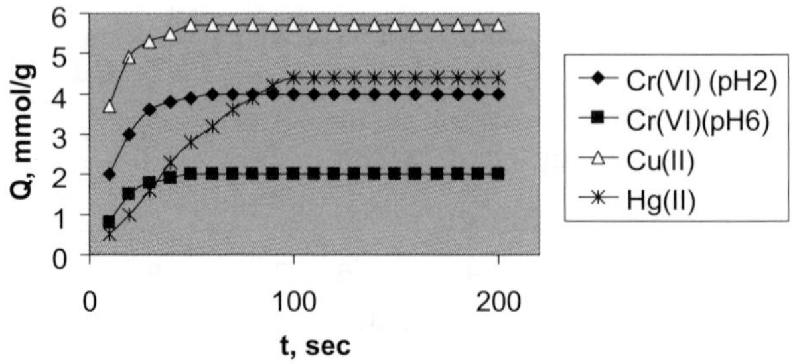

FIGURE 14.9

Adsorption of Cr(VI), Cu(II) and Hg(II) metal ions versus process time.

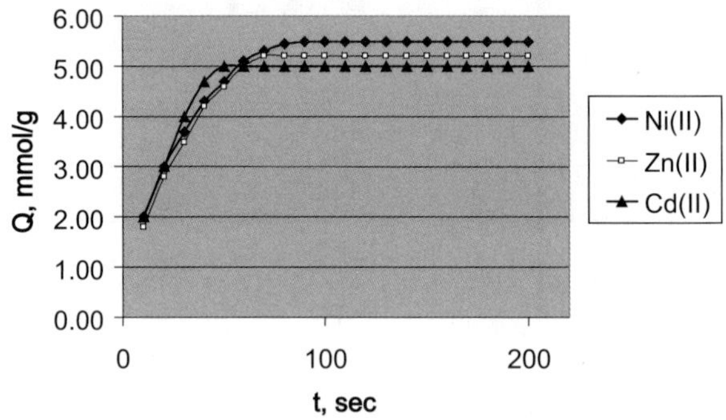

FIGURE 14.10

Adsorption of Ni(II), Zn(II) and Cd(II) metal ions versus process time.

3.3 Potentiality of NCPC in water treatment

Laboratory tests were conducted to reveal the efficiency of NCPC for purification of contaminated water. For this purpose, a 200 l. enamel container with a drainage pipe and a discharge valve at the bottom was connected through a water pump with a sand filter. The enamel container was filled with water containing metal ions Cd^{2+}, Cu^{2+}, Zn^{2+} with concentrations of 3 mg/l, 10 mg/l, and 5 mg/l, respectively; the temperature of the water was 24°C, pH = 6.8. A total of 4 liters of 0.5% NCPC solution with the ratio $C_{pei}/C_{ncc} = 0.78 \pm 0.15$ was prepared and added in the container with contaminated water. The water in the container was mixed thoroughly during 2 minutes, and 6 minutes after that the solution was discharged through the valve, water pump, and sand filter. The flow rate was about 1000 l/hr, and it was adjusted by the valve. Three samples of the water flowing out the sand filter were taken every 4 minutes after beginning the water discharge process to determine concentrations of Cd, Cu, and Zn.

Test results given in Table 14.3 confirm that (i) the lifetime of NCPC with the ratio $C_{pei}/C_{ncc} = 0.78 \pm 0.15$ is not greater than 8 minutes, (ii) the sediment after NCPC coagulation is easily removed even by a sand filtration, (iii) the filtered water does not contain NCPC and PEI (at least at lower

Table 14.3 Test Results of Proposed Water Purification Method

		Concentration in Water Flowing Out the Sand Filter, ug/l		
Metal Ions	**Initial Concentration, mg/l**	**After 4 Minutes**	**After 8 Minutes**	**After 12 Minutes**
Cu	10	5.1 ± 0.3	5.0 ± 0.3	4.9 ± 0.3
Zn	5	10.2 ± 0.5	9.8 ± 0.5	10.3 ± 0.5
Cd	3	5.3 ± 0.3	5.1 ± 0.3	5.2 ± 0.3
PEI	0.044	< 0.5	< 0.5	< 0.5

than the detectable levels), and (iv) metal ions are not washed off the sediment during the filtration process. These results show that the method is absolutely safe for the environment due to the possibility of controlling the lifetime of NCPC and the absence of NCPC and PEI in effluent water after filtration.

Thus the obtained results demonstrate that the proposed water purification method is highly efficient. It can be used for posttreatment of water to remove low concentrations of metal ions in plants making electronic plates, electrolysis shops, and so on.

3.4 Field tests of the soil remediation method

Field tests were conducted to reveal the efficiency of NCPC for remediation of soil and groundwater at a place that formerly housed a tannery where soil was contaminated by Cr(VI). The ground surface with the area of about 80 m^2 was divided in equal parts to compare the characteristics of soils treated (the first lot) and not treated (the second lot) by NCPC. Each of the lots was isolated by the wall with a height of 0.2 m.

First, the identity of soils in both lots was tested. Ground core samples with a diameter of 20 mm and a length of 70 mm were taken from five different places of each lot from the depth of 0–0.07, 0.2–0.27, 0.4–0.47, 0.6–0.67, and 0.75–0.82 m. Small stones were removed from the samples, and then each sample was mixed thoroughly, weighed, dried at 110°C during 15 minutes, and weighed again to determine the moisture of samples. A total of 100 mg of each sample was taken to determine Cr concentration C_{Cr} by NAA. The mean value of humidity and C_{Cr} of soil of each depth value and lot were calculated. Then each sample was put into a retort and filled with 200 ml of water and kept for 48 hours. After that the water was filtered through a paper filter and the concentration of extracted Cr was determined in water. The results of these tests are given in Table 14.4. As one can see from Table 14.4, C_{Cr} in the contaminated soil is about 60 mg/kg in both lots and the contamination extends for a depth over 0.8 m. Moreover, Cr concentration in extracted water is high, about 200 ug/l. The soil moisture is about 3% in both lots. These results show that the lots are identical and that one of them can be used as the experimental lot and another as the reference one.

In order to evaluate the efficiency of the proposed method of soil remediation, we have treated the first lot by NCPC. Eighty liters of 0.5% NCPC solution with the ratio $C_{pei}/C_{ncc} = 0.85$ was prepared. This concentrated solution was used to prepare the working solution with concentration of NCPC of 0.01%. The first lot was treated during 8 hours by 4 m^3 of working solution. The second lot was just watered without NCPC. Ground core samples from these two lots were taken from five different places and depths, treated, and tested as described above. Test results are given in Table 14.5.

Table 14.4 Comparison of Soil Characteristics of 2 lots

Parameter	Depth, m				
	0–0.07	0.2–0.27	0.4–0.47	0.6–0.67	0.75–0.82
	The First Lot				
C_{Cr} in soil, mg/kg	68 ± 20	54 ± 16	61 ± 18	50 ± 15	46 ± 15
Extracted C_{Cr} in water, ug/l	230 ± 50	200 ± 40	210 ± 40	190 ± 40	165 ± 35
Soil moisture, %	4 ± 1	3 ± 1	3 ± 1	3 ± 1	3 ± 1
	The Second Lot				
C_{Cr} in soil, mg/kg	64 ± 20	50 ± 16	69 ± 18	60 ± 15	54 ± 15
Extracted C_{Cr} in water, ug/l	200 ± 50	230 ± 40	230 ± 40	210 ± 40	185 ± 35
Soil moisture, %	4 ± 1	3 ± 1	3 ± 1	3 ± 1	3 ± 1

As shown in Table 14.5, the value of C_{Cr} in contaminated soil was not changed after watering. Over $t = 8$ hours the water penetrated into the soil up to a depth of $L_W = 0.7$ m. The soil moisture deeper than 0.7 m was not changed after watering. The Cr concentration in extracting waters of ground core samples taken from depths up to $L_{NCPC} = 0.5$ m is very low, about 1.4 ug/l, but it rises sharply from the depth of 0.5 m. It means that the lifetime t_{NCPC} of NCPC before coagulation

$$t_{NCPC} = \frac{L_{NCPC}}{L_W} * t \tag{14.18}$$

is about 5.5–6 hours and NCPC penetrated up to a depth of 0.5 m. This value of the NCPC lifetime is in good accordance with the data given in Figure 14.6.

The results obtained demonstrate a high efficiency of the proposed soil remediation method. Depth or area of remediation depends on NCPC life, that is, on the value of ratio C_{pei}/C_{ncc} and the penetration rate of NCPC solution. Accordingly, the depth of penetration of NCPC in soil or depth of remediation of soil can change from 1 to 100 cm, and the distance of NCPC moving with the groundwater or

Table 14.5 Comparison of Soil Characteristics of Lots After Treatment by NCPC

Parameter	Depth, m				
	0–0.07	0.2–0.27	0.4–0.47	0.6–0.67	0.75–0.82
	The First Lot				
C_{Cr} in soil, mg/kg	62 ± 20	59 ± 16	58 ± 18	60 ± 15	50 ± 15
Extracted C_{Cr} in water, ug/l	1.3 ± 0.2	1.5 ± 0.2	1.4 ± 0.2	190 ± 40	170 ± 35
Soil moisture, %	11 ± 3	13 ± 3	11 ± 3	10 ± 3	3 ± 1
	The Second Lot				
C_{Cr} in soil, mg/kg	60 ± 20	63 ± 16	63 ± 18	57 ± 15	54 ± 15
Extracted C_{Cr} in water, ug/l	230 ± 50	220 ± 40	200 ± 40	180 ± 40	190 ± 35
Soil moisture, %	12 ± 3	11 ± 3	12 ± 3	11 ± 3	3 ± 1

remediation zone of the ground can change from 1 to 100 meters. Due to the possibility of controlling the lifetime of NCPC, the method is safe for environment; nanoparticles are transformed into ordinary big-size particles after the required time. The method can be used for remediation of contaminated soil on industrial plants.

4 Conclusion

In order to improve the separation of carriers of metal ions from treated water, the metal ions can be bound to the polymeric molecules and carbon nanoparticles that form NCPC in water and precipitate rapidly.

The process of metal ion removal from water solutions involves the following stages: (i) sorption of metal ions by PEI and NCC, with simultaneous formation of NCPC, (ii) coagulation of NCPC containing metal ions, and (iii) removal of the coagulated NCPC by filtration or centrifugation in order to recover the metals. The contaminated soil is remediated due to the coagulation of NCPC forming insoluble compounds with metal ions and remaining in soil.

The metal ions interact with NCPC through the ion exchange and complexation mechanism. Studies of NCPC sorption properties show that the composition has high bonding capacity of 4.0–5.7 mmol/g at pH = 6 for most of the divalent metal ions. The percentage of sorption of Zn^{2+}, Cd^{2+}, Cu^{2+}, Hg^{2+}, Ni^{2+}, CrO_4^{2-} ions is higher than 99%, and the distribution coefficients are 10^1–10^3.

The lifetime of NCPCs before coagulation in the treated water and soil is 1 sec –1000 minutes and depends on the ratio of polymeric molecules and carbon nanoparticles concentrations. Accordingly, the depth of penetration of NCPC in a soil or the depth of remediation of soil can vary from 1 to 100 cm, and the distance of NCPC moving with groundwater or the remediation zone of ground can vary from 1 to 100 meters.

Laboratory tests have shown that when water is contaminated by metal ions at the level of 1–10 mg/l, the described method allows 100–1000 times decreasing the concentration of metal ions in water; the sediment after NCPC coagulation is easily removed by filtration; and the filtered water does not contain NCPC and PEI (at least lower than the detectable level).

Field tests have shown that when soil is contaminated by Cr at the level of 50–60 mg/kg, the described method allows 100 times decreasing concentration of metal ions in water passing through the soil.

The obtained results demonstrate a high efficiency of the proposed water purification and soil remediation methods and can be used in different industrial plants. Due to the opportunity of controlling the lifetime of NCPC, the method is safe for environment.

Glossary

Carboxyl group a set of four atoms bonded together and present in carboxylic acids, including amino acids; usually abbreviated as COOH

Equilibrium a state in which the concentrations of the reactants and products have no net change over time

Nanoparticle a particle having one or more dimensions of the order of 100 nm or less

Neutron activation analysis a nuclear process used for determining certain concentrations of elements in a vast amount of materials

Polyethyleneimine (PEI) a water-soluble polymer that has been found to have a great affinity to binding with metals, principally due to its polyelectrolyte character as well as their large macromolecular sizes, which make them very useful in sequestering metal ions from dilute solutions

Radionuclide an atom with an unstable nucleus; it undergoes radioactive decay and emits a gamma ray(s) and/or subatomic particles.

Remediation aremoval of pollution or contaminants from environmental media such as soil, groundwater, sediment, or surface water for the general protection of human health and the environment

Transmission electron microscopy (TEM) a microscopy technique whereby a beam of electrons is transmitted through an ultra-thin specimen, interacting with the specimen as it passes through

References

Abderrahim, O., Amine Didi, M., Moreau, B., Villemin, D., 2006. A new sorbent for selective separation of metal: polyethylenimine methylenephosphonic acid. Solvent Extraction Ion Exchange 24 (6), 943–955.

Chen, C., Wang, X., 2006. Adsorption of Ni(II) from aqueous solution using oxidized multiwall carbon nanotubes. Ind. Eng. Chem. Res. 45, 9144–9149.

Hsu, W.K., Terrones, M., Hare, J.P., Terrones, H., Kroto, H.W., Walton, D.R.M., 1996. Electrolytic formation of carbon nanostructures. Chem. Phys. Lett. 262, 161–166.

Hudson, M.J., Hunter-Fujita, F.R., Pecketta, J.W., Smithb, P.M., 1997. Electrochemically prepared colloidal, oxidised graphite. J. Mater. Chem. 7 (2), 301–305.

Juang, R.S., Chiou, C.H., 2000. Ultrafiltration rejection of dissolved ions using various weakly basic water-soluble polymers. J. Memb. Sci. 177, 207–214.

Juang, R.S., Chiou, C.H., 2001. Feasibility of the use of polymerassisted membrane filtration for brackish water softening. J. Memb. Sci. 187, 119–127.

Juang, R.S., Shiau, R.C., 2000. Metal removal from aqueous solutions using chitosan-enhanced membrane fltration. J. Memb. Sci. 165, 159–167.

Khaydarov, R.R., Khaydarov, R.A., Gapurova, O., 2010. Water purification from metal ions using carbon nanoparticle-conjugated polymer nanocomposites. Water. Res. 44, 1927–1933.

Li, Y.H., Wang, S., Wei, J., Zhang, X., Xu, C., Luan, Z., Wu, D., Wei, B., 2002. Lead adsorption on carbon nanotubes. Chem. Phys. Lett. 357, 263–266.

Li, Y.H., Ding, J., Luan, Z., Di, Z., Zhu, Y., Xu, C., Wu, D., Wei, B., 2003. Competitive adsorption of Pb2+, Cu2+ and Cd2+ ions from aqueous solutions by multiwalled carbon nanotubes. Carbon 41, 2787–2792.

Li, Y.H., Wang, S., Luan, Z., Ding, J., Xu, C., Wu, D., 2003. Adsorption of cadmium(II) from aqueous solution by surface oxidized carbon nanotubes. Carbon 41, 1057–1062.

Li, Y.H., Di, Z., Ding, J., Wu, D., Luan, Z., Zhu, Y., 2005. Adsorption thermodynamic, kinetic and desorption studies of Pb2+ on carbon nanotubes. Water. Res. 39, 605–609.

Li, Y.H., Zhu, Y., Zhao, Y., Wu, D., Luan, Z., 2006. Different morphologies of carbon nanotubes effect on the lead removal from aqueous solution. Diamond Relat. Mater. 15, 90–94.

Liang, P., Liu, Y., Guo, L., Zeng, J., 2004. Multiwalled carbon nanotubes as solid-phase extraction adsorbent for the preconcentration of trace metal ions and their determination by inductively coupled plasma atomic emission spectrometry. J. Anal. At. Spectrom 19, 1489–1492.

Lu, C., Chiu, H., 2006. Adsorption of zinc (II) from water with purified carbon nanotubes. Chem. Eng. Sci. 61, 1138–1145.

Lu, C., Liu, C., 2006. Removal of nickel (II) from aqueous solution by carbon nanotubes. J. Chem. Technol. Biotechnol. 81, 1932–1940.

Lu, C., Chiu, H., Liu, C., 2006. Removal of zinc(II) from aqueous solution by purified carbon nanotubes: kinetics and equilibrium studies. Ind. Eng. Chem. Res. 45, 2850–2855.

Lu, C., Chiu, H., Bai, H., 2007. Comparisons of adsorbent cost for the removal zinc (II) from aqueous solution by carbon nanotubes and activated carbon. J. Nanosci. Nanotechnol. 7, 1647–1652.

Molinari, R., Gallo, S., Pietro Argurio, P., 2004. Metal ions removal from wastewater or washing water from contaminated soil by ultrafiltration–complexation. Water. Res. 38, 593–600.

Peckett, J.W., Trens, P., Gougeon, R.D., Poppl, A., Harris, R.K., Hudson, M.J., 2000. Electrochemically oxidised graphite. Characterisation and some ion exchange properties. Carbon 38, 345–353.

Rao, G.P., Lu, C., Su, F., 2007. Sorption of divalent metal ions from aqueous solution by carbon nanotubes: A review. Separation Purif. Technol. 58, 224–231.

Rumeau, M., Persin, F., Sciers, V., Persin, M., Sarrazin, J., 1992. Separation by coupling ultrafiltration and complexation of metallic species with industrial water soluble polymers. Application for removal or concentration of metallic cations. J. Memb. Sci. 73, 313–322.

Steenkamp, G.C., Keizer, K., Neomagus, H.W.J.P., Krieg, H., 2002. Copper (II) removal from polluted water with alumina/chitosan composite membranes. J. Memb. Sci. 197, 147–156.

Vieira, M., Tavares, C.R., Bergamasco, R., Petrus, J.C.C., 2001. Application of ultrafiltration-complexation process for metal removal from pulp and paper industry wastewater. J. Memb. Sci. 194, 273–276.

CHAPTER

15

Quebracho colorado: A New Source for Water Coagulant—Performance on Dye Removal

J. Sánchez-Martín, J. Beltrán-Heredia, B. Coco-Rivero

Department of Chemical Engineering and Physical Chemistry, University of Extremadura, 06071 Badajoz, Spain

CHAPTER OUTLINE

The Role of Colloidal Systems in Environmental Protection. http://dx.doi.org/10.1016/B978-0-444-63283-8.00015-6

1 Introduction

Environmental concerns are growing in importance because the fragile Earth equilibrium is being broken by antropic actions. It is more than evident that the current rate of waste production, exhausting resources and producing contamination, is no longer sustainable in our finite planet (Dorf, 2001) where humankind must follow a cycle model rather than a linear vision of the environment (Schumacher, 1987). This is the philosophy of the well-known 3R: recycle, reuse, reduce. Modern life will no longer be possible if we treat the Earth like it had no limits. The presence of such limits is the basis of conservation and ecologism.

The scientific community is also worried about environmental sustainability, and every day more and more scientists are developing cleaner technologies to protect the legacy we will leave to our children. Many of these initiatives focus on water management and water and wastewater treatment even for developing countries in a technologically scarce scenario (Chinyama et al., 2012).

Water is a universal vector for the planetary chemistry. Almost everything can be dissolved in water, and aqueous effluents are the natural ways of disseminating nutrients and poisons (Buzier et al., 2011). Through aquatic paths pollutants can reach living organisms, either the simplest ones or the most complex animals, even humans (Ying, 2006; Halling-Sorensen et al., 1998; Shaul et al., 1991). Taking care of the environment, as well as of humankind, means caring for the water (WHO, 2008).

For this task, many remediation methods focus on polluted water or making drinking water safe for human consumption. One of the main uses is coagulation. Since it is a physical process, it does not involve chemical reactions (Andriamirado, 2007) intended to destabilize the suspended colloidal matter in the water matrix. This destabilization is due to electrical charge compensation, so positive-charged pollutants (cationic ones) usually respond to anionic coagulants, whereas negative-charged pollutants do the contrary (Bulatovic, 2007). Coagulants are therefore called anionic, cationic, or, in the most favorable cases, amphoteric.

Classical coagulants involved in water treatment are metallic salts such as $Al_2(SO_4)_3$ or $FeCl_3$. They were most widely used in the treatment of drinking water. However, some significant concerns have recently arisen, linking, for example, the use of alum to Alzheimer's disease (Flaten, 2001). This and other environmental problems have led scientists to conduct research on new coagulants, based mainly on natural products, in order to overcome such disadvantages (Sánchez-Martín and Beltrán-Heredia, 2012).

In this particular vision of sustainability, coagulants can be synthesized from many different raw substances. Among the most promising natural materials are tannins and specifically tannin extracts from *Schinopsis balansae*, that is, Quebracho colorado. This vegetal species was the main feedstock of tannin extract for a long time, until *Acacia mearnsii*, Black wattle, was used in industry. Tannin extracts are natural astringent products that are widely used for leather tanning (hence their name), but many other uses are being found (Pizzi, 2008). Their use as a coagulant precursor is one of tannin extracts' most interesting capacities present (Beltrán-Heredia et al., 2010).

With regard to natural tannin extracts, we focus our attention on vegetal water-soluble polyphenolic compounds with an exact chemical composition difficult to determine because of their very high impurity content. Economic reasons encourage scientists to work directly with these raw materials, without purifications. This is usually feasible, as we have reported previously (Beltrán-Heredia et al., 2011b). Namely, the nature of Quebracho tannins has been thoroughly characterized by specific

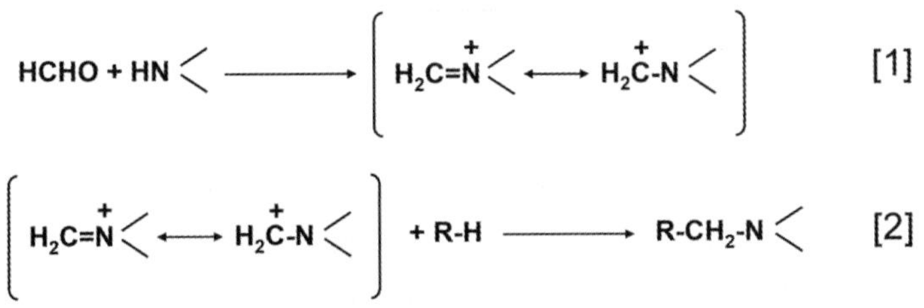

FIGURE 15.1

Reaction mechanism of aminomethylation for coagulant synthesis. (1) Formation of imine. (2) Formation of Mannich base.

researchers (Vivas et al., 2004), and it is said that it is mainly based on combinations of catechol, resorcinol, and pyrogallol-building blocks that provide a large structure where different operating changes can be performed. The feasibility of this raw material for coagulant synthesis has been fully demonstrated since it is used as feedstock for commercially available coagulants, such as Silvafloc (Beltrán-Heredia et al., 2009a).

Tannins have been used as a source of novel coagulants because they present interesting characteristics that make them useful according to sustainability principles (Starkl and Brunner, 2004; Muga and Mihelcic, 2008). They are affordable and ubiquitous compounds, being available in a wide variety of barks (Haslam, 1989). Alternatively, they are easy to extract and handle (Matthews et al., 1997) and their use does not affect environmental equilibrium because they can be employed in small amounts (Nakamoto et al., 2004). These and other sustainability premises make tannins optimum candidates for the production of new water treatment agents.

The polymerization of tannin extracts follows a Mannich reaction pattern. This chemical process is well known, and we have previously explained it (Beltrán-Heredia et al., 2011b). Figure 15.1 shows the mechanism of this reaction, which consists of two steps: (1) the initial formation of the imine and (2) the formation of the Mannich base. As a result of this chemical transformation, an electrically charged coagulant (cationic) is produced and therefore can be used in the destabilization of anionic compounds, suspended in water matrices.

We have synthesized an optimum coagulant at lab scale from *Schinopsis balansae* tannin extract. This chapter seeks to study three concomitant aspects: the efficiency of this coagulant in the removal of several dye compounds from water effluents; the optimization of operating conditions in the coagulation process; and the feasible theoretical model according to an adsorption-like model.

2 Materials and methods

2.1 Experimental installations

2.1.1 Coagulant synthesis reactor

For the synthesis of every coagulant, following the specific procedure, the experimental reactor consisted of:

- A thermostatic bath equipped with thermocontrol (SELECTA, Tectron Bio, Spain), which adjusted temperature within a range of ±0.5°C. A manual thermometer confirmed this point.

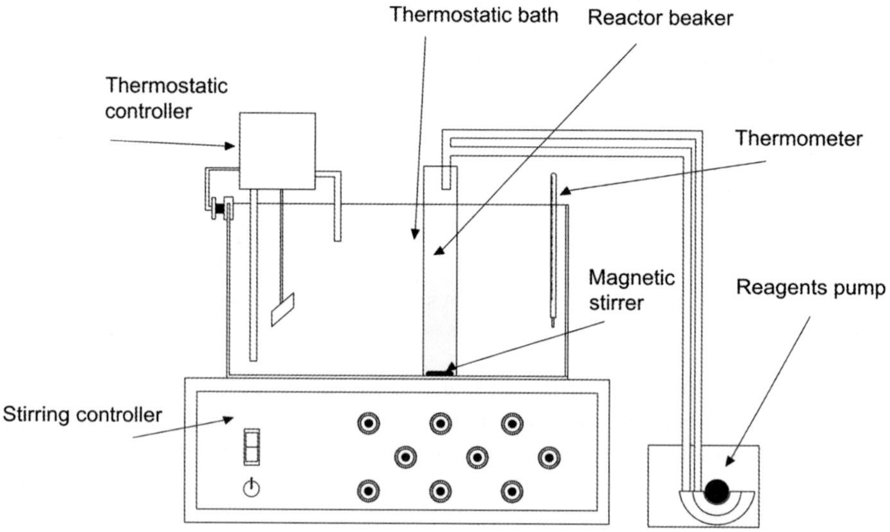

FIGURE 15.2

Synthesis reactor experimental installations. (For color version of this figure, the reader is referred to the online version of this book.)

- A multi-stirrer magnetic device (SELECTA, Multimatic 9N, Spain) that kept the reaction mixture homogeneous with the aid of a teflon-covered magnetic bar.
- A double-channel peristaltic hydraulic pump calibrated for the specific reagents (Cole Parmer, MasterFlex C/L).
- Glass beaker reactor (diameter 3.5 cm, height 18 cm).

The whole system is presented in Figure 15.2.

2.2 Buffered solutions

The trials with added pollutant were performed with pH-stable media. To this end, a standard pH-7 buffer solution was prepared of 1.2 g of NaH_2PO_4 and 0.88 g of Na_2HPO_4 in 1 L of distilled water. The pH was then adjusted to the desired level with HCl 1M or NaOH 1M. All reagents were analytical grade from PANREAC.

2.3 Reagents and model compounds

The synthesis of the tannin-based coagulants involved the following reagents:

- Tannin extract from *Schinopsis balansae* (Raoul-Duvall, France)
- Formaldehyde 37%, stabilized with methanol (Merck)
- Diethanolamine solution, analytical purity grade (Aldrich)

2.3.1 Model compounds

We considered up to nine different dyes: Carmine Indigo ($C_{16}H_8N_2Na_2O_8S_4$), Chicago Sky Blue 6B ($C_{34}H_{24}N_6Na_4O_{16}S_4$), Palatine Fast Black WAN ($C_{60}H_{36}N_9O_{21}Na_3S_3Cr_2$), Amaranth

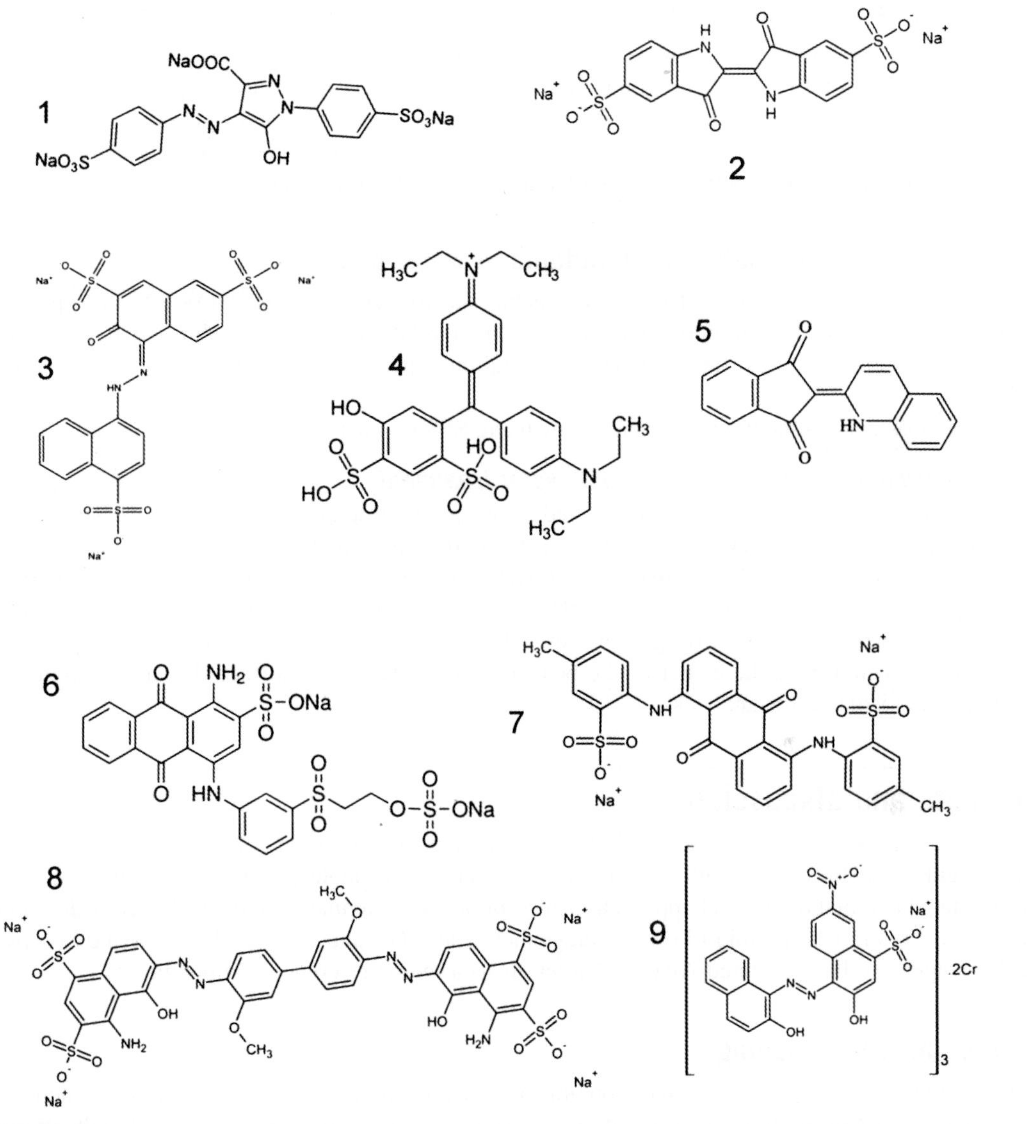

FIGURE 15.3

Chemical structures for dyes: (1)Tartrazine, (2) Carmine Indigo, (3) Amaranth, (4) Patent Blue V, (5) Quinoline Yellow WS, (6) Remazol Brilliant Blue, (7) Alizarin Violet 3R, (8) Chicago Sky Blue 6B, (9) Palatine Fast Black WAN.

($C_{20}H_{11}N_2Na_3O_{10}$), Patent Blue V ($C_{27}H_{31}N_2S_2O_7Na$), Quinoline Yellow ($C_{18}H_{13}NO_5SNa$), Remazol Brilliant Blue ($C_{22}H_{16}N_2Na_2O_{11}S_3$), Tartrazine ($C_{16}H_9N_4Na_3O_9S_2$), and Alizarin Violet 3R ($C_{28}H_{20}N_2Na_2O_8S_2$). Their chemical structures are shown in Figure 15.3.

2.4 Synthesis of the coagulant

Firstly, 2.5 g of the tannin extract were introduced in the reactor beaker. Subsequently, they were dissolved with 15.64 mL of diethanolamine (DEA) and 33.46 mL of distilled water. This solution is thermostated at a constant temperature of 30°C. The polymerization of the coagulant is achieved by adding 8.5 mL of formaldehyde drop by drop. The entire process usually takes 1 h, and the reaction is left in the same conditions of stirring and temperature for 24 h for ageing.

2.5 Analytical determination of pollutants in water matrix

Dye concentration was determined by direct visible spectroscopy in a HEλIOS UV/VIS spectrophotometer. According to Lambert-Beer's law, a direct relationship was established between the absorbance of the sample and the pollutant concentration. Different wavelengths were used for each colorant.

Nonlinear data adjustment was carried out by using SPSS 14.0.1 for Windows (SPSS Inc., 2005).

2.5.1 Experimental installation for wastewater treatment

Operating conditions of coagulation were studied by using a Jar test apparatus (JLT4, Velp Scientifica) as reported by Beltrán-Heredia et al. (2011b). A standard volume of 100 mL of contaminated water sample (usually with 100 $mg \cdot L^{-1}$ of dye content) were introduced in a 1-L beaker under strict thermal control. Experimental conditions, including pH, temperature, coagulant dosage, initial contaminant concentration, stirring rate, and stirring time, were adequately adjusted. After the coagulation period, a sample was collected from the center of the beaker, 3 cm under liquid surface. It was subsequently centrifuged and isolated for analysis.

3 Results and discussion

This section is presented in three subsections: a preliminary screening on colorant removal with this novel coagulant; a study on the influence of several operating variables, such as pH, temperature, or stirring rate; and a final theoretical approach to adsorption-like coagulation models. We consider three models: the Freundlich, Frumkin-Fowler-Guggenheim, and Gu and Zhu models. For the last two research stages, we have selected Alizarin Violet 3R as a model reference compound.

3.1 Preliminary screening

The preliminary screening was carried out on nine dyes, with completely different chemical structures. As can be observed from Figure 15.4, there exists a wide variety of efficiency levels within the group. All of them were treated under the same conditions: 1.5 mL of the lab-made coagulant and 100 $mg \cdot L^{-1}$ of dye concentration in a total volume of 100 mL. Anthraquinonic dyes, such as Alizarin Violet 3R or Remazol Brilliant Blue, reached significantly high removal efficiency levels—up to 60%—whereas the coagulation reached by other chemical groups was not so clear. In fact, azoic dyes such as Chicago Sky Blue 6B, Palatine Fast Black WAN, Amaranth, or Tartrazine did not present a clear tendency: The first of them were removed up to 90%, but the rest presented an almost null removal response. This high specificity makes the coagulant a selective treatment agent, and the reasons for this may be found in the chemical structure of the dyes or even in their molecular weight.

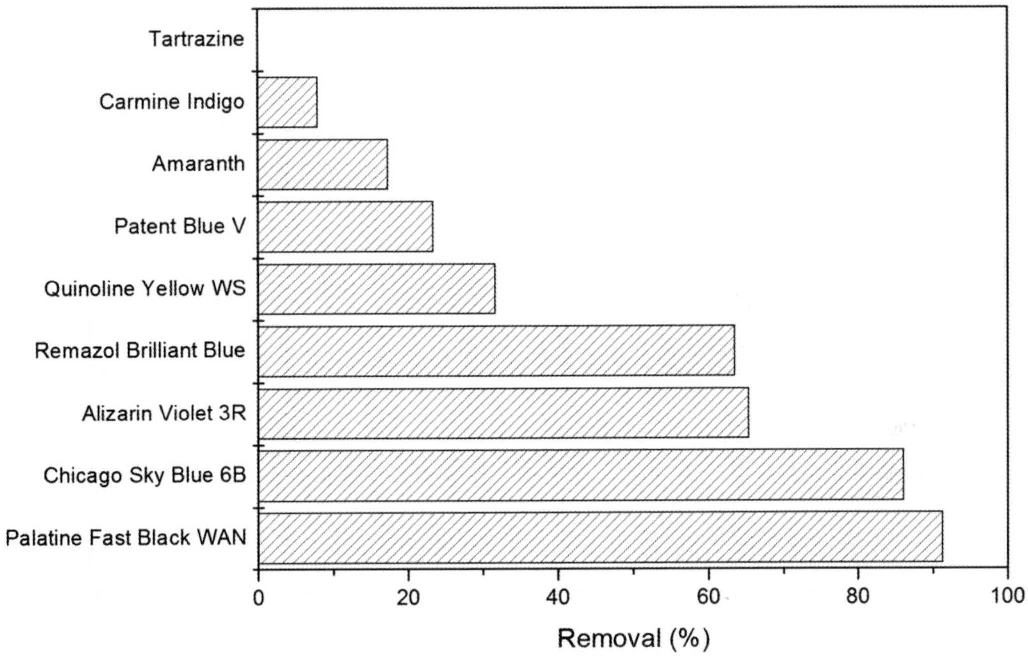

FIGURE 15.4

Preliminary screening for dye removal.

Amaranth and Tartrazine presented molecular mass of 500–600 g per mol, whereas Chicago Sky Blue 6B or Palatine Fast Black WAN presented 1000–1500 g per mol. For the rest of the chemical groups (triphenylmethane, indigoid, or quinoid), the coagulation efficiency was not very high.

3.2 Influence of operative conditions

The coagulation mechanism is quite complex and very difficult to model. However, it is obvious that the operating conditions such as the pH level or the stirring method are quite influential variables in determining the final efficiency of the entire process. Percentage dye removal is significantly affected by these physical and chemical parameters.

3.2.1 pH

In evaluating how pH level affects the removal of Alizarin Violet 3R, several experimental trials were performed with 1.5 mL of coagulant in 0.1 L of aqueous sample. An initial dye concentration of $100\ mg \cdot L^{-1}$ was fixed, and the reaction temperature was adjusted to 20°C. The standard protocol of 2 min, 100 rpm, plus 30 min, 30 rpm in the Jar test, was adopted for studying this variable.

As can be observed in Figure 15.5, there is a general tendency toward efficiency decrease with pH level. This effect can be explained if we note that the cationic character of the coagulant may be reduced as the medium becomes more alkaline. So the electrical and chemical interaction between amine groups and sulfonate groups (i.e., between coagulant and dye molecules) tends to be lower

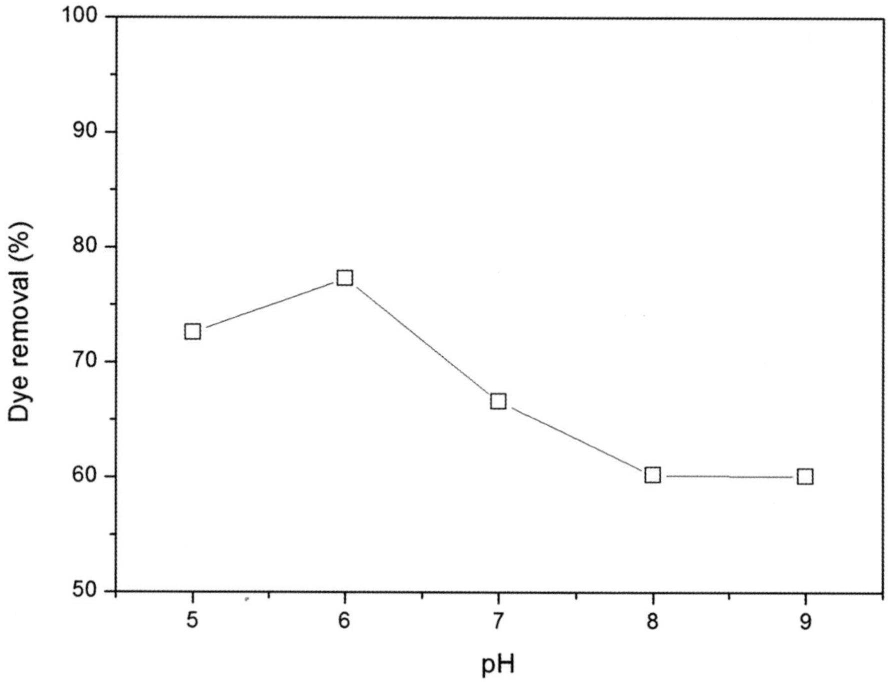

FIGURE 15.5

pH influence in Alizarin Violet 3R removal.

(Beltrán-Heredia et al., 2009, 2012). Consequently, pH levels in industrial scale-up installations should be kept at around 6–7.

3.2.2 Temperature

Similar experimental trials were performed by varying temperature in the thermostatic bath. Figure 15.6 depicts the decreasing tendency in the dye removal as temperature tends to be higher. pH level was set to 7, and the rest of the experimental conditions were exactly the ones presented in the previous section.

This behavior was already reported when dealing with natural coagulants (Sánchez-Martín et al., 2010) and might be linked to the destruction of flocs when the mobility of the suspended particles is increased due to thermal agitation (Bratby, 2006; Bulatovic, 2007).

3.2.3 Stirring conditions

It is more than obvious that the stirring conditions and the role of the stirring rate and period are significant variables in the efficiency of the coagulation ability (Rossini et al., 1999). This is why we have performed two experimental series regarding these parameters. Stirring time was varied between 10 and 45 min, and stirring rate was varied between 20 and 60 rpm. The rest of the experimental conditions were the same as before.

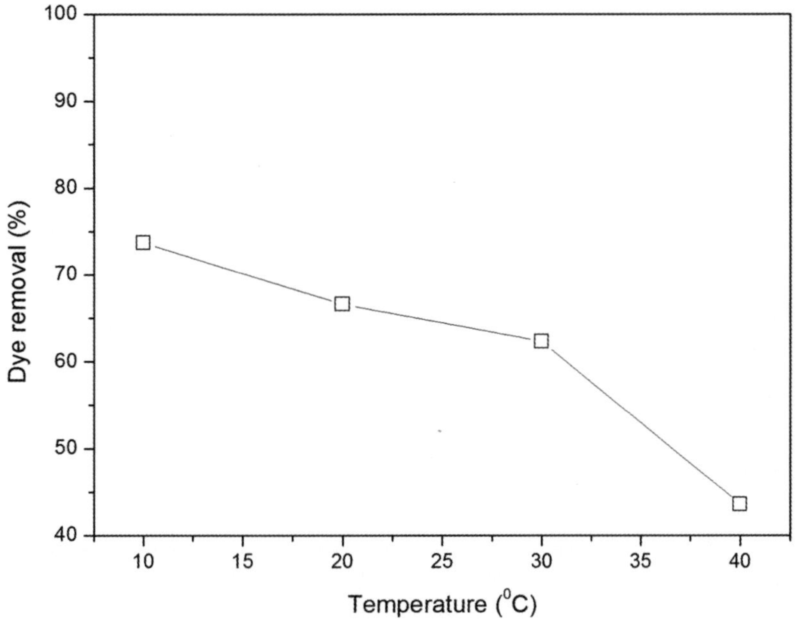

FIGURE 15.6

Temperature influence in Alizarin Violet 3R removal.

Figure 15.7 presents the results of these studies. As can be easily observed, this new tannin-based coagulant is efficient in the entire range of stirring conditions. A positive effect is shown as we increase the total stirring time, but there is no enhancement once the barrier of 20 minutes is reached. Almost no variation is presented when the stirring rate is varied. Every dye removal percentage is kept in high levels, around 70%.

3.2.4 Coagulant dosage

Increasing the coagulant dosage should be reported as an effective way of enhancing the efficiency of pollutant removal. However, it is important to evaluate the probable limit of this method (Beltrán-Heredia et al., 2012). This is why we planned an experimental series where pH was adjusted to 7, temperature was 20°C, a standard Jar test protocol was used, and initial dye content was 100 $mg \cdot L^{-1}$. Figure 15.8 presents the results.

It is clear that efficiency tends to grow as the coagulant dose increases. However, the experimental curve seems to grow rapidly up to 1.5 ml of coagulant; then there is no significant increase of dye removal percentage, which is around 70%—that is, the maximum level with this initial dye concentration.

This maximum removal percentage can be explained if we bear in mind that this dye (Alizarin Violet 3R) can present a surfactant-like behavior: High concentrations can lead to micellation (Holmberg et al., 2003) and, consequently, to higher removal efficiencies. This is reinforced by the apolar area of the molecule (nonsubstituted carbons in the aromatic rings) and the polar one (sulfonic groups). Section 3.3 explores this idea in greater depth.

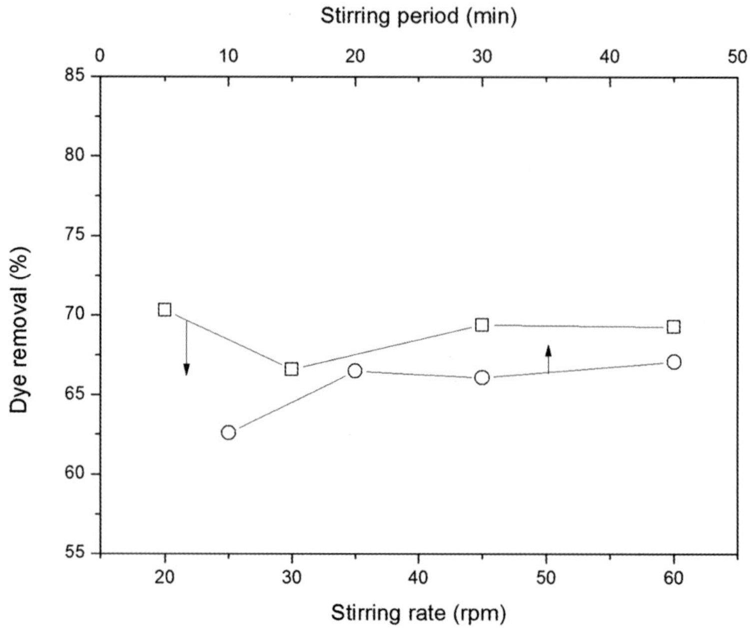

FIGURE 15.7

Stirring variables influence in Alizarin Violet 3R removal.

3.2.5 *Initial contaminant concentration*

In order to study the influence of the initial dye concentration, experimental trials from 20 to 400 $mg \cdot L^{-1}$ were performed. A fixed coagulant dosage of 1.5 mL was used for every experiment. The results of this study are shown in Figure 15.8. As this figure shows, increasing the initial dye concentration means increasing the efficiency of the dye removal, although the remaining concentration is higher as well. This result is in agreement with the previous statement, regarding the micellar behavior of this dye, so the percentage dye elimination tends to be higher due to the presence of bigger micelles that are significantly easier to remove (Esumi and Ueno, 2003).

3.3 Theoretical adsorption-like coagulation model

Coagulation is a very difficult process to model because a huge number of influential variables exist. However, it is possible to adopt a theoretical proposal based on the surfactant-polymer interaction models. These mathematical approaches are called *adsorption-like coagulation models,* and we have previously reported their use (Sánchez-Martín et al., 2011; Beltrán-Heredia et al., 2011a). Briefly, there are three such models: Freundlich, Frumnkin-Fowler-Guggenheim, and Gu and Zhu. They are extensively described elsewhere, including our own papers (Beltrán-Heredia et al., 2011c; Beltrán-Heredia and Sánchez-Martín, 2009). These models are based on the capacity variable (usually designed as q) which is given by Eq. 15.1.

$$q = \frac{(C_0 - C_1)V}{W} \tag{15.1}$$

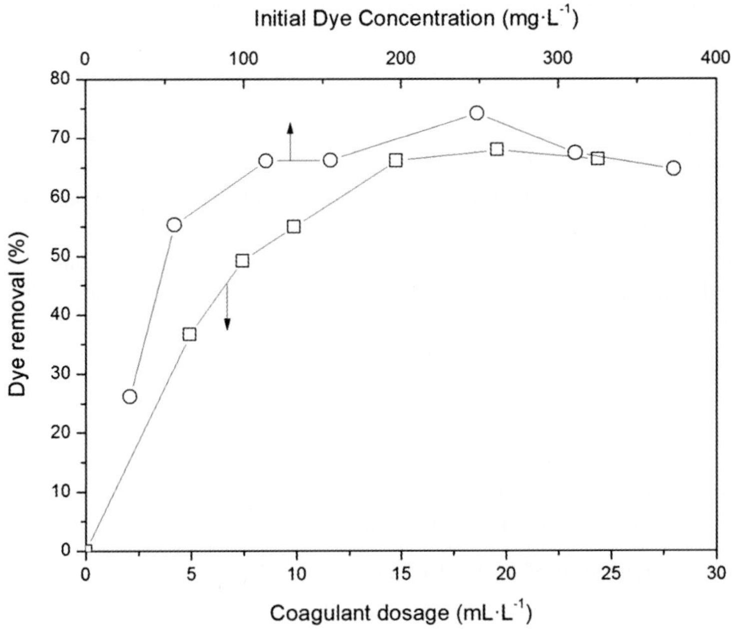

FIGURE 15.8

Coagulant dose and initial dye concentration influences in Alizarin Violet 3R removal.

where

C_0 is the initial contaminant concentration (mg L^{-1}),
C_l is the equilibrium contaminant concentration in bulk solution (mg L^{-1}),
V is the volume of solution (L), and
W is coagulant mass (g). This is calculated on the basis of tannin raw extract in the coagulant synthesis.

The Freundlich model is extensively used in the adsorption process; it presents q as a power function of the equilibrium dye concentration C_l according to Eq. 15.2:

$$q = k_f C_l^{nf} \tag{15.2}$$

where

n_f is the Freundlich adsorption order (dimensionless) and
k_f is the Freundlich adsorption constant ([L^{nf}]·[mg of coagulant]·[mg of removed dye^{nf-1}]).

For dilute solutions, a simple model ruled by Frumkin-Fowler-Guggenheim (FFG) Eq.15.3 can be applied to explain the affinity between coagulant and dye molecule (Fowler and Guggenheim, 1939; Frumkin, 1925).

$$\frac{\theta_l}{1-\theta_l} = C_l k_{12} \exp(\chi_{12}\theta_l) \tag{15.3}$$

where θ_l is the ratio between the current coagulation adsorption-like and maximum capacity.

$$\theta_l = \frac{q}{q_\infty} \tag{15.4}$$

k_{12} is a constant, being a measure of the interaction between dye and polymer surface, and χ_{12} is the Flory-Huggins parameter (Flory, 1953).

In this model, k_{12} and χ_{12} should be considered as adjustable parameters expressing the affinity for the surface and the lateral interactions in the adsorbed layer, respectively.

Finally, Zhu and Gu (1991) proposed a very simple model for polymer and suspended matter removal, assuming that the adsorbed layer is composed of contaminant aggregates. This model leads to the following Eq.15.5:

$$\frac{\theta_l}{1-\theta_l} = k_g C_l^{n_g} \tag{15.5}$$

where

n_g is the number of monomers in the aggregate (dimensionless), and
k_g is the Gu and Zhu constant for the studied model ($L^{ng} \cdot mg^{-ng}$).

Taking into account the definition of θ_l, Eq.15.5 becomes

$$q = q_\infty k_g \frac{C_l^{n_g}}{1 + k_g C_l^{n_g}} \tag{15.6}$$

Experimental data can be adequately modeled by using these mathematical approaches. This is shown in Figure 15.9 where the experimental and the predicted data are drawn. As can be seen, a classical

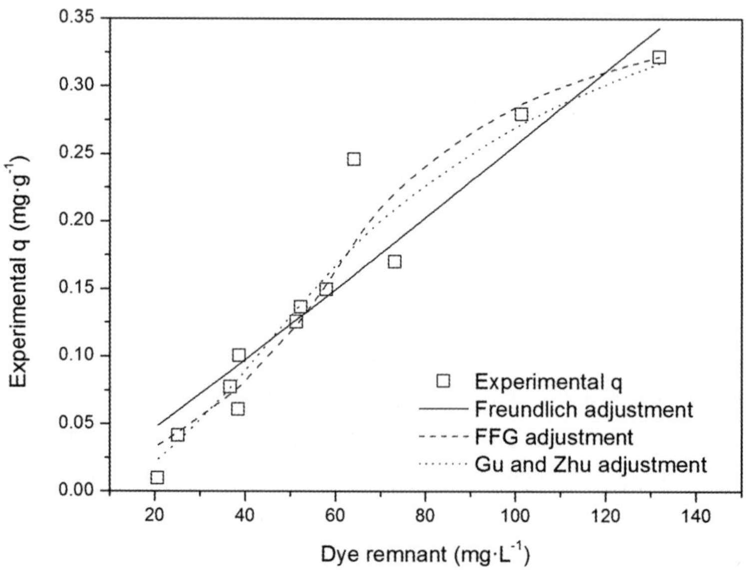

FIGURE 15.9

Theoretical model for dye removal.

Table 15.1 Theoretical Model Adjustments (Units in Text)

Model	Parameters	r^2
Freundlich	$k_l = 2.0 \cdot 10^{-3}$; $n_f = 1.05$	0.91
FFG	$q_\infty = 0.382$; $k_{12} = 3.7 \cdot 10^{-3}$; $\chi_{12} = 2.84$	0.95
Gu-Zhu	$q_\infty = 0.383$; $k_g = 6.08 \cdot 10^{-5}$; $n_g = 2.31$	0.93

S-shape curve is presented, with a slight increasing of q at low values of C_l. Then, q values increase rather quickly along the intermediate range of residual dye concentration and keep on increasing up to an asymptotic value, which corresponds to q_∞. This is shown in many previous papers (Somasundaran and Krishnakumar, 1997; Beltrán-Heredia et al., 2009b; Beltrán-Heredia et al., 2011a).

The experimental data fit reasonably well with the three proposed equations, with average r^2 above 0.90. The specific parameter values and the statistic summary for the systems and for each corresponding model are shown in Table 15.1. The graphical approach gave better data adjustment for Gu and Zhu and FFG models, since they predict adequately the shape of the curve. These adjustments were made according to nonlinear procedures, so homocedasticity and other hypothesis for linear fit can be disregarded (Kumar et al., 2008).

4 Conclusions

Quebracho colorado (*Schinopsis balansae*) tannin extract is a good precursor for coagulant agent. It can be synthesized from diethanolamine and formaldehyde, and the result is a feasible treatment agent for removing dyes from aqueous effluents. It is effective with several dyes of many types: anthraquinonic, azoic, or indigoid, among others. The coagulation process is fully working in a wide pH range, although the best working zone appears around neutrality. Temperature is negative-affecting, and the stirring variables are not very influent. The entire system can be modeled according to adsorption-like models, such as Freundlich, FFG, or Gu and Zhu, with a high r^2 correlation factor. Further studies must be carried out to scale up this coagulation process.

Glossary

Adsorbent whatever material that is used in the adsorption process; that is, the accumulation of atoms or molecules on the surface of a material. This process creates a film of the adsorbate (the molecules or atoms being accumulated) on the adsorbent's surface.

BOD_5 the acronym for biochemical oxygen demand after 5 days. It is a chemical procedure for determining the uptake rate of dissolved oxygen by the biological organisms in a body of water. It is a widely used and common technique for determining the quality of wastewater.

Cationization the chemical modification that involves conferring cationic character on a noncharged substance. Regarding tannin, cationization can be achieved by including a quaternary nitrogen in the polyphenolic net.

COD the acronym for chemical oxygen demand. COD is commonly used to indirectly measure the amount of organic compounds in water. Most applications of COD determine the amount of organic pollutants found in surface water (e.g., lakes and rivers), making COD a useful measure of water quality.

Design of experiments (also called DOE) the design of all information-gathering exercises where variation is present, whether under the full control of the experimenter or not. By using DOE, interactions between variables are shown, and conclusions can be obtained with a lower number of assays according to statistical principles.

Dye a colored substance that has an affinity to the substrate to which it is being applied. Dyes are generally applied in an aqueous solution and used either in the textile industry or in other specific applications, such as biochemistry, food production, or cosmetics.

Eutrophication an increase in the concentration of chemical nutrients in an ecosystem to an extent that produces an increase in the primary productivity of the ecosystem. Consequently, algae population can grow exponentially, so severe effects can appear in the aqueous medium such as anoxia and reduction of the biodiversity.

Frumkin-Fowler-Guggenheim model an adsorption isotherm that fits regular dilute solutions. It involves a theoretical consideration of the interaction between surfactants and polymer surfaces. The shape of the isotherm is S-type.

Gu-Zhu model an S-type adsorption isotherm that assumes that the adsorbed layer is composed of surfactant aggregates. This model considers these aggregates to be stabilized by the presence of the surface and can be reduced to the Langmuir model if the number of the monomers in the surfactant aggregate is equal to 1.

Mannich reaction an organic reaction that consists of an amino alkylation of an acidic proton placed next to a carbonyl functional group with formaldehyde and ammonia or any primary or secondary amine. The final product is a β-amino-carbonyl compound also known as a Mannich base.

Natural coagulants Apart from synthetic polyelectrolytes like aluminium sulfate or ferric chloride, there are several groups of coagulants that can destabilize colloids in suspended solutions. These are called natural coagulants due to their origin. Some of these can be found in clays, vegetal proteins (such as *Moringa oleifera* or *Opuntia ficus*), or polyphenol extracts (tannins).

Response surface methodology (RSM) a statistical procedure that explores the relationships between several explanatory variables and one or more response variables. The main idea of RSM is to use a sequence of designed experiments to obtain an optimal response. Box and Wilson suggest using a second-degree polynomial model to do this. They acknowledge that this model is only an approximation, but they use it because this model is easy to estimate and apply, even when little is known about the process.

Schinopsis balansae a hardwood tree that forms forests in the subtropical Gran Chaco ecoregion of northeastern Argentina, and Paraguay. It is also found in the wild Pantanal vegetation in Brazil. It is the second most important tannin feedstock, after *Acacia mearnsii* de Wild.

Surfactants wetting agents that lower the surface tension of a liquid, allowing easier spreading, and that lower the interfacial tension between two liquids. They are widely and commonly used in many chemical tasks, such as cosmetics, detergents, and high-tech devices.

Tannins astringent, bitter plant polyphenols that either bind and precipitate or shrink proteins. They are present in the large majority of plant barks, fruits, or pods, and they are used mainly in the tanning industry.

Acknowledgments

This investigation has been supported by the COMISIÓN INTERMINISTERIAL DE CIENCIA Y TECNOLOGÍA (CICYT) CTQ 2010-14823/PPQ project.

References

Andriamirado, L., 2007. In: Water treatment handbook, vol. 2. Degrémont, Rueil-Malmaison, France.

Beltrán-Heredia, J., Sánchez-Martín, J., 2009. Removal of sodium lauryl sulphate by coagulation/flocculation with Moringa oleifera seed extract. J. Hazard Mater. 164 (2–3), 713–719.

Beltrán-Heredia, J., Sánchez-Martín, J., Barrado-Moreno, M., 2011a. Long-chain an- ionic surfactants in aqueous solution. Removal by *Moringa oleifera* coagulant. Chem. Eng. J. 180, 128–136.

Beltrán-Heredia, J., Sánchez-Martín, J., Dávila-Acedo, M.A., 2011b. Optimization of the synthesis of a new coagulant from a tannin extract. J. Hazard Mater. 186 (2–3), 1704–1712.

Beltrán-Heredia, J., Sánchez-Martín, J., Delgado-Regalado, A., 2009. Removal of Carmine Indigo dye with *Moringa oleifera* seed extract. Ind. Eng. Chem. Res. 48 (14), 6512–6520.

Beltrán-Heredia, J., Sánchez-Martín, J., Frutos-Blanco, G., 2009a. Schinopsis balansae tannin-based flocculant in removing sodium dodecyl benzene sulfonate. Sep. Pur. Technol. 67 (3), 292–303.

Beltrán-Heredia, J., Sánchez-Martín, J., Gómez-Muñoz, C., 2012. Performance and characterization of a new tannin-based coagulant. Appl. Water Sci. 2 (3), 199–208.

Beltrán-Heredia, J., Sánchez-Martín, J., Gómez-Muñoz, M.C., 2010. New coagulant agents from tannin extracts: Preliminary optimisation studies. Chem. Eng. J. 162, 1019–1025.

Beltrán-Heredia, J., Sánchez-Martín, J., Jiménez-Giles, M., 2011a. Tannin-based coagulants in the depuration of textile wastewater effluents: elimination of anthraquinonic dyes. Water Air Soil Poll. 222 (1-4), 53–64.

Beltrán-Heredia, J., Sánchez-Martín, J., Martín-García, L., 2011b. Multiparameter quantitative optimization in the synthesis of a novel coagulant derived from tannin extracts for water treatment. Water Air Soil Poll. 223 (4), 2277–2286.

Beltrán-Heredia, J., Sánchez-Martín, J., Martín-Sánchez, C., 2011c. Remediation of dye-polluted solutions by a new tannin-based coagulant. Ind. Engin. Chem. Res. 50 (2), 686–693.

Beltrán-Heredia, J., Sánchez-Martín, J., Solera-Hernández, C., 2009b. Removal of sodium dodecyl benzene sulfonate from water by means of a new tannin-based coagulant: optimisation studies through design of experiments. Chem. Eng. J. 153 (1–3), 56–61.

Bratby, J., 2006. Coagulation and flocculation in water and wastewater treatment. IWA Publishing, London.

Bulatovic, S.M., 2007. Dispersion, Coagulation and Flocculation. In: Handbook of flotation reagents. Elsevier, Amsterdam.

Buzier, R., Tusseau-Vuillemin, M.H., Keirsbulck, M., Mouchel, J.M., 2011. Inputs of total and labile trace metals from wastewater treatment plants effluents to the Seine River. Phys. Chem. Earth A/B/C 36 (12), 500–505.

Chinyama, A., Chipato, P.T., Mangore, E., 2012. Sustainable sanitation systems for low income urban areas. A case of the city of Bulawayo, Zimbabwe. Phys. Chem. Earth A/B/C 50 (52), 223–228.

Dorf, R.C., 2001. Sustainable and appropriate technologies. Technology, humans, and society. Academic Press, San Diego.

Esumi, K., Ueno, M., 2003. Structure-performance relationships in surfactants. Marcel Dekker, New York.

Flaten, P., 2001. Aluminium as a risk factor in Alzheimer's disease, with emphasis in drinking water. Brain Res. Bull. 55 (2), 187–196.

Flory, P.J., 1953. Principles of polymer chemistry. Cornwell University Press, New York.

Fowler, R., Guggenheim, E.A., 1939. Statistical thermodynamics. Cambridge University Press, London.

Frumkin, A.N., 1925. Electrocapillary curve of higher aliphatic acids and the state equation of the surface layer. Int. J. Res. Phys. Chem. Chem. Phys. 116, 466–488.

Halling-Sorensen, B., Nielsen, S.N., Lanzky, P.F., Ingerslev, F., Holten Liitzhofl, H.C., Jorgensen, S.E., 1998. Occurrence, fate and effects of pharmaceutical substances in the environment—A review. Chemosphere 36 (2), 357–393.

Haslam, E., 1989. Plant polyphenols—Vegetables and tannins revisited. Cambridge University Press, Cambridge.

Holmberg, K., Jönsson, B., Kronberg, B., Lindman, B., 2003. Surfactants and polymers in aqueous solution. John Wiley & Sons, Chichester, England.

Kumar, V., Porkodi, K., Rocha, F., 2008. Isotherms and thermodynamics by linear and non- linear regression analysis for the sorption of methylene blue onto activated carbon: Comparison of various error functions. J. Hazard Mater. 151 (2–3), 794–804.

Matthews, S., Mila, I., Scalbert, A., Nommelly, D.M.X., 1997. Extractable and non- extractable tannins in barks. Phytochemistry 45 (2), 405–410.

Muga, H.E., Mihelcic, J.R., 2008. Sustainability of wastewater treatment technologies. J. Environ. Manag. 88 (3), 437–447.

Nakamoto, Y., Tsunoda, T., Ono, K., Kaeyama, N., Yazaki, Y., Jiang, F., Lawson, F., Uhlherr, P.H.T., 2004. Method for the production of tannins and its use. U.S. Patent 2004/0219234Al.

Pizzi, A., 2008. Tannins: Major sources, properties and applications. In: Belgacem, M., Galdini, A. (Eds.), Monomers, polymers and composites from renewable sources. Elsevier, Amsterdam.

Rossini, M., Garcia-Garrido, J., Galluzzo, M., 1999. Optimization of the coagulation- flocculation treatment: Influence of rapid mix parameters. Water Res. 33 (8), 1817–1826.

Sánchez-Martín, J., Beltrán-Heredia, J., Dávila-Acedo, M.A., 2011. Optimum coagulant from *Acacia mearnsii de Wild* for wastewater treatment. Chem. Eng. Technol. 34 (12), 2069–2079.

Sánchez-Martín, J., González-Velasco, M., Beltrán-Heredia, J., 2010. Surface water treatment with tannin-based coagulants from Quebracho (*Schinopsis balansae*). Chem. Eng. J. 165, 851–858.

Sánchez-Martín, J., Beltrán-Heredia, J., 2012. Nature is the answer. In: Sharma, S.K., Sanghi, R. (Eds.), Advances in water treatment and pollution prevention. Springer, Dordrecht, Germany.

Schumacher, E.F., 1987. Lo pequeño es hermoso (Small is beautiful). Hermann Blume. Hermann Blume, Hermann Blume, Madrid d.

Shaul, M.G., Holdsworth, T.J., Dempsey, R.C., Dostal, K.A., 1991. Fate of water soluble azo dyes in the activated sludge process. Chemosphere 22 (1–2), 107–119.

Somasundaran, P., Krishnakumar, S., 1997. Adsorption of surfactants and polymers at the solid-liquid interface. Coll. Surf. A. 123-124, 491–513.

SPSS Inc, 2005. SPSS 14.0 Developer's guide. Illinois, Chicago.

Starkl, M., Brunner, N., 2004. Feasibility versus sustainability in urban water management. J. Environ. Manag. 71 (3), 245–260.

Vivas, N., Nonier, M.F., Vivas de Gaulejac, N., Absalon, C., Bertrand, A., Mirabel, M., 2004. Differentiation of proanthocyanidin tannins from seeds, skins and stem of grapes *(Vitis vinifera)* and heartwood of quebracho (*Schinopsis balansae*) by matrix-assisted laser desorption/ionization time-of-flight mass spectrometry and thiacidolysis/liquid chromatography/electrospray ionization mass spectrometry. Anal. Chim. Acta. 513 (1), 247–256.

WHO, 2008. United Nations-Global Annual Assessment of Sanitation and Drinking Water. World Health Organization, Geneva.

Ying, C.G., 2006. Fate, behavior and effects of surfactants and their degradation products in the environment. Environ. Int. 32 (3), 417–431.

Zhu, B.Y., Gu, T., 1991. Surfactant adsorption at solid-liquid interfaces. Adv. Coll. Interf. Sci. 37 (1–2), 1–32.

CHAPTER

Biosorption of Uranium and Thorium by Biopolymers

16

C. Gok*, S. Aytas**

* Pamukkale University, Denizli Vocational School of Technical Sciences, Department of Electricity and Energy, 20070 Kinikli, Denizli, Turkey, ** Ege University, Institute of Nuclear Sciences, 35100 Bornova-Izmir, Turkey

CHAPTER OUTLINE

The Role of Colloidal Systems in Environmental Protection. http://dx.doi.org/10.1016/B978-0-444-63283-8.00016-8

1 Introduction

In recent years, the use of biotechnology to control and remove metal pollution has attracted much attention and gradually has become a hot topic in the field of metal pollution control because of its potential application. An alternative process is biosorption, which utilizes various natural biological materials, including bacteria, fungi, yeast, and algae. These biosorbents possess metal-sequestering properties and can be used to decrease the concentration of heavy metal ions in solution from the ppm to the ppb level. It can effectively sequester dissolved metal ions out of dilute complex solutions with high efficiency and quickly; therefore it is an ideal candidate for the treatment of high-volume and low-concentration complex wastewaters (Wang and Chen 2006).

Biosorption can be defined as the removal of metal or metalloid species, compounds, and particulates from solutions by biological materials. Large quantities of metals can accumulate through a variety of processes that are both dependent on and independent of metabolism. Both living and dead biomass as well as cellular products such as polysaccharides can be used for metal removal. Such algal biomass can be used in a natural form free in solution or immobilized by various techniques on to a solid support or to produce granules for metal removal/recovery. Algal biosorption techniques/processes can be used to remove toxic metals and/or radionuclides from liquid effluents before their safe discharge, in addition to their use as a recovery processes for metals of value. In the majority of cases of biosorption, metals can be considered as rapid metabolism-independent ionic and covalent binding to a particular structure of cell surfaces (Gadd and White 1990).

Biological materials have received increasing attention as a means of removal and recovery of heavy metals and radionuclides due to their high performance. When we look at the number of published paper on the biosorption-uranium and biosorption-thorium topic as listed in the ISI Web of Science database, an increasing number of published articles can be found, especially after 1995 (Figure 16.1). Until July 1, 2013, 462 and 70 articles had been issued, respectively, on biosorption uranium and biosorption thorium.

Thorium (Th) and uranium (U) are natural radionuclides and are widely distributed in nature due to nuclear power production as well as to a number of human activities (e.g., the mining, production, and use of phosphate fertilizers, copper metallurgy, and military activities). These elements are possible harmful pollutants in the environment (Sakamoto, Kano, and Imaizumi 2008; Harmsen and Haan 1980; Bituh

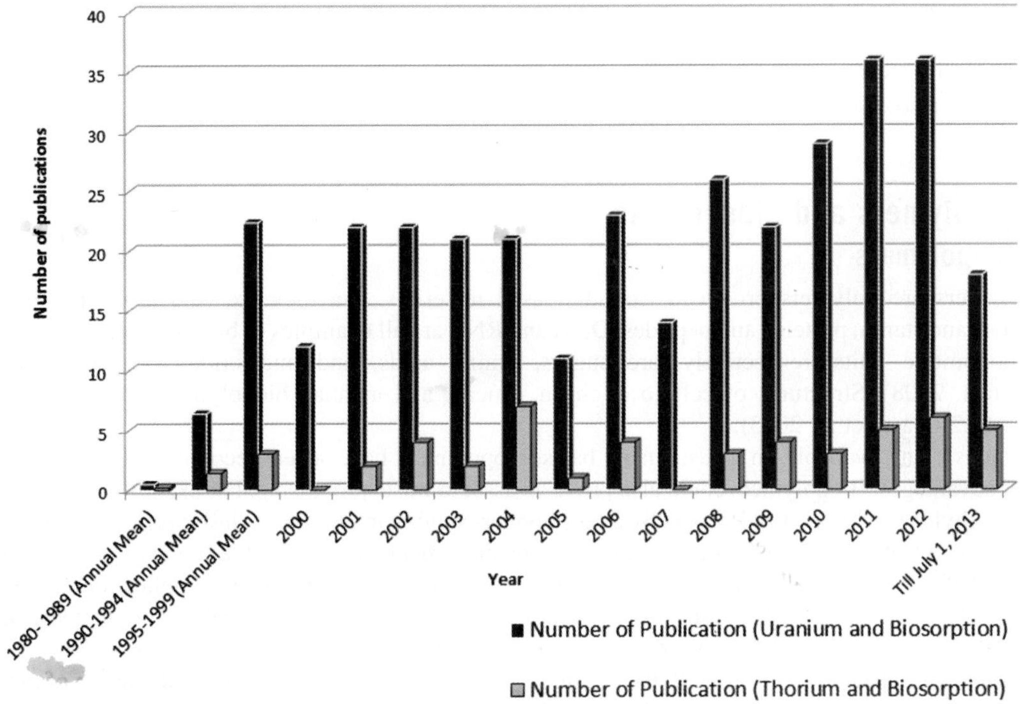

FIGURE 16.1

Number of papers appearing with biosorption-uranium and biosorption-thorium in the topic as listed in the ISI Web of Science database.

(Database searched at 20.12.2012).

et al. 2009). Uranium is one of the most serious contamination concerns because of its radioactivity and heavy-metal toxicity. Excessive amounts of uranium have found their way into the environment through activities associated with the nuclear industry. Uranium biosorption by various microorganisms and related biopolymers are reported in the literature (Genc et al. 2003; Gadd 1990, 1991).

Thorium is a naturally occurring radioactive element widely distributed over the Earth's crust. Some human activities such as exploitation of ores with associated thorium and nuclear fuel reprocessing can also concentrate this element (Salinas-Pedroza and Olguin 2004). Thorium is an important model element for tetravalent actinides in natural waters. It is also useful as a tracer when studying environmentally important processes (Osthols 1997). Since the last century, thorium has been extensively used in a variety of applications. These applications produce various gaseous, liquid, and solid wastes containing isotopes of uranium, thorium, and the daughter ions of Rn, Po, Bi, Ra. Liquid wastes are freed into the surface or the groundwaters of mines. Solid and liquid wastes are also produced during nuclear fuel production. Direct toxicity of thorium is low due to its stability at ambient temperatures; however, thorium fine powder is self-ignitable to thorium oxide (Metaxas 2003; Kaygun and Akyıl 2007).

This chapter will deal in general with biosorption and biosorption of radionuclides, as well as biopolymers used for biosorption of uranium and thorium in environmental protection. The modeling and mechanism of biosorption, process parameters, assessment, applications, and the future of biosorption are also discussed.

2 Biopolymers and biosorbents

2.1 Biopolymers

Biopolymers are polymers produced from biobased materials, and they are also biodegradable. Cellulose and starch, proteins and peptides, DNA and RNA are all examples of biopolymers, in which the monomeric units, respectively, are sugars, amino acids, and nucleotides (Encyclopædia Britannica, 2008). Structures of cellulose, chitin, glucan, and mannan biopolymers are shown in Figure 16.2 (Davis et al. 2003).

Besides being available on a sustainable basis, biopolymers have several economic and environmental advantages. Biopolymers could also prove an asset to waste processing. For example, replacing the polyethylene used in coated papers by a biopolymer could help eliminate plastic scraps occurring in compost. Whether the use of biopolymers will contribute to a more sustainable society is a question that will have to be studied by a representative life-cycle analysis (LCA) of each application and by comparison with existing applications. Representative LCAs are needed at material and product levels; they must make allowance for future developments in biopolymers and take account of all relevant environmental impacts associated with the complete product life cycle, including the depletion of raw materials, the agricultural use of fertilizers and pesticides, transportation, utilization, and waste disposal.

There are four main types of biopolymer based respectively on (1) starch, (2) sugar, (3) cellulose, and (4) synthetic materials. The ***starch-based polymer*** is a natural polymer that occurs as granules in plant tissue, from which it can easily be recovered in large quantities. It is obtained from potatoes, maize, wheat, tapioca and similar sources. Starch can be modified in such a way that it can be melted and reformed thermoplastically. The resulting material is thus suitable for conventional plastic-forming processes such as injection molding and extruding.

Sugar based biopolymers are the starting materials for polyhydroxibutyrate made from sucrose or starch through the process of bacterial fermentation. Varying the nutrient composition of the bacteria produces differences in the end product. This makes it possible to tune the properties of the material, for example, its moisture resistance. The polymer can be formed by injection, extrusion, blowing, and vacuum forming. Polylactides (lactic acid polymers) are made from lactic acid, which is in turn made from lactose (or milk sugar) obtained from sugar beet, potatoes, wheat, maize, and the like. Polylactides are water resistant and can be formed by injection molding, blowing, and vacuum forming.

The use of ***cellulose-based biopolymers*** for making packaging material such as cellophane is long established. The material is transparent and has good folding properties. Whether in the form of pure cellulose or of a nitrocellulose coating, the material is wholly biodegradable and can be composted by existing waste-processing plants (http://www.o2.org/ideas/cases/biopolymers.html). Cellulose is both the most common biopolymer and the most common organic compound on Earth. About 33% of all plant matter is cellulose (the cellulose content of cotton is 90%, and that of wood is 50% (Encyclopædia Britannica, 2008).

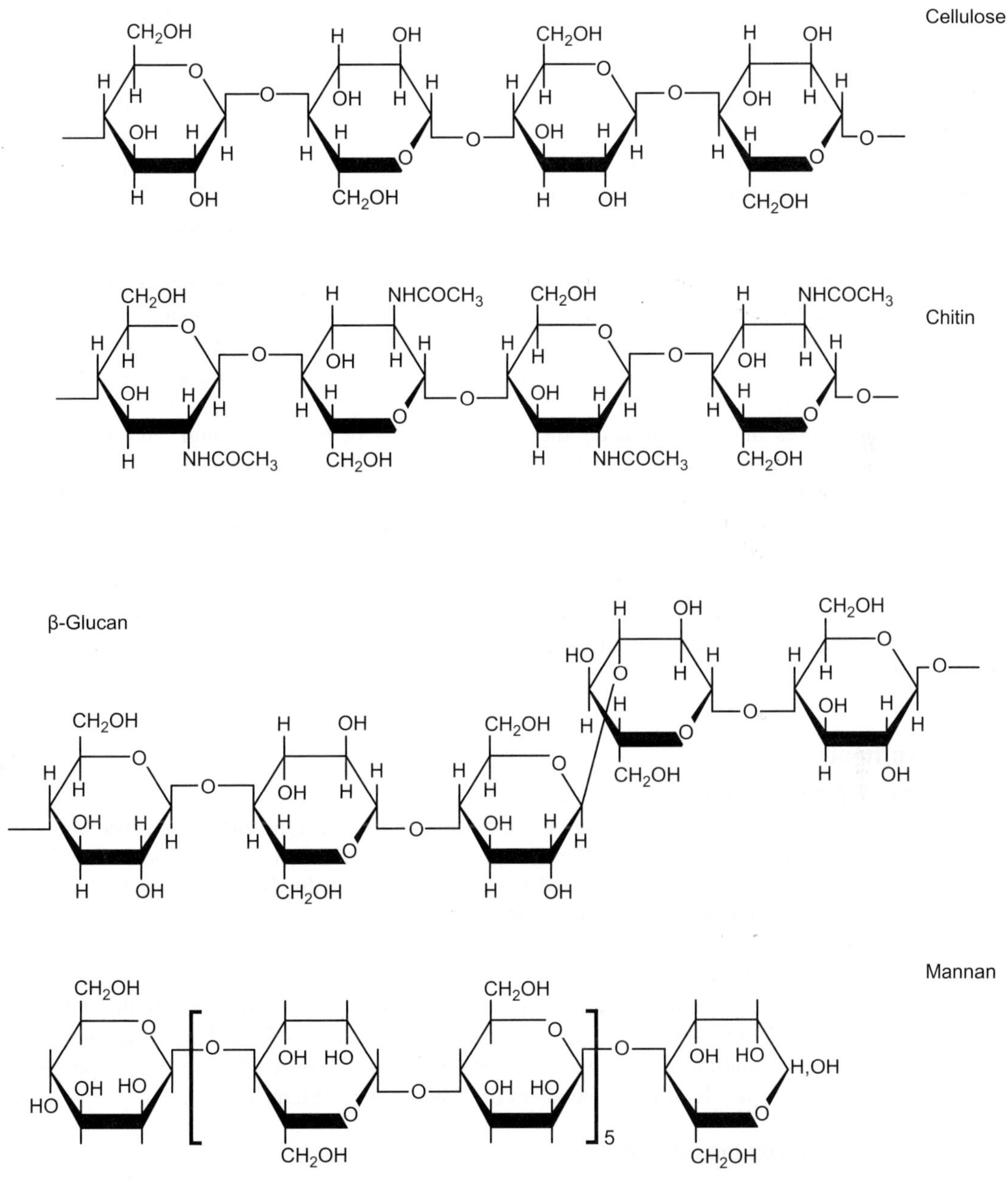

FIGURE 16.2 Structures of cellulose, chitin, glucan, and mannan.

(From Davis, T.A., Volesky, B., Mucci, A., Water Research, *37, 4311, 2003).*

Synthetic-based biopolymers are the relatively high price of biodegradable polymers of synthetic substances. For example, aliphatic aromatic copolyesters have prevented them from reaching a large-scale market. The best known application is for making substrate mats. Synthetic compounds derived from petroleum can also be a starting point for biodegradable polymers—for example, aliphatic aromatic copolyesters. These polymers have technical properties resembling those of polyethylene (LDPE). Although these polymers are produced from synthetic starting materials, they are fully biodegradable and compostable (http://www.o2.org/ideas/cases/biopolymers.html).

Many biopolymers such as chitosan, sodium alginate extracted from microalgae (Da Costa and Leite, 1991), shrimp, crab, and some fungi (Bosinco et al. 1997; Jang et al. 1991) are known to bind metal ions strongly and could be used for heavy metal adsorption. Biopolymers are nontoxic, selective, efficient and inexpensive, and thus are highly competitive with ion-exchange resins and activated carbon. Immobilizing biomass in a biopolymeric matrix may also improve biomass performance and biosorption capacity and facilitate the separation of biomass from metal-bearing solutions (Vijayaraghavan and Yun 2008).

One of the features that discriminates the microbial cells of bacteria, fungi, and algae from animal cells is the presence of a distinct barrier, the cell wall, on the outside of the cytoplasmic membrane of the cell. The wall protects the microorganisms against the injuries of the environment, controls the fluxes between the cytoplasm and the surrounding environment, and determines the shape and the rigidity of the cell. It is regarded as the external cellular skeleton of the cell. Obviously, the wall is directly in contact with the environment external to the cell and interacts especially with the substance soluble in the liquid medium. The interaction may be particularly pronounced in the case of metallic ions due to the anionic characteristics of the cell wall (Remacle 1990).

2.2 Biosorbents

2.2.1 Bacterial biosorbents

Bacterial cell walls have received much attention since the discovery of antibiotics. The walls of bacteria present common features, and three major groups of bacterial cell walls can be distinguished based on the importance of peptidoglycan and the adjacent layers: the Gram-positive bacteria, the Gram-negative bacteria, and the archaebacteria (Remacle 1990; Sleyter 1981).

Bacteria are the most abundant and versatile of microorganisms (Ferris and Beveridge 1985) and constitute a significant fraction of the entire living terrestrial biomass of $\sim 10^{18}$g. The high capacity for biosorption of heavy metals by microorganisms has been known for some time (Fairbridge 1972).

Knowledge of metal uptake by bacteria and other microorganisms is still in its infancy. The property of metal accumulation by cell walls will assume increasing importance due to problems stemming from the rising levels of toxic metals in the environment (Mann 1990).

2.2.2 Fungal and yeast biosorbents

Fungi are ubiquitous in natural environments and important in industrial processes. Morphologies range widely from unicellular yeasts to polymorphic and filamentous fungi, many of which have complex macroscopic fruiting bodies. Their most important roles are as decomposers of organic materials, with concomitant nutrients cycling, as pathogens and symbionts of animals and plants, and as spoilage organisms of natural and synthetic materials, for example, wood, paint, leather, food, and fabrics. They are also utilized as producers of economically important substances, for example, ethanol, citric acid, antibiotics, polysaccharides, enzymes, and vitamins (Gadd 993).

In the field of biosorption, the molds and yeast are of interest, and many research studies have been reported and reviewed. The molds are filamentous fungi. The yeasts are unicellular fungi, and most of them are classified with the Ascomycetes. The most important commercial yeasts are the baker's and brewer's yeasts, which are members of the genus Saccharomyces. The original habitats of these yeasts were undoubtedly fruits and fruit juices, but the commercial yeasts of today are probably quite different from wild strains because they have been greatly improved through the years by careful selection and genetic manipulation of eukaryotic cells, and they are thus excellent models for the study of many important problems in eukaryotic biology. Yeast cells are much larger than bacterial cells and can be distinguished microscopically from bacteria by their size and by the obvious presence of internal cell structures, such as the nucleus (Madigan et al. 1997).

2.2.3 Algae biosorbents

Algae, which abound in nature in aquatic habits, freshwater, marine environments, and moist soil, contain chlorophyll and carry out oxygenic photosynthesis. The algal groups include Chlorophyta (green algae), Euglenophyta (euglenoids, also considered with the protozoa), Chrysophyta (golden-brown algae, diatoms), Phaeophyta (brown algae), Pyrrophyta (dino-flagellates), and Rhodophyta (red algae). One of the key characteristics used in the classification of algal groups is the nature of the reserve polymer synthesized as a result of photosynthesis. Algae of the division Chlorophyta produce starch in a form very similar to that of higher plants. By contrast, algae of other groups produce a variety of reserve substances, some polymeric and some as free monomers (Wang and Chen 2009).

Algae are eukaryotic microorganisms that carry out the process of photosynthesis. In these organisms, as well as in green plants, an additional type of organelle is found: the chloroplast. The chloroplast is green and is the site where chlorophyll is localized and where the light-gathering functions involved in photosynthesis occur. Algae have been extensively studied due to their ubiquitous occurrence in nature. The term *algae* refers to a large and diverse assemblage of eukaryotic organisms that contain chlorophyll and carry out oxygenic photosynthesis. It should be noted that algae are distinct from cyanobacteria, which are also oxygenic phototrophs, but they are eubacteria (true bacteria) and are therefore evolutionarily distinct from algae. Although most algae are of microscopic size and hence are clearly microorganisms, a number of forms are macroscopic, some seaweeds growing to over 100 ft in length (Madigan et al. 1997).

2.2.4 Composite biosorbents

Composites may display more properties of the effective adsorbents than inorganic and organic components individually. To prepare composite ion exchangers or composite adsorbers, both inorganic and organic materials may be used as binding materials. Composite absorbers with inorganic binding matrices are based on the following binders: clay materials, cement, silica-gel, and alumina. Procedures for preparing cement or clay-based exchangers involve preparation of a binder-exchanger paste of suitable consistency, curing, and/or drying, followed by grinding and sieving (Sebesta and John 1995; Mahan and Holcombe 1992; Donat and Aytas 2005).

The algae-clay composite adsorbent was tested for its ability to recover U(VI) from diluted aqueous solutions by Donat and Aytas (2005). Macro marine algae (*Ulva sp.*) and clay (Na bentonite) were used to prepare composite adsorbent.

Boddu et al. (2003) thought that in order to provide physical support and increase the accessibility of the metal-binding sites for process, applications of composite chitosan biosorbent are also necessary. Hence, they attempted to prepare a biosorbent by coating chitosan on alumina.

2.3 Pretreatment of the biosorbents

Some researchers have found that biosorbents are singly inefficient to accumulate metal ions. So, some methods have been developed to solve this deficiency. As a result, chemical modification with different chemicals, immobilization or encapsulation methods, polymerization, and protonation techniques have been used to improve stability and capacity of biosorbents.

Alginate is one of the most frequently used biosorbents, and it was used either without pretreatment or with some modification techniques. Rangsayatorn et al. (2004) immobilized *Spirulina platensis* on alginate gel and silica gel for biosorption of cadmium (Rangsayatorn et al. 2004). Gok and Aytas (2009) used alginate beads polymerized with Ca^{2+} for the removal of uranium ions from aqueous solutions. Mimura et al. (2001; 2002) used alginate gel polymers to uptake americium and ruthenium. Also, freshwater algae were immobilized in calcium alginate for reduction of uranium. Ibañez et al. (2002) used protonated alginate beads for heavy-metal uptake.

Chen and Yang (2005) aimed to screen different surface modification methods on *Sargassum sp.* for the prevention of organic leaching and the enhancement of heavy-metal removal. Solisio et al. (2000) investigated the effect of acid pretreatment on the biosorption of chromium(III) by *Sphaerotilus natans* from industrial wastewater. Several biosorbents were prepared from orange peel by means of different chemical modifications, and Xuan et al. (2006) examined their adsorption characteristics for lead ion from aqueous solution).

2.4 Functional groups of biosorbents

The variety of structural components present in biomass means that many functional groups are able to interact with metal species (e.g., carboxyl, phosphate, hydroxyl, amino, thiol) to varying degrees and are influenced by physicochemical factors. For biosorption, defined as a physicochemical process independent of metabolism, such mechanisms as adsorption, ion exchange, and complexation/coordination may be important and, in these cases, biosorption can be rapid and reversible with biomass properties analogous to conventional ion exchange resins. Ion exchange is the replacement of an ion in a solid phase in contact with a solution by another ion. More specifically, it is the replacement of an absorbed, readily exchangeable ion by another (Sposito 1989).

Ralph Pearson (1969) introduced the hard-soft-acid-base (HSAB) principle. The HSAB principle indicates that hard acids prefer binding to hard bases (often forming bonds with substantial ionic character), while soft acids prefer binding to soft bases (often forming bonds with substantial covalent character). Although the HSAB principle is a foundational concept of the modern theory of acids and bases, the theoretical underpinnings of the HSAB principle remain murky.

Pearson (1969) classified metallic ions according to a "hardness scale" defined by their binding strength with F^- and I^-. The metallic ions forming strong binding with F^+ are called hard. Among them

Table 16.1 The Ligands Present in Biological Systems and Three Classes of Metals

Ligand Class	Ligands	Metal Classes
I: Ligands preferred to Class A	F^-, O^{2-}, OH^-, H_2O, CO_3^{2-}, SO_4^-, $ROSO_3^-$, NO_3^-, HPO_4^{2-}, PO_4^{3-}, ROH, $RCOO^-$, C=O, ROR	Class A: Li, Be, Na, Mg, K, Ca, Sc, Rb, Sr, Y, Cs, Ba, La, Fr, Ra, Ac, Al, Lanthanides, Actinides
II: Other important ligands	Cl^-, Br^-, N_3^-, NO_2^-, SO_3^{2-}, NH_3, N_2, RNH_2, R_2NH, R_3N, =N–, –CO–N–R, O_2, O_2^-, O_2^{2-}	Borderline ions: Ti, V, Cr, Mn, Fe, Co, Ni, Cu, Zn, Ga, Cd, In, Sn, Sb, As
III: Ligands preferred to Class B	H^-, I^-, R^-, CN^-, CO, S^{2-}, RS^-, R_2S, R_3As	Class B: Rh, Pd, Ag, Lr, Pt, Au, Hg, Tl, Pb, Bi

Source: Nierboer, E. and Richardson, D.M.S., Environ. Pollut. Ser. B, *1, 3, 1980. Pearson, R.G.,* Surv. Prog. Chem., *5, 1, 1969. Remacle, J., In* Biosorption of Heavy metals, *ed. B. Volesky, Boca Raton, FL: CRC Press, 1990.*

are Na^+, Mg^+, and Ca^{2+}, which are present in fairly high quantities in most living organisms. The soft ions are located at the other extremity of Pearson's scale, and among them are Hg^{2+}, Cd^{2+}, and Pb^{2+}. They are not known as biological nutrients and are usually toxic. Heavy metals are found among the soft ions. Metallic ions of an intermediate degree of hardness (e.g., Zn^{2+} an Co^{2+}), are less toxic and can even be detected in certain biomolecules where they assist in mediating specific biochemical reactions. In biological systems, hard ions form stable bonds with OH^-, HPO_4^{2-}, CO_3^{2+}, R-COO$^-$, and =C=O. All of these radicals include oxygen atoms. The soft ions form very strong bonds with CN^-, R-S$^-$, -SH$^-$, NH_2^-, and imidazol, that is, groups containing nitrogen and sulfur atoms. The hard ion binding is mainly of an ionic nature, whereas the soft ion binding displays a more covalent character. Nierboer and Richardson (1980) proposed a classification based on the atomic properties and the solution chemistry of the metal ions. They discriminated metal ions into three classes: oxygen-seeking, nitrogen- and sulfur-seeking, and a borderline or intermediate class.

According to the metal classification by Pearson (1969), as well as by Nieboer and Richardson (1980), metal affinity for ligands is indicated in Table 16.1 (Remacle, 1990).

2.5 Characterization of biosorbents

The biosorbents were characterized using some different techniques such as FTIR spectra, SEM micrograph, and thermogravimetric and surface analysis to understand the surface morphology (Vijayaraghavan and Yun 2008).

Chemical modeling of metal biosorption requires the characterization of the biomass used as biosorbent. In the study of Pagnanelli et al. (2000), to establish the nature and concentration of active sites on the cell wall, studies of ionic content, potentiometric titer, and IR spectrum in the solid phase of the lyophilized biomass were carried out. An IR analysis in solid phase was performed to give a qualitative and preliminary analysis of the main chemical groups present on the cell membrane (Pagnanelli et al. 2000).

Barbetta et al. (2009) prepared porous alginate hydrogels and characterized biopolymer with elemental analysis, surface area/pore-size distribution, particle-size determination, and nuclear magnetic resonance (NMR).

An alumina-supported biosorbent was characterized by high-temperature pyrolysis, scanning electron microscopy, and X-ray photoelectron spectroscopy. Boddu et al. (2003) determined the surface area, pore diameter, and pore diameter distribution with the nitrogen porosimeter on the basis of the Brunauer–Emmett–Teller (BET) adsorption isotherm.

2.6 Ca-alginate biopolymers

Alginate is an exopolymer extracted mainly from brown algae and various bacteria that have been used both as an immobilization material and as a biosorbent of several heavy metals. Moreover, alginate is one of the constituents of the cell walls of brown algae, and it consists of mannuronic and guluronic acid monomers. Oligopolymeric blocks of guluronic acid show a high selectivity for heavy metal ions (Seki and Suzuki 1998; Yang and Volesky 1999b).

Among these biosorbents, alginate is one of the most extensively investigated biopolymers for metal ion removal from aqueous solution as it is inexpensive, nontoxic, and efficient. Indeed, carboxylate functions of this polysaccharide are negatively charged in neutral and alkaline media and hence have greater affinity to cations (Ngomsika et al. 2006).

Alginate is a natural polysaccharide extracted from various species of brown seaweed, which constitutes 10–40% of the dry weight of all species of brown algae. It consists of linear copolymers composed of beta-D-Mannuronate (M) and alpha-L-Guluronate (G), linked by b-1,4- and a-1,4-glycosidic bonds. M and G units are organized in homopolymeric M–M and G–G blocks and heteropolymeric M–G blocks. The proportions of these blocks vary with the source of the alginate polymer. Alginate structure can be seen in Figure 16.3 (Smidsrod and Draget 1996). The carboxylate groups of the polymer give it the ability to undergo a sol-gel transition in the presence of multivalent cations and more specifically with calcium ions. The gelation properties of alginate can be attributed to the simultaneous binding of the divalent cations such as Ca^{2+} to different chains of α -L-guluronate blocks; regions of polyguluronate in one alginate are linked to a similar region in another alginate molecule by means of calcium binding the alginate polymers together, which gives rise to the famous "egg box structure" (Grant et al. 1973; Volesky 2003a; Nestle and Kimmich 1996).

Because of its ability to form stable structures, cross-linked alginate has been used for the removal of heavy metal from wastewater (Holan et al. 1993; Kuyucak and Volesky 1989; Romero-Gonzalez et al. 2001). Although it has been demonstrated that the alginates present in algae are capable of binding heavy metals through carboxyl groups (Chen and Yiacoumi 1997; Ibañez and Umetsu 2002; Konishi et al. 1993; Mimura et al. 2001), very few studies have been conducted on the potential use of alginates as a sorbent for heavy metal removal from aqueous solutions.

3 Biosorption

The term *biosorption* describes the removal of heavy metals by the passive binding to nonliving biomass from an aqueous solution. This implies that the removal mechanism is not metabolically controlled. In contrast, the term *bioaccumulation* describes an active process whereby removal of metals requires the metabolic activity of a living organism. In recent years, research on the mechanisms of biosorption has intensified since biomass can be employed to sequester heavy metals from industrial effluents (e.g., from the mining or electroplating industry) or to recover precious metals from processing solutions (Davis et al. 2003).

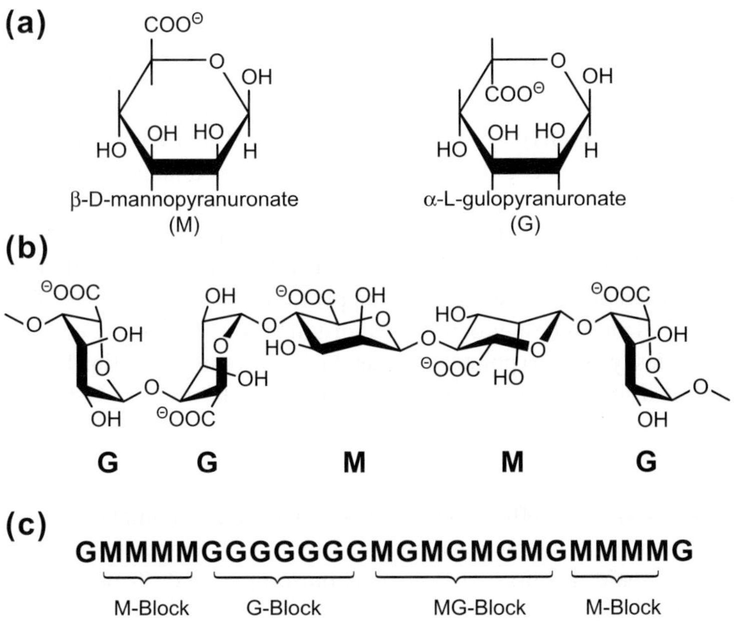

FIGURE 16.3

Alginate structural data: (a) alginate monomers; (b) the alginate polymer; (c) chain sequences of the alginate polymer.

(From Smidsrod, O. and Draget K.I., Carbohydrate European, *14, 6, 1996).*

3.1 Factors affecting biosorption

To identify the physiochemical conditions that are most effective for the removal of metal ions from wastewater, it is essential to have information about the effect of system parameters under varying process conditions. This is required for optimum system design criteria, because wastewater from different metallurgical and chemical process may contain mixtures of metals and other waste compounds (McKay and Allen 1997). Many factors can affect biosorption, and some important parameters are described in detail below.

3.1.1 pH

pH is one of the most important environmental factors for biosorption of heavy metals and radionuclides and is critical to the sorption process. It affects the solution chemistry of metals, the activity of functional groups in the biomass, and the competition of metallic ions (Veglio and Beolchini 1997). The pH value of solution strongly influences not only the site dissociation of the biomass surface, but also the solution chemistry of the heavy metals: hydrolysis, complexation by organic and/or inorganic ligands, redox reactions, precipitation, the speciation, and the biosorption availability of the heavy metals (Wang 2002).

For the majority of biosorbents the optimum pH seems to be slightly acid to around neutral (4–7) (Edyvean et al. 1997). The optimum pH for removal of U and Th sorption by *Pseudomonas* biomass was found to be pH 4.0 and pH 5.0, respectively (Pinaka et al. 2004). Thorium sorption by *Pseudomonas* biomass was strongly influenced by the pH of the solution, with optimum absorption at pH 4, compared to the Th^{4+} and $Th(OH)^{2+}{}_2$ ions formed at low pH that have been identified as a poor sorbate; the higher uptake at pH 4 could be correlated to the predominance of $[Th_2(OH)_2]^{6+}$ (Sar and D'Souza 2002).

Tsezos and Volesky (1981) indicated that solution pH affected the exhibited uptake. In general, lower biosorptive uptake was exhibited at pH 2 than at pH 4. No discernible difference in uptake was observed between pH 4 and pH 5 where the optimum pH for biosorption lies. The biomass of *Rhizopus arrhizus* at pH 4 exhibited the highest uranium and thorium biosorptive uptake capacity (g) in excess of 180 mg/g. Yang and Volesky (1999a) pointed out that the high uranium sorption by *Sargassum* biomass at pH 4.0 was attributed to the existence of hydrolysed uranyl ions, UO_2OH^+, $(UO_2)_3(OH)_5^+$, and $(UO_2)_2(OH)_2^{2+}$.

3.1.2 Temperature

Temperature can have a significant effect on biosorption. In general, optimal temperatures lie mostly in the 10–25°C range, with some interactions enhanced by higher temperatures. Any residual heat in industrial wastewaters would therefore not suppress biosorption and, if anything, might enhance it (Edyvean et al. 1997).

Temperature also has an influence on the biosorption of metal ions, but to a limited extent under a certain range of temperature, which indicates that the ion-exchange mechanism exists in biosorption to some extent. Biosorption process is usually not operated at high temperature because it will increase the operational cost (Wang 2002).

Ishikawa et al. (2002) investigated uranium and thorium recovery from dilute aqueous solutions by animal biopolymers. The uptake of U decreased with increasing temperature (30–50°C), whereas the Th uptake increased with increasing temperature in this study.

Biosorption of uranium by *Talaromyces emersonii* CBS 814.70 biomass was examined by Bengtsson et al. (1995). Studies on the effects of variation in temperature on the biosorptive capacity demonstrated no significant change in binding between 20°C and 60°C.

3.1.3 Contact time

The contact time of removal of heavy metals or radionuclides is critical to the design and economics of any biosorption process. In most of the biosorption studies, maximum equilibrium was reached within 2 h.

Ucun et al. (2002) found that the optimum biosorption time of chromium (VI) from aqueous solution by cone biomass of *Pinus sylvestris* was 2 h. At the biosorption of Cr (VI) from aqueous solution by *Rhizopus nigricans,* more than 75% of the ions were removed within 30 min of contact time, and maximum removal was obtained after 8 h (Bai R and Abraham 2001).

Batch sorption equilibrium of biosorption of uranium on *Sargassum* biomass was reached within 3 h (Yang and Volesky 1999a). *Aspergillus fumigatus* removed uranium (VI) very rapidly and reached equilibrium within 1 h of contact of biomass with the aqueous metal solution (Kuber et al. 1999). Lyophilized biomass of a *Pseudomonas* soilisolate adsorbed thorium (IV) optimally at pH 4, with 91% of equilibrium loading being reached in 1 min (Sar and D'Souza 2002).

3.1.4 Competing ions/co-ions

Real industrial effluent usually contains various ionic components, including metal cations and anions. Some studies indicate that cations and anions additional to the ions of interest have a generally detrimental impact on metal accumulation (Suh and Kim 2000). Usually, the biosorption capacity of one metal ion is interfered with and reduced by co-ions, including other metal ions and anions presenting in solution. However, the gross uptake capacity of all metals in solutions remains almost unchangeable (Wang and Chen 2006).

3.1.5 Initial concentration of metal ions and biomass concentration

The uptake rate of the metal ion will increase along with increasing the initial concentration if the amount of biomass is kept unchanged. Contrary to that, biosorptive capacity of the metal ions is inversely proportional to the initial concentration of the biomass when the initial concentration of metal ions is kept constant. Increase of the biomass concentration of the biosorption system could result in increasing the sorption site interactions. When the biomass concentration is low, metal ions in the solution would not only be adsorbed to the surface of the biomass, but also enter into an intracellular part through facilitating the concentration gradient of metal ion (Wang 2002).

3.2 Desorption studies

The critical cost factor for an adsorbent is often not its single exposure capacity but its lifetime exposure capacity, that is, the number of adsorption cycles the material can stand while remaining structurally sound and capable of adsorption. This presupposes that the metal adsorbed can be desorbed (Edyvean et al. 1997).

Desorption of loaded biomass enables reuse of the biomass and recovery and/or containment of sorbed materials, although it is desirable that the desorbing agent does not significantly damage or degrade the biomass (Gadd and White 1992). In some cases, desorption treatments may improve further sorption capacities, although in other cases there may be a loss of efficiency of the biomass. For operation of continuous flow systems, columns in parallel arrangements may allow sorption and desorption processes to occur without significant interruption. A variety of substances have been used as metal/radionuclide desorbents, including acids, alkalis, and complexing agents, depending on the substance sorbed, process requirements, and economic considerations. In addition, there may be a means of selective desorption (e.g., for certain metals). Combustion and subsequent recovery of metal/radionuclides from ash ("destructive recovery") may also be a possibility (Aksu 2005; Gadd 2009).

Fungus Lentinus sajor-caju could be regenerated using 10 mM sodium carbonate, with up to 93% recovery of uranium (Bayramoglu et al. 2006). The immobilized *T. conglobatum* cells adsorbed uranium almost quantitatively, and almost all uranium adsorbed was desorbed with 1 M Na_2CO_3 (Nakajima and Sakaguchi 1993). The uranium could be easily recovered from the metal-loaded *Sargassum biomass* by elution with 0.1 N HCl (Yang and Volesky 1999a). More than 90% of loaded thorium could be recovered from *Pseudomonas* biomass using 1 M $CaCO_3$ (Sar and D'Souza 2002). Desorption of adsorbed metals from *Pseudomonas aeruginosa* biosorbent released around 85% using citrate buffer (pH4, 0.2 M), whereas HCl could desorb 95% of the sorbed metal (Philip et al. 2000).

4 Mechanism of biosorption

Biosorption is a fast and reversible process for the removal of heavy metals or radionuclides from wastewater (Gavrilescu 2004). Adsorption and desorption studies invariably yield information on the mechanism of metal biosorption: how the metal is bound within the biosorbent. This knowledge is essential for understanding the biosorption process, and it serves as a basis for quantitative stoichiometric considerations, which constitute the foundation for mathematical modeling of the process (Volesky 1999).

The mechanism of metal biosorption is complicated and not fully understood. The status of biomass (living or nonliving), types of biomaterials, properties of metal-solution chemistry, and ambient/environmental conditions such as pH, will all influence the mechanism of biosorption (Wang and Chen 2006).

There are several mechanisms by which metal ions may be removed from a solution and attached to a sorbent particle surface. This is due largely to the complex chemistry of the sorbent surface. Metal sequestering by different parts of the cell can occur via various processes: complexation, chelation, coordination, ion exchange, precipitation, and reduction (McKay and Allen 1997).

Oxidation/reduction reactions may also be taking place in the biosorbent. Due to the complexity of the biomaterials used, it is quite possible that at least some of these mechanisms are acting simultaneously to varying degrees, depending on the biosorbent and the solution environment (Volesky 1999).

Extracellular polymeric materials are capable of binding and entrapment of metal and radionuclide species: Such phenomena may be significant in biofilms (White and Gadd, 1998). A variety of oxidoreductive transformations that alter metal or radionuclide speciation can lead to enhanced deposition in around cell walls, or even volatilization (Gadd 2000; Hockin and Gadd 2007).

5 Evaluation of biosorption studies

5.1 Sorption isotherms-equilibrium modeling

Sorption isotherms are mathematical models that describe the distribution of the adsorbate species among solid and liquid phases. They are important for the description of how sorbate will interact with a biosorbent and are critical in optimizing the use of sorbent.

Some empirical models for single-solute systems are listed in Table 16.2. Real wastewaters commonly contain a mixture of metal ions. Thus, multimetal biosorption models have been developed, and these models are listed in Table 16.3.

In the biosorption studies, three important sorption isotherm models were used mostly in order to investigate the biosorption isotherm—Langmuir, Freundlich, and Dubinin–Radushkevich (D–R) isotherm models—and are described below.

5.1.1 Langmuir isotherm

The Langmuir model suggests monolayer sorption on a homogeneous surface without interaction between sorbed molecules. In addition, the model assumes uniform energies of sorption onto the

Table 16.2 Frequently Used Single-Component Adsorption Models

Isotherm Types	Equations	Nomenclature
Langmuir (Langmuir 1918)	$q_e = \frac{Q_0 b C_e}{1+bC_e}$	q_e is equilibrium metal sorption capacity; C_e is equilibrium solute concentration in solution; Q_0 and b are Langmuir constants related to maximum sorption capacity (monolayer capacity) and bonding energy of adsorption (or "affinity"), respectively.
Freundlich (Freundlich 1906)	$q_e = K_F \times C_e^{1/n}$	K_F is a biosorption equilibrium constant, representative of the sorption capacity; and n is a constant indicative of biosorption intensity.
Langmuir–Freundlich (Sips 1948)	$q_e = \frac{Q_0 b C_e^{1/n}}{1+bC_e^{1/n}}$	Assuming that the surface is homogeneous, but that the sorption is a cooperative process due to adsorbate–adsorbate interactions.
Dubinin and Radushkevich (Dubinin et al. 1947)	$\ln C_{ads} = \ln x_m - \beta\varepsilon^2$ $\varepsilon = RT\ln\left(\frac{1}{1+C_e}\right)$ $E = \frac{1}{\sqrt{-2\beta}}$	C_{ads} is concentration of metal sorbed onto adsorbent, X_m is the maximum sorption capacity, β is a constant related to biosorption energy, ε is the Polanyi potential, R is a gas constant in kJ mol^{-1}, T is the temperature in Kelvin, E is mean biosorption energy.
BET model (multilayer sorption) (Brunauer et al. 1938)	$q_e = \frac{BQ^0 C_e}{(C_s - C_e)[1+(B-1)C_e/C_s]}$	Cs is the saturation concentration of the adsorbed component; B a constant indicating the energy of interaction between the solute and the adsorbent surface, and Q^0 is a constant indicating the amount of solute adsorbed forming a complete monolayer.
Redlich–Peterson (Redlich and Peterson 1959)	$q_e = \frac{K_{RP} C_e}{1+a_{RP} C_e^{\beta}}$	K_{RP}, a_{RP}, and β are the Redlich–Peterson parameters. The exponent β lies between 0 and 1. For $\beta = 1$ the model converts to the Langmuir form.
Radke–Prausnitz (Radke and Prausnitz 1972)	$q_e = \frac{arC_e^p}{a+rC_e^{p-1}}$	a, r, and p are related model constants.
Distribution coefficients model (Aksu 2005)	$q_e = K_d C_e$	K_d is distribution coefficient.

Table 16.3 Frequently Used Multicomponent Adsorption Models

Isotherm Types	Equations	Nomenclature
Langmuir (multi-component) (Langmuir 1918)	$q_{ei} = \frac{b_i q_{maxi} C_{ei}}{1+\sum_{i=1}^{N} b_i C_{ei}}$	C_{ei} and q_{ei} are the unadsorbed concentration of each component at equilibrium and the adsorbed quantity of each component per g of dried biomass at equilibrium, respectively. b_i and q_{maxi} are derived from the corresponding individual Langmuir isotherm equations.
Combined Langmuir–Freundlich (Sips 1948)	$q_{ei} = \frac{a_i C_{ei}^{1/n_i}}{1+\sum_{i=1}^{N} b_i C_{ei}^{1/n_i}}$	a_i,b_i, phase concentration of a single adsorbed component in equations.
Competitive Redlich–Peterson model (Bellot and Condoret 1993)	$q_e = \frac{K_{RPi} C_{ei}}{1+\sum_{i=1}^{N} a_{RPi} C_{ei}^{\beta i}}$	K_{RPi}, a_{RPi}, and β_i are the Redlich–Peterson parameters derived from the corresponding individual Redlich–Peterson isotherm equations.
IAST: Ideal Adsorbed Solution Theory (Radke and Prausnitz, 1972)	$\frac{1}{q_t} = \sum \frac{Y_i}{q_i^0}$	Y_i is the solute concentration of component i in the solid phase. q_i^0 is the phase concentration of a single adsorbed component in equations with C_i^0.

surface and no transmigration of the sorbate (Langmuir 1918). The nonlinear form of Langmuir equation can be expressed by the following equation:

$$q_e = \frac{Q_0 b C_e}{1 + bC_e} \tag{16.1}$$

where q_e is the amount of metal ions sorbed onto adsorbent, C_e is the equilibrium concentration of metal in solution, and Q_0 and b are Langmuir constants related to sorption capacity and sorption energy, respectively. A linear form of this equation can be given as follows:

$$\frac{C_e}{q_e} = \frac{1}{q_0^b} + \frac{C_e}{q_0} \tag{16.2}$$

Maximum sorption capacity (Q_0) represents monolayer coverage of sorbent with sorbate, and b represents enthalpy of sorption and should vary with temperature. Q_0 and b were calculated from the slope and intercept of linear plots of C_e/q_e versus C_e.

5.1.2 Freundlich isotherm

The Freundlich isotherm model proposes a monolayer sorption with a heterogeneous energetic distribution of active sites, accompanied by interactions among sorbed molecules (Freundlich 1906). The

Freundlich model stipulates that the ratio of solute adsorbed to the solute concentration is a function of the solution. The empirical model was shown to be consistent with exponential distribution of active centers, characteristic of heterogeneous surfaces. The amount of solute adsorbed, q_e, is related to the concentration of solute in the solution, Ce. The empirical Freundlich equation based on sorption is as follows:

$$q_e = K_F \times C_e^{1/n} \tag{16.3}$$

The constants in the Freundlich isotherm can be determined by plotting $\log q_e$ versus $\log C_e$ and making use of the above equation rewritten as

$$\log q_e = \log K_F + \frac{1}{n \log C_e} \tag{16.4}$$

Both K_f and n are empirical constants, being indicative of the extent of adsorption and the degree of nonlinearity between solution and concentration, respectively. Here the linear line obtained gives a slope that is the value of $1/n$, and the y-intercept is log K_f. The intercept is an indicator of adsorption capacity and the slope of adsorption intensity. A relatively slight slope (and hence a high value of n) indicates that adsorption is good over the entire range of concentrations studied, while a steep slope (and hence small n) means that adsorption is good at high concentrations but much less so at lower concentrations. A greater value of the intercept K_f indicates a higher capacity for adsorption than a smaller value.

5.1.3 Dubinin and Radushkevich isotherm

The equilibrium data can also be subjected to the Dubinin and Radushkevich (D–R) isotherm model to determine the nature of adsorption processes as physical or chemical (Dubinin et al. 1947). The sorption D-R isotherm model is applicable at low concentration ranges and can be used to describe sorption on both homogeneous and heterogeneous surfaces. This is postulated within an adsorption "space" close to sorbent surface. If the surface is heterogeneous and an approximation to a Langmuir isotherm is chosen as a local isotherm for all sites that are energetically equivalent, then the quantity $\beta^{1/2}$ can be related to the mean sorption energy, ε, which is the free energy of the transfer of 1 mol of metal ions from infinity to the surface of the sorbent. It can be represented by the general expression:

$$\ln C_{ads} = \ln x_m - \beta \varepsilon^2 \tag{16.5}$$

where X_m is the maximum sorption capacity, β is a constant related to energy, and ε is the Polanyi potential:

$$\varepsilon = RT \ln\left(\frac{1}{1 + C_e}\right) \tag{16.6}$$

where R is a gas constant in kJ mol^{-1} and T is the temperature in Kelvin. If ln C_{ads} is plotted against ε^2, β $(molK^{-1})^2$ and X_m (mmol g^{-1}) will be obtained from the slope and intercept, respectively. The mean biosorption energy can also be worked out using the following relationship:

$$E = \frac{1}{\sqrt{-2\beta}} \tag{16.7}$$

5.2 Kinetic modeling

Kinetics studies and dynamic continuous-flow investigations, offering information on the rate of the sorption metal uptake, together with the hydrodynamic parameters, are very important for biosorption process design (Volesky and Holan 1995). However, biosorption kinetics studies are insufficient according to literature published so far (Wand and Chen 2009).

Numerous kinetic models have been suggested to describe the reaction order of adsorption systems based on solution concentration. Kinetic models based on the capacity of the adsorbent have also been presented, such as Lagergren's first-order equation and Ho's second-order expression (Ho 2006). The first-order equation of Lagergren (Lagergren 1898) and the pseudo second-order equation are the most widely used kinetic models to describe the biosorption process; they are listed in Table 16.4. The pseudo second-order equation fitthe data very well in a large quantity of literature for biosorption (Ho 2006; Ho and McKay 1999). Ho (2006) presented a review on the application of second-order kinetic models to adsorption systems, including an earlier adsorption rate equation based on the solid capacity for a system of liquids and solids, the Elovich equation for adsorption of gases onto a solid, and applying a second-order rate equation for gas/solid and solution/solid adsorption systems, a second-order rate expression for ion-exchange reactions, and a pseudo-second-order expression.

5.3 Thermodynamic studies

The thermodynamic evaluation of a biosorption is necessary to conclude whether or not the process is spontaneous. The Gibbs free energy change, ΔG°, is an indication of the spontaneity of a chemical reaction and therefore is an important criterion for spontaneity. Also, both energy and entropy factors must be considered in order to determine the Gibbs free energy of the process.

Table 16.4 First-Order Equation of Lagergren and the Pseudo Second-Order Equation

Kinetic Models	Differential Equations	Integral Equations
First-order rate expression of Lagergren (Lagergren 1898)	$\frac{d_{q_t}}{d_t} = k_l(q_e - q_t)$	$\log(q_e - q_t) = \log q_e - \frac{k_l}{2.303}t$ $v_1 = k_l q_e V_l = k_l q_e$ where q is the amount of adsorbed pollutant on the biosorbent at time t; k_1 is the rate constant of Lagergren first-order biosorption; v_1 is the initial sorption rate.
Pseudo second-order equation (Ho and McKay 1999)	$\frac{d_{q_t}}{d_t} = k_2(q_e - q_t)^2$	$q_t = \frac{t}{\frac{1}{k_2 q_e^2} + \frac{t}{q_e}}$ $v_2 = k_2 q_e^2$ where k_2 is the rate constant of second-order biosorption; v_2 is the initial adsorption rate.

Thermodynamic parameters including the change in free energy (ΔG°), enthalpy (ΔH°) and entropy (ΔS°), can be used to describe the thermodynamic behavior of the biosorption of heavy metals. Thermodynamic parameters were calculated for the system by

$$\ln K_d = \frac{\Delta S^0}{R} - \frac{\Delta H^0}{RT} \tag{16.8}$$

where Kd is the distribution coefficient, ΔH°, ΔS° and T are the enthalpy, entropy, and temperature in kelvin, respectively, and R is the gas constant. The values of enthalpy (ΔH°) and entropy (ΔS°) can be obtained from the slope and intercept of log K_d versus 1/T plots, which were calculated by a curve-fitting program. Gibbs free energy (ΔG°) was calculated by:

$$\Delta G^0 = \Delta H^0 - T\Delta S^0 \tag{16.9}$$

In order to determine the thermodynamic parameters, experiments can be carried out at different temperature; for example, the temperature of a system can be changed from 293 to 323 K.

6 Assessment of biosorption

The biosorption of heavy metals by certain types of nonliving biomass is a highly cost-effective new alternative for the decontamination of metal-containing effluents. The mechanisms of metal biosorption allow the process to be scaled up and used in field applications, with packed-bed sorption columns being perhaps the most efficient for this purpose. Regenerating the biosorbents increases the process economy by allowing their reuse in multiple sorption cycles. The process results in metal-free effluents and small volumes of solutions containing concentrated metals, which can be easily recovered. Biosorption is a process that utilizes inexpensive dead biomass to sequester toxic heavy metals and is particularly useful for the removal of contaminants from industrial effluents. Biosorbents are prepared from the naturally abundant and/or waste biomass of algae, moss, fungi, or bacteria that have been killed, while the biomass is pretreated by washing with acids and/or bases before final drying and granulation (Kratochvil and Volesky 1998; Kratochvil et al. 1997; Fourest and Roux 1994; Brierley 1990; Brierley et al. 1988).

Although simple cutting and/or grinding of the dry biomass may yield stable biosorbent particles (Kratochhvil et al. 1997; Fourest and Roux 1994; Brierley 1990) and (Votapek et al. 1978; Kratochvil and Volesky 1997), some types of biomass have either to be immobilized in a synthetic polymer matrix (Greene and Bedell 1990; Jeffers et al. 1993) and/or rafted onto an inorganic support material such as silica (Bedell and Darnall 1990) in order to yield particles with the required mechanical properties (Mahan and Holcombe 1992). The biosorbent particles can then be packed in sorption columns, which are perhaps the most effective devices for the continuous removal of heavy metals. Biosorption columns operate on cycles consisting of loading, regeneration, and rinsing (Kratochvil et al. 1997; Kratochvil and Volesky 1997, 1998). The operation commences by loading the sorbent material, after which a metal-bearing effluent is passed through the packed bed and the heavy metals are taken up from the liquid by the biosorbent. When the metal-sorption capacity of the biosorbent is exhausted, the column is taken out of operation. Its bed can then be regenerated with solutions of acids and/or hydroxides; the regeneration produces small volumes of heavy-metal concentrates suitable for conventional metal-recovery processes (Brierley et al. 1986; Aldor et al. 1995). The cycle ends with rinsing and/or backwashing of the bed with water to remove the remains of the regenerants and suspended solids captured in the column. In order to make the biosorption process truly continuous, pairs of columns are employed in parallel so that, while one of the columns is undergoing regeneration and rinsing, the other is being loaded with heavy metals (Kratochvil and Volesky 1998).

Although many biological materials bind heavy metals, only those with sufficiently high metal-binding capacity and selectivity for heavy metals are suitable for use in a full-scale biosorption process. The first major challenge for the biosorption field was to select the most promising types of biomass from an extremely large pool of readily available and inexpensive biomaterials. Although this task is not complete, a large number of biomass types have been tested for their metal-binding capability under various conditions, and a summary of metal-biosorption results has been published (Volesky and Holan 1995; Wase and Forster 1997). Although several proprietary biosorption processes were developed and commercialized in the early 1990s, a lack of understanding of the mechanism underlying the metal-sorption process has hindered adequate assessment of process performance and limitations, and thus the expected widespread application of biosorption. Consequently, the selection of industrial effluents for pilot testing has remained largely intuitive, as has the task of scaling up the process. The next real challenge for the field of biosorption was to identify the mechanism of metal uptake by dead biomass (Kratochvil and Volesky 1998).

7 Application of biosorption

The application of biosorption for the sequestering of the lanthanides, the actinides, and related elements was primarily motivated by environmental concerns over the release to the environment and the subsequent fate of radioactive isotopes from the nuclear fuel power generation cycle (Tsezos 1997).

Over the past 25 years, intensifying research into metal biosorption elucidated the principles of this effective metal-concentration phenomenon. Biosorption can be cost-effective, particularly in environmental applications where low costs of the metal removal process are most desirable. Some efficient natural biosorbents have been identified that require little modification in their preparation. Engineering scale-up of the biosorption process seems fairly straightforward, based on experience from conventional sorption operations. While it seems that biosorption could hardly have any competition in many types of large-scale environmental metal removal applications, the process has not been applied as yet. Heavy metals (e.g., lead, copper, cadmium, zinc, and chromium) are toxic even at low concentrations. As they are nonbiodegradable, their threat is multiplied by their accumulation in the environment through the food chain.

The detoxification of metal-bearing wastewater is a pressing environmental concern. All metal-processing activities lose and/or discharge sometimes even large quantities of heavy metals. The development and implementation of a cost-effective process for removal/recovery of metals is essential in order to improve the competitiveness of industrial processing operations and to minimize the environmental hazard of toxic metal-containing effluents. The capacity of certain types of microbial and seaweed biomass to remove and concentrate heavy metals from solutions provides the basis for a cost-effective technology for detoxification of industrial effluents. New biosorbent materials can be extremely competitive and cost effective, particularly in this application. Further work with biosorbents identified for their high metal uptake could best be directed to derivation of engineering process scale-up parameters for application in the cleanup of two of the most ubiquitous types of metal-contaminated industrial effluents: acid mine drainage (AMD) and electroplating effluents.

The development and implementation of a cost-effective process for the removal and recovery of heavy metals is necessary on two fronts: first, to improve the competitiveness of mining (and electroplating) industries, which constitute a major industrial sector of manufacturing economies; and second, to eliminate toxic metals from effluents as required by environmental regulations. Biosorption technology removes heavy metals from wastewater without creating hazardous sludge at costs much lower than conventionally used ion exchange systems. Regeneration of the biosorbent and concentration of the metal solution for eventual recovery further increases the cost effectiveness of the process (Volesky and Naja 2005; Volesky 2003b).

8 Industrial perspective and aspects of biosorption

Biosorption is a proven cost-effective technique potentially for the treatment of high-volume low-concentration complex wastewaters containing metals or dyes. However, its performance under real industrial conditions is of concern. There have been few investigations examining the compatibility of the biosorbent for real industrial effluents (Vijayaraghavan and Yun 2008; Wang and Chen 2006).

On the other hand, several attempts to scale-up the biosorption process or to commercialize the process based on experiences from conventional sorption operations have not been successful so far. The biosorption has not been applied yet, while it seems that biosorption could hardly have any competition in many types of large-scale environmental metal removal applications (Volesky and Naja 2005).

Several attempts have been made to commercialize the biosorption in the fields of wastewater treatment and metal value recovery. The first patents for the biosorbents for wastewater treatment appeared in the early 1980s. In the early 1990s, other biomaterials were developed and commercialized, including AlgaSORB™ (*C. vulgaris*), AMT-BIOCLAIM™ (*Bacillus biomass*) (MRA), Bio-fix, and so on. B. V. SORBEX Inc., Advance Mineral Technologies Inc., Bio-recovery Systems Inc., and Visa Tech Ltd. developed different types of commercial metal-removal biosorbent (Tsezos 2001; Garnham 1997; Veglio and Beolchini 1997; Volesky 1990; Wang and Chen 2009). Even though some attempts have been made for biosorption application or commercialization, successful commercial application of biosorbents has not been carried out yet.

Biosorption is an economically feasible and a technically efficient technology for removing and recovering metals for solutions. The main attraction of biosorption is its cost effectiveness. It can comfortably fit into the metal treatment process where ion-exchange resins are applied. The major competing technology for the biosorbents is ion-exchange resins. While ion exchange can be considered a "mature" technology, biosorption is in its early developmental stages and further improvements in both performance and costs can be expected. However, the biosorption process is competitive in terms of capital and operational costs with the existing treatment technologies. While commercial ion-exchange resins are rather costly, the price of some biosorbents can be considerably cheaper (the ion-exchange resin price: $7–70/kg versus the estimated biosorbent cost: $ 3–7/kg). The market share of biosorbents is expected to grow exponentially in the next few years (Kuyucak 1990; Volesky 1990, 1999).

Eventually, the design and operation of the biosorption are similar to the established technologies for ion-exchange resin or activated carbon adsorption from the process of the application point of view. The existing wastewater treatment facilities using ion-exchange technologies can be easily converted to the biosorption process, with a good payback from savings produced by operating with an equally effective process using a more inexpensive sorbent material. The enormous potential of application for biosorption and its strong economic and technical advantages open considerable market opportunities that can actually be quantified through a responsible market analysis (Volesky and Naja 2007; Volesky 1990c). Furthermore, because of the difficulties existing for biosorption application urge, people should consider applying hybrid technologies, which involves combining various processes to treat real effluent (Wang and Chen 2009). We can say that there is a need to develop and propose to the market reliable, robust, simple, and effective process designs in order to arrive at a success story (Tsezos 2001).

9 The future of biosorption

Biosorption is basically at lab scale in spite of its development for many years. It cannot as yet be commercializedsufficiently. The incomplete understanding of the mechanism and the shortcomings of biosorption technology are two major factors that limit biosorption application. In future work several aspects may be considered.

The bacterial biomass represents an efficient and potential class of biosorbents for the removal of both dyes and metal ions. Unfortunately, the difficulties in reusing the microbial biomass, as well as the poor selectivity, hinder their applications under real conditions. Although some attempts have been made to commercialize biosorption for wastewater treatment, progress has been very modest considering that more than a decade of fundamental research has been done. The important features required for successful application of biosorption technology to real situations include, but are not limited to:

- Screening and selection of the most promising biomass, with sufficiently high biosorption capacity and selectivity.
- Optimizing the conditions for maximum biosorption, including optimization of pH, temperature, ionic strength and co-ion effects.
- Improving the selectivity and uptake via chemical and/or genetic modification methods.
- Examining the mechanical strength of biomass and if insufficient for reuse, improving rigidity by proper immobilization or other chemical methods.
- Testing the performance of biosorbents under different modes of operation.
- Analyzing the behavior of biosorbent for use with real industrial effluents and simultaneously analyzing the impact of water quality on the biosorption uptake of the specific pollutant of interest (Vijayaraghavan and Yun 2008).

For the future of biosorption, two trends of biosorption development for metal removal should be noted. One trend is to use hybrid technology for pollutant removal (Tsezos 2001), especially using living cells. Another trend is to develop good commercial biosorbents just like a kind of ion-exchange resin, and to exploit the market with great endeavor (Volesky 2007).

But the results are encouraging. Low-cost biosorbents will make the process economical and competitive against existing conventional technologies. Many waste biomass types have also been investigated for their biosorptive potential. These waste biomass types can be produced from the food and beverage industry, the pharmaceutical industry, or agriculture wastes and seaweed residues (Forster and Wase 1997; Volesky 1999; Wang and Chen 2009; Zouboulis 1997).

In addition to being cost effective, biosorption has huge potential, as many biosorbents are known to perform well, if not better than most conventional methods. Also, being aware of the hundreds of biosorbents able to bind various pollutants, sufficient research has been performed on various biomaterials to understand the mechanism responsible for biosorption. Therefore, through continued research, especially on the pilot and full-scale biosorption process, the situation is likely to change in the near future, with biosorption technology becoming more beneficial and attractive than currently used technologies (Vijayaraghavan and Yun 2008).

The researchers of various types of scientific background, from engineering to biochemistry, working together, will make a significant contribution to elucidating the biosorption mechanisms. Interdisciplinary efforts are mandatory and represent quite a challenge. Specific biosorption process applications have to be done in conjunction with industrial users/clients and require specific process engineering expertise and a serious commitment to development capital. A variety of investigation demonstrated that biosorption is a useful alternative to the conventional systems for the removal of heavy metal ions from aqueous solution. The development of the biosorption process requires further investigation in the direction of modeling, of regeneration and immobilization of biosorbents, and of treating the real industrial wastewater (Wang and Chen 2009).

10 Biosorption of radionuclides

The increasing demand for uranium has prompted a search for cheaper and more efficient methods for its recovery. Thorium, being a potential fuel for nuclear breeder reactors, is receiving similar attention. In addition to the ore processing recovery aspects, environmental problems resulting from intensified mining and ore processing activities call for the removal of radioactive elements from wastewater originating from these operations. Only about 15% of the total radioactivity in the ore leaves in the final uranium product, whereas the remaining 85% is discharged in the tailings. All 12 known isotopes of thorium are radioactive and, together with other unearthed not recovered radionuclides such as radium, polonium, and so on, find their way to the natural surface water bodies and environment. Uranium is also found in concentrations as high as 50 mg/liter in certain copper leach dumps and can be commercially extracted from the acidic wastewaters originating in the production of phosphoric acid from uranium bearing phosphate rock. A number of radionuclides and other heavy metal elements potentially can be extracted from seawater, provided that powerful and cheap sequestering agents are available. Malfunctions and leaks of nuclear reactor cooling fluids containing highly radioactive elements of different kinds further demonstrate the urgent need for efficient and element-selective materials of this sort. Biosorption potentially could be used for recovery as well as for cleanup purposes after we improve our understanding of the phenomenon and of the parameters that influence it (Tsezos and Volesky 1981).

10.1 Uranium biosorption

In examining the biosorptive uptake of uranium by microbial biomass, the equilibrium and rate of the process need to be defined. The equilibrium of biosorption has been successfully described by the use of the Langmuir and Freundlich relationships, which show the equilibrium distribution of the biosorbed element between the solution (liquid phase) and microbial biomass (soldi phase). Both models have been used and reported (Tsezos and Keller 1983; Tzesoz 1985, 1990; Tzesoz et al. 1995, 1996; Glombitza et al. 1984; Georgousis 1990). The effects of parameters such as solution pH, biomass growth conditions, and the solution ionic matrix have been investigated in recent years. The microbial biomass biosorptive uptake has been discussed in detail and has been presented in other publications by several authors (Tsezos 1985, 1990; Tsezos and McCready 1989; Ehlich and Brierley 1990). Most of the uranium biosorptive uptake studies have been conducted utilizing synthetic uranium solutions, that is, single-element solutions. Several different microbial biomass types have been examined for their uranium uptake capacity. (Tsezos and Volesky 1981; Tobin et al. 1994).The technical application potential of uranium biosorption is substantially dependent on the recovery of the uranium ,which has been sequestered by the microbial biomass as well as the potential for re-using the regenerated biomass in multiple biosorption-desorption cycles. Recovery of the adsorbed uranium can be achieved by the use of an appropriate elution solution capable of effectively stripping the adsorbed uranium from the exhausted biomass and bringing it back to a solution.

Studies on the engineering application of biosorption for the recovery of uranium from industrial process or waste solutions in batch form and at laboratory scale continuous pilot installations have been reported. The continuous laboratory pilot testing of uranium biosorption as a process for the removal/recovery of uranium from the complex waste or process solutions has confirmed that biosorption is a very selective process and that uranium can be selectively sequestered by microbial biomass out of complex leachate solution matrix.

Microbial biomass provides ligand groups on which metal species may bind by different mechanisms. Major classes of microbial biopolymers, such as proteins, nucleic acids, and polysaccharides, provide sites on which metal ions may bind. The metal ionic species should exhibit a preference for the ligand binding sites of the biomass based on their chemical coordination characteristics. Different ionic species of the same element can potentially exhibit preference for different binding sites (Tsezos 1990, 1997; Tsezos and McCready 1991).

10.2 Thorium biosorption

Thorium commonly exists along with uranium in nature, and from an environmental point of view the biosorption of thorium is of the interest. In general, thorium appears to be sequestered well by microbial biomass (Tsezos and Volesky 1981; Tobin et al. 1994). Although uranium and thorium are retained primarily by the fungal cell wall, uranium is localized within the cell wall chitin network, while thorium is localized on the external surface of the cell wall. This difference in the biosorptive loci enables the simultaneous biosorption of uranium and thorium from the same solution by the same biomass without immediate competition effects (Tsezos and McCready 1991; Tsezos 1997).

11 Biosorption of uranium and thorium by Ca-alginate biopolymer beads

The removal of uranium and thorium ions from aquatic solutions is very important because of the chemical and radiotoxicity of these elements. Ca-alginate biopolymer beads have high potentiality to uptake of U and Th. As a result of alginate's egg box structure, it has stable form, and cross-linked alginate has been used to remove heavy metal from wastewater with high capacity. Studies show that Ca-alginate beads can be successfully used for uranium and thorium recovery from aqueous solutions.

Ca-alginate can be used as a biosorbent without any treatment. Besides this, some kind of biosorbent such as algae, yeast, bacterial polymer or residual biomass, inorganic adsorbents can be immobilized in Ca-alginate.

Gok and Aytas (2009) used Ca-alginate beads to remove uranium from aqueous solution. The percent adsorption (%) and distribution constant (K_d) for uranium ions were found to be 91 ± 1 and 10043 ± 834 mL/g, respectively, under optimized experimental conditions. The theoretical capacity of biosorbent was calculated as 400 mg g^{-1} from the Langmuir isotherm.

Singhal et al. (2004) immobilized *Chlorella (Chlorella pyrendoidosa)*, a freshwater algae, into calcium alginate for reduction of uranium concentration in well water. In this study, well water samples were spiked with U(IV) ranging from 1000 to 2000 ppb. The contaminated solutions were passed through a glass column containing chlorella-impregnated beads of calcium alginate. Chlorella (*Chlorella pyrendoidosa*), a freshwater algae, was immobilized in sodium alginate in the form of beads by using 0.2 M calcium chloride solution. The solution was passed again through a charcoal solution to remove any trace of impurities. The concentration of uranium after treatment ranged from 10 to 20 ppb. Also, 99–100% of the adsorbed uranium on beads was recovered by 0.1M HNO_3.

Dilute calcium chloride brine solution was found to be effective in the solubilization of toxic heavy metals and long half-life radionuclides (Th-230, Ra-226 and Pb-210) from uranium ores and mill tailings. The recovery of heavy metals and radionuclides from uranium mill tailing effluents was studied with calcium alginate beads. The kinetic values, V_m and K, were calculated for uranium uptake by calcium alginate to be 96.2 mg/l/s and 0.125 g/l, respectively (Torma 1992).

The ability of Ca-alginate immobilized *Trichoderma harzianum* to remove and recover uranium from aqueous streams has been explored by Akhtar et al. (2009). The adsorption column containing 1.5 g dry weight of immobilized material purified 8.5 L of bacterial leach liquor (58 mg/L U) before breakthrough occurred and the biosorbent became saturated after 25 L of influent. Sorbed uranium was recovered in 200 ml of 0.1 N HCl, resulting in 98.1–99.3% elution by 0.1 N HCl.

Donat et al. (2009) discussed the sorption properties for U(VI) by alginate coated $CaSO_4 \cdot 2H_2O$ sepiolite and calcined diatomite earth (Kieselguhr) (ACSD). Cross-linked and alginate immobilized residual biomass from distillery spent wash was examined for its ability to function as a biosorbent for uranium, and the alginate-biomass matrix had a maximum biosorption capacity of 220 mg uranium/g dry weight of the matrix, even at low concentrations of biomass (Bustard and McHale 1997).

Hybrid composite films were prepared by incorporating 0.2, 0.4, and 0.6 g of natural and modified attapulgite clay particles into sodium alginate films by Guerra et al. (2009). The reactive centers attached to the pendant chains inside the sodium alginate films can coordinate thorium(IV) by presenting an isotherm saturated at 33.70×10^{-5} mol g^{-1} of this actinide.

12 Conclusion

Biosorption of uranium and thorium studies are rather weak as compared with studies of conventional techniques. Knowledge of biosorption of radionuclides is still in its infancy. Additionally, the waste products of food, beverages, or pharmaceutical s can be used as a biosorbent with high potentiality, and they are good candidates for the development of inexpensive biosorption processes.

Ca-alginate beads have a special place among biopolymers. These biopolymers have many advantages: They are natural, cheap, and biodegradable, and they can be successfully utilized in removal of uranium and thorium ions with high capacity. They can provide a useful alternative to conventional systems for the biosorption of radionuclides from waste streams. But these biopolymers have not yet attained adequate selectivity. Modification, immobilization, or encapsulation methods with different kinds of organic or inorganic materials would improve selectivity and stability of Ca-alginate beads.

There is a real potential in this area. The importance of biosorption in environmental protection suggests that further investigations should perhaps be directed toward interdisciplinary areas.

SYMBOLS AND TERMINOLOGY

a, r, and p constants of the Radke–Prausnitz isotherm model
a_i, b_i phase concentration of a single adsorbed component
b Langmuir constants related to sorption energy
B constant indicating the energy of interaction between the solute and the adsorbent surface
b_i and q_{maxi} constants derived from the corresponding individual Langmuir isotherm equations
C_{ads} concentration of metal sorbed onto adsorbent
C_e equilibrium concentration of metal in solution
C_{ei} unadsorbed concentration of each component at equilibrium
C_s the saturation concentration of the adsorbed component
E mean biosorption energy
k_1 rate constant of Lagergren first-order biosorption
k_2 rate constant of second-order biosorption
K_d distribution coefficient
K_f empirical constants of Freundlich isotherm
K_{RP}, a_{RP}, and β Redlich–Peterson parameters
K_{RPi}, a_{RPi}, and β_i Redlich–Peterson parameters derived from the corresponding individual Redlich–Peterson isotherm equations
n empirical constants of the Freundlich isotherm
q the amount of adsorbed pollutant on the biosorbent at time *t*
Q_0 Langmuir constants related to sorption capacity
Q^0 constant indicating the amount of solute adsorbed forming a complete monolayer
q_e amount of metal ions sorbed onto adsorbent
q_{ei} adsorbed quantity of each component per g of dried biomass at equilibrium
q_i^0 the phase concentration of a single adsorbed component in equations with C_i^0
R universal gas constant
T solution temperature
v_1 the initial sorption rate
v_2 the initial adsorption rate
X_m the maximum sorption capacity of D-R isotherm
Y_i the solute concentration of component *i* in the solid phase
β constant related to biosorption energy
ε the Polanyi potential
ΔH° the change in enthalpy
ΔS° the change in entropy
ΔG° Gibbs free energy change

Glossary

Adsorbent a material used to adsorb another substance onto its surface without chemically altering them

Adsorption the accumulation of substance onto the surface of adsorbent

Algae a large and diverse group of typically autotrophic, chiefly aquatic, eukaryotic organisms, ranging from unicellular to multicellular forms

Alginate natural heteropolysaccharide comprised of mannuronate and guluronate

Bacterial fermentation a process of fermentation carried out by bacteria through which organic compounds go through a chemical change caused by enzymes produced from bacteria

Bioaccumulation an intracellular metal accumulation from aqueous solution by a metabolically active process to living biomass

Biomass biological material derived from living, or recently living organisms

Biopolymer polymers produced from biobased materials

Biosorbent biological materials that is responsible for removing the undesirable substance in the biosorption process used for removing substance

Biosorption removal of substance from aqueous solution by the passive binding to nonliving biomass

Biotechnology technology based on biology, using living things to create products or to do tasks for human beings

Ca-alginate biopolymers matural biopolymer obtained by alginate gelation with calcium ions according to eggs-box model

Desorption the process of removing a sorbed substance by the reverse of sorption

Egg-box model interactions between polysaccharides and divalent cations like an egg box

Eukaryotic microorganisms an organism whose cells contain complex structures enclosed within membranes

Functional groups specific groups of atoms within molecules that have properties to bind metal ion species in solution such as phosphoryl, carboxyl, carbonyl, sulfydryl. and hydroxyl groups

Hydrogel an extremely hydrated water-insoluble polymer gel

Mechanism of biosorption explaining how the metal bound within the biosorbent as adsorption, ion exchange, coordination, and covalent bonding

Modification change the properties of adsorbent/biosorbent by chemically or physically with chemicals or temperatures to improve its capacity, selectivity, or stability

Polymer large molecules consisting of repeated chemical units

ppb parts per billion, a unit of concentration, also can be expressed as microgram per liter (μg/L)

ppm parts per million, a unit of concentration, also can be expressed as milligram per liter (mg/L)

Radionuclide or radioactive element an atom with unstable nucleus, which is a nucleus characterized by excess energy. These elements undergo radioactive decay through the emission of alpha particles, beta particles, or gamma rays to form more stable elements.

Seaweed a loose colloquial term encompassing macroscopic, multicellular, and benthic marine algae, such as various red, green, or brown algae

Sorption the term used for the action of both absorption and adsorption taking place simultaneously

Acknowledgment

The authors would like to acknowledge the Scientific and Technological Research Council of Turkey (TUBITAK) and Helmholtz Zentrum München, German Research Center for Environmental Health, Institute of Radiation Protection, Neuherberg, Munich.GERMANY for their support and for providing the facility for the study presented in this chapter.

References

Akhtar, K., Khalid, A.M., Akhtar, M.W., Ghauri, M.A., 2009. Removal and recovery of uranium from aqueous solutions by Ca-alginate immobilized *Trichoderma harzianum*. Bioresour. Technol. 100 (20), 4551–4558.

Aksu, Z., 2005. Application of biosorption for the removal of organic pollutants: A review. Process Biochem. 40, 997–1026.

Aldor, I., Fourest, E., Volesky, B., 1995. Desorption of cadmium from algal biosorbent. Can J. Chem. Eng. 73, 516–522.

Bai, R.S., Abraham, T.E., 2001. Biosorption of Cr (VI) from aqueous solution by *Rhizopus nigricans*. Bioresour. Technol. 79 (1), 73–81.

Barbetta, A., Barigelli, E., Dentini, M., 2009. Porous alginate hydrogels: Synthetic methods for tailoring the porous texture. Biomacromolecules 10 (8), 2328–2337.

Bayramoğlu, G., Çelik, G., Arica, M.Y., 2006. Studies on accumulation of uranium by fungus *Lentinus sajor-caju*. J. Hazard. Mater. 136 (2), 345–353.

Bedell, G.W., Darnall, D.W., 1990. Immobilization of nonviable, biosorbent, algal biomass for the recovery of metal ions. In: Volesky, B. (Ed.), Biosorption of Heavy Metals. CRC Press, Florida, USA, pp. 313–326.

Bellot, J.C., Condoret, J.S., 1993. Modeling of liquid chromatography equilibrium. Process Biochem 28, 365–471.

Bengtsson, L., Johansson, B., Hackett, T.J., McHale, L., McHale, A.P., 1995. Studies on the biosorption of uranium by *Talaromyces emersonii* CBS 814.70 biomass. Appl. Microbiol. Biotechnol. 42 (5), 807–811.

Bituh, T., Marovic, G., Franic, Z., Sencar, J., Bronzovic, M., 2009. Radioactive contamination in Croatia by phosphate fertilizer production. J. Hazard. Mater. 162 (2-3), 1199–1203.

Boddu, V.M., Abburi, K., Talbott, J.L., Smith, E.D., 2003. Removal of hexavalent chromium from wastewater using a new composite chitosan biosorbent. Environ. Sci. Technol. 37 (19), 4449–4456.

Bosinco, S., Dambies, L., Guibal, E., Roussy, J., Le Cloirec, P., 1997. Removal of chromium (VI) on chitosan gel beads: Kinetic modeling. Adv. Chitin Sci. 2, 453–461.

Brierley, J.A., 1990. Production and application of a *Bacillus*-based product for use in metals biosorption. In: Volesky, B. (Ed.), Biosorption of Heavy Metals. CRC Press, Boca Raton, FL, pp. 305–312.

Brierley, J.A., Brierley, C.L., Decker, R.F., Goyak, G.M., 1988. Metal recovery. U.S. Patent 4 789 481.

Brierley, J.A., Brierley, C.L., Goyak, G.M., 1986. AMT-BIOCLAIM: A new wastewater treatment and metal recovery technology. In: Lawrence, R.W., Branion, R.M.R., Ebner, H.G. (Eds.), Fundamental and applied biohydrometallurgy. Elsevier, Amsterdam, pp. 291–304.

Brunauer, S., Emmett, P.H., Teller, E., 1938. Adsorption of gases in multimolecular layers. J. Am. Chem. Soc. 60, 309–328.

Bustard, M., McHale, A.P., 1997. Biosorption of uranium by cross-linked and alginate immobilized residual biomass from distillery spent wash. Bioprocess Eng. 17 (3), 127–130.

Chen, J., Yiacoumi, S., 1997. Biosorption of metal ions from aqueous solution. Separation Sci. Technol. 32, 51–69.

Chen, J.P., Yang, L., 2005. Chemical modification of *Sargassum sp.* for prevention of organic leaching and enhancement of uptake during metal biosorption. Ind. Eng. Chem. 44 (26), 9931–9942.

Da Costa, A.C.A., Leite, S.G.F., 1991. Metals biosorption by sodium alginate immobilized chlorella homosphaera cells. Biotechnol. Lett. 13, 559–562.

Davis, T.A., Volesky, B., Mucci, A., 2003. A review of the biochemistry of heavy metal biosorption by brown algae. Water Res. 37, 4311–4330.

Donat, R., Aytas, S., 2005. Adsorption and thermodynamic behavior of uranium (VI) on *Ulva sp.* Na bentonite composite adsorbent. J. Radioanalytical. Nucl. Chem. 265 (1), 107–114.

Donat, R., Cilgi, G.K., Aytas, S., Cetisli, H., 2009. Thermodynamic parameters and sorption of U(VI) on ACSD. J. Radioanalytical. Nucl. Chem. 279 (1), 271–280.

Dubinin, M.M., Zaverina, E.D., Radushkevich, L.V., 1947. Sorption and structure of active carbons I. Adsorption of organic vapors. Zhurnal Fizicheskoi Khimii 21, 1351–1362.

Edyvean, R.G.J., Williams, C.J., Wilson, M.W., Aderhold, D., 1997. Biosorption using unusual biomasses. In: Wase, J., Forster, C. (Eds.), Biosorbents for Metal Ions. Taylor & Francis, London, pp. 165–182.

Ehrlich, H.L., Brierley, C.L., 1990. Microbial mineral recovery. McGraw-Hill, New York.

Encyclopædia Britannica, Cellulose. (2008). Retrieved January 11, 2008, from Encyclopædia Britannica Online.

Esposito, A., Pagnanelli, F., Veglio, F., 2002. pH-related equilibria models for biosorption in single metal systems. Chem. Eng. Sci. 57, 307–313.

Fairbridge, R.W., 1972. The encyclopedia of geochemistry and environmental science. Van Nostrand Reinhold, New York, p. 1321.

Ferris, F.G., Beveridge, T.J., 1985. Function of bacterial cell surface structure. BioScience 35, 172–198.

Forster, C.F., Wase, D.A.J., 1997. Biosorption: The future. In: Wase, J., Forster, C. (Eds.), Biosorbents for Metal Ions. Taylor & Francis Press, London, pp. 221–228.

Fourest, E., Roux, J.C., 1994. Improvement of heavy metal biosorption by mycelial dead biomasses (*Rhizopus arrhizus, Mucor miehei* and *Penicillium chrysogenum*): pH control and cationic activation. FEMS. Microbiol. Rev. 14, 325–332.

Freundlich, H.M.F., 1906. Über die adsorption in lösungen. Zeitschrift für Physikalische Chemie (Leipzig) 57 (A), 385–470.

Gadd, G.M., 1990. Heavy metal accumulation by bacteria and other microorganisms. Experientia 46, 834–840.

Gadd, G.M., 1991. Molecular biology and biotechnology of microbial interactions with organic and inorganic heavy metal compounds. Molecular Biology and Biotechnology of Extramophiles. Blackie, Glaskow, UK.

Gadd, G.M., 1993. Interaction of fungi with toxic metals. Phytologist 124, 25–60.

Gadd, G.M., 2000. Bioremedial potential of microbial mechanisms of metal mobilization and immobilization. Curr. Opin. Biotechnol. 11, 271–279.

Gadd, G.M., 2009. Biosorption: critical review of scientific rationale, environmental importance and significance for pollution treatment. J. Chem. Technol. Biotechnol. 84, 13–28.

Gadd, G.M., White, C., 1990. Biosorption of radionuclides by yeast and fungal biomass. J. Chem. Technol. Biotechnol. 49, 331–343.

Gadd, G.M., White, C., 1992. Removal of thorium from simulated acid process streams by fungal biomass: potential for thorium desorption and reuse of biomass and desorbent. J. Chem. Technol. Biotechnol. 55, 39–44.

Garnham, G.W., 1997. The use of algae as metal biosorbents. In: Wase, J., Forster, C. (Eds.), Biosorbents for metal ions. Taylor & Francis, London, UK, pp. 11–37.

Gavrilescu, M., 2004. Removal of heavy metals from the environment by biosorption. Eng. Life Sci. 4 (3), 219–232.

Genc, O., Yalcinkaya, Y., Buyuktuncel, E., Denizli, A., Arıca, M.Y., Bektas, S., 2003. Uranium recovery by immobilized and dried powdered biomass: characterization and comparison. Int. J. Miner. Process 68, 3–107.

Georgousi, Z., 1990. The loss of uranium uptake capacity of Rhzipus arrhizus immobilised biomass on recycle. M. Eng. Thesis, McMaster University, Canada.

Glombitza, Z., Iske, U., Gwenner, C.H.R., Krishnas, M.V., 1984. Biosorption of mercury. Acta. Biotechnol. 4, 281–284.

Gok, C., Aytas, S., 2009. Biosorption of uranium(VI) from aqueous solution using calcium alginate beads. J. Hazard. Mater. 168 (1), 369–375.

Grant, G.T., Morris, E.R., Rees, D.A., Smith, P.J.C., Thom, D., 1973. Biological interactions between polysaccharides and divalent cation: The egg-box model. FEBS Lett. 32, 195–198.

Greene, B., Bedell, G.W., 1990. In: Akatsuka, I. (Ed.), An introduction to applied phycology. SPB Academic Publishers, Amsterdam, Netherlands, pp. 137–149.

Guerra, D.L., Viana, R.R., da Costa, L.P., Airoldi, C., 2009. Sodium alginate films modified by raw and functionalized attapulgite for use of thorium(IV) adsorption: A thermodynamic approach. J. Phys. Chem Solids 70 (11), 1413–1421.

Harmsen, K., deHaan, F.A.M., 1980. Occurrence and behaviour of uranium and thorium in soil and water. Neth. J. Agric. Sci. 28, 40–62.

Ho, Y.S., 2006. Review of second-order models for adsorption systems. J. Hazard. Mater. 136, 681–690.

Ho, Y.S., McKay, G., 1999. Pseudo-second order model for sorption processes. Process Biochem. 34, 451–516.

Hockin, S., Gadd, G.M., 2007. Bioremediation of metals by precipitation and cellular binding, in *Sulphate-reducing bacteria: Environmental and engineered systems*. In: Barton, L.L., Hamilton, W.A. (Eds.). Cambridge University Press, Cambridge, pp. 405–434.

Holan, Z.R., Volesky, B., Prasetyo, L., 1993. Biosorption of cadmium by biomass of marine algae. Biotechnol. Bioeng. 41, 819–825.

http://en.wikipedia.org/wiki/Biopolymer#Biopolymers_as_materials (accessed 12.09.09.).

http://www.o2.org/ideas/cases/biopolymers.html, This material issue, About Biopolymers, is based on the report Tenminste houdbaar tot: verpakken met biopolymeren, compiled by het Nederlands Verpakkingscentrum, AgLink, KIEM and Proterra (accessed 08.09.09.).

Ibanez, J.P., Umetsu, Y., 2002. Potential of protonated alginate beads for heavy metals uptake. Hydrometallurgy 14, 89–99.

Ishikawa, S., Suyama, K., Arihara, K., Itoh, M., 2002. Selective recovery of uranium and thorium ions from dilute aqueous solutions by animal biopolymers. Biol. Trace. Elem. Res. 86 (3), 227–236.

Jang, L.K., Lopez, S.L., Eastmen, S.L., Pryfogle, P., 1991. Recovery of copper and cobalt by bio polymer gels. Biotechnol. Bioeng. 37, 266–273.

Jeffers, T.H., Bennett, P.G., Corwin, R., 1993. Biosorption of metal contaminants using immobilized biomass—Field studies. Report of Investigations, U.S. Bureau of Mines. Utah, Salt Lake City.

Kaygun, K.A., Akyil, S., 2007. Study of the behaviour of thorium adsorption on PAN/zeolite composite adsorbent. J. Hazard. Mater. 147 (1-2), 357–362.

Konishi, Y., Asai, S., Midoh, Y., Oku, M., 1993. Recovery of zinc, cadmium and lanthanum by biopolymer gel particles of alginic acid. Separation Sci. Technol. 28, 1691–1702.

Kratochvil, D., Volesky, B., 1997. Biominet Proceedings. In: Lortie, L., Bedard, C., Gould, W.C. (Eds.), CANMET, Natural Resources Canada. Ottawa, ON, Canada, pp. 41–60.

Kratochvil, D., Volesky, B., 1998. Advances in the biosorption of heavy metals. Trends Biotechnol. 16 (7), 291–300.

Kratochvil, D., Volesky, B., Demopoulos, G., 1997. Optimizing cu removal/recovery in a biosorption column. Water Res. 31, 2327–2339.

Kuyucak, N., 1990. Feasiblity of biosorbents application. In: Volesky, B. (Ed.), Biosorption of heavy metals. CRC Press, Boca Raton, FL, pp. 83–92.

Kuyucak, N.K., Volesky, B., 1989. Accumulation of cobalt by marine algae. Biotechnol. Bioeng. 33, 809–814.

Lagergren, S., 1898. Zur theorie der sogenannten adsorption geloester stoffe. Kungliga Svenska Vetenskapsakademiens. Handlingar 24, 1–39.

Langmuir, I., 1918. The adsorption of gases on plane surfaces of glass, mica and platinium. J. Am. Chem. Soc. 40, 1361–1403.

Madigan, M.T., Martinko, J.M., Parker, J., 1997. Brock biology of microorganisms, eighth ed. PrenticeHall, Upper Saddle River, NJ.

Mahan, C.A., Holcombe, J.A., 1992. Immobilization of algae cells on silica gel and their characterization for trace metal preconcentration. Anal. Chem. 64 (17), 1933–1939.

Mann, H., 1990. Biosorption of heavy metals by bacterial biomass. In: Volesky, B. (Ed.), Biosorption of heavy metals. CRC Press, Boca Raton, FL, p. 125.

McKay, G., Allen, S.J., 1997. Low-cost adsorbents in continuous process. In: Wase, J., Forster, C. (Eds.), Biosorbents for Metal Ions. Taylor & Francis, London, pp. 183–220.

Metaxas, M., Kasselouri-Rigopoulou, V., Galiatsatou, P., Konstantopolou, C., Oikonomou, D., 2003. Thorium removal by different adsorbents. J. Hazard. Mater. B. 97, 71–82.

Mimura, H., Ohta, H., Akiba, K., Onodera, Y., 2001. Uptake behavior of americium on alginic acid and alginate polymer gels. J. Radioanalytical. Nucl. Chem. 247 (1), 33–38.

Mimura, H., Ohta, H., Akiba, K., Onodera, Y., 2002. Uptake and recovery of ruthenium by alginate gel polymers. J. Nucl. Sci. Technol. 39 (6), 655–660.

Nakajima, A., Sakaguchi, T., 1993. Accumulation of uranium by *basidiomycetes*. Appl. Microbiol. Biotechnol. 38 (4), 574–578.

Nestle, N., Kimmich, R., 1996. Heavy metal uptake of alginate gels studied by NMR microscopy. Colloid Surf. A. 115, 141–147.

Ngomsika, A., Beea, A., Siauguea, J., Cabuila, V., Coteb, G., 2006. Nickel adsorption by magnetic alginate microcapsules containing an extractant. Water Res. 40, 1848–1856.

Nierboer, E., Richardson, D.M.S., 1980. The replacement of the nondescript term " heavy metals" by a biologically and chemically significant classification of metal ions. Environ. Pollut. Ser. B. 1, 3.

Osthols, E., Manccan, A., Farges, F., Charlet, L., 1997. Adsorption of thorium on amorphous silica: an EXAFS study. J. Colloid Interface Sci. 194, 10–21.

Pagnanelli, F., Petrangeli Papini, M., Toro, L., Trifoni, M., Vegliò, F., 2000. Biosorption of metal ions on *Arthrobacter sp.*: Biomass characterization and biosorption modeling. Environ. Sci. Technol. 34 (13), 2773–2778.

Pearson, R.G., 1969. Hard and soft acids and bases. Surv. Prog. Chem. 5, 1–52.

Philip, L., Iyengar, L., Venkobachar, C., 2000. Biosorption of U, La, Pr, Nd, Eu and Dy by *Pseudomonas aeruginosa*. J. Ind. Microbiol. Biotechnol. 25, 1–7.

Radke, C.J., Prausnitz, J.M., 1972. Thermodynamics of multi-solute adsorption from dilute liquid solutions. AIChE. J. 18, 761–769.

Rangsayatorn, N., Pokethitiyook, P., Upatham, E.S., Lanza, G.R., 2004. Cadmium biosorption by cells of *Spirulina platensis* TISTR 8217 immobilized in alginate and silica gel. Environ. Int. 30 (1), 57–63.

Redlich, O., Peterson, D.L., 1959. A useful adsorption isotherm. J. Phys. Chem., 1024.

Remacle, J., 1990. The cell wall and metal binding. In: Volesky, B. (Ed.), Biosorption of heavy metals. CRC Press, Boca Raton, FL, pp. 83–92.

Romero-Gonzalez, M.E., Williams, C.J., Gardiner, P.H.E., 2001. Study of the mechanism of cadmium biosorption by dealginated seaweed waste. Environ. Sci. Technol. 35, 3025–3030.

Sakamoto, N., Kano, N., Imaizumi, H., 2008. Biosorption of uranium and rare earth elements using biomass of algae. Bioinorganic Chem. Appl., 1–8.

Salinas-Pedroza, M.G., Olguin, M.T., 2004. Thorium removal from aqueous solutions of Mexican erionite and X zeolite. J. Radioanal. Nucl. Chem. 260 (1), 115–118.

Sar, P., D'Souza, S.F., 2002. Biosorption of thorium (IV) by a *Pseudomonas* biomass. Biotechnol. Lett. 24 (3), 239–243.

Sebesta, F., John, J., 1995. An overview of the development, testing and application of composite absorbers. LA-12875-MS, Los Alamos National Laboratory.

Seki, H., Suzuki, A., 1998. Biosorption of heavy metal ions to brown algae *Macrocystis pyrifera, Kjellmaniella crassiforia*, and Undaria pinnatifida. J. Colloid Interface Sci. 206 (1), 297–301.

Singhal, K., Joshi, S., Tirumalesh, K., Gurg, R.P., 2004. Reduction of uranium concentration in well water by *Chlorella* (*Chlorella pyrendoidosa*), a fresh water algae immobilized in calcium alginate. J. Radioanalytical. Nucl. Chem. 261 (1), 73–78.

Sips, R., 1948. On the structure of a catalyst surface. J. Chem. Phys., 490–495.
Sleyter, V.B., 1981. Morphopoietic and functional aspects of regular protein membranes present on prokaryotic cell wall. In: Kieermayer, O. (Ed.), Cell Biology Monographs, vol. 4,3. Springer-Verlag, New York.
Smidsrod, O., Draget, K.I., 1996. Chemistry and physical properties of alginates. Carbohydrate European 14, 6–13.
Solisio, C., Lodi, A., Converti, A., Borghi, M.D., 2000. The effect of acid pre-treatment on the biosorption of chromium(III) by *Sphaerotilus natans* from industrial wastewater. Water Res. 34 (12), 3171–3178.
Sposito, G., 1989. The chemistry of soils. Oxford University Press, New York.
Suh, J.H., Kim, D.S., 2000. Effects of Hg^{2-} and cell conditions on Pb^{2-} accumulation by *Saccharomyces cerevisae*. Bioprocess Eng. 23, 327–336.
Tobin, J.M., White, C., Gadd, G.M., 1994. Metal accumulation by fungi: Application in environmental biotechnology. J. Ind. Microbiol. 13, 126–130.
Torma, A.E., 1992. Remediation of uranium mill tailings by an integrated biological and chemical process, Conference: Winter annual meeting of the American Society of Mechanical Engineers (ASME). CA, Anaheim.
Tsezos, M., 1985. The selective extraction of metals from solution by microorganisms. A brief overview. Canadian Metallurgical Q. 24, 141–144.
Tsezos, M., 1990. Engineering aspects of metal binding by biomass. In: Ehrlich, H.L., Brierley, C.L. (Eds.), Microbial mineral recovery. McGraw Hill, New York, pp. 325–329.
Tsezos, M., 1997. Biosorption of lanthanides, actinides and related materials. In: Wase, J., Forster, C. (Eds.), Biosorbents for Metal Ions. Taylor & Francis, London, pp. 87–113.
Tsezos, M., 2001. Biosorption of metals. The experience accumulated and the outlook for technology development. Hydrometallurgy 59, 241–244.
Tsezos, M., Volesky, B., 1981. Biosorption of uranium and thorium. Biotechnol. Bioeng. 23, 583–604.
Tsezos, M., Keller, D., 1983. Adsorption of radium-226 by biological origin adsorbents. Biotecnol. Bioeng. 25, 201–215.
Tsezos, M., McCrady, R.G.L., 1989. The continuous recovery of uranium from biologically leached solution using immobilised biomass. Biotecnol. Bioeng. 34, 10–17.
Tsezos, M., McCready, R.G.L., 1991. The pilot plant testing of the continuous extraction of radonuclides using immobilised biomass. In: Sayler, G., Fox, R., Blackburn, J. (Eds.), Environmental biotechnology for waste treatment. Plenum Press, New York, pp. 249–260.
Tsezos, M., Remoudaki, E., Angelatou, V., 1995. A systematic study on equilibrium and kinetic of biosrptve accumulation. Int. Biodeterior Biodegradation 35 (1-3), 129–153.
Tsezos, M., Remoudaki, E., Angelatou, V., 1996. A study of the competing ions effects in the biosorptin of metal. Int. Biodeterior Biodegradation 36, 19–28.
Tsezos, M., 1985. The selective extraction of metals from solution by microorganisms. A brief overview. Canadian Metallurgical Q. 24, 141–144.
Ucun, H., Bayhan, Y.K., Kaya, Y., Cakici, A., Algur, O.F., 2002. Biosorption of chromium(VI) from aqueous solution by cone biomass of *Pinus sylvestris*. Bioresour. Technol. 85 (2), 155–158.
Veglio, F., Beolchini, F., 1997. Removal of metals by biosorption: A review. Hydrometallurgy 44, 301–316.
Vijayaraghavan, K., Yun, Y.S., 2008. Bacterial biosorbents and biosorption. Biotechnol. Adv. 26 (3), 266–291.
Volesky, B., 1990. Removal and recovery of heavy metals by biosorption. In: Volesky, B. (Ed.), Biosorption of heavy metals. CRC Press, Boca Raton, FL, pp. 8–43.
Volesky, B., 1999. Biosorption for the next century. A part of the Invited Lecture to be presented at the International Biohydrometallurgy Symposium. El Escorial, Spain.
Volesky, B., 2003a. Sorption and biosorption. (ISBN 0-9732983-0-8) BV-Sorbex, Inc. St. Lambert (Montreal). Canada, Quebec.

Volesky, B., 2003b. Biosorption process simulation tools. Hydrometallurgy 71, 179–190.
Volesky, B., 2007. Biosorption and me. Water Res. 41, 4017–4046.
Volesky, B., Naja, G., 2005. Biosorption: Application strategies. International Biohydrometallurgy Symposium. South Africa, Cape Town.
Volesky, B., Naja, G., 2007. Biosorption technology: Starting up an enterprise. Int. J. Technol. Transfer Commercialisation 6, 196–211.
Volesky, B., Holan, Z.R., 1995. Biosorption of heavy metals. Biotechnol. Prog. 11, 235–285.
Votapek, V., E. Marval, and K. Stamberg. 1978. U.S. Patent 4 067 821.
Wang, J., Chen, C., 2009. Biosorbents for heavy metals removal and their future. Biotechnol. Adv. 27 (2), 195–226.
Wang, J., Chen, C., 2006. Biosorption of heavy metals by *Saccharomyces cerevisiae*: A review. Biotechnol. Adv. 24 (5), 427–451.
Wang, J., 2002. Biosorption of copper (II) by chemically modified biomass of *Saccharomyces cerevisiae*. Process Biochem. 37, 847–850.
White, C., Gadd, G.M., 1998. Accumulation and effects of cadmium on sulphate-reducing bacterial biofilms. Microbiologica 144, 1407–1415.
Xuan, Z., Tang, Y., Li, X., Liu, Y., Luo, F., 2006. Study on the equilibrium, kinetics and isotherm of biosorption of lead ions onto pretreated chemically modified orange peel. Biochem. Eng. J. 31 (2), 160–164.
Yang, J., Volesky, B., 1999a. Biosorption of uranium on *Sargassum* biomass. Water Res. 33 (15), 3357–3363.
Yang, J., Volesky, B., 1999b. Cadmium biosorption rate in protonated *Sargassum* biomass. Environ. Sci. Technol. 33 (5), 751–757.
Zouboulis, A.I., Matis, K.A., Hancock, I.C., 1997. Biosorption of metals from dilute aqueous solutions. Separation Purif. Methods 26 (2), 255–295.

CHAPTER

Colloid-Mediated Transport and the Fate of Contaminants in Soils

17

Maria Gavrilescu

"Gheorghe Asachi" Technical University of Iasi, Faculty of Chemical Engineering and Environmental Protection; Department of Chemical Engineering; Department of Enviromental Engineering and Management, 71 Mangeron Blvd., 700050 Iasi, Romania; Academy of Romanian Scientists, 54 Splaiul Independentei, RO-050094 Bucharest, Romania

CHAPTER OUTLINE

The Role of Colloidal Systems in Environmental Protection. http://dx.doi.org/10.1016/B978-0-444-63283-8.00017-X

1 Introduction

Increased anthropogenic inputs of contaminants in terrestrial environments have caused considerable concern relative to their impact on soil and groundwater quality. The soil interacts with both the overlying atmosphere and the underlying strata, as well as with surface and underground bodies of water (Hillel 1998, Gavrilescu et al. 2009).

A pollutant in contact with water, soil, and air will partially dissolve into the water, partially volatilize into the air, and partially sorb to the soil surfaces, but the relative amounts of pollutant that are found in each phase depend on intermolecular attractive forces existing between pollutant, water, and soil molecules (Gavrilescu and Betianu 2008, Caliman and Gavrilescu 2009, Gavrilescu et al. 2009, Weiner 2000). Partitioning of a pollutant from one phase into other phases serves to deplete the concentration in the original phase and increase it in the other phases (Figure 17.1).

The soil has considerable but restricted capacity to process pollutants, so that when that capacity is reached, the soil may turn into simply a deposit or a vector in the transmission of pollutants to water resources as well as to the biological food chain by means of growing plants. Physical transport phenomena, as well as complex chemical and microbiological interactions that take place primarily in the upper layer of the soil, are the most noticeable processes occurring in soil in its relation with various contaminants (Hillel 1998).

It is essential that the mobility of applied chemicals in soils be investigated for the environmental assessment of their potential leaching to groundwater supplies. The movement of chemicals through soils to the groundwater is a major cause of degradation of soil and water resources, with great

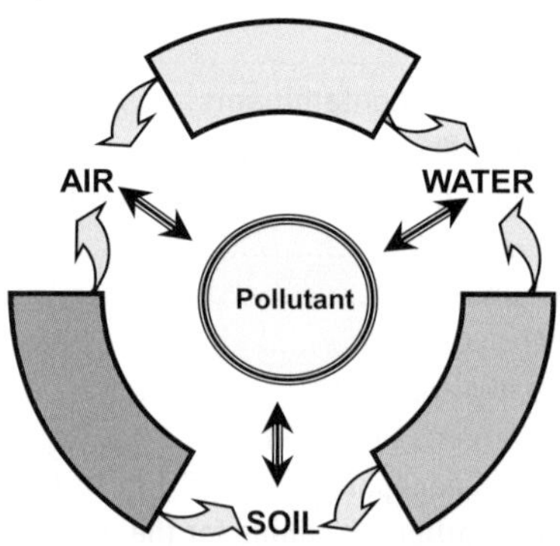

FIGURE 17.1

Partitioning of a pollutant among air, water, soil. (For color version of this figure, the reader is referred to the online version of this book.)

implications for human and livestock health, associated with this form of pollution (Caliman and Gavrilescu 2009, Hlihor et al. 2009, Gavrilescu et al. 2009).

Soil is important not only from an agronomic perspective but also from an environmental perspective for mitigating many of the potentially adverse effects of surface-applied contaminants on the quality of groundwater resources. A fundamental grasp of soil physical, chemical, and biological interfacial interactions at the molecular level is essential to understanding the behavior of metals, metalloids, and other compounds in the pedosphere and to restoring terrestrial ecosystem health globally.

A deep analysis of physical, chemical, and biological interfacial interactions in soil is essential in restoring, enhancing, and sustaining ecosystem productivity, services, and integrity and for establishing soil impact on environmental and human health (Huang, 2008).

With time, colloid migration in subsurface environments has attracted special attention because of its supposed role in assisting the transport of contaminants to groundwater. The interactions at physical-chemical-biological interfaces govern the mechanisms of transformations, speciation, dynamics, bioavailability, toxicity, and fate of various contaminants in soil and related environments (Huang, 2008). For the fate and transport of the contaminant to be altered from flow with the groundwater (for interactions between the subsurface environment and the contaminant to occur), it is necessary that both the contaminant property and the subsurface environment be favorable to these interactions (Knox et al. 1993).

Experimental evidence shows that many contaminants are not only transported in a dissolved state by water, but also sorbed to moving colloids, so that this colloid facilitated transport led to many studies published in the literature, related to numerous contaminants, such as heavy metals, radionuclides, pesticides, pharmaceuticals, hormones, and others (Von Gunten et al. 1988, Kan and Tomson 1990, Lindqvist and Enfield 1992, Grolimund et al. 1996, Tolls 2001, Hanselman et al. 2003, Thiele-Bruhn 2003, Mansfeldt et al. 2004, Gavrilescu 2004, 2005, Simunek et al. 2006, Gavrilescu et al. 2009). This is a very important achievement, since failure to account for colloid-facilitated solute transport can severely underestimate the transport potential and risk assessment for these contaminants (Simunek et al. 2006). In the presence of mobile subsurface colloids, contaminants may move faster and farther than in the absence of colloids, thereby bypassing the filter and buffer capacity of soils and sediments (Flury and Qiu 2008).

This chapter discusses some aspects of the current knowledge and recent developments regarding colloid-facilitated contaminant transport in soils, and examines theextent to which colloids may facilitate contaminant transport in porous media with respect to risk assessment, soil, and groundwater reclamation, and cleanup.

2 Soil structure and spatial variability

The combined effect of physical, chemical, or biological processes that operate with different intensities and at different scales confers to soils a high degree of spatial variability (Santra et al. 2008). Whereas water is always a potential transporter of contaminants, soil can be either a barrier to contaminant movement or a contaminant carrier, since the soil matrix provides solid surfaces to which contaminants can sorb, delaying or stopping their movement (Gavrilescu et al. 2009, Weiner 2000).

Table 17.1 Some Characteristics of the Soil Subsurface Zones (Sen and Khilar, 2007)

Zone	Characteristics
Vadose	is termed soil unsaturated zone is placed between the land surface and the phreatic zone or zone of saturation the unsaturated nature of the vadose zone introduces a third phase, air, which affects colloid partitioning between water and soil presence of oxygen presence of a gas–water interface presence of relatively high concentrations of particulate organic matter presence of relatively high microbial activity transients in flow and chemistry as a result of: dilution with rainwater concentration due to evaporation, large horizontal and vertical variations in the composition of the solid phase transport of contaminants occurs mainly vertical colloids can be trapped in immobile water, strained in thin water films and in the smallest regions of the pore space, or attached to the air–water interface itself.
Capillary	is a subsurface layer groundwater seeps up from a water table by capillary action to fill pores pores at the base of the capillary fringe are filled with water due to tension saturation is less than total capillary rise because of the presence of a mix in pore size
Saturated	the pressure head of the water is greater than atmospheric pressure lower organic matter contents lower oxygen content lower sorption capacity of the solid phase chemical conditions are less variable in time and space transport of contaminants occurs mainly horizontal

The soil subsurface is normally segregated into three regions, depending on their air and water content (Table 17.1). In the vicinity of the ground surface down to an aquifer water table, in the region called *vadose zone* (or *water unsaturated zone*), soil contains mostly air in the pore spaces, with some adsorbed and capillary-held water (Weiner 2000). From the top of the water table to the rock layer, soils contain mostly water in the pore spaces, so that this region is identified as the *saturated zone*.

Between the *vadose* and *saturated zones*, there is a transition region called the *capillary zone,* where capillary forces drain water upward from the water table. The depth of the capillary zone depends on the soil texture: The smaller the pore, the greater the capillary rising force.

Conventional ways of describing and predicting the movement of nonvolatile contaminants are to regard soil and groundwater as a two-phase system in which the partitioning of contaminants occurs between immobile solid constituents and the mobile water phase (de Jonge et al. 2004). In fact, the soil matrix, including the mineral solid phase, mobile or immobilized colloids, biofilms, and soluble substances, is a very complex system. Soil micromorphology is becoming more and more involved in studies on soils as porous media in relation to their physical properties. Soils are characterized by a high degree of spatial variability, due to the shared effect of physical, chemical, or biological processes that operate at different scales with different intensities.

The physical, chemical, and mineralogical properties of secondary clay minerals, hydrous oxides, carbonates, and humic substances, which are primary sorbents in soils especially for metals, largely dictate the kinetics of these reactions. These inorganic and organic soil constituents can exhibit both permanent and variable charges. A permanent charge arises primarily from ionic substitution when an ion substitutes for another of similar size in a crystal structure.

The principal source of a variable charge on soils is the protonation and deprotonation of functional groups on the colloid surfaces (hydroxyl: (–OH), carboxyl: (–COOH), phenolic: ($-C_6H_4OH$), and amine: ($-NH_2$)), while the net charge changes mainly with the pH and is positive at lower pH and negative at higher pH values (Sparks 1995). In most soils, there is a combination of both permanent and variable charge surfaces.

2.1 Soil—a three-phase system

Traditional description and prediction of the movement of volatile and nonvolatile contaminants in soils consider soil and groundwater as a two-phase system in which contaminants partition between immobile solid constituents and the mobile water phase occurs.

In fact, the soil system consists of three phases (Hillel 1998):

- the *solid phase*, which structures the soil matrix, consisting of particles that vary in chemical composition, size, shape, and orientation
- the *liquid phase* (water in the soil or soil solution), which contains dissolved matters
- the *gaseous phase* (the soil atmosphere)

The relative proportions of the three phases in the soil may change continuously, depending on variables such as weather, vegetation, and soil management (Hillel 1998).

2.1.1 Solid phase

Minerals, organic components, and organisms are among the major solid components of soils. The solid phase or the matrix consists of particles that vary in chemical composition, size, shape, and orientation, but also contains amorphous substances, particularly organic matter, which is attached to the mineral grains and may bind them in assemblies called aggregates (Hillel 1998).

The soil-solid phase affects the fate of pollutants, depending on the surface area of its constituents. For example, Yarón et al. (1996) showed that all the components of the soil-solid phase are more or less involved in the following processes:

- the transport of pollutants as solutes, as immiscible with water liquids or as vapors
- the degradation of the organic pollutants, due to the soil-solid-phase effect on the water/air ratio and induced biological activity
- retention and release of pollutants on and from the soil surface
- the surface-induced chemical degradation of pollutants

Strongly sorbing chemicals (e.g., heavy metals, radionuclides, pharmaceuticals, and explosives) in porous media are associated predominantly with the solid phase, which is commonly assumed to be stationary (Simunek et al. 2006).

The reactive constituents of the soil such as clay minerals, minerals others than clays, soil organic matters, as well as the soil-solid-surface net charge play a decisive role in the fate of pollutants in soil. The reactive components (nonliving and living), as well as the structural stability and associated nanoporosity of soils, control the transformation, transport, fate, and toxicity of metals and metalloids in the environment (Huang, 2008).

2.1.2 Liquid phase

The presence of the soil-liquid phase (the soil solution) is the consequence of a dynamic transformation of the dissolved constituents in various chemical species over a range of reaction time scales. In fact, the soil-liquid phase is an electrolyte, containing inorganic and organic ions and molecules (Yarón et al. 1996).

2.1.3 Air in soil

Air in soil has a different composition from atmospheric air because of the biodegradation of organic matter by soil organisms. In this context, O_2 content is decreased in soil air from its atmospheric value of 21% to about 15%, and CO_2 content is increased from its atmospheric value of about 0.03% to about 3%, since oxygen from soil pore space air is consumed and CO_2 is released by microbial metabolism. Much of the soil air is semi-trapped in pores and cannot readily equilibrate with the atmosphere (Weiner 2000).

The presence of air in soil as an oxygen vector promotes some processes and reactions. For example, aerobic biological processes are developed since oxygen is used for respiration by aerobic microorganisms, while anaerobic processes diminish the oxidation potential of water so that reducing conditions prevail, whereas oxidation conditions dominate in the unsaturated zone (Diels and Lookman 2007). As a result, certain metals (iron, manganese) can be mobilized by chemical reduction reactions changing them from insoluble to soluble forms, according to the reaction sequence illustrated by Eqs. 17.1–17.3 (Weiner 2000):

$$Fe(OH)_3(s) + 3H^+ + e^- \rightarrow Fe^{2+}(aq) + 3H_2O \tag{17.1}$$

$$Fe_2O_3(s) + 6H^+ + 2e^- \rightarrow 2Fe^{2+}(aq) + 3H_2O \tag{17.2}$$

$$MnO_2(s) + 4H^+ + 2e^- \rightarrow Mn^{2+}(aq) + 2H_2O \tag{17.3}$$

Equations 17.1–17.3 can be reversed and the metals redeposited as solid precipitates, mainly $Fe(OH)_3$ and MnO_2, if groundwater can transport dissolved Fe^{2+} and Mn^{2+} into zones where oxidizing conditions prevail (surface of a spring or lake).

2.1.4 Soil biota

The soil region where plants, microbes, and other soil organisms support one another (rhizosphere), includes thousands of species of bacteria, fungi, actinomycetes, worms, slugs, insects, mites, and the like, which can be as much as 100 times more numerous than in nonrhizosphere soil zones (Weiner 2000).

The microbial biomass constitutes only a very small proportion ($< 3\%$ of the total organic C in soil), although it is the most active and dynamic fraction of the living organic group in soil (Huang 2008b). Microbial activity that produces humic cements is promoted by root secretions and dead roots, which also produce a gelatinous substance that lubricates root penetration (mucigel) and various sugars, as well as aliphatic, aromatic, and amino acids.

2.2 Soil components with effect on contaminants immobilization or release

Some time ago, it was assumed that only the soil liquid and gaseous phases are mobile and could facilitate the transport of chemicals and nutrients through the vadose zone (de Jonge et al. 2004).

Nevertheless, part of the soil-solid phase is mobile under different geochemical conditions, and mobile organic and inorganic soil particles, colloidal in nature, may facilitate or impair the contaminant transport (Saiers 2002, Sen et al. 2002, Elimelech et al. 2002, Honeyman 1999, Kersting et al. 2009). The physical properties of soil that have the greatest effect on water and contaminant movement are *effective porosity*, *particle size range*, and *hydraulic conductivity*.

Major pathways and mechanisms for the transport of contaminants may differ in the vadose and saturated zone, due to different conditions (Sen and Khilar, 2006). Soil particles, such as minerals, organic matter, microorganisms, protozoans, or nematodes, are in permanent close association and interaction. They play key roles in affecting physical, chemical, and biological processes in soil environments. Soil components are not separate entities, but form an integrated system whose members are continuously in relationship with each other in the environment (Huang 2008). Interactions of soil minerals with organic matter and microorganisms exert a vital role in the formation of humic substances, mineral colloids, humic complexes, short-range-ordered metal oxides, aggregates, and the structure and porosity of soils (Huang 2008).

2.2.1 Organic matter

The properties of soil and its behavior in relation to the contaminants is strongly influenced by the amount of organic matter, which is a heterogeneous mixture of materials ranging from fresh plant and microbial residues to relatively inert humic compounds (Haynes 2008).

Two types of organisms can synthesize organic matter from nonorganic materials, namely, *autotroph* bacteria and chlorophyll-containing plants (Weiner 2000). Organic matter (OM) plays a fundamental role in the fate of contaminants in soil, since it can either adsorb trace elements and immobilize them inside organo-mineral complexes or travel with the soil solution and stimulate trace element transfer within the ecosystem toward groundwaters or surface waters (Pédrot et al 2008). Soil fungi metabolism produces good complexing agents, such as oxalate ion, and citric and other chelating organic acids.

The most important organic substance in soil is *humus*, a collection of variously sized polymeric molecules consisting of soluble fractions (humic and fulvic acids) and an insoluble fraction (humin) (Weiner 2000). Humic and fulvic acids are soluble organic acid macromolecules containing many carboxylic (–COOH) and hydroxyl (–OH) functional groups that ionize in water, releasing protons (H^+) and providing negative charge centers on the macromolecule to which cations are strongly attracted (Figure 17.2).

The amount of organic matter has a strong influence on the properties of soil and on the behavior of soil contaminants. Organic substances such as fulvic acid (FA), humic acid (HA), and many low-molecular-mass organic acids (LMMOAs) form stable complexes with aluminium and can perturb the crystallization of Al hydroxides and thus promote the formation of short-range-ordered (SRO) Al hydroxides (Violante et al. 2002; Colombo et al., 2004; Huang, 2008). Also, Huang (2008) highlighted the fact that interactions of soil minerals with organic substances and microorganisms could lead to a great impact on the formation and transformation of SRO Fe oxides, since they may influence Fe transformation through reduction and oxidation reactions. Metal cations may bind to humic and fulvic acids leading to structures illustrated in Figure 17.3.

These interactions strongly depend on physicochemical parameters such as pH, redox potential, temperature, pressure, and ionic strength, which also control both the amount and the speciation of waterborne elements (Grolimund et al. 1996, Tyler and Olsson 2001). Pedrot et al. (2008) found that

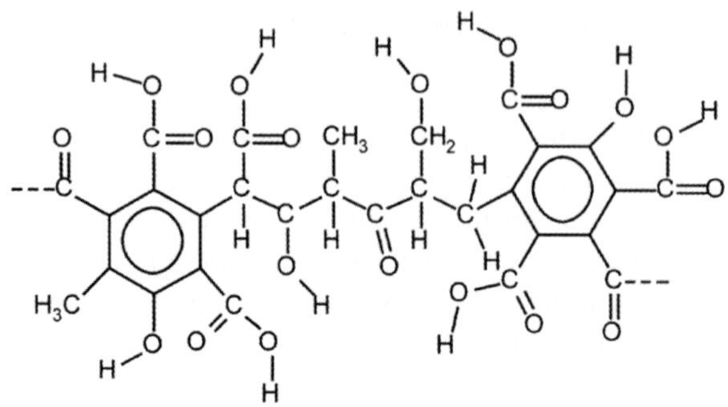

FIGURE 17.2

Fragment of a structural portion of an unionized humic or fulvic acid (Weiner 2000).

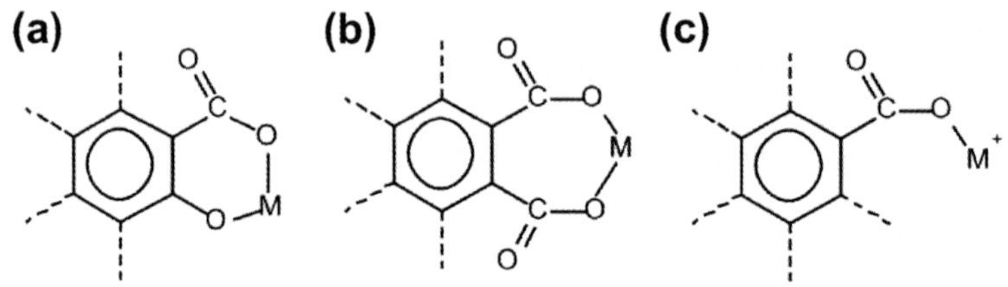

FIGURE 17.3

Structure of the fulvic and humic acids binding M^{2+} metals.

fractions higher than 5 kDa, and particularly the fraction ranging between 5 and 30 kDa which is very aromatic, are very efficient complexing agents for trace elements.

Depending on soil type, organic matter is typically less than 5% in most soils and represents the main factor in plant productivity. Peat soils can be 95% organic, while mineral soils can be less than 1% organic (Weiner 2000). Humic materials are not well defined chemically and have variable composition. It was found that the most abundant elements they contain are C: 45–55%, O: 30–45%, H: 3–6%, N: 1–5%, and S: 0–1%, but the accurate chemical structure depends on the source plant materials and the history of biodegradation (Weiner 2000).

The amount of organic matter in soil is usually expressed as either the weight fraction of organic carbon, f_{oc}, or the weight fraction of organic matter, f_{om}. Humic substances have the ability to sorb and sequester enzymes as well as retain their catalytic activity, although they are able to strongly inactivate enzyme activity depending on interaction mechanisms. On the other hand, soil organic matter can inhibit enzymes, while adsorption on humic polymers may reduce enzyme activity (Vuorinen and Saharinen 1996, Gianfreda et al. 1998, Huang 2008b).

2.2.2 Soil minerals

Soil minerals play a stabilizing role in organic matter through release of Al and Fe from soil minerals during soil formation, which complex and stabilize organic matter against microbial decomposition. In addition, soil minerals play an important catalytic role in accelerating the abiotic polymerization of phenolic compounds, polycondensation of phenolic compounds and amino acids, and subsequent formation of humic substances (Wang et al., 1986; Huang, 2000, 2008a,b).

Particle size is an important property that allows making distinctions among the different soil minerals: the coarse fraction and the fine earth fraction. In soils with little organic matter and in subsoils, mineral colloid-microorganism interactions are largely influenced by the mineralogical composition and pH of the soil. Mineral colloids have a stimulating effect on the activity of adhering bacteria by keeping the pH of microhabitats within the optimum physiological range for growth (Huang 2008).

The surface of mineral colloids provides a favorite environment for soil microbes, since mineral colloids are the most reactive fraction of inorganic components of soils for the reason that they have a large specific surface areas and high charge density characteristics (Theng and Orchard 1995). Besides the large surface areas of mineral colloids facilitating sorption, many mineral colloids also contain micropores, mesopores, or physical structures that can help entrap small organic compounds and exclude degradative enzymes (Mayer et al. 2004, Zimmerman et al. 2004a,b, Yu et al. 2006). Mineral colloids have a high affinity for enzymes, although their catalytic ability is not considered in this analysis.

Minerals in soils can be partially covered with organic materials, especially humic substances, which are to a large extent microbially resistant. Theng and Orchard (1995) illustrated the interactions of mineral colloid —organic material—microorganism as follows (17.4):

$$[\text{Mineral colloid} - \text{HS}]^{-} \ldots \text{M}^{n+} \ldots [\text{EPS} - \text{B}] \tag{17.4}$$

where HS is humic substances, M is divalent/polyvalent metal cation, EPS means extracellular polysaccharides, and B microbes and/or biofilm.

2.2.3 Colloids

Contaminants move in groundwater systems as dissolved species in the water, as flowing free-phase liquids, or combined with moving particulates. Particulates that can move through soils with groundwater must be small enough to move through the soil pore spaces. Such particulates are generally less than 2.0 mm in diameter and are called *colloids.* Colloids are a special class of matter, with properties that lie between those of the dissolved state and the solid or immiscible liquid states. Colloids are a dispersion of small particles (< 500 nm) from one material to another. Colloid particles are generally aggregate, made up of a large number of atoms or molecules that cannot be seen using a regular optical microscope. They can,however, be detected through light diffusion and sedimentation.

The presence of mobile colloids in the subsurface has been acknowledged for several decades. Interest in colloidal movement was propelled by the understanding that colloids can act as mobile phases for the transport of strongly sorbing contaminants.

Colloids, either organic or inorganic (clay mineral colloids—layered minerals containing Fe and Al oxides and amorphous minerals), are microscopic phases in a size range of 1 nm to 1 μm in at least one direction (Stumm and Morgan 1996, Kaplan et al. 1997, Kretzschmar and Sticher 1998, Shein and Devin 2007). Because of their small size, they do not tend to settle down out of suspensions, being influenced by Brownian motion and minor currents in the bulk solutions (Pédrot et al. 2008).

Based on their nature and composition, one can differentiate between three main classes of potentially hazardous colloidal particles: intrinsic colloidal particles, carrier colloidal particles, and biocolloids (Grolimund et al. 2007).

Colloids significantly affect most of the physical and chemical properties of soils. A contaminant can be found in three regions: dissolved, sorbed to mobile colloids, and sorbed to immobile surfaces (Fetter 1999).

2.2.3.1 Colloids in soils and groundwater

Naturally occurring colloidal particles are involved in many important processes in the subsurface zone. Colloids are often generated in the soil system by dispersion from soil aggregates in response to rainfall and infiltration (Ryan and Elimelech 1996).

The tendency of soil colloids to disperse from soil aggregates as a result of water infiltration is a natural phenomenon; it sometimes even leads to the development of illuvial subsurface horizons with higher contents of clay compared with the upper eluvial horizons. In the vadose zone, the source of mobile colloids is generally considered to be the in-situ release of water-dispersible colloids (de Jonge et al. 2004).

Role in colloid generation. Soils are penetrated by the atmosphere, hydrosphere, and geosphere (Figure 17.4). Soil's structure is expressed in its pore system, where the distribution of the pore radii shows a wide range from millimeters to micrometers. This system determines the inner surface of a soil, affecting many important soil parameters and processes such as turnover of nutrients, adsorption of toxicants, and bioavailability of these substances, water storage capacity, and water flow. Pores with radii of microns are the most frequent and have the greatest contribution to the inner surface of soil (de Jonge et al. 2009). These micropores are mainly built up by inorganic colloids such as clays, quartz particles, iron, and aluminium oxides, or organic colloids such as bacteria, fungi, or humic substances. To better understand parameters and processes. it is necessary to take a close look at the microstructure built up by these colloids. Because of their large specific surface area and their small size, many physical, chemical, and biological processes in soils are influenced by these particles (Thieme et al. 2009).

The chemical interactions of soils with the atmosphere and with water largely determine the composition of the Earth's atmosphere and groundwaters. These reactions are also important for cleaning the air and water after natural and human-induced pollution (Sparks and Sparks 2003). Dissolved ions in the soil solution are the immediate sources of the nutrients required by plants. Ions can enter into the soil solution in the following ways: mineral weathering; organic matter decay; rain; irrigation waters containing salts; fertilization; release of ions retained by the colloid or clay fraction of soils. The ions in the soil solution that are of primary interest are those essential for life or toxic to life. Essential elements for plants can be subdivided into those required in relatively large amounts (the *macronutrients*: hydrogen, carbon, nitrogen, oxygen, magnesium, phosphorus, sulfur, potassium, and calcium) and those required in small amounts (the *micronutrients*: boron, chlorine, vanadium, manganese, iron, copper, zinc, and molybdenum). The soil is involved in the supply of all essential elements. Although plants obtain their hydrogen, carbon, and oxygen from water and air, the soil supplies water-holding capacity and pore space for O_2 and CO_2.

Colloids and soil properties. Soil colloids are the finer size fractions of the soil (clay and organic matter), being also considered as the most chemically active portion of the soil because of their large surface area and the chemical structure of the materials involved. In addition, clays and organic matter impart to soils some colloidal properties, including retention of ions and molecules, water and gas adsorption, and shrinking and swelling, since colloidal particles react strongly with fluids due to their surface area (Hayes and Mingelgrin 1991).

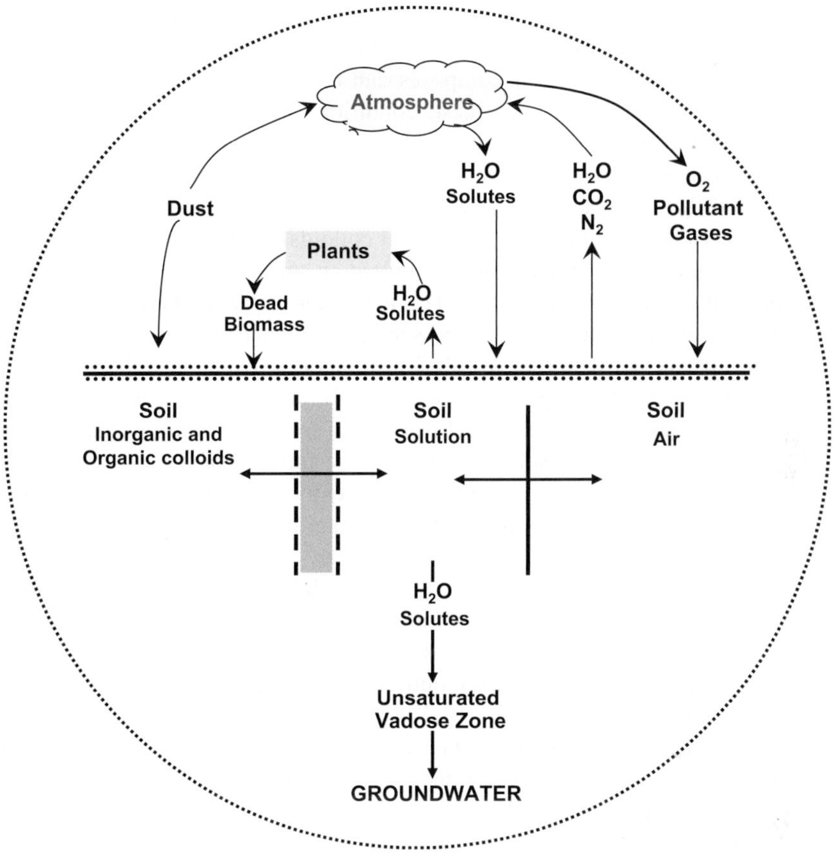

FIGURE 17.4

The cycle of mass transfers between the atmosphere, biosphere, the three soil phases, unsaturated (vadose) zone, and groundwater (the shaded region indicates the transition between soil particles and the soil solutions). (For color version of this figure, the reader is referred to the online version of this book.)

On the other hand, chemical pollution transfers and diverts chemicals from their natural cycles. The carbon, nitrogen, and phosphate in municipal wastes released to water streams are removed from the soil–plant cycle. When air and water are the disposal areas of wastes, pollution can result because air and water convert waste components back into their natural cycles only slowly. Soil has enormous surface area and catalytic activity, plus a supply of oxygen and water with which to deactivate pollutants, so that it is the appropriate location for degrading most wastes quickly and returns components to their natural cycles, thereby minimizing environmental disturbance by human activities (Bohn et al. 2001, Sparks and Sparks 2003).

The various cations adsorbed by soil colloids are subject to replacement by other cations through a process called *cation exchange*. Although cations may be adsorbed by soil colloids, they are not all held with equal tightness by the colloids. The order of strength of adsorption when they are present in equal quantities is Al > H > Ca > Mg > K > Na. Exchangeable ions can be released from the soil by a salt solution; soluble salts, on the other hand, can be removed by water alone. For example, hydrogen ions (H^+) generated as organic matter decomposes can displace Ca^{2+} and other cations from the colloidal complex (Adriano 2001). Under these conditions, the soil tends to become more acid (Bohn et al. 2001). Cation exchange reactions are reversible, so if limestone or another calcium compound is applied to an acidic soil, the reverse of the above exchange occurs. The Ca^{2+} ions replace H^+ and Al^{3+} ions, and the soil pH is raised. If sulfur or other acid-forming material is added to an alkaline soil, H^+ replaces the metal cations on the soil colloids and the soil pH decreases. If a water-soluble fertilizer (containing KCl, for example) is applied on soil, K^+ may be adsorbed on the colloid and will replace an equivalent quantity of Ca^{2+}, H^+, and other metals that appear in the soil solution. The K remains in an available condition but is less subject to leaching.

Cation exchange capacity (CEC) is a measure of the total of the exchangeable cations that a soil can adsorb (expressed as milliequivalents per 100 grams of soil, or centimoles of positive charge per kilogram of soil). CEC can range from as little as 1 meq/100 g soil for extremely coarse-textured soils to 50–60 meq/100 g soil for fine-textured soils containing large amounts of clay and organic matter. Soils can also adsorb negatively charged ions (anions) such as chloride, sulfate, or nitrate, and can adsorb uncharged molecules (e.g., pesticides). However, the magnitude of adsorption for these species is generally much less than for cations.

Exchangeable cations generally are available to both higher plants and microorganisms. Through cation exchange, hydrogen ions from the root hairs or microbe replace nutrient cations from the exchange complex. They enter the soil solution where they can either be absorbed by plant roots or be removed with drainage water. Factors affecting nutrient availability are relative amount of cation on the colloid surface (the higher the proportion of the cation in question, the easier it will be for the plant to obtain it); competition between nutrients; and influence of associated ions (for example: magnesium availability to plants has been shown to be limited by excessive quantities of exchangeable calcium) (Evans and Sorger 1966, Bengston et al. 2007).

Acidity, alkalinity, and pH are among the most important chemical properties of the soil. Since microorganisms and higher plants respond so markedly to their chemical environment, the importance of soil pH has long been recognized. Acid soils originate in regions where rainfall is high enough to leach exchangeable *basic cations* (e.g., Ca^{2+}, Mg^{2+}, K^+, Na^+) from the surface soil (Warfvinge and Sverdrup 1992, Adriano 2001, LeRoy Jackson and Barak 2005). Alkaline soils occur where basic cations have remained, or where calcareous (calcium carbonate) parent materials have dominated. In acid soils, hydrogen (H^+) and aluminum (Al^{3+}) dominate in the soil solution; in alkaline soils, hydroxyl (HO^-) ions dominate. Soil pH is in fact a measure of the concentration of hydrogen or hydroxyl ions in the soil solution. The pH scale runs from 0 to 14 and is logarithmic (i.e., a soil at pH 4 is 10 times more acidic than a soil at pH 5 and 100 times more acidic than a soil at pH 6). pH 7 is neutral; that is, the concentrations of hydrogen and hydroxide ions are equal. Lower than pH 7, acid conditions exist; higher than pH 7, alkaline conditions exist. Potential sources of acidity and alkalinity are the presence of aluminum oxides and hydroxides, sulfur or sulfides, fertilization with nitrogen in the ammonium (NH_4^+) form, decomposing plant residues and organic wastes, hydroxyl ions, carbonate ions from limestone, irrigation water pumped from limestone aquifers, calcareous parent material, and so on (Adriano 2001).

Colloids in the subsurface can significantly enhance the mobility and transport of contaminants (McCarthy and Zachara 1989) especially those that are comparatively insoluble (Grolimund et al.

1996, Gooddy et al. 2007). Dispersion of colloids is also suspected to be responsible for affecting soil physical properties such as surface crusting, surface erosion, water infiltration, and hydraulic conductivities (Miller and Baharuddin 1986, Shainberg et al. 1992, de Jonge et al. 2004).

Studies suggest that colloids are able to move through porous media even when the degree of saturation is very low (Harvey and Garabedian 1991, Wan and Tokunaga 1997). Over the last decade, there has been an increase in knowledge and understanding of the colloidal fines-associated contaminant transport in porous media (Sen and Khilar, 2007). Soil/water interactions generate various medium-dependent type colloids in the different environments.

Colloid behavior in natural soil–water systems is controlled by dispersion-flocculation processes, which are multifaceted phenomena, such as surface electrical potential, solution composition, shape of particles, initial particle concentration in suspension, and type and relative proportion of clay minerals (Evangelou 1998).

In groundwater a variety of inorganic and organic materials exist as colloids and small particles, including mineral precipitates (mainly iron, aluminum, calcium, and manganese oxides, hydroxides, carbonates, silicates, and phosphates), rock and mineral fragments, biocolloids (including viruses, bacteria, and protozoans) and macromolecular components of natural organic matter (Sen and Khilar 2007).

Carbon-rich colloids likely derive from the uppermost soil horizon, as well as many trace elements (e.g., heavy metals) usually associated with carbon colloid. Mineral colloids can interact strongly with negatively charged microbial cells coated with extracellular polysaccharides (EPS) in soil environments. They behave as a positively charged species or display amphoteric characteristics (Hung, 2009).

2.2.3.2 Types of colloids

Two colloidal substances, clay and humus, are present in soil essentially for its *cation exchange capacity* (CEC) since they serve as a cation warehouse or reservoir of the soil. Clay soils with high levels of organic matters (OM) have a much greater capacity to hold cations than sandy soils with very little organic matter. While colloids are subject to a similar subsurface fate and transport processes as chemical compounds, they are also subject to their own unique complexities (van Genuchten and Simunek 2004, Simunek 2006).

Mobile organic sorbents (MOS), such as dissolved and colloidal phase organic matter, control flow of water and transport of solutes in soils and determine sorption, partitioning, speciation, and ion exchange due to complexation, solubilization, carrier association, and the solvophobic effects (McCarthy and Zachara 1989, Totsche et al. 1997, Totsche and Kögel-Knabner 2004).

Cation exchange capacity is an important property of the soil, being a useful indicator of soil fertility, nutrient retention capacity, and capacity to protect groundwater from cation contamination (Grolimund and Borkovec, 2006).

2.2.3.3 Colloid generation and movement in subsurface environment

Colloids are formed in soil in some conditions, such as (Weiner 2000):

- when fragments of soil, mineral, or contaminant particles become detached from their parent solid because of weathering
- when dissolved minerals in groundwater undergo pH or redox potential changes (they result as fine precipitates)

- when introduced directly into groundwater from landfills, they can form as emulsions of small droplets from free-phase hydrocarbons or other immiscible liquids.

The most frequently occurring subsurface colloids are inorganic nanoparticles (clay minerals, metal oxides or oxyhydroxides, silica, carbonates), natural organic matter (NOM), bacteria and viruses (McCarthy and Zachara 1989, Kretzschmar et al. 1999, Metreveli and Frimmel 2007).

The need to examine colloid movement through geologic materials is determined by a variety of matters, including contaminant transport, soil-profile development, and subsurface migration of pathogenic microorganisms. For example, the movement of colloids through the vadose zone (illuviation) is an important process in soil genesis. In the vadose zone, colloid movement may facilitate the transport of some contaminants, as well as of pathogenic microbes (as biocolloids) during wastewater reclamation and aquifer recharge, which can induce a public health risk. Also, the deposition of mobile colloids may reduce soil permeability (Baveye et al. 1998, Redman et al. 2001, McGechan and Lewis 2002, DeNovio et al. 2004).

2.2.3.4 Entrapment and release phenomena

The processes of colloid transfer in soil pores (lessivage, input of fine fractions in suspensions, etc.) have been the focus of attention for soil scientists (Sen and Khilar 2006, Shein and Devin 2007). The migration of colloidal fines in porous media seems to be a reasonably well-understood topic, and their release and entrapment are thought to depend on various conditions (Table 17.2, Table 17.3). This effect could ultimately lead to plugging the porous media, which can drastically diminish the contaminant transport and subsequent containment under a constant pressure gradient flow as it occurs in groundwater flows (Khilar and Fogler 1998, Ramachandran and Fogler 1999, Sen et al. 2002, Sen and Khilar 2006). The entrapment phenomena depend on various factors as described in Table 17.3 (Pandya et al. 1998, Sen and Khilar 2006).

Migration of colloidal fines in porous media is a reasonably well-understood topic. Whether or not the particles get entrapped depends on various conditions, which could eventually lead to the plugging of the porous media (Sen and Khilar 2006). The plugging results in drastic reduction in flow rates and therefore, there exists a possibility of reduction of contaminant transport and subsequent containment under a constant pressure gradient flow as it occurs in groundwater flows.

Two behaviors of the released colloidal fines while flowing with the liquid phase through the porous medium may be determined by capture or by getting entrapped at the pore constrictions if the colloidal and hydrodynamic conditions that bring about their release are not likely to allow these particles to readhere back to the pore surface in the same conditions they were released (Sen et al. 2004, Sen and Khilar 2008). Entrapment can occur in three forms: size exclusion, multiparticle bridging, and surface deposition (conceptually depicted in Figure 17. 5).

Two important pore structure parameters that significantly affect the entrapment or piping of fines are the pore constriction size distribution and pore constriction to chamber interconnectivity, which can be described in terms of coordination numbers, as well as the number of pore constrictions per pore chamber (Sen and Khilar 2006). Media having small pore constriction size distributions and low coordination numbers show high probability of entrapment of fine particles, whose size or the ratio of size to that of pore constriction is a crucial parameter in determining the entrapment/plugging phenomenon. If the fine size is equal to or greater than the size of the pore constriction, then entrapment or plugging will inevitably take place. Figure 17.6 presents some qualitative results on whether plugging

Table 17.2 Fine Particles Releasing Mechanisms

Releasing Process	Characteristics of Mobilization	Factors Inducing Mobilization and Subsequent Transport	Examples	References
Chemically induced	In-situ particle released as a result of changes in solution chemistry may be an important transport mechanism for contaminant species expansion of the electrostatic double layers around like-charged colloids and grains	change in solution ionic strength (low ionic strength) critical salt concentration (CSC) critical total ionic strength (CTIS) high pH	noncalcareous silt loam soil sand soil highly weathered aquifer sand	Roy and Dzombak, 1996 Sen and Khilar, 2007 Saiers and Hornberger, 1999 Kaplan et al., 2006 Bergendahl and Grasso, 2003 McCarthy and Zachara, 1989
Hydrodynamically induced release	fine particles adhering to the pore surfaces of the porous media of relatively higher porosity and permeability can be released or be detached by the hydrodynamic forces hydrodynamic forces may act in more than one direction a lift force due to unsteady nature of viscous sublayer in the turbulent boundary layer, a hydrodynamic force along the surface causing the particle to slide a torque causing the particle to roll along the surface	fines are released only beyond a certain threshold value of a particular parameter: ''critical salt concentration'' (CSC), or "critical shear stress/critical velocity"		Khilar and Fogler, 1998, Sen and Khilar 2006

Table 17.3 Factors Affecting Plugging of the Porous Media and Effects on Contaminant Transport

Factor	Parameters Affected	Effects
Pore structure	pore constriction size distribution pore constriction to chamber interconnectivity	the probability of entrapment of fine particles is relatively high for porous media having small pore constriction size distributions and low coordination numbers
Size of colloidal fines	the ratio of size of fines to that of pore constriction the bead / particle ratio	if the fine size is equal to or greater than the size of the pore constriction, then entrapment or plugging will inevitably take place retention of particles due to surface deposition at sites other than the constriction sites also takes place if the bead / particle ratio is increasing for a fixed flow condition, then the extent of plugging by straining or convective jamming is decreasing
Concentration of fines	the ratio of size of fines to that of pore constriction	the concentration of fines is of high importance when the ratio of fines to size of pore constriction is in the range of 0.01 to 0.1
Hydrodynamic conditions	hydrodynamic forces acting on the released fine particles *flow velocity*	hydrodynamic forces affect the plugging phenomenon at the pore constrictions

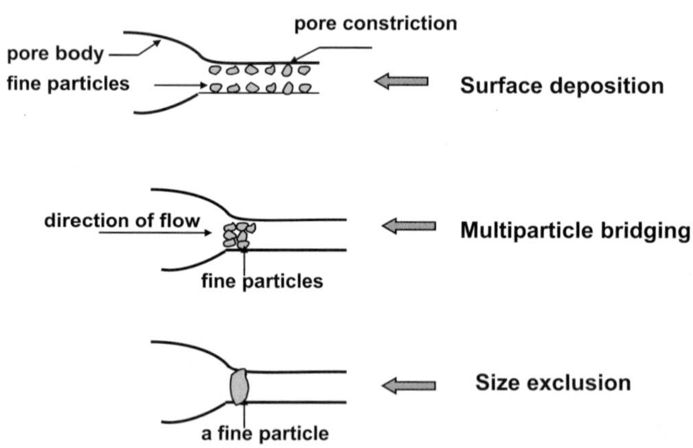

FIGURE 17.5

Conceptual diagrams for entrapment of fine particles at the pore constriction. (For color version of this figure, the reader is referred to the online version of this book.)

or other possible types of deposition could occur as determined by the ratio of the size of fines to size of the pore constrictions (Sen and Khilar 2008).

Khilar and Fogler (1998) found that, when the ratio of fine size to constriction size is between 0.1 and 0.01, there will be the possibility of multiparticle blocking, which depends on constriction of fines

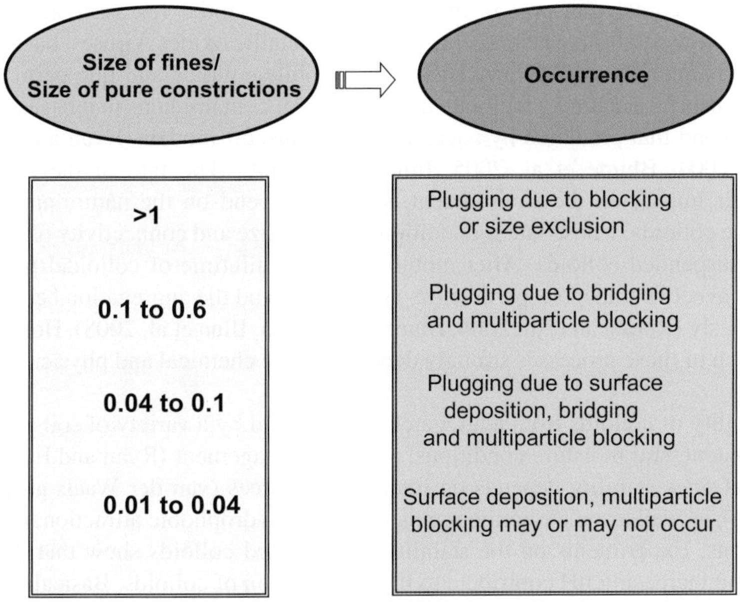

FIGURE 17.6

Dependence of plugging on the ratio of size of fines to size of pore constrictions. (For color version of this figure, the reader is referred to the online version of this book.)

and flow rate. Retention of particles due to surface deposition at sites other than the constriction sites also takes place. When the ratio of fines to size of pore constriction is in the range of 0.01 to 0.1, the concentration of fines assumes a high importance (Khilar and Fogler 1998).

The plugging phenomenon at the pore constrictions is affected by hydrodynamic forces acting on the released fine particles, in particular the flow velocity. At small values of the flow velocity, but for large particle density and size, the trajectories of the particles will be affected by sedimentation force, causing them to settle to the bottom wall of the pore space. Since the size of the fines is small in most cases of migration of fines in porous media, the fines will tend to follow the streamlines of the flow and form bridges at pore constrictions. The rate of formation and the structure of these bridges depend on the hydrodynamic conditions of the flow, so that at higher flow rates, the bridge will tend to break because of a higher drag force acting on them, while at lower flow rates, the number of bridges formed may be greater (Sen and Kilhar 2006).

2.2.3.5 Stability and transport

A comprehensive amount of literature exists on the topic of colloid mobilization and transport in model systems of packed sand demonstrating the importance of physical and chemical perturbations for the mobilization and transport of colloids (Kretzschmar and Sticher 1997, Grolimund et al. 1998, Grolimund and Borkovec 1999). However, in natural soils the processes of colloid mobilization and transport can be additionally complicated by the intense effects of pore structure on the active flow trail of water, thus affecting both in-situ colloid mobilization and subsequent transport.

Understanding colloidal transport in the subsurface is essential for assessing the migration of contaminants with low solubility, such as clay minerals, metallic oxides, viruses, bacteria, and organic macromolecules. Numerous studies have shown that mobile colloids and fine particulate matter are abundant in the subsurface and may facilitate the transport of contaminants that have a high affinity for their surfaces, beyond that predicted by aqueous-phase transport models (Ryan and Elimelech 1996; Rousseau et al. 2004, Blume et al. 2005, Ilina et al. 2008). The fate of the contaminants, and consequently their impact on the environment, strongly depend on the nature and behavior of the potentially mobile colloids, whose fate is conditioned by the size and connectivity of pores and the size and stability of suspended colloids. After mobilization, the lifetime of colloidal particles is mainly determined by convective transport, particle (re-) deposition and the aggregation behavior, which may occur simultaneously (Rousseau et al. 2004, Blume et al. 2005, Ilina et al. 2008). However, the relative importance of each of these processes strongly depends on the chemical and physical conditions of the system.

The dispersibility of colloids from aggregates is influenced by a variety of soil properties, such as pH, total clay content, soil moisture conditions, and soil management (Ryan and Elimelech 1996). In fact, the colloidal fines stability depends on interparticle forces (van der Waals attraction and electrolytic repulsion), as well as on hydration repulsive forces, hydrophobic attraction, polymer bridging, or steric repulsions. Experiments on the stability of dispersed colloids show that dissolved organic matter (DOM) and increasing pH contribute to the stabilization of colloids. Basically, in the saturated zone colloid transport is described by several filtration models, but in the unsaturated soil zone, the mobilization, retention, and migration of colloids is much less understood (Ryan and Elimelech 1996, Sen and Kilhar 2006).

Another class of colloidal fines says viruses and bacteria also fall into the colloidal size range and called biocolloids. The transport of biocolloids (viruses, bacteria, spores, and other microorganisms) through saturated and unsaturated porous media is of some significance, from the perspective of protecting groundwater supplies from contamination, assessing risk from pathogens in groundwater, natural and enhanced bioremediation, and design of better water treatment systems to remove biocolloids from drinking water supplies (Sen and Kilhar 2006).

The colloids' physicochemical, mineralogical, and surface charge characteristics, as well as solution chemistry influence the contaminant transport capacity of the readily dispersible colloids entering the aquatic system (Evangelou and Karathanasis 1991, Zhang et al. 2006). Colloid-associated contaminant transport can occur only if colloids are present in sufficient amounts to sorb a significant amount of the contaminant (Sen and Kilhar, 2006). This is dependent on the amount and nature of the colloidal particles present in the subsurface zone.

In water-saturated soils (Table 17.1), the mobilization and removal of colloids are usually accelerated with increasing amounts and duration of precipitation, so that the concentration peaks correspond to the lowest and highest values of water run (Rousseau et al. 2004, Shein and Devin 2007).

Studies in ideal porous media having uniform moisture content and steady pore-water velocity have shown that colloid deposition rates are sensitive to some soil physical and chemical properties, including volumetric moisture content, flow rate, pore-water ionic strength, as well as colloid size and composition (Wan and Tokunaga 1997, Gamerdinger and Kaplan 2001, Saiers and Lenhart 2003). DeNovio et al. (2004) attributed this fact to the interactions among three deposition mechanisms: mineral-grain attachment, air–water interface capture, and film straining. On the other side, the transport rates are dependent on three mechanisms of transport from the bulk fluid to the mineral grains: Brownian diffusion, interception, and sedimentation (Yao et al. 1971, DeNovio et al. 2004). They can be considered as functions of the physical properties of the porous medium–water–colloid system, including colloid diameter and density, grain size, and flow velocity (Yao et al. 1971, Logan et al., 1995, Tufenkji and Elimelech 2004).

Many colloids and microbes are electrostatically repelled by negatively charged solid surfaces since they are negatively charged (Simunek et al. 2006). This will lead to a slightly enhanced transport relative to fluid flow caused by an anion exclusion process. Also, size exclusion may enhance the advective transport of colloids, which limits their presence in the larger pores (Bradford et al. 2003, Bradford et al. 2006, Simunek et al. 2006).

3 Contaminants in soils

Soil contamination is caused by the presence of chemicals or other alterations in the natural soil environment, arising from the breaking of an underground storage tank, application of agrochemicals (in particular pesticides), percolation of contaminated surface water to subsurface levels, oil and fuel discarding, leaching of wastes from landfills, or direct discharge of industrial wastes to the soil. The most common chemicals involved are petroleum hydrocarbons, solvents, pesticides, and heavy metals.

3.1 Categories of contaminants in soil

3.1.1 Inorganic contaminants

3.1.1.1 Metals

Pollution of the environment with toxic metal is a result of many human activities such as mining and metallurgy. These metals have large economic and public health effects on the ecosystem (Gavrilescu 2004).

Most heavy metals are generally considered to be relatively immobile in most soils for short periods. However, it has been demonstrated that their mobility under certain solution and soil chemical conditions may exceed normal rates, and their presence in soil and groundwater may lead to significant threats for soil and water quality as well as human health (Maskall et al. 1995, Karathanasis, 1999).

Many potentially toxic trace metals and radionuclides are strongly adsorbed onto surfaces of mineral and organic compounds in soils and sediments, limiting their mobility in the environment. However, recent studies have shown that trace metals in soils, groundwater, rivers, and lakes can be carried by mobile colloidal particles (Kretzschmar and Schäfer 2005).

Bioaccumulation of metals in plants, followed by animal and human consumption of those plans, can lead to toxic effects. This mobility could be the consequence of water-dispersed colloidal particles migrating though soil macropores and fractures that can significantly enhance metal mobility, causing dramatic increases in transported metal load and migration distances (Mills et al. 1991, Liang and McCarthy 1995, Ouyang et al. 1996, Ryan and Elimelech 1996, Karathanasis 1999). For example, aluminum forms a series of highly reactive soluble species and poorly crystalline to noncrystalline mineral colloids, which include multiple reaction paths presented in details by Huang (2008).

Some metals—particularly As, Cd, Hg, Ni, Pb, Se, Te, Sn, and Zn—can form volatile metal-organic compounds in the natural environment by microbial mediation. Metal oxides have a significant role in influencing the physical, chemical, and biological properties of soils. They may exist as crystalline minerals, as short-range ordered (SRO) mineral colloids, or as surface coatings on clay minerals and organic matter.

3.1.1.2 Radionuclides

Environmental contamination caused by radionuclides, in particular by uranium and its decay products, is a serious problem worldwide. The development of nuclear science and technology has led to increasing release and disposal of nuclear waste containing uranium into the environment (Kretzschmar and Schäfer 2005, Gavrilescu et al. 2009).

Radionuclides are found in the soil as a consequence of some pathways (Todorov and Ilieva 2006, Gavrilescu et al. 2009)

- as part of Earth's layer (primordial radionuclides)
- generated and deposited by cosmic ray interactions
- as a result of anthropogenic activities

Artificial radionuclides are also introduced into the environment following nuclear power plant accidents or nuclear weapons tests, nuclear energy activities, scientific and other uses (Baca and Florkowski 2000, Todorov and Ilieva 2006).

The surprisingly rapid transport of cationic radionuclides that has been observed at hazardous waste sites has been attributed to colloid-facilitated transport (Rousseau et al. 2004). In soils, the mobility of radionuclides and other microelements is dependent on the binding mode of chemical elements to the clay–humus complex of the soil, since radionuclides are bound to various organic and inorganic solid phases of the soil. The migration potential of radionucide association with water-soluble, exchangeable oxide (Fe, Mn, etc.), and organic phases of the soil can be examined experimentally by single-stage and sequential extraction techniques using distilled water, variously concentrated acid and alkaline solutions, and various salt solutions (Sokolik et al. 2001, Goryachenkova et al. 2009).

3.1.2 Organic contaminants

Different sources such as the chemical industry, traffic, combustion of fossil fuels and other organics, water treatment by chlorination, and application of pesticides in agricultural areas release organic pollutants into the environment by direct input, transport, or precipitation processes (Kordel et al 1997, Widenfalk 2002). Contamination of soil, sediments, and groundwater by persistent and recalcitrant organic compounds such as chlorinated aromatics, polynuclear aromatics, heterocyclics and nitroaromatics is a widespread legacy of modern industrial, commercial efforts. The specific locations where chemicals were spilled or released, even decades ago, tend to remain the places with the highest concentrations of these contaminants (Kleka et al. 2001, Buccini 2004, Gavrilescu 2005).

Extensive scientific studies have shown that persistent organic pollutants (POPs) are some of the most dangerous pollutants released into the environment by humans (Tanabe et al. 1994, Csizer 2002, Buccini 2004). POPs are toxic substances released into the environment through a variety of human activities. They have adverse effects on the health of ecosystems, wildlife, and people (Kordel et al. 1997, Widenfalk 2002, Gavrilescu 2005).

The contamination of soil and groundwater by agrochemicals is a major environmental concern. The widespread use of pesticides over the past half century has led to their detection in many hydrologic systems (Barton and Karathanasis 2003). Numerous studies have been undertaken that show pesticide can sorb to colloids and potentially enhance their transport (Vinten et al. 1983, Seta and Karathanasis, 1997, Worrall et al. 1999, Kulikova and Perminova 2002).

Along with the dynamic of water flow, pesticide transport from the soil surface to groundwater is controlled by several processes, including plant uptake, sorption to solid phases, microbially mediated degradation, and dissolved and colloidal phase migration (Foster et al. 1991, Chilton et al. 2005).

Goody et al. (2007) have demonstrated that pesticides will form associations with colloids from a calcareous soil and produce pesticide-colloid complexes with a size generally in excess of 0.1μm.

3.2 Transport processes in soil involving contaminants

The movement of contaminants through soil is a process of continuous redistribution among the different phases they encounter. It is a process controlled by gravity, capillarity, sorption to surfaces, miscibility with water, and volatility (Weiner 2004). For predicting the environmental behavior of a pollutant, six properties are the most important (Weiner 2000), namely: *solubility in water, volatility, density, chemical reactivity, biodegradability, tendency to adsorb to solids.* Since environmental conditions are so varied, it is difficult to generate a simple set of properties that is always the most important to measure.

The distribution, mobility, bioavailability, and toxicity of chemicals depend not simply on their concentrations but critically, on their forms in soil (Huang 2008). Contaminants that are sparingly soluble in water and have a strong tendency to bind to the solid phase are assumed to be retarded in the soil (de Jonge et al. 2004).

The general categories of processes affecting subsurface fate and transport are hydrodynamic processes, abiotic processes, as well as biotic processes. Physical, chemical, and biological processes are not independent but rather, interactive with each other. Two basic elements affecting the transport and fate of contaminants in the subsurface are properties of the subsurface materials or the subsurface environment and physicochemical and biological properties of the contaminant (Knox et al. 1993) (Figure 17.7).

Pollutants have three possible naturally occurring fates (Weiner 2000, Gavrilescu 2004, Gavrilescu 2005) (Table 17.4):

- They remain unchanged in their current location.
- They are carried elsewhere by transport processes (through movement to other phases—air, water, or soil, volatilization, dissolution, adsorption, and precipitation; and/or movement within a phase under gravity, diffusion, and advection)
- They are transformed into other chemical species by natural chemical and biological processes (biodegradation, bioaccumulation, nonbiological chemical changes by processes such as oxidation-reduction, acid-base, hydration, hydrolysis, complexation, and photolysis reactions).

Contaminant transport is influenced by hydrodynamic processes (advection, dispersion, and preferential flow) through impacting the quantity and paths of flow of groundwater in the subsurface. Abiotic processes (sorption, ion exchange) affect contaminant transport as a result of interactions between the contaminant and the stationary subsurface material or due to chemical reactions (hydrolysis, redox reactions) that lead to contaminant degradation of transformation. Biotic processes can affect contaminant transport by metabolizing or mineralizing the organic contaminants, during aerobic, anoxic, and anaerobic biodegradation. Figure 17.7 presents a schematic representation of subsurface

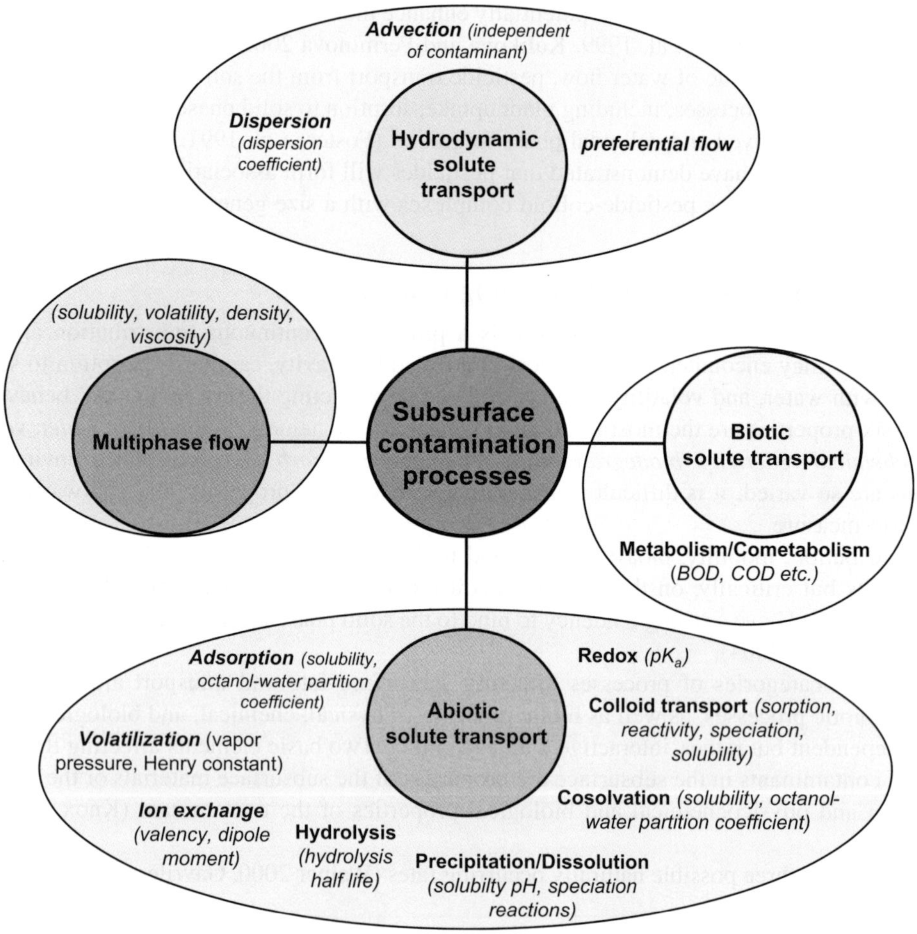

FIGURE 17.7

Schematic representation of some subsurface processes and corresponding subsurface and contaminant properties and interactions affecting the fate and transport of contaminants. (For color version of this figure, the reader is referred to the online version of this book.)

(adapted from Knox et al., 1993).

processes and corresponding subsurface and contaminant properties influencing these processes (Knox et al. 1993).

Nonreactive (conservative) chemicals will move through the subsurface environment with the groundwater (hydrodynamic processes) and will not be affected by abiotic or biotic processes that may be active in the subsurface (Knox et al. 1993). Contaminants that have the potential to be reactive (nonconservative) will not be affected during groundwater transport if the subsurface environment is not conducive to the reactions that affect the contaminants (Knox et al. 1993, de Jonge et al. 2004).

After infiltration of rainwater in urban areas, the pollutants, adsorbed or bound onto the mobile colloids, can discharge into the groundwater. Many contaminants, which are less soluble in water but

Table 17.4 Processes of Transport and Transformation the Pollutants Might Suffer in the Environment (Weiner 2000)

Process		Characteristics
Transport (contaminants can move to other phases)	*Sorption*	contaminants dissolved in liquid phase might be bound to solids by attractive chemical and electrostatic forces
	Precipitation	contaminants dissolved in liquid phase are transformed in compounds of low solubility (settable or not-settable) by changes in pH or oxidation-reduction potential, or by reaction with other species in water
	Sedimentation	small suspended solids in water grow large enough (by coagulation or flocculation) to settle to the bottom under gravity
	Volatilization:	dissolved contaminants move from water or soil into air in the form of gases or vapors
Transformations (by chemical reactions in the environment)	*Acido-basic reactions*	present in many environmental processes and influence the reactions of many pollutants, as a result of protons (H^+ ions) transfer between chemical species
	Complexation and chelation	electron-donor ligands bind polar or charged dissolved species to form *complex* or *coordination* compounds, often soluble and resisting removal by precipitation (common ligands include hydroxyl, carbonate, carboxylate, phosphate, and cyanide anions, as well as humic acids and synthetic chelating agents)
	Oxidation-reduction	chemical species transfer electrons, changing the oxidation states and the chemical properties of the electron donor and the electron acceptor
	Hydrolysis and hydration	in water a compound might form chemical bonds to water molecules or hydroxyl anions (all ions and polar compounds develop a hydration shell of water molecules, when the attraction to water is strong enough, a chemical bond can result: for example, hydration of dissolved carbon dioxide (CO_2) and sulfur dioxide (SO_2) forms carbonic acid, H_2CO_3 and sulfurous acid (H_2SO_3), respectively).
	Precipitation	if a solution of a salt becomes oversaturated, as in when the concentration of a salt becomes greater than its solubility limit, two or more dissolved species react to form an insoluble solid compound
	Photolysis	molecules that absorb solar radiation; exposure to sunlight can break chemical bonds and start chemical breakdown

Continued

Table 17.4 Processes of Transport and Transformation the Pollutants Might Suffer in the Environment (Weiner 2000)—cont'd

Process		Characteristics
Biological processes (aerobic or anaerobic)	Biodegradation	microbial metabolism (that is the biological reactions that convert organic compounds into energy and carbon for growth) involves the transfer of electrons from a pollutant molecule to other compounds present in the soil or water environment that serve as electron acceptors (molecular oxygen (O_2, involved in aerobic biodegradation), carbon dioxide (CO_2), nitrate (NO_3^-), sulfate (SO_4^{2-}), manganese (Mn^{2+}), and iron (Fe^{3+}) (all involved in anaerobic biodegradation, excepting oxygen)

have a strong tendency to bind to the solid phase, are assumed to be immobile and, therefore, present little danger to the groundwater (Zhang et al., 2005). However, readily dispersible colloidal particles in the solid phase may also be mobile in subsurface environments and thus transport significant amounts of contaminants to groundwater.

The transport of potentially harmful substances, such as heavy metals and pesticides, through the soil profile into the groundwater has been a major environmental concern.

4 Colloid-mediated transport of contaminants in soils

Contaminant fate and transport as well as risk assessment are based on an accurate representation of subsurface processes, and in the case of strongly sorbing contaminants, need to consider mobile colloids as potential contaminant carriers (Flury and Qiu 2008). The transport of contaminants, in particular colloid-mediated transport, is of fundamental interest in environmental sciences. Several investigations and experimental studies show that colloidal particles can facilitate the transport of different pollutants in soils, since field- and laboratory-scale observations have shown that in the presence of mobile colloidal particles (e.g., microbes, humic substances, clays, and metal oxides), colloids can act as pollutant carriers and thus provide a rapid transport pathway for strongly sorbing contaminants (Simunek et al. 2006).

An unexpected mobility of a low-solubility material may result when contaminants of low solubility can be transported as colloids or associated by adsorption or occlusion with colloids. When contaminants are sorbed to colloids, their transport behavior is determined by the properties of the colloid, not the sorbed contaminant (Weiner 2008).

The keys to assessment of the effects of colloids on subsurface transport are the colloid-contaminant interaction and the mobility of the colloid itself. Although characterized by small size, colloids are mobile and offer diverse surface functions allowing interactions with trace elements, playing a key role in the distribution of trace elements. The potential of enhanced transport of colloid-associated contaminants can be very significant, since mobile colloids often move at rates similar or faster as nonsorbing tracers (McCarthy and Zachara 1989, Simunek et al. 2006).

Due to their high specific surface area, colloids prove to have a high sorptive capacity and can be effective sorbents of low solubility, powerfully sorbing contaminants, acting as a transport vector of contaminants through the vadose zone, thus causing an increased risk of leaching of contaminants generally regarded as relatively immobile.

Major compounds that have been shown to increase the solubility and thus the mobility of nutrients and contaminants are surfactants, cosolvents, oxides and/or hydroxides, clay and minerals, humic substances, and humin-coated inorganic colloids (McCarthy and Zachara 1989, Brusseau et al. 1991, Kretzschmar and Sticher 1997, Liu and Chang 1997, Totsche et al. 1997, Degueldre et al. 2001, Totsche and Kögel-Knabner 2004). Their chemical properties, such as chemical diversity and high concentration of functional groups, are suspected to interact both with components of the immobile solid phase and with other dissolved and colloidal phase components of the liquid phase (Totsche and Kögel-Knabner 2004).

Colloid-mediated transport is environmentally relevant when three situations occur (Kretzschmar et al. 1999): first, mobile colloids are present in large concentrations; second the particles are transported over significant distances through uncontaminated zones of the porous medium; and third, the contaminant sorbs strongly to the mobile particles and desorbs only slowly.

4.1 Soil properties influencing in-situ colloid nobilization and transport

In-situ mobilization and transport of colloids in natural, structured soils depend on complex interactions among soil characteristics controlling the inherent dispersibility of colloids, and on dynamic properties determining the prevailing conditions for colloid mobilization and transport. De Jonge et al. (2004) showed that three main issues should be taken into account for evaluating the overall process of in-situ colloid mobilization and transport (Figure 17.8), namely:

- dispersion or release of in-situ colloids
- transport through the vadose zone
- stabilization of the dispersed colloids in the soil solution, depending on the actively conducting pore system size and geometry

Colloid mobilization, generation, and subsequent transport characterize a possible mechanism for structure formation, clogging, and structural changes of subsurface porous media, or enhanced mobilization of contaminants induced by mobile colloidal particles (Grolimund and Borkovec 2009).

It was demonstrated that a large number of soil physical and chemical factors together with colloid properties and soil structural heterogeneity determine the occurrence and magnitude of colloid- and colloid-facilitated transport in the vadose zone (de Jonge et al. 2004, Simunek et al. 2006, Flury and Qiu 2008).

Totsche and Kögel-Knabner (2004) consider that release, redistribution, transport, and retardation of mobile organic sorbents within the vadose zone are controlled by a complex interplay of biological and physicochemical interactions and processes.

4.2 Colloid properties

Colloids have a high surface area to mass ratio due to their very small size. It was found that groundwater concentrations of colloidal materials can be as high as 75 mg/L, corresponding to as

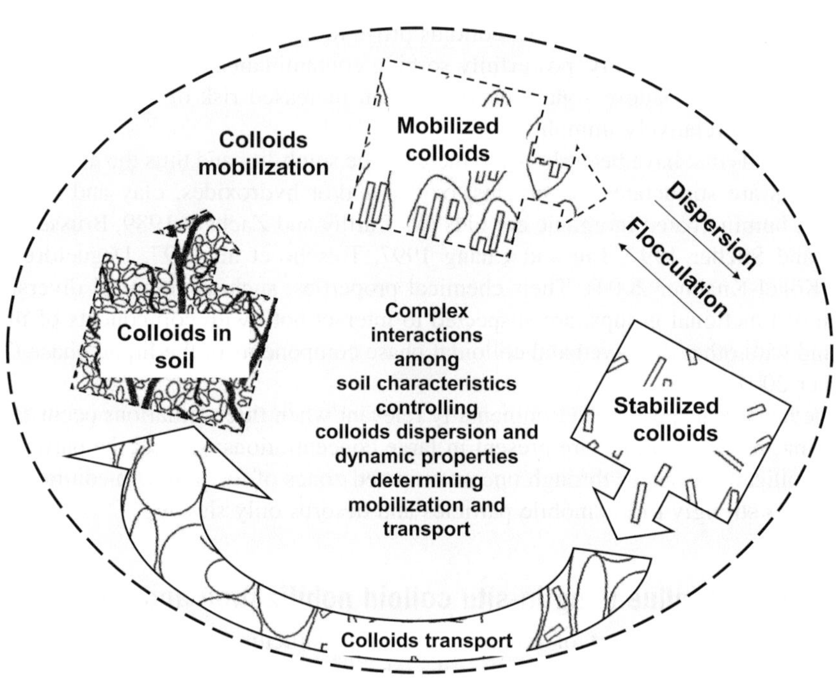

FIGURE 17.8

Main processes to be considered when investigating in situ colloid mobilization and transport. (For color version of this figure, the reader is referred to the online version of this book.)

(adapted from de Jonge et al., 2004).

many as 1012 particles/L, which represents a large surface area available for transporting sorbed contaminants (Weiner 2000).

The colloidal particles may coagulate. For coagulation to occur, two colloid particles must collide with enough energy to force past their repulsive double layers and approach close enough for London attractive forces to be effective. Brownian motion provides the energy for collision, as for overcoming the electrical repulsion of the double layer. The same is true for adsorption to soil particle surfaces (Weiner 2000, Sincero and Sincero 2003). Coagulation and adsorption of colloids are more efficient under high values of total dissolved solids (TDS).

Sorption of various compounds on mobile soil colloids affects leaching through facilitated transport mechanisms: Sorbed on soil colloids, metals, pesticides, and other organic contaminants can be transported through the soil to surface water or groundwater (Cornejo et al. 2005). Soil organic matter and organic particles from manure, slurry, and other organic wastes have a wide range of physical sizes, but the smaller ones provide the larger number of sites for sorption of pollutants (McGechan and Lewis, 2002).

4.3 Colloid-contaminant associations

The transport behavior of dissolved contaminant species has been studied for many years. Numerous data published in the literature have shown that colloids govern the fate of reactive elements and/or pollutants in soil, since they adsorb heavy metal ions and waterborne pollutants via their movements in aqueous systems and soils. Mobilization of the fine clay fraction from the solid matrix is considered to be the major source of mobile colloidal particles in subsurface systems. In this situation they can be the main vehicle for transport of pollutants (Grolimund and Borkovec 2006, Pédrot et al. 2008). Many contaminants can sorb onto colloids in suspension, thereby increasing their concentrations in solution beyond thermodynamic solubilities (Kim et al. 1992, Simunek et al. 2006).

Totsche and Kögel-Knabner (2004) considered that transport of reactive solutes in the presence of a mobile sorbent is conceptualized by assuming at least two interacting constituents, namely, the mobile organic sorbent and the freely dissolved solute, such as a contaminant.

4.3.1 Metals-colloid association

Heavy metals associated with colloids may show significantly different transport properties compared to dissolved heavy metals (Baumann, 2007). The transformation of metals and metalloids in the environment is influenced by physicochemical and biological interfacial interactions, with a significant role in controlling the transformation, transport, fate, and toxicity of metals and metalloids in soil and surrounding environments. Heavy metals are well known to bind to colloidal matter present in leachate and groundwater (Gounaris et al. 1993, Baun and Christensen 2004).

Pedrot et al. (2008) describe three groups of elements with regard to their mobilization via colloids:

- the dissolved group with alkaline, alkaline-earth metal elements, and other trace elements such as Mn, Si, B, and V, which are little mobilized by colloids and are quickly leached
- the intermediate group, which includes Cu, Zn, Co, and Ni, partly mobilized by colloids, thus evidencing different release behaviors
- trace element mobilities, depending on the involved colloid types (Al, Cr, U, Mo, Pb, Fe, Ti, Th).

Karathanasis (1999) demonstrates the important role of water-dispersible colloid particles in mediating and co-transporting heavy metals such as Cu and Zn in subsurface environments; he shows that colloid-mediated metal transport could be as high as 50 times that of control metal solutions, depending on solution conditions and colloid and soil properties. When subsurface environments show low ionic strength which enhances solid-phase dispersion, the transport of some heavy metals, such as copper and zinc, may be significantly enhanced by high surface-charge colloids mediation, if moderately high pH and organic carbon (OC) levels are also present (Karathanasis 1999). The magnitude of colloid-mediated transport is drastically reduced at high metal loadings or in the presence of low-surface-charge colloids with Fe- or Al-oxyhydroxide coatings. A specific metal sorption affinity for the colloid surface than for the soil matrix is thought to be the dominant mechanism facilitating metal transport.

For contaminants that are present in relatively low concentrations, it is generally accepted that association of the contaminants with colloids may control the processes in the subsurface environment. The cationic forms of metals (such as Cs^{+}, Cu^{2+}, and Ni^{2+}), as well as the anionic forms (such as CrO_4^{2-}, AsO_4^{3-}), can be in this situation, as well as some nonpolar organic compounds (polycyclic aromatic hydrocarbons, polychlorinated biphenyls, DDT) and polar organic compounds (Roy and Dzombak 1996, McCarthy and McKay 2004, Sen and Kilhar 2006). The basic mechanisms of contaminant association with colloidal particles are surface complexation, ion exchange, and hydrophobic partitioning. Also, the fate and behavior of colloid-contaminant associations in soil and water depend on pH, salinity, the presence of other competitive and/or complexing cations or anions, temperature, bedrock geology/ mineralogy, suspended particulate content, and water velocity, which in turn influence their sorption characteristics at solid/liquid interfaces (Sen and Kilhar 2006, 2008).

4.3.2 Organic compounds-colloid association

A large amount of organic chemicals, including xenobiotics, exhibit a strong affinity to humic substances, while the transformation of xenobiotics in terrestrial systems is greatly influenced by mineral components of soil (Huang 1998).

Both enzymes and mineral colloids are involved in the catalytic transformation of organic pollutants in soils. Surfacecatalyzed reactions of adsorbed pesticides could be induced by organic matter, but they could also hinder the degradation of some pesticides by diminishing both their accessibility to microbial action and their concentration in the soil solution (Huang and Bollag 1999, Huang 2008a, b).

Although microbial inhibitors can be used to repress biotic processes, their side effects on the surface reactivity of mineral colloids remain uncertain. It is therefore generally difficult to determine whether an organic pollutant is transformed abiotically or biotically, although they are thought to take place simultaneously (Huang and Bollag 1999).

Interaction of pesticide and other organic molecules with dissolved organic matter (DOM) affects leaching through facilitated transport mechanisms, since these interactions in solution may increase the solubility of the compound, resulting in an increase in leaching (Cox et al. 2000, Graber et al. 2001, Cornejo et al. 2005). Huang (2008b), who devoted many efforts to this subject, states that the processes of adsorption and abiotic degradation of xenobiotics through the action of the surfaces of soil minerals vary with the structural and surface properties of the minerals, saturating cations, and hydration status, molecular structures of xenobiotics, and associated environmental factors.

The bioavailability of sorbed xenobiotics to microorganisms is decisive in their transformation in soil. Alexander (1995) found that the degradation of organic pollutants may be considerably reduced when they are retained by soil colloids, as a result of the diminished bioavailability of chemicals involved in binding processes. Also, the bioavailability varies with the chemical properties of pollutants, the residence time of xenobiotics, the nature of the sorbent, the mechanisms of sorption, and the properties of the degradative organisms (Dec et al. 2002; Ehlers and Loibner 2006, Huang 2008b).

4.4 Transport and retention mechanisms and models

The transport of colloids in the soil pore space received growing attention because the transport of mobile colloids can cause the translocation of some contaminants in the soil profile (Amrhein et al. 1993, de Jonge et al. 1998, McGechan and Lewis 2002, Shein and Devin 2007). Also, the transport of living colloids or biocolloids (pathogenic microorganisms and viruses) by a water flow poses a threat tothe health and life of humans and biota (Redman 2001, Shein and Devin 2007).

Transport processes in subsurface environments are determined by complex interactions between the soil matrix and dissolved as well as particulate substances. Depending on environmental conditions, the transport of contaminants can be enhanced or retarded in the presence of colloids. Changes in the transport velocities and mass transfer rates of colloid-associated contaminants possess high environmental relevance.

Colloids are transported through the water-filled sections of the porous medium by advection and dispersion and are removed from the pore water by straining, air–water interface capture, and deposition onto soil–water interfaces.

Various interactions in colloid-facilitated contaminant transport, including colloid attachment to and detachment from the solid matrix and the air–water interface, contaminant adsorption to and desorption from colloids and transport with mobile colloids, and contaminant adsorption to and desorption from the solid matrix may be described quantitatively using a variety of models. Fate and transport models that neglect colloid-facilitated transport often underpredict contaminant movement (Flury and Qiu 2008).

The temporal and spatial variations of (bio)colloid concentration in a homogeneous, granular porous medium are described by the advection dispersion equation (shown here in one spatial dimension for simplicity) (Schijven and Hassanizadeh 2000, DeNovio et al. 2004) (Eq. 17.5), or in a equivalent form that includes parameters characterizing the water–air–soil–colloid system (Eq. 17.6).

$$\frac{\partial C}{\partial t} = D\frac{\partial^2 C}{\partial z^2} - v\frac{\partial C}{\partial z} \tag{17.5}$$

$$\frac{\partial C}{\partial t} + \frac{\partial \Gamma_{STR}}{\partial t} + \frac{\rho_c}{S_W}\left[f_{air}\frac{\partial \Gamma_{AWI}}{\partial t} + f_{soil}\frac{\partial \Gamma_{SWI}}{\partial t}\right] = A_{L}v\frac{\partial^2 C}{\partial z^2} - v\frac{\partial C}{\partial z} \tag{17.6}$$

where C is the pore-water colloid concentration at a distance z and time t; D is the hydrodynamic dispersion coefficient; Γ_{STR}, Γ_{AVI}, and Γ_{SVI} are immobile-phase colloid concentrations for removal by film straining (STR) air–water interface capture (AWI), and soil–water interface deposition (SWI); t is time; ρ_c is the ratio of colloid mass to its effective cross-sectional area; S_W is water saturation; f_{air} is the air–water interfacial area per unit void volume, f_{soil} is the soil–water interfacial area per unit void volume, AL is the longitudinal dispersivity; v is the average pore-water velocity; and z is the coordinate parallel to flow.

In Eq. 17.6, only the physical transport processes of advection and hydrodynamic dispersion are considered. In granular media, colloids or microorganisms are removed from the fluid phase by *physicochemical filtration* or attachment to sediment grain surfaces.

When an attachment mechanism is used to describe removal of particles from the liquid phase, the general equation for (bio)colloid transport and fate in a one-dimensional, homogeneous, granular porous medium becomes (Eq. 17.7) (Schijven and Hassanizadeh 2000).

$$\frac{\partial C}{\partial t} + \frac{\rho_b}{\varepsilon}\frac{\partial S}{\partial t} = D\frac{\partial^2 C}{\partial z^2} - v\frac{\partial C}{\partial z} \tag{17.7}$$

where S is the retained particle concentration, ρ_b is the dry bulk density of the porous medium; and ε is the volumetric water content. Specification of the kinetics models is required for the solution of Eqs. 17.5–17.7.

Perturbations in pore-water chemistry (such as ionic-strength reductions and pH increases) as well as increases in shear stress that accompany pore-water velocity increases will promote colloid release (DeNobio et al., 2004). The transport of colloids from mineral-grain surfaces may be facilitated by the movement of air–water interfaces associated with wetting and drying fronts (Gomez-Suarez et al. 2001; Saiers et al., 2003). Many investigators have developed different theoretical models on colloidal fines release and deposition which have been gathered by Ryan and Elimelech (1996), Sen and Killar (2006).

A number of different mathematical approaches have been applied to model the behavior of colloids and microorganisms in natural and engineered aqueous granular environments. The various colloid-facilitated solute transport models published in the literature consider the colloid transport and contaminant interactions in various manners: Some assume that colloids are nonreactive with the solid phase, others consider a first-order kinetic attachment of colloids, while several models consider an irreversible nonlinear kinetic attachment of colloids (Mills et al. 1991, Dunnivant et al. 1992, Corapcioglu and Jiang 1993, Saiers and Hornberger 1996, Simunek 2006).

Essential for improving predictions of colloid-facilitated transport of solutes in variably saturated porous media are the models that can accurately describe the various mechanisms controlling colloid and solute transport, and their mutual interactions and their interactions with the solid phase (Simunek, 2006). However, no effort to describe and predict contaminant transport can succeed if major pathways and mechanisms for transport are not well understood (Zhang et al. 2005).

Predicting the migration of colloids (e.g., clays, manufactured nanomaterials, etc.) or biocolloids (e.g., bacteria and viruses) in granular porous media is of significant interest in a number of environmental applications such as granular (deep-bed) filtration used in water and wastewater treatment, riverbank filtration (Medema et al. 2000), in-situ bioremediation (Steffan et al. 1999), and protection of groundwater supplies from microbial pathogens and (bio)colloid-associated pollutants (Ferguson et al. 2003).

Models that consider colloid-facilitated solute transport are based on mass balance equations for all colloid and contaminant species. Simunek et al. (2006) developed a one-dimensional numerical model based on the HYDRUS-1D software package that incorporates mechanisms associated with colloid and colloid-facilitated solute transport in variably saturated porous media, accounts for transient variably saturated water flow, and for both colloid and solute movement due to advection, diffusion, and dispersion, as well as for solute movement facilitated by colloid transport. Assuming that the porous medium consists of three phases—that is, solid phase, air, and water—the concept of two-site sorption to describe nonequilibrium adsorption–desorption reactions to the solid phase was considered, being further assumed that contaminants can be sorbed onto surfaces of both deposited and mobile colloids, fully accounting for the dynamics of colloid movement between different phases (Simunek et al. 2006).

Transient one-dimensional variably saturated water flow in porous media is described by Simunek et al. (2006) using the Richards equation (Eq. 17.8).

$$\frac{\partial \theta}{\partial t} = \frac{\partial}{\partial x}\left[K\left(\frac{\partial h}{\partial x} + \cos \alpha\right)\right] - S \tag{17.8}$$

where h is the water pressure head, θ is the volumetric water content, t is time, x is the spatial coordinate (positive upward), S is the sink term, α is the angle between the flow direction and the vertical axis, and $K(h)$ is the unsaturated hydraulic conductivity function. The authors proposed the analytical model of van Genuchten (1980) to evaluate the soil hydraulic properties, that is, the retention curve $\theta(h)$ and the unsaturated hydraulic conductivity function $K(h)$.

The van Genuchten's soil retention relationship was applied by Massoudieh and Ginn (2007) for calculating matrix potential and hydraulic conductivity based on saturation content (Eq. 17.9).

$$K(\theta) = K_S S_e^{1/2}\left[1 - \left(1 - S_e^{1/m}\right)^m\right]^2 \tag{17.9}$$

where K_S is the vertical saturation hydraulic conductivity, which is a function of concentration of colloids entrapped at the location it is being calculated, m is the van Genuchten retention parameter that is related to the uniformity of pore-size distribution and is usually specified by the soil type, and S_e is the effective saturation that is expressed by Eq. 17.10.

$$S_e = \frac{\theta - \theta_r}{\theta_S - \theta_r} \tag{17.10}$$

A more comprehensive equation for colloid transport describing both colloid/matrix and colloid/air–water interface mass partitioning in one-dimensional form is given by Simunek et al. (2006), as expressed by Eq. 17.11.

$$\frac{\partial \theta_W C_C}{\partial t} + \rho \frac{\partial S_C}{\partial t} + \frac{\partial A_{aW} \Gamma_C}{\partial t} = \frac{\partial}{\partial x}\left(\theta_W D_C \frac{\partial C_C}{\partial x}\right) - \frac{\partial q_C C_C}{\partial x} + R_C \tag{17.11}$$

where C_C is the colloid concentration in the aqueous phase, S_C and G_C are colloid concentrations adsorbed to the solid phase, and air–water interface, respectively; θ_w is the volumetric water content accessible to colloids (due to ion or size exclusion, θ_w may be smaller than the total volumetric water content u), D_C is the dispersion coefficient for colloids, ρ is the bulk density, A_{aW} is the air–water interfacial area per unit volume, q_C is the volumetric water flux density for colloids, while R_C represents various chemical and biological reactions. The second and third terms on the left side of Eq. 17.9 express the colloid mass transfer terms between the aqueous phase and the solid phase or the air–water interface, respectively. The first two terms on the right side of Eq. 17.9 represent the dispersive and advective colloid fluxes, respectively.

Simunek et al. (2006) defined all the other terms in Eq. 17.9 and then combined dissolved and colloid-facilitated contaminant transport equation (in one dimension) in an equation that sums the mass of contaminant associated with the different phases (contaminant in the liquid phase, contaminant sorbed instantaneously or kinetically to the solid phase, and contaminant sorbed to mobile colloids and immobile colloids attached to solid phase or air–water interface). Also, the various spatial mass fluxes are considered (dispersion and advective transport of the dissolved contaminant, as well as dispersion and advective transport of contaminant sorbed to mobile colloids), as well as sink/source reactions. Finally, the authors decided that for a complete description of colloid-facilitated solute transport, in addition to variably saturated flow equation, this mathematical development requires a total of nine coupled partial differential equations involving nine unknown variables, whose analytical solution is not easily found and necessitates a software package (Simunek et al. 2006).

Unfortunately, not all essential processes contributing to colloid and colloid-facilitated solute transport are included in the same model, such as the following: Colloid interactions with the air–water interface are not present in models for colloid-facilitated solute transport, which consider flow and transport only in fully saturated groundwater systems, usually for steady-state flow (Simunek et al. 2006). On the other hand, no colloid-facilitated transport model has considered straining and size exclusion as mechanisms of colloid retention and transport, respectively (Flurry and Qiu 2008).

Massoudieh and Ginn (2007) propose an advection–dispersion model with kinetic rate of capture and release to/from porous media, assuming that the source of colloid release in the porous media is separate from the captured colloids. This could be considered a reasonable hypothesis, since colloid reentrainment rates are often observed to be much smaller than filtration rates unless a significant change in the chemistry takes place. Massoudieh and Ginn (2007) formulated the bulk volumetric mass balance equation for mobile colloids, captured, attached available, and captured in a water–air interface, as expressed by (Eq. 17.12).

$$\frac{\partial(\theta G)}{\partial t}+\frac{\partial\theta v_p G}{\partial z}=\frac{\partial}{\partial z}\left(D_c\theta\frac{\partial G}{\partial z}\right)-B_d k_p G+B_d k_{rp} G_{si}-k_{fa}S_a G+k_{ra}S_a G_a \tag{17.12}$$

$$\frac{\partial G_{sf}}{\partial t}=k_p G \tag{17.13}$$

$$\frac{\partial G_{si}}{\partial t}=-k_{rp}G_{si} \tag{17.14}$$

$$\frac{\partial[S_a G_a]}{\partial t}=k_{fa}SG-k_{ra}S_a G_a \tag{17.15}$$

where G is the concentration of mobile colloids in pore water, v_p is the colloidal average velocity (calculated with Eq. 17.16, a_p is the colloid radius, and r_0 is the average pore radius.

$$v_p=\frac{Q}{\theta}\left[2-\left(1-\frac{a_p}{r_0}\right)\right] \tag{17.16}$$

This model is applicable when $a_p/r_0 << 1$.

D_c is the mechanical dispersion coefficient for colloids, B_d is the bulk density, k_p and k_{rp} are capture and release coefficients, respectively, k_{fa} and k_{ra} are capture and release rates to the air–water interface, G_{si} is the concentration of initially attached and available colloids expressed as mass of colloid per dry mass of solids, G_{sf} is the concentration of irreversibly captured colloids, and G_a is the concentration of colloids captured in the air–water interface expressed as mass of colloids over volume of air. S_a is the air–water interface area per unit volume that is a function of soil texture and saturation content, calculated using the Eq. 17.17.

$$S_a=\left(2\theta_s^b/r_0\right)\left[\frac{\lambda\theta_m}{b}\left(\theta^{-b}-\theta_0^{-b}\right)+\frac{1}{b-1}\left(\theta_s^{1-b}-\theta^{1-b}\right)\right] \tag{17.17}$$

where θ_m is the specific volume of a monolayer of water, λ is a constant, and b is another constant called porosity index. The authors (Massoudieh and Ginn, 2007) are assuming that the capture and release of colloids to the air–water interface takes place instantaneously and also that the surface area of the air–water interface is proportional to the air-phase volume, so that $G_a = K_a G$, where $Ka = k_{fa} / k_{ra}$ is the equilibrium air–water partitioning coefficient for colloidal particles, which depends on colloids hydrophobicity. By incorporating this equilibrium relationship into Eq. 17.12, it results in the expression 17.18.

$$\frac{\partial\{[\theta+S_a K_a]G\}}{\partial t}+\frac{\partial\theta v_p G}{\partial z}=\frac{\partial}{\partial z}\left(D_c\theta\frac{\partial G}{\partial z}\right)-B_d k_p G+B_d k_{rp} G_{si} \tag{17.18}$$

For solving Eq. 17.18, Cauchy boundary conditions are used at the top and the bottom (Eqs. 17.19, 17.20).

$$\left[-D_c\theta\frac{\partial G}{\partial z}+\theta v_p\right]_{x=0^+}=v_pG_0 \tag{17.19}$$

$$\left.\frac{\partial G}{\partial z}\right|_{z=L}=0 \tag{17.20}$$

In a similar manner, the transport equations for dissolved and colloid associated contaminants for the mobile phase can be written (Massoudieh and Ginn, 2007).

5 Anthropogenic colloids and engineered nanoparticles (ENP_s) in soil

Besides natural colloids, anthropogenic colloids resulting from various sources can be found in the environmental components, in particular in soil. Anthropogenic colloids are generated by physical, chemical, and/or biological processes "acting on materials created by humans" (EPRI 1999).

Anthropogenic colloids could originate from various sources (Figure 17.9) (EPRI 1999, Hassellov et al. 2008):

- degradation of the solid phase in the waste (degradation colloids)
- supersaturated solutions with respect to solid phase (precipitation colloids)
- attachment of nanoparticles to other colloids (pseudocolloids)
- manufactured (engineered) colloids (nanoparticles)

Colloidal nanoparticles (inorganic and organic) are prepared for a long time, while a number of processes allow control of their shape, particle size and distribution, structure, and surface properties (Pichot 2008). Therefore, the increased use of manufactured nanoparticle-containing products of the emerging nano-industry leads to the intensification of nanoparticle emissions into the environment, which could behave as colloids (EPRI 1999, Suijkerbuijk 2008, Triay et al. 1995).

Anthropogenic colloids/engineered nanoparticles can attend soil in the form of incidental nanoparticles, from grinding of primary or secondary minerals, near metal or mineral surface, combustion) or as engineered nanoparticles manufactured for specific applications, with a specific composition (NPCA 2008, Ostiguy et al. 2010).

Engineered nanoparticles are classified based on various criteria. The most relevant are chemical composition, size, and morphology characteristics (NPCA 2008, von den Kammer et al. 2008). Considering the chemical composition, engineered nanoparticles can be fullerenes, metals, oxides, complex compounds, quantum dots, or organic polymers (EPA 2007, von den Kammer et al. 2007).

Apart from the incidental release into the environment, the engineered nanoparticles can be deliberately introduced into the soil and groundwater for remediation purposes, due to their enhanced reactivity, surface area, subsurface transport and sequestration/sorptive characteristics (USEPA 2007). For example, zero-valent iron has been used successfully to remediate groundwater as a component of permeable reactive barriers (Quinn et al. 2005, USEPA 2007). Other nanosized materials, such as metalloporphyrinogens were used for the degradation of some chloroorganic pollutants (Dror 2005,

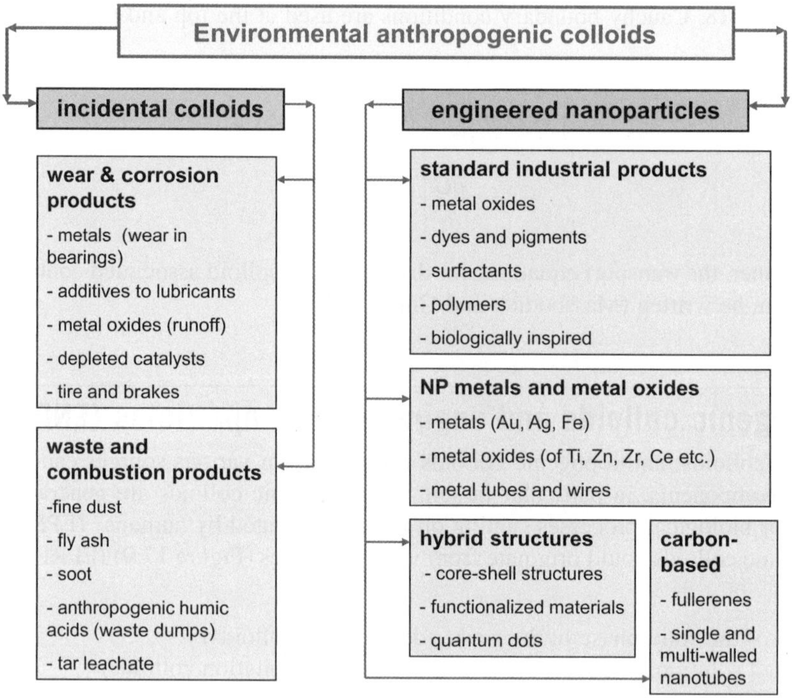

FIGURE 17.9

Sources of anthropogenic colloids/nanoparticles in the environment. (For color version of this figure, the reader is referred to the online version of this book.)

(adapted from von den Kammer et al., 2008).

USEPA 2007). Nanomaterials based on poly(ethylene)glycol modified urethane acrylate have been used to enhance the bioavailability of phenanthrene (Tungittplakorn et al. 2004, 2005). Table 17.5 describes succinctly some nanoparticle applications in environmental treatments (Mansoori et al. 2008).

The environmental nanoparticles can be involved in some processes that occur in soil, such as transport (mobilization-fixation), formation, transformation, dissolution, growth, aggregation-segregation, interaction with contaminants, nutrients, and nonorganic matters (co-transport), andelectron transfer (von den Kammer et al. 2008).

Soil is the environmental matrix where engineered nanoparticles will ultimately end up, and together with humic substances increase the porosity and surface areas. Being of an anthropogenic nature, the engineered nanoparticles can play a double role:

- one similar to that of natural colloids
- environmental contaminants.

Therefore, the fate of the engineered nanoparticles/colloids in the environment (in particular, in soil) should consider this dual behavior. The characterization of nanoparticles and their aggregates provides data for the environmental risk they can generate for both the environment and human health due to their contamination character (Hassellov et al. 2008, USEPA 2007).

Table 17.5 Summary of Environmental Treatments using Nanoparticles (Mansoori et al., 2008)

Type of Nanoparticle	Type of Treatment	Removal Target	Advantage	Disadvantage
Nanoparticle-based TiO_2	Photocatalyst oxidation	Organic pollutants	Nontoxicity, water insolubility under most conditions, photostability	High operation cost, hard to recover, sludge generation
Nanoparticle- based iron	Reduction adsorption	Heavy metals, anions, organic pollutants (dechlorination)	In-situ remediation, soil and water treatment, low cost, safe to handle	Hard to recover, sludge generation, cost for sludge disposal, health risk
Nanoparticle- based bimetallic	Reduction adsorption	Dechlorination, denitrification	Higher reactivity than the iron nanoparticle	Hard to recover, sludge generation
Nanoclay	Adsorption	Heavy metals, organic pollutants	Low-cost, unique structures, long-term stability, reuse, high sorption capacity, easy recovery, large surface and pore volume	Sludge generation
Nanotube and fullerene	Adsorption	Heavy metals, anions, organic pollutants	Treatment of pollution from air and water, exceptional mechanical properties, unique electrical properties, highly chemical stability	High capital cost, low adsorption capacity, hard to recover, sludge generation, health risk
Dendrimers	Encapsulation	Heavy metals, organic pollutants	Simple separation, renewable, large binding capacity, cost-effective, no sludge generation, reduce pollutant to the level of few ppb, treatment of pollution from soil and water	Costly
Micelles	Adsorption	Organic pollutants from soil	In-situ treatment, high affinity for hydrophobic organic pollutant	Costly
Metal-sorbing vesicles	Adsorption	Heavy metals	Reuse, high selective uptake profile, high metal affinity	
Magnetite nanoparticles	Adsorption	Heavy metals, organic pollutants	Simple separation, no sludge generation	External magnetical field required for separation, costly
Nanofiltration and nanosieve membranes	Nanofiltration	Organic and inorganic compound	LowER pressure than RO	Costly, prone to membrane fouling

6 Hazards and risks of engineered nanoparticles/colloids for the environment and human health

6.1 Health and safety risks linked to engineered colloids

Studies of processes that control in-situ mobilization and transport of colloids in soil are prerequisites for risk prediction and/or quantification, since colloids are seen as vectors for transport of contaminants.

Colloids present in soil can generate potential environmental and health risks as a result of their propensity to associate with inorganic, organic, and biological (pathogenic) contaminants (Patil 2012, Vendelboe et al. 2012). The association of strongly sorbing compounds with colloids enhances the mobility of contaminants (Kretzschmar et al. 1999, Siemens et al. 2004). The content of water-dispersible colloids can induce a significant risk potential for soil stability and functions, since "mobile colloids may increase the risk of colloidal facilitated transport of strongly sorbed environmental contaminants" (Vendelboe et al. 2012). Colloids in soil display a high sorptive capacity due to their high specific surface area. Therefore, they are considered good sorbents for contaminants with low solubility. This can facilitate the transport of contaminants through soil, especially through the vadose zone, generating an increased risk of leaching, in particular for contaminants that are usually considered to be relatively immobile. If macropores in soil are able to provide preferential pathways for water flow and suspended colloid, this facilitates the increase of flow velocities, increasing also the risk of rapid transport of contaminants to drain and groundwater (Kjaergaard 2003).

Some studies emphasize the alarming consequences of colloid-enhanced contaminants transport and the difficulties in predicting the real risks of soil and groundwater pollution due to the interactions of contaminants and colloids; too little is known about these interactions to permit risk assessment (Frimmel et al. 2007). These studies also deal with the leaching potential of contaminants sorbed on colloids under certain conditions and point out the incidence of ecological and health risks correlated with the physicochemical proprieties of colloids and contaminants, as well as their fate and environmental behavior (Klitzke et al. 2008, Sanchis et al. 2012, Siemens et al. 2004, Vendelboe et al. 2011).

Any risk assessment regarding the ecosystem and human health has to consider the properties of soil and natural colloids or engineered nanoparticles released in soils, environmental concentration and fate, transport, and effects (Simko and Mattsson 2010). Such unrestricted contaminant transport due to mobile colloids can lead to the rapid spread of contaminants in the subsurface and then to groundwater. This factor is of major relevance to risk assessment (Sen and Khilar 2006), which should therefore take into account both potential mobile and immobilized contaminants in judging the likelihood of contaminants leaching (Klitzke et al. 2008). Understanding the mobility, reactivity, ecotoxicity, and persistence of colloid contaminant associations is necessary in any assessment of therisks they pose to the environment. This issue gains in importance if colloids are introduced in the environment (soil, in particular) as nanoparticles originating from commercial products, as a result of their accidental release during production or transport (Hristozov and Malsch 2009, Nowack et al. 2007). In addition, engineered nanoparticles are intentionally released into the environment, accelerating the risk potential in soil or groundwater. The very properties that make nanoparticles technologically interesting may place them in a novel category of potentially toxic substances (Mueller and Nowack 2010,

Nowack et al. 2007). A number of studies concern the toxicity of nanoparticles, suggesting that smaller particles are more toxic than larger ones (Love et al. 2012, Oberdörster et al. 2007).

Therefore, exploring the sources and environmental pathways of colloids (natural and/or engineered) as transport vectors for contaminants helps identify hazards and associated risks (Nowack et al. 2007, Reijnders 2006, Zhang et al. 2011). Siemens et al. (2008) discussed the increased risk created by the accumulation of phosphorus in the soil and by the colloid-facilitated leaching of P when large volumes of electrolytes migrate through the soil matrix. Humic-rich forest topsoils may pose a high risk of colloid mobilization and a much larger one than that in arable soils at pH above 6 (Kaplan et al. 1997, Klitzke et al. 2008).

Methods to characterize and quantify natural colloids are an extremely helpful support in the decision-making process involved in risk management and control. This is true for both natural and engineered colloids (nanoparticles), where the lack of tools to detect and quantify them in soil or water can be a major constraint to describing fate exposure and risks associated with colloids behavior in the environment (Klitzke et al. 2008). The production, use, and waste treatment of some nanomaterials and their derivates could lead to losses, and nanomaterials are released into the environment (Nowack and Bucheli 2007, NPCA 2008). Here, they can induce relevant impacts due to their mobility, transport, ecotoxicity, modifications. Scientists, researchers, and regulatory agencies generally agree that the potential health and environmental risks of engineered nanoparticles should be evaluated over their entire life cycle (Grieger et al. 2012, Seager and Linkov 2008).

6.2 Assessing environmental risks generated by engineered nanoparticles

Risk assessment involves an established procedure for assessing whether a risk is associated with use of an agent, such as a chemical or other substance (Gavrilescu 2008, Grieger et al. 2012). When the ENP risk assessment follows the specific procedure of chemicals, current approaches to environmental risk assessment of various contaminants (in particular engineered nanoparticles) basically compare the actual or predicted environmental concentrations (PEC) to no adverse effect levels. The Norwegian Pollution Control Agency (NPCA, 2008) assumed that a similar strategy could be applied to evaluate the environmental risk of released contaminants existing as associations with natural or engineered colloids, taking into account the influence of environmental factors (temperature, pH, speciation, etc.) on colloid mobility and sorptive capacity.

In this context, risk assessment is employed as an internationally recognized process in which scientific and regulatory principles are applied in systematically in order to describe the hazards associated with the environmental and/or human exposure to chemical substances (Gavrilescu 2008, Grieger et al. 2012, Hristozov and Malsch 2009). Risk assessment consists of four basic steps: (1) hazard identification (which charts the inherent physicochemical and biological properties of a certain chemical and provides a uniform basis to evaluate the hazard potential); (2) dose-response assessment (developed so as to obtain a quantitative estimate of the chemical concentration, expected to not affect the human health or the ecosystems); (3) exposure assessment (which applies generic and/or specific scenarios for realistic exposure pathways for a chemical, considering, the so-called predicted environmental concentration (PEC) value for the specified scenarios or the measured values); and (4) risk characterization. Table 17.6 summarizes some of the most important scientific findings relevant to ENP risk assessment, addressing environmental and health issues.

Table 17.6 Framework for Assessing Environmental Risks Generated by Engineered Nanoparticles

Risk Assessment Steps	Literature and Experimental Data on:	Identified Issues	References
Hazard assessment	impact on atmospheric/stratospheric processes stability of soil bioavailability biological uptake and metabolism toxicity, ecotoxicity observed effects (mortality, mutations, inflammation, irritation, growth reduction, reproductive effects, stress) physiological, genetic, or functional effects end points of concern exposure mechanisms	**carbon nanotubes** are able to cause dose-dependent effects of interstitial inflammation and lesions in mice and rats (*in vivo* studies); grandulomas in rats; significant dose-dependent decrease in thiobarbituric acid reactive substances, especially in the gill, brain, and liver; dose-and time-dependent inhibition of cell proliferation and a decrease in cell-adhesive ability in human embryo kidney cells (in vitro studies); decrease of the viability of human osteoblastic lines and human epidermal keratinocytes after exposures **C_{60} fullerenes** tend to induce oxidative stress in living organisms; an increase in the numbers of bronchoalveolar lavage-recovered neutrophils (i.e., white blood cells) after intratracheal instillation in rats; a LD_{50} of 600 $mg \cdot kg^{-1}$ polyalkylsulfonated C_{60} in female rats after intraperitoneal administration (in vivo studies); a dose-dependent decrease in the viability of human epidermeal keratinocytes after exposure to C_{60}-phenylalanine, as no contribution to the effect was attributed to the phenylalanine groups (in vitro studies) **metal ENPs** induce more severe lung toxicity in *mice* than bulk particles from the same materials; Zn ENPs toxicity in humans is concentration-dependent and the most probable uptake path is	Beckett et al., 2005 Chen et al., 1998 Cui et al., 2005 Hishino et al., 2004 Hristozov & Malsch, 2009 Lam et al., 2004 Li et al., 1999 Lovric et al., 2005 Monteiro-Riviere et al., 2005 NPCA, 2008 Oberdörster et al., 1994 Robichaud et al., 2005, 2007 Rouse et al., 2006 Sayes et al., 2007 Warheit et al., 2007 Yang & Watts, 2005

		through the respiratory system; Al ENPs significantly inhibited the relative root growth in *Zea mays* (corn), *Glycine max* (soybean), *Brassica oleracea* (cabbage), *Daucus carota* (carrot) after administration of 2 mg·mL^{-1} for 24 h; TiO_2 (in vivo studiies)ENPs tend to cause more severe pulmonary damage in *mice* than larger particles; FeO ENPs significantly increased fetal skeletal malformations in rats and rabbits (*in vivo* studies) **quantum dots toxicity is influenced by** composition, size, surface charge, and coating; the cytotoxicity of QDs is influenced by the exposure to light and by temperature; adverse effects on mouse lymphocytes after exposure to CdSe/ZnS QDs, coated with albumin	
Dose-response assessment	dose based on surface area (more relevant than dose based on mass concentration) experiments in the laboratory or using mathematical models relevant end points (physiological, genetic and functional) epidemiological studies predictions of human effects, animal studies or other replacement tests	it is still largely unknown which properties influence the toxicity of most ENPs most of the current research on the properties of ENPs focuses on the identification of metrics and associated methods for the measurement of ENPs and their properties the toxicity of some ENPs is not mass-dependent, but influenced by other physicochemical characteristics (e.g., surface area, chemical composition, particle morphology) the toxicity of low-soluble ENPs is better described by their surface area than by their total mass the number of particles could be the most appropriate dose metrics the toxicity of some ENPs can be associated with the number of their surface functional groups	Hristozov & Malsch, 2009 NPCA, 2008 Oberdorster et al., 2007 Plata et al., 2008 Stoeger et al., 2006 a, b Wittmaack, 2007

Continued

Table 17.6 Framework for Assessing Environmental Risks Generated by Engineered Nanoparticles—cont'd

Risk Assessment Steps	Literature and Experimental Data on:	Identified Issues	References
Exposure assessment	scenarios for risk spreading known or anticipated figures for quantitative and qualitative parameters related to natural and engineered production, trade/transport and use possible routes for loss (accidents, leaks, disposal, wear, etc.) estimates of transformation and persistence for contaminants associated with natural and engineered colloids empirical data or procedures to predict the persistence and mobility in air, soil, and water	the environment may be exposed to ENPs during all stages of their life cycles: raw material production, transport and storage, industrial use, consumer use, waste disposal (including waste treatment, landfill, and recovery) the extent of environmental exposure to ENPs depends on their behavior in the environment some ENPs can strongly sorb to the soil particles and become completely inert and immobile properties of the soil, such as porosity and grain size, further influence the mobility of the particles surface photoreactions might induce photochemical transformations on the soil surface: some types of QDs were shown to degrade due to photolytic and oxidative reactions, as the degradation of their coatings can reveal toxic metalloid cores the biological processes in the environment can lead to the complete degradation of ENPs and sometimes they can only change their physical and/or chemical properties workers may be exposed to nanoscale materials while manufacturing these materials, formulating them into	Colvin, 2003 Hansen, 2009 Hardman, 2006 Hristozov & Malsch, 2009 Lecoanet & Wiesner, 2004 Lecoanet et al., 2004 Maynard et al., 2004 NPCA, 2008 Zhang, 2003

		products, transporting them or handling them in the storage facilities, through inhalation and/or dermal contact after the manufacturing process is complete the expected consumer exposure is highest for products in the categories "appliances," "health and fitness," and "home and garden" assessment of the consumer exposure to ENPs is significantly restricted by the lack of access to information about which commercially available products contain ENPs, the exact nanomaterial content of these products, and consumer behavior toward them	
Risk characterization	information from the hazard identification, dose-response, and exposure steps are considered together to determine and communicate the actual likelihood of risk to exposed populations comparison of the exposure concentration (PEC—predicted environmental concentration) with an exposure level that is assumed to have no effect (e.g., PNEC—predicted no-effect concentration) evaluation of the overall quality of data, the assumptions and uncertainties associated with each step, and the level of confidence in the resulting estimates	quantitative risk characterization compares the predicted environmental concentration (PEC) of a chemical agent with its predicted no-effect concentration (PNEC) each of the elements of risk assessment holds certain limitations and challenges toxicity has been reported on for many ENPs, but for most further investigation and confirmation are needed before hazard can be identified exposure assessment of ENPs must build on realistic environmental, occupational and consumer exposure scenarios exposure scenarios must be based on known or anticipated ENP production, trade/transport, use and disposal figures and coupled with a range of possible loss routes	Hansen, 2009 Hristozov & Malsch, 2009 Mueller & Nowack, 2008

Table 17.6 reveals that the environmental risk assessments of both natural and engineered colloids/nanoparticles require deep characterization of nanoparticles and their aggregates (Hassellov et al. 2008, Sanchis et al. 2012). The existing methods need optimization and development, but extensive data can be acquired from application of the specific methods and procedures of environmental chemistry of nanomaterials and from fundamental colloid chemistry (Hassellov et al. 2008).

An analysis of currently available data for risk characterization relative to the environment and human health shows that modeling of reliable exposure scenarios as an important step in risk assessment would be opportune (Frimmel et al. 2007, Sanchis et al. 2012).

The reactive transport models capable of predicting contaminant fate and behavior continue to be developed, driven by the need for more and comprehensive data for risk assessment, characterization, and management (Frimmel et al. 2007, Gavrilescu 2008, Gavrilescu et al. 2009). Some approaches considered useful in risk assessment are the Multi Criteria Decision Analysis, the Weight-of-Evidence, and the Expert Elicitation (Hristozov and Malsch 2009, Kandlikar et al. 2007, Kiker et al. 2005, Linkov et al. 2007, Seager and Linkov 2008, USEPA 2007). Application of some of these unconventional instruments to the risk assessment framework will generate a new perspective, diminishing uncertainties and delivering accurate risk characterization results.

7 Concluding remarks

Soils are characterized by a high degree of spatial variability owing to the combined effect of physical, chemical, or biological processes that operate with different intensities and at different scales. The composition of the soil, the stability of its structure, and the chemistry of the soil–water system are determinant factors in the detachment and mobilization of soil particles. Knowledge of the spatial variation of soil properties is important in several disciplines, for the development and application of methods for monitoring and characterizing soil–water–solute processes are among the most limiting factors in understanding the soil environment.

Colloid-facilitated transport is a potentially important and overlooked contaminant transport process, since the high specific surface area of colloidal particles with respect to the porous media can enhance their ability to absorb and transport solute contaminants. The effect of colloid-facilitated transport gains significance when the desorption rate of contaminants from colloids is relatively slow.

In many cases, colloids and contaminants associated with them may pass through the unsaturated zone to reach to the groundwater. The quantitative analysis of associated risk therefore requires reliable tools for modeling colloid-facilitated transport. Also, the main mechanisms involved in the retardation of colloids in the presence of an air phase should be the subject for future research.

The behavior and fate of chemicals in soil have proven to be complex topics inasmuch as the conditions involved are strongly influenced by time and space. Adequate characterization of the properties and behavior of anthropogenic/engineered colloids/nanoparticles is essential for understanding the correlations between their properties and the inherent toxicity. Because of their particular characteristics and behavior, accurate environmental and health risk assessment is required, serving as a support tool in risk management and control

Future research on this extremely important and exciting area of science should help sustain and enhance ecosystem productivity, services, and integrity, as well as evaluate the impact on human health and prosperity.

SYMBOLS AND ABBREVIATIONS

AL longitudinal dispersivity
AWI air–water interface capture
B biofilm
CEC cation exchange capacity
CSC critical salt concentration
CTIS critical total ionic strength
DOM dissolved organic matter
ENP engineered nanoparticle
EPS extracellular polysaccharides
FA fulvic acid
HA humic acid
HS humic substances
LMMOA low-molecular-mass organic acid
M metal cation
MOS mobile organic sorbents
NOM natural organic matter
OC organic carbon
OM organic matter
PEC predicted environmental concentration
PNEC predicted no effect concentration
POPs persistent organic pollutants
SRO short-range-ordered
STR film straining removal
SWI soil–water interface deposition
TDS total dissolved solids

Glossary

Adsorbtion the process by which molecules or colloids physically adhere to the surfaces of solids
Advection the process by which solutes are transported by the bulk motion of the flowing groundwater
Aerobic living, active, or occurring only in the presence of oxygen
Anaerobic living, active, or occurring in the absence of free oxygen
Anion a negatively charged ion
Aquifer water-bearing stratum of permeable rock, sand, or gravel
Bioavailability the availability of chemicals to microorganisms, which will influence the degradation of the chemicals
Biodegradation the breakdown of substances by microorganisms
Biota the animal and plant life of a region
Capillary fringe the subsurface layer in which groundwater seeps up from a water table by capillary action to fill pores. Pores at the base of the capillary fringe are filled with water due to tension saturation
Cation a positively charged ion
Coagulant a material that, when added to water, will combine with certain substances ordinarily present and form a precipitate comprising floc particles more or less gelatinous in character, having the capacity to remove colloids from water

Coagulation the agglomeration of colloidal or finely divided suspended matter by the addition to the liquid of an appropriate chemical coagulant, by biological processes, or by other means the process of adding a coagulant and other necessary reacting chemicals

Colloid any gas, solid, or liquid in a fine state of subdivision, with particles too small to be visible in an ordinary optical microscope, that is dispersed in a continuous gaseous, liquid, or solid medium and does not settle or settles very slowly; stable dispersion of fine particles in water that have a typical size less than 0.1 micron. Colloids containing iron (Fe), aluminum (Al) silica (SiO2), and organics are commonly found in water.

Complexing agent a dissolved ligand that binds with a simple charged or uncharged molecular species in a liquid solution to form a complex chemical species

Contamination a general term signifying the introduction into the environment of microorganisms, chemicals, wastes, or sewage, which render the environmental components unfit for their intended use

Degradation the process whereby a compound is transformed into simpler compounds; the changing of a soil to a more highly leached and a more highly weathered condition; usually accompanied by morphological changes such as development of an A2 horizon

Diffusion the migration of molecules, atoms, or ions from one fluid to another in a direction tending to equalize concentrations

Electron acceptor small inorganic or organic compound that is reduced to complete an electron transport chain; compound that is reduced in a metabolic redox reaction

Electron donor small inorganic or organic compound that is oxidized to initiate an electron transport chain; compound from which electrons are derived in a metabolic redox reaction

Engineered nanoparticle nanoparticles manufactured to have specific properties or a specific composition

Enzyme any of numerous proteins that are produced in the cells of living organisms and function as catalysts in the chemical processes of those organisms

Flow the movement of a fluid through the soil or over its surface

Fulvic acid the pigmented organic material that remains in solution after removal of humic acid by acidification. It is separated from the fulvic acid fraction by adsorption on a hydrophobic resin at low pH values.

Groundwater that portion of the water below the surface of the ground at a pressure equal to or greater than atmospheric

Heavy metals metallic elements with high molecular weights, generally toxic in low concentrations to plant and animal life. Such metals are often residual in the environment and exhibit biological accumulation. Examples include mercury, chromium, cadmium, arsenic, and lead.

Humic acid the dark-colored organic material that can be extracted from soil with dilute alkali and other reagents and that is precipitated by acidification to pH 1 to 2

In-situ in its original place; unmoved, unexcavated; remaining at the site or in the subsurface

Interfacial included between two plane surfaces or faces; relating to or situated at the boundary between two phases, such as liquid–liquid, liquid–solid, and gas–liquid interface

Ion exchange the process of purifying water by removing the ionized salts from solution. Most commonly, the process of ion exchange is used to achieve this by replacing hydrogen ions for cation impurities and hydroxyl ions for anion impurities.

Leaching the process or an instance of separating the soluble components from some material by percolation; the process or an instance of removing nutritive or harmful elements from soil by percolation

Microorganism an organism of microscopic or ultramicroscopic size

Mineralization the conversion of an element from an organic form to an inorganic state as a result of microbial activity

Modeling the use of mathematical equations to simulate and predict real events and processes

Percolation gravity flow of groundwater through the pore spaces in rock or soil

Permeability, soil the ease with which gases, liquids, or plant roots penetrate or pass through a bulk mass of soil or a layer of soil. Since different soil horizons vary in permeability, the particular horizon under question should be designated. The property of a porous medium itself that expresses the ease with which gases, liquids, or other substances can flow through it and is the same as intrinsic permeability.

Pollution the presence or introduction of a pollutant into the environment

Radionuclides nuclides that emit radioactivity (alpha, beta, or gamma particles) or fission into smaller nuclides

Remediation cleanup or other methods used to remove or contain a toxic spill or hazardous materials

Rhizosphere the zone of soil immediately adjacent to plant roots in which the kinds, numbers, or activities of microorganisms differ from those of the bulk soil

Risk the probability or likelihood of an adverse consequence from a hazardous situation or hazard, or the potential for the realization of undesirable adverse consequences from impending events. It is a measure of the probability and severity of an adverse effect to health, property, or the environment.

Saturated zone an underground geologic formation in which the pore spaces or interstitial spaces in the formation are filled with water under a pressure equal to or greater than atmospheric pressure

Soil the unconsolidated mineral or organic material on the immediate surface of the Earth that serves as a natural medium for the growth of land plants; the unconsolidated mineral or organic matter on the surface of the Earth that has been subjected to and shows the effects of genetic and environmental factors of climate (including water and temperature effects) and macro- and microorganisms, conditioned by relief, acting on parent material over a period of time. A product-soil differs from the material from which it is derived in many physical, chemical, biological, and morphological properties and characteristics.

Soil matrix an assemblage of mineral particles of various sizes, shapes, and chemical characteristics, together with organic materials, in various stages of decomposition and living soil populations

Sorption the removal of an ion or molecule from solution by adsorption and absorption. It is often used when the exact nature of the mechanism of removal is not known.

Specific surface the solid-particle surface area (of a soil or porous medium) divided by the solid-particle mass or volume, expressed in $m^2\ kg^{-1}$or $m^2\ m^{-3} = m^{-1}$, respectively

Subsurface the geologic zone below the surface of the Earth; rock and soil materials lying near, but not exposed, at the Earth's surface

Surface-charge density the excess of negative or positive charge per unit of surface area of soil or soil mineral

Transport the movement of mass from one point in the soil to another point in the soil

Vadose zone the aerated region of soil above the permanent water table

Volatility a measure of a compound's tendency to vaporize from the liquid state

Weathering the breakdown and changes in rocks and sediments at or near the Earth's surface produced by biological, chemical, and physical agents or combinations of them

Xenobiotic a compound foreign to biological systems; often refers to human-made compounds that are resistant or recalcitrant to biodegradation and/or decomposition

Acknowledgment

This work was partially supported by the Program IDEI, Grant ID_595, Contract No. 132/2007, and Program IDEI, Grant ID_559, Contract No. 265/2011, in the framework of the National Program for Research, Development and Innovation II—Ministry of Education and Research, Romania.

References

Adriano, D.C., 2001. Trace elements in terrestrial environments: Biogeochemistry, bioavailability, and risks of metals. Springer, New York.

Alexander, M., 1995. How toxic are toxic chemicals in soils. Environ. Sci. Technol. 29, 2713–2717.

Amrhein, C., Mosher, P.A., Strong, J.E., 1993. Colloid-assisted transport of trace metals in roadside soils receiving deicing salts. Soil Sci. Soc. Am. J. 57, 1212–1217.

Baca, T.E., Florkowski, T., 2000. The environmental challenges of nuclear disarmament. Springer, Berlin.

Barton, C.D., Karathanasis, A.D., 2003. Influence of soil colloids on the migration of atrazine and zinc through large soil monoliths. Water Air. Soil. Pollut. 143, 3–21.

Baumann, T., 2007. Colloid transport processes: Experimental evidence from the pore scale to the field scale. In: Frimmel, F.H., von der Kammer, F., Flemming, H.-C. (Eds.), Colloidal Transport in Porous Media. Springer, Berlin, pp. 57–85.

Baun, D.L., Christensen, T.H., 2004. Speciation of heavy metals in landfill leachate: A review. Waste Manag. Res. 22, 3–23.

Baveye, P., Vandevivere, P., Hoyle, B.L., DeLeo, P.C., de Lozada, D.S., 1998. Environmental impact and mechanisms of biological clogging of saturated soils and aquifermaterials. Crit. Rev. Environ. Sci. Technol. 28, 123–191.

Beckett, W., Chalupa, D., Pauly-Brown, A., Speers, D., Stewart, J., Frampton, M., Utell, M., Huang, L., Cox, C., Zareba, W., Oberdorster, G., 2005. Comparing inhaled ultrafine versus fine zinc oxide particles in healthy adults: A human inhalation study. Am. J. Respir. Crit. Care. Med. 171, 1129–1135.

Bengtson, P., Basiliko, N., Prescott, C.E., Grayston, S.J., 2007. Spatial dependency of soil nutrient availability and microbial properties in a mixed forest of *Tsuga heterophylla* and *Pseudotsuga menziesii*, in coastal British Columbia, Canada. Soil Biol. Biochem. 39, 2429–2435.

Bergendahl, J.A., Grasso, D., 2003. Mechanistic basis for particle detachment from granular media. Environ. Sci. Technol. 37, 2317–2322.

Blume, T., Weisbrod, N., Selker, J.S., 2005. On the critical salt concentrations for particle detachment in homogeneous sand and heterogeneous Hanford sediments. Geoderma 124, 121–132.

Bohn, H.L., McNeal, B.L., O'Connor, G.A., 2001. Soil Chemistry. John Wiley & Sons, New York.

Bradford, S.A., Simunek, J., Bettahar, M., van Genuchten, M.T., Yates, S.R., 2003. Modeling colloid attachment, straining, and exclusion in saturated porous media. Environ. Sci. Technol. 37, 2242–2250.

Bradford, S.A., Simunek, J., Bettahar, M., van Genuchten, M.T., Yates, S.R., 2006. Significance of straining in colloid deposition: Evidence and implications. Water Resources Res. 42, 12–15.

Brusseau, M.L., Wood, A.L., Rao, P.S.C., 1991. Influence of organic cosolvents on the sorption kinetics of hydrophobic organic chemicals. Environ. Sci. Technol. 25, 903–910.

Buccini, J., 2004. The global pursuit of the sound management of chemicals. The World Bank, Washington, DC.

Caliman, F.A., Gavrilescu, M., 2009. Personal care compounds, pharmaceuticals and endocrine disrupting agents in the environment—A review. CLEAN–Soil, Air, Water 34, 277–303.

Chen, H., Yu, C., Ueng, T., Chen, S., Chen, B., Huang, K., Chiang, L., 1998. Acute and subacute toxicity study of water-soluble polyalkylsulfonated C60 in rats. Toxicol. Pathol. 26, 143–151.

Chilton, P.J., Stuart, M.E., Gooddy, D.C., Williams, R.J., Johnson, A.C., 2005. Pesticide fate and behaviour in the UK Chalk aquifer, and implications for groundwater quality. Q. J. Eng. Geology Hydrogeology 38, 65–81.

Colombo, C., Ricciardella, M., Di Cerce, A., Maiuro, L., Violante, A., 2004. Effect of tannate, pH, sample preparation, ageing and temperature on the formation and nature of Al oxyhydroxides. Clays. Clay. Miner. 52, 721–733.

Colvin, V., 2003. The potential environmental impact of engineered nanoparticles. Nat. Biotechnol. 21, 1166–1170.

Corapcioglu, M.Y., Jiang, S., 1993. Colloid-facilitated groundwater contaminant transport. Water Resources Res. 29, 2215–2226.

Cornejo, J., Hermosin, M.C., Celis, R., Cox, L., 2005. Methods to determine sorption of pesticides and other organic compounds. In: Alvarez-Benedi, J., Munoz-Carpena, R. (Eds.), Soil-water-solute process characterization—an integrated approach. CRC Press, Boca Raton, FL, pp. 435–464.

Cox, L., Celis, R., Hermosin, M.C., Cornejo, J., Zsolnay, A., Zeller, K., 2000. Effect of organic amendments on herbicide sorption as related to the nature of the dissolved organic matter. Environ. Sci. Technol. 34, 4600–4605.

Csizer, Z., 2002. Persistent organic pollutants (POPs). UNIDO Programmes on Persistent Organic Pollutants (POPs), International Forum on Strategies and Priorities for Environmental Industries. June, Bratislava, 12–14.

Cui, D., Tian, F., Ozkan, C., Wang, M., Gao, H., 2005. Effect of single wall carbon nanotubes on human HEK293 cells. Toxicol. Lett. 155, 73–85.

Dec, J., Bollag, J.-M., Huang, P.M., Senesi, N., 2002. Impact of interactions between microorganisms and soil colloids on transformation of organic pollutants. In: Huang, P.M., Bollag, J.-M., Senesi, N. (Eds.), Interactions between soil particles and microorganisms. Impact on Terrestrial Ecosystem, vol. 8, IUPAC Series of Analytical and Physical Chemistry of Environmental Systems. John Wiley & Sons, Chichester, UK, pp. 323–378.

Degueldre, C., Bilewicz, A., Hummel, W., Loizeau, J.L., 2001. Sorption behaviour of Am on marl groundwater colloids. J. Environ. Radioact. 55, 241–253.

de Jonge, H., Jacobsen, O.H., de Jonge, L.W., Moldrup, P., 1998. Particle-facilitated transport of prochloraz in undisturbed sandy loam soil columns. Soil Sci. Soc. Am. J. 27, 1495–1503.

de Jonge, L.W., Kjaergaard, C., Moldrup, P., 2004. Colloids and colloid-facilitated transport of contaminants in soils: an introduction. Vadose Zone J. 3, 321–325.

de Jonge, L.W., Moldrup, P., Schjonning, P., 2009. Soil infrastructure, interfaces & translocation processes in inner space ("soil-it-is"): Towards a road map for the constraints and crossroads of soil architecture and biophysical processes. Hydrology Earth Syst. Sci. 13, 1485–1502.

DeNovio, N.M., Saiers, J.E., Ryan, J.N., 2004. Colloid movement in unsaturated porous media: Recent advances and future directions. Vadose Zone J. 3, 338–351.

Diels, L., Lookman, R., 2007. Microbial systems for in-situ soil and groundwater remediation. In: Marmiroli, N., Samotokin, B., Marmiroli, M. (Eds.), Advanced science and technology for biological decontamination of sites affected by chemical and radiological nuclear agents: NATO Science Series, vol. 75.Springer, The Netherlands, pp. 61–77.

Dror, I., Baram, D., Berkowitz, B., 2005. Use of nanosized catalysts for transformation of chloro-organic pollutants. Environ. Sci. Technol. 39, 1283–1290.

Dunnivant, F.M., Jardine, P.M., Taylor, D.L., McCarthy, J.F., 1992. Cotransport of cadmium and hexachlorobiphenyl by dissolved organic carbon through columns containing aquifer material. Environ. Sci. Technol. 26, 360–368.

Ehlers, G.A.C., Loibner, A.P., 2006. Linking organic pollutant (bio)availability with geosorbent properties and biomimetic methodology: A review of geosorbent characterization, and (bio)availability prediction. Environ. Pollut. 141, 494–512.

Elimelech, M., Ryan, J.N., 2002. The role of mineral colloids in the facilitated transport of contaminants in saturated porous media. In: Huang, P.M., Bollag, J.-M., Nicola, S. (Eds.), Interactions between soil particles and microorganisms: impact on the terrestrial ecosystem. John Wiley & Sons, New York, pp. 495–498.

EPRI, 1999. Colloids in saturated and partially saturated porous media: Approaches to the treatment of colloids in Yucca Mountain Total System Performance Assessment. TR-112135 Report. EPRI, Palo Alto, USA.

Evangelou, V.P., 1998. Environmental soil and water chemistry: Principles and applications. John Wiley & Sons, New York.

Evangelou, V.P., Karathanasis, A.D., 1991. Influence of pH, sodium adsorption ratio, and salt concentration on settling kinetics of suspended solids in coalmine ponds. J. Environ. Qual. 20, 783–788.

Evans, H.J., Sorger, G.J., 1966. Role of mineral elements with emphasis on the univalent cations. Annu. Rev. Plant. Physiol. 17, 47–76.

Fetter, C.W., 1999. Contaminant hydrogeology, second ed. Prentice Hall, New York.

Flury, M., Qiu, H., 2008. Modeling colloid-facilitated contaminant transport in the vadose zone. Vadose Zone J. 7, 682–697.

Frimmel, F.H., von der Kammer, F., Flemming, H.C., 2007. Colloidal transport in porous media. Springer, Amsterdam.

Foster, S.S.D., Chilton, P.J., Stuart, M.E., 1991. Mechanisms of groundwater pollution by pesticides. J. Inst. Water Environ. Manag. 5, 186–193.

Gamerdinger, A.P., Kaplan, D.I., 2001. Physical and chemical determinants of colloid transport and deposition in water-unsaturated sand and Yucca Mountain tuff material. Environ. Sci. Technol. 35, 2497–2504.

Gavrilescu, M., 2004. Removal of heavy metals from the environment by biosorption. Eng. Life Sci. 4, 219–232.

Gavrilescu, M., 2005. Fate of pesticides in the environment and its bioremediation. Eng. Life Sci. 5, 497–526.

Gavrilescu, M., 2008. Risk assessment and management (in Romanian). Iasi: Ecozone Press, Romania.

Gavrilescu, M., Betianu, C., 2008. Migration and fate of persistent organic pollutants in the atmosphere—a modelling approach. Environ. Eng. Manag. J. 7, 743–762.

Gavrilescu, M., Pavel, V.L., Cretescu, I., 2009. Characterization and remediation of soils polluted with uranium. J. Hazard. Mater. 163, 475–510.

Gianfreda, L., Sannino, F., Filazzola, M.T., Leonowicz, A., 1998. Catalytic behavior and detoxifying ability of a laccase from the fungal strain Cerrena unicolor. J. Mol. Catalysis B: Enzymatic 14, 13–23.

Gomez-Suarez, C., Busscher, H.J., van der Mei, H.C., 2001. Analysis of bacterial attachment from substratum surfaces by the passage of air-liquid interfaces. Appl. Environ. Microbiol. 67, 2531–2537.

Gooddy, D.C., Mathias, S.A., Harrison, I., Lapworth, D.J., Kim, A.W., 2007. The significance of colloids in the transport of pesticides through xhalk. Sci. Total Environ. 385, 262–271.

Goryachenkova, T.A., Kazinskaya, I.E., Kuzovkina, E.V., Novikov, A.P., Myasoedov, B.F., 2009. Association of radionuclides with colloids in soil solutions. Radiochemistry 51, 201–210.

Gounaris, V., Anderson, P.R., Holsen, T.M., 1993. Characteristics and environmental significance of colloids in landfill leachate. Environ. Sci. Technol. 27, 1381–1387.

Graber, E.R., Dror, I., Bercovich, F.C., Rosner, M., 2001. Enhanced transport of pesticides in a field trial with treated sewage sludge. Chemosphere 44, 805–811.

Grieger, K.D., Laurent, A., Miseljic, M., Christensen, F., Baun, A., Olsen, S.I., 2012. Analysis of current research addressing complementary use of life-cycle assessment and risk assessment for engineered nanomaterials: Have lessons been learned from previous experience with chemicals? J. Nanoparticles Res. 14, 958.

Grolimund, D., Borkovec, M., Barmettler, K., Sticher, H., 1996. Colloid-facillitated transport of strongly sorbing contaminants in natural porous media: A laboratory column study. Environ. Sci. Technol. 30, 3118–2132.

Grolimund, D., Borkovec, M., 2006. Release of colloidal particles in natural porous media by monovalent and divalent cations. J. Contam. Hydrol. 87, 155–175.

Grolimund, D., Barmettler, K., Borkovec, M., 2007. Colloid facilitated transport in natural porous media: Fundamental phenomena and modelling. In: Frimmel, F.H., von der Kammer, F., Flemming, H.-C. (Eds.), Colloidal transport in porous media. Springer, Berlin, pp. 3–27.

Hanselman, T.A., Graetz, D.A., Wilkie, A.C., 2003. Manure-borne estrogens as potential environmental contaminants: A review. Environ. Sci. Technol. 37, 5471–5478.

Hansen, S., 2009. Regulation and risk assessment of nanomaterials—Too little, too late? PhD Thesis Technical University of Denmark.

Hardman, R., 2006. A toxicological review of quantum dots: Toxicity depends on physicochemical and environmental factors. Environ. Health. Perspect. 114, 165–172.

Harvey, R.W., Garabedian, S.P., 1991. Use of colloid filtration theory in modeling movement of bacteria through a contaminated sandy aquifer. Environ. Sci. Technol. 25, 178–185.

Hassellov, M., Readman, J.W., Ranville, J.F., Tiede, K., 2008. Nanoparticle analysis and characterization methodologies in environmental risk assessment of engineered nanoparticles. Ecotoxicology 17, 344–361.

Hayes, M.H.B., Mingelgrin, U., 1991. Interactions between small organic chemicals and soil colloidal constituents. In: Bolt, G.H., de Boodt, M.F., Hayes, M.H.B., McBride, M.B., de Strooper, E.B.A. (Eds.), Interactions at the soil colloid—soil solution interface. Kluwer Academic Publishers, Dordrecht, pp. 323–408.

Haynes, R.J., 2008. Soil organic matter quality and the size and activity of the microbial biomass: Their significance to the quality of agricultural soils. In: Huang, Q., Huang, P.M., Violante, A. (Eds.), Soil mineral–microbe-organic interactions. Theories and applications. Springer, Berlin, pp. 3–37.

Hillel, D., 1998. Environmental soil physics. Elsevier, Academic Press, London.

Hlihor, R.M., Apostol, L.C., Smaranda, C., Pavel, V.L., Căliman, F.A., Robu, B.M., Gavrilescu, M., 2009. Bioavailability processes for contaminants in soils and their use in risk assessment. Environ. Eng. Manag. J. 8, 1199–1206.

Honeyman, B.D., 1999. Colloidal culprits in contamination. Nature 397, 56–59.

Hoshino, A., Hanaki, K., Suzuki, K., Yamamoto, K., 2004. Applications of t-lymphoma labeled with fluorescent quantum dots to cell tracing markers in mouse body. Biochem. Biophys. Res. Commun. 314, 46–53.

Hristozov, D., Malsch, I., 2009. Hazards and risks of engineered nanoparticles for the environment and human health. Sustainability 1, 1161–1194.

Huang, P.M., Bollag, J.-M., 1999. Minerals-organics-microorganisms interactions in the soil environment. In: Huang, P.M., Senesi, N., Buffle, J. (Eds.), Structure and surface reactions of soil particles. vol. 4. IUPAC series on analytical and physical chemistry of environmental systems. John Wiley & Sons, Chichester, UK, pp. 3–39.

Huang, P.M., 2002. Foreseeable impacts of soil mineral-organic matter-microorganism interactions on society: ecosystem health. In: Violante, A., Huang, P.M., Bollag, J.-M., Gianfreda, L. (Eds.), Soil mineral-organic matter-microorganism interactions and ecosystem health: Vol. 28A, Dynamics, mobility, and transformation of pollutants and nutrients. Elsevier, Amsterdam, The Netherlands, pp. 1–36.

Huang, P.M., 2008a. Impacts of physicochemical biological interactions on metal and metalloid transformations in soils: an overview. In: Violante, A., Huang, P.M., Gadd, G.M. (Eds.), Biophysico-chemical processes of heavy metals and metalloids in soil environments. John Wiley & Sons, Hoboken, NJ, pp. 4–52.

Huang, P.M., 2008b. Soil physical-chemical-biological interfacial interactions: An overview. In: Huang, Q., Huang, P.M., Violante, A. (Eds.), Soil mineral–microbe-organic interactions. Theories and applications. Springer, Berlin, pp. 3–37.

Ilina, T., Panfilov, M., Bues, M., Panfilova, I., 2008. A pseudo two-phase model of colloid transport in porous media. Transport Porous Media 71, 311–329.

Kan, A.T., Tomson, M.B., 1990. Ground water transport of hydrophobic organic compounds in the presence of dissolved organic matter. Environ. Toxicol. Chem. 9, 253–263.

Kandlikar, M., Ramachandran, G., Maynard, A., Murdock, B., Toscano, W., 2007. Health risk assessment for nanoparticles: a case for using expert judgment. J. Nanoparticle Res. 9, 137–156.

Kaplan, D.I., Sumner, M.E., Bertsch, P.M., Adriano, D.C., 1996. Chemical conditions conducive to the release of mobile colloids from ultisol profiles. Soil Sci. Soc. Am. J. 60, 269–274.

Kaplan, D.I., Bertsch, P.M., Adriano, D.C., 1997. Mineralogical and physicochemical differences between mobile and nonmobile colloidal phases in reconstructed pedons. Soil Sci. Soc. Am. J. 61, 641–649.

Karathanasis, A.D., 1999. Subsurface migration of copper and zinc mediated by soil colloids. Soil Sci. Soc. Am. J. 63, 830–838.

Kersting, A.B., Efurd, D.W., Finnegan, D.L., Rokop, D.J., Smith, D.K., Thompson, J.L., 1999. Migration of plutonium in groundwater at the Nevada test site. Nature 397, 56–59.

Khilar, K.C., Fogler, H.S., 1998. Migration of fines in porous media. Kluwer Academic Publishers, Dordrecht.

Kiker, G.A., Bridges, T.S., Varghese, A., Seager, P.T., Linkov, I., 2005. Application of multicriteria decision analysis in environmental decision making. Integrated Environ. Assess. Manag. 1, 95–108.

Kjaergaard, C., 2003. Colloid Mobilization and Transport in Structured Soils. The role of colloid dispersion, colloid stability and preferential flow. PhD Thesis. Aalborg University, Denmark.

Kleka, G.M., Rick, D.L., Will, M.E., Ritalahti, K., Marsh, T.L., 2001. Natural biological attenuation of phenoxy herbicides in groundwater: Dow Agroscience Paritutu Site, New Zealand. Bioremediation J. 5, 79–92.

Klitzke, S., Lang, F., Kaupenjohann, M., 2008. Increasing pH release colloidal lead in highly contaminated forest soil. Euro. J. Soil Sci. 59, 265–273.

Knox, R.C., Sabatini, D.A., Canter, L.W., 1993. Subsurface transport and fate processes. Lewis Publishers, Boca Raton, FL.

Kordel, W., Dassenakis, M., Lintelmann, J., Padberg, S., 1997. The importance of natural organic material for environmental processes in water and soils. Pure. Appl. Chem. 69, 1571–1600.

Kretzschmar, R., Sticher, H., 1997. Transport of humic-coated iron oxide colloids in a sandy soil: Influence of Ca^{2+} and trace metals. Environ. Sci. Technol. 31, 3497–3504.

Kretzschmar, R., Sticher, H., 1998. Colloid transport in natural porous media: Influence of surface chemistry and flow velocity. Phys. Chem. Earth 23, 133–139.

Kretzchmar, R., Borkovec, M., Grolimund, D., Elimelech, M., 1999. Mobile subsurface colloids and their role in contaminant transport. Adv. Agronomy 66, 121–193.

Kretzschmar, R., Schäfer, T., 2005. Metal retention and transport on colloidal particles in the environment. Elements 1, 205–210.

Kulikova, N.A., Perminova, I.V., 2002. Binding of atrazine to humic substances from soil, peat, and coal related to their structure. Environ. Sci. Technol. 36, 3720–3724.

Lam, C., James, J., McCluskey, R., Hunter, R., 2004. Pulmonary toxicity of single-wall carbon nanotubes in mice 7 and 90 days after intratracheal instillation. Toxicol. Sci. 77, 126–134.

Lecoanet, H., Wiesner, M., 2004. Velocity effects on fullerene and oxide nanoparticle deposition in porous media. Environ. Sci. Technol. 38, 4377–4382.

Lecoanet, H., Bottero, J., Wiesner, M., 2004. Laboratory assessment of the mobility of nanomaterials in porous media. Environ. Sci. Technol. 38, 5164–5169.

LeRoy Jackson, M., Phillip Barak, P., 2005. Soil chemical analysis: Advanced course: A manual of methods useful for instruction and research in soil chemistry, physical chemistry of soils, soil fertility, and soil genesis. Parallel Press, Madison, WI.

Li, X., Brown, D., Smith, S., MacNee, W., Donaldson, K., 1999. Short term inflammatory responses following intratracheal instillation of fine and ultrafine carbon black in rats. Inhal. Toxicol. 11, 709–731.

Liang, L., McCarthy, J.F., 1995. Colloidal transport of metal contaminants in groundwater. In: Allen, H.E., et al. (Eds.), Metal speciation and contamination of soils. Lewis Publishers, Boca Raton, FL, pp. 87–112.

Lindqvist, R., Enfield, C.G., 1992. Biosorption of dichlorodiphenyltrichloroethane and hexachlorobenzene in groundwater and its implications for facilitated transport. Appl. Environ. Microbiol. 58, 2211–2218.

Linkov, I., Satterstrom, K., Steevens, J., Ferguson, E., Pleus, R., 2007. Multicriteria decision analysis and environmental risk assessment for nanomaterials. J. Nanoparticle Res. 9, 543–554.

Liu, J.C., Chang, P.S., 1997. Solubility and adsorption behaviors of chlorophenols in the presence of surfactants. Water Sci. Technol. 35, 123–130.

Logan, B.E., Jewett, D.G., Arnold, R.G., Bouwer, E.J., O'Melia, C.R., 1995. Clarification of clean-bed filtration models. J. Environ. Eng. 121, 869–873.

Love, S.A., Maurer-Jones, M.A., Thompson, J.W., Lin, Y.-S., Haynes, C.L., 2012. Assesing nanoparticles toxicity. Anal. Chem. 5, 181–205.

Lovric, J., Bazzi, H., Cuie, Y., Fortin, G., Winnik, F., Maysinger, D., 2005. Differences in subcellular distribution and toxicity of green and red emitting CdTe quantum dots. J. Mol. Med. 83, 377–385.

Mansoori, G.A., Rohani Bastami, T., Ahmadpour, A., Eshaghi, Z., 2008. Environmental application of nanotechnology. Annu. Rev. Nano Res. 2, 1–73.

Maskall, J., Whitehead, K., Thornton, I., 1995. Heavy metal migration in soils and rocks at historical smelting sites. Environ. Geochem. Health 17, 127–138.

Massoudieh, A., Timothy, R., Ginn, T.R., 2007. Modeling colloid-facilitated transport of multi-species contaminants in unsaturated porous media. J. Contam. Hydrol. 92, 162–183.

Mayer, L.M., Schick, L.L., Hardy, K.R., Wagai, R., McCarthy, J., 2004. Organic matter in small mesopores in sediments and soils. Geochimica et Cosmochimica Acta. 68, 3863–3872.

Maynard, A., Baron, P., Foley, M., Shvedova, A., Kisin, E., Castranova, V., 2004. Exposure to carbon nanotube material: aerosol release during the handling of unrefined single-walled carbon nanotube material. J. Toxicol. Environ. Health. 67, 87–107.

McCarthy, J.F., Zachara, J.M., 1989. Subsurface transport of contaminants—mobile colloids in the subsurface environment may alter the transport of contaminants. Environ. Sci. Technol. 23, 496–502.

McCarthy, J.F., McKay, L.D., 2004. Colloid transport in the subsurface. Past, present, and future challenges. Vadose Zone J. 3, 326–337.

McGechan, M.B., Lewis, D.R., 2002. Transport of particulate and colloid-sorbed contaminants through soil. Part 1: General principles. Biosyst. Eng. 83, 255–273.

Medema, G.J., Juhasz-Holterman, M.H.A., Luijten, J.A., 2000. Removal of micro-organisms by bank filtration in a gravel-sand soil. In: Julich, W., Schubert, J. (Eds.), Proceedings of International Riverbank Filtration Conference. Düsseldorf, Germany: IAWR, pp. 161–168.

Metreveli, G., Frimmel, F.H., 2007. Influence of Na-bentonite colloids on the transport of heavy metals in porous media. In: Frimmel, F.H., von der Kammer, F., Flemming, H.-C. (Eds.), Colloidal Transport in Porous Media. Springer, Berlin, pp. 29–53.

Miller, W.P., Baharuddin, M.K., 1986. Relationship of soil dispersibility to infiltration and erosion of southeastern soils. Soil Sci. 142, 235–240.

Mills, W.B., Liu, S., Fong, F.K., 1991. Literature review and model (COMET) for colloid/metals transport in porous media. Ground Water 29, 199–208.

Monteiro-Riviere, N., Nemanich, R., Inman, A., Wang, Y., Riviere, J., 2005. Multi-walled carbon nanotube interactions with human epidermal keratinocytes. Toxicol. Lett. 155, 377–384.

Mueller, N., Nowack, B., 2008. Exposure modeling of engineered nanoparticles in the environment. Eviron. Sci. Technol. 42, 4447–4453.

Mueller, N.C., Nowack, B., 2010. Nanoparticles for remediation: Solving big problems with little particles. Elements 6, 395–400.

Nowack, B., Bucheli, T.D., 2007. Occurrence, behaviour and effects of nanoparticles in the environment. Environ. Pollut. 150, 5–22.

NPCA, 2008. Environmental fate and ecotoxicity of engineered nanoparticles. Report no. TA 2304/2007. In: Joner, E.J., Hartnik, T., Amundsen, C.E. (Eds.), Bioforsk. Norwegian Pollution Control Authority.

Oberdörster, G., Ferin, J., Lehnert, B., 1994. Correlation between particle size, *in vivo* particle persistence and lung injury. Environ. Health. Perspect. 102, 173–179.

Oberdorster, G., Oberdorster, E., Oberdorster, J., 2007. Concepts of nanoparticle dose metric and response metric. Environ. Health. Perspect. 115, A290.

Oberdörster, G., Stone, V., Donaldson, K., 2007. Toxicology of nanoparticles: A historical perspective. Nanotoxicology 1, 2–25.

Ostiguy, C., Roberge, B., Woods, C., Soucy, B., 2010. Engineered nanoparticles. Current Knowledge about OHS Risks R-656. Institute of Research for Health and Work Security. IRSST, Montreal, Canada.

Ouyang, Y., Shinde, D., Mansell, R.S., Harris, W., 1996. Colloid enhanced transport of chemicals in subsurface environments: A review. Crit. Rev. Environ. Sci. Technol. 26, 189–204.

Pandya, V.B., Bhuniya, S., Khilar, K.C., 1998. Existence of a critical particle concentration in plugging of a packed bed. AIChE. J. 44, 978–981.

Patil, S.P., 2012. Colloid and surfactant transport modeling through agricultural soil. PhD Thesis. Florida State University, USA.

Pédrot, M., Dia, A., Davranche, M., Bouhnik-Le Coz, M., Henin, O., Gruau, G., 2008. Insights into colloid-mediated trace element release at the soil/water interface. J. Colloid. Interface. Sci. 325, 187–197.

Pichot, C., 2008. Reactive nanocolloids for nanotechnologies and microsystems. In: Elasissari, A. (Ed.), Colloidal Nanoparticles in Biotechhnology. John Wiley & Sons, New York, pp. 1–30.

Plata, D.L., Gschwend, P.M., Reddy, C., 2008. Industrially synthesized single-walled carbon nanotubes: Compositional data for users, environmental risk assessments, and source apportionment. Nanotechnology 19. http://dx.doi.org/10.1088/0957-4484/19/18/185706.

Quinn, J., Geiger, C., Clausen, C., Brooks, K., Coon, C., O'Hara, S., Krug, T., Major, D., Yoon, W.-S., Gavaskar, A., Holdsworth, T., 2005. Field demonstration of DNAPL dehalogenation using emulsified zero-valent iron. Environ. Sci. Technol. 39, 1309–1318.

Ramachandran, V., Fogler, H.S., 1999. Plugging by hydrodynamic bridging during flow of stable colloidal particles within cylindrical pores. J. Fluid. Mech. 385, 129–156.

Redman, J.A., Grant, S.B., Olson, T.M., Estes, M.K., 2001. Pathogen filtration, heterogeneity, and potable reuse of wastewater. Environ. Sci. Technol. 35, 1798–1805.

Reijnders, L., 2006. Cleaner nanotechnology and hazard reduction of manufactured nanoparticles. J. Cleaner Production 14, 124–133.

Robichaud, C.O., Tanzil, D., Weilenmann, U., Wiesner, M.R., 2005. Relative risk analysis of several manufactured nanomaterials: An insurance industry context. Environ. Sci. Technol. 39, 8985–8994.

Robichaud, C.O., Tanzil, D., Wiesner, M.R., 2007. Assessing life cycle risks of nanomaterials. In: Wilsner Battero, M.R.,J. (Ed.), Environmental Nanotechnology. Applications and Impacts of Nanomaterials. McGraw-Hill, New York, pp. 481–524.

Rouse, J., Yang, J., Barron, A., Monteiro-Riviere, N., 2006. Fullerene-based amino acid nanoparticle interactions with human epidermal keratinocytes. Toxicol. Vitro 20, 1313–1320.

Rousseau, M., Di Pietro, L., Angulo-Jaramillo, R., Tessier, D., Cabibel, B., 2004. Preferential transport of soil colloidal particles: physicochemical effects on particle mobilization. Vadose Zone J. 3, 247–261.

Roy, S.B., Dzombak, D.A., 1996. Colloid release and transport processes in natural and model porous media. Colloids Surfaces A: Physicochemical Eng. Aspects 107, 245–262.

Ryan, J.N., Elimelech, M., 1996. Colloid mobilization and transport in groundwater. Colloids Surfaces 107, 1–56.

Saiers, J.E., Hornberger, G.M., 1996. The role of colloidal kaolinite in the transport of cesium through laboratory sand columns. Water Resources Res. 32, 33–41.

Saiers, J.E., Hornberger, G.M., 1999. The influence of ionic strength on the facilitated transport of cesium by kaolinite colloids. Water Resources Res. 35, 1713–1727.

Saiers, J.E., 2002. Laboratory observations and mathematical modeling of colloid-facilitated contaminant transport in chemically heterogeneous systems. Water Resources Res. 38, 3.1–3.14.

Saiers, J.E., Lenhart, J.J., 2003. Ionic-strength effects on colloid transport and interfacial reactions in partially saturated porous media. Water Resources Res. 39, 1256. http://dx.doi.org/10.1029/2002WR001887.

Sanchis, J., Farre, M., Barcelo, D., 2012. Analysis and fate of organic nanomaterials in environmental samples. In: Barcelo, D., Farre, M. (Eds.), Analysis and Risk of Nanomaterials in Environmental and Food Samples. Elsevier, , Amsterdam, pp. 131–168.

Santra, P., Chopra, U.K., Chakraborty, D., 2008. Spatial variability of soil properties and its application in predicting surface map of hydraulic parameters in an agricultural farm. Curr. Sci. 95, 937–945.

Sayes, C., Marchione, A., Reed, K., Warheit, D., 2007. Comparative pulmonary toxicity assessments of C60 water suspensions in rats: Few differences in fullerene toxicity in vivo in contrast to in vitro profiles. Nano Letters 7, 2399–2406.

Schijven, J.F., Hassanizadeh, S.M., 2000. Removal of viruses by soil passage: Overview of modeling, processes, and parameters. Crit. Rev. Environ. Sci. Technol. 30, 49–127.

Seager, T.P., Linkov, I., 2008. Coupling multicriteria decision analysis and life-cycle assessment for nanomaterials. J. Ind. Ecol. 12, 282–285.

Sen, T.K., Mahajan, S.P., Khilar, K.C., 2002. Colloid-associated contaminant transport in porous media: 1. Experimental studies. AIChE. J. 48, 2366–2374.

Sen, T.K., Shanbhag, S., Khila, K.C., 2004. Subsurface colloids in groundwater contamination: a mathematical model. Colloids Surfaces A: Physicochemical Eng. Aspects 232, 29–38.

Sen, T.K., Khilar, K.C., 2006. Review on subsurface colloids and colloid-associated contaminant transport in saturated porous media. Adv. Colloid. Interface. Sci. 119, 71–96.

Seta, T.K., Karathanasis, A.D., 1997. Atrazine adsorption by soil colloids and co-transport through subsurface environments. Soil Sci. Soc. Am. J. 61, 612–617.

Shainberg, I., Warrington, D., Laflen, J.M., 1992. Soil dispersibility, rain properties, and slope interaction in rill formation and erosion. Soil Sci. Soc. Am. J. 56, 278–283.

Shein, E.V., Devin, B.A., 2007. Current problems in the study of colloidal transport in soil. Eurasian Soil Sci. 40, 399–408.

Siemens, J., Ilg, K., Pagel, H., Kaupenjohann, M., 2008. Is colloid-facilitated phosphorus leaching triggered by phosphorus accumulation in sandy soils? J. Environ. Qual. 29, 97–105.

Siemens, J., Ilg, K., Lang, F., Kaupenjohann, M., 2004. Adsorption controls mobilization of colloids and leaching dissolved phosphorus. Euro. J. Soil Sci. 55, 253–263.

Simko, M., Mattsson, M.-O., 2010. Risks from accidental exposures to engineered nanoparticles and neurological health effects: A critical review. Particle Fibre Toxicol. 7, 42.

Simunek, J., He, C., Pang, L., Bradford, S.A., 2006. Colloid-facilitated solute transport in variably saturated porous media: numerical model and experimental verification. Vadose Zone J. 5, 1035–1047.

Sincero, A.P., Sincero, G.A., 2003. Physical-chemical treatment of water and wastewater. IWA Publishing, London.

Smith, C., Shaw, B., Handy, R., 2007. Toxicity of single walled carbon nanotubes on rainbow trout, (*Oncorhynchus mykiss*): Respiratory toxicity, organ pathologies, and other physiological effects. Aquat. Toxicol. 82, 94–109.

Sokolik, G.A., Ivanova, T.G., Leinova, S.L., Ovsiannikova, S.V., Kimlenko, I.M., 2001. Migration ability of radionuclides in soil-vegetation cover of Belarus after Chernobyl accident. Environ. Int. 26, 183–187.

Sparks, D.L., 1995. Kinetics of metal sorption reactions. In: Allen, H.E., Huang, C.W., Bailey, G.P., Bowers, A.R. (Eds.), Metal speciation and contamination of soil. CRC Press, Boca Raton, FL, pp. 35–58.

Sparks, D.L., Sparks, D.L., 2003. Environmental soil chemistry. Academic Press, London.

Steffan, R.J., Sperry, K.L., et al., 1999. Field-scale evaluation of in situ bioaugmentation for remediation of chlorinated solvents in groundwater. Environ. Sci. Technol. 33, 2771–2781.

Stoeger, T., Reinhard, C., Takenaka, S., Schroeppel, A., Karg, E., Ritter, B., Heyder, J., Schultz, H., 2006a. Instillation of six different ultrafine carbon particles indicates surface area threshold dose for acute lung inflammation in mice. Environ. Health Perspect. 114, 328–333.

Stoeger, T., Schmid, O., Takenaka, S., Schulz, H., 2006b. Inflammatory response to TiO_2 and carbonaceous particles scales best with BET surface area. Environ. Health Perspect. 115, A290–A291.

Stumm, W., Morgan, J.J., 1996. Aquatic chemistry. John Wiley–Interscience, New York.

Suijkerbuijk, M.P.W., 2008. Risk assessment of engineered carbonaceous nanoparticles in the aquatic environment. MSc Thesis. Utrecht University, The Netherlands.

Tanabe, S., Iwata, H., Tatsukawa, R., 1994. Global contamination by persistent organochlorines and their ecotoxicologcial impact on marine mammals. Sci. Total. Environ. 154, 163–177.

Theng, B.K.G., Orchard, V.A., 1995. Interactions of clays with microorganisms and bacterial survival in soil: a physicochemical perspective. In: Huang, P.M., Berthelin, J., Bollag, J.-M., McGill, W.G., Pages, A.L. (Eds.), Environmental impact of soil component interactions. vol. II. Metals, other inorganics and microbial activities. CRC/Lewis Publishers, Boca Raton, FL, pp. 123–143.

Thiele-Bruhn, S., 2003. Pharmaceutical antibiotic compounds in soils–A review. J Plant Nutr. Soil Sci. 166, 145–167.

Thieme, J., Sedlmair, J., Gleber, S.-C., Bertilson, M., von Hofsten, O., Takman, P., Hertz, H., 2009. High-resolution imaging of soil colloids in aqueous media with a compact soft X-ray microscope. J. Phys. (Conference Series) 186. http://dx.doi.org/10.1088/1742-6596/186/1/012107.

Todorov, P.T., Ilieva, E.N., 2006. Contamination with uranium from natural and anthropological sources. Rom. J. Phys. 51, 27–34.

Tolls, J., 2001. Sorption of veterinary pharmaceuticals in soils: A review. Environ. Sci. Technol. 35, 3397–3406.

Totsche, K.U., Danzer, J., Kögel-Knabner, I., 1997. DOM-enhanced retention of polycyclic aromatic hydrocarbons in soil miscible displacement experiments. J. Environ. Qual. 26, 1090–1100.

Totsche, K.U., Kögel-Knabner, I., 2004. Mobile organic sorbent affected contaminant transport in soil. Vadose Zone J. 3, 352–367.

Triay, I., Simmons, A., Levy, S., Nelson, S., Nuttal, H., Robison, B., Steinkampf, W., Viani, B., 1995. Progress report on colloid facilitated transport at Yucca Mountain. Los Alamos national Laboratory Report, LA 127779-MS, Los Alamos, USA.

Tufenkji, N., Elimelech, M., 2004. Correlation equation for predicting single-collector efficiency in physicochemical filtration in saturated porous media. Environ. Sci. Technol. 38, 529–536.

Tungittiplakorn, W., Cohen, C., Lion, L.W., 2005. Engineered polymeric nanoparticles for bioremediation of hydrophopic contaminants. Environ. Sci. Technol. 39, 1354–1358.

Tungittiplakorn, W., Lion, L.W., Cohen, C., Kim, J.K., 2004. Engineered polymeric nanoparticles for soil remediation. Environ. Sci. Technol. 38, 1605–1610.

Tyler, G., Olsson, T., 2001. Concentrations of 60 elements in the soil solution as related to the soil acidity. Euro. J. Soil Sci. 52, 151–165.

USEPA, 2007. Nanotechnology white paper. U.S. Environmental Protection Agency, EPA 100/B-07/001, Washington, DC.

van Genuchten, M.Th., Wagenet, R.J., 1989. Two-site/two-region models for pesticide transport and degradation: Theoretical development and analytical solutions. Soil. Sci. Soc. Am. J. 53, 1303–1310.

Vendelboe, A.L., Moldrup, P., Schjønning, P., Oyedele, D.J., Jin, Y., Scow, K.M., de Jonge, L.W., 2012. Colloid release from soil aggregates: Application of laser diffraction. Vadose Zone J. 11. http://dx.doi.org/10.2136/vzj2011.0070.

Vignati, D., 2004. Trace metal partitioning in freshwater as a function of environmental variables and its implications for metal bioavailability. PhD Thesis. University of Genève.

Vinten, A.J.A., Yaron, B., Nye, P.H., 1983. Vertical transport of pesticides into soil when adsorbed on suspended particles. J. Agric. Food. Chem. 31, 662–664.

Violante, A., Krishnamurti, G.S.R., Huang, P.M., 2002. Impact of organic substances on the formation and transformation of metal oxides in soil environments. In: Huang, P.M., Bollag, J.-M., Senesi, N. (Eds.), Interactions between soil particles and microorganism: Impact on the terrestrial ecosystem, IUPAC series on analytical and physical chemistry of environmental systems, vol. 8West Sussex, England: John Wiley & Sons, Chichester, pp. 133–188.

von den Kammer, F., Hofmann, T., Plathe, K., Hochella, M., Legros, S., Neubaner, E., Le Coustumer, P., 2008. Environmental colloids and nanoparticles: Occurrence, behaviour and relevance for engineered particles. Report. Center for Earth Science. University of Vienna.

von Gunten, H.R., Weber, V.E., Krahenbuhl, U., 1988. The reactor accident at Chernobyl: A possibility to test colloid-controlled transport of radionuclides in a shallow aquifer. J. Contam. Hydrol. 2, 237–247.

Vuorinen, A.H., Saharinen, M.H., 1996. Effects of soil organic matter extracted from soil on acid phosphomonoesterase. Soil Biol. Biochem. 28, 1477–1481.

Wan, J., Tokunaga, T.K., 1997. Film straining of colloids in unsaturated porous media: Conceptual model and experimental testing. Environ. Sci. Technol. 31, 2413–2420.

Wang, T.S.C., Huang, P.M., Chou, C.-H., Chen, J.-H., 1986. The role of soil minerals in the abiotic polymerization of phenolic compounds and formation of humic substances. In: Huang, P.M., Schnitzer, M. (Eds.), Interactions of Soil Minerals with Natural Organics and Microbes. Soil Science Society of America, Madison, WI, pp. 251–281.

Warfvinge, P., Sverdrup, H., 1992. Calculating critical loads of acid deposition with PROFILE—A steady-state soil chemistry model. Water. Air. Soil Pollut. 63, 119–143.

Warheit, D., Laurence, B., Reed, K., Roach, D., Reynolds, G., Webb, T., 2004. Comparative pulmonary toxicity assessment of single-wall carbon nanotubes in rats. Toxicol. Sci. 77, 117–125.

Weiner, E.R., 2000. Applications of environmental chemistry: A practical guide for environmental professionals. CRC Press, Boca Raton, FL.

Widenfalk, A., 2002. Pesticide bioavailability in aquatic sediments—A literature review. Inst. för miljöanalys, SLU, IMA Report. Department of Environmental Assessment Swedish University of Agricultural Sciences, Uppsala.

Wittmaack, K., 2007. In search of the most relevant parameter for quantifying lung inflammatory response to nanoparticle exposure: Particle number, surface area, or what? Environ. Health. Perspect. 115, 187–194.

Worrall, F., Parker, A., Rae, J.E., Johnson, A.C., 1999. A study of suspended and colloidal matter in the leachate from lysimeters and its role in pesticide transport. J. Environ. Qual. 28, 595–604.

Yang, L., Watts, D., 2005. Particle surface characteristics may play an important role in phytotoxicity of alumina nanoparticles. Toxicol. Lett. 158, 122–132.

Yao, K.M., Habibian, M.T., O'Melia, C.R., 1971. Water and waste water filtration: Concepts and applications. Environ. Sci. Technol. 5, 1105–1112.

Yarón, B., Calvet, R., Prost, R., 1996. Soil pollution: processes and dynamics. Springer, Berlin.

Yu, G., Saha, U.K., Kozak, L.M., Huang, P.M., 2006. Kinetics of cadmium adsorption on aluminum precipitation products formed under the influence of tannate. Geochim. Cosmochim. Acta. 70, 5134–5145.

Zhang, L., Tian, G., Liang, X., Liu, J., Peng, G., 2011. Effect of water dispersible colloids in manure on the transport of dissolved and colloidal phosphorus through soil column. Afr. J. Agric. Res. 6, 6369–6376.

Zhang, M., Li, W., Yang, Y., Chen, B., Song, F., 2005. Effects of readily dispersible colloid on adsorption and transport of Zn, Cu, and Pb in soils. Environ. Int. 31, 840–844.

Zhang, W., 2003. Nanoscale iron particles for environmental remediation: an overview. J. Nanoparticle Res. 5, 323–332.

Zimmerman, A.R., Chorover, J., Goyne, K.W., Brantley, S.L., 2004. Protection of mesopore-adsorbed organic matter from enzymatic degradation. Environ. Sci. Technol. 38, 4542–4548.

CHAPTER

18 Micellar Solutions to Recover Diesel in Sand

Leticia A. Bernardez, Luiz R.P. de Andrade Lima

Ingenium Consultoria em Engenharia Ltda, Avenida Estados Unidos, n. 528, Sala 1213, Salvador, Bahia, CEP 40010-020, Department of Materials Science and Technology, Federal University of Bahia, C.P. 6974, Salvador, BA, Brazil, 41810-971

CHAPTER OUTLINE

1 Introduction

1.1 Remediation of sites contaminated with nonaqueous phase liquids

Subsurface contamination by organic compounds represents a serious environmental problem. When released in the subsurface, most organic compounds form a separate phase commonly labeled NAPL (nonaqueous phase liquid). These NAPLs are complex mixtures containing a number of components such as aromatic hydrocarbons, several of which are suspected to be carcinogens. As those NAPLs

The Role of Colloidal Systems in Environmental Protection. http://dx.doi.org/10.1016/B978-0-444-63283-8.00018-1

migrate in the subsurface, they often reach residual saturation at which point their mobility is greatly reduced. Although most NAPL components have low solubility in groundwater, NAPLs can represent a long-term source of groundwater contamination because maximum concentration limits are much lower than the effective solubility of most NAPL components.

Widely used in the world, diesel fuel represents a permanent source of soil and water pollution. Resulting from the petroleum distillation, diesel fuel is a mixture of more than 2000 compounds composed of about 60% aliphatic and 40% aromatics hydrocarbons, which are toxic and recalcitrant to breakdown by microorganisms and are highly distributed in the environment (Bernardez, 2008). Remediation of these sites is of public interest because of the toxic and mutagenic potential of many diesel constituents.

When diesel leaks into soil, it will be located above the water table due to its density of 0.85–0.87. The mechanisms by which diesel moves in soil are by the saturated or unsaturated flow of liquid diesel, diesel dissolved in water and diffusion (volatilization). Due to the reduced mobility and the low volatility and solubility of NAPLs in groundwater, traditional methods designed to recover contaminants, such as pump-and-treat technology, air-sparging, or bio-venting, often fail when applied to NAPL-contaminated source zones. Excavation and ex-situ treatment is not always possible or economical when the contamination is located at great depths, is under buildings, or a large volume of soil has been contaminated. An alternative to conventional methods consists of injecting micellar solutions composed of surfactants and/or alcohols (Bernardez, 2005, 2008a,b, 2009; Bernardez et al., 2009).

1.2 Scope and objectives of the chapter

This chapter addresses the removal of diesel fuel by a washing process using micellar solution. Micellar solubilization is an important step for removal of organic contaminants from the soil matrix, especially for low aqueous solubility organic contaminants such as diesel. In addition to fundamental studies, laboratory and field studies on removal of diesel from contaminated soil and mathematical models applied are also reviewed to show the applicability of this technology. The combination of micellar solutions with other remediation techniques such as biodegradation applied to diesel recovery is presented.

2 Background

2.1 Surfactants

The term *surfactant* is a contraction of the phrase "surface-active agent." These compounds possess a wide range of applicability in the food, pharmaceutical, mineral processing, hydrometallurgy, cosmetics, and painting industries because they are among the most versatile chemical products. Surfactants are amphiphilic compounds possessing hydrophilic (polar) and hydrophobic (nonpolar) moieties. Due to this property, they tend to concentrate at the interfaces and modify interfacial tensions (Rosen, 1989).

Based on the nature of their hydrophilic moiety the surfactants are classified as four basic classes (Schramm, 1992):

(a) the non-ionic which has no charge groups in its head, (non-ionic surfactants are excellent detergents because they have moderate foaming properties and moderate wetting, have good emulsification and do not adsorb on charged surfaces, examples are polyoxyethylene alcohol and alkylphenol ethoxylate;
(b) the ionic which the head carries a net charge, if the charge is negative, the surfactant is more specifically called anionic (example: sodium dodecyl sulfate and sodium dodecyl benzene sulfonate), if the charge is positive, it is called cationic (example: laurylamine hydrochloride and cetyl trimethylammonium bromide) and
(c) zwitterioninc or amphoterics, which combine both a positive and a negative group (examples: lauramidopropyl betaine and cocoamido-2-hydroxypropyl sulfobetaine).

Efforts toward the formulation for new structures of surfactants have been done. These surfactants include the gemini surfactants, polymeric and polymerisable surfactants (Zana, 2002). These novel surfactants show synergistic interactions, enhanced surface and aggregation properties, less toxicity and high biodegradability. The last two characteristics are due to the more restricting regulations especially for personal care products and house detergents.

In aqueous systems, the addition of a surfactant enables two immiscible liquids to become soluble in one another. This striking phenomenon is attributed to the unique structure of the surfactant molecules.

In aqueous solutions the surfactants exist as monomers until a given concentration, termed the critical micellar concentration (CMC). As the surfactant concentration approaches the CMC, the lipophilic (nonpolar) parts of the surfactant monomers associate with each other to form micelles that consist of a hydrophobic core surrounded by a hydrophilic (polar) mantle. A micelle can be a sphere, an ellipsoid, a rod, or even a bi-layer. Their shape and size depend on the temperature, type of surfactant and its concentration, the presence of other ions in solution, or other water-soluble organic. Figure 18.1 shows the schematics of a phase diagram for cetyltrimethylammonium bromide (CTAB), which is a cationic surfactant, in water as a function of the solution temperature and surfactant concentration. At low surfactant concentration, the surfactant molecules are dispersed until the CMC. At this point, spherical micelles are formed. Increasing the concentration of the surfactant several transitions can be associated with the formation of micelles and the self-organization of micelles in several distinct morphologies (Brinker et al., 1999).

The shape of the micelle is of importance in determining various properties of the surfactant solution, such as its viscosity, its capacity to solubilize, and its cloud point. Changes in temperature, concentration of surfactant, additives in the liquid phases, and structural groups in the surfactant may cause change in the size, shape, and aggregation number of the micelle (Rosen, 1989). The number of surfactant molecules in a micelle is known as the aggregate number. Micelles are continually forming and dispersing. The half-life for micelles for ionic surfactants is about 0.01 seconds.

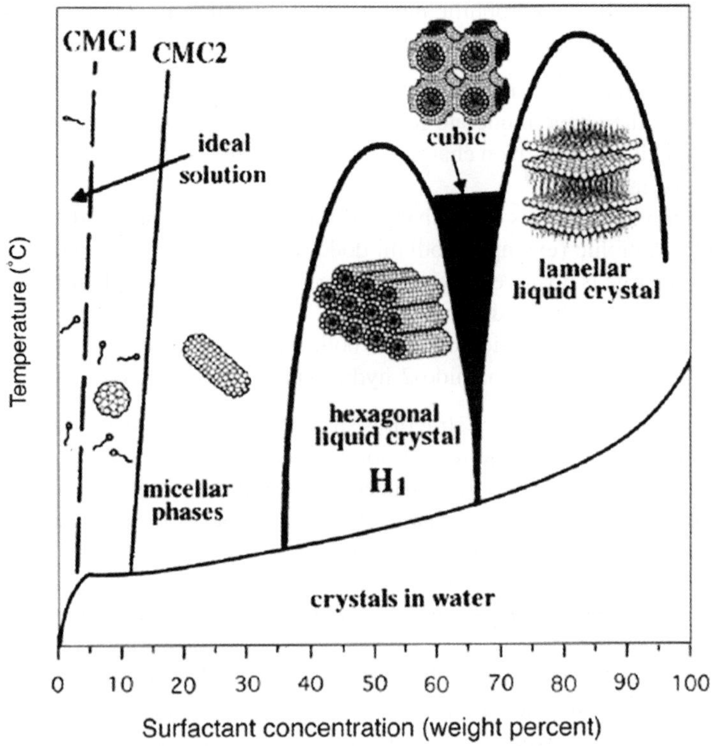

FIGURE 18.1

Schematics of a phase diagram for CTAB in water (Brinker et al., 1999).

2.2 Micellar solution

A micellar solution consists of a dispersion of micelles in a solvent (usually water) and forms when the concentration of amphiphile exceeds the critical micellar concentration (CMC). It persists until the amphiphile concentration becomes sufficiently high to form a lyotropic liquid crystal phase (Rosen, 1989). Micellar solutions are an isotropic phase.

The term *micellar solution* has alternative names such as microemulsion, transparent emulsion, swollen micelle, and solubilized oil. Micellar solutions are generally transparent in the absence of colored impurities, and when not transparent they are translucent, stable, isotropic liquid mixtures of oil, water, and surfactant, frequently in combination with a co-surfactant. The aqueous phase may contain salt(s) and/or other compounds, and the "oil" often is a complex mixture of different hydrocarbons. The two basic types of micellar solutions are direct (oil dispersed in water, o/w) and reversed (water dispersed in oil, w/o).

In ternary systems such as microemulsions, where two immiscible phases (water and oil) are present with a surfactant, the surfactant molecules may form a monolayer at the interface between the oil and water, with the hydrophobic tails of the surfactant molecules dissolved in the oil phase and the hydrophilic head groups in the aqueous phase. As in the binary systems (water/surfactant or oil/surfactant), self-assembled structures of different types can be formed, ranging, for example, from (inverted) spherical and cylindrical micelles to lamellar phases and bicontinuous microemulsions, which may coexist with predominantly oil or aqueous phases.

The microemulsion region is usually characterized by constructing ternary-phase diagrams. Three components are the basic requirement to form a microemulsion: an oil phase, an aqueous phase, and a surfactant. If a co-surfactant is used, it may sometimes be represented at a fixed ratio to surfactant as a single component, and treated as a single "pseudo-component." The relative amounts of these three components can be represented in a ternary-phase diagram. Each of the three components of the system is found at an apex of the triangle, where their corresponding volume fraction is 100%. Moving away from that corner reduces the volume fraction of that specific component and increases the volume fraction of one or both of the two other components. Each point within the triangle represents a possible composition of a mixture of the three components or pseudo-components, which may consist (ideally, according to the Gibbs phase rule) of one, two, or three phases. These points combine to form regions with boundaries between them, which represent the "phase behavior" of the system at constant temperature and pressure.

Figure 18.2 shows the schematics of a generic detergent (oil-water-surfactant) phase diagram (Brinker et al., 1999). It can be seen as the formation of micelles and the self-organization of micelles

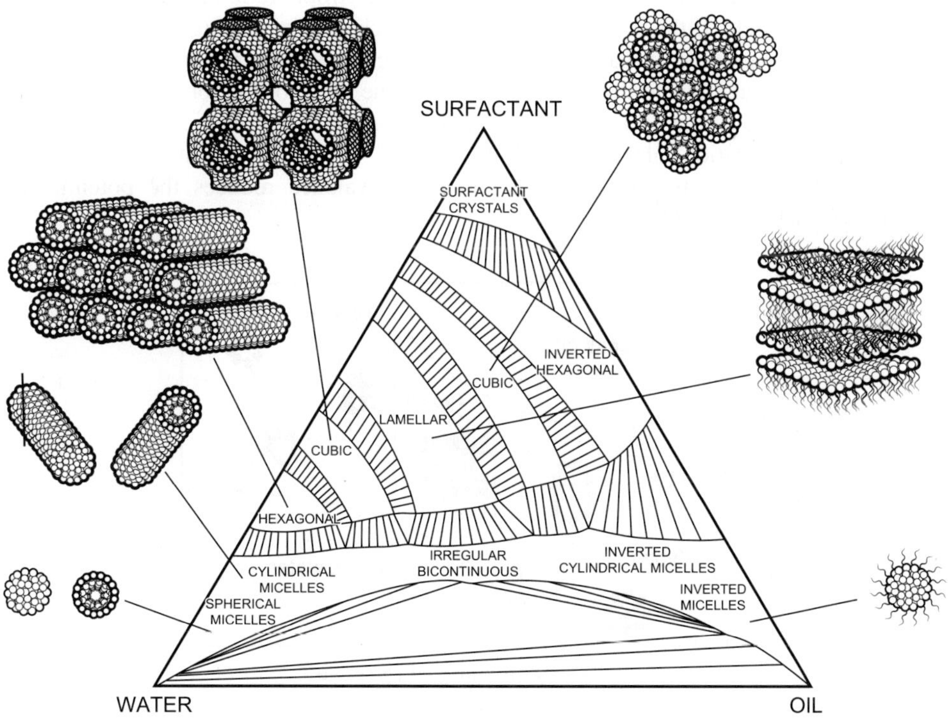

FIGURE 18.2

Schematics of oil-water-surfactant generic phase diagram (Brinker et al., 1999).

in several distinct morphologies, which has an impact on the solution rheology and NAPL mobilization (Brinker et al., 1999).

Figure 18.3 shows the phase behavior of microemulsions and the transition from oil solubilized in aqueous micelle (Windsor type I) to water solubilized in reverse micelle present in the oil phase (Windsor type II). The HLB *(hydrophile-lipohile-balance)* is a number that indicates a tendency of surfactant portioning between oil and water. A hydrophobic group has an HLB of 0.0, and the ethylene oxide head group has an HLB of 20. HLB decreases by increasing the salinity for ionic surfactants (Paria, 2008).

The stability of colloid explains the dispersion of the emulsion droplets. The electrostatic forces occurs due to the surface electrical charge exhibited when the droplets are put in contact with a polar medium, such as water, and are balanced with the attractive van der Waals forces. These attractive and repulsive forces can be represented as attractive and repulsive potentials, which decrease with the distance to the charged droplets. Figure 18.4 shows the schematics of the resulting potential energy of interaction of particles. Note in case 1 a potential barrier that will ensure system dispersion or stability. In case 2 the repulsive component is less important; therefore, it will not have a potential barrier and the particles will be brought together with or without large changes in the surface area, which are, respectively, the coalescence and the aggregation processes, which depend on the interfacial tensions.

The charge at the droplet surface can be evaluated with eletrophoretic mobility. The droplets are subjected to an electric field. If the droplet has an interfacial charge, it will migrate with a velocity that is proportional to the magnitude of that charge. In fact, the charge measured by this method gives the charge density or potential close to the droplet surface, at the plane of the ions shear, called the zeta potential. Figure 18.5 shows the zeta potential for bitumen in water emulsions as a function of electrolyte concentration and cation valence (Na^+, Ca^{2+}, and Al^{3+}). The zeta potential is negative, and the enhancement of the electrolyte concentration and cation valence reduces the potential and the emulsion stability (Schramm, 1992).

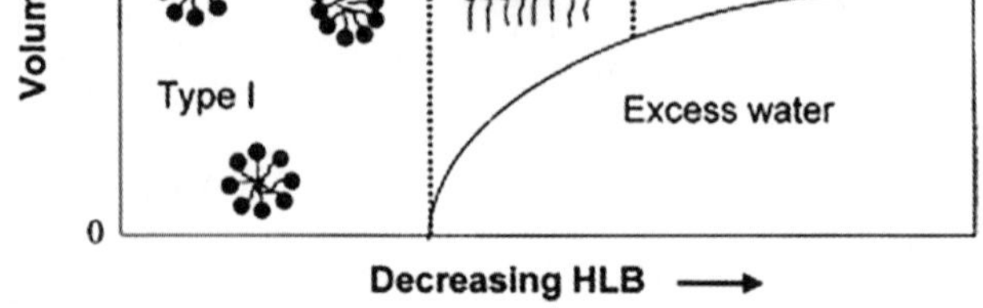

FIGURE 18.3

Phase behavior of microemulsions and transition from Windsor type I to Windsor type II system. HLB indicates a tendency of portioning between oil and water. (Paria, 2008).

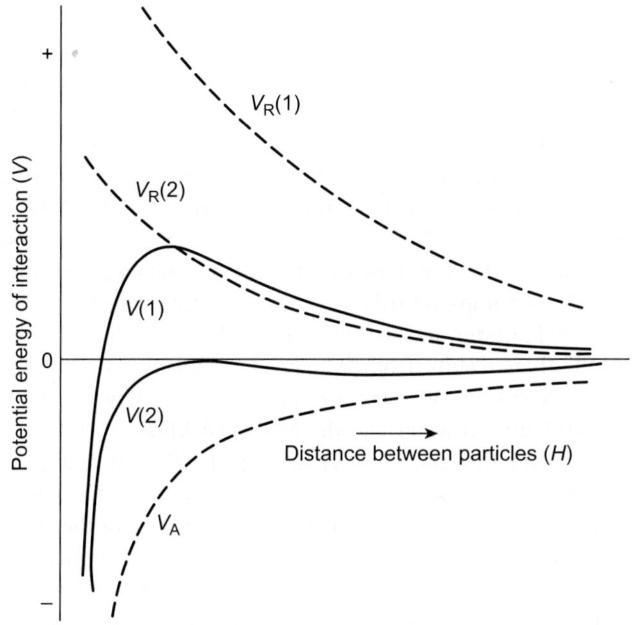

FIGURE 18.4

Effect of repulsive potentials ($V_R(1)$ and $V_R(2)$) and attractive potential (V_A) on the interaction energy (V(1) and V(2)) (Schramm, 1992).

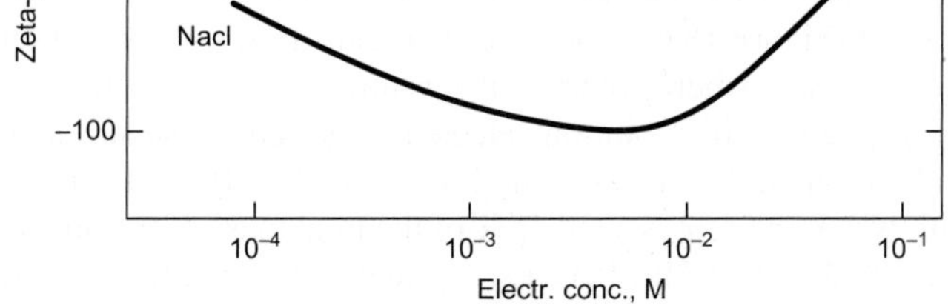

FIGURE18.5

Zeta potential for bitumen in water emulsions as a function of electrolyte concentration (Schramm, 1992).

3 Surfactant-enhanced site remediation

Surfactant-enhanced aquifer remediation (SEAR) has been proposed as a means of enhancing the removal of NAPL from contaminated sites. SEAR involves the injection of surfactant solutions to solubilize and mobilize NAPL constituents. The surfactant and contaminants are subsequently extracted through pumping wells. Various design approaches may be taken to SEAR, depending on the hydrogeological conditions and the physicochemical properties of the NAPL. The solubilization mechanism, which is at the heart of the SEAR process, is the formation in the groundwater of micelles, in which the molecules of the NAPL are dissolved and then transported by the groundwater. In the field, SEAR works similarly to pump-and-treat operations except that dilute surfactant solutions are injected into the contaminated aquifer and withdrawn together with the solubilized NAPLs.

The recovery of NAPL then results from enhanced solubility in the micellar solution and/or mobilization of the residual NAPL because the interfacial tension between the micellar solution and the NAPL is lower than that between water and the NAPL (AATDF, 1999).

Research directed at the evaluation and improvement of SEAR is currently active. Laboratory tests to screen various surfactant systems and to optimize the micellar solutions are being undertaken to determine whether solubilization or mobilization is most appropriate for the site. Several authors have successfully combined surfactants and surfactant and alcohols in a single micellar washing solution for different contaminants. Optimized micellar solutions have also been shown to recover diesel (Martel and Gélinas, 1996). In addition, column tests to assess the surfactant performance and the characterization of matrix-fluid interactions have been the focus of extensive research because they are essential prerequisites to field pilot testing.

4 Laboratory studies in columns of diesel recovery using micellar solutions

Most of the previous laboratory studies on surfactants have used large-volume (up to 19 pore volumes) and low-concentration surfactant solutions (0.1–4%) for NAPL recovery (Abdul et al., 1990, 1991, 1992; Fountain et al., 1991; Pennell et al., 1992, 1993). Previous studies suggested that the combinations of alcohols and surfactant to dissolve light oils and solvents are more effective than the use of either of them separately (Desnoyers et al., 1983; Martel et al., 1993).

Micellar solutions have been selected through the phase diagram. Pseudo-ternary-phase diagrams were used to select the optimal micellar solutions for the diesel recovery experiments in Martel and Gélinas (1996). The principle of the method is to create water-in-NAPL or NAPL-in-water microemulsions in Winsor type I (Figure 18.6a) and type II (Figure 18.6b) systems (Winsor, 1954). Winsor type III system (middle phase), where interfacial tension is ultra-low (10^{-3} – 10^{-4} mNm^{-1}) is not desirable (Figure 18.6c). The micellar solution selected consisted of the surfactant Hostapur SAS-60 (SAS) and two alcohols, n-butanol (n-BuOH), and n-pentanol (n-PeOH). The surfactant and the alcohols were mixed with water and represented 15% of the total mass of the aqueous micellar solution. The mass ratio of the alcohols n-BuOH/n-PeOH was equal to 1.5, and the mass ratio of alcohols over SAS was 1.0.

However, a very large number of surfactants, alcohols, and solvents are available, and choosing the composition of the micellar solution is not a trivial matter. Martel et al. (1993) and Martel and Gélinas

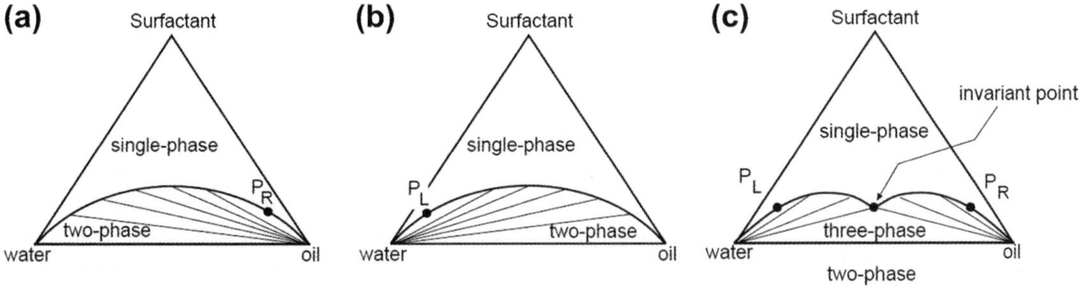

FIGURE 18.6

Schematic representation of ternary diagrams: (a) Type I, (b) Type II and (c) Type III.

(1996) have shown that the surfactant and alcohol can be efficiently selected, and their proportions in the micellar solution can be optimized, by combining phase diagrams and column tests that account for the properties of both the NAPL and the aquifer material. Sand column experiments with diesel present at residual saturation were used in Martel and Gélinas (1996) to optimize the micellar solution selected using the phase diagrams. The optimized washing solution was circulated through the column using a peristaltic pump. Complete diesel dissolution in a Winsor type I system was achieved in sand column with a concentrated alcohol–surfactant solution. A good agreement is observed between diesel dissolved in sand column experiments with downward injection and diesel dissolution predicted from phase diagrams. However, sand column experiments showed significant decrease in efficiency with a drop in temperature from 25 to 8°C (phase diagrams predict no change) and they are essential to check the efficiencies of washing solution selected by phase diagrams. Optimal injection concentration of solutions decreases as the alcohol ratio (n-AmOH:n-BuOH) increases and as the temperature decreases. In columns, optimal concentration of surfactant solutions has a maximum efficiency of 0.4 (dissolved diesel/active matter). This limited performance was explained by the authors due to the use of water to push a slug of washing solution that creates dilution in the column, the limited interfacial area between diesel and washing solution in the porous media, and the surface effects on the solid phase. Diesel extraction in sand columns was strongly influenced by the flow direction, the volume of the solution, and, to a lesser extent, the water salinity. Small modifications of flow velocity do not have a significant effect on diesel dissolution. The efficiency of diesel dissolution by surfactant solutions decreases as washing progresses because interfacial area between diesel and micellar solution decreases.

NAPL recovery is a function of displacement type (miscible or immiscible), but it also depends on a porous medium. Column experiments are also an effective screening tool for investigating the processes governing the NAPL recovery such as the identification of dispersion. It is also effective for testing the validity of the local equilibrium assumption for describing the solubilization of residual NAP and the mechanisms of interaction between the soil, micellar solution and contaminants (sorption, complexation, precipitation), essential parameters for the design of the remediation process.

Dwarakanath et al. (1999) analyzed the effect of numerous factors (precipitation and adsorption of surfactant, solubilization of contaminant, and soil hydraulic conductivity) on the removal of hydrocarbons by anionic surfactants during treatment of a soil column. In order to improve the removal of hydrocarbons from soil, different surfactant/additive systems have been proposed. Lee et al. (2002) studied the influence of soil texture and quality on the removal of organic contaminants, obtaining a removal ratio of 73 and 95% in batch and column experiments, respectively. The synergistic effect between anionic and non-ionic surfactants was investigated. In Khalladia et al. (2009), the removal of diesel fuel from soil in a glass column by a continuous washing process by an ionic surfactant sodium dodecyl sulfate (SDS) was investigated. The effects of the water flow rate and the contamination age have also been studied. The performance of water in the removal of diesel fuel was found to be nonnegligible. The effect of SDS is significant beyond a concentration of 8 mM, with removal equal to 97% of the diesel fuel. The presence of minerals in the soil composition improved the anionic surfactant activity on the surface. This result showed the effect of soil composition on the surfactant efficiency during the remediation of such a soil contaminated with diesel.

5 Application of micellar solution in situ to a diesel-contaminated site

During the last decade, numerous experiments involving micellar solutions for NAPLs remediation have been conducted; however, most of these experiments have been performed either in the laboratory scale with homogeneous mediums or under simple conditions (Powers et al., 1994; Pennell et al., 1996; Rao et al., 1997; Lee, 1998; Lee and Fountain, 1999; Lowe et al., 1999; Cowell et al., 2000; Dwarakanath and Pope, 2000; Fiorenza et al., 2000; Zhou and Rhue, 2000; Cheng et al., 2001; Chu and Kwan, 2003; Park and Bielefeldt, 2003).

One of the difficulties encountered in performing a real-site remediation and achieving a successful cleanup of a contaminated area is related to the assessment of the geological heterogeneities, which is an essential tool for determining NAPL distribution in a contaminated site and the most important factor controlling NAPL distribution in the subsurface (Kueper et al., 1993; Abdul and Ang, 1994; Fountain et al., 1996; Fetter, 1998; Pankow and Cherry, 1996). Another difficulty faced in developing a successful remediation scheme is the posttreatment system for the effluent solution. After surfactant flushing, the effluent solution has high concentrations of organic contaminants, and so it should be treated to prevent further contamination, before being disposed of. Separation of the contaminants from the surfactant solution has been technically difficult and extremely expensive; these limitations have hindered the application of surfactant flushing to NAPL remediation at genuinely contaminated sites.

Lee et al. (2005) investigated the capability of sorbitan monooleate (POE 20) to remove diesel fuel from a contaminated site, which had been used as a military vehicle repair area in Korea for several decades. A pilot-scale site was selected within the contaminated area for in-situ flushing; the selected site was composed of heterogeneous sandy and silt-sandy soils, with an average hydraulic conductivity (K) of 2.0×10^{-4} cm/s. In addition, an improved posttreatment procedure for the effluent surfactant solution was designed as a step in the surfactant-flushing process. The results of this study will provide meaningful information to those applying the in-situ surfactant flushing process to fuel-contaminated sites. About 88% of the total petroleum hydrocarbon was removed after three pore volumes of 2% sorbitan monooleate solution flushing; this amount was more than 75 times the amount removed when flushing with water alone.

6 Biodegradation of diesel-contaminated soil in soil column with synthetic surfactants and biosurfactants

The addition of surfactant in combination with other techniques has also been dsuggested as a possible way to increase the bulk concentration of the NAPLs and thus their bioavailability and biodegradation (Volkering et al., 1995).

Bioremediation of hydrocarbon-contaminated soils has been recognized as an efficient, economic, versatile, and environmentally sound treatment. However, contamination problems increase when ages of facilities, such as oil storage tanks and pipelines, increase over time. Also, when treating the oil contamination problems with bioremediation approaches, inadequate bioavailability of the hydrocarbons to microorganisms due to low solubility and volatility has been addressed as a limiting step in biodegradation (Bernardez, 2009). Therefore, introducing surfactants to the contaminated soil in order to solubilize hydrocarbons, to increase their water solubility, and to reduce surface tension has been found to be a useful approach.

Several reports in the literature on the effects of micellar solution on the rates of biodegradation of organic compounds present as solids or in sorbed phases have been presented. However, the effects of surfactants on biodegradation of hydrophobic organic compounds have been observed to be variable. Some studies have shown no change in biodegradation rates (Foght et al., 1989), others have shown decreases in biodegradation rates (Laha and Luthy, 1991), and yet other studies have shown an increase in biodegradation rates with the addition of non-ionic surfactant (Aronstein et al., 1991). Some of these studies have attributed the enhancement of the rates of biodegradation to the ability of the surfactant to increase the overall rate of dissolution of hydrocarbons to the bulk phase but note that the compounds present on the micelles were not available. Others concluded that the hydrocarbons in the micelle can be transported to the aqueous phase as soon as they are degraded (Liu et al., 1995) or be directly available for biodegradation (Guha and Jaffé, 1998) by interactions between bacteria and non-ionic surfactant.

More recently, the biosurfactants have been used to treat hydrocarbon contamination problems. Biosurfactants are a group of surface-active molecules produced mainly by hydrocarbon degrading microorganisms. Almost all surfactants currently in use are chemically derived from petroleum. Some of the advantages of biosurfactants over synthetic ones include lower toxicity, biodegradability, selectivity, specific activity at extreme temperatures, pH and salinity, the possibility of their production through fermentation, their potential applications in environmental protection and management, crude oil recovery, as antimicrobial agents in health care and food processing industries (Desai and Banat, 1997). These advantages have clearly put the biosurfactants ahead of their synthetic counterpart and have, as a result, elevated their commercial potential. Furthermore, the advantages have prompted a very high interest in the isolation of new microorganisms able to produce biosurfactants. Some of the organisms reported previously as biosurfactant producers include *Pseudomonas aeruginosa*, *Bacillus licheniformis*, *Arthrobacter spp.*, *Staphylococcus spp.*, and *Flavobacterium spp.* Strain MTN11.

Whole cells as well as extracellularly produced biosurfactants are the most common and most promising (Ilori and Amund, 2001). Most extracellularly secreted biosurfactants are growth associated and have been reported to be of importance in cellular architecture for a number of bacteria. These include the role of surfactin in *Bacillus subtilis* fruiting body formation and the role of *rhamnolipid* in *Pseudomonas aeruginosa* biofilm formation (Davey et al., 2003).

Some studies have addressed successful applications of biosurfactants to solubilized diesel from contaminated soil, and enhanced degradation has been demonstrated (Bai et al., 1997; Jain et al., 1991; Moran et al., 2000). Grace Liu et al. (2008) simulated in-situ bioremediation for diesel-contaminated sand by a column operation using biosurfactants. Several bioremediation approaches were adopted to examine the feasibility. Four lab-scale soil columns were operated specific to the following approaches: nutrient enhancement, bioaugmentation, biostimulation, and sterilized soil. The diesel-degradation efficiency was 67, 80, 45, and 24%, respectively.

7 Mathematical models

7.1 Numerical model description

To design a laboratory or field experiment of NAPL recovery, the dominant recovery process, which can be either dissolution or mobilization, must be identified. Knowing which process dominates is crucial when the NAPL is denser than water because, in that case, mobilization can lead to further downward migration of the NAPL, which can impede its recovery. Measuring the phase composition of the effluent during a column experiment is not sufficient to quantify dissolution and mobilization, and other techniques are required. One technique relies on the construction of phase diagrams, which have been shown by Falta (1998) and St-Pierre et al. (2004) to help predict the dominant recovery mechanism when co-solvents are used. Numerical models provide another method to quantify dissolution and mobilization by simulating either laboratory or field injection of micellar solutions. In addition, mathematical models for the description of surfactant-enhanced NAPL remediation allow a better understanding of the observed phenomena and the possibility of extending established knowledge to greater scales and more complex situations. Models enable a better design and performance of field remediation and a reduction of possible risks and are increasingly used as a basis for managing remediation problems.

There are a few examples of numerical simulations of NAPL recovery with micellar and/or co-solvent solutions, and most have limited applicability (one- or two-dimensional modeling, single species, and equilibrium mass transfer) and do not appropriately represent phase behavior. Abriola et al. (1993) developed a one-dimensional single-phase model to simulate, at the laboratory scale, the dissolution of dodecane in a micellar solution. They reproduced experimental results by incorporating rate-limited mass transfer of dodecane to the aqueous phase but neglected mobilization. Mason and Kueper (1996) also developed a one-dimensional model to simulate surfactant-enhanced dissolution of NAPLs, with rate-limited mass transfer between organic and aqueous phases. They calibrated their model with results from a laboratory experiment where tetrachloroethylene (PCE) was dissolved by injecting surfactants into a column. Reitsma and Kueper (1998 a,b) developed a one-dimensional compositional model to simulate alcohol flooding where phase diagrams are represented with Hand-plot equations, and they compared their results with measured laboratory data. Roeder and Falta (2001) simulated the recovery of an NAPL by alcohol in laboratory column experiments with an adapted version of a compositional multiphase flow simulator. Compositional models that rigorously simulate advection, dispersion, mass transfer of components, and phase behavior constitute powerful tools to understand the recovery of NAPLs.

The application of compositional simulators to reproduce either laboratory or field experiments of NAPL recovery, such as diesel, with micellar solutions is crucial to improve our understanding of controlling mechanisms for NAPL recovery and to test the reliability of the simulators.

The work presented by Bernardez et al. (2009) demonstrates the application of UTCHEM (Delshad et al., 1996; Pope et al., 1999) to simulate the recovery of diesel present at residual saturation in a column filled with sand by a micellar solution and the potential dissolution and mobilization of the diesel in a column filled with sand. This study demonstrates the potential of the compositional simulator, as opposed to models that simplify either the phase composition or the phase behavior, to simulate experimental results and to investigate the influence of various parameters on diesel recovery.

UTCHEM is a three-dimensional finite difference simulator, which was initially applied to enhanced oil recovery but has been since been adapted to simulate aquifer remediation (Brown et al., 1994; Delshad et al., 1996). The model solves pressure and energy balance equations, as well as mass conservation equations for up to 19 components (including water, oil, air, surfactant, polymer, alcohol), which can be present in four pseudo-phases (aqueous, oleic, air, and microemulsion). A brief description of the model is given here, and the reader is referred to Pope et al. (1999) for the detailed description of the mathematical and numerical formulation used in UTCHEM. The mass balance equation for each component k is given by:

$$\frac{\partial}{\partial t}(n\tilde{C}_k\rho_k) + \vec{\nabla} \cdot \left(\sum_{l=1}^{n_p} \left[\rho_k \left(C_{kl}\vec{u_l} - \vec{\vec{D}}_{kl} \right) \right] \right) = Q_k \tag{18.1}$$

where n is the porosity, $\tilde{C}_k$ is the total volume of component k per unit pore volume including the adsorbed phases (L^3/L^3PV), ρ_k is the density of the pure component at a reference pressure (M/L^3), and C_{kl} is the concentration of component k in phase l (L^3/L^3). The phase velocity, u_l (LT^{-1}), is given by Darcy's law, and the dispersive flux D_{kl} (L^2T^{-1}) is computed assuming that mechanical dispersion is Fickian. Sources and sinks are given by Q_k (L^3/T).

The pressure equation for the aqueous phase is obtained by using Darcy's law to represent phase velocities in Eq. 18.1, summing over all n_{cv} components and using the condition that $\sum_{k=1}^{n_{cv}} C_{kl} = 1$. The resulting equation has the following form (Delshad et al., 1996):

$$nC_t\frac{\partial P_1}{\partial t} + \vec{\nabla} \cdot \vec{\vec{k}} \cdot \lambda_{rTc} \vec{\nabla} P_1 = -\vec{\nabla} \cdot \sum_{l=1}^{n_p} \vec{\vec{k}} \cdot \lambda_{rlc} \cdot \vec{\nabla} z + \vec{\nabla} \cdot \sum_{l=1}^{n_p} \vec{\vec{k}} \cdot \lambda_{rlc} \cdot \vec{\nabla} P_{cl1} + \sum_{k=1}^{n_{cv}} Q_k \tag{18.2}$$

where C_t is the total compressibility $(ML^{-1}T^{-2})^{-1}$, P is the phase pressure ($ML^{-1}T^{-2}$), k is the permeability tensor (L^2), and λ_{rTc} and λ_{rlc} are total and relative mobilities, respectively $(ML^{-1}T^{-1})^{-1}$. The capillary pressure between the l and aqueous phases is given by P_{cll} ($ML^{-1}T^{-2}$), and z represents depth below a reference elevation (L).

UTCHEM (University of Texas Chemical Simulator) simulates the partition of alcohol between the microemulsion and the oleic pseudo-phases and accounts for changes in density, viscosity, and interfacial tension of the pseudo-phases resulting from alcohol partitioning. Empirical relationships are used to calculate saturations, phase densities and viscosities, relative permeabilities, and adsorption. Those relationships are then used in the governing equations. For each time step, the pressure

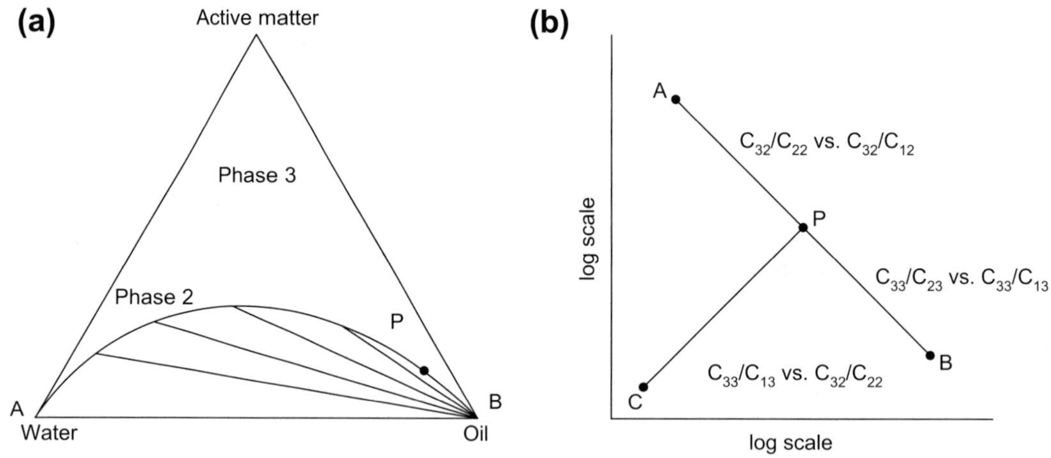

FIGURE 18.7

Representation of phase behavior with (a) pseudo-ternary diagram, and (b) Hand diagrams. P represents the plait point in both figures (Bernardez et al., 2009).

Eq. (2) is first solved implicitly and the mass conservation Eq. (1) for each component is then solved explicitly, following a procedure similar to the implicit pressure-explicit saturation method.

During NAPL recovery with micellar solutions, several phases flow simultaneously in the porous medium, and several components migrate in those phases. Take into account all components and phases in a numerical model is a complex problem, thus researchers proposed the use of pseudo-phases and pseudo-components (Vinatieri and Fleming, 1979; Prouvost et al., 1985) based on the fact that some species such as oil and amphiphilic species behave collectively and they can be considered a single pseudo-component in the phase behavior description.

These pseudo-components form three liquid mobile phases modeled with UTCHEM. The phases are aqueous (W′), oleic (O′) and microemulsion (M′). The aqueous phase consists of water with or without alcohol (s), the oleic phase consists of oil with or without alcohol(s) and the microemulsion phase consists of surfactant with or without alcohol(s).

The end components of the pseudo-ternary diagram thus correspond to active matter, water, and oil. The pseudo-ternary diagram is characterized by the miscibility curve also called binodal curve that separates the miscibility zone from the two-phase zone and by the tie lines within the two-phase zone that indicate the concentrations of the oleic and aqueous phases resulting from the partition of components between these two immiscible phases.

In the model, the phase representation obtained from the pseudo-ternary diagram is simplified by using Hand plots (Lake, 1989) where the concentration ratios are plotted on a log–log scale (Figure 18.7a, 18.7b).

In UTCHEM, the relationship that describes the binodal curve for the two-phase system is given by:

$$C_{31}/C_{21} = A(C_{31}/C_{11})^{B} \tag{18.3}$$

where A and B are empirical parameters and C_{kl} are the concentration of species k in phase l (L^3/L^3), and the equation for the tie lines is given by:

$$C_{31}/C_{21} = E(C_{33}/C_{13})^F \tag{18.4}$$

where F is equal to F=−1/B and E is equal to $E=C_{1P}/C_{2P}$.

The equilibrium phase behavior is then modeled using the pseudo-phase theory and Hand equations and the various types of phases are known to form the components.

Once the phase behavior is determined in UTCHEM, empirical relationships are used to calculate saturations, phase densities and viscosities, relative permeabilities, and adsorption. Those relationships are then used in the governing equations. For each time step, the pressure Eq. 18.2 is first solved implicitly, and the mass conservation Eq. (1) for each component is then solved explicitly, following a procedure similar to the implicit pressure-explicit saturation (IMPES) method.

Bernardez et al. (2009) reports the results of numerical simulations of laboratory experiments in which an environmentally significant compound, diesel, present at 18% residual saturation in a sand column, was recovered by injecting a micellar solution containing the surfactant Hostapur SAS-60 (SAS), and two alcohols, n-butanol (n-BuOH) and n-pentanol (n-PeOH). The micellar solution was developed and optimized for diesel recovery using phase diagrams and soil column experiments.

In this chapter numerical simulations of column experiments by Bernardez et al., 2009 using the compositional simulator UTCHEM is presented and discussed. The simulated results agree with the experimental results and show that the entire residual diesel in the sand column was recovered after the downward injection of 5 pore volumes of the micellar solution. Recovery of diesel occurs by enhanced solubility in the microemulsion phase and by mobilization. An additional series of simulations was performed by the authors to investigate the effects of phase transfer, alcohol partitioning, and component segregation on diesel recovery. These simulations suggest that diesel can be represented in the model by a single component, but that the pseudo-component approach for active matter and the assumption of local phase equilibrium leads to an underestimation of diesel mobilization.

7.2 Numerical simulations of diesel recovery

In Bernardez et al., 2009 diesel was assumed to be a pseudo-component and behaves as a pure component. The system considered in has 5 pseudo-components: water, (W), diesel as the "oil" (O), the surfactant SAS (S), and two alcohols, n-pentanol (A) and n-butanol (B). These pseudo-components form three liquid mobile phases modeled with UTCHEM. The phases, in this Winsor type I system are aqueous (W′), oleic (O′) and microemulsion (M′). The aqueous phase consists of water and alcohol (s), the oleic phase consists of oil and alcohol(s) and the microemulsion phase consists of surfactant and alcohol(s). For the diesel recovery experiment, the surfactant and the two alcohols were lumped together and labelled "active matter" in a pseudo-ternary diagram. Figure 18.8 shows the experimental phase diagram of the optimal surfactant solution for dissolving diesel. The end components of the pseudo-ternary diagram thus correspond to active matter, water, and diesel. The pseudo-ternary diagram was characterized by the miscibility curve also called binodal curve that separates the miscibility zone from the two-phase zone and by the tie lines within the two-phase zone that indicate the concentrations of the oleic and aqueous phases resulting from the partition of components between these two immiscible phases.

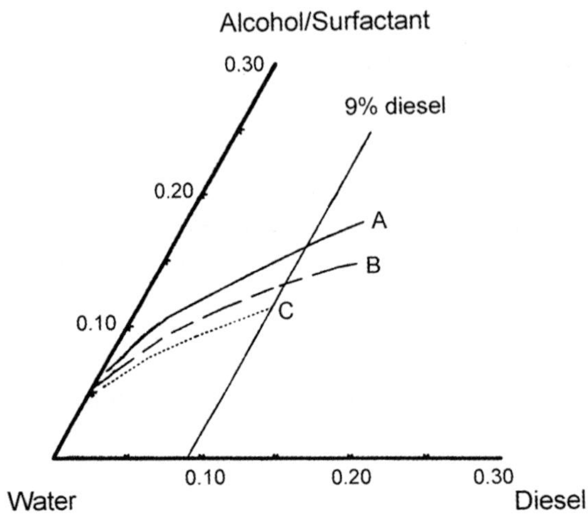

FIGURE 18.8

Experimental phase diagram of the optimal surfactant solutions for dissolving diesel at 25°C. (A) n-PeOH:n-BuOH(0,4:1,0)/SAS = 1; (B) n-PeOH:n-BuOH(1,5:1,0)/SAS = 1: CC) n-PeOH:n-BuOH(2,0:1,0)/SAS = 1 (Bernardez et al., 2009).

A base case simulation was first performed to ensure that the model could reproduce the experimental results (Figure 18.9). *Downward fluid injection and instantaneous mass transfer is assumed between the phases.* Instantaneous mass transfer is also assumed between the phases. In addition to the base case for downward injection, a second set of experimental results was available for upward

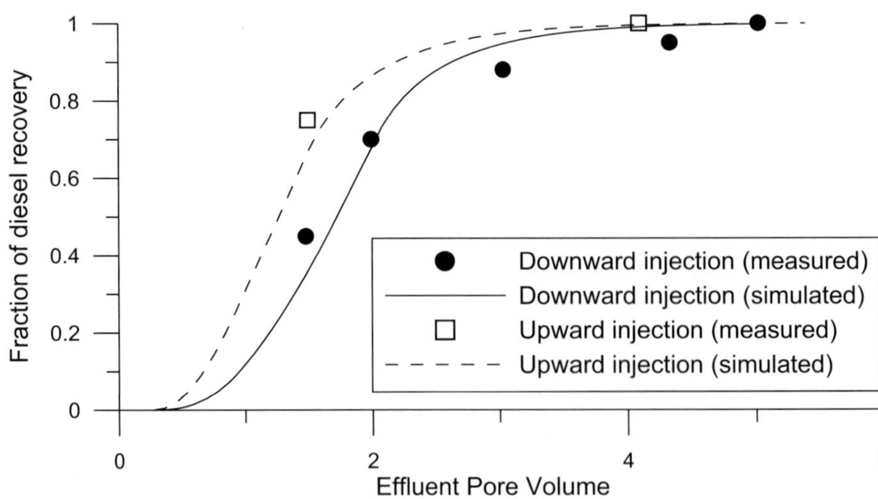

FIGURE 18.9

Simulated and measured cumulative diesel recovery fraction as a function of the effluent pore volume for downward and upward injection experiments (Bernardez et al., 2009).

injection of the micellar solution. This second data set compares the importance of diesel mobilization relative to dissolution by reversing the flow direction and allowing diesel buoyancy to play a significant role. Previous laboratory data had indicated higher displacement in the upward direction (Martel and Gélinas, 1996). This second case, along with the base case, served to test the model against experimental data.

A comparison of experimental and simulated diesel recovery for both downward and upward injection (Figure 18.9) indicates that the model adequately reproduces the experimental diesel recovery and thus provides a general validation of the simulation parameters, since diesel recovery is obtained for the same number of pore volumes compared to the experiment. The relative effects of dissolution and mobilization are also adequately represented for the different flow directions. However, for a larger number of pore volumes, there is always more pronounced tailing in the experimental diesel recovery curves compared to the simulated ones. *This phenomena was attributed by the authors to the small-scale heterogeneities, such as laminations observed in the sand columns and resulting from compaction during the column preparation.* These heterogeneities were not included in the numerical model because of the complexity in characterizing their geometry and properties. The extended tailing compared to the base case simulation and upflow experiment may also be due to displacement instability. Part of the mobilized diesel in the downflow experiment may be passed by the denser flushing solution, creating some instability.

The effects of nonequilibrium mass transfer and alcohol partition are shown in Figure 18.10 by comparing to the base case (Figure 18.9). When these two effects are included, initial recovery of diesel is slightly faster, which is closer to experimental results. In the case of alcohol partitioning, the maximum concentration of diesel is increased with respect to the base case because of the effect of alcohol on the height of the binodal curve. A *discret* chromatographic effect on active matter is observed because the alcohols no longer remain with the active matter but partitions into the diesel.

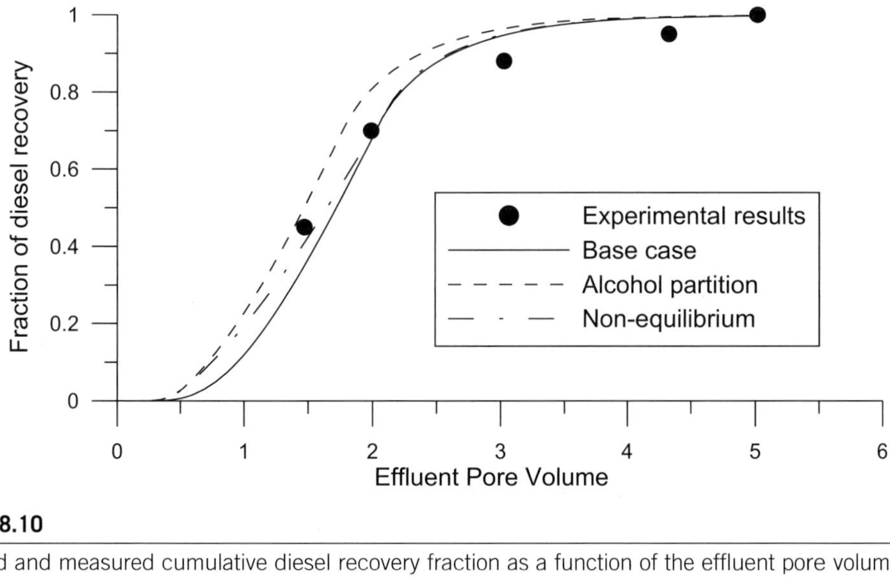

FIGURE 18.10

Simulated and measured cumulative diesel recovery fraction as a function of the effluent pore volume. Simulation of the effects of alcohol partition and nonequilibrium mass transfer (Bernardez et al., 2009).

The effects of the various processes are better seen in the composition of the fluid effluent in Figure 18.11, which shows the relative concentrations (C/Co) of the components as a function of the number of effluent pore volumes. For the base case, no component partitioning is considered, and all micellar solution components emerge simultaneously (Figure 18.11a). For the upward fluid injection case, diesel displacement leads to a faster release of diesel in the column effluent (Figure 18.11b). The

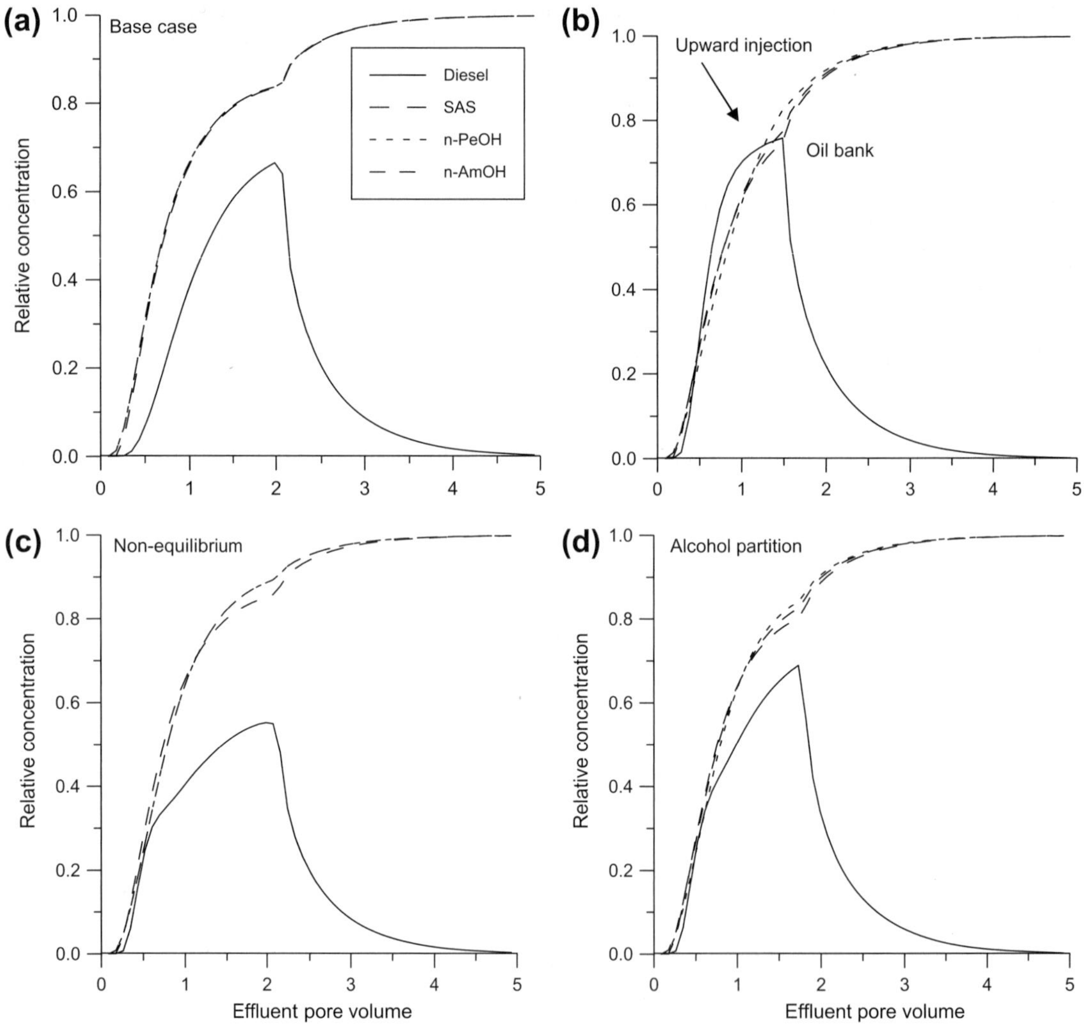

FIGURE 18.11

Simulated relative concentrations of diesel, surfactant (SAS), and alcohols (n-BuOH and n-PeOH) as a function of effluent pore volume produced: (a) base case, (b) upward injection, (c) nonequilibrium mass transfer, and (d) alcohol partitioning (Bernardez et al., 2009).

nonequilibrium mass transfer simulation also releases diesel earlier than for the base case, but the recovery rate is slightly lower afterward, even though the final recovery is similar (Figure 18.11c). This lower rate explains the diesel recovery behavior shown in Figure 18.10. A slight segregation of alcohols and surfactant is also observed for this case. For the case with alcohol partitioning, the results shown in Figure 18.11d are closer to the base case with the exception of earlier diesel recovery. Overall, the segregation of active matter components is not important and does not lead to significant chromatographic effects.

The chromatographic effects and conventional pump-and-treat simulations are also simulated. Diesel is represented in the model with multiple components, as opposed to a single component to test chromatographic effects. Thirteen components of diesel, including several aromatics and aliphatics, were considered in the simulations with equilibrium mass transfer assumed for each component. The mole fraction composition of the fuel assumed is the same as that used by Lee et al. (1992). For these laboratory experiments, available data did not suggest chromatographic effects (Martel and Gélinas, 1996). The case with multiple diesel components also did not produce chromatographic effects, which is consistent with analytical chromatograms from earlier experiments (Martel and Gélinas, 1996), and suggests that assuming pseudo-component behavior for diesel is adequate. The last case simulates flushing the residual diesel in the column only with water, to determine the gain in diesel recovery efficiency obtained by flushing with the micellar solutions. For conventional pump-and-treat simulations after injection of five pore volumes of water; only 0.3% of diesel is recovered, compared to 100% with the micellar solutions. The simulation suggests that the injection of 1500 pore volumes would be required to completely remove diesel by water flushing, thus highlighting the benefit of micellar solutions on recovery.

The comparison of experimental and simulated diesel recovery shows that recovery processes are well represented by the numerical model. The simulations suggest that recovery occurs by dissolution and mobilization, with dissolution being the prevalent mechanism in the base case. The surfactant and alcohols precede the chemical front as shown in Figure 18.11a.

Although simulated diesel recovery by Bernardez et al., 2009 generally matches the available experimental data, some technical challenges still exist in the laboratory when performing and simulating such column experiments. Micellar solutions and diesel fuel are complex mixtures, and the exact determination of their composition is difficult. Therefore, there can still be some uncertainty associated with experimental data used to test numerical models. Furthermore, some of the transfer coefficients needed for the simulations had to be estimated due to the absence of laboratory data, but the simulation results were not sensitive to variations of these coefficients within reasonable ranges of values (Bernardez, 1999).

The simulation of NAPL recovery with micellar solutions has numerous potential practical applications. As showed by Bernardez et al., 2009, simulations can be applied to understand processes observed experimentally and to investigate the effect of various parameters on NAPL recovery. Numerical models can also be applied in the design and interpretation of both field experiments and full site recovery systems. However, such simulations would require a very detailed site characterization, including the physical and chemical properties of the NAPLs, and would likely be time consuming and costly. Whereas diesel has mostly been used in experiments to study recovery mechanisms, other NAPLs are more likely to be recovered with micellar solutions, for example DNAPLs such as crude oil and PCBs.

8 Conclusions

Widely used in the world, diesel fuel represents a permanent source of soil and water pollution. Resulting from the crude distillation, diesel fuel is a mixture of more than 2000 compounds composed of 60% of aliphatic and 40% of aromatics hydrocarbons. The contamination at many sites by diesel contain a number of constituents, several of which pose a threat to public health. In this chapter, the removal of diesel fuel by a washing process using a micellar solution is discussed. Micellar solubilization is an important step for removal of organic contaminants from the soil matrix, especially for low aqueous solubility organic contaminants such as diesel. In addition to fundamental laboratory column studies on removal of diesel from contaminated porous media, phase diagrams and mathematical modes are essential tools for improving our understanding during diesel recovery in porous media and are also reviewed. The use of micellar solutions with or without combination with other techniques such as in-situ bioremediation is a potential cost-effective remediation measure to treat these sites. The addition of micellar solutions may increase the mass transfer of the contaminants from the diesel or surfaces and increase bulk concentration in the aqueous phase and thus the bioavailability.

GENERAL SYMBOLS

n porosity
$\tilde{C}_k$ total volume of component k per unit pore volume, including the adsorbed phases (L^3/L^3PV)
C_{kl} concentration of component k in phase l (L^3/L^3)
u_l phase velocity (LT^{-1})
D_{kl} dispersive flux (L^2T^{-1})
Q_k sources and sinks (L^3/T)
n_{cv} components
C_t total compressibility $(ML^{-1}T^{-2})^{-1}$
P phase pressure ($ML^{-1}T^{-2}$)
K permeability tensor (L^2)
P_{cl1} capillary pressure between the l and aqueous phases ($ML^{-1}T^{-2}$)
z depth below a reference elevation (L)
A and B empirical parameters
C_{kl} concentration of species k in phase l (L^3/L^3)

Greek symbols

λ_{rTc} **and** λ_{rlc} total and relative mobilities, respectively $(ML^{-1}T^{-1})^{-1}$
ρ_k the density of the pure component at a reference pressure (M/L^3)

Indices

k grid index in z direction
w aqueous phase
o oil phase

k components index: 1:water, 2:oil, 3 surfactant, 7:n-pentanol, 8:n-butanol alcohol
l phase index: 1:aqueous, 2: oil, 3:microemulsion

Glossary

Amphipathic in the case of surfactants having both lyophilic and lyophobic groups (properties) in the same molecule. Also referred to as being amphiphilic.

Amphoteric surfactant a surfactant molecule for which the ionic character of the polar group depends on solution pH

Anionic surfactant a surfactant molecule that can dissociate to yield a surfactant ion whose polar group is negatively charged

Air sparging in air sparging, air is injected into the ground below a contaminated area, forming bubbles that rise and carry trapped and dissolved contaminants to the surface where they are captured by a soil vapor extraction system.

Aromatics organic compounds that contain 6-carbon ring structures, such as creosote, toluene, and phenol, which often are found at dry cleaning and electronic assembly sites

Bioremediation treatment processes that use microorganisms (usually naturally occurring), such as bacteria, yeast, or fungi to break down hazardous substances into less toxic or nontoxic substances. Bioremediation can be used to clean up contaminated soil and water. In-situ bioremediation treats the contaminated soil or groundwater in the location in which it is found. For ex-situ bioremediation processes, contaminated soils must be excavated or groundwater pumped before they can be treated.

Bioventing an in-situ cleanup technology that combines soil vapor extraction methods with bioremediation. It uses vapor extraction wells that induce air flow in the subsurface through air injection or the use of a vacuum. Bioventing can be effective in cleaning up releases of petroleum products, such as gasoline, jet fuels, kerosene, and diesel fuel. See also Bioremediation.

Cationic surfactant a surfactant molecule that can dissociate to yield a surfactant ion whose polar group is positively charged

Contaminant any physical, chemical, biological, or radiological substance or matter present in any media at concentrations that may result in adverse effects on air, water, or soil

Electric double layer an idealized description of the distribution of free charges in the neighborhood of an interface

Emulsion a dispersion of droplets of one liquid in another, immiscible liquid, in which the droplets are of colloidal or near-colloidal sizes

Gibbs energy of interaction When two dispersed-phase species approach, they experience repulsive and attractive forces such as electrostatic repulsion and van der Waals attraction. The Gibbs energy of interaction may be thought of as the difference between Gibbs energies of the system at a specified separation distance and at infinite separation.

Hydrocarbon an organic compound containing only hydrogen and carbon, often occurring in petroleum, natural gas, and coal

Hydrophilic a qualitative term referring to the water-preferring nature of a species

Hydrophobic a qualitative term referring to the water-avoiding nature of a species. A functional group of a molecule that is not very water-soluble is referred to as a hydrophobe.

In situ term meaning "in its original place" or "on-site"; indicates unexcavated and unmoved. In -situ soil flushing and natural attenuation are examples of in-situ treatment methods by which contaminated sites are treated without digging up or removing the contaminants.

Interfacial tension See Surface tension.

Lipophilic the (usually fatty) organic-liquid-preferring nature of a species. The part of a molecule that is organic-liquid-preferring in nature is referred to as a lipophile.

Microemulsion a special kind of stabilized emulsion in which the dispersed droplets are extremely small (< 100 nm), and which is thermodynamically stable

Micelle an aggregate of surfactant molecules or ions in solution. Such aggregates form spontaneously at or above a surfactant concentration called the critical micelle concentration.

Nonionic surfactant a surfactant molecule whose polar group is not electrically charged

Nonaqueous phase liquid (NAPL) a chemical that does not dissolve well in water, typically an organic solvent. Dense nonaqueous phase liquids (DNAPLs) have a density greater than water. Light nonaqueous phase liquids (LNAPLs) have a density less than water. NAPLs, particularly DNAPLs, can contaminate groundwater, creating large underground plumes that persist for very long periods.

Polyaromatic hydrocarbon (PAH) a chemical compound that contains more than one fused benzene ring. They are commonly found in petroleum fuels, coal products, and tar. Pump-and–treat pump and treat is a general term used to describe cleanup methods that involve the pumping of groundwater to the surface for treatment. It is one of the most common methods of treating polluted aquifers and groundwater.

Surface tension The contracting force per unit length around the perimeter of a surface is usually referred to as surface tension if the surface separates gas from the liquid or solid phase, and interfacial tension if the surface separates two nongaseous phases. Surface tension can also be expressed in units of energy per unit surface area. For practical purposes, surface tension is frequently taken to reflect the change in surface free energy per unit increase in surface area.

Surface-active agent a compound, such as a detergent, wetting agent, or emulsifier, that helps to reduce surface tension among liquids and solids. See also Surfactant.

Surfactant a substance that lowers the surface or interfacial tension of the medium in which it is dissolved

Surfactant flushing an innovative treatment technology used to treat contaminated groundwater. Surfactant flushing of NAPLs increases the solubility and mobility of the contaminants in water, so that the NAPLs can be biodegraded more easily in an aquifer or recovered for treatment aboveground.

Tenside See Surfactant.

Zwitterionic surfactant a surfactant molecule that contains both negatively and positively charged groups.

References

Abdul, A.S., Gibson, T.L., Rai, D.N., 1990. Selection of surfactants for the removal of petroleum products from shallow aquifers. Ground Water 286, 920–926.

Abdul, A.S., Gibson, T.L., 1991. Laboratory studies of surfactant-enhanced washing of polychlorinated biphenyl from sandy material. Environ. Sci. Technol. 25, 665–671.

Abdul, A.S., Gibson, T.L., Ang, C.C., Smith, J.C., Sobczynski, R.E., 1992. In situ surfactant washing of polychlorinated biphenyls and oils from a contaminated site. Ground Water 302, 219–232.

Abriola, L.M., Dekker, T.J., Pennell, K.D., 1993. Surfactant enhanced solubilization of entrapped dodecane in soil columns: 2. Mathematical modeling. Environ. Sci. Technol. 27 (12), 2341–2351.

Aronstein, B.N., Calvillo, Y.M., Alexander, M., 1991. Effect of surfactants at low concentrations on the desorption and biodegradation of sorbed aromatic compounds in soil. Environ. Sci. Technol. 25, 1728–1731.

Bai, G., Brusseau, M.L., Miller, R.M., 1997. Biosurfactant-enhanced removal of hydrocarbon from soil. J. Contam. Hydrol. 25, 157.

Bernardez, L.A., 2005. Surfactant effects on solubilization, dissolution and biodegradation of polycyclic aromatic hydrocarbons from non-aqueous phase liquids. Ph.D Thesis. McGill University, Montreal, Canada, p. 205.

Bernardez, L.A., 2008a. Dissolution of polycyclic aromatic hydrocarbons from a non-aqueous phase liquid into a surfactant solution using a rotating disk apparatus. Colloids and Surfaces A: Physicochem. Eng. Aspects 320, 175–182.

Bernardez, L.A., 2008b. Investigation on the locus of solubilization of polycyclic aromatic hydrocarbons in non-ionic surfactant micelles with 1H-NMR spectroscopy. Colloids and Surfaces A: Physicochem. Eng. Aspects 324, 71–78.

Bernardez, L.A., 2009. A rotating disk apparatus for assessing the biodegradation of polycyclic aromatic hydrocarbons transferring from a non-aqueous phase liquid to solutions of surfactant Brij 35. Bioprocess and Biosyst. Eng. 32, 415–424.

Bernardez, L.A., Therrien, R., Lefebvre, R., Martel, R., 2009. Simulating the injection of micellar solutions to recover diesel in a sand column. J. Contam. Hydrol. 103 (3-4), 99–108.

Brinker, C.J., Lu, Y., Sellinger, A., Fan, H., 1999. Evaporation-induced self-assembly: nanostructures made easy. Adv. Mater. 11 (7), 579–585.

Brown, C.L., Pope, G.A., Abriola, L.M., Sepehrnoori, K., 1994. Simulation of surfactant enhanced aquifer remediation. Water Resour. Res. 30 (11), 2959–2977.

Chu, W., Kwan, C.R., 2003. Remediation of contaminated soil by a solvent/surfactant system. Chemosphere 53, 9–15.

Cowell, M.A., Kibbey, T.C.G., Zimmerman, J.B., Hayes, K.F., 2000. Partitioning of ethoxylated nonionic surfactants in water/NAPL systems: effects of surfactant and NAPL properties. Environ. Sci. Technol. 34, 1583–1588.

Davey, M.E., Caiazza, N.C., O_Toole, G.A., 2003. Rhamnolipid surfactant production affects biofilm architecture in. Pseudomonas aeruginosa PA01 J. Bacteriol. 185, 1027–1036.

Desai, J.D., Banat, I.M., 1997. Microbial production of surfactants and their commercial potential. Microbiol. Mol. Biol. Rev. 61, 47–64.

Delshad, M., Pope, G.A., Sepehrnoori, K., 1996. A compositional simulator for modeling surfactant enhanced aquifer remediation: 1. Formulation. J. Contam. Hydrol. 23, 303–327.

Desnoyers, J.E., Quirion, F., Héttu, D., Perron, G., 1983. Tar sand extractions with microemulsions. I. The dissolution of light hydrocarbons by microemulsions using 2-butoxyethanol and diethylmethylamine as cosurfactants. Can. J. Chem. Eng. 61, 672–679.

Dwarakanath, V., Kostarelos, K., Pope, G.A., Shotts, D., Wade, W.H., 1999. Anionique surfactant remediation of contaminated soil columns by nonaqueous phase liquids. J. Contam. Hydrol. 38, 465–488.

Dwarakanath, V., Pope, G.A., 2000. Surfactant phase behaviour with field degreasing solvent. Environ. Sci. Technol. 34, 4842–4848.

Falta, R.W., 1998. Using phase diagrams to predict the performance of cosolvent floods for NAPL remediation. Ground water Monit Remediation 18, 94–102.

Fetter, C.W., 1998. Contaminant hydrogeology, second ed. Prentice-Hall, Englewood Cliffs, NJ. 1–119.

Fiorenza, S., Miller, C.A., Oubre, C.L., Ward, C.H., 2000. NAPL removal, surfactants, foams, and microemulsions. CRC Press Boca Raton, FL, 249–291.

Foght, J.M., Gutnick, D.L., Westlake, D.W.S., 1989. Effect of emulsion on biodegradation of crude oil by pure and mixed bacterial cultures. Appl. Environ. Microbiol. 55 (1), 36–42.

Fountain, J.C., Klimek, A., Beikirck, M.G., Middleton, T.M., 1991. The use of surfactant for in situ extraction of organic pollutants from a contaminated aquifer. J. Hazard. Mater. 28, 295–311.

Fountain, J.C., Starr, R.C., Middleton, T., Beikirch, M., Taylor, C., Hodge, D., 1996. A controlled field test of surfactant-enhanced aquifer remediation. J. Ground water 34 (5), 910–916.

Grace Liu, P.W.G., Whang, L.M., Yang, M.C., Cheng, S.S., 2008. Biodegradation of diesel-contaminated soil: A soil column study. J. Chin. Inst. Chem. Eng. 39, 419–428.

Guha, S., Jaffé, P.R., Peters, C.A., 1998. Solubilization of PAH mixtures by a nonionic surfactant. Environ. Sci. Technol. 32, 930–935.

Ilori, M.O., Amund, O.O., 2001. Production of a peptidoglycolipid bioemulsifier by *Pseudomonas aeruginosa* grown on hydrocarbon. Z. Naturforsch 56C, 547–552.

Jain, D.K., Thompson, D.L.C., Lee, H., Trevors, J.T., 1991. A drop-collapsing test for screening surfactant producing microorganisms. J. Microbiol. Methods 13, 271–279.

Khalladia, R., Benhabilesa, O., Bentahara, F., Moulai-Mostefab, N., 2009. Surfactant remediation of diesel fuel polluted soil. J. Hazard. Mater. 164, 1179–1184.

Laha, S., Luthy, R.G., 1991. Inhibition of phenanthrene mineralization by nonionic surfactants in soil-water systems. Environ. Sci. Technol. 25 (11), 1920–1930.

Lake, L.W., 1989. Enhanced oil recovery. Prentice Hall, Englewood Cliffs, NJ, 550 pp.

Lee, M., 1998. The remediation of tetrachloroethylene by surfactant flushing in both the saturated and the unsaturated zone. Environ. Sci. Technol. 19, 1073–1083.

Lee, D.H., Cody, R.D., Kim, D.J., Choi, S., 2002. Effect of soil texture on surfactant-based hydrophobic organic remediation of contaminated soil. Environ. Int. 27, 681–688.

Lee, M., Kang, H., Do, W., 2005. Application of nonionic surfactant-enhanced in situ flushing to a diesel contaminated site. Water Res. 39, 139–146.

Liu, Z., Jacobson, A.M., Luthy, R.G., 1995. Biodegradation of naphthalene in aqueous nonionic surfactant systems. Appl. Environ. Microbiol. 61 (1), 145–151.

Lowe, D.F., Oubre, C.L., Ward, C.H., 1999. Surfactant and cosolvent for NAPL remediation. *A technology practices manual*. Lewis Publishers, New York, 205–277.

Martel, R., Gélinas, P.J., Desnoyers, J.E., Masson, A., 1993. Phase diagrams to optimize surfactant solutions for oil and DNAPL recovery in aquifers. Ground Water 55, 789–800.

Martel, R., Gélinas, P.J., 1996. Surfactant solutions developed for NAPL recovery in contaminated aquifers. Ground Water 34 (1), 143–154.

Mason, A.R., Kueper, B.H., 1996. Numerical simulation of surfactant flooding to remove pooled DNAPL from porous media. Environ. Sci. Technol. 30 (3), 3205–3215.

Moran, A.C., Olivera, N., Commendatore, M., Esteves, J.L., Sineriz, F., 2000. Enhancement of hydrocarbon waste biodegradation by addition of a biosurfactant from. Bacillus subtilis O9 Biodegradation 11, 65.

Paria, S., 2008. Surfactant-enhanced remediation of organic contaminated soil and water. Adv. Colloid Interface Sci. 138, 24–58.

Park, S., Bielefeldt, A.R., 2003. Aqueous chemistry and interactive effects on non-ionic surfactant and pentachlorophenol sorption to soil. Water Res. 37, 4663–4672.

Pennell, K.D., Abriola, L.M., Dekker, T.J., Weber, W.J., 1992. Surfactant enhanced solubilization of entrapped dodecane in soil columns. Am. Chem. Soc. Div. Environ. Chem. Preprints 32, 945–947.

Pennel, K.D., Abriola, L.M., Dekker, T.J., Weber, W.J., 1993. Surfactant enhanced solubilization of entrapped dodecane in soil columns: 1. Experimental investigation. Environ. Sci. Technol. 27 (12), 2332–2340.

Pankow, J.F., Cherry, J.A., 1996. Dense chlorinated solvents and other DNAPLs in groundwater: History, behavior, and remediation. Waterloo Press, Ontario, 337–506.

Pope, G.A., Sepehrnoori, K., Sharma, M.M., McKinney, D.C., Speitel, G.E., Jackson, R.E., 1999. Three-dimensional NAPL fate and transport model. EPA/600/R-99/011, 332 p.

Prouvost, L., Pope, G.A., Rouse, B., 1985. Microemulsion phase behavior: a thermodynamic modeling of the phase partitioning of amphiphilic species. Society of Petroleum Eng. J. (SPE), 693–703.

Rao, P.S.C., Annable, M.D., Sillan, R.K., Dai, D., Hatfield, K., Graham, W.D., Wood, A.L., Enfield, C.G., 1997. Field-scale evaluation of in situ cosolvent flushing for enhanced aquifer remediation. Water Resour. Res. 33 (12), 2673–2686.

Reitsma, S., Kueper, B.H., 1998a. Non-equilibrium alcohol flooding model for immiscible phase remediation: 1. Equation development. Adv. Water Resour. 21 (8), 649–662.

Reitsma, S., Kueper, B.H., 1998b. Non-equilibrium alcohol flooding model for immiscible phase remediation: 2. Model development and application. Adv. Water Resour. 21 (8), 663–678.

Roeder, E., Falta, R.W., 2001. Modeling unstable alcohol flooding of DNAPL-contaminated column. Adv. Water Resour. 24, 803–819.

Rosen, M.J., 1989. Surfactants and interfacial phenomena, second ed. John Wiley & Sons, New York. 431 pp.

Schramm, L.L., 1992. Emulsions: Fundamentals and applications in the petroleum industry. Advances in Chemistry Series, 231. American Chemical Society, 1–47.

St-Pierre, C., Martel, R., Gabriel, U., Lefebvre, R., Robert, T., Hawari, J., 2004. TCE recovery mechanisms using micellar and alcohol solutions: phase diagrams and sand column experiments. J. Contam. Hydrol. 71, 155–192.

Vinatieri, J.E., Fleming, P.D., 1979. The use of pseudocomponents in the representation of phase behaviour of surfactant systems. Society of Petroleum Eng. J. (SPE) Oct, 289–300.

Volkering, F., Breure, A.M., van Andel, J.G., Rulkens, W.H., 1995. Influence of nonionic surfactants on bioavailability and biodegradation of polycyclic aromatic hydrocarbon. Appl. Environ. Microbiol. 61 (5), 1699–1705.

Winsor, P.A., 1954. Solvent properties of amphiphilic compounds. Butterworths, London.

Zana, R., 2002. Dimeric and oligomeric surfactants. Behavior at interfaces and in aqueous solution: a review. Adv. Colloid and Interface. Sci. 97 (1-3), 203–251.

Zhou, M., Rhue, R.D., 2000. Screening commercial surfactants suitable for remediating DNAPL source zones by solubilization. Environ. Sci. Technol. 34, 1985–1990.

CHAPTER

Recent Advances on the Use of Surfactant Systems as Inhibitors of Corrosion on Metallic Surfaces

19

Alcides de Oliveira Wanderley Neto*, Tereza Neuma de Castro Dantas*, Afonso Avelino Dantas Neto†, Alexandre Gurgel**

** Universidade Federal do Rio Grande do Norte, Centro de Ciências Exatas e da Terra, Instituto de Química, Natal – RN, Brazil, † Universidade Federal do Rio Grande do Norte, Centro de Tecnologia, Departamento de Engenharia Química, Natal – RN, Brazil, ** Universidade Federal de Viçosa, Centro de Ciências Exatas e Tecnológicas, Departamento de Química, Viçosa – MG, Brazil*

CHAPTER OUTLINE

1 A review on the use of surfactant systems in the inhibition of corrosion on metallic surfaces

Technological and industrial activities widely involve the design and operation of metallic parts, especially surfaces. Corrosion on such surfaces is a major cause of concern, and much effort has been made to provide the means of inhibiting the development of corrosion phenomena. The action of electrolytes, acids, and other chemicals, together with variations in environmental conditions such as temperature, moisture ,and pressure, may contribute to the establishment of a corrosive situation. In the petroleum industry, for example, oil, its derivatives and water–oil emulsions have to be transported in

The Role of Colloidal Systems in Environmental Protection. http://dx.doi.org/10.1016/B978-0-444-63283-8.00019-3

long pipelines and stored in metallic tanks, which can be prone to corrosion. The sources of natural corrosion can be ascribed to the presence of gases dissolved in the formation water, like oxygen, carbon dioxide, and hydrogen sulfide, generating high salinity and/or acidity levels, despite the very diverse chemical nature of the materials that are handled. Normally, emulsions are formed, which require slightly higher temperatures to improve their transportation. This may be a further cause for the development of corrosion processes in the operation of such systems. Severe corrosion may lead to pipeline rupture, with obviously negative environmental impacts. The injection of corrosion inhibitors in different positions along the pipelines helps minime the phenomenon of internal corrosion. Also, solubilization of crude oil fractions, which are responsible for the formation of deposits in petroleum-processing operations or petroleum emulsions, must always be considered. The use of surfactant systems, such as micelle solutions and microemulsions, in particular, is an interesting option to solve such problems (Dantas et al., 2001, 2010; Dantas Neto et al., 2008; Moura et al., 2009).

Organic molecules are notably acknowledged as efficient corrosion inhibitors on metallic surfaces. In particular, the use of surfactants has been extensively and continuously investigated, with excellent results, and has been the focus of our research group for a number of years. We have aimed to examine the ability of surfactants to inhibit carbon-steel corrosion in both acidic and saline environments, using both commercial surfactants and compounds synthesized from renewable vegetable oils (Moura et al., 2009; Roberto et al., 2013; Wanderley Neto et al., 2012). The main reason to carry out investigations with carbon–steel surfaces is the extensive use of this metal alloy in the industry, with varying composition. In addition to petroleum pipelines, it can also be used in refrigeration and vapor generation systems, condensation and drinking water pipelines, and formation of protective coatings on industrial equipment. Enhanced corrosion resistance can be conferred to carbon–steel by incorporating other metals, such as chromium, copper, nickel, and aluminum, within its iron and carbon structure, which might increase the costs of equipment construction. An alternative approach, therefore, is the use of customized formulations containing corrosion inhibitors.

As far as chemical structure is concerned, the application of ammonium quaternary salts and other amine surfactants as inhibitors has confirmed that specific functional groups can be considered as reaction centers for the process of adsorption over charged metallic surfaces (El-Achouri et al., 1995, 1996; Hajjaji et al., 1993; Luo et al., 1998a,b). This has inspired studies on other chemical moieties that exhibit the same capacity. However, fewer reports can be found that are related to inhibition of corrosion on metallic surfaces, such as iron and steel, by surfactants in microemulsified media. This is a very promising research area, since there may be some difficulty in using binary surfactant solutions, or micelle systems, with regard to solubilization and stability over a wide range of ambient conditions.

Microemulsions are therefore considered a way to overcome such problems. They are thermodynamically stable, transparent, optically isotropic chemical systems, formed with the intimate mixture of polar and nonpolar solvents that can be stabilized by an interfacial surfactant layer. In many cases, co-solvents or co-surfactants are required. They possess a well-defined structure and, depending on the system composition and parameters like temperature, salinity, and pH, among others, the internal structure of microemulsions may range from aggregates consisting of droplets of oil dispersed in a continuous aqueous or polar microphase to droplets of water in a continuous oil phase, with various intermediate domains such as bicontinuous microemulsions. Also, depending on the composition of the microemulsion systems, the chemical nature of the compounds that form them, and how they interact with one another, the surfactant aggregates may be very stable or may easily destabilize upon specific changes in the environment to which they are applied. The possibility of tuning the properties of surfactant systems has been the main driving force for very strategic studies on interfacial phenomena, such as the controlled release of chemicals at specific targets and the search for alternative templates to optimize industrial processes and corrosion inhibition at metallic surfaces. The application of surfactant systems in the industry is advantageous in that important aspects affecting the industrial activities are accounted for—namely, the use of renewable materials, with reductions in cost and energy consumption, the recycling of chemicals, and optimization of disposal, treatment, and maintenance operations. However, previous knowledge of the interactions established by the various chemical substances that comprise these systems and the physical interfaces onto which they adsorb must be known, in order to avoid any negative effects that may lower the yield of the original procedures.

It is known that the inhibition of corrosive phenomena by surfactants is mainly due to the chemical structure of the surfactant molecules and their ability to adsorb onto surfaces and interfaces at relatively low concentrations, due to the presence of functional groups that are considered as reaction sites for the process of adsorption over neutral or charged metallic surfaces. They contain electronegative atoms like nitrogen, oxygen, and sulfur, which interact with anodic or cathodic sites on solid surfaces (Ali and Saeed, 2001). In that respect, it is interesting to highlight the huge variety of surfactant structures that are commercially available or can be synthesized in the laboratory, involving anionic, cationic, non-ionic, and zwitterionic structures.

Chemical moieties that render surfactants as good inhibitors of corrosion on metallic surfaces are heteroatoms and other chemical groups that are able to adsorb onto the surface, a phenomenon that is oriented according to existing polar sites. Groups like $-COOH$, $-NH_2$, $-OH$, $-SH$, and $-SO_3H$ are good polar sites that constitute the surfactant headgroup. When multifunctional groups are present, the ability of surfactants molecules to adsorb is further enhanced, forming a better protective film. The thickness of such film is mainly determined by the concentration of surfactant used and the amount of contacting time, allowing for the generation of monolayers or multilayers of surfactant molecules which impart resistance to the metal surface, thereby reducing material degradation. According to the nature of the technological activity, both solid and liquid inhibitors may be used, either chemically modified or obtained naturally. It is important to observe or envisage all differences in operation conditions, such as surface characteristics and the nature of the aggressive media. For example, amine surfactants are potential candidates for providing more intense adsorption interactions at solid–liquid interfaces with the formation of more efficient films that protect the metallic surface more effectively (Moura et al., 2009).

Some investigations have focused on the use of regional raw materials, like vegetable oils, to synthesize novel surfactant molecules at lower costs. In this chapter, we discuss some recent experimental advances provided by self-assembled chemical systems on the inhibition of corrosion on metallic surfaces. The works are based on the adsorption mechanism of micelle and microemulsion droplets at solid–liquid interfaces, altered by various parameters such as surfactant structure and operation conditions.

1.1 Mechanism of surfactant adsorption on metallic surfaces

The discussion on self-assembled systems in bulk solutions is widely available in the colloids literature, but description of surfactant adsorption on solid–liquid interfaces is less reported. Following Winsor's *R*-theory on energetic interactions at curved interfaces (Winsor, 1954), it is interesting to consider the different mechanisms for surfactant aggregation and adsorption when micelle or microemulsion systems are investigated, especially if additives like acids, bases, electrolytes, and nonionic co-surfactants are added. Therefore, concentration and composition are the main parameters to be considered when examining how surfactant molecules or surfactant aggregates interact with metallic surfaces. In particular, micelle solutions or even microemulsions are to be considered as self-assembled systems that effectively help inhibit any corrosive phenomena that may take place under certain circumstances. As pointed out above, the possibility of tuning the properties of the self-assemblies when the experimental conditions are altered is directly related to Winsor's *R*-theory. Protection occurs when a monolayer or multilayers of surfactant aggregates, either as micelles or microemulsion droplets, adsorb onto the metallic surfaces, and parameters like temperature, type of surfactant, and pH may affect the nature of the aggregates. Such films are formed from the point when the concentration of surfactant molecules is sufficiently high so as to saturate the medium, reaching the critical micelle concentration (CMC) of the compound being used at the experimental conditions employed (Love et al., 2005; Migahed et al., 2005; Shervedani et al., 2007; Ulman, 1991).

The formation of the so-called self-assembled monolayers (SAMs) is responsible for the protection of metallic surfaces. In surfactant systems, there is spontaneous adsorption promoted by interactions with the polar headgroups of the surfactant molecules and active sites on the surface of the metal. The nature of the surfactant hydrophobic chain also plays an important role in determining the extent of inhibition. Although corrosion inhibition may be effected under various surfactant concentrations and conditions, one can distinguish some possible mechanisms for surfactant adsorption, both in micelle and microemulsified media. According to Paria and Khilar (2004), the mechanism of surfactant adsorption on solid interfaces is a direct effect of surfactant concentration, with a range of different orientations and interaction properties. The following phenomena may be involved in the complete adsorption mechanism:

- Ionic exchange: replacement of adsorbed counter-ions with similar charged species from the surfactant molecules.
- Ion pairing: adsorption of surfactant ions from the bulk solution on oppositely charged sites of the solid surface that are not occupied by counter-ions.
- Hydrophobic adhesion: attraction between the hydrophobic groups of adsorbed molecules and species dispersed in the bulk solution.
- Adsorption induced by polarization of π electrons: interaction between electron-rich aromatic nuclei of surfactant molecules and strongly positive active sites on the solid surface.
- Adsorption induced by dispersion forces: attraction between species promoted by London and/or van der Waals forces that increase with increasing molecular weight of the adsorbed material.

In micelle systems at very low concentrations, below the CMC, surfactant monomers are sparsely adsorbed on the metal surface and protection may not be effective. Depending on the structure of the hydrophobic chain of the surfactant molecule, there may be some bending and accommodation of the molecule over the surface, allowing for slightly better coverage and protection. This is the particular case of compounds that also feature active sites on their hydrocarbon chains, such as epoxy groups or unsaturations. The hydrophobic chains may also be oriented toward the aqueous medium, due to stronger interactions between the polar headgroups and the surface, but this phenomenon is highly dependent on surfactant structure. The reverse situation may also be observed, whereby the polar headgroups remain in closer contact with the bulk aqueous solution and the hydrophobic chains interact with the metal surface. As concentration increases and the CMC is approached, adsorption enhances rapidly due to the formation of micelles in the bulk solution. At this point, hemi-micelles may form at the solid surface, providing complete surface coverage. At the CMC and above, however, more packed aggregates exist in the bulk solution and interact dynamically with the interfaces.

Recent theories, confirmed by some investigations of our research group, have indicated that completely formed micelles are adsorbed on the metal (Holmberg et al., 2002; Lindman et al., 1998; Wanderley Neto et al., 2009, 2012). A similar explanation had also been given by Tan and O'Haver (2004), when investigating the admicellization phenomena of styrene and non-ionic surfactant aggregates at water–silica interfaces. Interestingly, this may diminish the efficiency levels of corrosion inhibition, since packing at the solid–liquid interface is less pronounced than with hemi-micelles, allowing for penetration of corrosive agents that reach the metal surface. Depending on surfactant structure and operation conditions, the shape of the micelles may be distorted, and otherwise spherical aggregates turn into elongated structures that provide better protection. At sufficiently high-surfactant concentrations, much above the CMC, there may be the formation of multilayers, increasing once again the protection levels, which remain practically constant with further increases in surfactant concentration. At this point, water molecules and other weakly adsorbed ions that can trigger corrosive phenomena are displaced from the surface. This mechanism is depicted in Figure 19.1. Observe that, below the CMC, the surfactant molecules may lie on the metal surface or be oriented approximately perpendicularly to the surface in different ways, as explained earlier.

A similar mechanism is discussed with microemulsion systems. Now, because of their particular interfacial properties, different levels of interaction are established between the bulk solution and the surfaces. It is known that microemulsion droplets are highly curved and that mass transfer between the dispersed phase and the dispersing medium is more intense than in emulsions. Thus, the internal interfacial area of microemulsions is much higher, which is a very important property when applying surfactant systems as corrosion inhibitors. Microemulsions are formed at very distinct conditions that can be determined from phase behavior studies. Their efficiency in forming a protective film on the metal surface is highly altered by surfactant concentration and packing. For example, spherical or elongated droplets may form depending on the surfactant packing parameter, which is a more complete approach to the hydrophilic-lipophilic balance (Clint, 1992; Holmberg, 1998). Microemulsions can also remain stable at very high-surfactant concentrations. When used as corrosion inhibitors, the same mechanisms

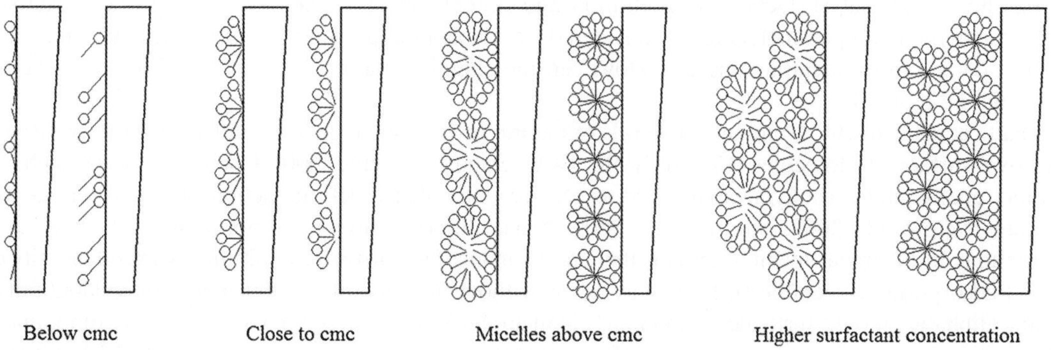

FIGURE 19.1

Evolution of surfactant adsorption at solid-liquid interfaces as a function of surfactant concentration in micelle systems.

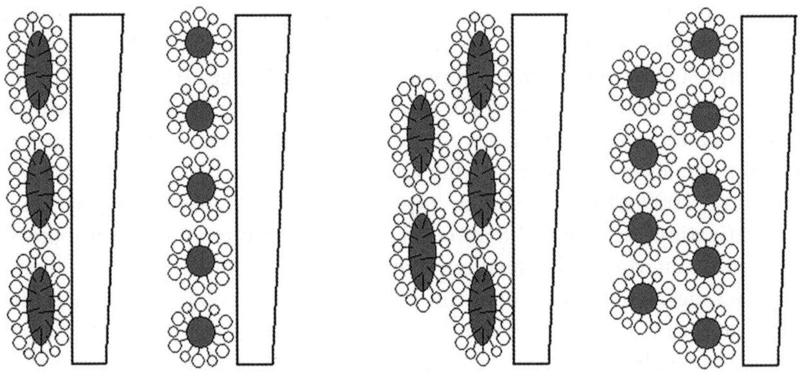

FIGURE 19.2

Adsorption of oil-in-water microemulsion droplets at solid-liquid interfaces with increasing surfactant concentration.

discussed above for micelle solutions may be observed when investigating how microemulsion droplets interact with the metal surfaces. Figure 19.2 is a schematic representation of the adsorption of microemulsion droplets on metal surfaces, for the particular case of oil-in-water (O/W) systems, at lower and higher surfactant concentrations. Note that when the surfactant concentration is relatively low, there may be somewhat poorer coverage and protection of the surface against corrosion agents, since the droplets remain fairly stable also at the solid–liquid interface as far as size and shape are concerned. Although several parameters are involved and simultaneously affect the final equilibrium established at the interfaces, this is probably the cause for the lower efficiency of some microemulsion systems that we have been examining as compared with micelle solutions formed with the same surfactant. This will be discussed later along with some experimental results.

1.2 Influence of surfactant structure and nature

It has been said that surfactant concentration and immersion or contacting time between reactive species are the main parameters to be considered when examining the ability of surfactant systems to inhibit corrosion on metallic surfaces. However, the chemical nature or structure of the surfactants must not be ignored.

Fatty acids extracted from vegetable oils or natural resins can be good candidates for the synthesis of novel surfactant molecules (El-Etre, 1998; El-Etre and Abdallah, 2000). Chemical groups such as amino, epoxy, and carboxylates are cited as reactive sites that block the action of corrosion agents (Dantas Neto et al., 2008; Moura et al., 2009; Nmai, 2004). Anionic and non-ionic surfactants are ubiquitous in chemical technology and their performance as corrosion inhibitors is excellent, since direct adsorption on metallic surfaces may occur without the need of counter-ions. Ethoxylated fatty acids, aliphatic anionic surfactants, and long-chain aromatic amines are other examples of compounds that have been investigated with success (Becerra et al., 2000; Deyab et al., 2007, 2009; El-Rehim et al., 2003; Shalaby and Osman, 2001, 2009; Soer et al., 2008).

The nature of the corrosive agent is also important. Some systems are better employed in either an acidic or a saline environment, depending on the solid material that must be protected. For instance, El-Sherbini and coworkers (2003 and 2005) reported that ethoxylated fatty acids are good corrosion inhibitors in sulfuric and hydrochloric acid solutions over zinc and aluminum surfaces. Acidic media can rapidly induce corrosion phenomena on metallic surfaces if no protective coatings are applied. Cationic surfactants have been frequently reported as effective corrosion inhibitors, particularly against sulfuric and hydrochloric acids (Branzoi et al., 2000; Migahed, 2005). Counter-ions are important for the adsorption of these surfactants onto hydrophilic, positively charged metallic surfaces. The superior ability of amine surfactants to inhibit corrosion phenomena is related to their adsorption mechanism, whereby interfacial films are formed by means of chelating effects that enhance the interaction between surfactant molecules and the metallic surface (Deyab, 2007; Nmai, 2004). In carbon–steel surfaces, for example, the corrosion velocity is affected by the evolution of gaseous hydrogen from the metallic surface. At $pH < 4$, this release is intensified and generates stress on the surface due to electrode discharges. As the concentration of H^+ ions is reduced, normally within a pH range between 4 and 10, the corrosion velocity is not much altered by pH, but is greatly affected by the amount of dissolved oxygen (Silverman et al., 1995). We are also particularly interested in salt-induced corrosion phenomena involved in the petroleum industry. For example, water injection is one technique used in primary and secondary oil recovery operations, and oil emulsions may be obtained. Because of Brazil's major offshore reserves, oil exploitation is widely performed on platforms above seawater. The water phase that is in contact with oil, or formation water, has various amounts of electrolytes such as chlorides, sulfates, and nitrates. This mixture must be pumped in metallic pipelines for processing, and the ability of surfactant systems to inhibit electrolyte action on metallic surfaces has also been investigated (Dantas Neto et al., 2008).

Several studies have been recently undertaken involving the use of natural products as potential sources of anticorrosion agents on various metallic surfaces under different ambient conditions (Bouyanzer et al., 2006; Chaieb et al., 2005; Ouachikh et al., 2009). Vegetable oils can be used as corrosion inhibitors due to the presence of fatty acids in their composition, such as oleic and maleic acids (Osman et al., 2003). The use of natural products is actually an important research area in our group, and we have used several vegetable oils with various purposes, in particular castor oil, soybean oil, cashew-nut shell liquid (CNSL), and coconut oil. Other researchers report the use of polymeric compounds, like acrylic esters grafted with copolymers (Patel and Dighe, 2007), polyaniline containing epoxy groups (Saravanan et al., 2007), or epoxy cardanol resins obtained from CNSL (Aggarwal et al., 2007). It has also been reported that several non-ionic polyethoxylated surfactants and ethoxylated fatty acid derivatives such as monooleates can be effective agents against corrosion on carbon–steel in the presence of naphthenic acids (Li et al., 2008, 2009; Mu and Li, 2005; Osman and Shalaby, 2003), with yields higher than 90%.

These are more representative works, showing that the ability of surfactants to inhibit corrosion is enhanced when their concentration in solution approaches the corresponding CMC. The adsorption of surfactant molecules on metallic surfaces can remarkably change their resistance properties, and this aspect has guided the elaboration of research projects to elucidate the relationship between the chemical nature of anticorrosion agents and the adsorption mechanisms that ultimately promote the protection of the metallic surfaces (Mandrino et al., 2008a, 2008b; Ramji, 2008). Thermodynamics obviously plays an important role in the description of the phenomena involved. It is noteworthy that adsorption results from the affinity between polar groups in the surfactant molecule and the solid surface, but self-assembly is the original mechanism in the overall phenomenon. Minimization of Gibbs free energy is the driving force that promotes lateral interactions between the hydrophobic chains of the surfactants, which are stronger than their own interactions with the surface. This is exactly the case of van der Waals forces that promote the hydrophobic effect in the self-assembly of surfactant-based aqueous systems (Tanford, 1980). Also, microemulsion phase behavior studies are valuable to tune experimental parameters such as the salinity and/or acidity of the aqueous phase, the nature of the organic solvent, and the type of surfactant, since these are the main components of such systems. Model assays can be carried out to mimic actual corrosive environments that help optimize future applications. These aspects are considered in the discussion of some of our own experimental results, in the following sections.

2 Current research on novel corrosion inhibitors

This section associates the subjects discussed in the previous section with actual, recent results on the use of surfactant systems as corrosion inhibitors on metallic surfaces. The influences of salt content, acidity, temperature, surfactant nature, and aggregation, as in micelle or microemulsion systems, have been investigated. In particular, we highlight the use of renewable raw material as precursors in the syntheses of surfactant molecules that can be used for such purpose.

A basic experimental procedure has been implemented in all cases to determine the CMC of the surfactants tested in distilled water or either hydrochloric acid or sodium chloride solution, at room temperature (27°C ± 1°C), when investigating acid- or salt-induced corrosion phenomena. Similarly, potentiometry measurements were carried out in a potentiostat connected with a microprocessor, working at a scanning rate of 2 mV/s. For each solution, the corrosion current (I_{corr}), corrosion potential (E_{corr}), and Tafel's anodic (β_a) and cathodic (β_c) curvatures were determined as a function of inhibitor concentration. Unless otherwise stated, the experiments were performed with an AISI 1010 carbon–steel electrode, which was initially kept at –0.2 V from the corrosion potential for 10 minutes, allowed to rest for 1 minute and then subjected to the open-circuit potential up to +0.08 V, during the experiments. More details about these procedures are given elsewhere (see Moura et al., 2009; Wanderley Neto et al., 2012).

2.1 Amine surfactants against acid- or salt-induced corrosion

2.1.1 Surfactants synthesized from dodecylamine

The inhibition of acid-induced corrosion phenomena on metallic surfaces has been investigated with aliphatic cationic surfactants, which are ammonium salts differing by their counter-ions, synthesized from dodecylamine with gaseous hydrochloric acid (HCl) and hydrogen bromide (HBr). A single acidification step is required to produce the surfactant salts with high purity and excellent yields, as generally indicated by Eq. 19.1, where X is either the chloride or bromide anion.

$$C_{12}H_{25}{-}NH_2\ (l)\ +\ HX\ (g) \rightarrow C_{12}H_{25} - NH_3^{+}X^{-}\ (s) \tag{19.1}$$

However, gaseous HCl or HBr are produced by reacting concentrated sulfuric acid (98%) with sodium chloride or potassium bromide. Purification of the chloride derivative was carried out with diethylether washings, and the bromide compound was recrystallized in distilled water. The pure surfactants dodecylammonium chloride ($C_{12}H_{25}NH_3^+Cl^-$, or DDAC, 80% yield) and dodecylammonium bromide ($C_{12}H_{25}NH_3^+Br^-$, or DDAB, 94% yield) have thus been obtained, and their ability to inhibit corrosion promoted by 0.1 M hydrochloric acid aqueous solutions has been examined by means of micelle or microemulsion systems.

The phase behavior of these surfactants (S) was investigated by constructing pseudoternary triangular-phase diagrams at room temperature. In this case, either pure distilled water or a 0.1 M HCl aqueous solution was used as the aqueous phase; kerosene was used as the oil phase. and butan-1-ol was the co-surfactant (C) tested. We have selected a co-surfactant/surfactant mass ratio of 2 in all experiments (C/S = 2), because of the more extensive microemulsion region formed under such circumstances. Figure 19.3 shows the phase diagrams for DDAC and DDAB when pure distilled water is the aqueous component of the mixtures. In Figure 19.4, the same systems are prepared by replacing pure water with the acid solution.

Both diagrams in Figure 19.3 show a very broad microemulsion region (WIV), demonstrating the ability of both surfactants to form single-phase, transparent mixtures with kerosene and butan-1-ol over a wide range of concentration. However, the presence of hydrochloric acid impairs the equilibrium established at the liquid–liquid interfaces between dispersed and dispersing microphases, as shown in Figure 19.4. A larger WIV region is observed in DDAC systems, which is an effect of the smaller ionic radius of chlorine counter-ions. Larger amounts of DDAB are therefore required to stabilize the aqueous and oil phases. Furthermore, in DDAC systems, the chlorine ions from both HCl and surfactant molecules induce a common-ion effect which helps enhance microemulsion formation and stabilization. However, a small WII region (microemulsion with excess acid water phase) is detected in the DDAC diagram, because of changes that occur in the geometric structure of the aggregates when the contents of aqueous and oil phases are altered. At equivalent concentrations, the surfactant molecules may be required to pack more closely, and this provokes system instability that generates the WII biphasic systems.

The CMC of DDAC and DDAB was determined by dissolving proper amounts of surfactant in a pure 0.1 M HCl solution (hereon referred to as system I) or preparing microemulsions with the same

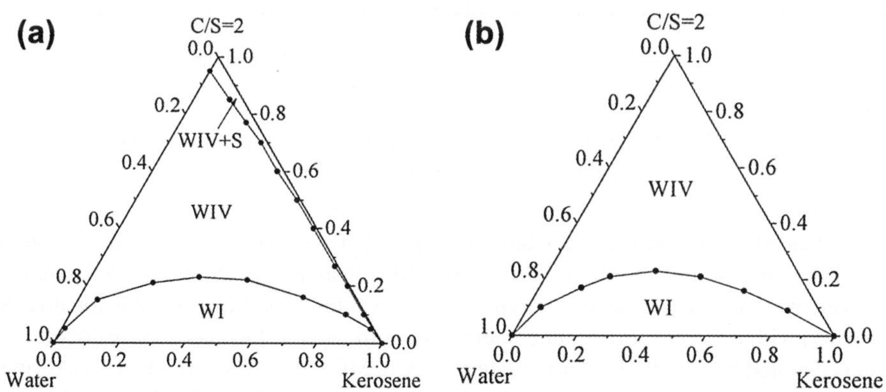

FIGURE 19.3

Pseudoternary phase diagrams for (a) DDAC/butan-1-ol/n-hexane/water and (b) DDAB/butan-1-ol/n-hexane/water systems.

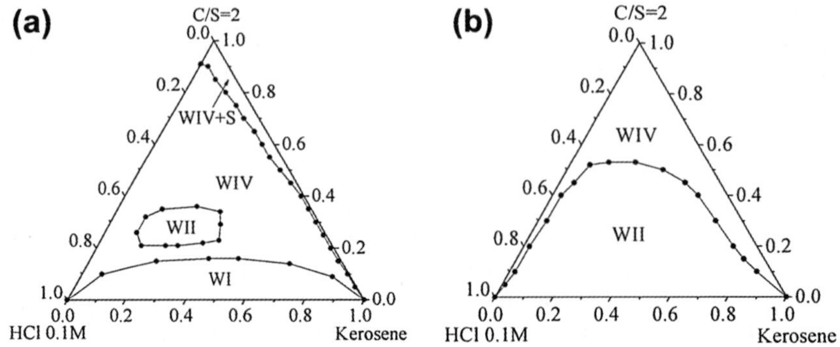

FIGURE 19.4

Pseudoternary phase diagrams for (a) DDAC/butan-1-ol/n-hexane/0.1 M HCl and (b) DDAB/butan-1-ol/n-hexane/0.1 M HCl systems.

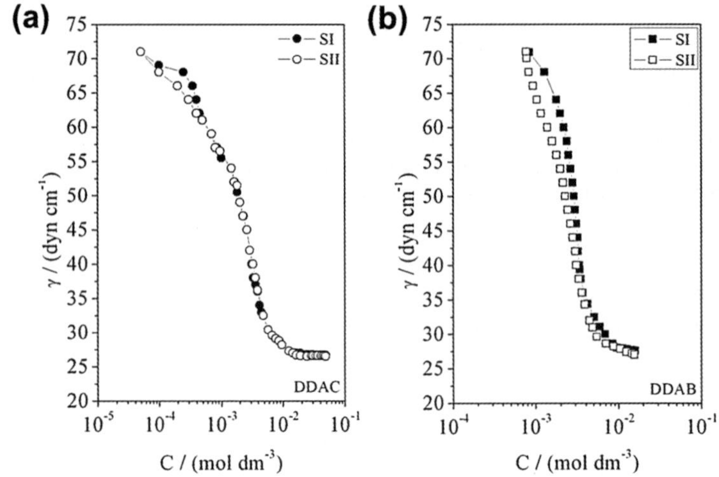

FIGURE 19.5

Surface tensions of liquid samples containing the amine surfactants (a) DDAC and (b) DDAB versus surfactant concentration at 27°C in micelle (S-I) or microemulsion (S-II) media.

components indicated above (surfactant/butan-1-ol/kerosene/0.1 M HCl, hereon called system II). The results are shown in Figure 19.5. System II is a microemulsion and could not be characterized by a CMC value. However, we have also observed variation in the values of surface tension for microemulsion samples prepared with different surfactant contents. An equivalent CMC has been ascribed to system II samples and is denoted here as P_C. Under these terms, the CMC values for surfactant DDAC in systems I and II are 4.1×10^{-3} mol/L and 6.8×10^{-3} mol/L, respectively. For DDAB, the CMC values are 6.4×10^{-3} mol/L and 5.1×10^{-3} mol/L, in micelle solution and microemulsified medium, respectively. The values are very close, with the same magnitude. In corrosion inhibition studies, however, it is important to know these concentrations in order to properly analyze the efficiency of the surfactant systems at each set of experimental conditions. In some cases, the surfactant system may be efficient even below the corresponding CMC. For example, Table 19.1 shows the main electrochemical parameters obtained from the polarization curves determined according to the experimental procedure presented above, for both surfactant systems. The last column of the table shows the efficiencies in corrosion inhibition for each system. The performance of some systems is excellent even below their CMC.

The data in Table 19.1 are also useful to acquire Tafel's current-potential curves by means of potentiokinetic assays. Typical curves for the systems investigated here are shown in Figure 19.6, in the presence and absence of surfactants, for both micelle and microemulsion media (systems S-I and S-II). For these curves, a surfactant concentration of 10^{-3} M was used based on the data of Table 19.1. Analysis of Figure 19.6 indicates that the corrosion potentials (E_{corr}) are more displaced toward positive values, while the corrosion currents (I_{corr}) decrease, showing that all systems can inhibit corrosion more effectively in anodic conditions.

Corrosion inhibition efficiencies can be assessed for each surfactant system in terms of the protection coverage on the metallic surface (θ) provided by the surfactant aggregates. This is shown in Figure 19.7 for DDAC and in Figure 19.8 for DDAB, in both micelle (system I) and microemulsion (system II) media. The data in Figures 19.7 and 19.8 are also fitted with the Frumkin mathematical model of adsorption isotherm. More details on this approach are also given elsewhere (Moura et al., 2009; Wanderley Neto et al., 2012). The fitting results are depicted by the full plotted curves, and the physicochemical parameters acquired from the model are listed in Table 19.2. These include the concentration and the Frumkin isotherm constants (K_F and A, respectively), the Frumkin object function ($\delta\Theta_F$), and the Gibbs free energy of adsorption ($\Delta_{ads}G$). The relatively low $\delta\Theta_F$ values indicate that the Frumkin model appropriately fits the experimental data, and the negative $\Delta_{ads}G$ values characterize the spontaneity of the adsorption phenomena. The lack of functional groups or polar sites on the surfactants' hydrophobic chains is responsible for a more effective lateral contact between the molecules, thereby forming a more compact film that provides a satisfactory coverage efficiency.

It is convenient to refer to Figures 19.1 and 19.2 again to discuss these results more fully. It can be understood that a complete, protective layer is formed on the metal surface when the surfactant concentration reaches the CMC in micelle systems, and the critical P_c value in microemulsion samples. At this point, maximal efficiency in surface coverage is reached for each surfactant. At higher concentrations, the adsorption of surfactant molecules at the solid–liquid interface only generates more layers, and a plateau of constant corrosion inhibition level is reached. The adsorption mechanism depicted in Figures 19.1 and 19.2 helps explain such experimental observations. In particular, the ammonium headgroups of DDAC and DDAB molecules are responsible for their interactions with the positively charged metal surface that becomes negatively charged due to counter-ion adsorption. For systems I and II, with both surfactants, direct, spherical aggregates are expected, and, for this reason, the ammonium groups are oriented toward the continuous aqueous medium. While there may be changes in the shape of the aggregates at the solid–liquid interface in system I, at surfactant concentrations below and above the CMC, which induce surfactant rearrangement and occasionally improve the protection levels, there is only accommodation of typical microemulsion droplets on the metallic surface in system II. In summary, both DDAC and DDAB are good corrosion inhibitors on carbon–steel in 0.1 M HCl media. The efficiencies are higher than 80%

Table 19.1 Electrochemical Parameters for Carbon–Steel In 0.1 M HCl Solution in the Presence of The Corrosion Inhibitors DDAC and DDAB for Systems I And II

Electrolytic Medium	[S] / (mol L^{-1})*	I_{corr} / (μA cm^{-2})	β_a / (mV dec^{-1})	β_c / (mV dec^{-1})	E_{corr} / mV	E %
0.1 M HCl	**0**	**363**	**75**	**386**	**-404**	**-**
DDAC in S-I (CMC = 4.1 × 10^{-3} mol L^{-1})	2.6 × 10^{-6}	196	79	396	-377	46
	2.0 × 10^{-5}	186	68	300	-385	49
	2.6 × 10^{-4}	46	58	128	-212	87
	7.8 × 10^{-4}	13	68	236	-384	96
	1.6 × 10^{-3}	9	88	207	-376	98
DDAC in S-II (P_c = 6.8 × 10^{-3} mol L^{-1})*	6.0 × 10^{-5}	127	75	249	-476	65
	2.0 × 10^{-4}	69	76	307	-476	81
	2.0 × 10^{-3}	35	82	379	-89	90
	1.0 × 10^{-2}	15	49	213	-398	96
	5.0 × 10^{-2}	7	67	229	-90	98
	1.0 × 10^{-1}	16	84	183	-292	96
DDAB in S-I (CMC = 6.4 × 10^{-3} mol L^{-1})	1.0 × 10^{-6}	98	82	264	-506	73
	6.5 × 10^{-6}	96	77	268	-494	74
	3.0 × 10^{-5}	84	70	133	-485	77
	1.3 × 10^{-4}	56	83	416	-493	85
	1.0 × 10^{-3}	15	74	346	-442	96
DDAB in S-II (P_c = 5.1 × 10^{-3} mol L^{-1})*	5.0 × 10^{-6}	190	77	203	-244	47
	2.0 × 10^{-5}	155	54	196	-368	57
	1.0 × 10^{-3}	63	42	118	-157	83
	5.0 × 10^{-3}	26	64	386	-29	93
	1.0 × 10^{-2}	22	80	474	-24	94

(*)[S] = surfactant concentration; P_c = critical surfactant concentration in microemulsion systems.

for systems with surfactant concentrations just below the CMC or P_c, and they are as high as 98% at higher surfactant concentrations. Their micelle solutions are slightly more efficient than equivalent microemulsion formulations under a specific range of surfactant concentration. This effect is more evident for DDAB and directly reflects the surfactant reorganization at solid interfaces. The complete efficiency results are shown in Figure 19.9 for both surfactants in micelle and microemulsion formulations.

The relevance of microemulsions is confirmed in applications where surfactant dilution is a possibility, as in oil transportation in pipelines and protection of internal surfaces of metallic reactors. Since microemulsions are renowned for dissolving larger amounts of surfactants than emulsions and micelle solutions, they can be used in more diverse or unpredictable situations without any impairment in corrosion efficiency. For DDAC and DDAB, however, the surfactant concentration must not be lower than 5 × 10^{-3} M, which is the approximate critical point of the formulations tested in a 0.1 M HCl environment (see Table 19.1).

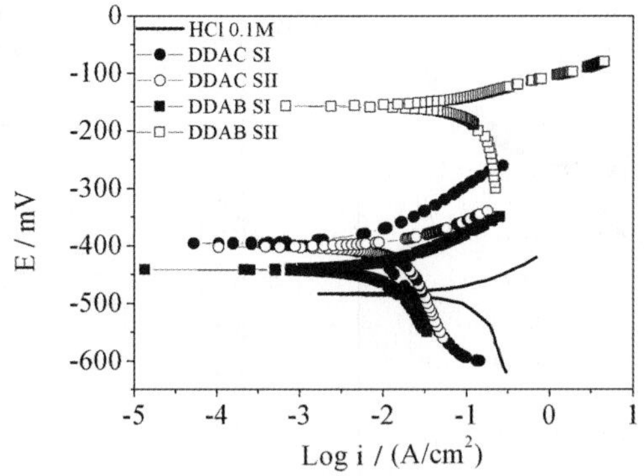

FIGURE 19.6

Potentio-kinetic polarization curves for carbon-steel in 0.1 M HCl in the absence and presence of inhibitors DDAC and DDAB as micelle (S-I) and microemulsion systems (S-II), at a constant surfactant concentration of 10^{-3} M.

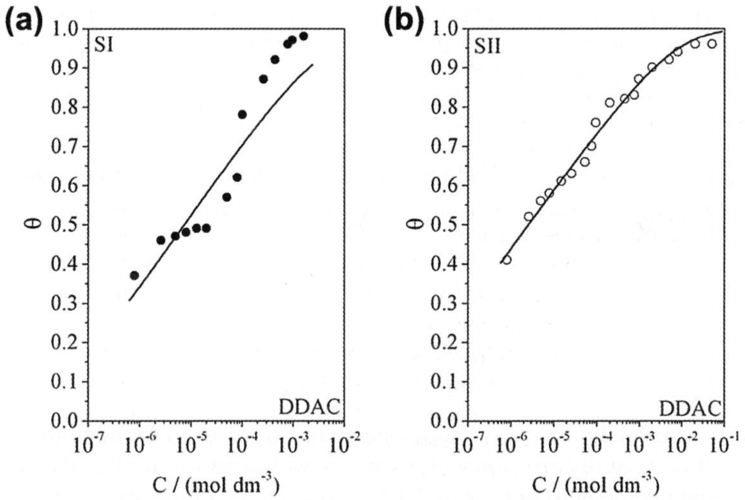

FIGURE 19.7

Surface covered fraction (θ) obtained from electrochemical measurements for DDAC systems: (a) micelle solutions; (b) microemulsion systems. The full curves are fittings to the experimental circular points according to the Frumkin adsorption isotherm model.

*2.1.2 Sodium 12-*N,N*-diethylamino-octadec-9-enoate, a castor oil derivative*

In a recent article, we discussed the ability of amine surfactants synthesized from ricinoleic acid, a major component of castor oil (Moura et al., 2009), to inhibit salt-induced corrosion phenomena. In that investigation, we were mainly concerned with assessing the differences in surfactant structure at fixed temperature and salinity conditions. In another departure, we selected the unsaturated derivative, named sodium 12-*N,N*-diethylamino-octadec-9-enoate, or AR1S, to be used in micelle and microemulsion

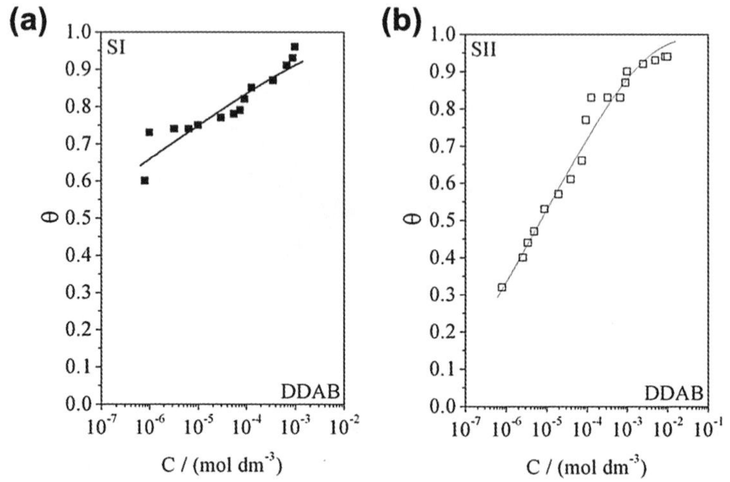

FIGURE 19.8

Surface covered fraction (θ) obtained from electrochemical measurements for DDAB systems: (a) micelle solutions; (b) microemulsion systems. The full curves are fittings to the experimental square points according to the Frumkin adsorption isotherm model.

Table 19.2 Physicochemical Parameters from the Frumkin Model in DDAC and DDAB Systems

Surfactant	System	K_F	A	$\delta\Theta_F$	$\Delta_{ads}G$ / (kJ/mol)
DDAC	I	9.66×10^6	−8.54	1.09×10^{-1}	−39.8
DDAC	II	1.20×10^8	−11.5	2.63×10^{-2}	−46.1
DDAB	I	1.65×10^{12}	−20.8	4.22×10^{-2}	−69.7
DDAB	II	6.54×10^6	−7.72	4.06×10^{-2}	−38.9

systems tested at different salinity levels and temperatures, in order to examine the influence of such parameters on the efficiencies of corrosion inhibition. The surfactant was used as the main component of micelle (system I) and microemulsion (system II) formulations, in a procedure similar to that described above for DDAC and DDAB. Phase diagrams were constructed at 27°C. In this case, 0.5 M and 1.0 M NaCl aqueous solutions were used as aqueous phases in the preparation of the samples and as corrosion agents. Again, butan-1-ol was the co-surfactant, and kerosene was the oil used in preparing the samples. A co-surfactant/surfactant mass ratio of 1.0 was employed. Tensiometry and potentiokinetic measurements were made at 30°C and 60°C. The work electrode was made of API5LX Gr X42 steel, which is another common material used in pipelines in the petroleum industry (Dantas Neto et al., 2008).

Figure 19.10 shows the pseudo-ternary-phase diagrams for the AR1S systems at the higher saline concentrations used. There is a rich variety of equilibrium systems (from WI to WIV), but the composition of the WIV microemulsion regions favors the formation of oil-in-water dispersions. The main purpose in varying the salt concentration is to investigate the formation of microemulsion structures under different corrosive environments, because changes in electrolyte content alter the

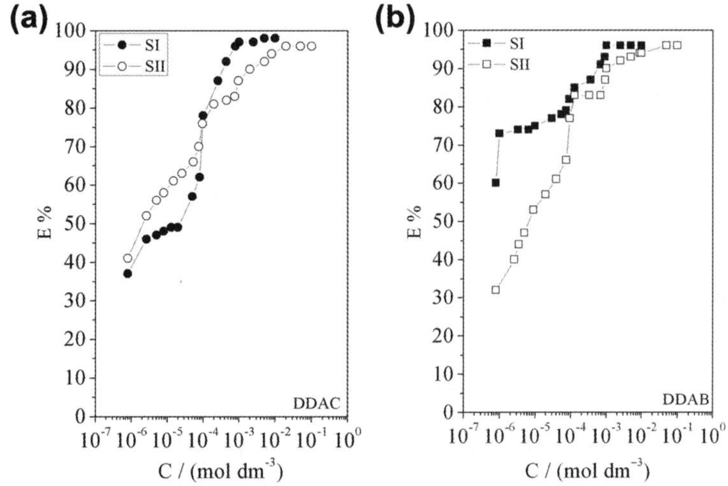

FIGURE 19.9

Corrosion inhibition efficiency data (*E*%) for (a) DDAC and (b) DDAB dissolved in systems I and II.

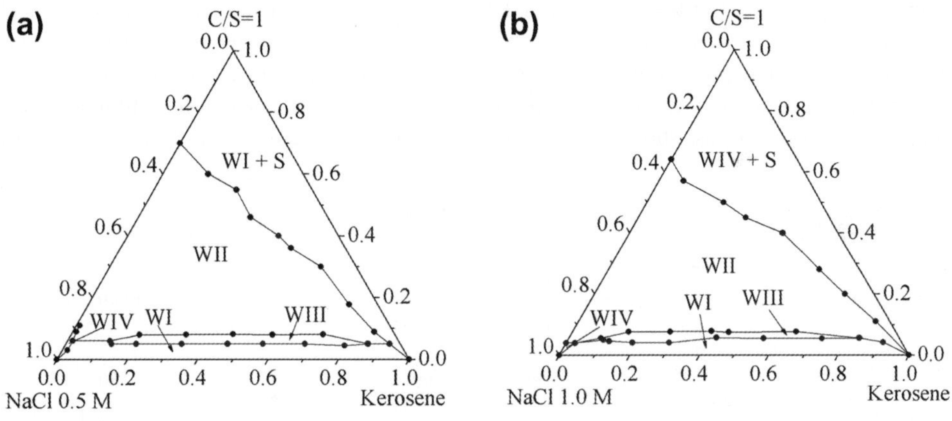

FIGURE 19.10

Pseudoternary phase diagrams for AR1S systems with (a) 0.5 M NaCl and (b) 1.0 M NaCl as aqueous phases. Butan-1-ol is used as cosurfactant at a C/S ratio = 1 and kerosene is the oil phase. The diagram was constructed at 27°C.

microemulsion properties. This is particularly important in the transportation of petroleum and its derivatives, since in many cases water is the major component of the oil–water mixtures. By knowing the extent of the microemulsion region, one can establish the range of concentrations within which they can still be efficient as corrosion inhibitors. Direct micelles or microemulsion droplets should then be formed, and, although the concentration of corrosion inhibitor is reduced because of dilution during pumping, the aggregates are not broken and the inhibition mechanism is maintained.

Table 19.3 Physicochemical Parameters from the Frumkin and Langmuir Models in AR1S Systems

System	Salt Concentration	Temperature	K_F	A	$\delta\Theta_F$	K_L	$\delta\Theta_L$
I	0.5 M	30 °C	7.86×10^{10}	−23.4	8.31×10^{-3}	425	0.01
I	0.5 M	60 °C	3.06×10^{20}	−53.9	1.80×10^{-2}	245	0.18
I	1.0 M	30 °C	3.61×10^{13}	−32.7	2.16×10^{-2}	286	0.15
I	1.0 M	60 °C	1.90×10^{22}	−60.2	6.07×10^{-3}	206	0.20
II	0.5 M	30 °C	7.78×10^{15}	−35.7	1.25×10^{-2}	572	0.01
II	0.5 M	60 °C	3.37×10^{15}	−36.5	1.40×10^{-2}	399	0.12
II	1.0 M	30 °C	7.82×10^{8}	−18.2	7.97×10^{-3}	388	0.01
II	1.0 M	60 °C	8.41×10^{12}	−30.1	1.23×10^{-2}	333	0.13

The adsorption phenomena at the solid–liquid interface for AR1S systems was examined by carrying out corrosion current assays, as described above for other systems. In this case, both Frumkin and Langmuir isotherms were used to model the experimental results (Dantas et al., 2002). The results are shown in Table 19.3. Frumkin and Langmuir parameters are indicated by the subscripts *F* and *L*, respectively. The lower $\delta\Theta_F$ values indicate that Frumkin's adsorption isotherm is more suitable than the Langmuir's approach to fit the experimental data. The adsorption phenomena generate a very broad charge field that involves the interface. As a consequence, the values of the lateral parameter suggest repulsive interactions. The predominant effect is the interaction between surfactant molecules and the metallic surface, and not lateral interactions. The high values of corrosion inhibition efficiency corroborate this proposal, as shown in Figure 19.11 (surface coverage, θ, and corresponding

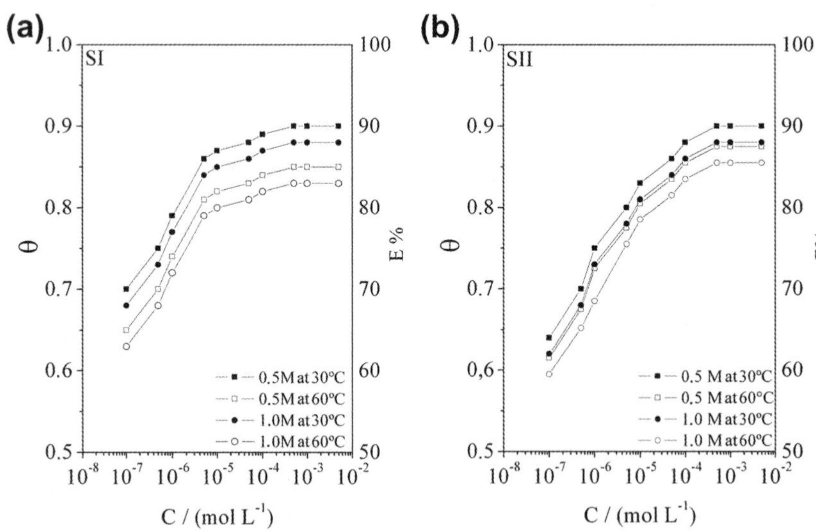

FIGURE 19.11

Surface covered fraction (θ) obtained from electrochemical measurements and corrosion inhibition efficiency data (*E*%) for AR1S systems in (a) Micelle and (b) Microemulsion medium, as a function of NaCl concentration and temperature.

percentage efficiency, *E*%), for the salinity and temperatures tested. Both systems I and II are good corrosion inhibitors, with efficiencies as high as 90%. In this case, microemulsions are slightly more efficient, especially at lower surfactant concentrations. Although the efficiency increases with increasing surfactant concentration, the opposite occurs at higher temperatures. In chloride-rich media, pitting corrosion is the main observed mechanism that is strongly affected by temperature, because of enhanced ionic mobility. More uniform corrosion on the metallic surface is an effect of free chloride ions in solution at higher temperatures.

2.2 Epoxy surfactants against acid- or salt-induced corrosion

Epoxy surfactants can be synthesized by some routes employing vegetable oils. Epoxy-containing organic compounds have important properties that can be used in many promising, technological applications, such as preparation of plasticizers, stabilizers for polyvinyl chlorides, and precursors to polyether polyalcohols and polyurethanes (Lligadas et al., 2006; Metzger and Bornscheuer, 2006). We have carried out an investigation on the use of some epoxy derivatives from castor oil and soybean oil as corrosion inhibitors. The synthetic routes involved epoxidation of ricinoleic, linoleic, and linolenic acids via the initial formation of peracetic acid from peroxidation of acetic anhydride, followed by reaction with each fatty acid and final saponification with aqueous sodium hydroxide. Three different compounds have been obtained after purification, named as follows: sodium 12-hydroxy-octadec-9-enoate (SAR), sodium 9,10-epoxy-12-hydroxy-octadecanoate (SEAR), and sodium 9,10,11,12-diepoxy-octadecanoate (SEAL). Their chemical structures are shown in Figure 19.12. Note that SAR does not have any epoxy group, but its carbon chain is unsaturated. It has been included in this study to compare its performance against that of its epoxy counterparts. The aggregation behavior of all surfactants in distilled water has been confirmed with tensiometry measurements. Table 19.4 shows the CMC's of the surfactants at 27°C and the surface tension (γ) of each solution at the CMC.

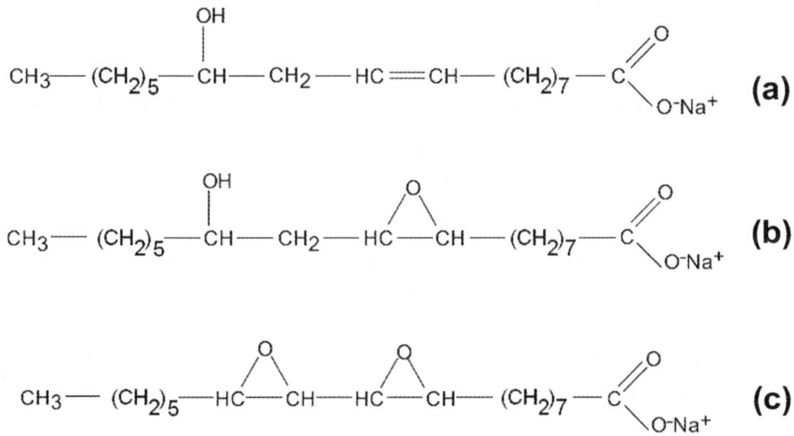

FIGURE 19.12

Chemical structures of epoxy surfactants synthesized from vegetable oils: (a) sodium 12-hydroxy-octadec-9-enoate (SAR), (b) sodium 9,10-epoxy-12-hydroxy-octadecanoate (SEAR), and (c) sodium 9,10,11,12-diepoxy-octadecanoate (SEAL).

Table 19.4 Critical Micelle Concentration and Surface Tension at CMC for Epoxy Surfactants

Surfactant	CMC / (mol L^{-1})	γ / (mN m^{-1})
SAR	8.00×10^{-3}	52.05
SEAR	9.20×10^{-3}	52.70
SEAL	9.40×10^{-3}	54.05

The potential of the surfactants as corrosion inhibitors was assessed in a number of experiments, on API5LX Gr X42 steel. The surface coverage (θ) and percentage efficiency ($E\%$) were determined as a function of surfactant concentration, temperature, salinity, and acidity, aiming at possible applications as inhibitors of acid- and salt-induced corrosion phenomena.

The influence of surfactant concentration on θ and $E\%$ is shown in Figure 19.13, for experiments carried out in 0.5 M NaCl aqueous solution at 30°C. The experimental data have been fitted with the Frumkin and Langmuir isotherms, and they are more satisfactorily correlated with the Frumkin model. For this reason, Frumkin's isotherm is used in all subsequent investigations, as discussed below. It has been observed that insertion of epoxy moieties on the surfactant hydrocarbon chains causes repulsive interactions between the molecules, which are enhanced by increasing the epoxy content (SAR → SEAR → SEAL). In liquid micelle systems that interact with solid, metallic surfaces, an opposite behavior to that of bulk micelles is expected, due to the high polarity of the metallic surfaces. In bulk micelles, the repulsive interactions of the aggregates are diminished with the attachment of epoxy groups to the chains. Micelles of epoxy surfactants may have a slightly flat shape, as opposed to SAR micelles, which tend to be reasonably spherical. This greatly affects their performance in covering metallic surfaces. For SAR, adsorption of spherical micelles is expected only above the CMC, while at lower concentrations (typically between 10^{-4} and 10^{-3} M, there can be formation of hemi-micelles on the surface. As a result, this surfactant can be more effective only above the CMC. On the other hand, the epoxy surfactants SEAR and SEAL behave differently than SAR, in that even below their CMC's there can be some protection of the surface, as indicated in Figure 19.13 by their higher efficiency values. This is probably due to changes in the conformation of the molecules, whereby more intense interactions are established between the surfactant chains and the metal surface. The molecules can then cover the surface more effectively, as monomers (below the CMC) and impair the action of corrosion agents.

The effect of higher salinity levels on θ and $E\%$ curves has been determined for each surfactant at 30°C in 1.0 M NaCl aqueous media. The results are shown in Figure 19.14, where the results for 0.5 M NaCl are reproduced for comparison. In a 1.0 M NaCl environment, a small reduction in the efficiency is observed, but again the epoxy surfactants cover the surface more efficiently even below their CMCs. Above the CMC, more elongated epoxy aggregates are expected at the solid–liquid interface. This reduction in the θ values is associated with minor decreases in the intensity of lateral interactions and adsorption constant in the Frumkin model, which is an indication that adsorption at the interface occurs by means of hemi-micelles or admicelles, rather than fully formed micelles.

The effect of temperature has also been examined. Figure 19.15 shows the aspects of the θ and $E\%$ curves obtained for the surfactants at a fixed salinity level of 0.5 M NaCl. Experiments have been made at 60°C, and the results are compared with those obtained at 30°C. Corrosive phenomena are greatly

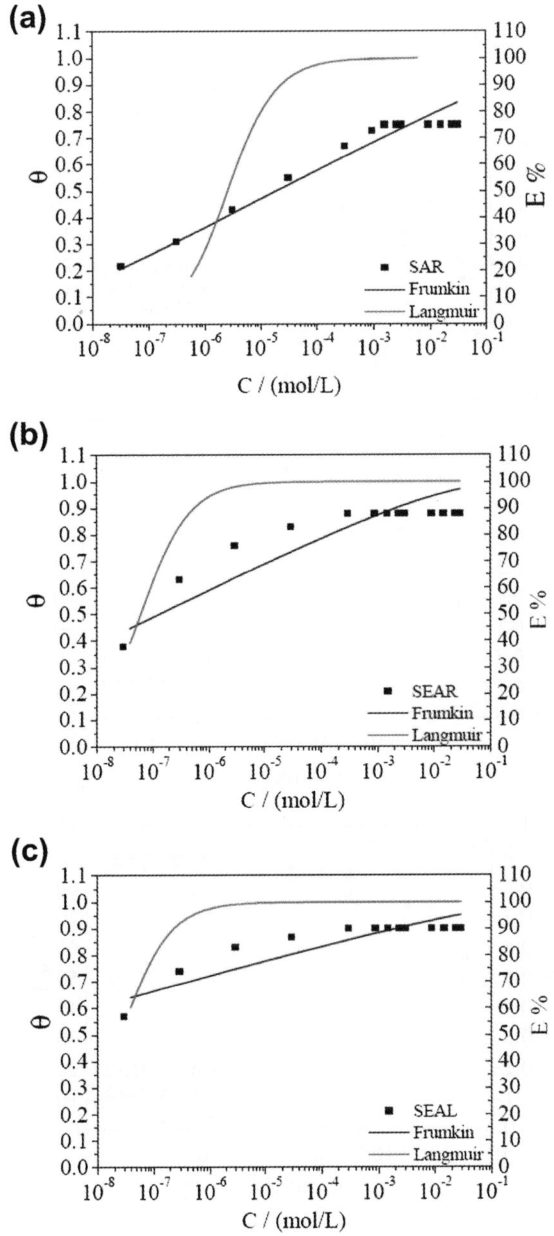

FIGURE 19.13

Influence of structure and concentration of epoxy surfactants on the surface coverage and efficiency of corrosion inhibition in 0.5 M NaCl media at 30 °C: (a) SAR, (b) SEAR and (c) SEAL. (For color version of this figure, the reader is referred to the online version of this book.)

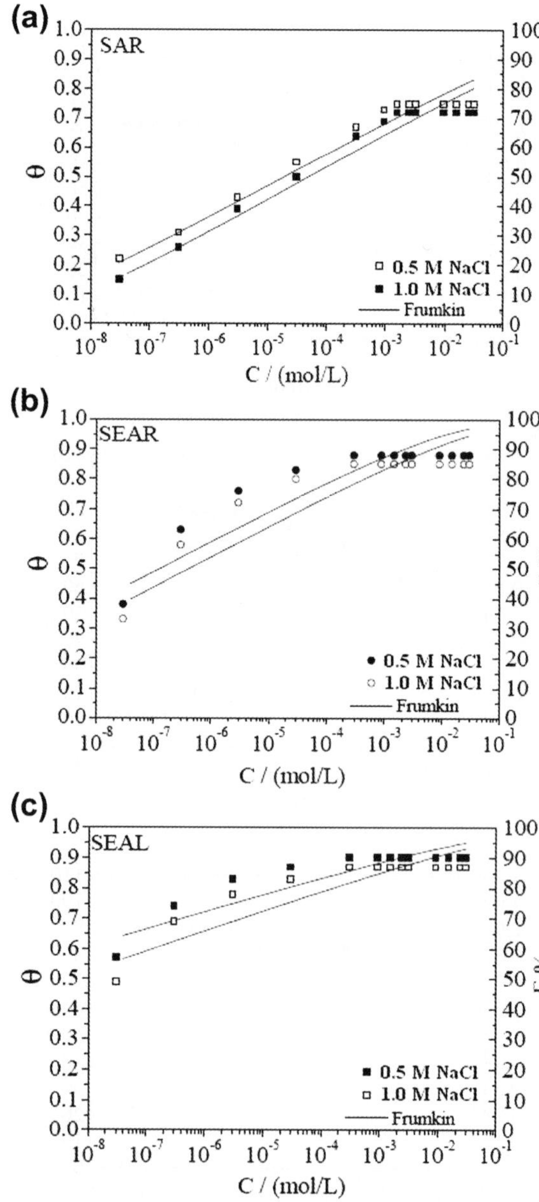

FIGURE 19.14

Influence of salinity on the surface coverage and efficiency of corrosion inhibition at 30 °C: (a) SAR, (b) SEAR and (c) SEAL.

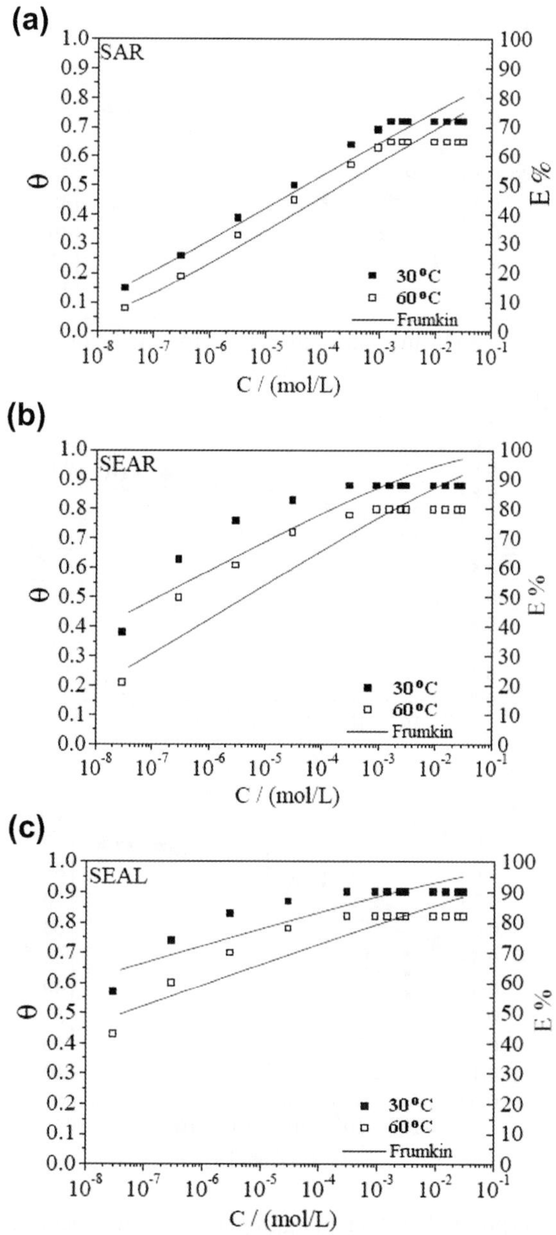

FIGURE 19.15

Influence of temperature on the surface coverage and efficiency of corrosion inhibition in a 0.5 M NaCl environment: (a) SAR, (b) SEAR and (c) SEAL.

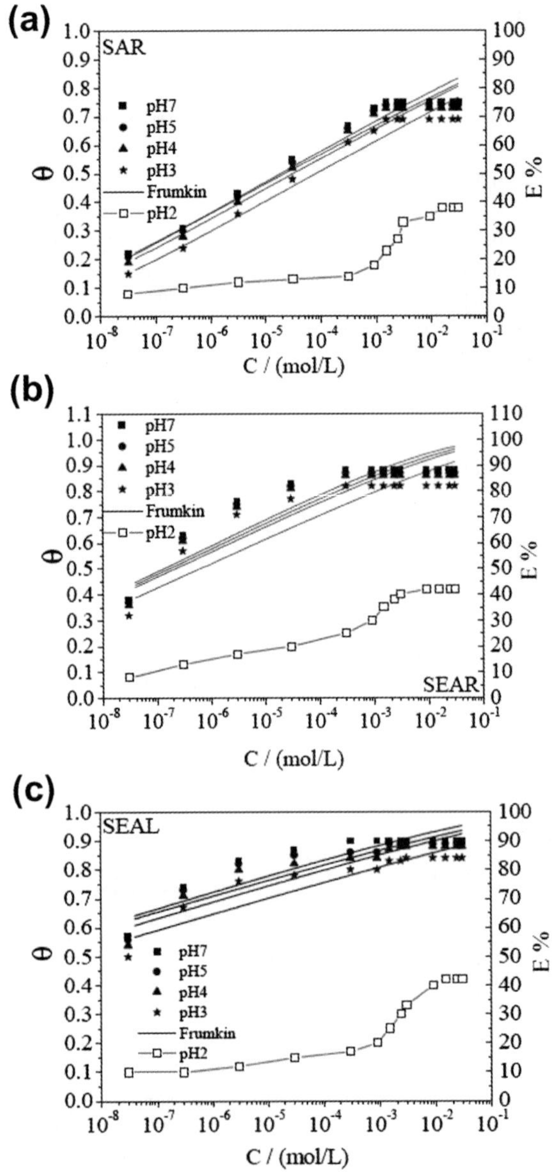

FIGURE 19.16

Influence of pH on the surface coverage and efficiency of corrosion inhibition in a 0.5 M NaCl environment kept at 30°C: (a) SAR, (b) SEAR and (c) SEAL. (For color version of this figure, the reader is referred to the online version of this book.)

enhanced by increasing temperature, because of desorption of surfactant aggregates from the metallic surface. Temperature also seems to cause the same effects in surfactant behavior and performance as those observed in the concentration and salinity studies, in that SAR has a slightly lower efficiency than SEAR and SEAL at 60°C.

Finally, the performance of the surfactants has been tested at different acidity levels, in 0.5 M NaCl solutions at 30°C. The results for pH values between 2 and 7 are shown in Figure 19.16.

Acidity is an important parameter to be considered in the prevention of corrosion, affecting the behavior of surfactant systems and their ability as corrosion inhibitors. Figure 19.16 shows that the surfactants' efficiencies decrease with decreasing pH, because of demicellization promoted by free H^+ ions in solution, which protonate the anionic surfactant molecules, thereby regenerating the corresponding fatty acids. Furthermore, it is possible that the epoxy groups are cleaved at lower pH, with the formation of hydroxyl groups that render an even more acidic character to the molecule. However, OH groups are also active sites that interact with the metallic surface, which could explain why the efficiency values do not drop dramatically at lower pHs. Only at pH = 2 can one provoke changes that negatively affect the efficiency of the surfactant systems, as indicated by the curves in Figure 19.16.

In summary, the effects of salinity and the surfactant's structure and concentration are more pronounced than temperature effects, at least within the levels established for such variables. Differences in behavior and performance are remarkable for SAR systems as compared to both SEAR and SEAL systems, providing evidence of the influence of epoxy groups on surfactant aggregation. These observations can be confirmed by carrying out radiation scattering experiments. Small-angle X-ray scattering (SAXS) is a useful technique that provides some preliminary information on size and shape of aggregates. Some measurements were made at the Brazilian Synchrotron Light Laboratory (LNLS), by running specific samples in the new SAXS2 beam line installed at LNLS. The sample-to-detector distance was fixed at 707.106 mm, allowing for a scattering vector amplitude (q) ranging between 0.18 nm^{-1} and 4.756 nm^{-1} at a wavelength of 0.1488 nm. Although this is part of another ongoing research project, a few interesting results have been obtained that permit establishing a correlation between the size and shape of aggregates, the nature of interactions between them, and parameters such as salinity and temperature. Overall, spherical aggregates are formed with these surfactants with repulsive interactions, which help corroborating the potentiokinetic data presented above. In Table 19.5, some representative data on particle size are given at different salinities, surfactant concentrations, and temperatures for the SAR surfactant. These are average particle sizes, but the SAXS curves (not shown here) indicated that reasonably spherical particles were formed above the CMC, with relatively low polydispersity. By increasing salinity, more aggregation seems to be induced, as suggested by the larger particles formed. The pictorial representation of micelles in saline media is shown in Figure 19.17 as salinity increases. The distinct sizes of the bulk micelles is a result of different concentrations of

Table 19.5 Average Diameter of SAR Micelles in Different Solutions from SAXS Measurements

	Average Particle Diameter / nm		
Solution	**T = 30°C (1) [SAR] = 1 % m/v**	**T = 60°C (1) [SAR] = 1 % m/v**	**T = 60°C (1) [SAR] = 2 % m/v**
SAR in water	36.94	36.94	33.94
SAR in 0.1 M NaCl	39.25	Not defined	38.06
SAR in 0.5 M NaCl	43.31	48.31	43.31
SAR in 1.0 M NaCl	48.31	57.09	52.33

(1) *T = temperature; [SAR] = surfactant concentration in the medium, in mass/volume percentage.*

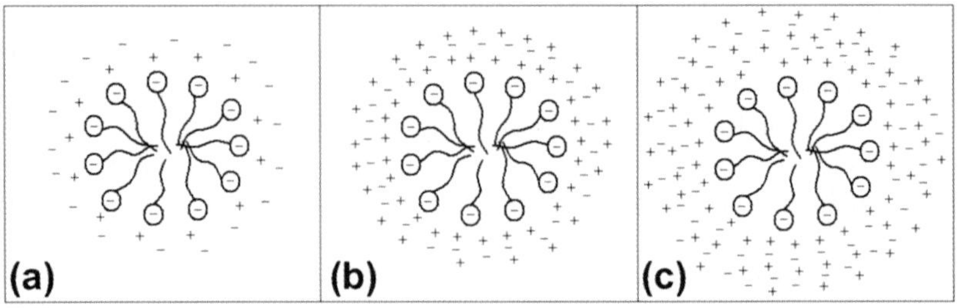

FIGURE 19.17

Schematic representation of micelles formed in aqueous solution with increasing salinity levels: (a) 0.1 M NaCl; (b) 0.5 M NaCl; (c) 1.0 M NaCl.

counter-ions surrounding the surfactant's polar headgroups. As salinity increases, a cloud of electrical charges is formed, which induces the lateral separation of the surfactants and consequently increases micelle size.

If the concentration of surfactant in the medium is increased, especially above the CMC, the same behavior is observed, particularly at higher salinity levels. Although temperature promotes surfactant desorption, which is responsible for the decrease in the efficiency of corrosion inhibition, the SAXS data have shown that the systems can be stable at higher temperatures under different conditions of salinity and composition, in spite of the experimental difficulties encountered when examining the 0.1 M NaCl systems at 60°C with 1% m/v surfactant. This is a very important observation because, in real applications, the confirmation that temperature gradients do not seem to materially affect the stability of the aggregates is an advantage that must be pointed out.

3 Conclusions

The improvement of modern technology and the development of research on novel materials have generated a large number of customized metallic parts, with particular properties, in several industrial applications. Metal corrosion has always been a major problem in the industry, and there have always been continuous requirements to expand the offer of potential, low-cost corrosion inhibitors. This can be accomplished both on practical and fundamental levels, by optimizing the industrial processes and understanding how the inhibition mechanism is affected at the interfaces.

However, many commercial corrosion inhibitors exhibit both high efficiency and toxicity, which has led some environmental agencies to prohibit their use. Consequently, there has been growing interest in carrying out scientific investigations on the use of environmentally friendly compounds; surfactants and biocompatible corrosion inhibitors are special categories of chemicals that are suitable for sustainable industrial applications. Molecules with very distinct hydrophilic and lipophilic characters may be used in a variety of applications, overcoming solubilization problems and enhancing extraction procedures.

The use of surfactant-based formulations in the inhibition of corrosion on metallic surfaces is an innovative approach in chemical technology. Studies have focused on the use of several surfactant solutions and, more recently, microemulsions have been cited as potential inhibitors, with efficiencies higher than 90% in various corrosive situations. In some applications microemsulsions are highlighted as excellent transport media that are able to dissolve higher amounts of active matter, or surfactants, remaining stable over a wide range of experimental conditions. In particular salinity, acidity, and temperature are important variables to be considered in the investigations. It is important to understand how these parameters influence the adsorption mechanisms of surfactant molecules and aggregates at solid–liquid interfaces. Another interesting aspect is the use of renewable resources, such as vegetableoils, to synthesize novel surfactant molecules that can be used in the formulations with many advantages. We have demonstrated the efficiency of some ionic surfactants derived from castor and soybean oils in inhibiting corrosion on steel surfaces by means of tensiometry and potentiokinetic measurements. Also, association of experimental techniques is valuable to provide a complete description of all mechanisms involved in the development of corrosive phenomena and to optimize the most adequate inhibition processes. The results that our research group have been obtaining are very promising and have inspired more comprehensive investigations on the nature of the corrosive phenomena and the performance of micelle and microemulsion systems as templates for corrosion inhibitors with regard to surfactant structure.

SYMBOLS AND TERMINOLOGY

A lateral interaction parameter of Frumkin's isotherm model
AR1S sodium 12-*N*,*N*-diethylamino-octadec-9-enoate
CMC critical micelle concentration
CNSL cashew-nut shell liquid
C/S co-surfactant/surfactant ratio
dec decade (used in potentiokinetic measurements to quantify Tafel's curvatures)
DDAB dodecylammonium bromide
DDAC dodecylammonium chloride
E_{corr} corrosion potential
I_{corr} corrosion current
K_F concentration constant of Frumkin's isotherm model
LNLS Brazilian Synchrotron Light Laboratory
O/W oil-in-water
P_c critical surfactant concentration in microemulsion systems
q scattering vector amplitude
SAMs self-assembled monolayers
SAR sodium 12-hydroxy-octadec-9-enoate
SAXS small-angle X-ray scattering
SEAL sodium 9,10,11,12-diepoxy-octadecanoate
SEAR sodium 9,10-epoxy-12-hydroxy-octadecanoate
WI Winsor I system (microemulsion and excess oil in equilibrium)
WII Winsor II system (microemulsion and excess water in equilibrium)
WIII Winsor III system (microemulsion and excess oil and water in equilibrium)
WIV Winsor IV system (single-microemulsion phase)

Greek letters

β_a Tafel's anodic curvature
β_c Tafel's cathodic curvature
γ surface tension
$\delta\Theta_F$ Frumkin's object function
$\Delta_{ads}G$ Gibbs free energy of adsorption

Glossary

Admicelles pre-micelle adsorbed surfactant aggregates that can form at a concentration below the CMC; known as critical admicelle concentration (or cac) in some situations, typically as two-dimensional clusters

Cardanol common name given to a mixture of phenols extracted from cashew-nut shell liquid (*Anacardium occidentale*), with a long hydrocarbon chain attached to the benzene ring at a meta position that can be saturated or unsaturated

Castor oil the vegetable oil obtained from the beans of the castor plant (*Ricinus communis*)

CMC critical micelle concentration, or the minimal amount of surfactant molecules required to form bulk micelles in solution at fixed conditions, like temperature

Counter-ion the ion associated with the larger part of an ionic surfactant molecule, like sodium ions in the sodium dodecylsulfate molecules

Co-surfactant normally a low-molecular-weight organic solvent, typically an alcohol or amine, that helps stabilizd microemulsion formulations under specific experimental conditions

Electrolyte normally an inorganic salt, like sodium chloride, added to surfactant systems to examine the effect of salinity on their performance and stability

Epoxy group a saturated cyclic ether composed of two carbon atoms and one oxygen atom forming an approximately equilateral triangle

Headgroup the main polar site or moiety of a surfactant molecule, like the carboxylate or sulfate groups in anionic surfactants, and the polyethylene oxide group in nonionic molecules

Hemi-micelles distorted or incomplete three-dimensional surfactant clusters that can form at some surfaces, like solid-liquid interfaces, because of preferential interactions between surfactant molecules and the surface

Hydrophilic-lipophilic balance an empirical or theoretical correlation that establishes comparative standards for the classification of surfactant molecules according to their polar or nonpolar character, which ultimately defines the range of applications for the compounds

Hydrophobic effect a phenomenon whereby compounds that feature regions with different polarities on their moleucles, like surfactants, tend to form clusters or aggregates in a polar solvent in order to avoid unfavorable interactions with the solvent, being mostly responsible for the micellization of surfactant molecules in several circumstances

Micelle solutions homogeneous mixture obtained when single or polydisperse surfactants are dissolved in a pure aqueous or nonaqueous solvent above the CMC

Micelle aggregation of surfactant molecules formed when they are added to a given solvent in a concentration above their CMC at the experimental conditions imposed, characterizing the micellization phenomenon

Microemulsions optically isotropic, transparent, thermodynamically stable chemical systems, formed with the intimate mixture of polar and nonpolar solvents that can be stabilized by an interfacial layer of surfactant molecules, sometimes requiring additives like co-surfactants or co-solvents

Monomer in surfactant systems, corresponds to a single, nonaggregated surfactant molecule, dissolved in a given solvent at concentrations below its CMC

Packing parameter: an extension of the hydrophilic-lipophilic balance concept useful to characterize the type or shape of the clusters formed when surfactants are in solution

Polydispersity the state of being polydisperse; in surfactant systems, polydispersity is characterized by a mixture of similar molecules with different molecular weights, like commercial nonionic polyethoxylated surfactants, or nonuniform microdispersions like micelles or microemulsion droplets that feature a distribution of particle sizes

Saponification chemical treatment or neutralization of fatty acids with bases like sodium hydroxide, whereby the original nonionic molecules are converted to anionic surfactants

Self-assembly the spontaneous aggregation of surfactant molecules when dissolved in solution at sufficiently high concentrations, forming a variety of packed structures depending on a number of factors, like surfactant structure, temperature, and additives

Surfactants organic molecules that feature chemical moieties or groups with opposite polarities that render them active at surfaces and interfaces, reducing interfacial tensions by adsorption

Zwitterionic surfactants molecules that feature positive and/or negative charges when dissolved in solution depending on the pH

Acknowledgments

The authors gratefully acknowledge financial support from CNPq / Brazil (National Council for Scientific and Technological Development), ANP-PRH 14 / Brazil (Human Resources Program 14 of the Petroleum National Agency), and the support granted by the Brazilian Synchrotron Light Laboratory (LNLS) under proposal D11A - SAXS1 7021, with the allocation of SAXS beam time and contribution with consumables and travel expenses.

References

Aggarwal, L.K., Thapliyal, P.C., Karade, S.R., 2007. Anticorrosive properties of the epoxy-cardanol resin based paints. Prog. Org. Coatings 59 (1), 76–80.

Ali, S.A., Saeed, M.T., 2001. Synthesis and corrosion inhibition study of some 1,6-hexanediamine-based N,N-diallyl quaternary ammonium salts and their polymers. Polymer 42 (7), 2785–2794.

Becerra, H.Q., Retamoso, C., Macdonald, D.D., 2000. The corrosion of carbon steel in oil-in-water emulsions under controlled hydrodynamic conditions. Corrosion Sci. 42 (3), 561–575.

Bouyanzer, A., Majidi, L., Hammouti, B., 2006. Effect of eucalyptus oil on the corrosion of steel in 1M HCl. Bull. Electrochem. 22 (7), 321–324.

Branzoi, V., Branzoi, F., Baibarac, M., 2000. The inhibition of the corrosion of Armco iron in HCl solutions in the presence of surfactants of the type of N-alkyl quaternary ammonium salts. Mater. Chem. Phys. 65 (3), 288–297.

Chaieb, E., Bouyanzer, A., Hammouti, B., Benkaddour, M., 2005. Inhibition of the corrosion of steel in 1 M HCl by eugenol derivatives. Appl. Surface. Sci. 246 (1–3), 199–206.

Clint, J.H., 1992. Surfactant aggregation. Chapman and Hall, New York.

Dantas, T.N.C., Dantas Neto, A.A., Moura, E.F., 2001. Microemulsion systems applied to breakdown petroleum emulsions. J. Petroleum. Sci. Eng. 32 (2–4), 145–149.

Dantas, T.N.C., Moura, E.F., Scatena Júnior, H., Dantas Neto, A.A., Gurgel, A., 2002. Micellization and adsorption thermodynamics of novel ionic surfactants at fluid interfaces. Colloids and Surfaces A: Physicochemical and Engineering Aspects 207, 243–252.

Dantas, T.N.C., Dantas Neto, A.A., Rossi, C.G.F.T., Gomes, D.A.A., Gurgel, A., 2010. Use of microemulsion systems in the solubilization of petroleum heavy fractions for the prevention of oil sludge waste formation. Energy Fuels 24, 2312–2319.

Dantas Neto, A.A., Dantas, T.N.C., Moura, M.C.P.A., Barros Neto, E.L., Gurgel, A., 2008. Microemulsions as potential interfacial chemical systems applied in the petroleum industry. In: Monzer Fanum (Ed.), Microemulsions: Properties and applications, vol. 144. CRC Press, Boca Raton, FL, pp. 411–449.

Deyab, M.A., 2007. Effect of cationic surfactant and inorganic anions on the electrochemical behavior of carbon steel in formation water. Corrosion Sci. 49 (5), 2315–2328.

Deyab, M.A., Dief, H.A.A., Eissa, E.A., Taman, A.R., 2007. Electrochemical investigations of naphthenic acid corrosion for carbon steel and the inhibitive effect by some ethoxylated fatty acids. Electrochimica Acta. 52 (28), 8105–8110.

Deyab, M.A., Abd El-Rehim, S.S., Keera, S.T., 2009. Study of the effect of association between anionic surfactant and neutral copolymer on the corrosion behaviour of carbon steel in cyclohexane propionic acid. Colloids and Surfaces A: Physicochemical and Engineering Aspects 348, 170–176.

El-Achouri, M., Hajji, M.S., Salem, M., et al., 1996. Some nonionic surfactants as inhibitors of the corrosion of iron in acid chloride solutions. Corrosion 52 (2), 103–108.

El-Achouri, M., Hajji, M.S., Kertit, S., Essassi, E.M., Salem, M., Coudert, R., 1995. Some surfactants in the series of 2-(alkyldimethylammonio) alkanol bromides as inhibitors of the corrosion of iron in acid chloride solution. Corrosion Sci. 37 (3), 381–389.

El-Etre, A.Y., 1998. Natural honey as corrosion inhibitor for metals and alloys. I. Copper in neutral aqueous solution. Corrosion Sci. 40 (11), 1845–1850.

El-Etre, A.Y., Abdallah, M., 2000. Natural honey as corrosion inhibitor for metals and alloys. II. C-steel in high saline water. Corrosion Sci. 42 (4), 731–738.

El-Rehim, S.S.A., Hassan, H.H., Amin, M., 2003. The corrosion inhibition study of sodium dodecyl benzene sulphonate to aluminium and its alloys in 1.0 M HCl solution. Mater. Chem. Phy. 78 (2), 337–348.

El-Sherbini, E.E.F., Abd-el-Wahab, S.M., Deyab, M.A., 2003. Studies on corrosion inhibition of aluminum in 1.0 M HCl and 1.0 M H_2SO_4 solutions by ethoxylated fatty acids. Mater. Chem. Phy. 82 (3), 631–637.

El-Sherbini, E.E.F., Wahaab, S.M.A., Deyab, M.A., 2005. Ethoxylated fatty acids as inhibitors for the corrosion of zinc in acid media. Mater. Chem. Phy. 89 (2–3), 183–191.

Hajjaji, N., Rico, I., Srhiri, A., Soufiaoui, M., Benbachir, A., 1993. Effect of N-alkylbetaines on the corrosion of iron in 1-M HCl solution. Corrosion 49 (4), 326–334.

Holmberg, K., 1998. Quarter century progress and new horizons in microemulsions. In: Dinesh O, Shah (Ed.), Micelles, microemulsion, and monolayers: Science and technology. Marcel Dekker, New York, pp. 161–192. Chapter 7.

Holmberg, K., Jönsson, B., Kronberg, B., Lindman, B., 2002. Surfactants and polymers in aqueous solution, second ed. John Wiley & Sons, Chichester, England.

Li, X.H., Deng, S.D., Mu, G.N., Fu, H., Yang, F.Z., 2008. Inhibition effect of nonionic surfactant on the corrosion of cold rolled steel in hydrochloric acid. Corrosion Sci. 50 (2), 420–430.

Li, X.H., Deng, S.D., Fu, H., Mu, G.N., 2009. Inhibition by tween-85 of the corrosion of cold rolled steel in 1.0 M hydrochloric acid solution. J. Appl. Electrochem. 39 (7), 1125–1135.

Lindman, B., Tiberg, F., Piculell, L., Olsson, U., Alexandridis, P., Wennerström, H., 1998. Surfactant self-assembly structures at interfaces, in polymer solutions, and in bulk: micelle size and connectivity. In: Dinesh O, Shah (Ed.), Micelles, microemulsion, and monolayers: Science and technology. Marcel Dekker, New York, pp. 101–126. Chapter 4.

Lligadas, G., Ronda, J.C., Galia, M., Biermann, U., Metzger, J.O., 2006. Synthesis and characterization of polyurethanes from epoxidized methyl oleate based polyether polyols as renewable resources. J. Polym. Sci. Part A-Polym. Chem. 44 (1), 634–645.

Love, J.C., Estroff, L.A., Kriebel, J.K., Nuzzo, R.G., Whitesides, G.M., 2005. Chem. Rev. 105 (4), 1103–1169.

Luo, H., Guan, Y.C., Han, K.N., 1998a. Inhibition of mild steel corrosion by sodium dodecyl benzene sulfonate and sodium oleate in acidic solutions. Corrosion 54 (8), 619–627.

Luo, H., Guan, Y.C., Han, K.N., 1998b. Corrosion inhibition of a mild steel by aniline and alkylamines in acidic solutions. Corrosion 54 (9), 721–731.

Mandrino, D., Godec, M., Torkar, M., Jenko, M., 2008a. Study of oxide protective layers on stainless steel by AES, EDS and XPS. Surface and Interface Analysis 40 (3–4), 285–289.

Mandrino, D., Godec, M., Kocijan, A., Lamut, M., Torkar, M., Jenko, M., 2008b. Oxide and nitride protective layers formed on stainless steel by thermal treatment: SEM, AES, WDS and corrosion measurements. Meta 47 (2), 119–123.

Metzger, J.O., Bornscheuer, U., 2006. Lipids as renewable resources: current state of chemical and biotechnological conversion and diversification. Appl. Microbiol. Biotechnol. 71 (1), 13–22.

Migahed, M.A., 2005. Electrochemical investigation of the corrosion behaviour of mild steel in 2 M HCl solution in presence of 1-dodecyl-4-methoxy pyridinium bromide. Mater. Chem. Phys. 93 (1), 48–53.

Migahed, M.A., Abd-El-Raouf, M., Al-Sabagh, A.M., Abd-El-Bary, H.M., 2005. Effectiveness of some non ionic surfactants as corrosion inhibitors for carbon steel pipelines in oil fields. Electrochimica Acta. 50 (24), 4683–4689.

Moura, E.F., Wanderley Neto, A.O., Dantas, T.N.C., Scatena Júnior, H., Gurgel, A., 2009. Applications of micelle and microemulsion systems containing amine surfactants synthesized from ricinoleic acid as carbon-steel corrosion inhibitors. Colloids and Surfaces A: Physicochemical and Engineering Aspects. 340, 199–207.

Mu, G.A., Li, X.H., 2005. Inhibition of cold rolled steel corrosion by Tween-20 in sulfuric acid: Weight loss, electrochemical and AFM approaches. J. Colloid. Interface Sci. 289 (1), 184–192.

Nmai, C.K., 2004. Multi-functional organic corrosion inhibitor. Cement and Concrete Composites 26 (3), 199–207.

Osman, M.M., Shalaby, M.N., 2003. Some ethoxylated fatty acids as corrosion inhibitors for low carbon steel in formation water. Materials Chem. Phys. 77 (1), 261–269.

Osman, M.M., El-Ghazawy, R.A., Al-Sabagh, A.M., 2003. Corrosion inhibitor of some surfactants derived from maleic-oleic acid adduct on mild steel in 1 M H_2SO_4. Mater. Chem. Phys. 80 (1), 55–62.

Ouachikh, O., Bouyanzer, A., Bouklah, M., et al., 2009. Application of essential oil of artemisia herba alba as green corrosion inhibitor for steel in 0.5 M H_2SO_4. Surface Rev. Lett. 16 (1), 49–54.

Paria, S., Khilar, K.C., 2004. A review on experimental studies of surfactant adsorption at the hydrophilic solid–water interface,. Adv. Colloid Interface. Sci. 110 (3), 75–95.

Patel, C.J., Dighe, A., 2007. Novel isocyanate-free self-curable cathodically depositable epoxy coatings: Influence of epoxy groups on coating properties. Prog. Org. Coatings 60 (3), 219–223.

Ramji, K., Cairns, D.R., Rajeswari, S., 2008. Synergistic inhibition effect of 2-mercaptobenzothiazole and Tween-80 on the corrosion of brass in NaCl solution. Appl. Surface Sci. 254 (15), 4483–4493.

Roberto, E.C., Wanderley Neto, A.O., Martínez-Huitle, C.A., Fonseca, J.L.C., Dantas, T.N.C., Gurgel, A., 2013. The effect of type of self-assembled system and pH on the efficiency of corrosion inhibition of carbon-steel surfaces. Prog. Org. Coatings 76 (10), 1308–1315.

Saravanan, K., Sathiyanarayanan, S., Muralidharan, S., Azim, S.S., Venkatachari, G., 2007. Performance evaluation of polyaniline pigmented epoxy coating for corrosion protection of steel in concrete environment. Prog. Org. Coatings 59 (2), 160–167.

Shalaby, M.N., Osman, M.M., 2001. Synergistic inhibition of anionic and nonionic surfactants on corrosion of mild steel in acidic solution. Anti-Corrosion Methods and Mater 48 (5), 309–318.

Shalaby, M.N., Osman, M.M., 2009. Protection of copper surface against corrosion by cationic surfactant in seawater. J. Dispersion Sci. Technol. 30 (5), 677–683.

Shervedani, R.K., Mehrjardi, A.H., Babadi, M.K., 2007. Comparative electrochemical study of self-assembled monolayers of 2-mercaptobenzoxazole, 2-mercaptobenzothiazole and 2-mercaptobenzimidazole formed on polycrystalline gold electrode. Electrochim. Acta. 52 (24), 7051–7060.

Silverman, D.C., Kalota, D.J., Stover, F.S., 1995. Effect of pH on corrosion inhibition of steel by polyaspartic acid. Corrosion Sci. 51 (11), 818–825.

Soer, W.J., Ming, W.H., Koning, C.E., van Benthem, R.A.T.M., 2008. Towards anti-corrosion coatings from surfactant-free latexes based on maleic anhydride containing polymers. Prog. Org. Coatings 61 (2-4), 224–232.

Tan, Y., O'Haver, J.H., 2004. Lipophilic linker impact on adsorption of and styrene adsolubilization in polyethoxylated octylphenols. Colloids and Surfaces A: Physicochemical and Engineering Aspects. 232, 101–111.

Tanford, C., 1980. The hydrophobic effect: formation of micelles and biological membranes, second ed. John Wiley & Sons, New York.

Ulman, A., 1991. An introduction to ultrathin organic films: From Langmuir–Blodgett to self-assembly. Academic Press, New York.

Wanderley Neto, A.O., Gurgel, A., Dantas, T.N.C., Moura, E.F., Dantas Neto, A.A., 2009. Micelle solutions and microemulsions formed with ionic surfactants applied as corrosion inhibitors: Characterization by small-angle X-ray scattering. In: Proceedings of the XIX Brazilian Synchrotron Light Laboratory's Annual Users Meeting, Campinas (SP), Brazil, 204–204.

Wanderley Neto, A.O., Moura, E.F., Scatena Junior, H., Dantas, T.N.C., Dantas Neto, A.A., Gurgel, A., 2012. Preparation and application of self-assembled systems containing dodecylammonium bromide and chloride as corrosion inhibitors of carbon-steel. Colloids and Surfaces A: Physicochemical and Engineering Aspects 398, 76–83.

Winsor, P.A., 1954. Solvent properties of amphiphilic compounds. Butterworth, London.

CHAPTER

20 The Role of Corrosion Inhibitors in Protecting Metallic Structures against Corrosion in Harsh Environments

El-Sayed M. Sherif

Center of Excellence for Research in Engineering Materials (CEREM), College of Engineering, King Saud University, Al-Riyadh, Saudi Arabia; Electrochemistry and Corrosion Laboratory, Department of Physical Chemistry, National Research Centre (NRC), Cairo, Egypt

CHAPTER OUTLINE

1 Introduction

Corrosion is a natural process and can be defined as the undesirable deterioration of a material, usually a metal, through a reaction with its environment (Riggs 1973). This reaction is electrochemical in nature, as in the case of metals and alloys due to their high conductivity. The corrosion reaction can

The Role of Colloidal Systems in Environmental Protection. http://dx.doi.org/10.1016/B978-0-444-63283-8.00020-X

also be chemical deterioration as in the case of nonconducting materials, such as ceramics, plastics, and concrete. The driving force that causes metals to corrode is a natural consequence of their temporary existence in metallic form. To reach this metallic state from their occurrence in nature in the form of various chemical compounds (ores), it is necessary for them to gain and store the energy that is later returned through the corrosion process. The amount of energy required and stored varies between metals. It is relatively high for active metals such as aluminum, magnesium, and iron, while relatively low for noble metals, such as copper, silver, and gold. Most metals corrode upon contact with water or moisture, acids, bases, salts, oils, aggressive metal polishes, and other solid and liquid chemicals. Corrosion is often insidious and hidden until striking at the worst moment of a system operation (Roberge and Revie 2007, Sherif 2010, Sherif and Ahmed 2010, Sherif and Almajid 2010, Sherif, Erasmus and Comins 2010).

Corrosion can cause dangerous and expensive damage to everything from automobiles, home appliances, and drinking water systems to pipelines, bridges, and public buildings. The importance of corrosion studies is threefold. First, economically: that includes the cost of material losses resulting from the wasting away or sudden failure of piping, tanks, ships, hulls, marine structures, and the like. The most commonly quoted figure for corrosion costs in the United States is \$276 billion (or about 3.1% of the U.S. gross domestic product (GDP)) in 1998, which was reported in the NACE Corrosion Costs Study. This estimate is a bit outdated now, and closer examination of the NACE data indicates that the costs of corrosion in the United States in 2012 totaled more than \$1 trillion. The corrosion costs also include the losses of energy and water reserves accompanying the production of metals and the fabrication of new structures as replacements for the corroded ones. In addition, a human effort is involved in the re-design and rebuilding of corroded equipment, which otherwise would be available for other useful purposes. Another important consideration for corrosion study is the conservation of mineral and metal resources, of which the world's supply is limited. Third is the environmental factor, which involves the reduction of potentially damaging and toxic metallic ions into the soil and water sources (McEwan 2004).

A variety of methods have been used to protect metallic structures against corrosion, where the harsh environment attacks the exposed metal surfaces. One of the most important methods in that regard is the use of corrosion inhibitors (Sherif and Park 2005, Sherif 2006, Sherif and Park 2006, 4065, Sherif and Park 2006, 6556, Sherif, Erasmus, and Comins 2007, Sherif 2011, 961, and Sherif 2013). By definition, a corrosion inhibitor is a chemical substance or combination of substances that, when added in small concentration to an environment, prevents or reduces corrosion without significant reaction with the components of the environment. Under this condition, the corrosion rate of the exposed metals and/or alloys decreases by a reaction at the metal–environment interface (Roberge 2000). Inhibitors are added to many systems—notably, coolants, fuels, hydraulic fluids, boiler water, engine oil, and many other fluids used in industry. (Sanyal 1981). For example, water treatment manufacturers combine many different inhibitors to produce a strong product that will inhibit a number of different metals in a wide range of field conditions. As well as in the oil extraction and processing industries, inhibitors have always been considered to be the first line of defense against corrosion (Gad Allah et al. 1989, Altoé et al. 1996, Al-Sayegh, Carew, and Al-Hashem 2001, Prokhorenkov et al. 2002, Makarenko et al. 2003, and Khalifa et al. 2003).

A great number of scientific studies have been devoted to the subject of corrosion inhibitors (Khaled and Hackerman 2004, Sherif and Park 2005, Sherif and Park 2006, 1313, Sherif and Park 2006, 4665, El-Etre 2006, Sherif, Erasmus, and Comins 2007, 144, Sherif, Erasmus. and Comins 2007,

470, and Sherif et al. 2007). Also, the scientific and technical corrosion literature has descriptions and lists of numerous chemical compounds that exhibit inhibitive properties. Of these, only a few are actually used in practice. This is partly because the desirable properties of an inhibitor usually extend beyond those simply related to metal protection. Considerations of cost, toxicity, availability, and environmental friendliness are of considerable importance. However, most of what is known has grown from trial-and-error experiments, both in the laboratory and in the field. Rules, equations, and theories to guide inhibitor development or use are very limited (Roberge 2000).

The present review brings together the definition, classification, inhibition efficiencies, and industrial applications of corrosion inhibitors in controlling the corrosion of metals and alloys in harsh environments such as aqueous media and atmosphere. In addition, the aim is to shed more light on the most frequently used techniques to obtain the inhibition efficiencies of inhibitors on surfaces of metallic structures.

2 Classification of corrosion inhibitors

Inhibitors are chemicals that react with a metallic surface, giving it a certain level of protection. Inhibitors often work by adsorbing themselves to form a film on the metallic surface and thus protect it through slowing the corrosion processes through one of these ways: (1) Increasing the anodic or cathodic polarization behavior (Tafel slopes); (2) reducing the movement or diffusion of ions to the metallic surface; and (3) Increasing the electrical resistance of the metallic surface. For this reason, inhibitors have been classified differently by various investigators. Some authors, for example, prefer to group inhibitors by their chemical functionality—for example, passivators, precipitators, vapor phase, cathodic, anodic, mixed, neutralizing, film-forming type, and absorbents (Bregmann 1963, Bockris and Reddy 1970, Riggs 1973, Rozenfield 1981, Sanyal 1981, Uhlig and Revie 1985, Talbot D. and J. Talbot 1998, and Roberge and Revie 2007). Putilova et al. (Putilova, Balezin, and Barannik 1966) have also described three types of inhibitors; type A forms a protective film on the metal surface; type B reduces the aggressiveness of the corrosive media; and type ABforms a protective film and also reduces the aggressiveness of the corrosive agents. However, by far the most popular organization scheme consists in regrouping corrosion inhibitors in a functionality scheme that is normally divided into five groups: (1) passivating inhibitors; (2) cathodic inhibitors; (3) organic inhibitors; (4) precipitation inhibitors; and (5) volatile corrosion Inhibitors (Roberge 2000):

2.1 Passivating inhibitors

Passivating inhibitors are the most widely used and the most effective of all inhibitors (Pryor and Cohen 1953 and Williams and McMurray 2001). This type of inhibitor reacts with the metal being corroded to form a very thin film that prevents any further metal being lost. The film is only maintained by the continued presence of inhibitor in the corrosive medium. They are called anodic inhibitors because they passivate the anodic reaction and cause a large anodic shift of the corrosion potential, forcing the metallic surface into the passivation range (Robertson 1951, Konno, Kobayashi and Nagayama 1982, Aramaki 2001, and Kending and Buchheit 2003). In order for an anodic corrosion inhibitor to be effective, it must accomplish a sufficient dose to "coat" the internal surfaces of the system plus some excess to maintain this film. If the dose is markedly below the level required to achieve protection, exposed areas of metal will continue to corrode. This is because if all of the anode sites are not passivated, the cathodic reaction will be supported by the remaining anodic sites and pitting will occur. For this reason, anodic inhibitors are also termed as unsafe inhibitors as "dangerous." On the other hand, overdosing with anodic inhibitors is unlikely to have any detrimental effect. Passivation inhibitors themselves are classified into two types: (1) oxidizing anions and (2) nonoxidizing ions.

2.1.1 Oxidizing passivation inhibitors

Oxidizing anions, such as chromate, nitrite, and nitrate, work through passivation of metals and alloys in the absence of oxygen (Schweitzer 1989). They are incorporated into the oxide film of the metal, thus stabilizing it and stifling further dissolution. In the case of chromate, it is the least expensive inhibitor for use in water systems and is widely used in the recirculation-cooling systems of internal combustion engines, rectifiers, and cooling towers. Chromates work by reacting with ferrous ions to produce a mixture of chromium and iron oxides. Therefore, they form a protective oxide film over the anodic sites on the metal surface (Schweitzer 1989, Schweitzer 1996, and Zein El Abedin 2001). Sodium chromate, typically in concentrations of 0.04–0.1% was used for these applications. At higher temperatures or in freshwater with chloride concentrations above 10 ppm, higher concentrations are required. If the concentration of chromate falls below a concentration of 0.016%, corrosion will be accelerated (Alexander and Moccari 1993 and Ilevbare and Burstein 2003).

Nitrite is considered an anodic inhibitor and provides passivation corrosion protection because it reacts and bonds with the metal at the point of potential metal loss by forming a thin, protective, monomolecular film, thus minimizing the anodic corrosion reaction and reducing metal loss. The monomolecular film formed from nitrites can develop over either clean or rusted surfaces. Applying it over clean surfaces minimizes corrosion, and applying it over rusted surfaces inhibits further corrosion. Nitrite-based corrosion inhibitors are easy to test, and they provide excellent corrosion protection; they are also economical to use when there is very little air ingression into the system. However, excessive or continuous air ingression into the system will oxidize the nitrite to the less effective nitrate form, thus resulting in corrosion of iron material in the system. The most common application for nitrites is in antifreeze-type cooling water systems because these systems have little tendency to react with alcohols or ethylene glycol. Nitrites are also applied in the inhibition of the internal surfaces of pipelines used to transport petroleum products and gasoline. Nitrite enters the water phase and effectively eliminates the corrosion process (Schweitzer 1989).

2.1.2 Nonoxidizing passivation inhibitors

Nonoxidizing ions, such as phosphate, molybdate, and tungstate, work successfully and have shown a powerful efficiency as passivating corrosion inhibitors for metals and alloys in the presence of oxygen (Robertson 1951, Alexander and Moccari 1993, and Zein El Abedin 2001). Accordingly, low concentrations of these anions in the absence of oxygen accelerate the corrosion of metals. On the other hand, the passivation of the metal surface is evidently associated with the presence of a sufficient concentration of the inhibitor ions at the surface in the presence of oxygen (Kuznetsov 1996).

Molybdate-based corrosion inhibitors usually consist of an aqueous solution of sodium molybdate, a pH buffer, possibly a dispersing agent, and an azole. Corrosion protection is provided by a protective film that is formed by a chemical reaction between molybdate and metal. In this case, the metal surface must be reasonably clean and free of corrosion products. Molybdate-based corrosion inhibitors are easy to test and economical to use; they also provide excellent corrosion protection regardless of the amount of air ingression into the system. Obviously, most molybdate-based inhibitors were almost used in neutral or approximate neutral media and the effect of inhibition was evident when the concentration of molybdate was very high (thousands of mg/L). The corrosion inhibition of tungstate was more fully explored, but most of the studies also focused on neutral or approximately neutral solutions (Mu et al. 2006). The toxicity of tungstate compared to chromate is also very low.

2.2 Cathodic inhibitors

Cathodic inhibitors, such as zinc oxide and a volatile amine present in steam, either decrease the cathodic reaction or selectively precipitate on cathodic areas to increase the surface impedance and limit the diffusion of reducible species to these areas (Riggs 1973, McEwan 2004, and Roberge and Revie 2007). Zinc oxide, for example, retards corrosion by inhibiting the cathodic reaction. Under this condition, the cathodic reaction is either the reduction of water to hydrogen gas, as follows:

$$H_2O = OH^- + H^+, \tag{20.1}$$

$$2H^+ + 2e = H_2, \tag{20.2}$$

or oxygen reduction to hydroxide ion,

$$2H_2O + O_2 + 4e^- = 4OH^-. \tag{20.3}$$

The volatile amine present in steam, as another example, is used in the boilers that drive turbines in order to protect the pipework in which the condensed water passes. Here, the amine is moved by the steam through steam distillation to the remote pipework. In general, cathodic inhibitors are less effective than the anodic type. In contrast, they often form a visible film along the cathode surface, which polarizes the metal by restricting the access of dissolved oxygen to the metal substrate. The film also acts to block hydrogen evolution sites and prevent the resultant depolarizing effect.The amine increases the pH, thereby making proton reduction less favorable. However, cathodic inhibitors can provide inhibition through three different mechanisms: (1) cathodic poisons; (2) cathodic precipitates; and (3) oxygen scavengers.

2.2.1 Cathodic poisons

Cathodic poisons are substances that interfere with the cathodic reduction processes that must balance the anodic corrosion reaction. When corrosion occurs in an acid (low-pH) solution, some of the reduced hydrogen does not form gaseous hydrogen, but instead diffuses into the metal as atomic hydrogen. This can happen during the acidic pickling and electroplating of metals. Substances such as arsenic, antimony, sulfur, selenium, tellurium, and cyanide ions prevent the hydrogen atoms from forming hydrogen gas and are called cathodic poisons. Cathodic poisons facilitate contamination by keeping hydrogen in atomic form, in which hydrogen more readily diffuses into the metal. Environments with hydrogen sulfide contain both hydrogen and a cathodic poison (i.e., sulfur); they are especially dangerous for alloys and metals. Hydrogen sulfide is often encountered in the petroleum industry during the drilling and completion of oil and gas wells, and during the storage and piping of petroleum products containing hydrogen sulfide (Lyons and Plisga 2005). However, cathodic poisons can also increase the susceptibility of a metal to hydrogen-induced cracking since hydrogen can also be absorbed by metal during aqueous corrosion or cathodic charging.

2.2.2 Cathodic precipitates

Cathodic precipitate inhibitors block the cathodic reaction by depositing at the cathode sites as a result of a local increase in pH value. The most widely used cathodic precipitation inhibitors are carbonates of calcium and magnesium. They occur in natural waters and are used as inhibitors, usually requiring only an adjustment of pH. Ionic cathodic inhibitors, such as calcium or magnesium, may be precipitated as oxides to form a protective layer on the metal. In the case of zinc salts, zinc hydroxide is precipitated at the cathodic site, suppressing the reduction reaction of the dissolved oxygen (McEwan 2004). This is similar to the natural precipitation of ferric hydroxide, except that in the case of zinc hydroxide, the precipitate is nonporous. Pitting is usually prevented even if complete coverage does not occur because the current density generated over the large number of anodes, when supported by relatively few cathodes, is insufficient to cause a rapid local dissolution. Cathodic inhibition does not interfere with the anodic reaction but largely shifts the corrosion potential to the more negative values. These inhibitors are therefore, often classed as safe, since they reduce the corrosion rate at all concentrations (McEwan 2004).

2.2.3 Oxygen scavengers

The oxygen scavenger is a chemical substance that reacts with dissolved oxygen to reduce corrosion; this inhibitor type prevents cathodic depolarization. Oxygen scavengers are added to water either alone or with another corrosion inhibitor to retard corrosion. The most common effective oxygen scavengers used at ambient temperature are sodium sulfite (Na_2SO_3) and sulphur dioxide (SO_2). Sulfite can be used for low- to medium-pressure boilers. Quinones and cobalt salts are typically used as catalysts added with the sodium sulfite to overcome its low reaction with oxygen at low temperatures. This is also because the oxygen scavenger process is a redox reaction and requires a catalyst. Although sulfite is an effective oxygen scavenger, it is nonvolatile and does not leave the boiler with the steam, thus providing no protection in the condensate system. Sulfite also does not reduce hematite to magnetite and is ineffective in repassivating boilers with existing rust. Sulfite is the least expensive and most active (when catalyzed) for lower and medium pressure boilers. In its reaction with oxygen, sodium sulfite produces sodium sulfate, which contributes solids to the circulating boiler system (McEwan 2004 and Schweitzer 2005):

$$2Na_2SO_3 + O_2 = 2Na_2SO_4 \tag{20.4}$$

Another oxygen scavenger is hydrazine (N_2H_4), which is almost invariably used at elevated temperatures for high-pressure boilers to remove oxygen from the system. Hydrazine does not contribute solids to the system, thereby reducing boiler blow-down, or the mechanical removal of solids from the after-boiler section as sludge. It also promotes the formation of the protective magnetite film on the boiler tubes and drum, and converts red iron dust (hematite) to magnetite. It is because of these passivation effects that an excess of scavenger to oxygen is required when changing a boiler system form a nonpassivating scavenger to one that passivates. Hydrazine does not produce corrosive gases at high temperatures and pressures, and in application, reacts with oxygen to form nitrogen and water: (McEwan 2004 and Schweitzer 2005):

$$N_2H_4 + O_2 = 2H_2O + N_2 \tag{20.5}$$

Other compounds that can be used as oxygen scavenger corrosion inhibitors include diethylhydroxylamine, carbohydrazide, and hydroquinone (Walker and Zupanovich 1987). Removal of air from mud by defoaming and mechanical degassing is an essential first step before a scavenger can lower the dissolved oxygen content. According to Veldman and Trahan (2001), corrosion of iron in aqueous alkanolamine, physical solvents, or a combination of alkanolamine and physical solvent solutions used to remove acid gases from natural gas, synthetic gas, or light hydrocarbon streams can be reduced by adding mixtures of oxygen scavengers or mixtures of oxygen scavengers and sodium molybdate. The oxygen scavengers serve as corrosion inhibitors by reducing the iron oxides and hydroxides to the more corrosion-resistant magnetite form. Vacuum deaeration and boiling the solution help get rid of the oxygen and thus can also facilitate the role of the oxygen scavenger in inhibiting the cathodic reduction reaction and hence decreasing the corrosion rate.

2.3 Organic inhibitors

Both anodic and cathodic effects are sometimes observed in the presence of organic inhibitors, but, as a general rule, organic inhibitors affect the entire surface of a corroding metal when present in sufficient concentration. Organic inhibitors protect the metal by forming a hydrophobic film on the metal surface. The chemical structure of organic molecules plays a significant role and determines their effectiveness as corrosion inhibitors. Riggs (1973) has presented a fairly comprehensive generic classification of organic compounds used as inhibitors. Accordingly, some factors that contribute to the effectiveness of organic inhibitors are: (a) the size of the organic molecule; (b) the aromaticity and/or conjugated bonding; (c) carbon chain length; (d) bonding strength to metal substrate; (e) the type and number of bonding atoms or groups in the molecule (either π or σ); (f) the ability for a layer to become compact or cross linked; (g) the ability to form a complex with the atom as a solid within the metal lattice; and (h) adequate solubility in the environment. Because film formation is an adsorption process, the temperature and pressure in the system are also important factors.

Organic compounds containing polar groups including nitrogen, sulfur, and oxygen (Gomma and Wahdan 1994, Zucchi, Trabanelli and Fonsati 1996, Kendig and Jeanjaquet 2002, Ma et al. 2003, Ma et al. 2004, Christy et al. 2004, Zhang, Gao, and Zhou 2004, Otmacic et al. 2004, and Wang, Chen and Zhao 2004) and heterocyclic compounds with polar functional groups and conjugated double bonds (Al-Kharafi, 1988, Elmorsi and Hassanein 1999, Otmacic and Stupnisek-Lisac 2002, Dafali et al. 2002, Scendo, Poddebniak and Malyszko 2003, Sherif 2011, and Sherif 2013) have been reported to be good corrosion inhibitors. The inhibiting action of these organic compounds is usually attributed to their interactions with the metal surface via their adsorption. Adsorption of organic molecules on the surface of a metal usually takes place by physical or chemical adsorption. The physical adsorption (physisorption) process is one in which there are van der Waals (electroattractive) forces of interaction between gas or liquid molecules and electrically charged metal surface. On the other side, chemical adsorption (chemisorption) occurs due to a charge transfer, where the adsorbed species must have a dipole in the molecule. Chemisorption usually occurs through electron donation to a metal surface. However, polar functional groups are regarded as the reaction center that stabilizes the adsorption process (Fontana and Staehle 1970). In general, the adsorption of an organic inhibitor on a metal surface depends on the nature and surface charge of the metal, the adsorption mode, the inhibitor's chemical structure, and the type of electrolyte solution (Riggs 1973). Amines and azole derivatives such as benzotriazole, mercaptobenzothiazole, benzimidazole, and imidazole are well-known corrosion inhibitors for metals and alloys in different corrosive media (Fontana and Staehle 1970,

Srivastafa, Mukerjee and Agarwal. 1979, Al-Hajjar and Al-Kharafi 1988, and Lakshminarayanan, Kannan and Rajagopalan 1994).

Another category of inhibitors, called ohmic inhibitors and also referred to as general filming inhibitors, reduces the corrosion rate by decreasing the mobility of ionic species between anodes and cathodes on the corroding metal surface through decreasing the ionic conductivity of the solution (Davis 2000). Because it is usually impractical to increase resistance of the bulk electrolyte, increased resistance is practically achieved by the formation of a microfilm on the metal surface. If the film is deposited selectively on anodic areas, the corrosion potential shifts to more positive values. If the deposition occurs on cathodic areas, however, the corrosion potential will be shifted to the more negative values. In case of the inhibitor film covering both anodic and cathodic areas, the corrosion potential will shift slightly in either direction. These inhibitors include amines, which are cationic, and sulfonates, which are anionic (Nunez, 2007).

2.4 Precipitation inhibitors

Precipitation-inducing inhibitors are film-forming compounds that have a general action over the metal surface, blocking both anodic and cathodic sites indirectly. This effect is achieved through the formation of a bulky precipitation film over the entire surface. For example, hard water that is high in calcium and magnesium is less corrosive than soft water because of the tendency of the salts in the hard water to precipitate on the surface of the metal and form a protective film. The most common inhibitors of this category are the silicates and the phosphates (Armstrong, Peggs, and Walsh 1994). Many investigators (e.g., Lahodny and Kastelan 1981) have found sodium silicate to be very effective in controlling corrosion of carbon steel, and the optimum quantity of silicate required to suppress the corrosion has been cited as less than 10 ppm. Sodium silicate is also used in many domestic water softeners to prevent the occurrence of rust water. In aerated hot water systems, sodium silicate protects steel, copper, and brass. However, protection is not always reliable and depends heavily on pH and a saturation index that relies on water composition and temperature. Phosphates also require oxygen for effective inhibition. Silicates and phosphates do not afford the degree of protection provided by chromates and nitrites; however, they are very useful in situations where nontoxic additives are required.

2.5 Volatile corrosion inhibitors

In the early 1900s, it was discovered that a chemical could protect metal from corrosion by vaporizing and depositing on the metals. These chemicals were nitrites, and the first useful one was DICHAN (dicyclohexylammonium mitrite). This compound was developed after World War II to protect military equipment. DICHAN was adapted by some companies as a powder and as a paper coating to protect metal parts. In general, volatile corrosion inhibitors (VCIs), also called vapor phase inhibitors (VPIs), are compounds transported in a closed environment to the site of corrosion by volatilization from a source. They are similar to the organic adsorption-type inhibitors, and they possess a high-vapor pressure (Bastidas, Cano and Mora 2005). The VCI method is used primarily for articles made from carbon steel, stainless steel, cast iron, galvanized steel, nickel, chromium, aluminium, and copper (Wachter, Skei, and Stillman 1951 and Skinner 1993). As a consequence,

VCI compounds can be used to inhibit atmospheric corrosion of metals without coming into direct contact with the metal surface. VCI is placed in the vicinity of the metal to be protected and is transferred by sublimation (the process whereby a solid or gas changes state without becoming a liquid) and is then condensed onto the surface to be protected. For a compound to be classified as a VCI, it must possess two basic properties: It must be volatile, and the vapors alone must be able to retard the corrosion rate.

All known VCIs are organic compounds, with at most a minor inorganic component. They can be broadly divided into two groups: (1) highly volatile compounds and (2) solid compounds at room temperature (Mercer 1985 and McEwan 2004). Highly volatile compounds generally have a vapor pressure greater than 2 kPa (0.02 bar) at 20°C. These compounds are normally amines, and typical examples are ammonia, cyclohexylamine, morpholine, octadecylamine, and butyleamine. The most common use of these materials is in boilers. The VCI is transported with the steam to prevent corrosion in condenser tubes by neutralizing acidic carbon dioxide. Being volatile, they are carried over with the steam and subsequently condense in the water produced in the steam-return lines, where they serve a dual purpose. Due to their alkaline nature, they maintain a high pH value in the condensed water, effectively neutralizing the carbon dioxide that would otherwise produce corrosive acidic water. They also form a protective film on the steam pipes and act as normal film-forming corrosion inhibitors.

The second category (solid compounds at room temperature) is typically solid compounds that volatilize by sublimation and provide corrosion protection by adsorption on the metal surface. Commercially, they come in a variety of forms and compositions, but they all provide corrosion protection for metals in enclosed spaces, such as shipping containers. Numerous solid compounds can protect metals against corrosion at room temperature such as the salts of dicyclohexylamine, cyclohexylamine and hexamethylene-amine. For example, in the case of steel-corrosion inhibition, the majority of the commercial inhibitors are of the amine salt type, such as dicyclohexylamine nitrate (Dichan^{R}), dicyclohexylamine carbonate (CHC^{R}), cyclohexylamine m-nitro benzoate, and cyclohexylamine benzoate. In addition to these organic salts, there are a number of organic compounds that are not salts, but are also widely used as VCIs. The most common of these are triazoles (e.g., benzotriazole) and imidazole-derivatives.

Volatile corrosion inhibitors are available commercially in a number of different forms designed to meet different application requirements. The most common forms are as follows (Wachter, Skei, and Stillman 1951, Prenosil 2001, McEwan 2004): *powders*: the most common form of VCIs and used for the inhibition of relatively large container volumes; *tablet*: usually made by compressing the VCI powder mixed with a small quantity of binder material; *foam pads*: usually provided with adhesive backings so that they can be attached to the inside of containers (polyurethane foam pads have been used extensively in the past, but polyethylene is now recommended in order to avoid the decomposition problem associated with polyurethane foam pads); *impregnated wrappings*: such as cellulose paper and polyethylene plastic. They are used mostly to protect components in storage and during transit; and *coatings*: provide a suitable temporary protective coating for components that are too large to be conveniently wrapped or boxed. Typically, the VCI coatings are clear or opaque formulations that form very thin films. The most common application for the coatings is on machined surfaces, where it is important to maintain tolerances. The films are very thin and contain no abrasive pigments.

2 Inhibition efficiencies of corrosion inhibitors

The percentage of the efficiency of an inhibitor, *IE* (%), can be obtained using the equation (Sherif and Park 2005, Sherif 2006, Sherif and Park 2006, 4065 and Sherif and Park 2006, 6556):

$$IE\ (\%) = \frac{K_{Corr}^{Un} - K_{Corr}^{In}}{K_{Corr}^{1}}\ x\ 100 \tag{20.6}$$

where K_{Corr}^{Un} and K_{Corr}^{In} are the corrosion rates of the metal in the uninhibited and inhibited media, respectively.

There are numerous ways to calculate the values of K_{Corr} for metals and/or alloys in an electrolyte (or even in the atmosphere) depending on the corrosion test method. K_{Corr} can be obtained gravimetrically from the weight-loss method as follows (Sherif and Park 2005, Sherif 2006, and Sherif and Park 2006, 4065, Sherif and Park 2006, 6556 and Sherif, et al. 2009, 1364):

$$K_{Corr} = \frac{\Delta W}{At} \tag{20.7}$$

where ΔW is the weight loss *(mg)*, *A*, the area of the tested coupons *(dm^2)*, and *t*, the immersion time *(day)*. The values of K_{Corr} can also be calculated from weight loss using different expressions as fellows (Sherif, Erasmus, and Comins 2007, 144 and Sherif et al. 2009, 1364):

$$K_{Corr} = \frac{543\ \Delta W}{ADt} \tag{20.8}$$

Here, ΔW represents the weight loss (gm) of metal coupons, *A*, the total area of the coupon (square inches), *D*, the density of metal (g/cm^3), and *t*, the exposure time (h).

The corrosion rate can be also calculated from polarization curves of the metal in the aggressive environment in the absence and presence of the inhibitor molecules as follows (Metikoŝ–Huković, Babić and Grubać 1998, Sherif and Park 2005, Sherif 2006, and Sherif and Park 2006, 4065, Sherif and Park 2006, 6556 and Sherif et al. 2009, 1364):

$$K_{Corr} = \frac{j_{Corr}\ k\ E_W}{D\ A} \tag{20.9}$$

where j_{Corr} is the corrosion current, *k* is a constant that defines the units for K_{Corr} ($k = 128{,}800$ milli inches (amp-cm-year)), E_W is the equivalent weight in grams/equivalent of the metal, *D* is the density of the metal in grams/cm^3, and *A* is the area of electrode that is exposed to the corrosive media in cm^2. The inhibition efficiency, Equation (20.6), will then be obtained from the values of K_{Corr}, which in turn were obtained as shown in Eq. 20.8. Since, the corrosion current is directly proportional to the corrosion rate of the metal; the inhibition efficiency can be calculated from polarization data by the following relation (Metikoŝ–Huković, Babić and Grubać 1998):

$$IE\ (\%) = \frac{j_{Corr}^{un} - j_{Corr}^{in}}{j_{Corr}^{un}} \times 100 \tag{20.10}$$

In this case, j_{Corr}^{in} and j_{Corr}^{un}, represent the corrosion current in the presence and absence of the inhibitor molecules, respectively.

Electrochemical impedance spectroscopy (EIS) is a powerful technique that has been successfully used in studying the corrosion and passivation phenomena of metals and alloys (Macdonald 1987, Ma et al. 2002, Sherif and Park 2005, Sherif 2006, and Sherif and Park 2006, 4065, Sherif and Park 2006, 6556, Sherif and Park 2005, Sherif and Park 2006, 1313, Sherif and Park 2006, 4665, Sherif, Erasmus and Comins 2007, 144, Sherif, Erasmus, and Comins 2007, 470, Sherif et al. 2007, Sherif et al. 2009, 1364, Sherif, Erasmus, and Comins 2008, 2009, and Sherif et al. 2009, 1385). The inhibition efficiency obtained can be obtained by EIS as follows (Ma et al. 2002):

$$IE\ (\%) = \frac{R_P - R_p^o}{R_P} \times 100 \tag{20.11}$$

R_p and R_p° represent the charge transfer resistances in the electrolytic solution without and with inhibitor, respectively.

The adsorption and/or precipitation of a corrosion inhibitor onto a metal surface can be evaluated perfectly by the quartz crystal analysis (QCA) method (Kanungo, Kumar, and Contractor 2002, Park and Yoo 2003, Losito et al. 2005, Sherif and Park 2005, Sherif and Park 2006, 1313, and Akrout, et al. 2007). By using this technique, the increase in mass at the QCA electrode is calculated from the decrease in frequency according to the Sauerbrey equation (Park and Yoo 2003),

$$\Delta W = -\left[\Delta f / C_f\right], \tag{20.12}$$

where ΔW is the mass change, Δf is the change in frequency, and C_f is the sensitivity constant of the electrochemical quartz crystal microbalance (EQCM) electrode. This relation enables us to calculate the mass of the inhibitor molecules adsorbed or precipitated on the surface even if it was in the range of a nanogram as well as the number of layers of the adsorbed inhibitor molecules onto the metal surface.

3 Uses of corrosion inhibitors in industry

The major industries using corrosion inhibitors are oil and gas exploration and production, petroleum refining, chemical manufacturing, heavy manufacturing, water treatment, and the product additive industries. This is because corrosion inhibitors are commonly added to boiler water, coolants, fuels, hydraulic fluids, engine oil, and many other fluids used in industry. For example, the inhibition process is used internally with carbon–steel pipes and vessels as an economic corrosion control alternative to stainless steels and alloys, coatings, or nonmetallic composites, and can often be implemented without disrupting any other process. Volatile amines form a protective film on the steel surface and, at the same time, act as an anodic inhibitor and thus are used in the boilers to minimize the effects of acid. For fuels, various corrosion inhibitors can be used. Some components include zinc dithiophosphates. In oil refineries, hydrogen sulfide can corrode steels so that it is removed, often using air and amines by conversion to polysulfides. Water and water/glycol solutions are common heat transfer fluids used in cooling systems and recirculating chillers. Silicate-based corrosion inhibitors are often used for the treatment of cooling water systems. For closed water systems, a mixture of molybdate and phosphate corrosion inhibitor is best utilized. A combination of film-forming corrosion inhibitor and biocide is most often used to protect metals in open cooling systems, tanks, marine, and seawater applications.

4 Summary

The role of corrosion inhibitors in protecting metallic structures against corrosion in harsh environments has been reported. Corrosion inhibitors, when added to an environment, reduce the corrosion rate of metals. Corrosion inhibitors can be classified into different categories depending on how they protects the structure from being corroded in the aggressive media. These are passivating, cathodic, http://www.corrosionsource.com/technicallibrary/corrdoctors/Modules/Inhibitors/organic.htm organic, precipitation, and volatile corrosion inhibitors. Passivating inhibitors work by inhibiting the anodic sites on the surface in either the absence or presence of oxygen. Cathodic inhibitors decrease the cathodic reaction and therefore decrease the corrosion rate. Organic compounds can also be used as inhibitors and usually decrease the corrosion rate of metals via their adsorption onto the entire surface. Precipitation inhibitors work by precipitation or deposition on certain places or on the whole metal surface to block the active sites and prevent corrosion. Volatile corrosion inhibitors (VCIs) belong to a different category and are used mainly to protect metallic structures exposed to atmospheric corrosion, where protection is achieved by sublimation of the VCI and then its adsorption onto the surface to be protected to form an invisible layer. This layer isolates the metallic structure from being exposed to the harsh environment and hence prevents it from corrosion. The inhibition efficiency of corrosion inhibitors can also be obtained either gravimetrically from the loss in weight technique or electrochemically by using different methods such as potentiodynamic polarization, electrochemical impedance spectroscopy, and electrochemical quartz crystal microbalance.

SYMBOLS AND TERMINOLOGY

A area of the tested metal in a certain environment
C_f sensitivity constant of an electrochemical quartz crystal microbalance electrode
CHC^R dicyclohexylamine carbonate
D density of the tested metal
Δf change in frequency
$Dichan^R$ dicyclohexylammonium nitrite
ΔW weight loss of the metal in a certain environment
EIS electrochemical impedance spectroscopy
EQCM electrochemical quartz crystal microbalance
E_W equivalent weight in grams/equivalent of the metal
GDP gross domestic product
IE (%) percentage of the efficiency of an inhibitor
j_{Corr} corrosion current of the tested metal
j^{in}_{Corr} corrosion current of the tested metal in presence of inhibitor
j^{un}_{Corr} corrosion current of the tested metal in absence of inhibitor
K a constant that defines the units for the corrosion rate
K_{Corr} corrosion rate of the metal in a certain environment
K^{In}_{Corr} corrosion rate of the metal in an inhibited environment
K^{Un}_{Corr} corrosion rate of the metal in an uninhibited environment
QCA quartz crystal analysis

R_p charge transfer resistance of a metal in an inhibited electrolytic solution
R_p° charge transfer resistance of a metal in an uninhibited electrolytic solution
t immersion time of the metal in a certain environment
VCIs volatile corrosion inhibitors
VPIs vapor-phase inhibitors

Glossary

Acid a chemical substance that yields hydrogen ions (H^+) when dissolved in water
Active a state in which a metal tends to corrode; referring to the negative direction of electrode potential (opposite of passive or noble)
Active metal a metal ready to corrode, or being corroded
Active potential the potential of a corroding material
Additive a substance added in a small amount, usually to a fluid, for a special purpose, such as to reduce friction or corrosion
Adsorption the surface retention of solid, liquid, or gas molecules, atoms, or ions by a solid or liquid
Anion an ion or radical that is attracted to the anode because of the negative charge
Anode the electrode at which oxidation or corrosion of some component occurs. Electrons flow away from the anode in the external circuit.
Anode corrosion the dissolution of a metal acting as an anode
Anodic inhibitor a chemical substance or combination of substances that prevent or reduce the rate of the anodic or oxidation reaction by a physical, physicochemical or chemical action
Anodic polarization the change in the initial anode potential resulting from current flow effects at or near the anode surface. Potential becomes mode noble (more positive) because of anodic polarization.
Anodic reaction an oxidation process by which a transfer of positive charge from the electronic to the ionic conductor occurs
Atmospheric corrosion the gradual degradation or alteration of a material due to contact with the atmosphere
Base an aqueous substance that can accept hydrogen ions
Cathodic polarization the change of the electrode potential in the active (negative) direction due to current flow; a reduction from the initial potential resulting from current flow effects at or near the cathode surface. Potential becomes more active (negative) because of cathodic polarization.
Cation a positively charged ion that migrates through the electrolyte toward the cathode under the influence of a potential gradient
Chemisorption the binding of an adsorbate to the surface of a solid by forces whose energy levels approximate those of a chemical bond
Conductivity the ratio of the electric current density to the electric field in a material
Corrosion the chemical or electrochemical reaction between a material, usually a metal, and its environment that produces a deterioration of the material and its properties
Corrosion inhibitor any substance or collection of substances that, when added in small concentration to an environment, effectively decreases the corrosion rate
Diffusion the spontaneous movement of atoms or molecules to new sites within a material
Environment the surroundings or conditions (physical, chemical, mechanical) in which a material exists
Noble metal a metal whose potential is highly positive relative to the hydrogen electrode
Electrical resistivity the electrical resistance offered by a material to the flow of current, times the cross-sectional area of current flow and per unit length of current path; the reciprocal of the conductivity

Electrochemical corrosion corrosion that is accompanied by a flow of electrons between cathodic and anodic areas on metallic surfaces

Electrochemical impedance the frequency-dependent complex-valued proportionality factor (A£/Ai) between the applied potential or current and the response signal. This factor is the total opposition (0 or 0 – cm^2) of an electrochemical system to the passage of charge. The value is related to the corrosion rate under certain circumstances.

Film a thin, not necessarily visible, layer of material

Harsh environment any condition of extremes relative to the human condition applies, so that temperature, humidity, atmosphere (including pressure), radiation, and shock are included whether indoors or not

Hydrolysis decomposition or alteration of a chemical substance by water

Inorganic being or composed of matter other than hydrocarbons and their derivatives, or matter that is not of plant or animal origin

Ion an atom, or group of atoms, that has gained or lost one or more outer electrons and thus carries an electric charge

Neutralization a chemical reaction whereby an acid and a base react to form water and a salt

Noble metal a metal whose potential is highly positive relative to the hydrogen electrode

Noble potential a potential more cathodic (positive) than the standard hydrogen potential

Ore a type of rock that contains minerals such as gemstones or metals that can be extracted through mining and refined for use

Organic being or composed of hydrocarbons or their derivatives, or matter of plant or animal origin

Oxidation the loss of electrons, gain of oxygen, or loss of hydrogen by an atom, ion, or molecule during a chemical reaction

Passivation the process in metal corrosion by which metals become passive

Passivators a type of inhibitor that appreciably changes the potential of a metal to a more noble (positive) value

Passive the state of the metal surface characterized by low corrosion rates in a potential region that is strongly oxidizing for the metal

Passivity a condition in which a piece of metal, because of an impervious covering of oxide or other compound, has a potential much more positive than that at the metal in the active state

Physisorption the binding of an adsorbate to the surface of a solid by forces whose energy levels approximate those of condensation

Pitting localized corrosion of a metal surface, confined to a point or small area, that takes the form of cavities or pits

Polarization a change in the potential of an electrode during electrolysis, such that the potential of an anode becomes more noble, and that of a cathode more active, than their respective reversible potentials

Potentiodynamic the technique for varying the *potential* of an electrode in a continuous manner at a preset rate

Potentiostatic the technique for maintaining a constant electrode potential

Reduction a half-reaction in which a chemical species decreases its oxidation number, usually by gaining electrons

Resistance the opposition that a device or material offers to the flow of direct current, equal to the voltage drop across the element divided by the current through the element

Salt an ionic compound, which can result from the neutralization reaction of acids and bases

Sublimation the transition of an element or compound from the solid to gas phase with no intermediate liquid stage

Tafel slope the relationship between current and potential of a polarized electrode over a definite region that can be approximated by: $E = \pm B \log (j/j_o)$, where, E = change in open circuit potential, j = the current density, B and j_o = constants. B is known as the Tafel slope.

References

Akrout, H., Maximovitch, S., Bousselmi, L., Triki, E., Dalard, F., 2007. Evaluation of corrosion non toxic inhibitor adsorption for steel in near neutral solution: L(+) ascorbic acid. Materials and Corrosion 58 (3), 202–206.

Alexander, D.B., Moccari, A.A., 1993. Evaluation of corrosion inhibitors for component cooling water systems. Corrosion (NACE) 49, 921–928.

AL-Hajjar, F.H., Al-Kharafi, F.M., 1988. 2-amino-thiazole and 2-amino-4,6-dimethylpyrimidine as corrosion inhibitors for copper. Corros. Sci. 28 (2), 163–171.

Al-Sayegh, A., Carew, J., Al-Hashem, A., 2001. Evaluation of oil field chemicals for secondary oil recovery systems in Kuwait. Anti-Corros. Methods & Mater. 48 (4), 245–251.

Altoé, P., Pimenta, G., Moulin, C.F., Diaz, S.L., Mattos, O.R., 1996. Evaluation of oilfield corrosion inhibitors in CO_2 containing media—A kinetic study. Electrochim. Acta 41 (7-8), 1165–1172.

Aramaki, K., 2001. The inhibition effects of chromate-free, anion inhibitors on corrosion of zinc in aerated 0.5 M NaCl. Corros. Sci. 43 (3), 591–604.

Armstrong, R.D., Peggs, L., Walsh, A., 1994. Behaviour of sodium silicate and sodium phosphate (tribasic) as corrosion inhibitors for iron. J. Appl. Electrochem. 24 (12), 1244–1248.

Bastidas, D.M., Cano, E., Mora, E.M., 2005. Volatile corrosion inhibitors—a review. Anti-Corros. Meth. Mater. 52 (2), 71–77.

Bockris, J.O.'M., Reddy, A.K.N., 1970. Modern electrochemistry, second ed. Plenum Press, New York.

Bregmann, J.I., 1963. Corrosion inhibitors. Macmillan Co, New York.

Christy, A.G., Lowe, A., Otieno-Alego, V., Stoll, M., Webster, R.D., 2004. Voltammetric and Raman microspectroscopic studies on artificial copper pits grown in simulated potable water. J. Appl. Electrochem. 34 (2), 225–233.

Dafali, A., Hammouti, B., Touzani, R., Kertit, S., Ramdani, A., El Kacemi, K., 2002. Corrosion inhibition of copper in 3 per cent NaCl solution by new bipyrazolic derivatives. Anti-Corros. Meth. Mater. 49 (2), 96–104.

Davis, R. Joseph, 2000. Corrosion: Understanding the basics. ASM International, Materials Park, Ohio, USA.

El-Etre, A.Y., 2006. Khillah extract as inhibitor for acid corrosion of SX 316 steel. Appl. Surf. Sci. 252 (24), 8521–8525.

Elmorsi, M.A., Hassanein, A.M., 1999. Corrosion inhibition of copper by heterocyclic compounds. Corros. Sci. 41 (12), 2337–2352.

Fontana, M.G., Staehle, K.W., 1970. Advances in corrosion science and technology. Plenum Press, New York vol. 1.

Gad Allah, A.G., Badawy, M.W., Rehan, H.H., Abou-Romia, M.M., 1989. Inhibition of corrosion of α-brass (Cu-Zn, 67/33) in acid chloride solutions by some amino pyrazole derivatives. J. Appl. Electrochem. 19 (6), 928–932.

Gomma, G.K., Wahdan, M.H., 1994. Effect of temperature on the acidic dissolution of copper in the presence of amino acids. Mater. Chem. Phys. 39 (2), 142–148.

Ilevbare, G.O., Burstein, G.T., 2003. The inhibition of pitting corrosion of stainless steels by chromate and molybdate ions. Corros. Sci. 45 (7), 1545–1569.

Kanungo, M., Kumar, Anil, Contractor, A.Q., 2002. Studies on electropolymerization of aniline in the presence of sodium dodecyl sulfate and its application in sensing urea. J. Electroanal. Chem. 528 (1-2), 46–56.

Kending, M.W., Buchheit, R.G., 2003. Corrosion inhibition of aluminum and aluminum alloys by soluble chromates, chromate coatings, and chromate-free coatings. Corrosion (NACE) 59 (5), 379–400.

Kendig, M.W., Jeanjaquet, S., 2002. Cr(VI) and Ce(III) Inhibition of oxygen reduction on copper. J. Electrochem. Soc. 149 (2), B47–B51.

Khaled, K.F., Hackerman, N., 2004. *Ortho*-substituted anilines to inhibit copper corrosion in aerated 0.5 M hydrochloric acid. Electrochem. Acta 49 (3), 485–495.

Khalifa, M.A., El-Batouti, M., Mahgoub, F., Bakr Aknish, A., 2003. Corrosion inhibition of steel in crude oil storage tanks. Materials and Corrosion 54 (4), 251–258.

Konno, H., Kobayashi, S., Nagayama, M., 1982. The hydration of barrier oxide films on aluminium and its inhibition by chromate and phosphate ions. Corros. Sci. 22 (10), 913–923.

Kuznetsov, Yurii I., 1996. Organic inhibitors of corrosion of metals. Plenum Press, New York.

Lahodny, O., Kastelan, L., 1981. The influence of pH on the inhibition of corrosion of iron and mild steel by sodium silicate. Corros. Sci. 21 (4), 265–271.

Lakshminarayanan, V., Kannan, R., Rajagopalan, S.R., 1994. Cyclic voltammetric behavior of certain copper-azole systems using carbon paste electrodes. J. Electroanal. Chem. 364 (1–2), 79–86.

Losito, I., Malitesta, Cosimino, De Bari, Isabella, Calvano, Cosima-Damiana, 2005. X-ray photoelectron spectroscopy characterization of poly(2,3-diaminophenazine) films electrosynthesised on platinum. J. Thin Solid Films 473 (1), 104–113.

Lyons, William C., Plisga, Gary J., 2005. Standard handbook of petroleum and matural gas engineering, second ed. Gulf Publishing Co, New York.

Ma, H., Yang, C., Yin, B.S., Li, G.Y., Chen, S.H., Luo, J.L., 2003. Electrochemical characterization of copper surface modified by *n*-alkanethiols in chloride-containing solutions. J. Appl. Surf. Sci. 218 (1-4), 144–154.

Ma, H., Chen, S., Niu, L., Zhao, S., Li, S., Li, D., 2002. Inhibition of copper corrosion by several Schiff bases in aerated halide solutions. J. Appl. Electrochem. 32 (1), 65–72.

Macdonald, J.R., 1987. Impedance spectroscopy. John Wiley & Sons, New York.

Makarenko, V.D., Petrovskii, V.A., Galiullin, M.F., Chernov, V.Yu., Makarenko, I.O., 2003. Methods of improving the corrosion resistance in oil well lining and compressor pipes. Chemical and Petroleum Engineering. 39 (7–8), 429–432.

McEwan, J.J., 2004. Corrosion control in Southern Africa, second ed. Corrosion Institute of South Africa.

Mercer, A.D., 1985. Test methods of corrosion inhibitors. Brit. Corros. J. 20 (2), 61–70.

Metikoŝ–Huković, M., Babić, R., Grubać, Z., 1998. Corrosion protection of aluminium in acidic chloride solutions with nontoxic inhibitors. J. Appl. Electrochem. 28 (4), 433–439.

Mu, Guannan, Li, Xianghong, Qu, Qing, Zhou, Jun, 2006. Molybdate and tungstate as corrosion inhibitors for cold rolling steel in hydrochloric acid solution. Corros. Sci 48, 445–449.

Nunez, Magdalena, 2007. Prevention of metal corrosion: New research. Nova Science Publishers, New York.

Otmacic, H., Stupnisek-Lisac, E., 2002. Copper corrosion inhibitors in near neutral media. Electrochim. Acta 48 (8), 985–991.

Otmacic, H., Telegdi, J., Papp, K., Stupnisek-Lisac, E., 2004. Protective properties of an inhibitor layer formed on copper in neutral chloride solution. J. Appl. Electrochem. 34 (5), 545–550.

Park, S.-M., Yoo, J.-S., 2003. Electrochemical impedance spectroscopy for better electrochemical measurements. Anal. Chem. 75 (21), 455A–461A.

Prenosil, M., 2001 (Jan.). Volatile corrosion inhibitor coatings. Supplement to Materials Performance 40, 14–17.

Prokhorenkov, V.D., Ostrikov, V.V., Knyazeva, L.G., Chernyshova, I.Yu., 2002. Producing Corrosion Inhibitors from Spent Lubricating Oils. Chemical and Petroleum Engineering 38 (9–10), 630–633.

Pryor, M.J., Cohen, M., 1953. The inhibition of the corrosion of iron by some anodic inhibitors. J. Electrochem. Soc. 100, 203–215.

Putilova, I.N., Balezin, S.A., Barannik, V.P., (Translated from Russian), 1966. Metallic corrosion inhibitors. Pergamon Press, London.

Riggs Jr., O.L., 1973. Corrosion Inhibitors, second ed. C.C. Nathan, Houston, TX.

Roberge, Pierre R., 2000. Handbook of corrosion engineering. McGraw-Hill, New York.

Roberge, Pierre R., Winston Revie, R., 2007. Corrosion inspection and monitoring. John Wiley & Sons, New York.

Robertson, W.D., 1951. Molybdate and tungstate as corrosion inhibitors and the mechanism of inhibition. J. Electrochem. Soc. 98, 94–100.

Rozenfield, I.L., 1981. Corrosion inhibitors. McGraw-Hill, New York.

Sanyal, B., 1981. Organic compounds as corrosion inhibitors in different environments—a review. Progress in Organic Coatings 9 (2), 165–236.

Scendo, M., Poddebniak, D., Malyszko, J., 2003. Indole and 5-chloroindole as inhibitors of anodic dissolution and cathodic deposition of copper in acidic chloride solutions. J. Appl. Electrochem. 33 (3–4), 287–293.

Schweitzer, Philip A., 2005. Corrosion engineering handbook, second ed. Marcel Dekker, New York.

Schweitzer, Philip A., 1989. Corrosion and corrosion protection handbook. Marcel Dekker, New York.

Schweitzer, Philip A., 1996. Corrosion engineering handbook. Marcel Dekker, New York.

Sherif, E.M., Park, S.-M., 2005. Effects of 1,5-naphthalenediol on aluminum corrosion as a corrosion inhibitor in 0.50 M NaCl. J. Electrochem. Soc. 152, B205–B211.

Sherif, E.M., Park, S.-M., 2005. Inhibition of Copper Corrosion in 3.0% NaCl Solution by *N*-Phenyl-1,4-phenylenediamine. J. Electrochem. Soc. 152 (10), B428–B433.

Sherif, E.M., Park, S.-M., 2006. 2-Amino-5-ethyl-1,3,4-thiadiazole as a corrosion inhibitor for copper in 3.0% NaCl solutions. Corros. Sci. 48, 4065–4079.

Sherif, E.M., Park, S.-M., 2006. Effects of 1,4-naphthoquinone on aluminum corrosion in 0.50 M sodium chloride solutions. Electrochem. Acta 51 (7), 1313–1321.

Sherif, E.M., Park, S.-M., 2006. Effects of 2-amino-5-ethylthio-1,3,4-thiadiazole on copper corrosion as a corrosion inhibitor in aerated acidic pickling solutions. Electrochim. Acta 51, 6556–6562.

Sherif, E.M., Park, S.-M., 2006. Inhibition of copper corrosion in acidic pickling solutions by *N*-phenyl-1,4-phenylenediamine. Electrochem. Acta 51 (22), 4665–4673.

Sherif, E.-S.M., 2011. Effects of 5-(3-aminophenyl)-tetrazole on the inhibition of unalloyed iron corrosion in aerated 3.5% sodium chloride solutions as a corrosion inhibitor. Mater. Chem. Phys. 129, 961–967.

Sherif, E.-S.M., 2013. Comparative study on the inhibition of inhibition of iron corrosion in aerated stagnant 3.5 wt.% sodium chloride solutions by 5-phenyl-1Htetrazole and 3-amino-1,2,4-triazole. Ind. Eng. Chem. Res. 52, 14507–14513.

Sherif, E.-S.M., 2006. Effects of 2-amino-5-(ethylthio)-1,3,4-thiadiazole on copper corrosion as a corrosion inhibitor in 3% NaCl solutions. Appl. Surf. Sci. 252, 8615–8623.

Sherif, E.-S.M., Almajid, A.A., 2010. Surface protection of copper in aerated 3.5% sodium chloride solutions by 3-amino-5-mercapto-1,2,4-triazole as a copper corrosion inhibitor. J. Appl. Electrochem. 40, 1555–1562.

Sherif, E.-S.M., Ahmed, A.H., 2010. Synthesizing new hydrazone derivatives and studying their effects on the inhibition of copper corrosion in sodium chloride Solutions. Synth. React. Inorg. Metal-Organic Nano-Metal Chem. 40, 365–372.

Sherif, E.-S.M., El-Shamy, A.M., Ramla, M.M., El-Nazhawy, A.O.H., 2007. 5-(Phenyl)-4H-1,2,4-triazole-3-thiol as a corrosion inhibitor for copper in 3.5% NaCl solutions. Mater. Chem. Phys. 102 (2–3), 231–239.

Sherif, E.-S.M., Potgieter, J.H., Comins, J.D., Cornish, L., Olubambi, P.A., Machio, C.N., 2009. Effects of minor additions of ruthenium on the passivation of duplex stainless-steel corrosion in concentrated hydrochloric acid solutions. J. Appl. Electrochem. 39 (8), 1385–1392.

Sherif, E.-S.M., Potgieter, J.H., Comins, J.D., Cornish, L., Olubambi, P.A., Machio, C.N., 2009. The beneficial effect of ruthenium additions on the passivation of duplex stainless steel corrosion in sodium chloride solutions. Corros. Sci. 51 (6), 1364–1371.

Sherif, E.-S.M., Erasmus, R.M., Comins, J.D., 2007. Corrosion of copper in aerated acidic pickling solutions and its inhibition by 3-amino-1,2,4-triazole-5-thiol. J. Colloid Interface Sci. 306, 96–104.

Sherif, E.-S.M., Erasmus, R.M., Comins, J.D., 2007. Corrosion of copper in aerated synthetic sea water solutions and its inhibition by 3-amino-1,2,4-triazole. J. Colloid Interface Sci. 309, 470–477.

Sherif, E.-S.M., Erasmus, R.M., Comins, J.D., 2007. Effects of 3-amino-1,2,4-triazole on the inhibition of copper corrosion in acidic chloride solutions. J. Colloid Interface Sci. 311, 144–151.

Sherif, E.-S.M., Erasmus, R.M., Comins, J.D., 2008. Inhibition of copper corrosion in acidic chloride pickling solutions by 5-(3-aminophenyl)-tetrazole as a corrosion inhibitor. Corros. Sci. 50 (12), 3439.

Sherif, E.-S.M., Erasmus, R.M., Comins, J.D., 2009. Inhibition of corrosion processes on copper in aerated sodium chloride solutions by 5-(3-aminophenyl)-tetrazole. J. Appl. Electrochem. 39, 83.

Sherif, E.-S.M., Erasmus, R.M., Comins, J.D., 2010. In situ Raman spectroscopy and electrochemical techniques for studying corrosion and corrosion inhibition of iron in sodium chloride solutions. Electrochim. Acta 55, 3657–3663.

Skinner, William, 1993. A new method for quantitative evaluation of volatile corrosion inhibitors. Corros. Sci. 35 (5–8), 1491–1494.

Srivastafa, R.D., Mukerjee, R.C., Agarwal, A.K., 1979. Corrosion of electroplated Cu—Cd alloys and its inhibition. Corros. Sci. 19 (1), 27–33.

Talbot, David, Talbot, James, 1998. Corrosion science and technology. CRC Press, New York.

Uhlig, H.H., Revie, R.W., 1985. Corrosion and corrosion control, third ed. John Wiley & Sons, New York.

Veldman, Ray R., Trahan, David O., 2001. Gas treating solution corrosion inhibitor. U.S. Patent no. 6299838 issued on Sept. 10.

Wachter, A., Skei, T., Stillman, N., 1951. Dicyclohexylammonium nitrate, a volatile inhibitor for corrosion preventive packaging. Corrosion 7, 284–294.

Walker, Jerry L., Zupanovich, John D., 1987. Stabilized sodium erythorbate boiler corrosion inhibitor compositions and methods. United States Patent no. 4681737 issued on July 21.

Wang, C., Chen, S., Zhao, S., 2004. Inhibition effect of AC-treated, mixed self-assembled film of phenylthiourea and 1-dodecanethiol on copper corrosion. J. Electrochem. Soc. 151 (1), B11–B15.

Williams, G., McMurray, H.N., 2001. Chromate inhibition of corrosion-driven organic coating delamination studied using a scanning Kelvin probe technique. J. Electrochem. Soc. 148 (10), B377–B385.

Zein El Abedin, S., 2001. J. Appl. Electrochem. 31 (6), 711–718.

Zhang, D.-Q., Gao, L.-X., Zhou, G.-D., 2004. Inhibition of copper corrosion by bis-(1-benzotriazolymethylene)-(2,5-thiadiazoly)-disulfide in chloride media. Appl. Surf. Sci. 225 (1-4), 287–293.

Zucchi, F.G. Trabanelli, Fonsati, M., 1996. Tetrazole derivatives as corrosion inhibitors for copper in chloride solutions. Corros. Sci. 38 (11), 2019.

CHAPTER

Petroleum Products Transporting Pipeline Corrosion—A Review

21

N. Muthukumar

Corrosion Protection Division, Central Electrochemical Research Institute, Karaikudi—630 006, India; Institute of Advanced Energy, Kyoto University, Gokasho, Uji, Kyoto-611-0011, Japan; ***Present Address:*** *Water & Process Technologies, GE (General Electric) India Technology Centre, Bangalore - 560066, India.*

CHAPTER OUTLINE

The Role of Colloidal Systems in Environmental Protection. http://dx.doi.org/10.1016/B978-0-444-63283-8.00021-1

1 Corrosion

Corrosion is simply defined as the destruction of materials under chemical or electrochemical action by the surrounding environment. Corrosion entails the conversion of a metal from the atomic to ionic state, with the loss of one or more electrons. Rusting of iron with the formation of iron oxide is a typical example of corrosion. Corrosion of metal and alloys lead to enormous losses to any country. The growing need of metal in every walk of life and loss of it by corrosion has put a big question mark before humankind regarding how to prevent corrosion so that it can be used to its maximum extent. Scientists have provided many preventive methods. Corrosion can be prevented to a large extent, if not stopped completely. A survey carried out by the U.S. Minerals Management Service (Atlas 1981) indicates that most problems are due to internal corrosion, which occur more frequently than problems due to external corrosion.

During normal pipeline operations, warm petroleum products with small amounts of salt and water travel through the pipeline. Factors viz. nutrients (nitrite; phosphate, etc), oxygen, chloride and bacteria are the causative factors for internal corrosion in petroleum transporting pipelines. In oil-transporting pipelines, stagnation of water occurs due to the slopes in the landscape, and this acts as a breeding ground for bacteria. Generally, degradation starts at the interface between diesel and water. Under this condition, a good inhibitor is needed for preventing pipeline corrosion. It has also been estimated that 40% of all internal corrosion of pipelines in the gas and oil industry can be attributed to microbial corrosion. Different methods are used for protecting petroleum product pipelines from corrosion. Application of a protective surface coating is one of these methods. This is an expensive procedure, and it is impossible to coat the pipeline when it is already corroded and filled with petroleum product. One of the efficient and cost-effective approaches to mitigate this problem is application of corrosion inhibitors. Thus there is an urgent and immediate need to develop am efficient corrosion inhibitor.

1.1 Economic loss due to corrosion

In a study of corrosion cost conducted jointly by C.C. Technologies Inc., USA, Federal Highway Agencies (FHWA), USA, and the National Association of Corrosion Engineers (NACE), direct corrosion cost was estimated to be around $US276 billion, which represents approximately 3.1% of the national gross domestic product (GDP). Based on an extensive survey conducted by Battelle Columbus Laboratories in Columbus, Ohio, and the National Institute of Standards and Technology (NIST), in 1975, the cost was estimated to be $US82 billion, which would have exceeded $US350 billion in view of price inflation over the last 25 years. Because of the long time involved in conducting the cost structure, it is not possible to update the information every year. However, both studies show that corrosion costs are staggering and that a figure of about US$350 billion appears to be a reasonable estimate for another two to three years. At least 35% of the above amount could have been saved by taking appropriate corrosion control measures. In the UK, the corrosion cost was estimated to be 4–5% of GDP (Uhlig, 1985). In Japan, the cost of corrosion was estimated at about 5258 trillion yen per year. For most industrialized nations, the average corrosion cost is 3.5–4.5% of GDP. Below are some startling figures on corrosion losses:

- The corrosion cost of gas and liquid transmission pipelines in the United States exceeds US$7 billion. The figures for the major oil-producing countries in the Gulf region are not known; however, the cost is expected to be very high because of the highly corrosive environment in the region (NACE).
- The corrosion-free life of automobiles in the coastal regions of the Arabian Gulf is only about six months.
- Nearly 95% of the concrete damage in the Arabian Gulf coastal region is caused by reinforcement corrosion and consequent spalling of concrete (Rashid-uz-Zafar et al. 1992).
- It is estimated that 10% of all aircraft maintenance in the United States is spent on corrosion remediation (Tullmin et al. 1990).
- Major annual corrosion losses to the tune of 350 million euros in transport, 280 million euros in marine, 250 million euros in buildings and construction, and 180 million euros in oil and chemical industries have been reported in the UK (Hoar 1971). These are uncorrected 1971 figures.
- About $120 billion is spent on maintenance of aging and deteriorating infrastructures in the United States.

- Automotive corrosion costs US$23.4 billion annually in the United States (NACE 2002).
- Every newborn baby in the world now has an annual corrosion debt of $40.

The National Association of Corrosion Engineers–International India Section (NACE) states that the loss of materials due to corrosion was about Rs.35,000 crores each year, nearly 2.5–4% of India's GDP. It was estimated that the annual cost of all forms of corrosion to the oil and gas industries was $13.4 billion in the United States for the year 2001, of which microbially influenced corrosion accounted for about $2 billion (Koch 2001). The importance of internal corrosion as a cause of leakage of steel tanks was documented by various sources. A comprehensive U.S. Environmental Protection Agency report documents that 6–10% of tank failures were caused by internal corrosion. In France, a major oil company estimated that 10% of underground storage tanks leaked due to internal corrosion, while other data from France indicated that 8.5% of the leakages were caused by internal corrosion. Switzerland has reported that internal corrosion was the cause for 5% of its tank leakages. The percentage of internal leaks was lower than that of external leakages, and Sweden reported that half of its leaking tanks were only due to internal corrosion.

1.2 Principles of corrosion

Among the principles of corrosion are the following:

1. Thermodynamic principles
2. Electrochemical principles

1.2.1 Thermodynamic principles

Thermodynamic (Bockris and Reddy 1970) and electrochemical principles play a major role in determining the corrosion behavior of materials. Thermodynamics indicates the spontaneous direction of a chemical reaction. It is used to determine whether or not the corrosion is theoretically possible.

1.2.2 Electrochemical principles

Electrochemical principles are extensively used to determine the corrosion behavior of materials. The corrosion reaction can be represented by partial reactions such as metal oxidation and reduction of some reducible species of the environment, both occurring simultaneously at equal rates at the mixed potential of the reaction. Corrosion reaction mainly occurs at the metal–environment interface.

The electrochemical nature of corrosion can be illustrated by the attack of iron in hydrochloric acid. When iron is dipped in acid, a vigorous action occurs; as a result hydrogen gas is evolved and iron gets dissolved.

Hence the reaction is

$$Fe + 2H^{+} \rightarrow Fe^{2+} + H_2 \tag{21.1}$$

The above reaction can be divided into two partial reactions:

$$Fe \rightarrow Fe^{2+} + 2e^{-} \quad \text{oxidation (anodic reaction)} \tag{21.2}$$

$$2H^{+} + 2e^{-} \rightarrow H_2 \quad \text{reduction (cathodic reaction)} \tag{21.3}$$

1.3 Theories of corrosion

Corrosion theories may be classified into (1) homogeneous and (2) heterogeneous theories.

1.3.1 Homogeneous theory

Surface heterogeneous of the metal causes the corrosion process. Wagner and Traud (1938) emphasized that the necessary condition for corrosion is the dissolution of the metal and electronation reaction that takes place at the metal–environment interface. Metals become unstable owing to the charge transfer reaction taking place at the interface. Therefore it is necessary that the potential difference across the interface be more negative than the equilibrium potential for the metal dissolution (anodic) reaction, or more positive than the equilibrium potential for the electronation (cathodic) reaction. This theory is helpful in explaining the corrosion of pure metals and uniform corrosion, as the cathodic and anodic sites interchange their characters quite frequently.

1.3.2 Heterogeneous theory

According to this theory, the presence of impurities on the metal surface is necessary, and thus a local cell is set up between cathodic and anodic areas. Because of the electrochemical reactions taking place at the interface between the metal and ionically conducting films or actual electrolyte, the metal becomes unstable and hence corrodes.

Thus a corroding metal consists of

1. An electron sinks area where a deelectronation reaction occurs.
2. An electron source area where an electronation reaction occurs.
3. An ionic conductor to keep the ion current flowing.

This model is the basis for the local cell theory of corrosion or heterogeneous corrosion.

2 Corrosion of petroleum production and transportation pipelines

Pipelines play an extremely important role throughout the world as a means of transporting gases and liquids over long distances from their sources to the ultimate consumers. At present, there are approximately 460,000 km (285,000 miles) of common carrier pipelines transporting about 46% of all the crude oil and refined products moved in the United States. In 1984, there were more than 1.6×10^6 km (10^6 miles) of natural gas pipelines in service in the United States, where about 25% of them were in interstate service. At that time, there were approximately 2,800,000 km (174,000 miles) of pipeline carrying liquids in interstate service in the United States. Construction continues to add more miles of pipelines. In 1986, 9800 km (6090 miles) of natural gas pipeline were constructed, along with 5740 km (3567 miles) of crude oil pipelines and 2660 km (1652 miles) of pipelines in the United States for refined products. In India, the Indian Oil Corporation's cross-country crude oil and product pipelines network spanning over 9000 km meets the vital energy needs of the country.

Enormous amounts of money are being spent every year on prevention, monitoring, inspection, and land repair of corrosion-related damages. Internal corrosion control programs usually involve chemical treatment with corrosion inhibitors. Corrosion inhibitors are marketed in several basic types, such as oil-soluble, water-dispersible, water-soluble, limited-solubility (gunkers), and volatile, and each performs uniquely in different pipeline conditions. They can be applied in a batch procedure where the

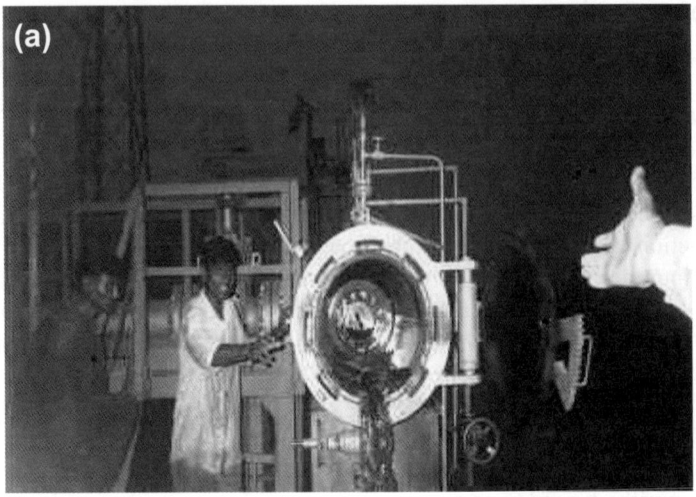

FIGURE 21.1a

Pigging program in a diesel transporting. (For color version of this figure, the reader is referred to the online version of this book.)

FIGURE 21.1b

PIG instrument Pipeline. (For color version of this figure, the reader is referred to the online version of this book.)

persistent nature of a heavy protective film may last weeks or months. Or they can be continuously metered into the pipeline in low concentrations through a continuous injection programmer, in which a thin film is gradually laid down and maintained over time. The chemicals work very well, provided that they can reach the wall of the pipe and form an effective film. If deposits of dirt, corrosion products, and bacteria are inside the pipe, they can both consume chemicals meant to treat the walls of the pipe and prevent the chemical from contacting the walls of the pipe beneath the deposits. For these reasons, pipelines should be as clean as possible when corrosion inhibitors are applied. A pigging program (Figures 21.1a–21.1b) is the most cost-effective and efficient method of cleaning the pipe surface for the application of liquid inhibitors.

2.1 Forms of corrosion in petroleum production

In the petroleum industry, many types of corrosion occur in both the production system and petroleum products transporting pipelines.

2.1.1 Sulfidation

When the sulfur activity (partial pressure, concentration) of the gaseous environment is sufficiently high, sulfide phases can be formed instead of oxide phases. The mechanisms of sulfide formation in gaseous environments and beneath molten-salt deposits have been determined in recent years. In the majority of environments encountered in practice by oxidation-resistant alloys, Al_2O_3 or Cr_2O_3 should form in preference to any sulfides, and destructive sulfidation attack occurs mainly at sites where the protective oxide has broken down. The role of sulfur, once it has entered the alloy, appears to tie up the chromium and aluminium as sulfides, effectively redistributing the protective scale-forming elements near the alloy surface and thus interfering with the process of formation or re-formation of the protective scale. If sufficient sulfur enters the alloy so that all immediately available chromium or aluminium is converted to sulfides, then the less stable sulfides of the base metal may form for morphological and kinetic reasons. It is these base metal sulfides that are often responsible for the observed accelerated attack because they grow much faster than the oxides or sulfides of chromium or aluminum. In addition, they have relatively low melting points, so that molten slag phases are often possible (Garverick 1994). The oxygen, sulfur and temperatures are major characteristics factors for sulfidation process in coal gasification system.

2.1.2 Pitting corrosion

Pitting of metals is extremely localized corrosion that produces sharply defined holes. The attack on the interior walls of the hole is usually reasonably uniform, but may be irregular where the specific conditions introduce a secondary intergranular attack. Every engineering metal or alloy is susceptible to pitting. Pitting occurs when one area of a metal surface becomes anodic with respect to the rest of the surface or when highly localized changes in the corrodent in contact with the metal, as in crevices, cause accelerated localized attack. There is a high ratio of cathode to anode area, and penetration progresses more rapidly than when pits are numerous and close together. Buried pipelines that fail because of corrosion originating on the outer side surface usually fail by pitting corrosion (Evans 1981). Another form of corrosion affecting pipelines involves a combination of pitting and erosion.

2.1.3 Crevice corrosion

The presence of narrow openings or spaces (gaps) between metal-to-metal or nonmetal-to-metal components may give rise to localized corrosion at these sites. Similarly unintentional crevices such as cracks, seams, and other metallurgical defects could serve as sites for corrosion initiation.

Resistance to crevice corrosion can vary from one alloy–environment system to another. Passive alloys, particularly those in the stainless steel group, are more prone to crevice corrosion than materials that exhibit more active behavior. Stainless steel may be susceptible to crevice corrosion in other chloride-containing natural water. In seawater, localized corrosion of copper and its alloys at crevices is different from that of stainless steel-type materials because the attack occurs outside of the crevice rather than within.

2.1.4 Stress corrosion cracking (SCC)

The cracking of metal (or) alloy due to the simultaneous presence of tensile stress and a specific corrosive environment is known as stress corrosion cracking (Garverick 1994). Several theories have been advanced to explain the mechanisms of SCC. Two major theories are the electrochemical and stress sorption theories.

Electrochemical Theory: According to the electrochemical theory, galvanic cells are set up between regions of exposed metal surfaces, for example, metal grains and heterogeneous phases; anodic paths are thus established. For example, the precipitation of $CuAl_2$ from an Al-4Cu alloy along grain boundaries produces copper-depleted paths in the edges of the grains. When the alloy, stressed in tension, is exposed to a corrosive environment, the ensuing localized electrochemical dissolution of metal, combined with localized plastic deformation, opens up a crack. With sustained tensile stress, protective films that form at the tip of the crack rupture, causing fresh anodic material to be exposed to the corrosive medium, and the SCC is propagated.

Stress-Sorption Theory: According to the stress-sorption theory, SCC generally proceeds by weakening of the cohesive bonds between surface-metal atoms through adsorption of damaging substances in the environment. Because chemisorption is specific, damaging components are also specific. The surface energy of the metal is said to be reduced, increasing the probability that the metal will form a crack under tensile stress. Adsorption of any kind that reduces surface energy should favor crack formation.

The presence of hydrogen sulfide and carbon dioxide in hydrocarbon reservoirs can create a significant materials problem. Material degradation can take many forms, including sulfide-stress cracking, hydrogen-induced cracking, and stress corrosion cracking. Each must be understood if appropriate measures are to be taken to minimize its effect in filed operations. Sulfide–stress cracking can occur when H_2S is present in the reservoir and is in contact with high-strength steels commonly used in drilling, completing, and producing wells. Sulfide-stress cracking is an embrittlement phenomenon in which failures can occur at stresses well below the yield strength of the material. For SCC to occur, three conditions must be met. First, a surface tensile stress must be present. It is important to remember that tensile stresses can be both applied and residual. The second requirement is that the particular material must be susceptible. Third condition is environment like H_2S concentration, pH, temperature, strength level and cold work. In general, lower H_2S concentration take longer to promote cracking than higher concentrations.

2.1.5 Stray current corrosion

If an extraneous direct current (DC) in the earth is traversed by a conductor, part of the current will transfer to the lower-resistance path thus provided. Direct currents are much more destructive than alternating currents (AC); an equivalent AC current causes only about 1% of the damage of a DC current (Ewing 1948). Regions of current arrival (where electrons depart) will become cathodic, and those regions where the current departs will become anodic. With corrodible metals such as carbon and low-alloy steels, corrosion in the anodic areas is the result.

2.1.6 Microbiologically influenced corrosion (MIC)

Macro- and microscopic organisms influence corrosion in two principal ways: (1) by creating mats or obstructions on the surface, which produce differential aeration cells, or (2) by absorbing hydrogen from the surface of steel and thus removing the hydrogen as a resistance factor in the corrosion cell. Certain sulfate-reducing bacteria operate this way, creating sulfurous acid in the vicinity of cathodic areas of steel, thereby accelerating the corrosion. Most of the documented cases in which biological organisms are the sole cause of, or an accelerating factor in, corrosion involve localized forms of attack. (Miller 1970). MIC occurs in secondary oil wells, petroleum industry pipeline cooling water system of oil refinery, the fire protection system, and so on.

2.1.7 Erosion corrosion

When movement of a corrodent over a metal surface increases the rate of attack due to mechanical wear and corrosion, the attack is called erosion corrosion. It is encountered when particles in a liquid impinge on a metal surface, causing the removal of protective surface films, such as air-formed protective oxide films or adherent corrosion products, and exposing new reactive surfaces that are anodic to uneroded neighboring areas on the surface. Nearly all-flowing or turbulent corrosive media can cause erosion corrosion. The attack may exhibit a directional pattern related to the path taken by the corrodent as it moves over the surface of the metal.

2.1.8 Effects in slow-moving and stagnant waters

Slow-moving and stagnant waters allow loosely adherent solid corrosion products to form on metal surfaces and aggravate corrosion. In addition, in closed systems where a corrosion inhibitor is used, the effectiveness of the corrosion inhibitor is reduced where the water is stagnant or nearly so. In designing for corrosion control, stagnant zones should be eliminated by the following methods.

- Allowing free drainage of water and suspended solids.
- Installing baffling to eliminate stagnant liquid zones.
- Increasing the frequency of cleaning
- Providing gas vent lines
- Providing strainers or separators to remove foreign material

2.1.9 Hydrogen damage

Hydrogen damage is a form of environmentally assisted failure that results most often from the combined action of hydrogen and residual or applied tensile stress. Hydrogen damage to specific alloys or groups of alloys manifests itself in many ways, such as cracking, blistering, hydride formation, and loss in tensile ductility. For many years, these failures have been collectively termed *hydrogen embrittlement*; this term persists even though it is improperly used to describe a multitude of failure modes involving hydrogen, several of which do not demonstrate the classical features of embrittlement (that is, reduced load carrying capability or fracture below the yield strength). As may be appreciated from the numerous classes of hydrogen damage, there are many explanations or theories for these various forms of degradation. The preeminent theories for hydrogen damage are based on pressure, surface adsorption, decohesion, enhanced plastic flow, hydrogen attack, and hydride formation (Garverick 1994).

2.2 Factors influencing corrosion in petroleum production

2.2.1 Oxygen

Although oxygen is not normally present at depths more than approximately 100 m below the surface, it is nevertheless responsible for a great deal of the corrosion encountered in oil and gas production. However, oxygen-induced internal corrosion problems tend to be greater in oil production, where much of the processing and handling occurs at near ambient pressure. This makes oxygen contamination through leaking pump seals, casing and process vents, open hatches, and open handling (as in mud pits during drilling, trucking, and so on) highly likely. Also, the failure of oxygen-removal processes (gas stripping and chemical scavenging) is a relatively common occurrence in water flood systems.

At ambient conditions, water equilibrated with air will contain on the order of 7 to 8 ppm of oxygen. Under such conditions, mass transport-limited rates of general corrosion of steel range from about 0.25 mmpy in a stagnant system to 15 mmpy in a highly turbulent one. However, by chemically scavenging the oxygen concentration down to the order of 0.007 to 0.0008 ppm, the corresponding rates are reduced to less than about 0.01 mmpy. Such rates are acceptable. However, under these conditions, magnetite forms as a stable protective corrosion product film and further lowers the corrosion rate by introducing a slower, anodically controlling step.

2.2.2 Hydrogen sulfide, polysulfides, and sulfur

Hydrogen sulfide, when dissolved in water, is a weak acid and is therefore corrosive because it is a source of hydrogen ions. In the absence of buffering ions, water equilibrated with 1 atm of H_2S has a pH of about 4. However, under high-pressure formation conditions, pH values as low as 3 have been calculated. Hydrogen sulfide can also play other roles in corrosion in oil and gas production. It acts as a catalyst to promote absorption by steel of atomic hydrogen formed by the cathodic reduction of hydrogen ions. This accounts for its role in promoting sulfide stress cracking of high-strength steels (yield strength greater than approximately 690 Mpa, or 100 ksi). Hydrogen sulfide also reacts with elemental sulfur. In a gas phase with a high H_2S partial pressure, sulfanes (free acid forms of a polysulfide) are formed so that elemental sulfur is rendered mobile and is produced along with the remaining gaseous constituents. However, as the pressure decreases traveling up the production tubing, the sulfanes dissociate and elemental sulfur precipitates. Various solvent treatments are used to avoid plugging by such sulfur. In the aqueous phase, under acidic conditions, sulfanes are also largely dissociated in H_2S and elemental sulfur. However, enough strongly oxidizing species can remain either as polysulfide ions or as traces of sulfanes to play a significant role in corrosion reactions.

Oxygen contamination of sour (H_2S- containing) systems can also result in polysulfide formation. At high temperatures (150 to 230°C, or 300 to 450°F) and H_2S partial pressures (thousands of pounds per square inch) encountered in deep sour gas wells, a so-called barnacle type of localized corrosion (Rhodes 1976) can occur, resulting in corrosion rates of several hundred miles per year (Place 1979). This type of attack is strongly promoted by polysulfide-type species and requires the presence of some minimum chloride concentration. Although initially recognized in deep sour well environments, this same mechanism may operate at lower rates under much milder conditions. In the barnacle mechanism, corrosion can be sustained beneath thick but porous iron sulfide deposits (primarily pyrrhotite, FeS) because the FeS surface is an effective cathode. The anodic reaction beneath the FeS deposit is dependent on the presence of a thin layer of concentrated $FeCl_2$ at the Fe/FeS interface. This intervening $FeCl_2$ layer is acidic due to ferrous ion hydrolysis, thus preventing precipitation of FeS directly on the corroding steel surface and enabling the anodic reaction to be sustained by the cathodic reaction on the external FeS surface.

2.2.3 Carbon dioxide

Carbon dioxide, like H_2S, is a weakly acidic gas and becomes corrosive when dissolved in water. However, CO_2 must first hydrate to carbonic acid (H_2CO_3) (a relatively slow reaction) before it is acidic. There are other marked differences between the two systems. Velocity effects are very important in the CO_2 system; corrosion rates can reach very high levels (thousands of miles per year), and the presence of salts is often important. Whether or not corrosion in a CO_2 system is inherently controlled or uncontrolled depends critically on the factors governing the deposition and retention of protective iron carbonate (siderite, $FeCO_3$) scale. On the other hand, there are factors that determine the rate of corrosion on bare steel. These factors govern the importance of maintaining corrosion control. Conditions favoring the formation of a protective iron carbonate scale are:

- Elevated temperature (decreased scale solubility, decreased CO_2 solubility, and accelerated precipitation kinetics)
- Increased pH, as occurs in bicarbonate-containing waters (decreased solubility)
- Lack of turbulence

Iron carbonate lacks conductivity and therefore does not provide an efficient cathode surface. Thus, the types of pitting mechanism found in oxygenated and in H_2S-containing systems do not occur.

2.2.4 Concentrated brines

Dense halide brines of the cations of calcium, zinc, and, more rarely, magnesium are sometimes used to balance formation pressures during various production operations. All can be corrosive because of dissolved oxygen or entrained air. In addition, such brines may be corrosive because of the acidity generated by the hydrolysis of the metallic ions:

$$Zn^{2+} + H_2O \rightarrow ZnOH + H^+ \quad (21.4)$$

Corrosivity due to acidity is worst with dense zinc brines. More expensive calcium bromide brines are now often used at densities above 1.7 g/cm^3 (14 lb/gal) (attainable with $CaBr_2$ brines) to avoid long–term exposure to $ZnCl_2$ brines.

2.3 Factors influencing oil degradation

2.3.1 Bacteria

These organisms can either be autotrophic or heterotrophic and either aerobic or anaerobic. Heterotrophic bacteria derive their energy and carbon requirements from organic sources. Autotrophs are bacteria, which obtain their energy from light or from the oxidation of inorganic materials and the carbon by assimilation. Anaerobic bacteria do not require oxygen for their growth. It is the major group in moist soil and water contributing to the corrosion process. Sulphate-reducing bacteria (SRB) and sulfur-oxidizing bacteria are the major groups involved in microbial corrosion.

2.3.2 Nitrogen and phosphorus

There is some confusion and considerable apparent conflict in the literature regarding the limitation of petroleum biodegradation by available concentrations of nitrogen and phosphorus. Lepetit and Barthelemy (1968) reported that the concentrations of available nitrogen and phosphorus in water are severe limiting factors for microbial hydrocarbon degradation. Kinney (1968), however, proposes that nitrogen and phosphorous are not the limiting factor, since microorganisms require nitrogen and phosphorus for their incorporation into biomass and that the availability of these nutrients within the same area as that of hydrocarbon is critical. Colwell et al. (1975) concluded that Metula oil was degraded slowly in the marine environment, most probably because of the limitations imposed by the relatively low concentrations of nitrogen and phosphorus availability. The culture studies have indicated that the available concentrations of nitrogen and phosphorus severely limit the extent of hydrocarbon degradation of the most major oil spills. Rates of nutrient replenishment generally are inadequate to support rapid biodegradation of large quantities of oil. The addition of nitrogen and phosphorus-containing fertilizers can be used to stimulate the microbial degradation of hydrocarbon.

2.3.3 Oxygen

As with nutrients, a controversy has arisen over whether oxygen is absolutely required for hydrocarbon degradation or whether hydrocarbons are subject to anaerobic degradation. Grishchenkov (2000) and Zinjarde (2002) support the view that anaerobic degradation by microorganisms at best proceeds at negligible rates in nature. The existence of microorganisms, which are capable of anaerobic hydrocarbon metabolism, has not been excluded (Atlas 1981). During storage of oil products, underground conditions are usually limited in oxygen and it is of special concern if hydrocarbon degradation is possible under anaerobic conditions. The initial attack on hydrocarbon requires oxygen, but for the subsequent steps, anaerobic processes may degrade partially oxygenated intermediates further. Under oxygen limited conditions and accumulation of degradation products in the form of fatty acids occur. It has been shown that with the sulphate reducing in the fuel, and even in a relatively static system, the microbes themselves can produce membranes, biosurfactants, and gaseous by-products that carry up into fuel, increasing the surface area available for microbial activity. In fact, there have been several reports of isolated microorganisms, which are capable of alkane dehydrogenation under anaerobic conditions (Atlas 1981). These organisms have enzymatic mechanisms, which should permit the addition of water across the double bond, forming a secondary alcohol, and therefore permit anaerobic growth.

In the case of the *Pseudomonas* sp. strain studied by Senez and Azoulay (1961), the organisms consumed oxygen when grown on heptanes, even though it had an n-heptane dehydrogenase enzyme. The importance of oxygen for hydrocarbon degradation is evidenced by the fact that the major degradative pathways for both saturated and aromatic hydrocarbons involve oxygenases and molecular oxygen. Degradation of hydrocarbons has been reviewed by Atlas (1981). SRB can use a wide range of fatty acids, including acetate, lactate, propionate, butyrate, and fatty acids up to C_{18} and some phenyl substituted acids (Widdel 1980). The generally accepted view is that anaerobic degradation of hydrocarbons is a very slow process as compared to aerobic degradation. Over very long periods, however, it might be significant. Reports on the anaerobic degradation in natural ecosystems have suggested that nitrate or sulphate could serve as an electron acceptor in place of oxygen (Atlas 1981).

2.3.4 Temperature and pressure

Schwarz et al. (1945) examined the growth and utilization of hydrocarbons at ambient and in-situ pressure for deep-sea bacteria. The rate of hydrocarbon utilization at high pressure and ambient temperatures were found to be significantly lower than the rates found under conditions of ambient temperatures and atmospheric pressure, whereas 94% hexadecane was utilized within 8 weeks at 1 bar while at 500 bars it took 40 weeks for similar degradation. It appears that oil, which enters deep-ocean environments, will be degraded very slowly and persist for long periods of time. Biodegradation can occur over a wide range of temperatures from below 0°C to 70°C (Atlas 1981, 1984). Most hydrocarbon degrading microorganisms prefer a near-neutral pH, while source organisms can tolerate extreme values.

3 Corrosion control and prevention

3.1 Materials selection

Carbon steel is the almost exclusive choice of pipeline designers. This is true for pipeline systems that are used to gather or collect the natural gas, crude oil, or water; it is also true for those pipelines used to transport substances over distances from hundreds of feet to hundreds of miles. It is also the case for piping systems that are used to distribute natural gas, water, water-refined liquids, and so on, to the end user (Garverick 1994). Table 21.1 lists the composition of some standard pipe steels for the petroleum production system and petroleum-transporting pipelines. Metallic corrosion can be prevented by either changing the metal or altering the environment, or by separating the metal from the environment or changing the electrode potential of the metal. A large number of corrosion failures are due to improper design of equipment. and the corrosion control can therefore be warranted at the design stage itself. Corrosion protection mostly involves bulk alloying or surface coatings. Surface coatings may pose problems related to adhesion, thermal expansion compatibility, and the like. Surface processing of metals has been improved by the ion-implantation technique (Prjor and Keir, 1957) and laser treatment (Tan 1994, Oldfield 1987), which results in a homogeneous and single-phase surface layer. Recently, electron beam surface area glazing has been found to increase the clear life of iron-base-tool materials (De Vivekanand and Banerjee 1997). Corrosion can be prevented by changing the electrode potential and taking the metal to the immune region (or) passive region. According to the Pourbaix diagram, this can be accomplished by making the potential of the cathode equal to the open-circuit potential of the anode.

3.2 Cathodic protection

Cathodic protection is defined as the reduction or prevention of corrosion by making it a cathode in the electrolytic cell. There are two methods of applying cathodic protection to metallic structures: the

Table 21.1 Typical Compositions of ASTM and API Pipe Steels

Steel	Composition, %											
	C	Mn	Si	P	S	Cr	Mo	Nb	V	Ti	Al	B
A106, Grade A	0.25 max	0.27 -0.93	0.10 min	0.048 max	0.058 max	...	...	...	...	...	...	...
A106, Grade B	0.30 max	0.29-1.06	0.10 min	0.048 max	0.058 max	...	...	...	...	...	...	...
A335, Grade P2	0.10-0.20	0.30 -0.61	0.10 -0.30	0.045 max	0.045 max	0.50-0-0.81	0.44-0.65	...	...	...	...	...
A335, Grade P5	0.15max	0.30-0.60	0.50 max	0.030 max	0.030 max	4-6	0.45-0.65	...	...	...	...	...
A335, Grade P7	0.15max	0.30 -0.60	0.50-1	0.030 max	0.030 max	6-8	0.44-0.65	...	...	...	...	...
A335, Grade P11	0.15max	0.30-0.60	0.50-1	0.030 max	0.030 max	1-1.50	0.44-0.65	...	...	...	...	...
A335, Grade P22	0.15max	0.30 -0.60	0.50 max	0.030 max	0.030 max	1.90-2.60	0.87-1.13	...	...	...	...	...
A335, ClassY52	0.26 max	1.40 max	...	0.040 max	0.050 max	...	...	...	...	...	...	...
API 5L – X46	0.30 max	1.35 max	...	0.040 max	0.050 max	...	...	...	...	...	...	...
API 5L – X60	0.26 max	1.35 max	...	0.040 max	0.050 max	...	...	0.05 min	0.02 min	0.03 min **(a)**	...	...
API 5L, Grade X52	0.21	0.90	0.26	0.015 max	0.015 max	...	...	...	0.09	...	0.030	...
API 5A, Grade X55	0.45	1.30	0.26	0.015 max	0.015 max	...	...	...	...	...	0.007	...
API 5AX, Grade N-80	0.28	1.48	0.26	0.015 max	0.015 max	0.20	0.10	...	...	...	0.007	...
API 5AX, Grade P-110	0.28	1.48	0.26	0.015 max	0.015 max	0.22	0.23	...	...	...	0.007	...
API 5AC, Grade C-90	0.29	0.50	0.26	0.015 max	0.015 max	1.08	0.33	...	0.03	...	...	0.0015
API 5L, Grade A	0.17	0.50		0.020	0.020	...	...	...	...	...	...	...
API 5L, Grade X60	0.05	1.11	0.017	0.007	0.006	...	...	0.045	...	...	0.045	...

Niobium, vanadium, and titanium are used at manufacturer's option.

galvanic or sacrificial anode and the impressed current method. In each method, a dc supply is more available for the protection of metal structures. The choice of the method to be used depends on a number of economic and technical considerations. An active metal, generally, zinc is used as a sacrificial anode in contact with the corroding material. The two metals in contact form a galvanic cell, the terminals of which have been short-circuited. In the impressed current method of cathodic protection, the external source of direct current is connected to the structure to be protected and the anode. The negative terminal of the power source is connected to the structure to be protected. Some important impressed current anodes are graphite, scrap iron, platinum, and lead–silver alloys. Power sources used in these systems are rectifiers, dc batteries, and so on. Sacrificial anode systems and impressed current anode system are complementary to each other. Cathodic protection can be applied to buried pipelines, underground cables, equipment for handling and storing chemicals, steel structures in marine atmosphere, hulls of ships, and oil-cargo-ballast tanks.

3.3 Anodic protection

Anodic protection is defined as the protection of a metal by maintaining it in the passive condition. This technique is based on the phenomenon of passivity. The metal to be protected is given a fixed potential so as to produce a passive film on it, and the structure is protected from the corrosive environment.

3.4 Use of coating

Corrosion can be prevented by separating the metal from the corrosive environment by using protective coatings. The two types of coatings are metallic and nonmetallic. Their characteristics are:

Good resistance to corrosion
Perfect adherence to the underlying metal
Continuity so as to cover the metal surface completely.

Cathodic and anodic metallic coating provide a physical barrier between the metal surface and environment, and discontinuity in the coating will result in localized attack (e.g., brass, chromium. or gold on steel as cathodic coating). Zinc and aluminium on steel as anodic coating. Nonmetallic coatings are such as vitreous enamels and portland cement coating. The chemical conversion coatings are such as varnishes, paints, enamels, and lacquers.

4 Corrosion inhibitors

4.1 Introduction

An inhibitor is a chemical substance which, when added to an environment in small concentration, effectively checks, decreases, or prevents the reaction of the metal with the environment (Riggs 1973). Corrosion inhibitors are added to many systems including cleaning pads, cooling systems, refinery units, chemical operations, steam generators, ballast tanks, and oil and gas production units. Corrosion inhibitors are used to protect metals from corrosion, including temporary protection during storage or transport as well as localized protection required, for example, to prevent corrosion that may result from accumulation of small amounts of an aggressive phase. One example is brine, in a nonaggressive phase, such as oil. An efficient inhibitor is compatible with the environment, is economical for application, and produces the desired effect when present in small concentrations.

4.2 Classification of inhibitors

Inhibitor selection is based on the metal and the environment. A qualitative classification of inhibitors is presented in Figure 21.2. Inhibitors can be classified as environmental conditioners and interface inhibitors.

4.2.1 Environmental conditioners (scavengers)

Corrosion can be controlled by removing the corrosive species from the medium. Inhibitors that decrease corrosivity of the medium by scavenging the aggressive substances are called environmental conditioners or scavengers. In near-neutral and alkaline solutions, oxygen reduction is a common cathodic reaction. In such situations, corrosion can be controlled by decreasing the oxygen content using scavengers (Noack 1982).

4.2.2 Interface inhibitors

Interface inhibitors control corrosion by forming a film at the metal/enviroment interface. Interface inhibitors can be classified into liquid and vapor-phase inhibitors.

4.2.2.1 Liquid-phase inhibitors

Liquid-phase inhibitors are classified as anodic, cathodic, or mixed inhibitors, depending on whether they inhibit the anodic, cathodic, or both electrochemical reactions.

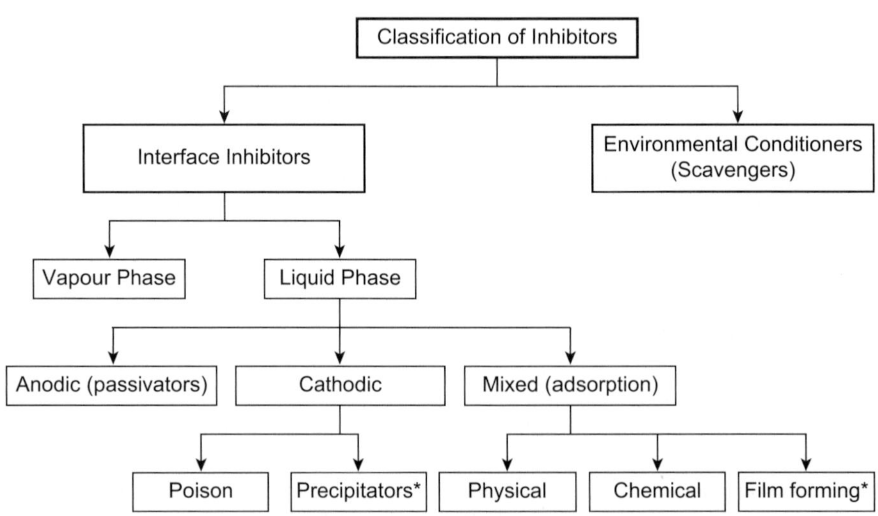

FIGURE 21.2

Classification of inhibitors.

4.2.2.2 Anodic inhibitors

Anodic inhibitors are most often used in near-neutral solutions where sparingly soluble corrosion products, such as oxides, hydroxides, or salts, are formed. They form or facilitate the formation of passivating films that inhibit the anodic metal dissolution reaction. Anodic inhibitors are often called passivating inhibitors.

4.2.2.3 Cathodic inhibitors

Cathodic inhibitors control corrosion by either decreasing the reduction process (cathodic poison) or precipitating selectively on the cathodic areas (cathodic precipitators). Cathodic poisons, such as sulfides and selenides, are adsorbed on the metal surface; whereas in near-neutral and alkaline solutions, inorganic anions, such as phosphates, silicates, and borates, form protective films that decrease the cathodic reaction rate by limiting and diffusing oxygen to the metal surface. Cathodic precipitators increase the alkalinity at cathodic sites and precipitate insoluble compounds on the metal surface. The most widely used cathodic precipitators are the carbonates of calcium and magnesium.

4.2.2.4 Mixed inhibitors

About 80% of inhibitors are organic compounds that cannot be designated specifically as anodic or cathodic and are known as mixed inhibitors. The effectiveness of organic inhibitors is related to the extent to which they adsorb and cover the metal surface. Adsorption depends on the structure of the inhibitor, surface charge of the metal, and type of electrolyte (Frignani et al. 1980). Mixed inhibitors protect the metal in three possible ways: physical adsorption, chemisorption, and film formation. Physical (or electrostatic) adsorption is a result of electrostatic attraction between the inhibitor and metal surface. When the metal surface is positively charged, (anionic) inhibitors are facilitated (Figure 21.3). Positively charged molecules acting in combination with a negatively charged intermediate can inhibit a positively charged metal. Anions such as halide ions in solution adsorb on the positively charged metal surface, and organic cations subsequently adsorb on the dipole (Figure 21.4a–21.4b).

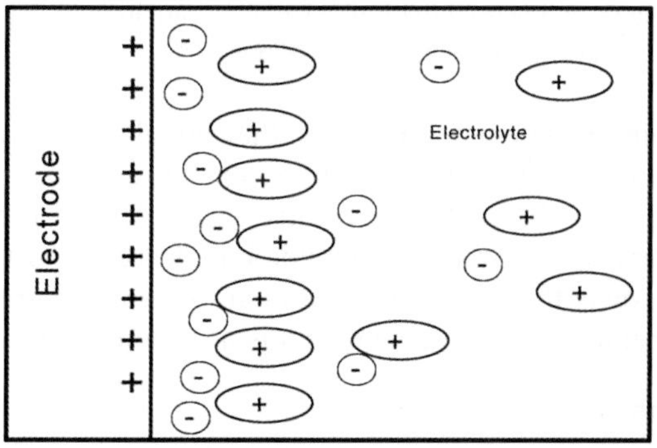

FIGURE 21.3

Adsorption of negatively charged inhibitor on a positively charged metal surface.

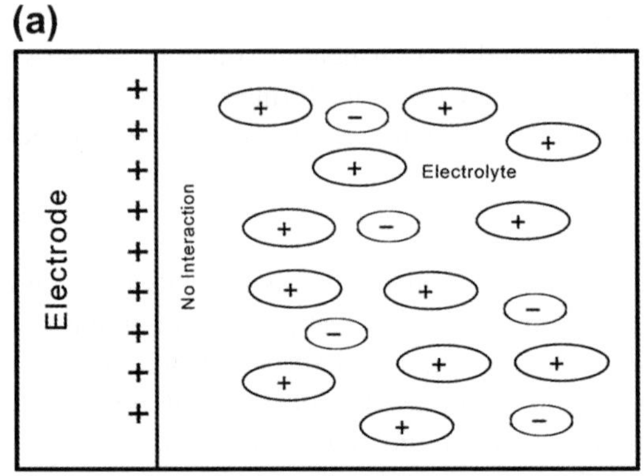

FIGURE 21.4a

Positively charged inhibitor molecule does not interact with positively charged metal surface.

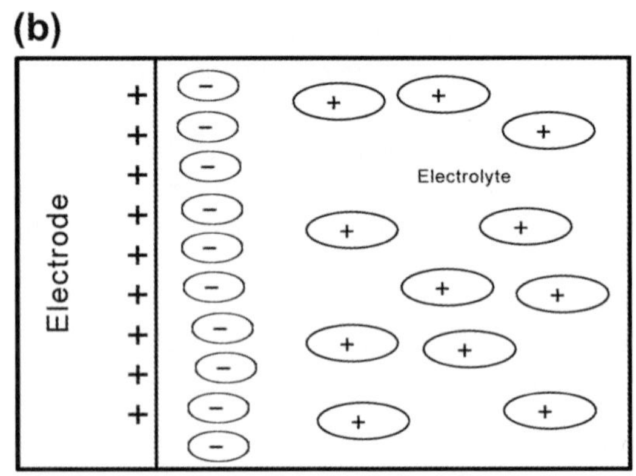

FIGURE 21.4b

Synergistic adsorption of positively charged inhibitor molecule and anion on a positively charged metal surface. List of some corrosion inhibitors for oilfield applications [14].

Physically adsorbed inhibitors interact rapidly, but they are also easily removed from the surface. Increase in temperature generally facilitates desorption of physically adsorbed inhibitor molecules. The most effective inhibitors are those that chemically adsorb (chemisorb), a process that involves charge sharing or charge transfer between the inhibitor molecules and the metal surface. Chemisorption takes place more slowly than physical adsorption. As temperature increases, adsorption and inhibition also increase. Chemisorption is specific and is not completely reversible (Sastri 1998).

4.2.2.5 Vapor-phase inhibitors

Temporary protection against atmospheric corrosion, particularly in close environments, can be achieved using vapor-phase inhibitors (VPI). Substances having low but significant pressure of vapor with inhibitive properties are effective (Levin et al. 1965).

4.3 Mechanistic aspects of corrosion inhibition

4.3.1 Environmental conditioners (scavengers)

In near-neutral solution, the common cathodic reaction is oxygen reduction:

$$O_2 + 2H_2O + 4e^- \leftrightarrow 4OH^- \tag{21.5}$$

Scavengers deplete the oxygen by chemical reaction: for example, hydrazine removes oxygen by the following reaction:

$$5O_2 + 2(NH_2 - NH_2) \leftrightarrow 2H_2O + 4H^+ + 4NO_{2^-} \tag{21.6}$$

4.3.2 Anodic inhibitors (passivators)

The mechanism of anodic inhibition can be explained using the polarization diagram of an active-passive metal (Fontana 1986). In the absence of inhibitors, the metal corrodes in the active state at a rate corresponding to point A in Figure21.4. As the concentration of inhibitor is increased, the corrosion rate also increases until a critical concentration and a critical corrosion rate are reached. At the critical concentration, there is a rapid transition of the metal to the passive state, and the corrosion rate is decreased.

4.3.3 Cathodic inhibitors

In acid solution, the cathodic reaction is typically the reduction of hydrogen ions to hydrogen, which combine and form hydrogen molecules:

$$H^+ + e^- \leftrightarrow H \tag{21.7}$$

$$2H \leftrightarrow H_2 \tag{21.8}$$

In alkaline solution, the cathodic reaction is typically oxygen reduction (Eq. 21.5).

Cathodic inhibitors impede reduction reactions. Substances with high overpotential for hydrogen and those that form precipitates at the cathode are effective in acid and alkaline solution, respectively. In this case the slope of the anodic polarization curve is unaffected, but the slope of the cathodic polarization curve is changed (Hackerman and Snavely 1984).

4.3.4 Mixed inhibitors (adsorption)

Adsorption occurs as a result of electrostatic forces between the electric charge on the metal and the ionic charges or dipoles on the inhibitor molecules. The potential at which there is no charge on the metal is known as the zero-charge potential (ZCP) (Table 21.2) (Trasatti 1971). The charge on a metal surface in a given medium can be determined from the corrosion potential (E_{corr}) and zero-charge potential. When the difference, E_{corr}-ZCP, is negative, the metal is negatively charged and adsorption of cations is favored; when E_{corr}-ZCP is positive, the metal is positively charged and adsorption of anions is favored.

The charge on inhibitors depends on the presence of loosely bound electrons, lone pair electrons, π-electron cloud, aromatic (e.g., benzene) ring systems, and functional groups containing elements of groups V and VI of the periodic table. Most organic inhibitors possess at least one functional group, regarded as the reaction center or anchoring group. The strength of adsorption depends on the charge

Table 21.2 Values of Zero Charge Potentials

Metal	Zero Charge Potentials, mV (SHE)
Ag	-440
Al	-520
Au	+180
Bi	-390
Cd	-720
Co	-450
Cr	-450
Cu	+90
Fe	-350
Ga	-690
Hg	-190
In	-650
Ir	-40
Nb	-790
Ni	-300
Pb	-620
Pd	0
Pt	+20
Rh	-20
Sb	-140
Sn	-430
Ta	-850
Ti	-1050
Tl	-750
Zn	-630

on this anchoring group [rather than on the hetero atom (i.e., atoms other than carbon, including nitrogen and sulfur) present in the anchoring group]. The structure of the rest of the molecule influences the charge density on the anchoring group (Sastri 1998). Water molecules adsorb on the metal surface immersed in an aqueous phase. Organic molecules adsorb by replacing the water molecules:

$$[\text{Inhibitor}]_{\text{soln}} + [\text{nH}_2\text{O}]_{\text{absorbed}} \leftrightarrow [\text{Inhibitor}]_{\text{absorbed}} + [\text{nH}_2\text{O}]_{\text{soln}} \quad (21.9)$$

where n is the number of water molecules displaced by one inhibitor molecule.

The ability of the inhibitor to replace water molecules depends on the electrostatic interaction between the metal and the inhibitor. On the other hand, the number of water molecules displaced depends on the size and orientation of the inhibitor molecule. Thus, the first interaction between inhibitor and metal surface is nonspecific and involves low-activation energy. This process, called physical adsorption, is rapid and in many cases, reversible (Lorenz 1962). Under favorable conditions, the adsorbed molecules involved in chemical reaction (chemisorption) form bands with the metal surface.

Chemisorption is specific and is not reversible. The bonding occurs with electron transfer or sharing between metal and inhibitor. Electron transfer is typical for transition metals having vacant, low-energy electron orbitals. Inhibitors having relatively loosely bound electrons transfer charge easily. The inhibition efficiency of the homologous series of organic substances differing only in the heteroatom is usually in the following sequence: P > S > N > O. On the other hand, electronegativity, that is, the ability to attract electrons, increases in the reverse order.

Adsorption strength can be deduced from the adsorption isotherm, which shows the equilibrium relationship between concentrations of inhibitors on the surface and in the bulk of the solution. Various adsorption isotherms to characterize inhibitor efficiencies are presented in Table 21.3 (Smialowska and Wieczorek 1971; Trasatti 1974; Bockris and Swinkels 1964; Langmuir 1947). To evaluate the nature and strength of adsorption, the experimental data (e.g., corrosion rate) are fitted to the isotherm, and from the best fit, the thermodynamic data for adsorption are evaluated.

4.3.5 Vapor-phase inhibitors (VPIs)

The process of vapor-phase inhibition involves two steps: transport of inhibitor to the metal surface and interaction of inhibitor on the surface. A VPI may vaporize in the undissociated molecular form, or it may first dissociate and then vaporize. Amines vaporize in the undissociated molecular form. Subsequently, reaction with water, present as moisture at the surface, dissociates the inhibitor. On the other hand, dicyclohexylamine nitrite dissociates liberating amine and nitrous acid, which deposit on the metal surface (Rosenfeld et al. 1964). Both in molecular and in dissociated forms, VPIs adsorb either physically or chemically on the metal surface to inhibit corrosion. Some of the cationic and anionic corrosion inhibitors used in petroleum oil wells and transporting pipeline are listed in Figures 21.5 and 21.6. Volatile inhibitors, such as diethylamine, can also be injected at the same time in order to prevent corrosion in condensing liquids at the top of the pipe. This can provide protection to the initial portion in the line where condensation rates are high, although volatile inhibitors are generally considered for their limited travel ability not going very far down the line.

4.4 Biocides

Biocides are the most underused, misunderstood, and misapplied chemical products in the petroleum industry for many reasons. They are used to combat a problem that is subtle and difficult to detect. In

Table 21.3 Adsorption Isotherm[a]

Name	Isotherm[b]	Verification Plot
Langmuir	$\theta / 1 - \theta = \beta . c$	$\theta / (1 - \theta)$ vs. log c
Frumkin	$[\theta / (1 - \theta)]\, e^{f\theta} = \beta . c$	θ vs. log c
Bockris - Swinkels	$\theta / (1 - \theta)^n . [\theta + n(1-\theta)]^{n-1} / n^n = c . e^{-\beta} / 55.4$	$\theta / (1 - \theta)$ vs. log c
Temkin	$\theta = (1/f) \ln K . c$	θ vs. log c
Viral Parson	$\theta . e^{2f\theta} = \beta . c$	θ vs. $\log(\theta / c)$

[a]*See [46-49].*
[b]*θ. % P/100. surface coverage; β, ΔG, free energy of adsorption; R, gas constant; T. temperature: c, bulk inhibitor concentration; n, number of water molecules replaced per inhibitor molecule: f, inhibitor interaction parameter (+, attraction; and -, repulsion); K, constant; and %P = 1-inhibited corrosion rate / uninhibited corrosion rate.*

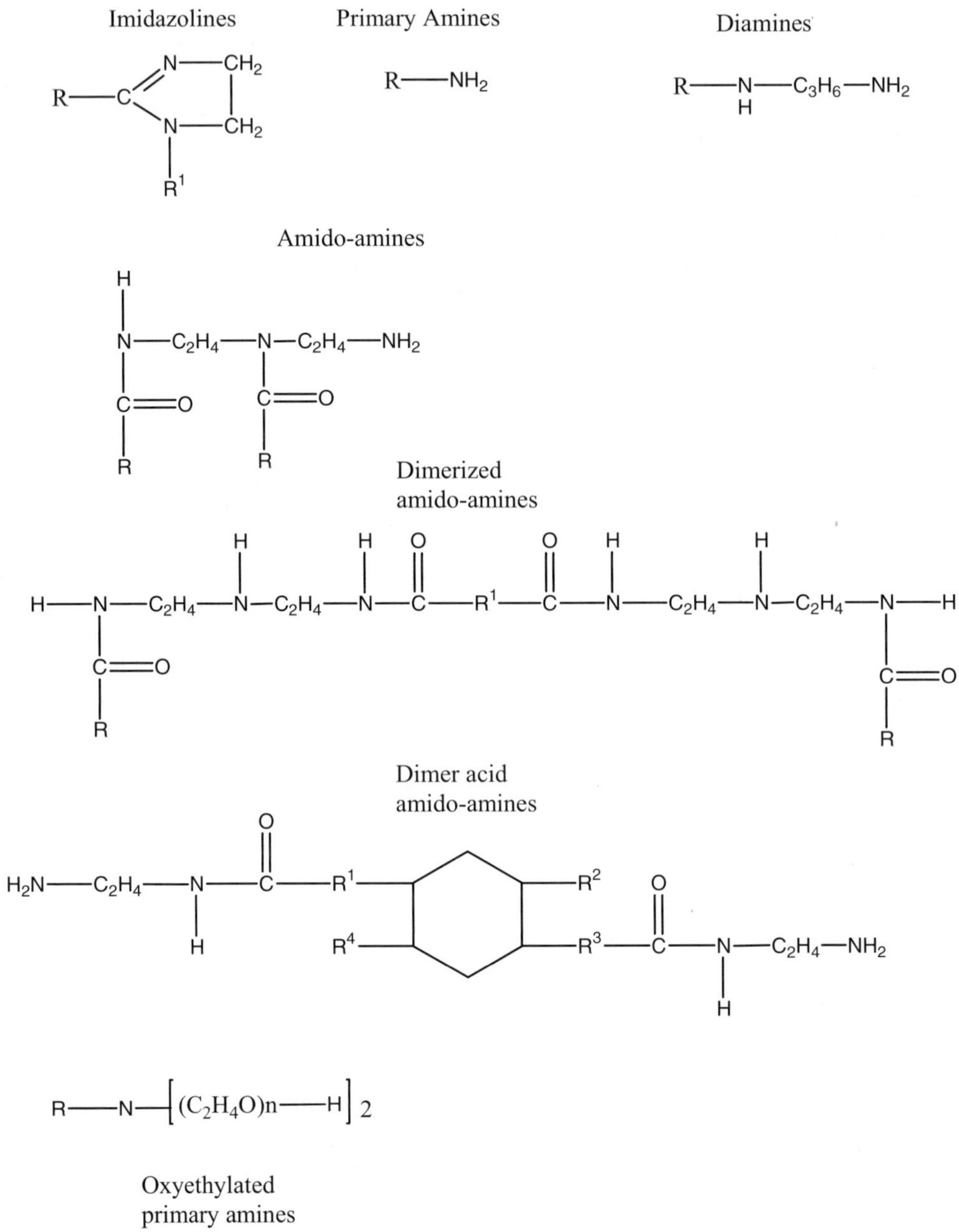

FIGURE 21.5

Cationic molecular structures in commercial oilfield inhibitors.

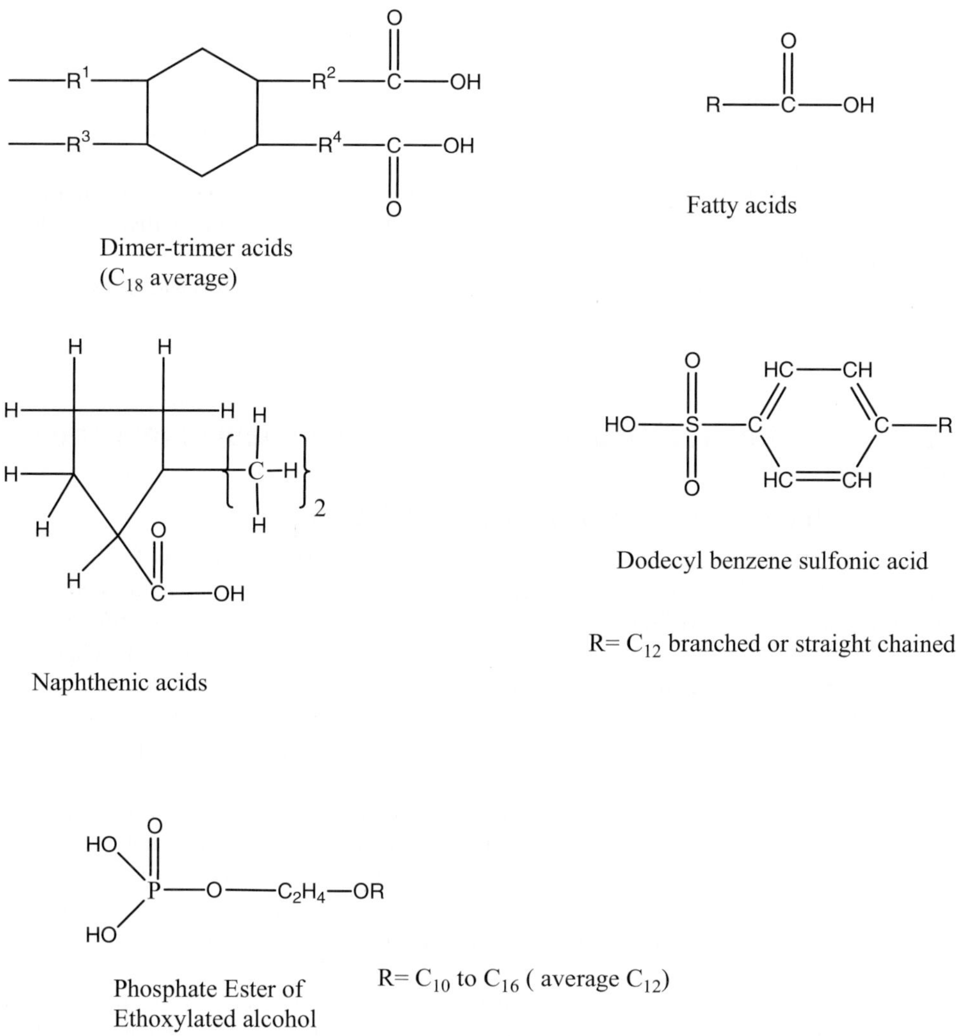

FIGURE 21.6

Anionic molecular structures in commercial oilfield inhibitors.

general, biocides are needed to control the activity of the bacteria in a system. However, biocides alone cannot solve a microbiological problem. Very few sample and reliable means of monitoring are available to the supplier or the end user, and the benefits of biocide use take a long time to become evident. Finally, once bacteria are well established they are nearly impossible to control without drastic measures (Widdel 1980). Hence, identification of microbes and the role of inhibitors on microbial growth need to be studied in detail.

Examples of chemicals used as fuel system biocides (fuel+/or water soluble)

- Methylenebisthiocyanate
- Hexahydro-1, 3,5-tris (2-hydroxyethyl)-S-triazine
- Mixture of 5-chloro-2-methyl-4-isothiazolin-3-one and 2-methyl-4-isothiazoline-3-one
- Ethylene or diethylene glycol monomethyl ether
- Mixture of 4-(2-nitrobutyl) morpholine and 4.4'-(2-ethyl-2-nitromethylene_-dimorpholine
- Mixture of 2's-oxy-bis(4,4,6-trimethyl-1, 3,2-dioxaborinane) and 2, 2 –(1-methyl-trimethylene di-oxy) bis – (4-methyl – 1,3,2 – dioxaborinane)
- Methyl-1-(butyl carbomoyl)-2-benzimidazole carbamate-1-(2-hydroxyethyl-)-2-alkyl-(-C-18)-2-imidazoline, etc.

5 An overview of corrosion and its control in petroleum production pipeline

5.1 Corrosion and its control in oil and gas production

Dougherty et al (2004) identified some protective films when the corrosive gases H_2S, CO_2 are present in the oil and gas system. The corrosion mechanisms in oil, gas, gas-condensate fields, and gas and oil pipelines were compared by Moiseeva (2005). Altoe et al. (1996) studied the commercial oilfield corrosion inhibitor (propyl amine) on carbon dioxide containing media. Their active mechanism in the presence of CO_2 is still little understood as far as the kinetics are concerned. Hausler and Schmitt (2004) studied the hydrodynamic and flow effects on corrosion inhibition. Chemical corrosion inhibition of steel in CO_2 and H_2S environments under high flow rates, hence high shear stresses, has long been a puzzle from a mechanistic point of view. Systematic progress, however, showed that a relationship existed between fluid velocity and inhibitor concentration for equal corrosion rate, thereby opening the possibilities of corrosion inhibition at even higher flow rates, although with higher inhibitor concentrations. Tung et al. (2001) reported the study of corrosion control effect of H_2S scavengers in multiphase systems. Aggravated corrosion conditions caused by H_2S in mixtures of oil, brine, and gas were responsible for some of the highest equipment failure rates in the petroleum industry.

Corrosion control is the major concern in the petroleum industry, specifically in environments like., CO_2, NaCl solution. It will result different corrosion products e.g., $FeCO_3$ and iron oxide. Many organic amine-based inhibitors were used to control corrosion in pipelines and production equipment (Galicia et al. 2004). These products were prepared from high-molecular-weight film-forming petroleum products, with the addition of corrosion inhibitors acting through different mechanisms, along with fillers, plasticizers, and a solvent. After the film forming inhibited petroleum compositions (FIPC) is applied to a metal surface, evaporation of the solvent leaves a film of the product with a thickness of 20–200 μm, depending on the type of FIPC. Gonik (2005) also reported the effect of diphilic corrosion inhibitor on the iron passivity in oilfield electrolytes and described the film forming inhibitors. Singh et al. (1990) studied the effects of some amines as inhibitors for the corrosion of API-45 grade steel in NACE sour solution (5% NaCl and 0.5% CH_3COOH solution saturated with H_2S), using weight-loss, hydrogen-estimation, electrochemical, and tensile strength of materials techniques. The selection of amines chosen was made up of ethanol amine (EA), triethyl amine (TEthA), dimethyl amine (DMA), diethyl amine (DEthA), triethanol amine (TFA), n-tributyl amine (TBA), cyclohexyl amine (CHA), dicyclohexyl amine (DCHA), pyridine (Py), ethylene diamine (EDA), furfuryl amine (FA), octyl amine (OA), and tri-iso-octylamine (TIOA). It was found that TIOA affords maximum protection to the metal under all the conditions tested. The inhibition efficiency of all the amines except that of TIOA decreased with rise in temperature from 25 to 70°C.

Hasheem et al. (2004) studied the synergistic effect of water treatment chemicals on the corrosion rate of L-80 steel in seawater, formation water, and a combination of both. The corrosion rate of tubular grade L-80 carbon steel under downhole conditions of a northern oilfield of Kuwait was investigated using injection seawater, formation water, and a 50:50 mixture of both waters in the presence of commercially available corrosion inhibitor (a water-soluble filming amine), scale inhibitor (aqueous blend containing polyacrylate and ethylene glycol), and biocide (blend of formaldehyde/quaternary ammonium and phosphonium sulfate) products separately and in combination with each other. The corrosion inhibitor at the manufacturer's recommended dosage level of 5 ppm concentration was unable to control corrosion. The addition of the scale inhibitor at the manufacturer's dosage level of 5 ppm led to reduction of corrosion in the seawater and mixed water systems but not in the formation water where the opposite, increased corrosion, was observed.

Sorkhabi et al. (2002) studied the inhibition effects on carbon steel by amines in petroleum-water mixtures. The corrosion inhibition efficiency afforded by ethylene diamine (EDA), hexyl amine (HA), butyl amine (BA), *tert*-butyl amine (*t*-BA), propyl amine (PA), and *iso*–propyl amine (i-PA) have been studied by polarization, electrochemical impedance spectroscopy (EIS), and optical micrography studies. Kirkov (1987) studied the influence of drag reducers and inhibitors on the rate and mechanism of corrosion in petroleum white products in the presence of moisture. Bojes et al. (2002) described the batch corrosion inhibitors for petroleum product (diesel). They examined the effect of contact time, the inhibitor/diluent ratio, and the nature of diluent on the corrosion inhibition effectiveness of a batch corrosion inhibitor. Powell and Wende (1997) studied selection of chemical products for oilfield applications in arctic environments.

Jeyaraman and Saxena (1996) reported the corrosion inhibitors for hydrocarbon systems. They provided directions to select the various corrosion inhibitors for different areas of petroleum industry based on their available structure–activity relationship. The present review highlights the utility of most of the inhibitors (used in production and distillation of crude oils), which are either N-containing compounds or their derivatives. Jayaraman et al. (1991) studied the control of internal corrosion of petroleum pipelines by inhibitors. The present studies were carried out (1) to generate corrosion-rate data under accelerated laboratory test conditions for two different crudes and (2) to investigate the corrosion-inhibiting behavior of various known amines and aminoamides prepared from nonedible oils that are available in India. Laboratory results showed that low-sulfur Bombay high crude did not give a measurable corrosion rate for API 5LX-52-pipeline steel, even under accelerated test conditions. However, high-sulfur Dubai crude severely corroded the same steel under identical test conditions. The high wax content (10.8 wt%) in Bombay High Crude might be one of the important factors for its protective ability and hence for the inability to generate measurable corrosion rate data.

The corrosion-inhibiting performance of three known primary amines of C12 to C18 carbon atoms in Dubai crudes from potentiodynamic polarization data showed an increase in inhibitor efficiency with carbon atoms. Al-Sabbagh et al. (1996) reviewed some of the organic corrosion inhibitors for steel pipelines in oilfields. N-benzyl triethanol ammonium chloride (BTAC) and ethoxylated n-benzyl triethanol ammonium chloride (EBTAC) were synthesized. The prepared compounds were characterized and then tested for their evaluation as corrosion inhibitors for steel pipelines in typical crude oil and residual fuel oil (RFO) media through weight-loss determination. In addition, the surface properties of the investigated compounds were studied through surface tension measurements in a hydrocarbon system. Results showed that the two additives achieved a substantial decrease in the corrosion rate and that EBTAC exhibited improved surface and thermodynamic properties and attained more inhibition effect.

Imidazolines have been used as corrosion inhibitors in the oil and gas industry for many years (Wang et al. 1999; Cheng and Zhan 1993; Hong et al. 2002; Sastri and Perumareddi 1994; Tan et al. 1996; McCormack et al. 2002; Martin and Valone 1985). However, the mechanism of corrosion inhibition, the interrelationship between surfactant adsorption and corrosion inhibition, and the main characteristic factors determining surfactant adsorption in aqueous medium have been rarely reported. This article (Muthukumar et al. 2007) uses recently developed methodology and equations to describe the interrelationship between the surfactant adsorption and corrosion inhibition. The main factors, such as surfactant concentration, functional group, hydrophobic chain length, and environment, affecting corrosion inhibition, were discussed as well. A reasonable prediction model was developed to predict corrosion inhibition based on this method, which can be applied to relevant industrial uses of corrosion inhibition. Ramachandran et al. (1997, 1999) also studied oleic imidazoline class molecules extensively for corrosion inhibitor oilfield applications, and they described the model for understanding how imidazoline works. The development and testing of a new corrosion inhibitor (IMP-ICCA-9710) were studied by Benitez Aguilar et al. (2002). The inhibitor was an imidazoline type and was dispersible in water and soluble in hydrocarbons. It demonstrated its effectiveness in pipeline protection against corrosion agents in crude oil, and it displayed a protective film that adheres to the metal surface for long periods. The self-assembly monolayer mechanism for corrosion inhibition in iron in imidazolines was reported by Ramachandran et al. (1996). Khalifa et al. (2003) studied the corrosion inhibition of steel in crude oil storage tanks. Metsik et al. (1986) studied the mechanism of action of oil-soluble corrosion inhibitors and petroleum-based protective compositions. They investigated the retardation of carbon–steel corrosion by these protective films when the specimens were exposed in a salt fog chamber or immersed in a 3% sodium chloride solution.

In order to evaluate the relative importance of the insulating (shielding) component in retarding the diffusion of oxygen, and also to determine the factors involved in the protective effect of certain corrosion inhibitors, film-forming inhibited petroleum-based compositions (FIPCs), and other protective products, measurements were taken to determine the current of oxygen diffusion through films of these materials, in air and in distilled water. Andreev et al. (1983) studied cyclohexylamine as a new inhibitor, under conditions simulating carbon dioxide corrosion of industrial pipeline steel. Castaneda et al. (2004) studied the film stability for API 5LX-52 line pipe steel in CO_2(aq) and Cl-(aq) solutions in the presence of amine-based inhibitors under hydrodynamic conditions. Additions of CO_2 (g) to API 5L X-52 steel exposed in brine solutions promoted the formation of $FeCO_3$ on active sites left by dissolution of Fe_xO_y films under hydrodynamic conditions. When the rotating rate was introduced to the steel-electrolyte interface, competition for the active sites was shown during electrochemical experiments because of the diffusion phenomenon of the ionic species to the surface of the metal interface. CO_2 additions showed kinetics and formation of iron carbonate, which was the protective film formed because of the free sites or porous regions at the interface electrode-electrolyte. Addition of filming inhibitor showed the equivalence of having a protecting (corrosion product) layer and a chemical formed layer that was more stable due to an outer layer formed by the absorption and interaction between active sites and ionic species in solution. The effects of some fatty acids on the inhibiting properties of nitrated petroleum products (white petrol) were studied by Arkuszewaska (1983). Durnie et al. (1999) studied the performance of a broad range of carbon dioxide corrosion inhibitors. Data for the thermodynamics of adsorption of these compounds have provided important information about the adsorption mechanism for corrosion inhibitors. The fundamental constants of the Temkin adsorption isotherm, that is, molecular interaction and adsorption equilibrium constants, were used to glean important information about the behavior of adsorbed corrosion inhibitors. Adsorption data were also used to calculate $\Delta_a dH$, $\Delta_a dG$, and $\Delta_a dS$, and the results demonstrated that some compounds were chemisorbed on the electrode.

Papavinasam et al. (2002) studied the evaluation of green inhibitors for oil. The current test methods, their reproducibility and effectiveness, and their accuracy in relation to the selection of green inhibitors were discussed. Hedges and Lockledge (1996) studied the continuing development of environmental-friendly corrosion inhibitors for petroleum production. Increasing concern has arisen regarding toxicity, biodegradability, and bioaccumulation of oilfield chemicals, which are discharged into the environment from offshore production platforms. The development and application of these inhibitors were discussed with particular reference to the European North Sea. Martin et al. (1999) studied the development of environmentally acceptable corrosion inhibitors. The influences of alkoxylation, starting amine type, and acid chain length on imidazoline properties were studied in addition to the properties of blends. The properties considered were corrosion inhibition, biodegradability, marine organism toxicity, octanol/water partitioning, and vapor pressure. From the overall literature, it can be concluded that no mechanism has been proposed on corrosion inhibition for diesel, kerosene, and petrol transporting pipelines. Although some products (inhibitors) are available in the market, there is no mechanistic study has been carried by scientists.

5.2 Acidizing inhibitors

A survey of the literature on high-temperature acid–corrosion inhibitors indicates the following compounds to be effective either alone or in combination with one another. These are: (1) acetylenic alcohols (Foster et al. 1959; Frunkhouser 1961; Poling 1967; Tedeschi 1975; Brindsi et al. 1981; Muzyezko et al. 1972; Mago 1987; Rosenfeld 1981), (2) alkenyl phenones (Frenier et al. 1987), (3) cinnamaldehyde derivatives (Growcock et al. 1989), (4) organic nitrogen compounds such as amines, amides, heterocyclics, and quaternary ammonium salts (Meyers 1984; Annand and Woodson 1986; Frenier et al. 1988; Redmore 1976; Woodson 1976), (5) condensation products of carbonyls and amines (Monroe 1963) and inorganic compounds such as arsenic and cuprous halides (Stern 1955; Matsushima and Uhlig 1966). Some of the patented inhibitor formulations contained compounds with most of the active groups in a single molecule. Most of the formulations were proprietary items and were not indigenously available.

Recommended inhibitor formulations were mixtures of N-containing compounds, acetylenic compounds, and surfactants. The N-containing compounds were frequently condensation products of amines with aldehydes. The amines included fatty, cycloaliphatic, aryl, and heterocyclic amines. Prominent examples were $C_{12}-C_{18}$ primary amines, cyclohexylamines, aniline and methyl substituted anilines, alkylpyridines, benzimidazole and high-molecular-weight amines like rosin amines. The condensating aldehyde was in most cases formaldehyde. Condensation products of formaldehyde with ammonia or aniline, condensation products of urotropin with aniline, and mixtures of pyridine bases gave only poor protection at temperature above 100°C (Keeney and Johnson 1973; Migahed et al. 2005). The efficiency of amine-type inhibitors in acidizing solutions based on hydrochloric acid was markedly improved by adding metal halides, especially copper (I) iodide (Quinlan and McBride 1977). The high-temperature performance was generally improved by introducing acetylenic compounds into the inhibitor package. The compound mostly used was propargyl alcohol; however, other derivatives such as 1-hexyne-3-ol were also applied (Schmitt 1984). Protection of acetylenic inhibitors was generally enhanced by the presence of iron ions. It was explained by the participation of iron ions in the condensation and polymerization reactions leading to the protection of surface films.

Two heterocyclic compounds, namely, QJ-1 and QJ-2 were synthesized from indigenous chemicals, and their corrosion-inhibiting action was evaluated by the weight-loss method and electrochemical technique in 15% HCl under boiling conditions. The performance of synthesized compounds was compared with that of propargyl alcohol, a commercially available acidizing inhibitor. Both QJ-1 and QJ-2 showed better performance than that of propargyl alcohol for mild steel (Quraishi et al. 1996). The corrosion inhibitors used in acidizing were reviewed by Cizek (1984). Neemla et al. (1992, 1989) have studied the performance of quarternary ammonium iodides and some additives in 15% HCl for J-55 and N-80 steels. α-alkenyl phenones gave excellent protection for oilfields. The effect of propargyl alcohol on corrosion and hydrogenation of steel in hydrochloric acid solution has been studied by Pati et al. (1990). Growcock et al. (1988) studied the inhibition of steel corrosion in hydrochloric acid with 3-phenyl-2-propyn – 1-ol and trans cinnamaldehyde. The effect of the mixture of amine, ester, and quaternary ammonium compound has been studied by Jeyaprabha et al. (1998). Selected condensation products of aromatic aldehydes and p-phenylenediamine were synthesized (Quraishi and Jamal 2000) and evaluated as corrosion inhibitors for mild steel and oil well steel in 15% HCl at 105 ± 2°C by the weight-loss method.

Two new organic inhibitors, namely, dibenzylidene acetone (DBA) and Di-N-dimethyl-amino-benzylidene acetone (DDABA), were synthesized in the laboratory, and their inhibiting action was evaluated on corrosion of N-80 steel and mild steel (MS) in HCl by the weight-loss technique (Suroor Athar 2001). Some new triazole-based corrosion inhibitors (Quraishi et al. 1998) were synthesized, and their influence on corrosion inhibition of mild steel in 15% HCl hydrochloric acid solution under boiling conditions was studied using the weight-loss method. Three new organic inhibitors, namely, dicinnamylidene acetone (DCA), disalicylidene acetone (DSA), and divanillidene acetone (DVA), were synthesized in the laboratory, and their inhibiting action was studied on corrosion of N-80 alloy in boiling hydrochloric acid (HCl) by the weight-loss method (Quraishi 2002). A new corrosion inhibitor, namely, 4-salicylideneamino-3-hydrazino-5-mercapto-1,2,4-triazole (SAHMT), was synthesized, and its influence on the corrosion inhibition of oil-well tubular steel (N-80) and mild steel in 15% hydrochloric acid (HCl) solution under boiling condition was studied (Quraishi and Jamal 2001).

5.3 Monitoring

Development of an integrity management program to control the internal corrosion of such pipelines depends on our ability to monitor the efficiency of the inhibitor performance. Controlling the electrode potential in low-conductivity solutions was a major difficulty because of the drift of the corrosion potential that progressively polarized the electrode. Papavinasam et al. (2003) studied the comparison of techniques for monitoring corrosion inhibitors in oil and gas pipelines. The reliability of weight loss, linear polarization resistance (LPR), electrochemical impedance spectroscopy

(EIS), electrochemical noise (EN), and externally mounted hydrogen probes for monitoring inhibitor performance in oil and gas pipelines were investigated. Reliable EIS measurements could not be obtained with regularity in operating oil and gas pipelines. Electrochemical noise was a reliable method of monitoring the effects of corrosion inhibitors on general corrosion rates in oil and gas pipelines. Noise parameters, pitting index, pitting factor, and pit indicator may be used with caution to obtain information regarding pitting corrosion. EN appears to be a promising monitoring method in oil and gas pipelines, but better theoretical understanding is still required. Chen et al. (2004) investigated monitoring of corrosion and flow characteristics in oil/brine mixtures of various compositions. The feasibility of monitoring corrosion processes by means of electrochemical impedance and noise measurements in paraffin oil/brine mixtures of various compositions between 0 and 80 vol% of oil were done. Hence many investigators evaluated the various electrochemical techniques in oil–water systems.

In contrast, electrolyte resistance (ER) measurements, performed with a homemade device delivering an analog signal allowing the mean value as well as the fluctuations of the ER to be measured in real time, were not sensitive to the corrosion potential drift. Bouazaze et al. (2005) reported a new approach for monitoring corrosion and flow characteristics in oil-brine mixtures. Monitoring of the corrosion process of various compositions between 0 and 80% in a volume of oil was investigated by means of electrochemical impedance and noise measurements. Hong et al. (2000, 2001) studied the monitoring of corrosion in multiphase pipelines. They investigated the corrosion inhibition efficiency of carbon steel in multiphase flow by electrochemical impedance spectroscopy (EIS) and electrochemical noise methods. They found that EIS and EN (electrochemical noise) techniques are good methods to study the corrosion and inhibition of pipeline steel under multiphase flow conditions.

Caunter (1999) investigated the remote corrosion monitoring of pipelines and plant sites. A newly developed remote corrosion monitoring system was used in monitoring oil and gas pipelines in remote and hostile environments, as well as process plants in hazardous locations. Silva et al. (2004) studied the effect of flow on the corrosion mechanism of different API pipeline steels grades in NaCl solutions containing CO_2. Electrochemical results were obtained during study of the corrosion of X52, X60, X65, and X70 pipeline steel samples, immersed in a 3 wt% NaCl solution saturated with CO_2 at 20°C under static and controlled turbulent flow conditions. Electrochemical techniques (e.g., linear polarization resistance) and polarization curves were also used to determine the effect of turbulent flow on the corrosion kinetics of the different steels studied. Dougherty et al. (1999) reported the criteria for selecting corrosion inhibitors for arctic and subsea high-velocity flow lines. They also described the different methodologies for evaluating corrosion inhibitors such as the flow loop test, rotating cylinder electrode (RCE), and emulsion tendency test. Becerra et al. (2000) studied the corrosion of carbon steel in oil–water emulsions under controlled hydrodynamic conditions. They found that the effect of the oil content on the electrochemical activity of carbon steel varies with the internal phase relationship (IPR). For emulsion with low IPR (oil contents up to 20 wt%), the electrochemical activity was slightly higher than that of the base surfactant solution. Schiapparelli et al. (1980) conducted their electrochemical study by employing contaminated water. In the present study, electrochemical behavior has been investigated by using contaminated water. Impedance and polarization techniques were employed. The study focuses on biodegradation of diesel/inhibitor and its effect on corrosion. Generally, water and bacteria contribute to corrosion. Hence, to find the mechanism whereby corrosion occurs, the contaminated water was used for electrochemical studies.

5.4 Risk assessment and safety management

Corrosion-related failures of processing facilities are a major source of risk to offshore oil and gas installations. Corrosion can be a life-limiting cause of deterioration by general wastage and/or pitting and/or environmentally assisted cracking to plant items, which in turn can lead to loss of containment of hydrocarbon fluids. Identifying hazards and assessing risks are fundamental for any management process. A hazard has the potential to cause harm or damage. Risk is the combination of the severity of the effect and the likelihood of it happening. Industrial risk assessment is a careful examination of potential hazards that may affect the operation of a business; these risks may be associated with the safety and integrity of physical assets, risks to the environment, and financial risks from various decisions, as well as risks from corrosion or poor corrosion mitigation procedures. Most offshore processing equipment (pipelines) is fabricated from carbon steel. This is an economic choice, based on life-cycle costing at the design stage of a project. Use of carbon steels means that potential hazards are present due to internal corrosion damage from aqueous produced fluids that contain acidic gasses, carbon dioxide, and hydrogen sulphide. The iron corrosion product films are only partially protective and are particularly susceptible to localized erosion corrosion under highly turbulent conditions. All offshore equipment and systems that contain seawater and injection water systems are at risk from corrosion. In addition, corrosion-related failures can result in hydrocarbon releases and significant loss of production, as well as increased costs of maintenance, repair, or replacement. Management of corrosion is therefore a major driver for safety, environmental, and economic issues within the industry.

Internal corrosion and the integrity of installations are typically monitored by recognized inspection procedures. In many pipelines and processing systems, the rate of corrosion is controlled by injection of inhibitor chemicals. External surfaces and the internals of vessels are usually protected by corrosion control coatings, whil the water wet internals of some vessels will also have cathodic protection systems installed. Some production systems can become infected with sulphate-reducing bacteria that then generate hydrogen sulphide as a by-product of respiration. This toxic chemical also increases the risk of some forms of corrosion damage.

Selection of appropriate construction materials for sulphide-containing fluids and/or production systems with high partial pressures of carbon dioxide is vital for some installations. Ensuring inherent safety means that corrosion-resistant alloys have to be employed in some systems. Safety-critical items such as downhole safety valves have traditionally been fabricated from such materials. Installations fabricated in carbon steels often rely on chemical treatment packages for inhibition and biological control; the availability of correct dosage levels and monitoring of performance are crucial for a successful corrosion management program. Procedures adopted to achieve the required objectives are then subject to independent verification. The same inspection techniques are employed to assess the condition of static equipment, but the emphasis should now be on using the data to provide the basis for continuous improvement by means of predictive strategies. A further objective should be the integration of health and safety management and corrosion management into the day-to-day overall asset management system. A major concern for offshore safety is hydrocarbon releases. Separation equipment and processing plant on installations are densely packed and exposed to both internal and external corrosive environments. The hazards associated with releases are well recognized within the industry, and considerable resources are directed toward managing these corrosion risks.

The above discussion provides a high-level introduction to safety and corrosion issues on offshore installations. In particular, it identifies the requirements of corrosion management with reference to management of offshore pipeline safety and asset integrity. An outline of the typical procedures that should be adopted in order to implement the risk-based strategies presented in this chapter provide the basis of current practice and also identifies how further improvements can be made by means of risk control systems. Corrosion management has a key role to play in ensuring asset integrity, control of hydrocarbon releases, and safety. Successful management of corrosion also influences the economic outcome by ensuring cost-effective selection of materials, chemical treatments, coatings, cathodic protection systems, and appropriate designs.

Corrosion management is that part of the overall management system that is concerned with the development, implementation, review, and maintenance of corrosion policy. A corrosion policy includes establishment of organizational structures with defined responsibilities, reporting routes, practices, procedures, processes and resources. This requires the demonstration of responsibility and accountability for corrosion performance, managing risks, decreasing costs, controlling compliance, and motivating personnel.

5.5 Risk assessment methodologies

A number of key elements are characteristic of all effective risk assessments:

- Resources—Sufficient personnel and funding must be allocated to implement the desired approach to risk assessment. Frequent data integration and reassessment of risks must be performed to continually account for changing conditions on the pipeline.
- Commitment—The operator must commit to the approach and must be willing to take appropriate action to mitigate identified risks.
- Quality of Data—The quality of the data used in the models must be continually assessed and improved. Conservative assumptions must be used when data do not exist or are determined to be of poor quality. Trending analysis must be used to identify data that must be obtained to more accurately reflect risk to a particular segment.
- Weighting Factors—A structured set of weighting factors must be identified and consistently applied. Sensitivity analysis should be performed to ensure that there is a clear understanding of how each of the factors will drive the relative risk score. Additionally, risk modifiers or "spikers" can be incorporated into the models to elevate assumed risks, if specific conditions are known to exist.
- Segmentation—A structure that clarifies how the weighting factors are assigned relative to individual pipeline segments must be established. Initially, entire pipeline systems may be unfairly "penalized" with a high-risk rating, even though a specific deleterious condition exists within only one segment. The model must clearly define how this situation is to be accounted for and the approach required to limit the risk rating to the area of the segment for which the risk drivers actually apply.
- Documentation—Documenting risk decisions and all of the data used to arrive at the decisions is very important, as it supports the technical basis for assignment of risk. All of the procedures used to facilitate the process, including the source for each of the data elements, must be documented.
- Continuous Improvement—It is expected that each risk assessment approach will be continually refined, based on lessons learned in data-gathering and risk-mitigation efforts. Commitment to continuous improvement is an extremely important part of a comprehensive risk assessment approach.

6 Recommendations and summary

During transporting the petroleum products through the pipeline, small amounts of salt and water travel through it. The salt and water can accumulate in areas of low flow such as bypass piping and future provisions meant for connections. Factors such as nutrients (nitrite, phosphate, etc.), oxygen, chloride, and bacteria are the causative factors for internal corrosion in the petroleum-transporting pipeline. The types and ability of microorganisms that degrade petroleum hydrocarbons have been widely documented. It has also been estimated that 40% of all internal corrosion of pipeline in the gas and oil industry can be attributed to microbial corrosion. In order to prevent the effects of chemical impact and microbial growth, several lines of approach may be used, namely, good housekeeping practices, treatment with biocides to limit growth, use of special tank linings, and good inhibitors. In oil-transporting pipelines, stagnation of water occurs due to the slopes in the landscape, and this acts as a breeding ground for bacteria (Figure 21.7). The solubility of water in the hydrocarbon system is mentioned in Table 21.4. Due to the solubility of water in the hydrocarbon system, preservative is needed to avoid the degradation of oil in the petroleum product pipeline. Today it is well established that bacterial species exist in oil pipelines and degrade petroleum products in Indian pipelines also (Schiapparelli and Meybaum, 1980; Muthukumar et al. 2006). Generally, degradation starts at the interface between diesel and water. Under this condition, a good inhibitor is needed ti prevent pipeline corrosion. Different methods are used to protect petroleum product pipelines from corrosion. Application of a protective surface coating is one such method. This is an expensive procedure; moreover, it is impossible to coat the pipeline, which is already corroded and filled with petroleum product.

Addition of an inhibitor or biocide avoids the degradation process at the petrol–water interface and metal loss due to corrosion. An efficient and cost-effective approach to mitigate this problem is application of corrosion inhibitors. Corrosion could be controlled by inhibitor injection in the transportation pipeline by adding oil-soluble, water-dispersible, and filming amine-type corrosion inhibitors that can disperse sufficiently into stratified water layers. Many misapplications of inhibitors

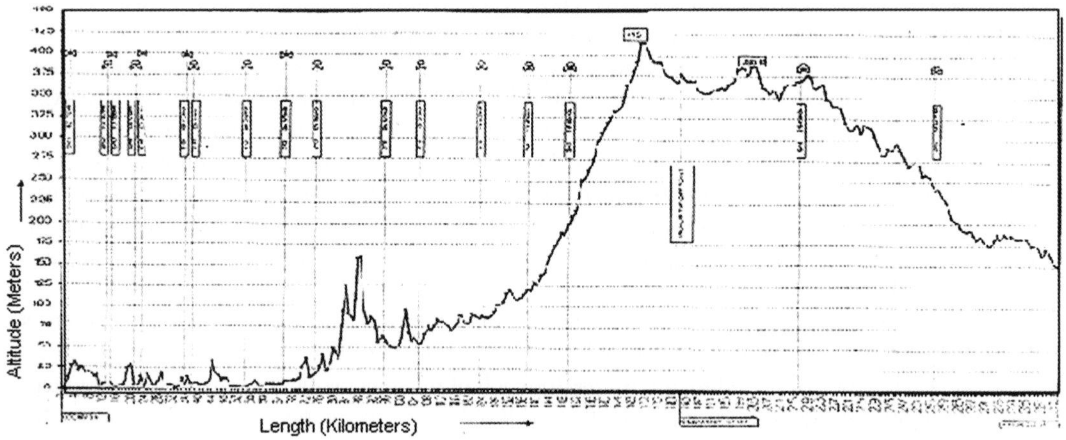

FIGURE 21.7

Land topography of existing pipeline in southwest India.

Table 21.4 Solubility of Water in Hydrocarbon System

Hydrocarbon	°C	°F	Solubility	
			mg/100g	Gal/1000 Bbl
n-Butane	20	68	6.5	1.6
Isobutane	19	66	6.9	1.7
n-pentane	15	59	6.1	1.6
	25	77	12.0	3.2
Isopentane	20	68	9.4	2.4
n-Hexane	20	68	11.1	3.1
Cyclohexane	20	68	10.0	3.3
n-Heptane	20	68	12.6	3.6
n-Octane	20	68	14.2	4.2
Benzene	20	68	43.5	16.1
Heptane-1	21	70	104.7	30.8
Butene-1	20	68	39.7	11.1
Gasoline	4	40	6.0	1.8
	10	50	7.2	2.1
	16	60	8.2	2.4
	21	70	9.2	2.7
	27	80	10.2	3.0
	32	90	11.3	3.3
	38	100	12.3	3.6
	43	110	13.6	4.0

occur because, the characteristics of the inhibitors are not considered before use, and it plays an important role in microbial corrosion (Muthukumar et al. 2007; Maruthamuthu 2005). Thus there is an urgent and immediate need to develop an efficient corrosion inhibitor.

Organic compounds used as typical oilfield corrosion inhibitors function by forming a film or protective barrier between metal and the corrosive fluids which are either anodic, cathodic, or mixed-type behavior. Biocides are used to combat the activity of bacteria in a system, a problem that is subtle and difficult to detect. Biocides are the most underused, misunderstood, and misapplied chemicals used in the petroleum industry for many reasons. However, biocides alone cannot solve a microbial problem. Once bacteria are well established it is nearly impossible to effect control without drastic measures. Hence it is the right time to develop a multifunctional inhibitor for the tropical pipeline.

In a recent tender call made by the Indian oil corporation Limited to procure inhibitors/biocides for a pipeline, the entry fee alone accounted for $11,000, which highlights the availability of a big market for new inhibitors. Hence, identification of good inhibitors/biocides as well as the role of inhibitors in microbial growth/degradation needs to be studied in detail in the oil–water system. It is also important to select the best inhibitors/biocides.

A cross-country pipeline in India transports petroleum products such as kerosene, petrol, and diesel. This pipeline has intermittent product delivery with pressure-booting stations at different locations. Severe corrosion and microfouling problems have been observed in the pipeline, even though a corrosion inhibitor was added. About 200 to 400 kg of muck (corrosion product) was received from a 20- km stretch of the pipeline within 30 days. Since carboxylic acid and ester-based corrosion inhibitor was added in the pipeline at one point for the entire 1400-km stretch, the formation of a micro-acidic environment was created by carboxylic acid. In the laboratory, 100 ppm and 200 ppm of inhibitor in water show a pH of about 5.5 and 4.2, respectively (Maruthamuthu et al. 2005). Water can stratify in the pipeline when the velocity is too slow, typically 3-4 ft/sec or if the pipeline is operated in "stop–start" mode. Hence it is possible that the solubility of inhibitor reduces the pH at various areas of the pipeline. It can be assumed that the formation of low pH is the reason for the bacterial proliferation (Rajasekar et al. 2005), which creates severe corrosion problems.

Another cross-country pipeline is located in southwest India, which is transporting white petrol (Naphtha) for generating electric power. The length of the pipeline is 5.5 km, and corrosion products of about 10 kg were collected from the pipeline every 2 months (Rajasekar et al. 2005). The white petrol had 400 ppm of organic sulfur, and it got adsorbed on the material. The adsorbed sulfur layer becomes cathode, and the remaining part of the material becomes anode. Small cathode and large anode create a pitting type of corrosion. A large quantity of sludge was noticed in the naphtha storage tank where the disposal of sludge had to be cleared by a pollution control board. These two case studies are the origin of the present study and encourage developing a novel inhibitor for the diesel and naphtha pipeline.

Organic compounds used as typical oilfield corrosion inhibitors function by forming a film or protective barrier between metals and the corrosive fluids because of their anodic, cathodic, or mixed-type behavior. Linear or cyclic amines, fatty acids, or quaternary amines are commonly used as inhibitors. Corrosion inhibitors used in the petroleum industry are the amino group containing imidazolines, amidoamines, and polythiol compounds. The basic types, all of which have long-chain hydrocarbons (C18) as a part of the structure, include (a) aliphatic fatty-acid derivatives, (b) imidazolines, (c) quartnaries, and (d) rosin derivatives. Oil-soluble, water-dispersible, film-forming amine-type organic compounds are a type of inhibitor that can disperse sufficiently into stratified water layers and prevent corrosion.

As the literature reveals, until today almost all the organic molecules used as corrosion inhibitor have not contained a redox active unit. As part of our synthetic study, we were interested in the development of a novel and efficient corrosion inhibitor for petroleum product pipeline applications.

An efficient corrosion inhibitor should be able to bind irreversibly to the metal surface, provide good surface coverage, and at the same time withstand microbial degradation. Quinones are an intriguing class of molecules that act as terminal electron acceptors toward anaerobic microbial oxidation in photosynthesis. In certain biomembrane assemblies, they are part of electron transfer chains; for example, in mitochondrial respiration, this role is played by ubiquinone or coenzyme Q, and they have an antimicrobial activity in petroleum products.

With this background in mind, a redox active organic molecule 1-aminoanthraquinone was identified for our study. The presence of a polar amine functional group is amenable for further functional group manipulations. Different 1-aminoanthraquinone derivatives were synthesized by a simple coupling reaction with a long carbon chain acid chloride. Introduction of a lengthy alkyl chain has led to a much-desired solubility in petroleum products. All the derivatives were evaluated for their corrosion inhibition efficiency using the weight-loss method under field conditions generated in the laboratory. Polarization and impedance measurements were used to study the surface binding characteristics (Muthukumar et al. 2007).

The mechanism of action for redox active AAQ derivatives may be very different from that of conventional inhibitors. The developed compounds act as a corrosion inhibitor as well as a biocide

(Muthukumar et al. 2009). Six biocides (cationic and non-ionic) were also employed to study the biodegradation of diesel in the diesel-water interface. In addition, the impact of micelle characteristics on biocidal efficiency has been investigated (Muthukumar et al. 2007). The impact of corrosion inhibitors on the environment has received increased attention in recent years. The new generation of environmental regulations requires the replacement of toxic chemicals with so-called Green chemicals. In the present study, neem oil, castor oil, and punga oil were used for corrosion inhibition evaluation in the petroleum product (diesel), along with 2% water containing 120 ppm chloride. Weight-loss and polarization studies were employed to find out the corrosion inhibition efficiency (IE). Weight-loss study revealed that castor oil gave higher efficiency (89%) when compared to neem oil and punga oil (Muthukumar et al. 2007).

The following conclusions have been derived from Muthukumar et al. (2003, 2005, 2006, 2007, 2009) and Rajasekar et al. (2005).

1. In the present chapter, a detailed review has been presented with scope of the present work. Factors influencing corrosion in petroleum production and transporting pipelines have been discussed in detail, with scanning of the literature. Factors such as oxygen, temperature, nitrogen, phosphorus, and sulphides influence corrosion in the petroleum industry, and their impact on metals has been described. Corrosion control and prevention have been discussed with material selection, cathodic protection, and use of coatings. In addition, a detailed study on corrosion inhibitors for petroleum production and transporting pipelines are screened and presented in detail. Literature collected on corrosion and its control of petroleum production and transporting pipelines have been carried out and presented.
2. The role of bacteria on oil degradation and its influence on corrosion has been studied. Two systems (biotic and abiotic) with and without inorganic content and bacteria were employed for studying degradation and corrosion. The aerobic heterotrophic bacterial population (HB) was found to be higher in the presence of inorganic medium than its absence. Oil degradation by microbes was characterized by Fourier transform infrared spectroscopy (FTIR) and nuclear magnetic resonance (NMR). The corrosion studies were carried out by the gravimetric method. It was found that *Gallionella* sp. degraded aliphatic protons -CH2 -CH2 to -O-CH2, whereas *Brucella* sp. converted only aromatic ring to aliphatic protons.
   The following inferences have been made from this study: (a) inorganic contents in contaminated water determine oil degradation in storage tanks and transporting pipelines; AND (b) the degraded product may adsorb on pipeline, which would enhance the rate of microbial corrosion. *Brucella* sp. is the major aromatic degrader, which breaks the benzene ring to aliphatic. *Gallionella* sp helps for the addition of oxygen, converts only the aliphatic group, and accelerates corrosion by ferric formation (Fig.21.8).
   The presence of inorganic content also determines the oil degradation and microbiologically influenced corrosion of materials. This investigation clearly points to diesel degradation and its impact on corrosion as a new research area in corrosion science and engineering (Muthukumar et al. 2003).
3. The effect of the water-soluble (WSCI) and oil-soluble (OSCI) corrosion inhibitor on the growth of bacteria and its corrosion inhibition efficiency were investigated. Corrosion inhibition efficiency was studied by rotating the cage test and flow loop techniques. The nature of biodegradation of corrosion inhibitors was also analyzed by using FT-IR, NMR, and GC–MS.

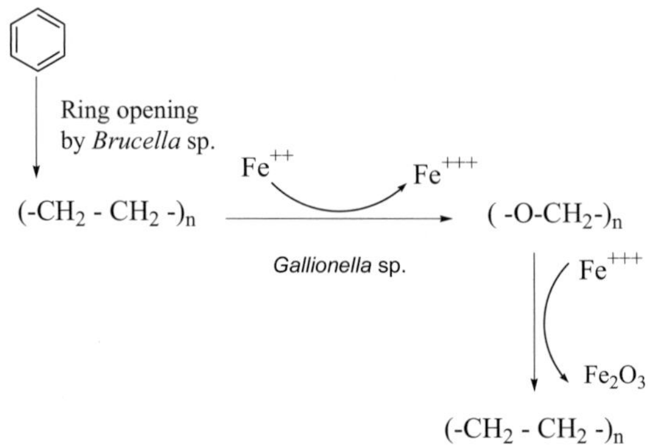

FIGURE 21.8

Schematic diagram of oil degradation by bacterial species.

The bacterial isolates (*Serratia marcescens* ACE2 and *Bacillus cereus* ACE4) have the capacity to degrade the aromatic and aliphatic hydrocarbons present in the corrosion inhibitor. The degraded products of corrosion inhibitor and bacterial activity determine the electrochemical behavior of API 5LX steel.

The weight-loss data and electrochemical studies reveal that bacterial culture reduces the efficiency of inhibitors.

The inhibition efficiency of WSCI in the flow loop system was 36%, while upon adding OSCI the efficiency was 31%. Bacteria reduced the efficiency of WSCI to about 12 to 21% within 7days. The efficiency of OSCI was reduced to the range between 31 and 58% in the presence of bacteria. It indicates that WSCI is better than OSCI under laboratory conditions. FTIR results indicate that ester and carboxylic acid adsorb on the metal surface in the presence of WSCI, whereas in the presence of OSCI, amide (CO-NH) adsorbed on the metal surface. Generally, bacteria prefers a water–oil interface for utilizing energy from inhibitor/diesel. Since ester dissolves in oil, it kills most of the bacteria in oil. Subsequently, carboxylic acid adsorbs on the pipeline. Hence, it can be assumed that WSCI gives good inhibition efficiency in the laboratory, but at the same time the carboxylic acid encourages the chemolithotrophic bacteria in the pipeline. The OSCI adsorbs on the metal surface and forms as film. Since bacteria prefer inorganic and organic nutrients on the metal surface, it encourages corrosion. Hence, it can be concluded that the inhibitor should kill the bacteria in both water and the oil system; subsequently, it adsorbs on the metal surface for better efficiency. The present study reveals that a nondegradable oil-soluble corrosion inhibitor is needed for the petroleum product pipeline to avoid the microbial degradation of corrosion inhibitors (Muthukumar et al. 2006, 2007).

4. Biodegradation occurs at the interface between diesel and water. The microbial contamination can result in inhibitor/fuel degradation that leads to the unacceptable level of turbidity, filter plugging, corrosion of storage tanks and pipeline, and souring of stored products. Hence, the selection of biocides/inhibitors is an important aspect of the petroleum product transporting pipeline. Six biocides (cationic and non-ionic) were employed to study the biodegradation of diesel in the diesel–water interface. The biocidal efficiency of the biodegradation of diesel was

examined using FTIR, NMR, and GC-MS. But the cationic biocides (CTAB, CPB, and DDAB) gave good biocidal efficiency at the interface. The data are explained in terms of a model that postulates the formation of "micelle" at the diesel–water interface.
Micelle formation at the diesel/water interface by cationic biocides (CTAB, CPB, and DDAB) is an important factor in controlling oil degradation (Figure 21.9).
Cationic biocides (CTAB, CPB, and DDAB) are efficient in controlling the oil degradation and show good biocidal activity at the diesel/water interface when compared to non-ionic biocides (BNP, BRIJ-35, and TRITON-X-100). Non-ionic biocides have better bactericidal efficiency than cationic biocides individual systems of oil medium, but it does not work well at the diesel/water interface (Muthukumar et al. 2007).

5. Accordingly, the main objective of the present invention is to provide a simple process for the synthesis of different alkyl chain substituted 1-aminoanthraquinone derivatives of fatty acids, such as oleic acid, stearic acid, palmitic acid, lauric acid, deconoic acid, and octonoic acid and their evaluation as effective corrosion inhibitors in petroleum products transporting pipelines.
Preparation of 1-aminoanthraquinone derivatives can be carried out in a simple one-step transformation. The preparation method is quite suitable for large-scale preparation of the compounds. 1-aminoanthraquinone, the starting material for the synthesis of this new class of inhibitors, is cheap and easily available. The solubility of 1-aminoanthraqunone was completely reversed after the introduction of the lengthy alkyl chain.
The redox potentials of newly prepared amide derivatives 8–13 were much less than 1-aminoanthraquinone. AAQ derivatives can be used as a biocide (Muthukumar et al. 2007).
6. This is the first report showing that quinone derivatives 8-13 are useful corrosion inhibitors in diesel as well as the white petrol transporting pipeline.
Polarization and electrochemical impedance spectroscopy results suggest that these compounds act as anodic inhibitor. AFM and FTIR results confirm the film formation on the metal surface. All the studied 1-aminoanthraquinone derivatives show the inhibition efficiency of 50–90% in the concentration range of 25 ppm, 50 ppm, and 100 ppm. It is obvious from the impedance table that as the inhibitors' concentration increased; resistance of inhibitor layer increased (Muthukumar et al. 2007, 2009). The order of inhibitor efficiency is as follows:

Compound 8 (C_8) < Compound 9 (C_{10}) < Compound 10 (C_{12}) < Compound 11 (C_{15}) < Compound 12 (C_{18}) < Compound 13 (C_{18})

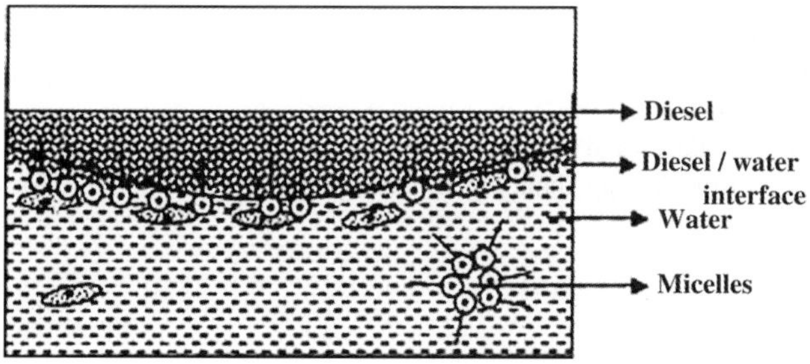

FIGURE 21.9

Cationic biocides in water and oil phase.

7. Punga oil, castor oil, and neem oils can be used as corrosion inhibitors in the petroleum product pipeline. They act as mixed inhibitors. Among the three compounds, castor oil showed higher corrosion inhibition efficiency. All Green inhibitors offered good inhibition against the corrosion of carbon steel (API 5LX) in the diesel/water mixture. The main components of Green inhibitors are fatty acids (oleic acid, stearic acid, palmitic acid) and amino acids (tyrosine, phenylalanine, histidine). The inhibitive action of fatty acids were attributed to the formation of a passive layer of fatty acids on the metal surface. At the corroding surface, these acids (fatty acids and aminoacids) may perform the following functions. (a) Chemisorptions on the surface, (b) forming a more or less stable complex, with corrosion intermediate from the dissolution sequence, and (c) forming a complex with a final oxidative propensity. The electron distribution throughout the molecule will have some effect on the above three functions (Muthukumar et al. 2007).

   At anodic areas, iron dissolves as Fe^{2+}. This reaction is rapid in most media.

$$Fe \rightarrow Fe^{2+} + 2e^-$$

In deaerated solutions, the cathodic reaction is given as

$$2H_2O + 2e^- \rightarrow H_2 + 2OH^-$$

The cathodic reaction is accelerated by dissolved oxygen.

$$O_2 + 2H_2O + 4e^- \rightarrow 4OH^-$$

Therefore, the over all reaction can be expressed by

$$\begin{array}{l} Fe^{2+} + 2\,[R_1R_2\,COO^-] \longrightarrow Fe\,(COOR_1R_2)_2^{2-} \\ \qquad\qquad\qquad\qquad\qquad \Big\downarrow e^- \\ \qquad\qquad\qquad Fe\,(COO^-R_1R_2) + COO^-R_1R_2 \xrightarrow{e^-} Fe + COO^-R_1R_2 \end{array} \tag{21.10}$$

DESCRIPTION OF TERMINAL SYMBOLS

Terminal Symbol Description
GDP gross domestic product
FHWA Federal Highway Agencies
NIST National Institute of Standards and Technology
>, = =, < numerical comparison; greater than, equal, smaller than (also applicable for character comparisons)
θ surface coverage
ΔG free energy of adsorption
R gas constant
T temperature
c bulk inhibitor concentration
n number of water molecules replaced per inhibitor molecule
f inhibitor interaction parameter (+, attraction; and −, repulsion)

K constant
CTAB N-cetyl-N,N,N-trimethylammonium bromide
CPB cetylpyridinum bromide
DDAB dodecylethyldimethylammonium bromide
BNP 2-Bromo-2-nitro-propane-1, 3-diol (Bronopol)
BRIJ-35 polyoxyethyleneglycol dodecyl ether
TRITON-X-100 polyethylene glycol-p-isooctyl phenyl ether
WSCI water-soluble corrosion inhibitor
OSCI oil-soluble corrosion inhibitor
HB heterotrophic bacterial population
SRB sulphate-reducing bacteria
AAQ aminoanthraquinone
IPR internal phase relationship
RCE rotating cylinder electrode
SCE standard calomel electrode
API American Petroleum Institute
EN electrochemical noise
EIS electrochemical impedance spectroscopy
ER electrolyte resistance
SAHMT 4-salicylideneamino-3-hydrazino-5-mercapto-1,2,4-triazole
DVA divanillidene acetone
DCA dicinnamylidene acetone
DSA disalicylidene acetone
MS mild steel
DDABA di-N-dimethyl-aminobenzylidene acetone
DBA dibenzylidene acetone
RFO residual fuel oil
BTAC N-benzyl triethanol ammonium chloride
EBTAC ethoxylated n-benzyl triethanol ammonium chloride
EDA ethylene diamine
i-PA *iso* – propyl amine
PA propyl amine
***t*-BA** *tert*-butyl amine
BA butyl amine
HA hexyl amine
EA ethanol amine
TEthA triethyl amine
DMA dimethyl amine
DEthA diethyl amine
TFA triethanol amine
TBA n-tributyl amine
CHA cyclohexyl amine
DCHA dicyclohexyl amine
Py pyridine
FA furfuryl amine
OA octyl amine
TIOA tri-iso-octylamine
NACE National Association of Corrosion Engineers

FIPC Film-forming inhibited petroleum compositions
DC direct current
AC alternating current
SCC stress corrosion cracking

Glossary

Anodic inhibitors usually used in near-neutral solutions where sparingly soluble corrosion products, such as oxides, hydroxides, or salts, are formed. They form, or facilitate, the formation of passivating films that inhibit the anodic metal dissolution reaction. Anodic inhibitors are often called passivating inhibitors.

Anodic protection the protection of a metal by maintaining it in the passive condition. This technique is based on the phenomenon of passivity. The metal to be protected is given a fixed potential so as to produce a passive film on it and the structure is protected from corrosive environment.

Atomic force microscope (AFM) or scanning force microscope (SFM) a very high-resolution type of scanning probe microscopy, with demonstrated resolution of fractions of a nanometer, more than 1000 times better than the optical diffraction limit

Biocide a chemical substance capable of killing living organisms, usually in a selective way

Cathodic inhibitors control corrosion by decreasing the reduction process (cathodic poison) or by precipitating selectively on the cathodic areas (cathodic precipitators)

Cathodic protection the reduction (or) prevention of corrosion by making it a cathode in the electrolytic cell. There are two methods of applying cathodic protection to metallic structures: galvanic or sacrificial anode and impressed current method.

Coatings Corrosion can be prevented by separating the metal from the corrosive environment by using protective coatings.

Corrosion the disintegration of a material into its constituent atoms due to chemical reactions with its surroundings

Corrosion inhibitors a chemical substance that, when added to an environment in small concentration, effectively checks, decreases, or prevents the reaction of the metal with the environment

Electrochemical theory According to the electrochemical theory, galvanic cells are set up between regions of exposed metal surfaces, for example, metal grains and heterogeneous phases; anodic paths are thus established.

Environmental conditioners (scavengers) Corrosion can be controlled by removing the corrosive species from the medium. Inhibitors that decrease the corrosivity of the medium by scavenging the aggressive substances are called environmental conditioners or scavengers.

Erosion corrosion occurs when movement of a corrodent over a metal surface increases the rate of attack due to mechanical wear and corrosion

Fourier transform infrared (FTIR) spectroscopy measurement technique for collecting infrared spectra. Instead of recording the amount of energy absorbed when the frequency of the infrared light is varied (monochromator), the IR light is guided through an interferometer.

Gas chromatography-mass spectrometry (GC-MS) a method that combines the features of gas-liquid chromatography and mass spectrometry to identify different substances within a test sample

Gross domestic product (GDP) one of the basic measures of a country's overall economic performance

Hydrogen damage a form of environmentally assisted failure that results most often from the combined action of hydrogen and residual or applied tensile stress

Infrared spectroscopy (IR spectroscopy) the subset of spectroscopy that deals with the infrared region of the electromagnetic spectrum

Interface inhibitors control corrosion by forming a film at the metal/environment interface. Interface inhibitors can be classified into liquid- and vapor-phase inhibitors.

Liquid-phase inhibitors classified as anodic, cathodic or mixed inhibitors depending on whether they inhibit the anodic, cathodic, or both electrochemical reactions.

Microbial corrosion also called bacterial corrosion, bio-corrosion, microbiologically influenced corrosion, or microbially induced corrosion (MIC); is caused or promoted by microorganisms, usually chemoautotrophs. It can apply to both metals and nonmetallic materials.

Mixed inhibitors About 80% of inhibitors are organic compounds that cannot be designated specifically as anodic or cathodic and are known as mixed inhibitors.

Nuclear magnetic resonance used to study molecular structures

Pigging program the most cost-effective and efficient method of cleaning the pipe surface for the application of liquid inhibitors.

Stress-sorption theory According to the stress-sorption theory, stress corrosion cracking (SCC) generally proceeds by weakening the cohesive bonds between surface-metal atoms through adsorption of damaging substances in the environment.

Vapor-phase inhibitor (VPI) the process of vapor-phase inhibition involves two steps: transport of inhibitor to the metal surface and interaction of inhibitor on the surface. A VPI may vaporize in the undissociated molecular form, or it may first dissociate and then vaporize.

Acknowledgments

I would like to thank my research guides Dr. N.Palaniswamy and Dr. S.Maruthamuthu & my research colleague Dr. A.Rajasekar, Central Electrochemical Research Institute, Karaikudi, India for their constant guidance and valuable help towards my research program on petroleum product pipeline corrosion.

References

Al-Sabbagh, A.M., Osman, M.M., Omar, A.M.A., El-Gamal, I.M., 1996. Anti-corrosion Meth. Mater. 43, 11.

Altoe, P., Pimenta, G., Moulin, C.F., Diaz, S.L., Mattos, O.R., 1996. Electrochim. Acta. 41, 1165.

Annand, R.R., Woodson, A.E., 1986. US Patent, 3982894. p. 10128.

Arkuszewaska, I.N., 1983. Studies of the effect of some fatty acids on the inhibiting properties of nitrated petroleum products, Corros. Sci. 23, 75.

Atlas, R.M., 1981. Microbial degradation of petroleum hydrocarbons: an environmental perspective. Microbial. Rev. 45, 180.

Atlas, R.M., 1984. Petroleum microbiology. Macmillan, New York.

Becerra, H.Q., Retamoso, C., Macdonald, D.D., 2000. The corrosion of carbon steel in oil-in-water emulsions conditions, Corros. Sci. 42, 561–575.

Benitez Aguilar, J.L., Martinez Ortiz, C., Roldan, R., 2002. Oil Gas J. 100, 66.

Bockris, J.O.M., Reddy, A.K.N., 1970. Modern electrochemistry. Plenum Press, 1091.

Bockris, J.O.M., Swinkels, D.A.J., 1964. Adsorption of n-Decylamine on Solid Metal Electrodes, J. Electrochem. Soc. 111, 736.

Bojes, J., Girgis, M., Goerz, K., Rippon, I., 2002. A laboratory evaluation of the variables that affect the application of batch corrosion inhibitors–Phase–I, CORROSION 2002. NACE International, Houston, TX. Paper no. 02289.

Bouazaze, H., Huet, F., Nogueita, R.P., 2005. A new approach for monitoring corrosion and flow characteristics in oil/brine mixtures, Electrochim. Acta. 50, 2081.

Brindsi, F.A. Blecks T.W., Sullivan, T.E., 1981. U.S. Patent, 4302246.

Castaneda, H., Zambrano, G., Angeles, C., Genesca, J., 2004. Film stability for API 5L X-52 line pipe steel in CO_2(aq) and Cl-(aq) solutions in presence of amine based inhibitor under hydrodynamic conditions. Corrosion, 15p.

Caunter, C., 1999. Oil Gas J. 97, 4.

Chen, Y., Ruschau, G.R., Koch, G.H., Nour, M.M., Bouazaze, H., Huet, & Nogueira, R.P., 2004. Monitoring corrosion and flow characteristics in oil/brine mixtures of various compositions. Corrosion, 2004, p. 20.

Cheng, J.X., Zhan, J.M., 1993. Corrosion 49, 170.

De Vivekanand, P.K. Banerjee, S., 1997. Proc. International Conference on Corrosion. p. 1258.

Dougherty, J.A., 2004. Aggressive agents as corrosion inhibitors in gas and oil production. Corrosion 13.

Dougherty, J.A., Ahn, Y.S., 1999. Criteria for the selection of corrosion inhibitors for arctic and sub sea high velocity flow lines, Corrosion 9. NACE International, Houston, TX. Paper No.4.

Durnie, W., De Marco, R., Jefferson, A., Kinsella, B., 1999. Development of a structure - Activity relationship for oil field corrosion inhibitors. J. Electrochem. Soc., 146, 1751–1756.

Evans, U.R., 1981. The corrosion and oxidation of metals. Edward Arnold, London.

Ewing, S.P., 1948. Corrosion by stray current. In: Uhlig, H.H. (Ed.), Corrosion Handbook. John Wiley & Sons, New York, p 601.

Federal Highway Administration (FHWA), Office of the Infrastructure and Development 2001. Report FHWA-RD-01–156.

Fontana, M.G., 1986. Modern theory—Principles in corrosion engineering. McGraw-Hill, New York p.445.

Foster, G.L., Oakes, B.D., Kucera, C.H., 1959. Acetylenic Corrosion Inhibitors, Ind. Eng. Chem. 51, 825.

Frenier, W. Growcock, F. Dixon, B., Lopp, V.R., 1987. Corrosion, European patent Applied, BP212,752, C.A.107 (1987) p.81472.

Frenier, W.W., Growcock, F.B., Lopp, V.R., 1988. Prod. Eng. 3, 584.

Frignani, A., Trabanelli, G., Zucchi, F., Zucchini, M., 1980. Proceeding of Fifth European Symposium of Corrosion Inhibitors. Ferrara, Italy, p. 1185.

Frunkhouser, J.G., 1961. Acid Corrosion Inhibition With Secondary Acetylenic Alcohols. Corrosion 17, 283.

Galicia, M., Castañeda, H., Perez, R., Barraza, I., 2004. Film kinetics of X-65 steel in the presence of amine based inhibitors and H_2S for the system Fe-CO_2-Cl under hydrodynamic conditions. Corrosion, 10p.

Garverick, L. (Ed.), 1994. Corrosion in the petrochemical industry. ASM international, Materials Park. OH 44073–0002. p.13.

Gonik, A.A., 2005. Effect of a diphilic corrosion inhibitor on the iron passivity in oil-field electrolytes. Prot. Metals 41, 173.

Grishchenkov, V.G., Townsend, R.T., McDonald, T.J., Autenrieth, R.L., Bonner, S., Boronin, A.M., 2000. Degradation of petroleum hydrocarbons by facultative anaerobic bacteria under aerobic and anaerobic conditions, Process Biochem. 35, 889.

Growcock, F.B., Frenier, W.W., Andreozzi, P.A., 1989. Corrosion 31, 130.

Growcock, F.B., Lopp, V.R., 1988. Inhibition of Steel Corrosion in HCl by Derivatives of Cinnamaldehyde: Part II. Structure–Activity Correlations, Corros. Sci. 28, 817.

Growcock, F.B., Freiener, W.W., 1988. Kinetics of Steel Corrosion in Hydrochloric Acid Inhibited with *trans*-Cinnamaldehyde, J. Electrochem. Soc. 135, 817.

Hackerman, N., Snavely, E.S., 1984. in Delinder, L.S.V (ed.)., Corrosion Basics, NACE. Houston, TX, p.127.

Hausler, R.H., Schmitt, G., 2004. Hydrodynamic and flow effects on corrosion inhibition. Corrosion, 20 p.

Hedges W.M., Lockledge, S.P., 1996. The continuing development of environmental friendly corrosion inhibitors for petroleum production, *Corrosion* 96, Houston, TX, Paper No.151.

Hoar, T.P., 1971. Report of the committee on Corrosion and Protection. HMSO, London.

Hong, T., Sun, Y.H., Jepson, W.P., 2002. Study on corrosion inhibitor in large pipelines under multiphase flow using EIS, Corros. Sci. 44, 101.

Hong, T., Chen, Y., Sun, Y.H., Jepson, W.P., 2001. Monitoring corrosion in multiphase pipelines, Mater. Corros. 52, 590.

Hong, T., Gopal, M., Jebson, Mater, 2000. Effect of oil in multiphase flow on corrosion product film in large horizontal pipeline, Corros. 51, 439.

Jayaraman, A., Neemla, K.D., Saxena, R.C., 1991. Corrosion inhibitor studies on steels in hydrochloric acid, Corros. Prevent. Cont. 38, 8.

Jeyaprabha, C., Muralidharan, S., Jeyaperumal, D., Venkatachari, G., Rengaswamy, N.S., 1998. Anti-Corros. Meth. Mater 45, 148.

Jeyaraman, A., Saxena, R.C., 1996. Corrosion inhibitors in hydrocarbon systems, CORROSION 96, Paper No.221. NACE International Annual Conference and Exposition, Houston, TX.

Keeney, B.R., Johnson, J.W., 1973. U.S. Patent, 3773465.

Khalifa, M.A., El-Batouti, M., Mahgoub, F., Bakr Aknish, A., 2003. Corrosion inhibition of steel in crude oil storage tanks, Mater. Corrosion 54, 251.

Kinney, D., 1968. Biodegradation of hydrocarbons in the marine environment, Appl. Environ. Microbiol. 27, 125.

Kirkov, P., 1987. The influence of drag reducers and inhibitors on the rate and mechanism of corrosion in petroleum white products in the presence of moisture, Electrochim. Acta. 32, 921.

Koch, G.H., Brongers, M.P.H., Thompson, N.G., Virmani, Y.P., Payer, J.H., 2001. Corrosion costs and preventive strategies in the United States. FHWA-RD-01–156 [Online]. Federal Highway Administration, Washington, DC. http://www.corrosioncost.com.

Langmuir, I., 1917. J. Am. Chem. Soc. 39, 1848.

Lepetit, J., Barthelemy, H., 1968. Ann. Ins. Pasteur Paris 114, 149.

Levin, S.A., Gintzbergy, S.A., Dinner, I.S., Kuchinsky, V.N., 1965. Proceedings of Second European Symposium on Corrosion inhibitors. Ferrara, Italy p.765.

Lorenz, W. Z. *Phys.Chem.* 219(1962) 421. and 244 (1970) 65.

Mago, B.F. 1987. U.S. Patent, 4263167.

Martin, J.A., Valone, F.W., 1985. Corrosion 41, 281.

Martin, R.L., Alink, B.A., Mcmahon, J.A., Weare, R., 1999. Further advance in the Development of environmentally acceptable corrosion inhibitors, *Corrosion* 99, paper No. 98, Houston, TS.

Maruthamuthu, S., Mohanan, S., Rajasekar, A., Muthukumar, N., Ponmarriappan, S., Subramanian, P., Palaniswamy, N., 2005. Role of corrosion inhibitor on bacterial corrosion in petroleum product pipelines, Ind. J. Chem. Tech. 12, 567.

Matsushima, I., Uhlig, H.H., 1966. Protection of Steel from Hydrogen Cracking by Thin Metallic Coatings, J. Electrochem. Soc. 113, 555.

McCormack, P., Jones, P., Rowland, S.J., 2002. Rapid Communications in Mass. Spectrometry 16, 705.

Metsik, R.E., Khallik, A.A., 1986. Chemist. Tech. Fuel. Oils 22, 394.

Meyers, K.O. 1984. U.S. Patent, 485874.

Migahed, M.A., Abd-El-Raouf, M., Al-Sabagh, A.M., Abd-El-Bary, H.M., 2005. Effectiveness of some non ionic surfactants as corrosion inhibitors for carbon steel pipelines in oil fields, Electrochim. Acta. 50, 4683.

Miller, J.D.A., Tiller, A.K., 1970. Microbial corrosion of buried and immersed metal, in Microbial Aspects of Metallurgy. In: Miller, J.D.A. (Ed.). American Elsevier, pp. 61–106.

Moiseeva, L.S., 2005. Prot. Metals 41, 76.

Monroe, R.F. Kucera, C.H., Oates, B.D., 1963. U.S. Patent 3007454.

Muthukumar, N., Maruthamuthu, S., Palaniswamy, N., 2006. Water-soluble inhibitor on microbiologically influenced corrosion in diesel pipeline, Colloids. Surf. B. Biointerfaces. 53, 260.

Muthukumar, N., Maruthamuthu, S., Mohanan, S., Palaniswamy, N., 2007. Influence of an oil soluble inhibitor on microbiologically influenced corrosion in a diesel transporting pipeline, Biofouling 23, 395.

Muthukumar, N., Maruthamuthu, S., Ilangovan, A., Palaniswamy, N., 2007. Surface analysis of inhibitor films formed by 1-aminoanthraquinones on API 5L-X60 steel in diesel—water mixtures, Electrochim. Acta. 52, 7183.

Muthukumar, N., Maruthamuthu, S., Ilangovan, A., Palaniswamy, N., Kimura, A., 2009. 1-Aminoanthraquinone derivatives as a novel corrosion inhibitor for carbon steel API 5L-X60 in white petrol—water mixtures, Material. Chem. Phys. 115, 444.

Muthukumar, N., Maruthamuthu, S., Palaniswamy, N., 2007. Role of cationic and nonionic surfactants on biocidal efficiency in diesel-water interface, Colloids. Surf. B. Biointerfaces. 57, 152.

Muthukumar, N., Maruthamuthu, S., Palaniswamy, N., 2007. Impact of micelles on the biocidal efficiency in a diesel—water interface, Surface and Interface Analysis 39, 533.

Muthukumar, N., Maruthamuthu, S., Palaniswamy, N., 2007. Green Inhibitors for petroleum product pipelines, Electrochemistry 75, 1.

Muthukumar, N., Mohanan, S., Maruthamuthu, S., Subramanian, P., Palaniswamy, N., Raghavan, M., 2003. Role of Brucella sp. and Gallionella sp. in oil degradation and corrosion, Electrochem. Comm. 5, 421.

Muzyezko, T.M. Share, S., Martin, J.A., 1972. U.S. Patent, 3105106.

Neemla, K.D., Saxena, R.C., Jeyaraman, A., 1992. Corros. Prevent. Cont. 39, 69.

Neemla, K.D., Jayaraman, A., Saxena, R.C., Agarwal, A.K., Krishna, R., 1989. Bull. Electrochem. 5, 250.

Noack, M.G., 1982. Mater. Perform. 21, 26.

Oldfield, J.W., 1987. Bull. Electrochem. 3, 597.

Papavinasam, S., Doiron, A., Revie, R.W., 2002. Evaluation of green inhibitors. In: Proceedings of the International Pipeline Conference. IPC. B, p 1997.

Papavinasam, S., Revie, R.W., Attard, M., Demoz, A., Michaelian, K., 2003. Corrosion 59, 1096–1111.

Pati, B.B., Chatterjee, P., Singh, T.B., Singh, D.D.N., 1990. Corrosion 46, 354.

Place Jr., M.C., 1979. Corrosion Control-Deep sour gas production paper presented at the 54th Annual fall technical conference and exhibition of the society of petroleum engineers of AIME. Society of Petroleum Engineers, Las Vegas, NV.

Poling, G.W., 1967. J. Electrochem. Soc. 114, 1209.

Powell, D.E., Wende, E.V., 1997. Mater. Perform. 36, 58.

Prjor, M.J., Keir, D.S., 1957. J. Electrochem. Soc. 104, 269.

Quinlan, P.M. McBride, R.D. 1977. U.S. Patent 4026807.

Quraishi, M.A., Jayaperumal, D., Subramanian, P., Natesan, M., Venkatachari, G., Balakrishnan, K., 1996. Bull. Electrochem. 12, 100.

Quraishi, M.A., Jamal, D., 2000. New and Effective Corrosion Inhibitors for Oil-Well Steel (N-80) and Mild Steel in Boiling Hydrochloric Acid, Corrosion (NACE) 56, 156.

Quraishi, M.A., Qasim Ansari, Mohd, Ahmad, Shamim, Venkatachari, G., 1998. Bull. Electrochem. 14, 302.

Quraishi, M.A., Sardar, N., Ali, H., 2002. Corrosion (NACE) 58, 317.

Quraishi, M.A., Jamal, D., 2001. Mater. Chem. Phys. 68, 283.

Rajasekar, A., Maruthamuthu, S., Muthukumar, N., Mohanan, S., Subramanian, P., Palaniswamy, N., 2005. Corros. Sci. 47, 257.

Ramachandran, S., Tsai, B.L., Blanco, M., Chen, H., Tang, Y., Goddard III, W.A., 1997. J. Phys. Chem. A. 101, 83.

Ramachandran, S., Jovancicevic, V., 1999. Corrosion 55, 259.

Ramachandran, S., Tsai, B.L., Blanco, M., Chen, H., Tang, Y., Goddard, W.A., 1996. Langmuir 12, 6419.

Rashid-uz-Zafar, S., Al-Sulaiman, G.J., Al-Gahtani, A.S., 1992. Symp. Corrosion and the control. Saudi Arabia, Riyadh. May, 110.

Redmore, D. 1976. U.S. Patent, 3965264. p. 109012.

Rhodes, P.R. 1976. Corrosion mechanism of carbon steel in aqueous H_2S solutions, Abstract 107, in extended abstracts, Vol. 76-2, The Electrochemical Society, p 300.

Riggs, O.L. Jr., 1973. in Nathan, C.C. (Ed.), Corrosion inhibitors, NACE, Houston, TX, p. 11.
Rosenfeld, I.L., Persiantseva, V.P., Terentief, P.B., 1964. Corrosion 20, 222.
Rosenfeld, I.L., 1981. Corrosion inhibitors. McGraw Hill, New York, p. 98.
Sastri, V.S., 1998. Corrosion inhibitors: Principles and applications. John Wiley & Sons, New York, p. 39.
Sastri, V.S., Perumareddi, J.R., 1994. Corrosion 50, 432.
Schiapparelli, E.R., Meybaum, B.R., 1980. Mater. Perform. 19, 47.
Schmitt, G., 1984. Br. Corros. J 19, 165.
Schwarz, J.D., Walker, R.R., 1945. Colwell. Can. J. Microbiol. 21, 682.
Senez, J.C., Azoulay, E., 1961. Biochem. Biophys. Acta 471, 307.
Silva, L.M., Hernandez, J., Genesca, J., Duran, R., Mendoza, J., 2004. Effect of flow on the corrosion mechanism of different API pipeline steels grades in NaCl solutions containing CO_2. Corrosion, 2004, p. 23.
Singh, D.D.N., Singh, T.B., Chatterjee, P., Paty, B.B., Mukherjee, K.P., 1990. Corrosion Prev. Control 37, 37.
Smialowska, Z.S., Wieczorek, G., 1971. Corros. Sci. 11, 843.
Sorkhabi, H.A., Nabavi-Amri, S.A., 2002. Electrochim. Acta. 47, 2239–2244.
Stern, M., 1955. J. Electrochem. Soc. 102, 663.
Suroor Athar, S.M., Ali, H., Quraishi, M.A., 2001. Anti-Corr. Meth. Mater 48, 251.
Tan, Y.J., 1994. Corrosion 50, 226.
Tan, Y.J., Bailey, S., Kinsella, B., 1996. Corros. Sci. 38, 1545.
Tedeschi, R.J., 1975. Corrosion 31, 130.
Trasatti, S., 1971. J. Electroanal. Chem. Inter. Electrochem. 33, 351.
Trasatti, S., 1974. J. Electroanal. Chem. 53, 335.
Tullmin, M.A.A., Roberge, P.R., Grenier, L., Little, M.A., 1990. Can. Aeronautics Space J. 42, 272.
Tung, N.P., Hung, P.V., Tien, H.D., Loi, C.M., 2001. Study of corrosion control effect of H_2S scavengers in multiphase systems. In: Proceedings—SPE International Symposium on Oilfield Chemistry. P. 747.
Uhlig, H.H., 1985. Corrosion and corrosion control, third ed. John Wiley & Sons, New York.
Uhlig, H.H., 1963. Corrosion and Corrosion Control. John Wiley & Sons, New York.
Wagner, C., Traud, W., 1938. Zeit. Electrochem. 44, 391.
Wang, D., Li, S., Ying, Y., Wang, M., Xiao, H., Chen, Z., 1999. Corros. Sci. 41, 1911.
Wang, W., Free, M.L., Horsup, D., 2005. Metallurgical and materials transactions B. Process Metallurgy Mater. Processing Sci. 36, 335.
Widdel, E., 1980. J. Appl. Bacterial, 271.
Woodson, A.E. 1976. U.S. Patent, 3982894.
Zinjarde, S.S., Pant, A.A., 2002. Mar. Pollut. Bull. 44, 118.

CHAPTER

22

Aerogels in the Environment Protection

D.R. Chen[*], **X.H. Chang**[†], **X.L. Jiao**[*]

[*] *School of Chemistry and Chemical Engineering, Shandong University, Jinan, 250100, P. R. China,*

[†] *School of Chemistry and Chemical Engineering, Luoyang Normal University, Luoyang, 471022, P. R. China*

CHAPTER OUTLINE

1 Introduction

As a kind of typical porous solid materials composed of colloidal particles or polymer molecules, aerogels have high specific surface areas (e.g., 3200 m^2/g), low bulk density, and electric resistivity. Their porosity usually is 80∼99.8%, and the aperture is between 1 and 100 nm. Thus they exhibit

The Role of Colloidal Systems in Environmental Protection. http://dx.doi.org/10.1016/B978-0-444-63283-8.00022-3

many special properties compared with conventional solid materials, and recently aerogels have attracted great interest. Since Kistler first reported on the SiO_2 aerogel in 1931 (Kistler 1931), the development of aerogels has been complicated by the difficulty in removing the solvent from the gel networks. In the last decade organic, inorganic, and inorganic/organic aerogels such as silica, carbon, and resorcinol-formaldehyde (RF) aerogels have been studied extensively s (Cantin et al. 1974). Typically, the preparation process for aerogels includes the sol-gel process and subsequent drying. One view is that only the supercritical drying can obtain aerogels, while others believe that all dried gels with high porosity are aerogels. Aerogels can exhibit granular, monolithic, and film forms with many excellent properties. Normally, the wet-gels composed of nanoparticles with the size of 1 ~ 100 nm and a large amount of solvent existing in their networks have been obtained through the traditional sol-gel process. Then the drying process is applied to remove the solvent from the networks to result in porous structure with low density. The porous microstructures of the aerogels can be modified by both the sol-gel parameters and the drying process. Moreover, the abundant functional groups such as the -OH, -NH_2, and -COOH groups on the surface of sol particles composing the aerogels can modify the aerogels' properties, which is an important consideration for their applications.

Before 1974, no practical application is had been developed for the aerogels until Cantin's group reported the use of silica aerogels as detectors in Cerenkov radiation (Cantin et al. 1974). Subsequently, a series of applications of aerogels weare developed. Up toTo date, the aerogels have been applied in the areas of thermal insulation, catalysts, catalyst carriers, acoustic impedance coupled materials, drug carriers, environmental protection, and so on. In the past decades, several reviews on the preparation, microstructure, and application of the aerogels have been given published (Akimov 2003; Cantin et al. 1974; Gesser and Goswami 1989; Moreno-Castilla and Maldonado-Hódar 2005; Pajonk 2003; Schneider and Baiker 1995). Particularly in 1999, Pajonk summarized the catalytic applications of aerogels for environmental purposes (Pajonk 1999). In the last decade, many reports about theaerogels and their focused on the environmental applications have been issued; in this chapter, we give a mini-review of these correlative researches.

2 Preparation of aerogels

The preparation of aerogels mainly includes the sol-gel process and supercritical drying.

2.1 Sol-gel process

The sol-gel process is the preparation of wet-gel, regardless of whether the precursors are inorganic salts or metal alkoxides. The main steps in this process are as follows: the precursors dissolve in solvents (water or organic solvents) to form a uniform solution and hydrolyze or alcoholyze; then condensation usually occurs to form sol particles with size of ca. 1.0 ~ 10.0 nm and is gradually transformed to a gel; and finally the aerogels are obtained through evaporation or drying. The basic principles are as follows (Hench and West 1990):

① **Solvation.** The metal cations (M^{Z+}) attract water molecules to form a solvent-unit $M(H_2O)_n^{Z+}$ (Z is the valence number of M ions) and strongly trend to release the H^+ cations in order to maintain its coordination number:

$$M(H_2O)_n^{Z+} \rightarrow M(H_2O)_{n-1}(OH)^{Z-1} + H^+ \tag{22.1}$$

As other ions are introduced into the reaction system, the polymerization reaction might arise, but the reaction is extremely complicated.

② **Hydrolysis.** The non-ionizing molecular precursors such as metal alkoxides $M(OR)_n$ (n is the valence of metal M) reacted with water:

$$M(OR)_n + H_2O \rightarrow M(OH)(OR)_{n-1} + ROH \tag{22.2}$$

Reaction could be extended to form $M(OH)_n$.

③ **Condensation.** Dehydration condensation:

$$-M-OH + HO-M- \rightarrow M-O-M- + H_2O \tag{22.3}$$

Dealcohol reaction:

$$-M-OH + HO-M- \rightarrow M-O-M- + ROH \tag{22.4}$$

2.2 Supercritical drying

Wet-gel is wrapped in a large number of organic solvents and water, and the drying process is often accompanied by volume shrinkage. Because the surface tension always is greater than zero, the surface tension of the liquid/vapor easily leads to shrinkage and pore collapse. In order to avoid the contraction and collapse of the pores, the supercritical drying usually is used. The so-called critical state is a coexistence of the gaseous and liquid state in which the liquid density and saturated vapor pressure in the interface of the same phase disappear. When the wet-gels are heated to the temperature and pressure higher than the critical ones of organic solvents or water contained in the gel, the capillary force in the gel disappears. According to the critical temperature level of extraction medium, the supercritical drying can be divided into high-temperature and low-temperature supercritical drying (Hauthal 2001).

Currently, the supercritical drying method is commonly used to prepare the ultra-light and crack-free aerogels ($\leq$ 100 kg/m^3). The atmospheric pressure drying makes the wet-gel strongly contract because the gel contains the hydrophilic surface groups such as a large number of -OH to make the microporous gel have a very strong capillary force. In the evaporation process of water or ethanol, with the evaporation proceeding, the liquid reduces and a strong capillary action pulls all the nanoscale pores, which tend to disappear and destroys the nanopore structure, which leads to the bulk density rapidly increasing, and the distorted, cracking, and large-density xerogels or nanoporous aerogels are obtained.

The basic principle behind supercritical drying is that the gas and liquid become a homogeneous fluid in the supercritical state; because the interface between the gas and liquid no longer exists, this fluid gradually releases from the gel. Because of the disappearance of the gas–liquid interface, the capillary action does not exist. Therefore the gel contraction and structural damage will not appear until all the liquid is removed from the gel. Finally, the ultra-light aerogels with nanopores are obtained.

2.2.1 High-temperature supercritical drying

In the high-temperature supercritical drying process, the commonly used solvents are alcohols such as CH_3OH, C_2H_5OH, and C_3H_7OH; their critical temperatures are more than 220°C. Due to the high critical temperature and pressure of water (Tc = 374.1°C, Pc = 217.6 MPa), the supercritical conditions easily cause a solvation. Thus for the hydrogel, the water must be replaced with an organic solvent before the supercritical drying (Marr and Gamse 2000).

2.2.2 Low-temperature supercritical drying

Commonly used low-temperature supercritical drying medium is supercritical liquid CO_2; its critical temperature is only 31°C. It is necessary to replace the water with anhydrous ethanol to obtain the alcohol gel before supercritical drying. The gel is added into the extraction autoclave, and high-pressure CO_2 gas is cooled to form liquid in the pipeline after filling into the autoclave, which is slowly warmed up to a supercritical temperature. Then the CO_2 medium is released slowly with heating until the pressure inside the reactor is balanced to the external atmospheric pressure. In the process of alcohol gel in contact with liquid CO_2, the ethanol in pores of gels gradually dissolves into liquid CO_2 to form a single- alcohol/ CO_2-based solution system, which discharges slowly from the separation autoclave. Investigations on some subcritical and ambient pressure drying processes have made a breakthrough in the laboratory. When the separation autoclave has no alcohol to be observed, the temperature and pressure slowly decrease to the atmospheric temperature and pressure.

For the hydrogel, the organic medium dissolved in the liquid CO_2 replaces the water within the hydrogel, and then the liquid CO_2 is reused to substitute the organic media. Compared with the high-temperature supercritical drying, the CO_2 supercritical drying is therefore both energy-saving and safe because CO_2 is inert gas as well as nonburning and nonexplosive. At present, the CO_2 supercritical drying is used to prepare aerogels on a large-scale, in which a suitable supercritical temperature and pressure as well as appropriate drying rate must be chosen to obtain high-quality aerogels (Chen and Mao 2007).

2.3 Formation parameters of aerogels

In the preparative process of aerogels, the concentration of precursors, pH value of solution, solvent and water consumption, metal oxide loading, aging time and temperature of wet gel, supercritical drying parameters, and other factors have affected the properties of aerogels. To prepare the aerogels with special structures and properties, the relationships between them must be studied. Boonstra found that in acid conditions the hydrolysis accelerated, accompanied with a corresponding increase in gelation time; the pH value of solution determined the final particle size and surface area of the composition. In the high concentrated alkali solution, the particle size is ca. 1.0 nm, and the gel has meso-/macropores. In the neutral condition, the particle size is 2.5 ~ 20 nm, and the gel shows irregular pore structure, but in acidic conditions the particle size is 0.5 nm ~ 3 μm, and the gel has good porosity (Gonzalez, Lopez, and Gomez 1997). Chiara and coworkers studied the preparation of a series of microporous silica/zirconia composite aerogels using the reaction of cyclohexene peroxide by the sol-gel method, found to change the hydrolysis rate of tetraethyl orthosilicate (TEOS) by adjusting temperature and pH value to control the specific surface area, pore structure, and surface acidity of the aerogels (Palazzi et al. 2000).

2.4 Preparation of multicomponent aerogels

It is important to improve the structures and optimize the quality of aerogels by adding other components in the formation of aerogels. The multicomponent aerogels can be divided into three types

(Chen and Gan 1997): (1) two or three kinds of metal alkoxides hydrolyzed to form the mixed gel and then to transform them into the aerogels such as Al_2O_3/SiO_2 and TiO_2/SiO_2 aerogels through supercritical drying process, (2)other substances are added in a certain formation stage of the alcohol gel to make it fully dispersed and then to transform them into aerogels such as (C_{60}/C_{70})-SiO_2 and SiC-SiO_2 aerogels by supercritical drying, and (3)to enable the metal or metal oxide to be deposited on the oxide aerogels to form multicomponent aerogels such as Fe/SiO_2 and Pd/SiO_2 aerogels.

Although the preparation processes of aerogels have been studied for many years and their applications are increasingly widespread, further research is still needed.

(1) There are fewer studies on the multicomponent and nonoxide aerogels than on SiO_2 and single-metal oxide aerogels.
(2) Besides the polar solvents such as H_2O and C_2H_5OH, the inert solvent sol systems for the preparation of aerogels are scarcely reported.
(3) The atmospheric pressure and subcritical drying processes have lower-cost; however, the aerogels prepared by supercritical drying exhibit good quality. Furthermore, it is difficult to prepare large pieces of aerogels due to the mass transfer limitations during the supercritical drying process, so the industrialization of aerogels is a slow process.
(4) The formation mechanisms of aerogels, especially the physical and chemical changes of gel in the aging process, are not clear; many mechanisms are limited to speculation.

3 Adsorption applications of aerogels

3.1 Adsorption of heavy metal ions in wastewater

In recent years, the increasing wastewater containing heavy metal ions has caused great harm to environmental safety and human health with industrial development. The use of adsorption materials, including aerogels, to remove the heavy metal ions in water is a hot topic in current research. In particular, carbon aerogels due to a high specific surface area and abundant functional groups have demonstrated a good application in the removal of heavy metal ions. So some research has focused on its use for the adsorptive removal of metal ions. Goel et al studied the adsorption of carbon aerogels for Pb^{2+} (Goel et al. 2005a), and found the adsorption mechanism following the Langmuir model, in which the kinetic data would be useful for the fabrication and design of wastewater treatment plants. They also indicated that the carbon aerogels were feasible for the removal of Hg^{2+} from solution and also had high adsorption ability to Ni^{2+} and Cd^{2+} in aqueous solution (Goel et al. 2005b, 2005c, 2006). Moreover, the adsorption rate of Pb^{2+}, Hg^{2+} and Ni^{2+} ions onto the aerogels were also determined using pseudo-first order and pseudo-second order rate equations. It was clearly found that the pseudo-second order equation was able to describe the adsorption of all the three metal ions as evidence from the correlation coefficient values. According to the coefficiency of correlation, it is deduced that the model of Langmuir correlated to the experimental data well and better than the Freundlich model for all the three metal ions.

Bag and coworkers studied the adsorption of heavy metal cations on the chalcogenide-based clusters aerogels prepared via sol-gel process and CO_2 supercritical drying (Bag et al. 2007). The chalcogels showed a remarkably high capacity for removing heavy metals from the contaminated water. For example, starting with water contaminated by 645 ppm Hg^{2+}, 10 mg of chalcogel would remove Hg^{2+} down to the concentration levels of ~0.04 ppm. The chalcogels worked directly as potential adsorbents for Hg^{2+} without prior modification of their surface, although the detailed mechanism of metal ion removal was not known.

Lee and Bae introduced the novel reactive V_2O_5/TiO_2 aerogels for the removal of Hg^0. Aerogel catalyst contained reactive monovanadates on its surfaces, resulting in higher reactivity for the Hg^0 removal than xerogel and impregnated catalysts. XPS analyses on the surfaces of nanosized catalysts after the removal of Hg^0 suggested that the adsorbed Hg^0 oxidatively transformed HgO by surface vanadates (mono- and poly- vanadates), which was consistent with the Mars-Maessen mechanism (Wu, Wu, and Fu 2007).

The removal of Cd^{2+}, Pb^{2+}, Hg^{2+}, Cu^{2+}, Ni^{2+}, Mn^{2+}, and Zn^{2+} cations by carbon aerogels has been found to be dependent on the ion concentrations, pH value of solution, contact time, adsorbent dose and temperature. Surface complexation and ion exchange are the major removal mechanisms, and the adsorptive behavior satisfies not only the Langmuir assumptions but also the Freundlich assumptions, that is, multilayer formation on the surface of the adsorbent with an exponential distribution of site energy. The carbon aerogel showed nearly 100% adsorptive removal of heavy metal ions under optimal conditions of dosage 10.0 mg/L for aqueous solutions containing 3.0 mg/L metal ions in 48 h (Daniel, Sannino, and Guerra 2008).

3.2 Adsorption of organic pollutants in wastewater

Alumina aerogels have a very good adsorption performance for pyridine molecules, which are found to be coordinated to the Al^{3+} cations as well as through hydrogen bonds to relatively acidic surface -OH groups (Khaleel and Klabunde 2002). Modified hydrophobic silica aerogels exhibit excellent adsorption for different toxic organic compounds (Standeker, Novak, and Knez 2007), which could efficiently remove toluene, benzene, ethylbenzene, chloroform, xylene, chlorobenzene, 1,2-dichloroethan and trichloroethylene from wastewater. The hydrophobic degrees were dependent on the variations in a molar ratio of methyltrimethoxysilane (MTMS)/tetramethoxysilane (TMOS) or trimethylethoxysilane (TMES)/tetramethoxysilane (TMOS) in original precursors, and the molar ratios of the precursors, water, solvent and catalyst, as well as a type of co-precursor significantly also affects the hydrophobicity. The aerogels are shown to be very efficient adsorbents for a variety of volatile organic compounds in water; the adsorption capacity exceeds that of comparable active carbon, for all toxic organic compounds measured in ranges of factors from 15 to 400.

Activated carbon aerogels can also adsorb organic pollutants such as phenol, methylene blue, I_2, and VB_{12} in aqueous solution (Yang, Wu, and Fu 2007). It has been found that the carbon aerogels after CO_2 activation have high surface areas and exhibit excellent absorption ability to the pollutants. The increase in pore-size volume is conducive to adsorb the small-molecule pollutants such as phenol and I_2; however, the mesopores favor the adsorption of large molecule pollutants such as methylene blue and VB_{12}. Moreover, the studies on the adsorption of several basic dyes such as methylene blue (MB), crystal violet (CV), rhodamine B (RB), and acridine orange (AO) on the RF carbon aerogels showed that the major impact factors on the adsorption process are the surface acidic groups and pore size distribution of aerogels (Yang, Wu, and Fu 2008). The electrostatic interaction between the cationic dye and surface negative charge of aerogels could increase the adsorption amount of aerogels to dyes. After the aerogels were treated with HNO_3, H_2O_2, and other oxidants, the amount of surface acidic groups such as acidic carboxyl and phenolic increased, resulting in an increase of adsorption capacity of carbon aerogels on basic dyes. In addition, for MB and AO, with the increase of specific surface areas and micropore volume of carbon aerogels, the adsorption capacity increased. For the large molecular CV and RB, the increase in micropore volume of carbon aerogels has no influence on the adsorption of dyes on the carbon aerogels, which indicates that the pore-size distribution of carbon aerogels has an obvious influence on adsorption. In short, both the pore structure and surface acidic groups affect the adsorption ability of carbon aerogels. Furthermore, it has been found that the adsorption efficiency of carbon aerogels for reactive brilliant red X-3B dye is higher than that of RF aerogels. The pH value, R/C ratio, as well as the adsorbent dosage in the preparation process and particle size of RF adsorbent, have an important role in the adsorption (Lee and Bae 2009). For the adsorption removal of 1,2-dichloroethane (DCH) from aqueous solution with concentration of 1.0 ppm, the adsorption capacity of the crystalline syndiotactic polystyrene aerogels is up to 50 mg/g; however, amorphous aerogels almost do not adsorb any DCH. It is believed that only the crystalline nanoporous phase can adsorb DCH (Meena et al. 2005).

The oil removal from water by inverse fluidization of aerogels was investigated, in which surface-treated hydrophobic aerogel (nanogel) granules of 500 and 850 μm, 1.7 and 2.3 mm, and 0.5 and 2.3 mm were fluidized by a downward flow of oil-contaminated water in an inverse fluidization mode. Aerogel particles are sufficiently robust to be fluidized; their hydrophobic surface gives them a strong affinity for oil and other organic compounds, with the exclusion of water. Oil concentrations of about 2000 ppm in water could be reduced to less than 10 ppm by the inverse fluidization process. Their advantages are extremely low-energy consumption (low-pressure drop) and the large absorption capacity during oil removal (Quevedo, Patel, and Pfeffer 2009).

The MgO aerogels for the aerosil OT (AOT) have higher adsorption amounts than those for didodecyl dimethyl-ammoniumbromide (DDAB) in aqueous solution, and the adsorption amounts are higher than those of cationic surfactants. This suggests that the interaction is the weak Lewis acid sites (surface Mg^{2+}), with the weak Lewis base SO_3^- site of the surfactant and the Na^+ of the surfactant with a nearby surface O^{2-} site (Jeevanandam and Klabunde 2002). However, the chitosan-based aerogels exhibited high adsorption capability for anionic surfactant sodium dodecyl benzene sulfonate (SDBS) molecules and could remove them in aqueous solution. The adsorption is mainly due to the electrostatic attraction; the pore size of aerogels also affects the adsorption kinetics. The larger average pore size is, the faster the adsorption rate is (Chang, Chen, and Jiao 2008).

3.3 Adsorption of organic molecules in air

Activated carbon aerogels have high adsorption ability to some organic molecules such as methanol, acetone, cyclohexane, and benzene; the maximum adsorption amounts are 1605, 1435, 1584, and 1750 mg/g, respectively, which are about 2∼3 times larger than those of the sulfonic-groups on those polyacrylonitrile- based or pitch-based activated carbon fibers,; the adsorption speeds for organic vapors are much higher than those for granular-activated carbon but closed to those on activated carbon fibers. Moreover, the activated carbon aerogels exhibit outstanding desorption and regeneration properties (Wu, Sun, and Fu 2006).

Benzene, toluene, and xylenes (known as BTX) are volatile organic air pollutants that play a key role in the formation of tropospheric ozone and other oxidants in the atmosphere. For the adsorption of BTX on activated carbon aerogels (Fairén-Jiménez, Carrasco-Marín, and Moreno-Castilla 2007), the micropores of ca.1.05 nm contributed most to the adsorption of BTX. So the pore-size control of carbon aerogels is important for adsorption applications. The maximum adsorption amount of toluene on the activation carbon aerogels is 1180 mg/g under static conditions at 25°C. Moreover, as the aerogels were heated to 400°C, the adsorbed toluene would desorb completely. The adsorption amount of toluene remained constant after repeated adsorption-desorption, showing that the carbon aerogels have good removal ability and repeated adsorption-desorption performance (Maldonado-Hódar et al. 2007).

The adsorption of SiO_2 aerogels for benzene vapor showed the saturation adsorption amount to be more than 3000 mg/g (Zhang et al. 2008), but the saturation adsorption amount of benzene on activated carbon powder and activity carbon fiber (ACF) was less than 500 mg/g. The adsorption equilibrium time for benzene on SiO_2 aerogels was 10 h, and those of activated carbon and ACF were less than 40 mins, indicating that the pore structure has a significant influence on the adsorption rate. All the pores of ACF were in the fiber surface, the adsorption rate was the fastest, and SiO_2 aerogels exhibited that with the three-dimensional porous network structure a longer time was required for the benzene molecules to reach the surface of holes.

In summary, the research on the adsorption properties of aerogels has made great progress, but further studies on the semiconductive and magnetic aerogels with excellent adsorption properties are still needed. Moreover, the relationship of the microstructure of aerogel to its adsorption properties, especially the mechanism of selective adsorption performance, should be investigated.

4 Catalytic applications of aerogels

4.1 Catalytic decomposition of organic pollutants

In the 1930s, Kistler pointed out that the aerogels have potential application in the catalyst and catalyst carrier for its high porosity and specific surface area. For example, the photo-degradation rate of salicylic acid on TiO_2 aerogels photocatalyst under UV irradiation is about 10 times more than that on TiO_2 (Degussa) powder (Tomkiewicz, Dagan, and Zhu 1994).

Silica aerogels are excellent adsorbents for some volatile organics, and the titania aerogels exhibit photocatalytic activity for the degradation of organics; thus their combination is expected to be able to remove volatile organic pollutants. For example, silica-titania aerogels showed excellent adsorption and photocatalytic decomposition for benzene in air. The photocatalytic activity of titania aerogels to decompose benzene increased with the increase in crystallinity, in which these differences were not so much affected by the adsorption capacity. A combination of silica and titania would be advantageous in improving the adsorption capacity and probably completely decompose the intermediates to form CO_2 gas (Yoda, Suh, and Sato 2001).

The titania aerogels for the photocatalytic decomposition of three phenol para-derivatives—p-chlorophenol, p-nitrophenol ,and 4-hydroxybenzoic acid—were more efficient in comparing to the titania-silica aerogels. Although all the aerogels exhibited photocatalytic activity, the presence of two separated phases in titania-silica aerogels does not favor photocatalysis. The identified degradation products of p-chlorophenol and p-nitrophenol are hydroquinone and benzoquinone as well as chloride ions and nitrate ions, respectively. However, the photodegradation course did not depend on the surface area and pore volumes of the catalysts but on the kind of pollutant (Malinowska et al. 2003).

Cao et al. compared the catalytic activity of two titania-silica aerogels, one containing anatase TiO_2 nanocrystal domains (ATND) and the other none for the photocatalytic oxidation of iso-propanol and trichloroethylene in air (Cao, Yeung, and Yue 2006). The aerogel catalysts not containing anatase TiO_2

domains did not exhibit activities for two pollutants. The removal efficiency of ATND is 9.7% for trichloroethylene and 10.1% for isopropanol. However, the highest removal efficiency of P25 TiO_2 for trichloroethylene and isopropanol were 1.5% and 1.3%, respectively. They authors also found that the photocatalytic oxidation reaction of trichloroethylene on the aerogel catalysts exhibited an average conversion of 30% at a trichloroethylene concentration of 56 ppm (Cao, Yeung, and Yue 2007). Padilla-Serranos indicated that the activities of Pt catalysts supported carbon aerogels in the combustion of o-xylene, and m-xylene increased with the increase of Pt particles (Padilla-Serrano, Maldonado-Hódar, and Moreno-Castilla 2005). In addition, the catalysts were activated during the consecutive combustion runs, which depended on the Pt particle size and type of xylene isomer. The lower activity of catalysts with smaller Pt particles was due to the stronger Pt-O bonds formed during the combustion process of xylene.

The vanadia-titania aerogel catalysts prepared via the sol-gel method and CO_2 supercritical drying showed a very high activity and thermal stability for gas-phase oxidation of 1,2-dichlorobenzene (DCB) in the range of 150–600°C. The selectivity for CO_2 gas was kept over 95% unless the reaction temperature was relatively low (< 200°C). The conversion efficiency was maintained over 90% at above 350°C, with a residence time of less than 0.05 s; deactivation was not observed during the 48 h of operation at 450°C. The surface structure of vanadia played a critical role in catalytic activity, but the structure of crystalline titania appeared to have exclusive dependency. For the vanadia-titania catalysts with the inherent characteristics of aerogels such as high specific surface area and chemical homogeneity, little bulk vanadia was observed and catalytic activity was maintained or improved because of the formation of surface mono- and poly-vanadates after repeated reactions (Choi et al. 2006).

At present, the catalytic conversion of harmful NO_x to harmless substances is a key area in environment protection. Aerogels of $Fe_2O_3/Cr_2O_3/Al_2O_3$, MgO/Fe_2O_3, and Cr_2O_3/Al_2O_3 were applied to make the NO in styrene-chloroprene rubber to transform the harmless substances. It was found that at $T \approx 327$°C as the Mg/Fe ratio changed from 1/2 to 1/1 the conversion rate increased. When the Mg/Fe ratio is 0.287 the conversion rate is very high (Willey, Lai, and Peri 1991). However, when using Fe_2O_3/SiO_2 aerogels as catalyst the selectivity of the reduction of NO to N_2 is higher than 98% below 460°C. The catalytic activity is found to be related to the density of Lewis acid sites (Fabrizioli, Bürgi, and Baiker 2002).

The feasibility of using carbon aerogels as bed electrodes in a three-dimensional electrode reactor for the electrochemical catalytic oxidation of simulated phenol wastewaters was studied. It was found that the removal efficiency depended on the airflow, the feed phenol concentration, pH value, ratio of aerogel mass to solution volume, and repeated times. Compared with commercial carbon particle electrodes, the carbon aerogels exhibited stronger and longer-term activity to remove phenol-simulated wastewater. For 250 mg/L of phenol solution, 98% of the initial phenol chemical oxygen demand can be removed after treatment for 20 min. Moreover, high removal efficiency can be kept basically after being reused for many times (e.g. 93% and 82% after 20 and 50 times, respectively) (Lv, Wu, and Fu 2009).

4.2 Catalytic decomposition of inorganic pollutants

The catalytic properties of mesoporous iron oxide-silica aerogels prepared by a sol-gel process combined with supercritical CO_2 drying were investigated in the selective oxidation (SCO) of ammonia and the selective reduction (SCR) of NO in a continuous fixed-bed reactor in the temperature range of 210–550°C at ambient pressure. The catalytic activity of the aerogels in SCO correlated with the abundance of strongly bound ammonia adsorbed on Lewis sites (low coordinated iron). High selectivity to nitrogen (97%) could be reached up to 500°C, whereas at higher temperatures the formation of N_2O and NO became significant. For all aerogels, the selective catalysis to nitrogen in SCR was >98% at T<460°C, and the activation energies varied from 38 to 53 kJ/mol. The catalytic activity for SCR did not correlate with the population density of Lewis sites. It is proposed that SCO predominantly occurs on Lewis sites consisting of highly dispersed iron atoms of low coordination, whereas in SCR these sites do not play an important role (Wojciechowska and Lomnicki 1999).

The TiO_2-SiO_2 aerogels as photocatalyst prepared via sol-gel technique and C_2H_5OH supercritical drying are used to degrade the cyanide pollutant in aqueous solutions. As the molar ratio of Ti:Si is equivalent to 6:1 in TiO_2-SiO_2 aerogel, the molar ratios of TEOS:H_2O:C_2H_5OH:HNO_3 is taken as 1:8: 10:0.25, and the chemical reaction conducts for 90 min at 25°C (supercritical drying at 10^5 bar and 295°C), TiO_2-SiO_2 aerogel with a specific surface area of 850 m^2/g can be obtained, which has the best removal efficiency of 98.74% for cyanide (Ismail et al. 2004).

The selective oxidation of H_2S in the presence of excess water and ammonia over the V_2O_5-TiO_2 aerogel catalysts prepared via the sol-gel method with subsequent CO_2 supercritical drying showed very high conversion of H_2S without harmful emission of SO_2. Conversion of H_2S over aerogels is from 93.8% to 98.0%, higher than those of the xerogel (78.1%) and impregnated V_2O_5-TiO_2 (79.7%). The high catalytic performance of the aerogel originated from their highly dispersed VOx species and high reducibility (Kim et al. 2006).

Ultra-low loading Pt nanocatalysts on carbon aerogels prepared by atomic layer deposition (ALD) exhibited high catalytic activity (Wojciechowska and Lomnicki 1999), although the Pt loading in the CAs is extremely low, with the density of Pt estimated to be ~0.047 mg/cm^2 in the sample prepared with 2 ALD cycles. The oxidation studies of CO demonstrated near 100% conversion of CO to CO_2 at the temperatures ranging from 150 to 250°C (King et al. 2008).

The Au/titania-coated silica aerogels showed high oxidation activity for the CO conversion, in which Au loadings ranged from 0.14 to 20 wt%. All the Au nanoparticles as small as 2 nm were well dispersed in the samples, preserving their original diameters after heat treatment at 673 K. The oxidation activity of the catalyst for CO increased monotonously with the increase of Au loading. The Au nanoparticles on this support were resistant to thermal sintering: The diameter of the particles remained well below 5 nm even after calcination at 973 K for 2 h, and the CO conversion was up to near 100% (Tai and Tajiri 2008).

The preferential oxidation of CO gas on the Co-promoted Pt catalysts supported on silica (Co/Pt-SiO_2, PtCo-SiO_2 and Pt-SiO_2) aerogels in a fixed-bed reactor system under atmospheric was investigated, in which all the as-prepared catalysts were reduced at 400°C for 2 h prior to the oxidation reaction. The total flow rate was 100 mL/min containing 1.1% H_2, 2500 ppm CO and 2500 ppm O_2, and the reaction temperature was 25–300°C with a 25°C increment and held for 1.5 h at each temperature. The overall catalytic activities were in order of Co/Pt-SiO_2 aerogels ~ PtCo-SiO_2 aerogels>>Pt-SiO_2 aerogels>commercial Pt/Al_2O_3, suggesting that the catalytic activity for the preferential oxidation of CO was significantly influenced by the reducible Pt sites at low temperature. The major roles of Co were not only to make Pt easily reduced, but also to stabilize Pt in its reduced state, as inferred from the high catalytic activity in a wide range of temperatures (Choi, Shin, and Suh 2008).

The alumina aerogel catalyst with a surface area of 550 m^2/g destroys carbon tetrachloride with an efficiency of > 99% at 400°C. Its reactivity to carbon tetrachloride is remarkably higher than that of commercial alumina, which has a surface area of 155 m^2/g. This alumina aerogel adsorbed and decomposed large quantities of carbon tetrachloride through surface interactions that drove further interactions with the bulk of the particles as the $AlCl_3$ forming on the surface sublimed (Khaleel and Dellinger 2002).

The titania-silica aerogel-like microspheres (TSAMs) (with mean diameters of around 100 μm, bulk density of 0.3 g/cm^3, specific surface area of 415 m^2/g, pore-size distribution between 2 and 35 nm, and average pore size of 15 nm) exhibited excellent photocatalytic activity for the degradation of methylene blue (MB), which acted as a simulated aqueous organic pollutant in wastewater (Liu et al. 2008).

5 Other applications of aerogels for environmental purposes

5.1 Storage of hydrogen

Kabbour and coworkers studied the hydrogen storage property of high-surface-area carbon aerogels synthesized through the sol-gel polymerization of resorcinol with formaldehyde in aqueous solution to produce organic gels that are supercritically dried and subsequently pyrolyzed in an inert atmosphere. Theyfound that after the carbon aerogels were activated with CO_2 at 950°C, the maximum amount of adsorbed hydrogen was 5.3 wt%. The carbon aerogels could offer engineering viability for hydrogen sorption applications by controlling surface area, pore volume, and substituent/dopant levels, where both the gravimetric and volumetric density are important parameters (Kabbour et al. 2006).

Zhou prepared the carbon aerogels derived from resorcinol and furfural by a sol-gel technique using ambient pressure drying, followed by carbonization and activation. Addition of a KOH precursor prior to the sol-gel polymerization process significantly increased the microporosity of carbon aerogels and enhanced hydrogen storage capacity. A KOH-catalyzed carbon aerogel exhibits hydrogen uptake of 5.2 wt% and micropore volume of 1.06 ± 0.02 cm^3/g. The KOH catalyst plays an important role in accelerating the gelation, strengthening a cross linking, solidifying the carbon network, and increasing microporosity. A linear dependence of hydrogen uptake maximum with surface area and micropore volume was obtained. The optimal micropore volume and surface area of carbon aerogels could be obtained by controlling the synthesis process (Tian et al. 2009).

Vajo et al. found that the MgH_2 nanoparticles incorporated in a mesoporous carbon aerogel demonstrated accelerated hydrogen exchange kinetics but no thermodynamic change in the equilibrium hydrogen pressure. The improved hydrogen sorption kinetics is due to both the aerogel limiting the maximum MgH_2 particle diameter and a catalytic effect from the Ni and Cu wetting layers. At 250°C, MgH_2 filled Ni decorated and Cu decorated carbon aerogels released H_2 at 25 wt%/h and 5.5 wt%/h, respectively, while a MgH_2 filled aerogel without catalyst desorbed only 2.2 wt%/h (all wt %/h values are with respect to MgH_2 mass). At the same temperature, MgH_2 ball milled with synthetic graphite desorbed only 0.12 wt%/h, which demonstrated the advantage of incorporating nanoparticles in a porous host (Gross et al. 2009). They also developed a new approach to the incorporation of MgH_2 in the nanometer-sized pores of a carbon aerogel scaffold. The loading of MgH_2 was determined as 15 ~ 17 wt%, of which 75% is reversible over ten cycles. Incorporated MgH_2 had >5 times faster dehydrogenation kinetics than ball-milled activated MgH_2, which may be attributed to the particle size of the former being smaller than that of the latter. Cycling tests of the incorporated MgH_2 showed that the dehydrogenation kinetics is unchanged over four cycles. The results demonstrate that confinement of metal hydride materials in a nanoporous scaffold is an efficient way to avoid aggregation and improve cycling kinetics for hydrogen storage materials (Zhang et al. 2009). Moreover, they also have successfully incorporated pure $NaAlH_4$ into the carbon aerogel. The $NaAlH_4$ melt filled the aerogel pores to 94% capacity by capillary action. $NaAlH_4$@aerogel displayed a greatly reduced dehydrogenation temperature, with an onset at ~140°C, compared to both bulk $NaAlH_4$ and extensively ball-milled $NaAlH_4$ (>230°C). Unlike uncatalyzed bulk sodium alanate, $NaAlH_4$@aerogel was readily rehydrogenated to full capacity at ~60°C under 100 bar H_2. Although the first dehydrogenation of catalyzed bulk $NaAlH_4$ was superior to that of $NaAlH_4$@aerogel, the dehydrogenation kinetics of $NaAlH_4$@aerogel rivaled that of catalyzed bulk $NaAlH_4$ after the first cycle. No enhancement was observed for $NaAlH_4$ mixed with graphite, clearly demonstrating that the presence of carbon alone was insufficient for improved kinetics. Rather, the confinement of $NaAlH_4$ to nanoscale dimensions within aerogel pores, which both limited diffusion distances for phase segregation and recombination and blocked agglomeration into larger particles with cycling, was responsible for the dramatic improvement of cycling kinetics even in the absence of a catalyst (Stephens et al. 2009).

Kang et al. synthesized nitrogen-doped (1.2~4.5 wt%) carbon xerogels from carbonization of resorcinol-formaldehyde polymer in an ammonia atmosphere at various temperatures. The maximum hydrogen uptakes were measured to be 3.2 wt% at –196°C and 0.28 wt% at 35°C. Hydrogen adsorption had a stronger correlation with specific surface area than nitrogen content at the low temperature of –196°C. At the higher temperature of 35°C, optimal nitrogen doping enhanced hydrogen adsorption by electronic modification of carbon in agreement with previous theoretical predictions (Kang, Lee, and Lee 2009).

Meisner synthesized the spherical nanoporous carbon particles from carbon precursor solutions of sucrose with either silica sols or colloidal silica particles, or both, in a direct one-step aerosol-assisted process followed by carbonization and then removal of the silica template, and measured hydrogen adsorption at various temperatures and pressures. The carbon particles show very high porosity with narrow pore-size distributions, surface areas up to 2000 m^2/g, and pore volumes up to 4.0 cm^3/g. The hydrogen sorption of > 4.0 wt% at 77 K and > 20 bar was found for the unimodal nanoporous carbon particles (Hu, Lu, and Meisner 2008).

Aerogel catalysts contained high active components, which can be very evenly dispersed in the carrier. The catalysts have excellent thermal stability and high catalytic activity, selectivity, and long life, exhibiting extensive applications. The studies on the sol-gel cost and production process are still in the development stage, which has limited their catalytic applications. The research on the photocatalytic degradation of organics by aerogel catalysts has been developed. It is believed that with further simplification of the preparation process, the significant decrease of production costs, and deep research on the law of light catalysis, aerogels could play a unique role in the governance of environmental pollutants. In addition, to improve the catalytic activity of aerogels, the surface modification of aerogels particles is another focus of the study.

5.2 Thermal insulation

The problems of coal and petroleum energy depletion and global warming have become a main challenge and threat to human survival and have thus greatly boosted research efforts to develop clean alternative energy and emission control of global warming gases. These efforts, however, will prove insufficient if they are not accompanied by energy-saving practices. One of the major energy-saving practices is thermal insulation (e.g., reducing fuel and electricity usage through better insulating boilers and buildings, respectively).

Silica aerogels, a class of mesoporous materials with extremely high porosities and specific surface areas, exhibit potential applications in high-temperature insulation. Their applications as thermal insulators are particularly promising because of the existence of highly interconnected pores, with apparent thermal conductivities commonly lower than 0.1 W/m·K at room temperature (Hüsing and Schubert 1998; Pierre and Pajonk 2002). Native silica aerogels however have found difficulties in commercialization because of their fragility, hydrophilicity and demand of supercritical drying in production. Fragility makes it very difficult to apply the silica aerogels in monolithic form while hydrophilicity restricts silica aerogels from long-term usage since cracking can readily occur through moisture adsorption at ambient conditions. Furthermore, supercritical drying limits silica aerogels from large scale production. Because of these drawbacks, silica aerogels have been found only very limited use in highly specific situations such as in NASA's Stardust program for capturing particles in space and in nuclear reactors as radiation detectors. Consequently, researchers in this area devote efforts to improve the mechanical strength of, introduce hydrophobicity to, and avoid use of supercritical drying in production of silica aerogels, while trying to maintain their low thermal conductivities, particularly at high temperatures. For example, Wei and co-workers indicated that silica-poly(vinylpyrrolidone) composite aerogels have high hydrophobicity (static contact angle $>120°$), good mechanical strength (Young's modulus of bending >30 MPa), and low high-temperature thermal conductivity (0.063 W/m·K at 300°C), which are critical characteristics for practical applications of aerogels, particularly in energy saving areas for long-term usage and large scale production (Wei, Lu, and Chang 2008). They also found a new class of opacified monolithic aerogels of ultra-low high-temperature thermal conductivities, in which the carbon nanofibers were successfully incorporated into the mesoporous network of silica aerogels at the concentration as high as 20 wt % through an accelerated-gelation sol-gel process to enhance the dimensional stability of the silica aerogels and to suppress the thermal radiations that become dominant at high temperatures. The composite aerogels have a ultra low thermal conductivity of 0.050 W/m·K at 500°C, whereas maintaining a thermal stability above 500°C (much better than the conventional high-temperature thermal insulation materials: 0.3 W/m·K at 500°C for glass fibers, 0.1 W/m·K at 527°C for alumina fused brick and 1.7 W/m·K at 527°C for sillimante) (Wei, Lu, and Chang 2009).

So far, all nanoporous insulation materials are based on SiO_2 aerogels as a carrier of nanopore. But all the ultra-light aerogels have low strength and poor toughness. Although the SiO_2 aerogels have a relatively high intensity, and still can not serve as a single block material for insulation projects, thus the SiO_2 aerolgels with practical value should be enhanced and toughened by a variety of approaches. For example, Bostaph and co-workers prepared the hydrolytic silica aerogels using the optimum reaction mixture composition by weight is 20% $Si(OCH_3)_4$, 75% C_3H_7OH, 5% H_2O, trace NH_4OH under nonsupercritical conditions for thermal insulation, and found that even though the tendency to fracture in this system is minimized, the gel nevertheless fractures into 2 ~ 5 cm size pieces during the drying, making it useless to be used as large panels for thermal insulation. However, micronized powder of this aerogel (3 ~ 10 μm size particles) exhibits thermal insulation characteristics which are twice that of precipitated silicas with similar porosity, when used in evacuated pouches (Yoldas, Annen, and Bostaph 2000).

Chang and coworkers successfully incorporated the carbon nanofibers into the mesoporous network of silica aerogels at concentrations as high as 20 wt% through an accelerated-gelation sol-gel process to enhance the dimensional stability of the silica aerogels and to suppress the thermal radiations that become dominant at high temperatures, to achieve an ultralow thermal conductivity of 0.050 W/m·K at 500°C, whereas maintaining a thermal stability above 500°C (much better than the conventional high-temperature thermal insulation materials: 0.3 W/m·K at 500°C for glass fibers, 0.1 W/m·K at 527°C for alumina fused brick, and 1.7 W/m·K at 527 °C for sillimante) (Yim, Kim, and Yoo 2002).

Up to date, aerogels are mainly used for scientific purposes such as particle collector and detectors in high-energy physics, and dust collector in the universe. The aerogels materials have extensive applications in the fields of military and aerospace, the most successful application as a super-insulation material is the field of aerospace. In the civil context, aerogels can effectively improve the energy efficiency and reduce energy consumption, and significantly reduce the consumption of plant-based fuels, which is important to the ecological and environmental protection. Their applications are now irreplaceable and its application areas will become an increasingly widespread in the civil, aerospace, military and other high-tech fields. The lightweight, composite and flexible aerogels for the insulation materials would be the main direction of development.

5.3 Acoustic absorption

Noise pollution is a kind of environmental pollution, which has become a major hazard to human. Noise pollution and water pollution, air pollution was seen as the three major environmental issues worldwide. Although noise control technology has matured, but because of modern industrial, transport a large scale, we should adopt noise control large number of places, so research and development of new high-performance materials, the elimination of noise pollution is also necessary. The aerogels with excellent sound absorption (low sound velocity down to 90 m/s) have been widely studied for different frequency ranges. Hooley et al found that the acoustic transmission losses in the silica aerogel granules of sizes 80 μm and 3.5 mm are at least 10 dB higher than in the glass wool sample of the same thickness (4 cm) for the frequency range 300 ~ 1700 Hz (Forest, Gibiat, and Hooley 2001). Matsumoto and co-workers investigated sound velocities and attenuation in an aerogel sample of 3 mm length and 3 mm diameter squeezed between a 10 MHz transmitter and receiver quartz using sound transmission techniques. Pulsed measurements were performed during several cool down cycles (300 ~ 0.7 K) in vacuum as well as while filling the aerogel with gaseous and liquid. The striking features of all experiments were that a profound change of the sound mode in the 3 He-filled aerogel occur around 5, 8, and 12 bar, and that neither sound velocity nor absorption in the aerogel has been found very reproducible from cool-down to cool-down (Putselyk et al. 2003). Aerogels and alcogels display very different acoustic behaviors, since, for alcogels, longitudinal wave velocity is always around the velocity in alcohol, for aerogels, it is often significantly lower than the velocity of air. Biot's model is appropriate for the description of sound propagation in both alcogels and aerogels, quantitatively in term of velocity and qualitatively in term of absorption. Because it distinguishes the solid, fluid and skeletal contributions, it allows prediction of acoustical behavior if the aerogel is filled with gas other than air (Forest, Gibiat, and Woignier 1998).

6 Conclusions

Aerogels containing a large number of pores with a typical size of 1 ~ 100 nm are nanostructured porous materials and have shown unique properties in mechanics, acoustics, thermal, optical, and other aspects. In particular, the aerogels demonstrate extensive applications for environmental purposes. Significantly different from the micron-/millimeter-porous materials, the aerogels have very low thermal conductivity because of their nanopores. For example, the thermal conductivity of silica aerogels is lower than that of the corresponding glassy material with low 2 ~ 3 orders of magnitude, because the fine network structures of aerogels effectively limit the spread of the local thermal excitation. Nanopores inhibit the heat conduction of the gas molecules. The light and sound scattering of aerogels are much smaller than those of traditional porous materials. These unique properties not only attract much attention in basic research, but also have a wide range of applications in many areas.

The aerogels also serve as a new and efficient gas filter material for their uniform pore-size distribution and high porosity. Their nanoporous network, large specific surface area, and controlled meso-scale structure make them the best candidate material for the adsorption of heavy metal ions and organic molecules pollutants. Particularly, they can be used as a new type of catalyst or catalyst carrier.

With their high specific surface area and rich surface groups, controlled pore structure, and modified surface, the aerogels have potential applications in terms of environmental protection such as the adsorption of toxic and harmful substances and catalytic degradation of toxic and harmful substances. However, the tedious preparation process and expensive cost limit their applications.

Glossary

Acoustic absorption As the sound waves in the medium in the transmission or reflection during the interface process, the energy decreases. The sound absorption process is a dissipation process. Sound absorption is mainly caused by heat conduction, medium viscosity, and molecular relaxation processes.

Activated carbon aerogel The carbon aerogel is heated at a certain temperature for different times in CO_2 atmosphere, and their specific surface areas can reach to the original multiple with the pore size changing.

Aerogel a kind of unique solid materials that have extremely high surface areas (up to 3200 m^2/g), low bulk densities (up to 95% of their volume is air) and large open pores. The ordinary preparation process of areogel includes several steps: the sol-gel process, the aged step, the exchange of solvent, and the drying of gel. Supercritical drying is the ordinary process; the freeze-drying and ambient drying have also been applied in the preparation of aerogels. The aerogels exhibit extremely low thermal conductivity and low sound velocity.

Storage of hydrogen Hydrogen is a clean and efficient energy carrier. The hydrogen storage and transportation is a key factor that limits the application of hydrogen. There are three main methods: high-pressure gas storage, cryogenic liquid storage, and solid-state storage; solid-state hydrogen storage by a chemical reaction or physical adsorption is considered the most promising method because its energy density and safety. Porous solid adsorption mechanism is based on physical adsorption, of which hydrogen storage capacity is proportional to surface area, while the capacities of materials increase with the micropores increasing.

Supercritical drying a drying process of wet-gel to aerogels. In this procedure the liquid solvent achieves the critical point to transit to the supercritical gas phase, in which there are no liquid/gas interfaces in the pores during the drying process, so the cytoskeleton will not collapse.

Thermal insulation There are three main thermal transportations: the gas phase, the solid phase, and radiation transportation. The aerogel possesses high porosity more than 80%, and the size of pores composed by the nanoskeleton connecting with each other distributes between 2 and 50 nm. The thermal property of aerogel is different from that of other porous materials attributed to its particular architecture, and aerogel is becoming the super insulating material.

References

Akimov, Yu K., 2003. Fields of application of aerogels (review). Instrum. Exp. Tech. 46 (3), 287–299.

Bag, Santanu, Pantelis, N., Trikalitis, Peter J., Chupas, Gerasimos S., Armatas, Kanatzidis, Mercouri G., 2007. Porous aemiconducting gels and aerogels from chalcogenide clusters. Science 317 (5837), 490–493.

Cantin, M., Casse, M., Koch, L., Jouan, R., Mestreau, P., Roussel, D., Bonnin, F., Moutel, J., Teichner, S.J., 1974. Silica aerogels used as Cherenkov radiators. Nucl. Instrum. Methods 118 (1), 177–182.

Cao, Shengli, Yeung, King Lun, Yue, Po-Lock, 2006. Preparation of freestanding and crack-free titania-silica aerogels and their performance for gas phase, photocatalytic oxidation of VOCs. Appl. Catalysis B: Environ. 68 (3-4), 99–108.

Cao, Shengli, Yeung, King Lun, Yue, Po-Lock, 2007. An investigation of trichloroethylene photocatalytic oxidation on mesoporous titania-silica aerogel catalysts. Appl. Catalysis B: Environ. 76 (1-2), 64–72.

Chang, Xinhong, Chen, Dairong, Jiao, Xiuling, 2008. Chitosan-based aerogels with high adsorption performance. J. Phys. Chem. B 112 (26), 7721–7725.

Chen, Longwu, Gan, Lihua, 1997. Aerogels. Bull. Chem. (in Chinese) 8, 21–27.

Chen, Xiaobo, Mao, Samuel S., 2007. Titanium dioxide manomaterials: Synthesis, properties, modifications, and applications. Chem. Rev. 107 (7), 2891–2959.

Choi, Jinsoon, Shin, Chee Burm, Park, Tae-Jin, Suh, Dong Jin, 2006. Characteristics of vanadia-titania aerogel catalysts for oxidative destruction of 1,2-dichlorobenzene. Appl. Catalysis A: Gen. 311, 105–111.

Choi, Jinsoon, Shin, Chee Burm, Suh, Dong Jin, 2008. Co-promoted Pt catalysts supported on silica aerogel for preferential oxidation of CO. Catalysis Commun. 9 (5), 880–885.

Daniel, Christophe, Sannino, Diana, Guerra, Gaetano, 2008. Syndiotactic polystyrene aerogels: Adsorption in amorphous pores and absorption in crystalline nanocavities. Chem. Mater. 20 (2), 577–582.

Fabrizioli, P., Bürgi, T., Baiker, A., 2002. Environmental catalysis on iron oxide-silica aerogels: Selective oxidation of NH_3 and reduction of NO by NH_3. J. Catalysis 206 (1), 143–154.

Fairén-Jiménez, D., Carrasco-Marín, F., Moreno-Castilla, C., 2007. Adsorption of benzene, toluene, and xylenes on nonolithic carbon aerogels from dry air flows. Langmuir 23 (20), 10095–10101.

Forest, L., Gibiat, V., Hooley, A., 2001. Impedance matching and acoustic absorption in granular layers of silica aerogels. J. Non-Crystalline Solids 285 (1-3), 230–235.

Forest, L., Gibiat, V., Woignier, T., 1998. Biot's theory of acoustic propagation in porous media applied to aerogels and alcogels. J. Non-Crystalline Solids 225 (1), 287–292.

Gesser, H.D., Goswami, P.C., 1989. Aerogels and related porous materials. Chem. Rev. 89 (4), 765–788.

Goel, Jyotsna, Kadirvelu, K., Rajagopal, Chitra, Garg, Vinod Kumar, 2005b. Investigation of adsorption of lead, mercury and nickel from aqueous solutions onto carbon aerogel. J. Chem. Technol. Biotechnol. 80 (4), 469–476.

Goel, Jyotsna, Kadirvelu, K., Rajagopal, C., Garg, V.K., 2005a. Removal of lead(II) from aqueous solution by adsorption on carbon aerogel using a response surface methodological approach. Ind. Eng. Chem. Res. 44 (7), 1987–1994.

Goel, Jyotsna, Kadirvelu, K., Rajagopal, C., Garg, V.K., 2005c. Removal of mercury(II) from aqueous solution by adsorption on carbon aerogel: Response surface methodological approach. Carbon 43 (1), 197–200.

Goel, Jyotsna, Kadirvelu, K., Rajagopal, C., Garg, V.K., 2006. Cadmium (II) uptake from aqueous solution by adsorption onto carbon aerogel using a response surface methodological approach. Ind. Eng. Chem. Res. 45 (19), 6531–6537.

Gonzalez, Richard D., Lopez, Tessie, Gomez, Ricardo, 1997. Sol—Gel preparation of supported metal catalysts. Catalysis Today 35 (3), 293–317.

Gross, Adam F., Ahn, Channing C., Van Atta, Sky L., Liu, Ping, Vajo, John J., 2009. Fabrication and hydrogen sorption behaviour of nanoparticulate MgH_2 incorporated in a porous carbon host. Nanotechnology 20, 204005.

Hauthal, Werner H., 2001. Advances with supercritical fluids [review]. Chemosphere 43 (1), 123–135.

Hench, Larry L., West, Jon K., 1990. The sol-gel process. Chem. Rev. 90 (1), 33–72.

Hu, Qingyuan, Lu, Yunfeng, Meisner, Gregory P., 2008. Preparation of Nanoporous Carbon Particles and Their Cryogenic Hydrogen Storage Capacities. J. Phys. Chem. C. 112 (5), 1516–1523.

Hüsing, Nicola, Schubert, Ulrich, 1998. Aerogels-airy materials: Chemistry, structure, and properties. Angew. Chem. Int. Ed. 37 (1-2), 22–45.

Ismail, A.A., Ibrahim, I.A., Ahmed, M.S., Mohamed, R.M., El-Shall, H., 2004. Sol-gel synthesis of titania-silica photocatalyst for cyanide photodegradation. J. Photochem. Photobiol. A: Chem. 163 (3), 445–451.

Jeevanandam, P., Klabunde, K.J., 2002. A study on adsorption of surfactant molecules on nagnesium oxide nanocrystals prepared by an aerogel route. Langmuir 18 (13), 5309–5313.

Kabbour, Houria, Baumann, Theodore F., Satcher, Joe H., Saulnier, Angelique, Ahn, Channing C., 2006. Toward new candidates for hydrogen storage: High-surface-area carbon aerogels. Chem. Mater. 18 (26), 6085–6087.

Kang, Kyung Yeon, Lee, Burtrand I., Lee, Jae Sung, 2009. Hydrogen adsorption on nitrogen-doped carbon xerogels. Carbon 47 (4), 1171–1180.

Khaleel, A., Dellinger, B., 2002. FT-IR investigation of adsorption and chemical decomposition of CCl_4 by high surface-area aluminum oxide. Environ. Sci. Technol. 36 (7), 1620–1624.

Khaleel, Abbas A., Klabunde, Kenneth J., 2002. Characterization of aerogel prepared high-surface-area alumina: In situ FT-IR study of dehydroxylation and pyridine adsorption. Chem.—A Euro. J. 8 (17), 3991–3998.

Kim, M.I., Park, D.W., Park, S.W., Yang, X., Choi, J.S., Suh, D.J., 2006. Selective oxidation of hydrogen sulfide containing excess water and ammonia over vanadia-titania aerogel catalysts. Catalysis Today 111 (3-4), 212–216.

King, Jeffrey S., Wittstock, Arne, Biener, Juergen, Kucheyev, Sergei O., Wang, Yinmin M., Baumann, Theodore F., Giri, Sandeep K., Hamza, Alex V., Baeumer, Marcus, Bent, Stacey F., 2008. Ultralow loading Pt nanocatalysts prepared by atomic layer deposition on carbon aerogels. Nano Lett. 8 (8), 2405–2409.

Lee, Woojin, Bae, Gwi-Nam, 2009. Removal of elemental mercury (Hg(0)) by nanosized V2O5/TiO2 catalysts. Environ. Sci. Technol. 43 (5), 1522–1527.

Liu, Mingxian, Gan, Lihua, Pang, Yingcong, Xu, Zijie, Hao, Zhixian, Chen, Longwu, 2008. Synthesis of titania-silica aerogel-like microspheres by a water-in-oil emulsion method via ambient pressure drying and their photocatalytic properties. Colloids Surfaces A: Physicochem. Eng. Aspects 317 (1-3), 490–495.

Lv, Guifen, Wu, Dingcai, Fu, Ruowen, 2009. Performance of carbon aerogels particle electrodes for the aqueous phase electro-catalytic oxidation of simulated phenol wastewaters. J. Hazard. Mater. 165 (1-3), 961–966.

Maldonado-Hódar, Francisco J., Moreno-Castilla, Carlos, Carrasco-Marín, Francisco, Pérez-Cadenas, Agustín F., 2007. Reversible toluene adsorption on monolithic carbon aerogels. J. Hazard. Mater. 148 (3), 548–552.

Malinowska, B., Walendziewski, J., Robert, D., Weber, J.V., Stolarski, M., 2003. The study of photocatalytic activities of titania and titania-silica aerogels. Appl. Catalysis B: Environ. 46 (3), 441–451.

Marr, R., Gamse, T., 2000. Use of supercritical fluids for different processes including new developments–a review. Chem. Eng. Processing 39 (1), 19–28.

Meena, Ajay Kumar, Mishra, G.K., Rai, P.K., Rajagopal, Chitra, Nagar, P.N., 2005. Removal of heavy metal ions from aqueous solutions using carbon aerogel as an adsorbent. J. Hazard. Mater. 122 (1-2), 161–170.

Moreno-Castilla, C., Maldonado-Hódar, F.J., 2005. Carbon aerogels for catalysis applications: An overview. Carbon 43 (3), 455–465.

Padilla-Serrano, M.N., Maldonado-Hódar, F.J., Moreno-Castilla, C., 2005. Influence of Pt particle size on catalytic combustion of xylenes on carbon aerogel-supported Pt catalysts. Appl. Catalysis B: Environ. 61 (3-4), 253–258.

Pajonk, G.M., 1999. Some catalytic applications of aerogels for environmental purposes. Catalysis Today 52 (1), 3–13.

Pajonk, Gerard Marcel, 2003. Some applications of silica aerogels. Colloid Polym. Sci. 281 (7), 637–651.

Palazzi, Chiara, Oliva, Liliana, Signoretto, Michela, Strukul, Giorgio, 2000. Microporous zirconia-silica mixed oxides made by dol-hel as vatalysts for the Liquid-Phase oxidation of olefins with hydrogen peroxide. J. Catalysis 194 (2), 286–293.

Pierre, Alain C., Pajonk, Gerard M., 2002. Chemistry of aerogels and their applications. Chem. Rev. 102 (11), 4243–4266.

Putselyk, S., Eska, G., Abe, S., Matsumoto, K., 2003. Acoustic properties of 97% porous aerogel at low temperatures. Phys. B-Condensed Matter 329, 309–310.

Quevedo, Jose A., Patel, Gaurav, Pfeffer, Robert, 2009. Removal of oil from water by inverse fluidization of aerogels. Ind. Eng. Chem. Res. 48 (1), 191–201.

Kistler, S.S., 1931. Coherent expanded aerogels and jellies. Nature 127, 741–741.

Schneider, Michael, Baiker, Alfons, 1995. Aerogels in catalysis. Catalysis Rev.: Sci. Eng. 37 (4), 515–556.

Standeker, Suzana, Novak, Zoran, Knez, Zeljko, 2007. Adsorption of toxic organic compounds from water with hydrophobic silica aerogels. J. Colloid Interface Sci. 310 (2), 362–368.

Stephens, Robert D., Gross, Adam F., Van Atta, Sky L., Vajo, John J., Pinkerton, Frederick E., 2009. The kinetic enhancement of hydrogen cycling in $NaAlH_4$ by melt infusion into nanoporous carbon aerogel. Nanotechnology 20, 204018.

Tai, Yutaka, Tajiri, Koji, 2008. Preparation, thermal stability, and CO oxidation activity of highly loaded Au/titania-coated silica aerogel catalysts. Appl. Catalysis A: Gen. 342 (1-2), 113–118.

Tian, H.Y., Buckley, C.E., Wang, S.B., Zhou, M.F., 2009. Enhanced hydrogen storage capacity in carbon aerogels treated with KOH. Carbon 47 (8), 2128–2130.

Tomkiewicz, Micha, Dagan, Geula, Zhu, Zhu, 1994. Morphology and photocatalytic activity of TiO_2 Aerogels. Res. Chem. Intermediates 20 (7), 701–710.

Wei, Te-Yu, Lu, Shih-Yuan, Chang, Yu-Cheng, 2008. Transparent, hydrophobic composite aerogels with high mechanical strength and low high-temperature thermal conductivities. J. Phys. Chem. B 112 (38), 11881–11886.

Wei, Te-Yu, Lu, Shih-Yuan, Chang, Yu-Cheng, 2009. A new class of opacified monolithic aerogels of ultralow high-temperature thermal conductivities. J. Phys. Chem. C 113 (17), 7424–7428.

Willey, R.J., Lai, H., Peri, J.B., 1991. Investigation of iron oxide-chromia-alumina aerogels for the selective catalytic reduction of nitric oxide by ammonia. J. Catalysis 130 (2), 319–331.

Wojciechowska, M., Lomnicki, S., 1999. Nitrogen oxides removal by catalytic methods. Clean Technol. Environ. Policy 1 (4), 237–247.

Wu, Dingcai, Sun, Zhuoqi, Fu, Ruowen, 2006. Structure and adsorption properties of activated carbon aerogels. J. Appl. Polymer Sci. 99 (5), 2263–2267.

Wu, Xinbo, Wu, Dingcai, Fu, Ruowen, 2007. Studies on the adsorption of reactive brilliant red X-3B dye on organic and carbon aerogels. J. Hazard. Mat. 147 (3), 1028–1036.

Yang, Wenjin, Wu, Dingcai, Fu, Ruowen, 2007. Porous structure and liquid-phase adsorption properties of activated carbon aerogels. J. Appl. Polymer Sci. 106 (4), 2775–2779.

Yang, Wenjin, Wu, Dingcai, Fu, Ruowen, 2008. Effect of surface chemistry on the adsorption of basic dyes on carbon aerogels. Colloids Surfaces A: Physicochemical Eng Aspects 312 (2-3), 118–124.

Yim, Tae-Jin, Kim, Sun, Yoo, Ki-Pung, 2002. Fabrication and thermophysical characterization of nano-porous silica-polyurethane hybrid aerogel by sol-gel processing and supercritical solvent drying technique. Korean J. Chem. Eng. 19 (1), 159–166.

Yoda, Satoshi, Suh, Dong Jin, Sato, Tsugio, 2001. Adsorption and photocatalytic decomposition of benzene using silica-titania and titania aerogels: Effect of supercritical drying. J. Sol-Gel Sci. Technol. 22 (1), 75–81.

Yoldas, B.E., Annen, M.J., Bostaph, J., 2000. Chemical engineering of aerogel morphology formed under non-supercritical conditions for thermal insulation. Chem. Mater. 12 (8), 2475–2484.

Zhang, hu, Gross, Adam F., Van Atta, Sky L., Lopez, Maribel, Liu, Ping, Ahn, Channing C., Vajo, John J., Jensen, Craig M., 2009. The synthesis and hydrogen storage properties of a MgH_2 incorporated carbon aerogel scaffold. Nanotechnology 20, 204027–204032.

Zhang, Zhihua, Shen, Jun, Ni, Xingyuan, Li, Yang, Wang, Bo, Wu, Guangmin, Zhou, Bin, 2008. Benzene adsorption properties of silica aerogel-fiber composites. Paper presented at the 2nd IEEE International Nanoelectronics Conference (NEC), March 24–27, in Shanghai, China.

CHAPTER

Effective Utilization of Solid Waste from Leather Industry

23

N. Nishad Fathima, J. Raghava Rao, Balachandran Unni Nair

Chemical Laboratory, Central Leather Research Institute, Council of Scientific and Industrial Research, Adyar, Chennai, India

CHAPTER OUTLINE

The Role of Colloidal Systems in Environmental Protection. http://dx.doi.org/10.1016/B978-0-444-63283-8.00023-5

1 Origin of pollution from the leather industry: an overview

Leather making is several centuries' old technology, which attracts people even today because of the versatile properties leather possesses over a synthetic material. Leather is a unique material for several applications like footwear and garments, as it possesses good vapor permeability, strength, and thermal insulation because of its natural fiber matrix. However, the wastes generated from this industry make it fall in the "red" category. With present pollution control and treatment technologies, it is possible to reduce a considerable amount of liquid waste that is generated without causing serious damage to the environment. But the major problem the industry is facing is disposal of the solid wastes generated by the various operations extending from tanning to finished products. A part of these wastes such as trimmings splits and shavings are used in the manufacture of leather boards, while a major part of the wastes are dumped. The unavailability of dumping sites is now increasing the pressure on the tanning industries to come up with new ways to address these wastes.

2 Solid waste: new source for creating wealth

Industrial solid waste is an important part of the total waste problem. The amount of industrial solid waste is growing in many countries; therefore waste management and waste treatment methods are of utmost importance. Waste management is all about preventive methods, whereas waste treatment is about methods for end-of-pipe applications. Solid waste primarily falls into two categories: hazardous waste and nonhazardous waste. Nonhazardous waste consists of building and demolition waste (gloss, wood, tar, scrap, asbestos, etc.), flammable waste (office waste, active coal, etc.), nonflammable waste (sludge, shredder waste, salts, etc.), coating used in the metal industry, and asbestos. Hazardous waste consists of contaminated soil, mud, sludge, and residues from treated waste; as the hazardous waste poses a greater potential environmental threat, it is managed more strictly than other categories. Like municipal waste, most hazardous waste is managed in accordance with national governmental requirements. Requirements are set for issues such as operating permits for the treatment, storage, or disposal of hazardous waste. There are several groups of treatment methods: first, reuse of waste, second, methods of treatment, third, storage, and finally, disposal methods.

Many by-products from various industrial activities that have traditionally been treated as waste are now being viewed as "new raw materials." The solid wastes produced by different industries are being categorized according to its composition. Although various treatment methods are being practiced as per norms of the legislation, the disposal costs continue to rise. The following paragraphs provide insight into the currently practiced disposal methodologies for the problematic solid wastes (World Leather 2005).

3 Treatment of solid waste: state of the art

Treatment is any process that changes the physical, chemical, or biological character of a waste to make it less of an environmental threat. Treatment can neutralize the waste, recover energy from a waste, render the waste less hazardous, or make the waste safer to transport, store, or dispose of.

3.1 Incineration

Incineration is the high-temperature burning (rapid oxidation) of a waste. It is also known as controlled–flame combustion or calcination and is a technology that destroys organic constituents in waste materials. New techniques are developed for this burning process, used as energy-generating methods. Also, incineration is considered a volume-reducing step to decrease the cost of landfilling. The commonly employed methods of incineration include industrial furnaces and boilers. The industrial furnace is a type of hazardous waste treatment device that uses thermal energy to recover energy or materials from wastes. It includes cement kilns, lime kilns, aggregate kilns, phosphate kilns, coke ovens, blast furnaces, and smelting furnaces; a boiler is a type of device that can be used to treat hazardous waste. Boilers use controlled flame combustion and recover thermal energy in the form of steam or heated gases.

3.2 Gasification

Gasification provides a major route for treating solid wastes of many types. Often termed the ultimate solution, the system operates by an automatic feed of solid waste into a gasification unit. In general, briquettes of solid waste are fed into the top of the reactor and heated in the absence of air. This causes a breakdown of the organic matter (distillation) and the production of various combustible gases, mainly carbon monoxide, methane, and hydrogen.

As the solids, now in the form of a charcoal containing inorganic matter, pass down the reactor toward the base, they are finally combusted to produce an ash. Part of the heat from this stage is used for heating water or for fuel to generate electricity.

3.3 Biological treatment

This method usually involves treatment of waste by bacteria, fungi, or algae to remove and degrade the hazardous constituents. Composting and anaerobic digestion of organic waste, such as food waste, garden (yard), and park waste and sludge, is common in both developed and developing countries. The advantages of the biological treatment include reduced volume in the waste material, stabilization of the waste, destruction of pathogens in the waste material, and production of biogas for energy use. The end products of the biological treatment can, depending on its quality, be recycled as fertilizer and used for soil treatment. Composting is an aerobic process, and a large fraction of the degradable organic carbon (DOC) in the waste material is converted into carbon dioxide (CO_2). Anaerobic treatment is usually linked with methane (CH_4) recovery and combustion for energy. Mechanical-biological (MB) treatment of waste is becoming popular in Europe. In MB treatment, the waste material undergoes a series of mechanical and biological operations that aim to reduce the volume of the waste as well as stabilize it to reduce emission from the final disposal. The operations vary by application. Typically, the mechanical operations separate the waste material into fractions that will undergo further treatment (composting, anaerobic digestion, combustion, recycling). These may include separation, shredding, and crushing of a material. The biological operations include composting and anaerobic digestion (Beck-Friis 2001, Binner 2002, Hellebrand 1998).

3.4 Storage of solid waste

Storage is the holding of waste for a temporary period of time. At the end of the storage period, the waste is treated, disposed of, or stored elsewhere. Solid waste in many cases is temporarily stored at waste transfer stations. At the transfer station, waste is offloaded from the local collection routes and in some cases sorted according to type. The waste is then loaded onto larger trucks or rail cars for transport to either a municipal waste treatment or a disposal facility. Hazardous waste, in many cases, is stored prior to treatment or disposal. The most common hazardous waste storage practices are container storage, storage in tanks, and storage in containment buildings.

3.5 Deactivation

Deactivation is a process that removes the hazardous nature of the waste by neutralizing the characteristics of ignitability, corrosivity, and reactivity.

3.6 Micro-encapsulation

This technique coats the surface of the waste material with a thin layer of plastic or resin to prevent the material from leaching hazardous waste constituents.

3.7 Landfilling

Landfilling is a very common disposal method for solid waste. Landfilling used to be a fast, inexpensive, easy method for disposing of waste. The drawbacks of landfilling are becoming more and more obvious; think of environmental impacts, health risks, and landfill space. The solid waste often takes up valuable space in the landfills. The main problem associated with landfilling is the potential for leachate contamination of ground- and surface water. This has created governmental regulations requiring leachate control and groundwater monitoring. Disposal in areas such as wetlands, floodplains, permafrost zones, watersheds, and critical habitats should be prohibited. Nowadays both municipal and hazardous waste landfills usually have liner systems and leachate collection systems to prevent contamination of the groundwater under the landfill. Groundwater monitoring is also generally required, along with corrective action, if releases of hazardous pollutants occur. Once the landfills stop receiving waste, they are required to close by putting an impermeable cover on the landfill to prevent rainwater from entering.

3.8 Dumping

Dumping is an illegal method but unfortunately is practiced on a worldwide scale. Dumping can be done in an uncontrolled or a controlled manner. Uncontrolled dumping takes place through leakages. Otherwise, waste is dumped willfully. While costs for waste treatment like incinerations are rising, dumping practices occur more often too.

3.9 Land application

Another disposal option is land application. Land application involves applying the waste directly onto the surface or mixing it into the soil. The results of various studies on land application reflect positive gains in this type of disposal. The one problem with land application is its environmental and health impact.

3.10 Reuse of solid waste

In many cases solid waste, when collected separately, can be reused in the production process, for example, spillages. Other solid waste can be reused in another industry; for instance, slag (waste) from the metal industry can be used as input for concrete or other waste to be reused as agricultural fertilizer. Often reuse needs active management in order to develop alternatives for using solid waste. In many European countries, industries look for alternative uses for solid waste, in order to decrease the costs of disposal (which are rising heavily). Of course, there is also waste that cannot be reused again, for example, toxic waste. Again, waste management is important.

4 Significant characteristics of solid waste from tanneries

The most significant characteristics of solid waste from tanneries are

- The untanned collagenous wastes and fats (raw and soaked trimmings, often mixed with fleshing) produce odors. However, provided they are not heavily contaminated with process chemicals, they are certainly a most suitable material for agricultural fertilizers.
- The sulfide-bearing wastes (trimming and fleshing) from the limed hide and beam house residues and sludge may lead to malodorous hydrogen sulfide fumes; however, the concentration in this form is unlikely to be hazardous.
- The chrome-bearing wastes (shavings, buffing dust, and sludge from equalized effluents). Sludge from effluent treatment may contain up to 3.5% chromium, leather fragments (buffing dust, shaving, scrap splits, etc.), and 2.0–5.0% Cr_2O_3.

4.1 Fleshing: source and availability

Fleshing is an important by-product of tannery and is a good source of collagen protein. It is one of the wastes available in wet conditions and also in huge quantities. Fleshing is adipose tissue, which is loose connective tissue dominated by fat cells, or adiposities, and is removed either manually or mechanically from plumped hide or skin after the liming process. The fleshing operation results in a clean, uniform surface. The main constituents of fleshing are provided in Table 23.1. The flesh

Table 23.1 Composition of Fleshing

Component	Fleshing	
	Kg	%
Water	240	80
Collagen	24	8
Salts	24	8
Fats	12	4
Total	300	100

Source: United Nations Industrial Development Organization (UNIDO).

removed from the hide prior to unhairing and liming (green fleshing), particularly if the hides have a large amount of tissue, is less damaging to the environment. The fleshing in that case is not contaminated by sodium sulfide and lime, and it can be made into saleable and edible products at a rendering plant. Limed fleshing can be processed into glue. In India, a major recycling plant now dries and pulverizes both raw and limed fleshing for use as a high-protein supplement for chicken feed. In Europe, fleshing is being compressed under high pressure for grease extraction. This can be purified for tallow and can also be used as boiler fuel due to its high calorific value. Hair and fleshing have also been used for the production of bio-gas; both hair and fleshing have boosted gas production. The gas is then burned to provide hot water for heating greenhouses or to run generators with electricity sales to the national grid. Solid residues in a coarse, dry powder form are used for ground conditioning, and the liquid can be applied directly as a nitrogen-high fertilizer.

4.1.1 Preparation of feed ingredient from green fleshing

Green fleshing comprises protein and fatty tissues. They are also mixed with soft offals like lungs, stomach meat, and condemned organs obtained from slaughterhouses and then transferred to rendering equipment. The proteinous wastes get coagulated, and the fatty tissues arw split into fats above 60°C. The fat melts and the proteinous tissues get cooked. The total mass in the form of slurry is loaded into a centrifuge to separate tallow. The protein is completely dried and pulverized into a protein meal. The product is termed meat and bone meal.

In subtropical countries, the green fleshing contains 10–15% fat, the rest being collagen protein. The trend to install rendering plants has been facilitated because of the high return obtained through tallow. Some tanneries sell green fleshing to rendering plants.

4.1.2 Preparation of hydrolysates from fleshing

The microbiological method for utilization of green and limed fleshings for the production of value-added products such as enzymes has been studied extensively (Kumar et al. 2008, Taylor et al. 2006). Delimed fleshings, mixed with *Pediococcus acidolactici* culture, sugar, and common salt, on fermentation yield silage, which can be used in animal feeds. This can also be used in aquaculture and animal feeds to supplement arginine, phenylalanine, and tyrosine (Bhaskar et al. 2007). Hydrolysis of fleshing with alkaline proteinase resulted in the isolation of three fractions: fat—cattle tallow, collagen hydrolysate—glue, and protein concentrate for fodder. Further purification of the collagen hydrolysate fraction into edible gelatin was achieved (Simeonova et al. 1996).

4.1.3 Biomethanation and production of biodiesel of fleshing

Biomethanation of fleshing yields biogas, which can be used for various purposes and is also economical (Aloy 1987). The efficiency of liquefied tannery fleshing and sludge in combination with cow dung was studied for biogas production and found to be effective (Vasudevan et al. 2007). The possible use of fleshing fat in the production of biodiesel (fatty-acid methyl esters) by trans-esterfication for use as a replacement for fossil fuels was investigated (Colak et al. 2005). Fleshing with a high-fat content, fluidized by heating, was treated with sulphuric acid, and the fats were extracted with biodiesel (biofuel based on fatty-acid methyl esters). The fats were converted by the catalytic transesterification process into fatty-acid methyl esters after suitable purifying, as biodiesel or heating biofuels (Stepan et al. 2009).

4.2 The drawbacks of present disposal methods

The handling and any recovery of products contained in limed fleshing in the tannery have long been a serious expense. Moving this material for eventual dumping incurs labor costs, even when the specialized equipment developed for the purpose is employed. Another factor against dumping is the restrictions in force, with more coming into force, regarding the disposal of organic materials. Although various systems of recovering fat in the form of tallow and the protein contained in the resulting sludge have been adopted, most of them require prodigious amounts of energy for their operation, making the process, at best, marginally economic. Hence, finding an effective and economical solution for disposal of fleshing is the need of the hour.

5 Chromium—an insight into the metal and its toxicological evidences

Chromium is a vital element found in the Earth's crust that finds its way into many materials and products used in industrialized countries. Important uses of chromium are given in Table 23.2 (Petura et al. 1998).

Discharge of chromium containing tannery wastes is of concern to the environment due to the toxic nature of certain chromium(III) salts (Rajaram et al. 1995, Ramasami 1996). Toxicity of chromium(III)

Table 23.2 Various Uses of Chromium (Petura et al. 1998)

Various Uses of Chromium	%
Stainless steel	10–18
Wood preservation	38
Metal finishing, corrosion resistant	15
Leather tanning	10
Pigment	8
Oil drilling mud	4
Textile mordant	3
Magnetic tapes	2
Others	2–4

ion has been the subject of much debate (Bianchi et al. 1983, Snow 1991). Although the implication of chromium(III) in glucose and lipid metabolism has been considered beneficial (Jeejeebhoy et al. 1977, Anderson et al. 1997, Mertz 1998), epidemiological and animal studies have firmly established hexavalent chromium compounds as potent carcinogens (Costa 1991). Chromium(III) species have been shown to be mutagenic and genotoxic (Ke and Costa 2007). Although trivalent chromium compounds are kinetically inert to ligand substitution reactions, they do react strongly with DNA and proteins (Levina et al. 2003, Zhitkovich 2005). Trivalent chromium leads to the formation of different chromium–DNA adducts such as DNA protein, interstrand DNA, and DNA-amino acid cross links (Zhitkovich et al. 2000, 2005, O'Brien et al. 2003). Salen-based chromium(III) complexes have been found to induce apoptosis in lymphocyte cell cultures (Rajaram et al. 1995).

Chromium(III) complexes in specific ligand environments have been proposed to bring about oxidative damage to plasma proteins, glycoproteins, and DNA (Voitkun et al. 1998, Vijayalakshmi et al. 2000, Shrivastava and Nair 2000, 2001). The effects of soil pollution by the metal ion are observed when soil contains large amounts of soluble and biologically assimilatable forms of chromium (Bartlett 1991).

Exposure to Cr(III) (as a mixture of Cr(III) aqua and hydroxo complexes) in leather tannery workers led to a pattern of DNA damage in lymphocytes, including DNA-protein cross links and an increased incidence of micronuclei, which was similar to genotoxicity markers found in stainless steel welders exposed to Cr(VI)-containing fumes (Medeiros et al. 2003, Levina et al. 2007). These results point to a similarity in the mechanisms of DNA damage, promoted by either Cr(III) or Cr(VI) compounds. Chronic use of Cr picolinate has been reported to cause nephrotoxicity in humans if ingested in excess (Cerulli et al. 1998). In leather processing, the effect of heat on the formation of Cr(VI) has been studied (Fathima et al. 2001).

5.1 Present disposal methods for removal of chromium(VI)

Classical treatment technologies for the removal of heavy metals include the (i) chemical precipitation (addition of chemicals, e.g., caustic soda, sulfide), (ii) ion exchange, (iii) electrodialysis, and (iv) adsorption (Rich and Cherry 1982, Fathima et al. 2006). However, there are significant disadvantages, such as incomplete metal removal and high-energy requirements associated with these methods (Trujillo et al. 1991). The advantages and disadvantages of each method are presented in Table 23.3.

5.1.1 Chemical precipitation

The recovery of metals as hydroxides and their reuse is one of the methods for combating metal pollution (Prasad and Nair 1994). The widely practiced method of removal of chromium(VI) from wastewaters is by reduction followed by hydroxide precipitation and sedimentation. In general, the traditional reduction–precipitation treatment has limitations such as difficulty in complying with the discharge limits through this method, especially when inorganic and organic ligands are present in the wastewater as the residual chromium(III) concentration in the treated water tends to be significantly high. Also, the land disposal of chromium hydroxide and sulfide sludge is expensive.

5.1.2 Ion exchange

The ion-exchange process has been a strong alternative to precipitation for the removal of Cr(VI) ions because of its ability to reduce dissolved chromate concentration to almost nondetectable levels in the

Table 23.3 Advantages and Disadvantages of Conventional Metal Removal Technologies from Wastewaters (Fathima et al. 2006)

Method	Advantages	Disadvantages
Chemical precipitation	Simple	Difficulty in separation
	Relatively cheap	Disposal of resulting toxic sludge
		Not very effective
		(iv) Metals held in solution by complexing agents (e.g., cyanide or EDTA) are difficult to precipitate
		(v) The presence of multiple metal species may lead to removal difficulties as a result of amphoteric nature of different compounds
		(vi) Process can be costly, depending on reagents used, required system controls, and required operator involvement in system operation
Electrochemical treatment	Metal recovery possible	Applicable for high metal concentrations
		Sensitive to specific conditions, such as the presence of certain interfering compounds
Reverse osmosis	Pure effluent/permeate (available for recycling)	Membrane scaling/fouling
	Bacteria and other particles are removed well	Expensive
	Low effluent concentrations possible	Reject stream is 20–90% of feed flow
	(iv) Small floor space requirements	(iv) Limited concentration range of operation
	v) Less energy intensive than evaporation	(v) Membrane sensitive to pH
	(vi) Can return valuable additives	
Ion exchange	Effective	Sensitive to the presence of particles
	Possible metal recovery	Expensive resins
	Relatively insensitive to flow variations	Usually not feasible at high level of TDS
	(iv) Low-energy demands	(iv) Requires tight operation and maintenance—equipment complex
	(v) Returns metal as metal salt solution	(v) Limited concentration ability
	(vi) Handles dilute feed	(vi) May require evaporation to increase concentration
		(vii) Excess regenerate required: 3-5X stoichiometric quantity, therefore waste
		(viii) Feed concentration must be closely monitored

Continued

Table 23.3 Advantages and Disadvantages of Conventional Metal Removal Technologies from Wastewaters (Fathima et al. 2006)—cont'd

Method	Advantages	Disadvantages
Adsorption	Conventional sorbents	Not very effective for certain metals
	Low effluent contaminant level possible	Slow adsorption kinetics
		Spent regenerant must be disposed off
Electrodialysis	(i) Achieves higher concentration than reverse osmosis or ion exchange	(i) Feed must be filtered
	(ii) Energy efficient	(ii) Membrane sensitive to flow distribution, Ph, and suspended solids
	(iii) Organics not concentrated	(iii) Equipment uses multicell stacks, incurs leakage
	(iv) Inorganic salts transport at different rates minimizes return of unwanted inorganics	(iv) Chemical adjustment of return material
		(v) New technology—membrane life uncertain

treated water. Several studies have been carried out to study the potential application of anion exchange processes for chromate's selective removal and possible recovery of chromate-laden wastewaters (Newman and Reed 1980, Bolto and Pawlowski 1987, Sengupta et al 1988). Loss in anion exchange capacities due to oxidative action of Cr(VI) with an increased number of cycles was a major problem, which has been resolved with a more chemically and thermally stable anion exchangers. In the presence of competing sulfate and chloride anions, anion exchange resins offer much greater chromate removal capacity at acidic pH compared to alkaline pH.

5.1.3 Electrodialysis

The possibility of selectively removing neutral salts contained in spent chromium tanning solutions has been studied by the electrodialysis technique (Rao et al. 1989). The effluent from rinsing stainless steel, which had been etched in nitric acid and hydrofluoric acid, was studied using electrodialysis and ion exchange (Wisniewski and Suder 1995). Electrodialysis has both advantages and disadvantages. For instance, the process requires very little energy and can recover highly concentrated solutions. Also, the equipment is available as a preassembled unit. On the other hand, electrodialysis membranes are susceptible to fouling and must be regularly replaced.

5.1.4 Adsorption/biosorption

Various studies have been published documenting the adsorption of chromium on activated carbon (Bansal et al. 1979, Leyva-Ramos et al. 1995). The high cost of activated carbon has stimulated interest in examining the feasibility of using cheaper raw materials. The metal-sequestering ability of microorganisms, such as yeast, bacteria, and fungi, that are generated in large quantities as waste by-products from fermentation industries have been investigated and reported (Volesky and Holan 1995). The use of low-cost natural resources for the removal of chromium is actively pursued by researches in preference to the other conventional methods. The applications of water hyacinth weeds (Singaram 1994), treated sawdust (Daiyan et al. 1995), barks of *Acacia arabica*, and *Artocarpus heterophyllus* (Swami et al. 2002), brown seaweed biomass (David et al. 1998, Yun et al. 2001), and coconut shell (Sivasamy et al. 2000) as materials for removal of chromium from the synthetic and industrial wastewater have been reported.

6 Dye: a ubiquitous pollutant

Dye pollution is a serious environmental issue. Dyes, often toxic, will cause biological magnification in marine or aquatic organisms when released, disrupting the food chain and ecology of habitats.

Many industries generate a considerable amount of colored wastewater. Color is the first contaminant to be recognized in wastewater (Banat et al. 1996). The presence of very small amounts of dyes in water (less than 1 ppm for some dyes) is highly visible and undesirable (Robinson et al. 2001).

Over 100,000 commercially available dyes exist, and more than 7×10^5 tons per year are produced annually (McMullan et al. 2001). Due to their good solubility, synthetic dyes are common water pollutants, and they may frequently be found in trace quantities in industrial wastewater. An indication of the scale of the problem is given by the fact that 2% of dyes that are produced are discharged directly in aqueous effluent (Guthrie et al. 2003, Robinson et al. 2001). Many of these dyes are also toxic and even carcinogenic, and this poses a serious hazard to aquatic living organisms (Delée et al. 1999,

Bianchi et al. 1998). However, wastewater-containing dyes is very difficult to treat, since the dyes are recalcitrant organic molecules, resistant to aerobic digestion, and are stable to light, heat, and oxidizing agents (Sun and Yang 2003, Bhavani et al. 1998).

The presence of dyes in water reduces light penetration and has a derogatory effect on photosynthesis. Dyes may also be problematic if broken down anaerobically in the sediment leading to the production of toxic amines. The effluents from the dyestuff manufacturing and some similar industries are generally highly colored with a large amount of suspended organic solids and hence are the important sources of water pollution (Gregory 2007). From an environmental point of view, the removal of synthetic dye is of great concern, since some dyes and their degradation products may be carcinogens and toxic and consequently their treatment cannot depend on biodegradation alone (Pagga and Braun 1986, Reife 1993).

Many studies have been conducted on the toxicity of dyes and their impact on the ecosystem (Shenair 1996, Hunger 2003), as well as the environmental issues associated with the manufacturing and subsequent usage of dyes (Saha 1996, Phillips 1996). The disposal of colored wastewater without sufficient treatment is toxic to aquatic life (Heuper 1969). They also pose a problem because of their carcinogenicity and toxicity (Hunger 2003).

6.1 Present disposal methods for dye removal

Various physical, chemical, and biological pre- and posttreatments can be employed to remove color from dye containing wastewater (Sandhya et al. 2005, Fathima et al. 2008). Table 23.4 provides a list of processes employed for decolorization.

6.1.1 Physical methods

Methods that do not require the use of chemicals for treatment, for example, membrane-filtration techniques (ultrafiltration, nanofiltration, and reverse osmosis) have been used for water reuse and

Table 23.4 Technologies Used for Decolorization

Methods	Processes	Materials Used	References
Physical	Adsorption	Activated carbon and low-cost materials	McKay 1982; Singh and Srivastava 2001
	Irradiation	Gamma rays and electron beams	Hosono et al. 1993
	Membrane filtration	Nanofiltration membranes	Chakraborty et al. 2003
Chemical and Electrochemical methods	Coagulation/precipitation	Iron/alumina/lime or without lime	Lin and Liu 1994, Lin and Chen 1997
	Ozonation	Cl_2, NaOCl, Ozone	Muthukumar et al. 2001
	Fenton oxidation	High temperature, Iron/H_2O_2	Hassan and Hawkyard 2002
	Ion exchange	Ion exchange resins	Raghu and Basha 2007
	Oxidation/reduction	Electrocoagulation, electrooxidation, electroflotation	Gutiérrez and Crespi 1999
Photocatalytic	Advanced oxidation	UV/Solar/H_2O_2/TiO_2	Kang et al. 2000
Biological	Aerobic/anaerobic	Activated sludge, biodegradation	Banat et al. 1996

chemical recovery. The major disadvantage of the membrane processes is that they have a limited lifetime due to membrane fouling and the cost of periodic replacement must thus be included in any analysis of their economic viability (Crossley 2002). Adsorption of synthetic dyes on inexpensive and efficient solid supports was considered as a simple and economical method for their removal from water and wastewater (Slokar and le Marechal 1998). The excellent adsorption properties of carbon-based supports have been exploited for the decolorization of dyes in the industrial effluents (Dussoubs et al. 1991, Low et al. 1995, Danis et al. 1999). Adsorption capacity of silica and alumina has been exploited for removal color from wastewater (Ahmed and Ram 1992, Saleem et al. 1994, 2007).

In recent decades, researchers have focused their attention on alternative low-cost materials. Various low-cost sorbents such as peat, charred/carbonated rice husk, coal, charcoal, fly ash (Khare et al. 1987, Singh and Rawat 1994), wood, silica gel, clay materials (Theng and Wells 1995, Juang et al. 1997), and agriculture wastes have been tried with varying success for color removal. The adsorbents used include alginate beads (Aravindhan et al. 2006, 2007), sawdust (Asfour et al. 1985), anaerobic digester slurry (Namasivayam and Yamuna 1994), paddy straw (Deo and Ali 1993), and industrial solid waste such as bentonite clay (Anirudhan and Ramachandran 2006) and Fe(III)/Cr(III) hydroxide (Namasivayam et al. 1994). The disadvantages of the adsorption process are that the adsorbent needs to be regenerated, which adds to the cost of process and is sometimes a very-time consuming procedure.

6.1.2 Chemical and electrochemical methods

Chemical methods used for discoloration of wastewater include reduction, oxidation, complexometry, ion exchange, and neutralization. Coagulation or flocculation combined with flotation and filtration, precipitation-flocculation with Fe(II)/$Ca(OH)_2$, electroflotation, electrokinetic coagulation, conventional oxidation methods by oxidizing agents (ozone), and irradiation have been demonstrated (Forgacs et al. 2004). The electrochemical process for the decolorization of wastewater is also followed (Muthukumar et al. 2007). Other emerging techniques, known as advanced oxidation processes (AOPs), which are based on the generation of very powerful oxidizing agents such as hydroxyl radicals, have been applied for dye degradation (Haag and Hoigne 1983, Slokar and le Marechal 1998, Vonguriten and Oliveras 1997, Aravindhan et al. 2006, Fathima et al. 2008, Zou and Zhu 2008). Ion exchange is yet another chemical process available to decolorize the colored wastewater containing ionic dye stuffs (Liu et al. 2007, Wu et al. 2008). Although these methods are efficient for the treatment of water contaminated with pollutants, they are very costly and commercially unattractive.

6.1.3 Biological treatment method

Biodegradation methods such as fungal decolorization, microbial degradation, adsorption by (living or dead) microbial biomass, and bioremediation systems are commonly applied to the treatment of industrial effluents because many microorganisms such as bacteria, yeasts, algae, and fungi are able to accumulate and degrade different pollutants (Banat et al. 1996, McMullan et al. 2001, Fu and Viraraghavan 2001, Banat et al., 1996). However, their application is often restricted because of technical constraints such as operational parameters, the composition of industrial wastewater, and the structure and substituents of dye molecule. Although many organic molecules are degraded, many others are recalcitrant due to their complex chemical structure and synthetic organic origin (Banat et al. 1996).

7 Solid wastes to remove toxic liquid wastes: new paradigm

The solid waste emanating from tanneries is a major concern among tanners and environmentalists alike, as scarcity of land is a major constraint of the land disposal method. Also, these solid wastes are rich in protein matter, which also have economical value. Hence, evaluation of the existing practices for waste disposal and development of more effective alternatives to reuse and final disposal of solid waste are the need of the hour. Recently, attempts have been made to minimize the adverse environmental affects caused by the indiscriminate disposal of solid wastes, especially hazardous waste like lime fleshing (containing sulfides) and to maximize the returns from solid waste by investigating the feasibility of removal of toxic liquid wastes using solid waste.

Thus, this study explores the possibility of a holistic solution to both the solid waste and the liquid waste problem. Also, the pollutant-loaded fleshing has been used in the manufacture of a tanning salt, thereby providing complete solution to the above-mentioned problems.

8 Removal of chromium(VI) using fleshing

The removal of toxic Cr(VI) using iron-treated fleshing has been attempted (Fathima et al. 2005). For the removal of chromium(VI), the adsorbent has to be modified so as to enable it to bind with the anionic chromium(VI). The modification of sorbent by the grafting of ligands has been proposed as a way to improve sorption properties: For example, Denizli et al. (1997) incorporate a dye-ligand onto synthetic polymers to enhance its metal ion-sorption properties. Such metal-treatment can proceed by adsorption and/or precipitation steps of the corresponding salts into the porous structure of the adsorbent. Fleshing contains proteins such as collagen, which have carboxyl groups that can coordinately complex with metal ions such as Fe^{3+} to give a cationic Fe(III) complex (Ramos et al. 1994, Raji and Anirudhan 1998). This cationic complex can bind to the anionic $Cr_2O_7^{2-}$ (Eq. 23.1). To establish the role of iron, the Cr(VI) removal experiments were carried out with both raw and iron-treated fleshing with an initial Cr(VI) concentration of 50 mg/L. It was observed that the raw fleshing has 8% as against the 98% Cr(VI) removal efficiency of the iron-treated fleshing after a treatment time of 5 h. This confirms that it is the cationic Fe^{3+} complex that binds the anionic Cr(VI).

$$2[R - Fe]^{n+}[Cl^-]_n + n[K_2^+][Cr_2O_7]^{2-} \rightleftarrows [R - Fe]_2^{n+}[Cr_2O_7]_n^{2-} + 2nKC1 \\ R \; - \; \text{protein carboxyl group} \tag{23.1}$$

The effect of pH and the initial concentration of chromium(VI) on the removal of Cr(IV) by iron-treated fleshing have also been carried out. The ultimate adsorption capacity of iron-treated fleshing is 51 mg of chromium(VI) per gram of fleshing. The capacity of untreated fleshing is 9 mg/g, such that iron treatment increases the adsorption capacity of fleshing by 10-fold. The measured adsorption kinetics is well described by a pseudo-second-order kinetic model. The uptake of chromium(VI) by fleshing is best described by the Langmuir adsorption isotherm model. X-ray photoelectron spectroscopic (XPS) studies show that the iron is incorporated into the protein matrix. Shifts in XPS spectra suggest that dichromate binding occurs with iron at active adsorption sites and that iron-treated fleshing removes chromium(VI) without reducing it to chromium(III).

9 Removal of dye using fleshing

Removal of dye using fleshing has been attempted recently (Fathima et al., 2012). Since fleshing is a biodegradable material, glutaraldehyde cross linking has been done to provide stability to the adsorbent. The various parameters such as effect of pH, role of salt concentration, adsorbent dosage, and initial concentration of dye have been standardized. About 90% dye removal has been achieved by treating 100 mg/L dye solution with 12 g/L of glutaraldehyde crosslinked fleshing. The dye removal has been found to follow the Langmuir type of adsorption isotherm. In order to study the potential of the prepared adsorbent on the removal of dyes from real-world wastewaters, commercial studies were carried out. These studies are essential as they may throw light on the feasibility of the adsorbent to perform in conditions where other chemicals (syntans and fatliquors used in leather manufacture) present could compete for the same available sites. Wastewater from a sectional stream employing the same dye as is used for batch experiments was collected from a tannery. The optimized parameters from batch experiments were used for the commercial wastewater. From the results, it could be seen that the % reduction of dye and the dye uptake capacity of the adsorbent have not been reduced considerably. The % reduction of dye from the commercial wastewater was found to be 91% as compared to 95% dye reduction in the case of synthetic dye solution under optimum conditions. The slight reduction in the uptake could be attributed to the presence of neutral salts and other constituents in the commercial wastewater. Also, the uptake of dye from the commercial liquor was found to be 33 mg/g for 12 g/L of the adsorbent.

10 Manufacture of tanning salt from pollutant-loaded solid wastes: a holistic solution

The above-mentioned studies prove the efficiency of fleshing as a potential adsorbent for both chromium(VI) and dye removal from wastewaters. However, the disposal of pollutant-loaded fleshing poses a similar solid waste disposal challenge. Hence, unless a solution to the disposal of pollutant-loaded fleshing is presented, this method will not be practicable. The dye-loaded adsorbent has been used as a reducing agent in the manufacture of a tanning salt, basic chromium sulfate (BCS), which is the most widely used tanning agent in leather industry. The scheme showing the utilization of raw fleshing for the adsorption of dye from aqueous solution and using the same for the preparation of basic chromium sulfate is depicted in Figure 23.1. The chrome content of the fleshing reduced BCS was estimated to be 21%. The shrinkage temperature of leathers is found to be above 110°C. The chromic oxide content in the wet blue leathers is in the range of $2.85 \pm 0.5\%$. The color of the wet blue made from fleshing reduced BCS is comparable to commercial tanned leathers. This shows that the pollutant-loaded adsorbent could be effectively reused in the same industry.

11 Conclusions

The present review explored the potential of cross-linked fleshing as an adsorbent for removal of dyes and chromium(VI) from wastewaters. Various parameters required for maximum removal efficiency have been described. The efficiency of fleshing for dye removal in real-life conditions has been tested

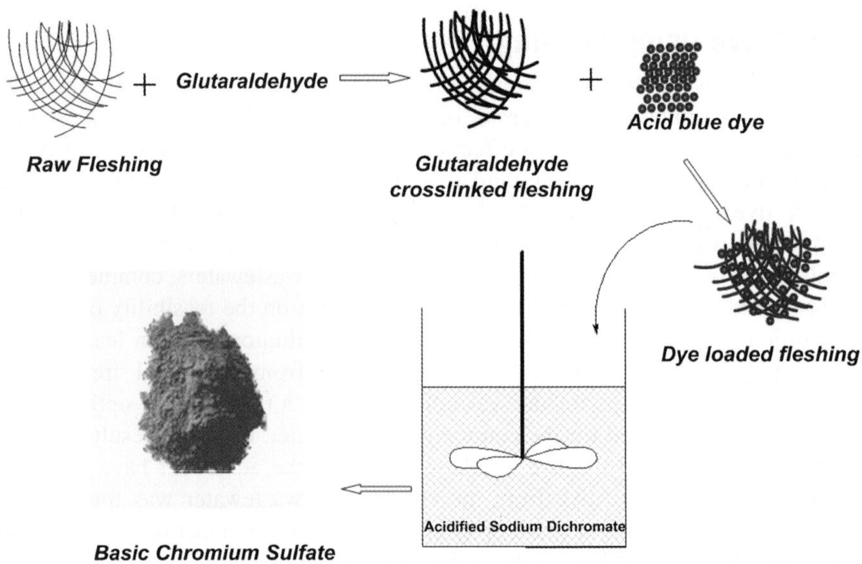

FIGURE 23.1

Scheme showing various steps involved in the use of raw fleshing for the removal of acid blue dye from aqueous solution and the utilization of same for the preparation of basic chromium sulfate. (For color version of this figure, the reader is referred to the online version of this book.)

by carrying out trials on commercial wastewaters. Also, the disposal problem related to the dye-loaded adsorbent has been tackled by using it as a reductant in the manufacture of a tanning salt. Thus, cross-linked fleshing can be used judiciously as a potential adsorbent for removal of toxic dyes and chromium(VI).

SYMBOLS AND TERMINOLOGY

BCS basic chromium sulfate
MB mechanical-biological treatment
DOC degradable organic carbon
DNA deoxy ribonucleic acid
XPS X-ray photoelectron spectroscopy

Glossary

Basic chromium sulfate (BCS) salt used for tanning
Fleshing inside part of the skin composed of proteins and fats; is removed during leather processing
Green fleshing flesh removed before "liming" operation during leather processing
Hide skin of a large animal, such as cow or buffalo
Langmuir isotherm monolayer adsorption of pollutant on the adsorbent

Leather a general term for hide or skin with its original fibrous structure more or-less intact, and that has been treated so as to be imputrecible. The hair or wool might or might not have been removed

Shrinkage temperature temperature at which the leather shrinks to one-third its original length

Skin skin of smaller animals, such as goat or sheep

Tannery place where leather processing is done

Wet blue chrome-tanned leather in wet condition

References

Ahmed, M.N., Ram, R.N., 1992. Removal of basic dye from wastewater using silica as adsorbent. Environ. Pollut. 77, 79–87.

Aloy, M., 1987. Methane gas from tannery effluent. Leather 189, 53–54.

Anderson, R.A., Cheng, N.Z., Bryden, N.A., Polansky, M.M., Cheng, N.P., Chi, J.M., 1997. Elevated intakes of supplemental chromium improve glucose and insulin variables in individuals with type 2 diabetes. Diabetes 46, 1786–1791.

Anirudhan, T.S., Ramachandran, M., 2006. Adsorptive removal of tannin from aqueous solutions by cationic surfactant-modified bentonite clay. J. Colloid Interface Sci. 299, 116–124.

Aravindhan, R., Fathima, N.N., Rao, J.R., Nair, B.U., 2006. Utilization of calcium alginate beads as adsorbent for removal of dyes from tannery wastewaters. J. Am. Leather Chem. Ass. 101, 223–230.

Aravindhan, R., Fathima, N.N., Rao, J.R., Nair, B.U., 2006. Wet oxidation of acid brown dye by hydrogen peroxide using heterogeneous catalyst Mn-salen-Y zeolite: A potential catalyst. J. Hazard. Mater. 138, 152–159.

Aravindhan, R., Fathima, N.N., Rao, J.R., Nair, B.U., 2007. Equilibrium and thermodynamic studies on the removal of basic black dye using calcium alginate beads. Coll. Surf. A. Physicochem. Eng. Aspects 299, 232–238.

Asfour, H.M., Nassar, M.M., Fadali, O.A., el Guendi, M.S., 1985. Colour removal from textile effluents using hardwood sawdust as an adsorbent. J. Chem. Technol. Biotechnol. 35, 28–35.

Banat, I.M., Nigam, P., Singh, D., Marchant, R., 1996. Microbial decolorization of textile-dye-containing effluents: A review. Bioresour. Technol. 58, 217–227.

Bansal, R.C., Donnet, J.B., Stoeckli, F., 1979. Active carbon. Marcel Dekker, New York, pp. 389–391.

Bartlett, R.J., 1991. Chromium cycling in soils and water: Links, gaps and methods. Environ. Health. Perspect. 92, 17–24.

Beck-Friis, B.G., 2001. Emissions of ammonia, nitrous oxide and methane during composting of organic household waste. Swedish University of Agricultural Sciences, Uppsala, 331 p. (doctoral thesis).

Bhaskar, N., Sakhare, P.Z., Suresh, P.V., Gowda, L.R., Mahendrakar, N.S., 2007. Biostabilization and preparation of protein hydrolysates from delimed leather fleshings. J. Sci. Ind. Res. 66, 1054–1063.

Bhavani, K.D., Dutta, P.K., Ravi Kumar, M.N.V., Sridhari, T.R., 1998. Trends in color removal from textile mill effluents. Colorage 40, 25–34.

Bianchi, R., Vandevivere, P.C., Verstraete, W., 1998. Treatment and reuse of wastewater from the textile wet-processing industry: review of emerging technologies. J. Chem. Technol. Biotechnol. 72, 289–302.

Bianchi, V., Celotti, L., Lanfranchi, G., Majone, F., Marin, G., Montaldi, Sponza, G., Tamino, G., Venier, P., Zantedeschi, A., Levis, A.G., 1983. Genetic effects of chromium compounds. Mutat. Res. 117, 279–300.

Binner, E., 2002. The impact of mechanical-biological pretreatment on the landfill behaviour of solid wastes. Workshop Bio Wastes. Brussels, 8-10.04.2002. In: Brandrup, J., Immergut, E.H., Grulke, E.A. (Eds.). John Wiley& Sons, New York, p. 16.

Bolto, B.A., Pawlowski, L., 1987. Wastewater treatment by ion exchange. Spon Publ. London, 57–62.

Cerulli, J., Grabe, D.W., Gauthier, I., Malone, M., McGoldrick, M.D., 1998. Chromium picolinate toxicity. Ann. Pharmacother. 32, 428–431.

Colak, S., Zengin, G., Ozgunay, H., Sari, O., Sarikahya, H., Yuceer, L., 2005. Utilization of leather industry prefleshings in biodiesel production. J. Am. Leather Chem. Assoc. 100, 137–141.

Costa, M., 1991. DNA-protein complexes induced by chromate and other carcinogens. Environ. Health Perspect. 92, 45–52.

Crossley, C., 2002. How the dye industry is benefiting from membrane technology. Filtr. Separ. 39, 36–38.

Daiyan, S.K.A., Basu, J.K., Biswas, A.K., 1995. An effective adsorbent for the removal of hexavalent chromium from aqueous solution. Chem. Environ. Res. 4, 261–267.

Danis, U., Gurses, A., Canpolat, N., 1999. Removal of some azo dyes from wastewater using PAC as adsorbent. Fresen. Environ. Bull 8, 358–365.

David, K., Pinhentel, P., Volesky, B., 1998. Removal of trivalent and hexavalent chromium by seaweed biosorbent. Environ. Sci. Technol. 32, 2693–2698.

Delée, W., Hawkes, F.R., O'Neill, Hawkes D.L., Lourenco, N.D., Pinheiro, H.M., 1999. Colour in textile effluents—sources, measurement, discharge consents and simulation: A review. J. Chem. Technol. Biotechnol. 74, 1009–1018.

Denizli, A., Salih, B., Arica, M.Y., Kesenci, K., Hasirici, V., Piskin, E., 1997. Cibacron blue F3GA-incorporated macroporous poly(2-hydroxyethyl methacrylate) affinity membranes for heavy metal removal. J. Chromatogr. A. 758, 217–226.

Deo, N., Ali, M., 1993. Dye adsorption by a new low cost material congo red-1. Indian J. Environ. Prot. 13, 496–508.

Dussoubs, M.D., Mazet, M., Pronost, J., 1991. Removal of dyestuffs by wood charcoal. Environ. Technol. 12, 625–634.

Fathima, N.N., Aravindhan, R., Rao, J.R., Nair, B.U., 2005. Solid waste removes toxic liquid waste: Adsorption of chromium(VI) by iron complexed protein waste. Environ. Sci. Technol. 39, 2804–2810.

Fathima, N.N., Aravindhan, R., Rao, J.R., Nair, B.U., 2008. Dye house wastewater treatment through advanced oxidation process using Cu-exchanged Y zeolite: A heterogeneous catalytic approach. Chemosphere 70, 1146–1151.

Fathima, N.N., Aravindhan, R., Rao, J.R., Nair, B.U., 2008. Dye containing wastewaters: Sources, concerns and mitigation. In: Trivedi, P.C. (Ed.), Industrial Pollution and Bioremediation, 2008 Aavishkar Publications, Jaipur, pp. 56–75.

Fathima, N.N., Rao, J.R., Nair, B.U., June 2012. Tannery solid waste to treat toxic liquid wastes: A new holistic paradigm. Environ. Eng. Sci. 29, 363–372. http://dx.doi.org/10.1089/ees.2010.0445.

Fathima, N.N., Rao, J.R., Nair, B.U., 2001. Chromium(VI) formation: Thermal studies on chrome salt and chrome tanned hide powder. J. Amer. Leather Chem. Ass. 96, 444–449.

Fathima, N.N., Rao, J.R., Nair, B.U., 2006. Management of chromium in industrial wastewaters. In: Trivedi, P.C. (Ed.), Industrial Pollution and Its Management, 2006 Aavishkar Publications, Jaipur, pp. 1–32.

Forgacs, E., Cserhati, T., Oros, G., 2004. Removal of synthetic dyes from wastewaters: a review. Environ. Int. 30, 953–971.

Fu, Y., Viraraghavan, T., 2001. Fungal decolorization of dye wastewaters: A review. Bioresour. Technol. 79, 251–262.

Gregory, P., 2007. Toxicology of textile dyes: Environmental aspects of textile dyeing (ed.) Christie R.M. Woodhead Publishing UK, Woodhead Textiles Series No. 66, pp. 44–73.

Guthrie, J.T., Lloyd, J.R., Pearce, C.I., 2003. Removal of colour from textile wastewater using whole bacterial cells: A review. Dyes Pigments 58, 179–196.

Haag, W.R., Hoigne, J., 1983. Ozonation of bromide-containing waters: Kinetics of formation of hypobromous acid and bromate. Environ. Sci. Technol. 17, 261–267.

Hellebrand, H.J., 1998. Emissions of nitrous oxide and other trace gases during composting of grass and green waste. J. Agr. Eng. Res. 69, 365–375.

Heuper, W.C., 1969. Occupational and environmental cancers of the urinary system. Yale University Press, New Haven, CT, p. 216.

Hunger, K., 2003. Health and safety aspects, Industrial dyes: Chemistry, properties, applications. In: Hunger, K. (Ed.), Wiley-Vch Verlag GmbH & Co. KGaA, Weinheim, pp. 625–641.

Jeejeebhoy, K.N., Chu, R.C., Marliss, E.B., Greenberg, G.R., Bruce-Robertson, A., 1977. Chromium deficiency, glucose intolerance, and neuropathy reversed by chromium supplementation, in a patient receiving long-term total parenteral nutrition. Am. J. Clin. Nutr. 30, 531–538.

Juang, R.S., Tseng, R.L., Wu, F.C., 1997. The ability of activated clay for the adsorption of dyes from aqueous solutions. Environ. Technol. 18, 525–531.

Ke, Q., Costa, M., 2007. Overview of chromium(III) toxicity, The nutritional biochemistry of chromium(III). In: Vincent, J.B. (Ed.). Elsevier Publications, Amterdam, The Netherlands, pp. 257–263.

Khare, S.K., Panday, K.K., Srivarstara, R.M., Singh, V.N., 1987. Removal of victoria blue from aqueous solution by fly ash. J. Chem. Technol. Biotechnol. 38, 99–104.

Kumar, G.A., Swarnalatha, S., Sairam, B., Sekaran, G., 2008. Production of alkaline protease by pseudomonas aeruginosa using proteinaceous solid waste generated from leather manufacturing industries. Bioresour. Technol. 99, 1939–1944.

Levina, A., Codd, R., Dillon, C.T., Lay, P.A., 2003. Chromium in biology: Toxicology and nutritional aspects. Prog. Inorg. Chem. 51, 145–250.

Levina, A., Mulyani, I., Lay, A.P., 2007. Redox chemistry and biological activities of chromium(III) complexes, The nutritional biochemistry of chromium(III). In: Vincent, J.B. (Ed.). Elsevier Publications, Amterdam, The Netherlands, pp. 224–256.

Leyva-Ramos, R., Rubio, L.F., Coronado, R.M.G., Barron, J.M., 1995. Adsorption of trivalent chromium from aqueous solutions onto activated carbon. J. Chem. Technol. Biotechnol. 62, 64–67.

Liu, C.H., Wu, J.S., Chiu, H.C., Suen, S.Y., Chu, K.H., 2007. Removal of anionic reactive dyes from water using anion exchange membranes as adsorbers. Water Res. 41, 1491–1500.

Low, K.S., Lee, C.K., Wong, A.M., 1995. Carbonized spent bleaching earth as a sorbent for some organic dyes. J. Environ. Sci. Health. Part A. Environ. Sci. Eng. Toxic. Hazar. Subst. Control A. 31, 673–685.

McMullan, G., Meehan, C., Conneely, A., Kirby, N., Robinson, T., Nigam, P., Banat, I.M., Marchant, R., Smyth, W.F., 2001. Microbial decolourisation and degradation of textile dyes. Appl. Microbiol. Biotechnol. 56, 81–87.

Medeiros, M.G., Rodrigues, A.S., Batoreu, M.C., Laires, A., Zhitkovich, A., Rueff, J., 2003. Biomarkers of chromium exposure and cytogenetic damage in leather tanning and welding industry workers. NATO Sci. Ser. Ser.-I 351, 132–141.

Mertz, W., 1998. Chromium research from a distance: From 1959 to 1980. J. Am. Coll. Nutr. 17, 544–547.

Muthukumar, M., Karuppiah, M.T., Bhaskar Raju, G., 2007. Electrochemical removal of CI acid orange 10 from aqueous solutions. Separ. Purif. Technol. 55, 198–205.

Namasivayam, C., Yamuna, R.T., 1994. Utilising biogas residual slurry for dye adsorption. Amer. Dyestuff Rep. 83, 22–28.

Namasivayam, C., Yamuna, R.T., Jayakumar, R., 1994. Dye removal from waste water by adsorption on waste Fe(III)/Cr(III) hydroxide'. Waste. Manag. 4, 643–648.

Newman, J., Reed, L., 1980. Weak-base vs strong base anion exchange resins for the recovery of chromate from cooling tower blowdown. AIChE Sym. Ser. Water-1979 197, 76.

O'Brien, T.J., Ceryak, S., Patierno, S.R., 2003. Complexities of chromium carcinogenesis: Role of cellular response, repair and recovery mechanisms. Mutat. Res. 533, 3–36.

Pagga, U., Braun, D., 1986. The degradation of dye stuffs: Part II. Behaviour of dye stuffs in aerobic biodegradation tests. Chemosphere 15, 479–491.

Petura, J.C., James, B.R., Vitale, R.J., 1998. Chromium (VI) in soils. In: Meyers, R.A. (Ed.), Wiley Encyclopaedia series in Environmental Science. Environmental Analysis and Remediation, vol. 2, pp. 1142–1158.

Phillips, D.A.S., 1996. Environmentally friendly, productive and reliable priorities for cotton dyes and dyeing process. J. Soc. Dyers Colourists 112, 183–187.

Prasad, B.G.S., Nair, B.U., 1994. Recovery and recycling of regenerated chromium from spent chrome liquor. Indian J. Environ. Health 36, 267–271.

Rajaram, R., Nair, B.U., Ramasami, T., 1995. Chromium(III) induced abnormalities in human lymphocyte cell proliferation: Evidence for apoptosis. Biochem. Biophys. Res. Commun. 210, 434–440.

Raji, C., Anirudhan, T.S., 1998. Batch Cr(VI) removal by polyacrylamide-grafted saw dust. Kinetics and thermodynamics. Water Res. 32, 3772–3780.

Ramasami, T., 1996. Greening of chrome tanning in Indian leather industry. ILIFO J. Cleaner Tanning 1, 12–14.

Ramos, R.C., Martinez, J., Coronado, R.M.G., 1994. Adsorption of chromium (VI) from aqueous solutions on activated carbon. Water. Sci. Technol. 30, 191–197.

Rao, J.R., Prasad, B.G.S., Narasimhan, V., Ramasami, T., 1989. Electrodialysis in the recovery and reuse of chromium from industrial effluents. J. Memb. Sci. 46, 215–224.

Reife, A., 1993. Dyes, environmental chemistry. In: Kirk-Othmer encyclopedia of chemical technology, fourth ed. John Wiley & Sons, New York, p. 753.

Rich, G., Cherry, K., 1982. In: Hazardous wastes treatment technologies. Pudran Publishing Co., New York, pp. 1–169.

Robinson, T., McMullan, G., Marchant, R., Nigam, P., 2001. Remediation of dyes in textile effluent: A critical review on current treatment technologies with a proposed alternative. Bioresour. Technol. 77, 247–255.

Saha, C., 1996. Eco-textile: A novel concept of cleaner product. Tex. Dye. Prin. 29, 13–18.

Saleem, M., Afzal, M., Mahmood, F., Hameed, A., 1994. Thermodynamics of adsorption of rhodamin B and nile blue sulfate on alumina from aqueous solutions. J. Chem. Soc. Pak. 16, 83–86.

Saleem, M., Pirzada, T., Qadeer, R., 2007. Sorption of acid violet 17 and direct red 80 dyes on cotton fiber from aqueous solutions. Colloids Surf. A. Physicochem. Eng. Asp. 292, 246–250.

Sandhya, S., Padmavathy, S., Swaminathan, K., Subrahmanyam, Y.V., Kaul, S.N., 2005. Microaerophilic-aerobic sequential batch reactor for treatment of azo dyes containing simulated wastewater. Process Biochem. 40, 885–890.

Sengupta, A.K., Roy, T., Lessen, D., 1988. Modified anion exchange resins for improved chrome selectivity and increased efficiency of regeneration. React. Polym. 9, 293.

Shenair, A., 1996. Azo dyes on textiles vs. German ban: An objective assessment—Part III Another study. Colourage 43, 41–44.

Shrivastava, H.Y., Nair, B.U., 2000. Cleavage of human orosomucoid by a chromium(V) species: Relevance in biotoxicity of chromium. Biochem. Biophys. Res. Commun. 279, 980–983.

Shrivastava, H.Y., Nair, B.U., 2001. Chromium(III) mediated structural modification of glycoprotein: Impact of the ligand and the oxidants. Biochem. Biophys. Res. Commun. 285, 915–920.

Simeonova, L.S., Dalev, P.G., 1996. Utilization of a leather industry waste. Waste. Manag. 16, 765–769.

Singaram, P., 1994. Removal of chromium from tannery effluent by using weeds. Indian J. Environ. Health 36, 197–199.

Singh, B.K., Rawat, N.S., 1994. Comparative sorption equilibrium studies of toxic phenols on fly ash. J. Chem. Technol. Biotechnol. 61, 307–317.

Sivasamy, P., Muthumari, R., Kumari, V.J.T., Shanmugam, L.P.N., Shanthi, S., 2000. EDTA assisted removal of heavy metal ions by adsorption on activated charcoal and coconut shell carbon. Indian J. Environ. Prot. 20, 761–771.

Slokar, Y.M., Le Marechal, M.A., 1998. Methods of decolorization of textile wastewaters. Dyes. Pigments 37, 335–356.

Snow, E.T., 1991. A possible role for chromium(III) in genotoxicity. Environ. Health Perspect. 92, 75–81.

Stepan, E.S., Velea, S., Gaidau, C., Filipescu, L., Niculescu, D., Radu, A.C., 2009. Surfactants and Biofuels obtained from leather industry wastes, International Leather Engineering Symposium held in Turkey during April 29–May 1, pp. 1–11.

Sun, Q., Yang, L., 2003. The adsorption of basic dyes from aqueous solution on modified peat-resin particle. Water Res. Vo 37, 1535–1544.

Swami, A., Ramteke, D.S., Sarin, R., 2002. Evaluation of natural polymers for wastewater treatment. J. Indian Ass. Environ. Manag. 29, 90–95.

Taylor, M.M., Marmer, W.N., Brown, E.M., 2006. Effect of fillers prepared from enzymatically modified proteins on mechanical properties of leather. J. Am. Leather Chem. Assoc. 101, 128–137.

Theng, B.K.G., Wells, N., 1995. Assessing the capacity of some Newzeland clays for the decolourising vegetable oil and butter. Appl. Clay Sci. 9, 321–326.

Trujillo, E.M., Jeffers, T.H., Fergusan, C., Stevenson, H.Q., 1991. Mathematically modelling the removal of heavy metals from a wastewater using immobilised biomass. Environ. Sci. Technol. 25, 1559–1560.

Vasudevan, N., Ravindran, A.D., 2007. Biotechnological process for the treatment of fleshing from tannery industries for methane generation. Curr. Sci. 93, 1492–1494.

Vijayalakshmi, R., Kanthimathi, M., Subramanian, V., Nair, B.U., 2000. DNA cleavage by a chromium(III) complex. Biochem. Biophys. Res. Commun. 271, 731–734.

Voitkun, V., Zhitkovich, A., Costa, M., 1998. Cr(III)-mediated crosslinks of glutathione or amino acids to the DNA phosphate backbone are mutagenic in human cells. Nucleic. Acids. Res. 26, 2024–2030.

Volesky, B., Holan, Z.R., 1995. Biosorption of heavy metals. Rev. Biotechnol. Prog. 11, 235–250.

Vonguriten, U., Oliveras, Y., 1997. Kinetics of the reaction between hydrogen peroxide and hypo bromous acid: Implication on water treatment and natural systems. Water Res. 31, 900–906.

Wisniewski, J., Suder, S., 1995. Water recovery from etching effluents for the purpose of rinsing stainless steel. Desalination 101, 245–253.

World Leather, Waste creates new growth, December 2004/ January 2005.

Wu, J.S., Liu, C.H., Chu, K.H., Suen, S.Y., 2008. Removal of cationic dye methyl violet 2B from water by cation exchange membranes. J. Memb. Sci. 309, 239–245.

Yun, Y.S., Park, D., Park, J.M., Volesky, B., 2001. Biosorption of trivalent chromium on the brown seaweed biomass. Environ. Sci. Technol. 35, 4353–4358.

Zhitkovich, A., 2005. Importance of chromium-DNA adducts in mutagenicity and toxicity of chromium(VI). Chem. Res. Toxicol. 18, 3–11.

Zhitkovich, A., Shranger, S., Messer, J., 2000. Metabolism of Cr(VI) by cysteine leads to the formation of binary and ternary Cr-DNA adducts in the absence of oxidative DNA damage. Chem. Res. Toxicol. 13, 1114–1124.

Zou, L., Zhu, B., 2008. The synergistic effect of ozonation and photocatalysis on color removal from reused water. J. Photochem. Photobiol. A. Chem. 196, 24–32.

CHAPTER

Applications of Porous Media Combustion Technology

24

M. Abdul Mujeebu*, A.A. Mohamad†, M.Z. Abdullah**

*Department of Building Engineering, College of Architecture and Planning, University of Dammam, 31451 Al-Dammam, Saudi Arabia, †Department of Mechanical and Manufacturing Engineering, CEERE, The University of Calgary, Calgary, Alberta T2N 1N4, Canada, **Porous Media Combustion Laboratory, School of Mechanical Engineering, Universiti Sains Malaysia, 14300 Nibong Tebal, Penang, Malaysia*

CHAPTER OUTLINE

1 Introduction

Porous media combustion (PMC), also known as filtration combustion, pertains to the heterogeneous interaction between two different media, usually a solid and a gas. The theory of filtration combustion involves a new type of flame with exothermic chemical transformation during fluid

The Role of Colloidal Systems in Environmental Protection. http://dx.doi.org/10.1016/B978-0-444-63283-8.00024-7

motion in a porous matrix. The term *filtration combustion* was introduced by Russian scientists for combustion phenomena in which gas flow through porous media (in Russian—filtration flow) plays a principal role. This term does not correspond to Western scientific terminology tradition, yet it can be found in special literature as a synonym for combustion within porous media. In the PMC terminology, a porous medium (PM) is defined as a solid medium with interconnected pores (high permeability), available as either a packed bed of discrete solids (mainly ceramic material or metal) or foam.

The stationary and the transient systems are the two major design approaches commonly employed in filtration combustion. The first approach (stationary) is widely used in radiant burners and surface combustor-heaters (for glass tempering and industrial drying processes, such as pain drying) due to high radiant emissivity of the solid. Here, the combustion zone is stabilized near the exit of the burner. The second approach(transient) to filtration combustion leads to the "excess enthalpy" flame theory where an unsteady reaction zone possesses the freedom to freely propagate as a combustion wave in the downstream or upstream direction due to positive and negative energy fluxes (Mujeebu et al., 2009a). Porous medium burners exhibit the feature of combusting subnormal lean mixtures due to the intense heat transfer across the solid to preheat the mixture to the temperatures that sustain chemical reactions. With such energy recirculation, the porous burners have wider flame-stability limits and can hold an extended range of firing capacities, a feature that cannot be obtained by conventional burners especially at their lean operation. Therefore, the ultra-low CO and NOx emissions, associated with the low gas temperatures at such lean conditions, characterize the performance of the porous burners at significantly higher turndown ratios (Mohamad et al., 1994). The relatively higher thermal emission with the lower exhaust gas emissions and the excellent stability behavior are all mandatory requirements for industrial applications. This testifies to the widespread porous media in combustors of boilers, gas turbines, heat recovery units, and incinerators for fuel saving and higher energy utilization (Kamal and Mohamad, 2006).

A comprehensive review on this topic, covering the history of developments up to the year 2009, has been reported in our previous work (Mujeebu et al., 2009a). However, the present chapter covers an exclusive outline of the applications of PMC.

2 Applications of PMC technology

2.1 Household heating systems and burners

The high-power modulation and physical compactness offered by PM burners makes them interestingly useful for household heating applications. As the highest emissions occur during the warm-up phase, a higher power modulation automatically leads to a decrease of waste gas emissions. Also in stationary operation, emissions of porous media burners are minimized in comparison to many domestic gas appliances working with free flames because it is possible to control the combustion temperature with the porous material in the combustion region. Solid materials have good thermal properties in comparison to gases; thus heat can be efficiently transported out of the combustion region leading to a significant decrease of the NO formation, which is strongly temperature dependent (Mohamad et al., 1994). It has been reported that the emissions of nitrous oxides and carbon monoxide from a PM burner are clearly below the most stringent European emission standards. Figure 24.1 shows a schematic sketch and a photograph of a complete heating system for household applications

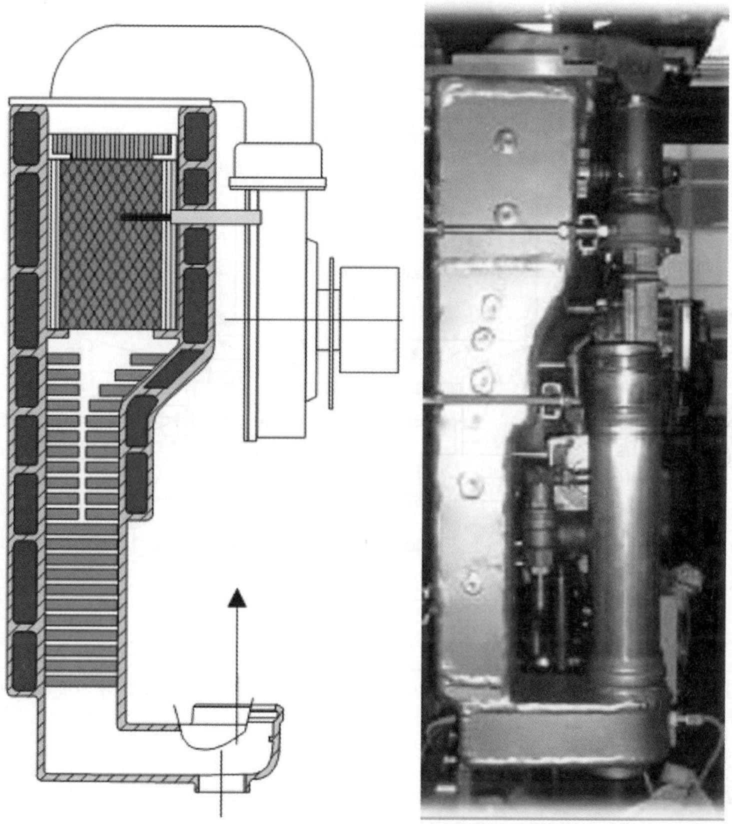

FIGURE 24.1

Porous burner and integrated heat exchanger unit for household applications (Mößbauer et al., 1999). (For color version of this figure, the reader is referred to the online version of this book.)

on a porous burner basis. It is about half as large as a conventional heating system for the same nominal power output (Mößbauer et al., 1999).

Development of a porous radiant recirculated burner (PRRB) has been reported by Jugjai and Rungsimuntuchart (2002). They used the heat recirculation system based on the PMC technology in combination with a swirling central flow burner to improve the efficiency of the domestic gas burners. A porous medium was used to preheat the combustion air in a liquefied petroleum gas (LPG) cooking stove, and an increase in the thermal efficiency by about 30% from its normal value has been found. A maximum preheating effect on the combustion air was reported to be 300°C. The schematic of the PRRB is shown in Figure 24.2.

2.2 Air-heating systems

In the process industry, dryers are used to remove moisture from agricultural products, food, textiles, chemicals, leather, wood, paper, and many other goods. For this purpose, hot-air streams are often applied to evaporate the water and to remove the moisture. The combustion chambers needed for the

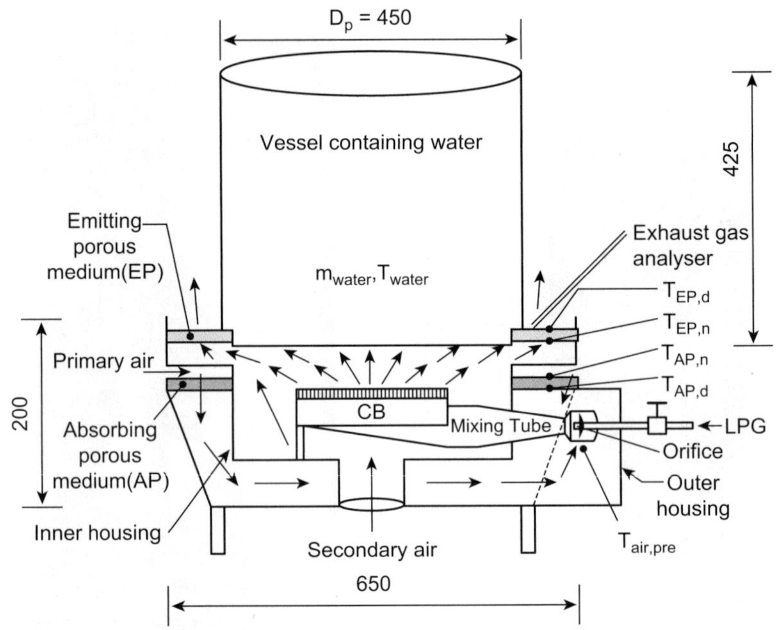

FIGURE 24.2

Schematic of the PRRB in combination with the conventional burner developed by Jugjai and Rungsimuntuchart (2002).

industrial gas burners used in these applications are large in size, which are often undesirable. Small and compact hot-air-heating systems can be realized if porous media combustors are used. The porous medium burners have a burning length in the range of 10 cm. Thus, very small hot-air-heating systems for industrial drying are possible, as indicated in Figure 24.3 (Mößbauer et al., 1999).

2.3 Vehicle heating systems

The trend in vehicles for the near future will go in the direction of more comfort, life span, and safety, particularly in cars and caravans. Therefore, new heating systems for cars and caravans become more and more important. The first heaters for these applications had the problem that the pollutants were too high for operation in closed garages, particularly in subterranean parking garages. The aim of a better acceptance of these heating systems could be achieved with the porous medium burner technology. With this principle, it is possible to obtain low-pollutant emissions with even better space usage. Figure 24.4 shows an enhanced view of the complete independent vehicle heating system on a porous burner basis (Mößbauer et al., 1999).

2.4 Steam generation

Steam generators are used for steam turbines, industrial processes, hospitals, laundries, or even in ships. Conventional devices are usually very large in size due to the necessary length of the free flames

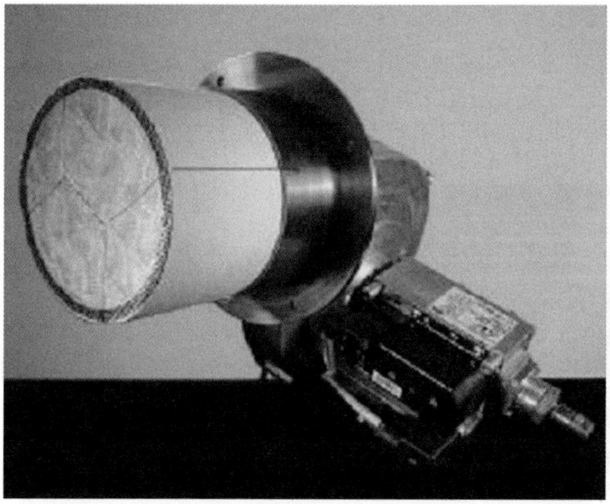

FIGURE 24.3

Air-cooled porous medium burner for use in air-heating systems (Mößbauer et al., 1999). (For color version of this figure, the reader is referred to the online version of this book.)

for a complete conversion of the fuel. This fact requires that burners have long flame tubes for steam generation. Additionally, a very large diameter of the combustion chamber is required because of the limited radiation heat transfer from the combustion region to the working fluid. Both the length and the diameter of combustion chambers for steam generation devices could be reduced by applying the porous burner technology. Furthermore, industrial boilers produce extensive amounts of NOx due to high flame temperature; combustion in PM is an ultimate method for reducing NOx without sacrificing combustion efficiency (Mohamad et al., 1994). Figure 24.5 shows a conception for the generation of steam on a porous medium basis. After preheating the feed water, the water is evaporated in two evaporation stages. By using the high temperature level in the superheating stage, the steam could be heated up to temperatures of 400°C.

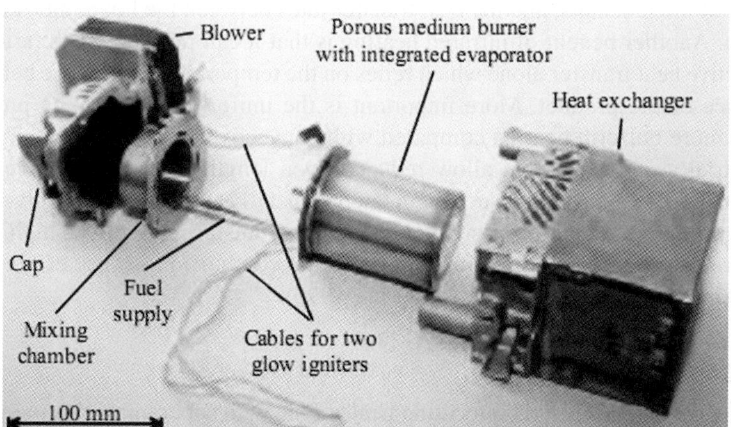

FIGURE 24.4

Independent vehicle heating system on porous burner basis (Mößbauer et al., 1999). (For color version of this figure, the reader is referred to the online version of this book.)

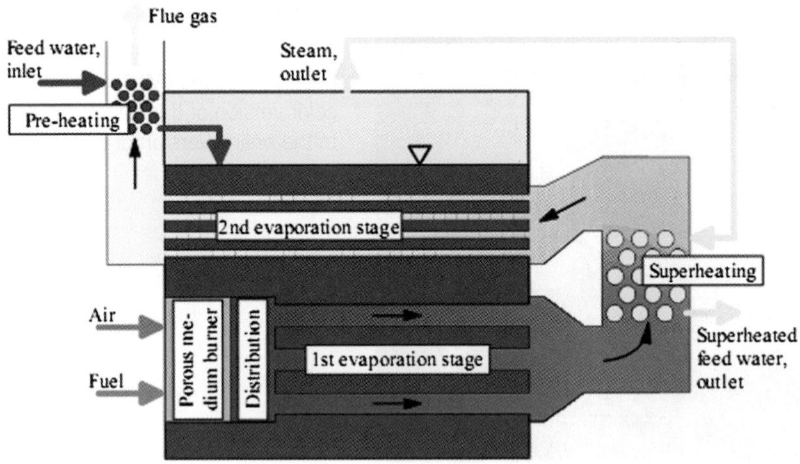

FIGURE 24.5

Schematic of steam generator using PMC (Mößbauer et al., 1999). (For color version of this figure, the reader is referred to the online version of this book.)

2.5 Infrared heating

Infrared heating is widely used by industry for manufacturing processes, such as glass tempering and paint drying. The heat transfer rates for infrared heaters are determined by the temperature of the source and the absorption characteristics of the product being heated. Because the radiation output is proportional to the fourth power of the temperature, the temperature of the source dominates the performance of the device. This temperature depends on the emissivity of the radiating matrix, the rate of energy release of the reactants, and the heat transfer rates between the hot combustion products and the porous media. Another benefit of infrared heating is that it can provide much faster heating times than from convective heat transfer alone which relies on the temperature difference between the heated air and the surface of the product. More important is the uniformity of heating process. Radiative heating provides more uniform heating compared with convective heating process. These advantages applied to industrial applications can allow reduced oven lengths, increased conveyor speeds, and improved surface finish of products such as coated paper that might be sensitive to air currents associated with convective drying. While the radiative heat transfer rates can be high, the combustion temperatures of radiant burners are relatively low and the combustion is very efficient, allowing the emissions of NOx and CO to be low (Howell et al., 1996).

2.6 Internal combustion engines

Hanamura et al. (1997) proposed a reciprocating heat engine with superadiabatic combustion in porous media. The system consisted of two pistons and a thin porous medium in a cylinder, one being a displacer piston and the other a power piston. These created reciprocating motions with a phase relation, similar to those of an ordinary Stirling engine. By means of the reciprocating flow system, the residual combustion gas enthalpy was effectively regenerated to induce enthalpy increase in the mixture through the porous medium, which provided heat storage. The results showed that due to heat recirculation, the maximum temperature was higher than the theoretical one in free space.

The thermal regeneration concept for diesel engines was introduced by Park and Kaviany (2002) for the roles of the porous insert motion and the fuel injection strategies on the fuel evaporation and combustion and on the engine efficiency. They claimed that a regenerative diesel engine using an in-cylinder reciprocating porous regenerator as shown in Figure 24.6 has the potential to improve fuel-air mixing and combustion. The porous insert is attached to a rod and moves in the cylinder, synchronized, but out of phase with the piston. During the regenerative heating stroke, the regenerator remains just beneath the cylinder head for most of the period and moves down to the piston (as it approaches the top dead center). During the regenerative cooling stroke, the regenerator moves up and remains in the original position, until the next regenerative heating stroke. Following the combustion and expansion, the products of combustion (exhaust gases) retain an appreciable sensible heat. During the regenerative cooling stroke, the hot exhaust gas flows through the insert and stores part of this sensible heat by surface-convection heat transfer in the porous insert (with large surface area). For the proposed engine, a thermal efficiency of 53% was claimed, compared to 43% of the conventional diesel engines.

The prototype of a compact zero emission steam engine using porous burner technology, suitable for automotive engine applications, was made and tested by Mößbauer et al. (2001). The proposed engine had a power turndown ratio of about 1:20, which is roughly 5 times greater than the modulation range of competitive burner technologies. The porous burner yielded a very low emission at any state

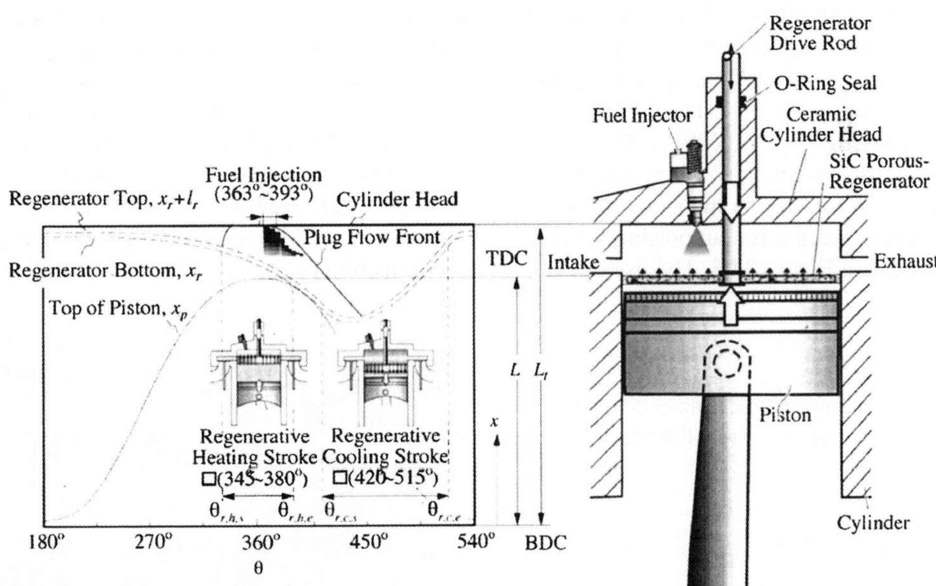

FIGURE 24.6

Sequence of motion of the regenerator and piston and physical rendering of fuel injection and air blowing during the regenerative heating stroke (Park and Kaviany, 2002).

FIGURE 24.7

Multifuel burner for a one-piston ZEE introduced by Mößbauer et al. (2001). (For color version of this figure, the reader is referred to the online version of this book.)

of operation, clearly below the most stringent super-ultra-low emission vehicle standards. The power density of about 3000 kW/m^2 was obtained, which is about 10 times greater than that of other premixed, low-emission burner technologies, resulting in very small burner units. Further, the porous burner technology allowed complex combustion chamber geometries to be realized, which is of special interest in automobiles. The experimental results obtained from the prototypes clearly indicated that porous medium burners together with a modern steam engine concept could be successfully applied to car engines and offer significant advantages compared to the conventional state-of-the-art engines. The multifuel burner used for a one-piston zero emission engine (ZEE) is shown in Figure 24.7.

Novel concepts for internal combustion engines based on the application of PMC technology were proposed by Weclas (2001, 2004, 2005). His study proved that gas flow, fuel injection, and its spatial distribution, vaporization, mixture homogenization, ignition, and combustion could be controlled or positively influenced with the use of porous media reactors. The key features of the highly porous medium for supporting the mixture formation, ignition and combustion in IC engines, are illustrated in Figure 24.8. The use of PMC in direct injection (diesel or gasoline) IC engines was

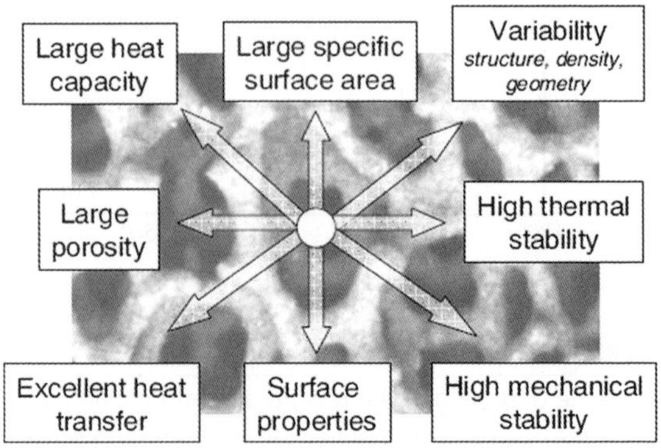

FIGURE 24.8

Main features of porous structures to be utilized to support engine processes (From the work of Weclas).

proposed by Durst and Weclas (2001). Polasek and Macek (2003) presented the simulation of IC engine equipped with a PM to homogenize and stabilize the combustion of compression ignition engines. The purpose of the PM matrix was to ensure reliable ignition of lean mixture and to limit maximum in-cylinder temperature during combustion.

2.7 Gas turbines and propulsion

As an excellent breakthrough in the application of PMC, a porous burner with matrix stabilized combustion for gas turbines was built by Ellzey and William Jr. (2003). The evaporation processes in porous media for gas turbine applications were numerically analyzed by Periasamy et al. (2004). Evaporation of a pointwise-injected kerosene spray in a carbon–carbon porous medium was considered. The effects of porous medium temperature, fuel flow rate, air inlet temperature, and porous medium geometry on the evaporation of spray had been analyzed. For the size under consideration, a porous medium heat source of 642 W was required to achieve 97% complete evaporation for an air inlet temperature of 473 K. Evaporation characteristics were not found to vary much with porous medium geometry, as the porous medium was modeled as a momentum sink. But thermal effects of porous media were found to be more dominant.

The characteristics of combustion within porous media, which are attractive in a propulsion context, are the ability to burn leaner and hotter than a free flame with low emissions, no cooling requirement for the combustor itself, and the potential to operate free from combustion-induced noise. The performance of a PM combustor applicable for gas turbines, at elevated pressures and inlet temperatures, has been studied by Noordally et al. (2004). The combustor was formed of reticulated porous ceramics, untreated to augment or sustain chemical reaction. The results showed that the combustor could operate in a "superadiabatic" mode, with low emissions. Porous materials are often used for the injector face plate of liquid propellant rocket engines (LPRE). A multiscale modeling strategy was developed by Sozeret al. (2006) to develop predictive capabilities for such porous media fuel flow problems.

2.8 Oil and gas recovery

Gas-phase combustion in porous media has many potential applications for the oil and gas industry. Such applications include enhanced oil and gas recovery, formation heat treatment for remediation of near wellbore formation damage, downhole steam generation for heavy oil recovery, in-situ preheating of bitumen for improved pumping, increased temperatures in gas condensate reservoirs, and improved gas production from hydrate reservoirs. Sanmiguel et al. (2003) studied the various processes and relevant controlling mechanisms associated with gas-phase combustion in porous media, especially at elevated pressures. A novel apparatus has been designed, constructed, and commissioned in order to evaluate the effects of controlling parameters such as operating pressure, gas-flow rate, type and size of porous media, and equivalence ratio on combustion characteristics. It was shown that the velocity of the combustion front decreases as the operating pressure of the system increases, and during some test operating conditions, the apparent burning velocities are over 40 times higher than the open-flame laminar burning velocities. The sustained propagation of combustion fronts in porous media is a necessary condition for the success of in-situ combustion for oil recovery. Compared to other recovery methods, in-situ combustion involves the complexity of exothermic reactions and temperature-dependent chemical kinetics. In the presence of heat losses, the possibility of ignition and extinction also exists. Akkutlu and Yortsos (2003) addressed some of these issues by studying the properties of forward combustion fronts propagating at a constant velocity in the presence of heat losses.

2.9 Lighting

Qiu and Hayden (2006) had developed a new radiant burner based on PMC for a lighting application. The porous medium used in the burner was an oxide ceramic fiber felt with the desired porosity and fiber diameter (on the order of 10 mm). It was composed of a mixture of rare earth oxides, Yb_2O_3 (93.5 wt%) and CeO_2 (3.0 wt%), with minor additives (3.5 wt%) as hardening agents. The burner emitted the desired spectral radiation and generated fairly intense light under conditions of heat recuperation. The light radiant burner can be used as an alternative low-glare light source in an integrated heat and light system in which the light is distributed through light pipes. The schematic of the proposed burner assembly is shown in Figure 24.9. Their study was extended (Qiu and Hayden, 2009) to analyze the performance of radiant burners by using oxygen-enriched air, which was produced by means of polymer membranes. The membranes were shown to be an effective means of obtaining an oxygen-enriched environment for gas combustion in the radiant burners. Two different porous radiant burners, a reticulated ceramic burner and a ceramic fiber burner, were investigated. The experimental results showed that the radiation output and the radiant efficiency of these burners could be increased noticeably with oxygen enrichment in the combustion air. The schematic of the oxygen enrichment membrane module used in their study is shown in Figure 24.10, and the structure of the hollow fiber membrane is illustrated in Figure 24.11. The thin, permeable perfluoropolymer membrane on the outer surface of the porous polymer hollow fiber wall forms a separating layer. Pressurized air feed enters the inside of hollow fibers. Oxygen-enriched air is obtained outside the fibers and exits the module through an outlet port. Meanwhile, the air stream flowing through the inside of the hollow fibers becomes enriched in nitrogen and is exhausted to the atmosphere through a retentive port.

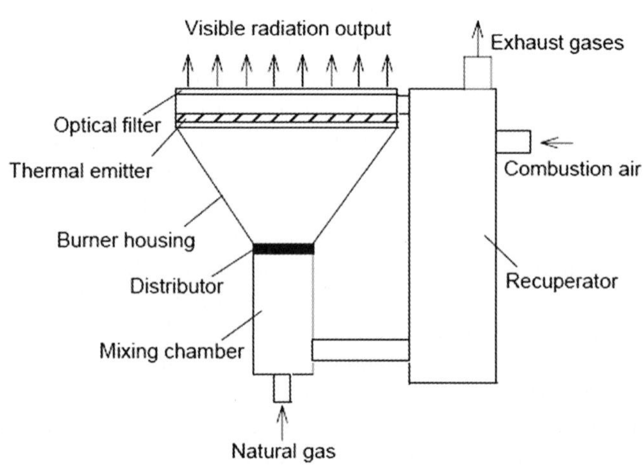

FIGURE 24.9

Structure of the radiant burner for lighting application introduced by Qiu and Hayden (2006).

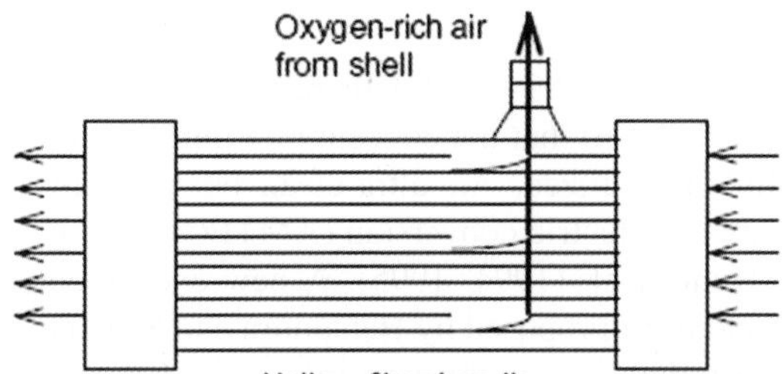

FIGURE 24.10

Configuration of the countercurrent hollow-fiber membrane module for oxygen-enrichment (Qiu and Hayden, 2009).

2.10 Hydrogen production

Production of hydrogen from gases such as methane and hydrogen sulfide is another potential application of PMC. Filtration combustion waves formed in an inert porous media were analyzed comparatively for methane-air and diluted hydrogen sulfide (20% H_2S + 80% N_2)-air mixtures by Bingue et al. (2002). Temperature, velocity, and chemical products of the combustion waves were studied experimentally in the range of equivalence ratios (φ) from 1 to 2.5 for methane and from 1 to 5.5 for hydrogen sulfide, at a filtration velocity of 12 cm/s. The practical goal of the study in the rich and ultra-rich region was to explore the extent of conversion of these reactants into commercially viable products such as hydrogen (H_2), syngas (H2 + CO), and sulfur (S_2). For both methane and hydrogen sulfide combustion, upstream (under adiabatic) propagation corresponded to the range of equivalence ratios from stoichiometry to 1.7, and downstream (superadiabatic) wave propagation was observed for ultra-rich ($\varphi \geq 1.7$) mixtures. The products of methane partial oxidation, dominant for ultra-rich waves, were H_2, CO, and C_2 hydrocarbons where up to 60% of the methane was converted to CO and H_2. Similar observation for H_2S partial oxidation products revealed a maximum conversion rate of 20% to H_2 and 50% for S_2. Later (Bingue et al., 2004), they performed another study to optimize the H_2 production from methane. The contributions of Shabunya (2001), Slimane et al. (2004), Pedersen-Mjaanes (2005), and Raviraj and Ellzey (2006) are also significant in this area.

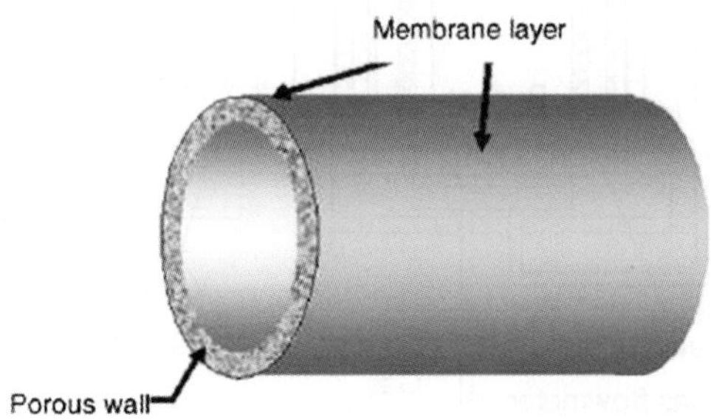

FIGURE 24.11

The hollow fiber membrane structure used by Qiu and Hayden (2009) for oxygen-enrichment. (For color version of this figure, the reader is referred to the online version of this book.)

2.11 CHP and thermoelectric conversion

As suggested by Mohamad (2005) and Echigo et al. (1993, 1994, 1995), there is an attractive scope in the application of PMC for the production of electricity, so that the concept of a mini-combined heat and power (CHP or cogeneration) system could be realized. An ultra-clean burner with cellular ceramic porous medium for an alkali metal thermal to electric conversion (AMTEC) system suitable for hybrid electric vehicles was developed by Hunt et al. (1994, 1995). A new energy conversion method from combustion heat to electric power had been proposed by Hanamura and Kumano (2003) and Hanamura et al. (2005) on the basis of superadiabatic combustion in a pair of thermoelectric porous elements. A mixture of about 1% methane in air was introduced into the elements, where the flow direction changed regularly. By energy recirculation through the heat storage of the porous elements, the combustion temperature became higher than the theoretical flame temperature by a factor of about two. A sharp temperature gradient was automatically established in the elements. The electric power was generated with a total thermal efficiency of 0.2%.

Recently, Qiu and Hayden (2007) focused on thermophotovoltaic (TPV) power generation using natural gas-fired radiant burners. These burners were used as radiation sources to establish and test two TPV prototype systems. It was found that for a nonsurface combustion radiant burner, the radiation output could be enhanced using a thermal radiator with a porous structure. They also developed a cascaded radiant burner that generated two streams of radiation output. One stream illuminated silicon

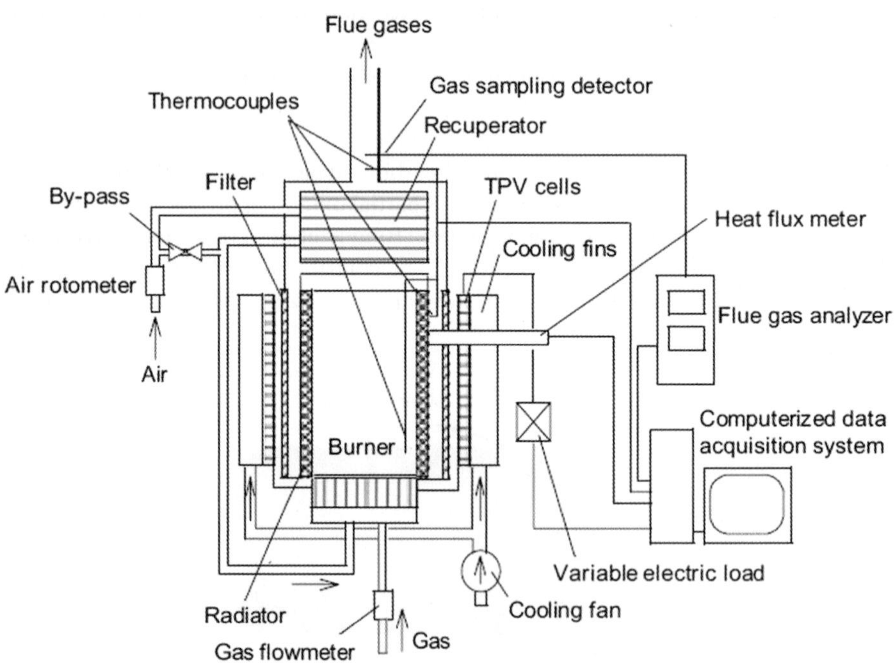

FIGURE 24.12

Schematic of the TPV cogeneration setup using porous radiant burner and GaSb TPV cells introduced by Qiu and Hayden (2007).

concentrator solar cells, while the other drove GaSb cells of low band-gap. In this way, useful radiation output and thus TPV system efficiency were significantly increased due to the cascaded utilization of combustion heat and optimized thermal management. The schematic of one of the proposed cogeneration schemes using TPV is shown in Figure 24.12. Recently this concept has been introduced to micro- and mesoscale applications, which will be discussed in section 2.15 of this chapter. More research is needed in this area to develop technically feasible and energy efficient CHP systems.

2.12 Combustion of low CV fuels

PMC is a promising technique for burning low calorific gases (LCG). The effective combustion of LCG offers a significant saving in primary energy reduction of hazardous emissions. Nevertheless, the fluctuating composition, calorific value, and high inert content of such mixtures result in technical difficulties concerning combustion stability. The intensive internal heat recuperation and the high-flame stability of porous media burners offer great chances for LCG utilization. An interesting study on this area was reported by Echigo et al. (1983). Later on, Al-Hamamre (2006) investigated the combustion behavior of LCG mixtures emitted from landfills and pyrolysis processes, both numerically and experimentally with SiC and Al_2O_3 porous structures. The influence of the combustion zone properties, the thermal load, and the duct temperature were investigated. Experimentally, the operational range of the burners was specified at different power rates and different mixture temperatures.

2.13 VOC oxidation

The highly polluting volatile organic components (VOC), such as phenol, formaldehyde, acetone, benzole, and others, can be eliminated by oxidation in inert porous media or filtration combustion. A remarkable contribution to this area has been provided by Dobrego et al. (2005, 2006, 2007). They introduced a new regenerator–recuperator scheme for VOC oxidation using PMC. Its main characteristics, such as maximum temperature of reactor, temperature of exhaust gases, and NO emission were calculated and compared with the same characteristics of other types of reactors (co-flow filtration combustion wave reactor, counterflow heat exchange recuperative reactor, and reverse flow regenerative reactor). It was shown that the regenerator–recuperator scheme provides important advantages (wide range of flow rate, and equivalence ratio of the combustible, low temperature of the exhaust gases) compared to conventional schemes.

2.14 Micro- and mesoscale applications

Combustion at small scales presents several challenges, including high heat loss and short-flow residence times. Heat recirculation is an effective method to limit heat loss and improve combustion performance. However, new methods of achieving heat recirculation in a small volume must be developed for practical devices. In the recent past, a few researchers found that PMC was a promising option to tackle this issue.

A mesoscale lean premixed PMC system that employed PM in the combustion chamber and in the preheating annulus around the combustor, with methane as fuel, was developed and tested by Marbach and Agrawal (2006). Results showed that the reactants were preheated in excess of 600 K by recirculating thermal energy from the reaction zone. Heat loss to the surroundings decreased, and heat recirculation to the reactants increased with PM in the annulus and with insulation of exterior surfaces of the system.

Later, this concept was utilized together with the flame stabilized on the surface of a silicon-carbide-coated carbon foam by Marbach et al. (2007). The effectiveness of the proposed concept was demonstrated experimentally for methane combustion in a chamber volume of 0.364 cm^3 and an overall system volume of 1.5 cm^3. Experiments were conducted for reactant flow velocities varying from 0.25 to 1.0 m/s in the equivalence ratio range of 0.50 to 0.80. Results showed excellent agreement between measured and computed temperature profiles at different reactant flow rates. It was claimed that the modified system achieved a significant reduction in heat loss as compared to the baseline design tested experimentally.

A mesoscale liquid fuel film combustor with a central porous inlet was devised by Li et al. (2008). The effects of porous material type and bead size on the flame structures and combustion characteristics were examined. Porous media made of stainless steel and bronze were tested with different fuel and air-flow rates, equivalence ratios, and bead sizes. The flame structure and its corresponding stabilization mechanism were different between the stainless steel and the bronze porous media combustor. In the stainless steel case, the high specific heat capacity enhanced fuel vaporization and fuel–air mixing, and the flame anchor located on the surface of the porous cap. In the bronze case, due to its low heat capacity, the flame was swept downstream where the recirculation zone above the porous cap offered a low-velocity field to help anchor the flame.

Very recently, a novel mesoscale combustion concept as shown in Figure 24.13, comprising a flow-blurring injector, a counterflow heat exchanger to preheat reactants using product gas enthalpy,

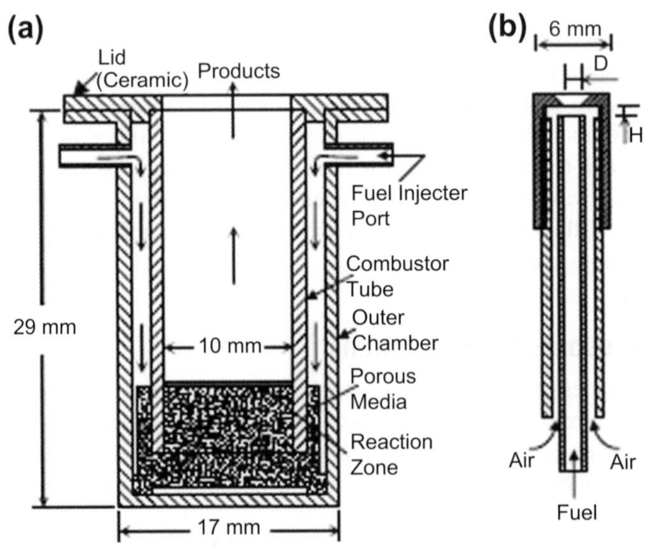

FIGURE 24.13

Schematic of the mesoscale combustion system of Sadasivuni and Agrawal (2009) (a) combustor system and (b) flow-blurring injector.

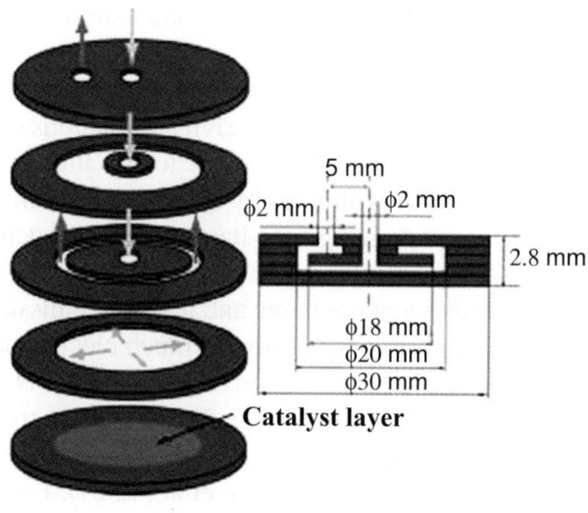

FIGURE 24.14

The radial-flow-type combustor developed by Kamijo et al. (2009).

and a PM for combustion, was introduced by Sadasivuni and Agrawal (2009). The overall system was 30 mm long and 17 mm in diameter, with combustor diameter of 10 mm. The system, with kerosene as fuel, achieved a heat release rate of up to 460 W, pertaining to energy density of 90 MW/ m^3 based on total volume and 230 MW/m^3 based on combustor volume. A numerical model incorporating conjugate heat transfer, radiation heat exchange, flow and heat transfer in PM, and heat release by combustion was developed and validated.

Kamijo et al. (2009) developed and tested a microscale catalytic combustor using high-precision ceramic tape-casting technology, as shown in Figure 24.14, for high-temperature applications like TPV power generation. Pd catalyst/nanoporous alumina support fabricated through anodic oxidation was employed for the catalyst layer, having good adhesion to the substrate. The ignition temperature of the catalyst was found to be around 225°C, and its activity was on a par with that of the catalyst prepared with the conventional sol-gel methods. The wall temperature as high as 850°C and extremely high heat generation density of $2 - 5 \times 10^8$ W/m^3 have been achieved. It was also found that the wall temperature gradient was significantly reduced with the proposed catalyst arrangement.

2.15 Combustion of liquid fuels

The application of porous media to liquid fuel combustion has been introduced only in the recent past. Typical applications of this technology are quite new and have had a strong impact on industrial and domestic applications. A potential application of a liquid fuel porous ceramic burner is the incineration of liquid hazardous waste. Liquid hazardous wastes are difficult to incinerate in conventional burners because they are often low-energy-content fuels and contain chlorinated species. These wastes can be efficiently burned in porous ceramic burners owing to the higher volumetric heat release. Also, the soot formed in conventional incinerators often acts as condensation sites for hazardous materials, becoming hazardous itself when emitted to the atmosphere. This problem can be eliminated by porous ceramic burners as the liquid waste is vaporized before combustion. The radiant preheating of the fuel–air mixture, which is inherent to its design, may also aid the combustion of mixed fuels containing less reactive species (Howell et al., 1996).

Liquid fuel-fired porous combustors are classified into (1) fuel-vaporizing type and (2) fuel-spraying type. Commonly, spraying type is used for industrial and civil applications such as boilers and furnaces. In the spraying type, homogeneous combustion in large-range load is little obtained because the flame is relatively large and the stability of the flame is affected by aerodynamics among the air and the droplets. Compared with the spraying type, the vaporizing type has advantages such as a high turndown ratio, compactness, and the easy proportional control that matches a gaseous fuel-fired boiler. Its combustion becomes close to homogeneous and has a smaller flame and less soot emission by mixing the vaporized fuel with air. But it has the disadvantage of consumption of much electric power for fuel vaporization. For instance, the power consumption of a kerosene vaporizing-type boiler is 800W at ignition and 470W at combustion on the main burner for a 55kW class boiler. Therefore, the total thermal efficiency of the boilers may be decreased due to the high amount of energy loss during the conversion process from fossil fuel energy to electricity in power generation (Fuse et al., 2003). For more details on liquid fuel combustion in PM, readers are encouraged to consult our recent review (Mujeebu et al., 2009c), which provides fundamentals, modeling aspects, investigations, and applications.

2.16 Miscellaneous applications

Apart from the aforementioned, a lot of other works on the application of PMC to various processes such as thermal recovery of sorbents (Futko et al., 2007), metal hydride heat conversion (Fateev and Rabinovich, 1997, 1998), heat exchangers, HCl synthesis, oxidization of volatile organic compounds (VOC), and powder combustion (Mujeebu et al., 2009b) have been reported.

3 Conclusion

Applications on porous media combustion have been made. Yet many of the works have been left without citation due to page limitation but without any intention. More investigations have to be done on the application of PMC in gas turbines, IC engines, combustion of gaseous fuels with low heating value such as the producer gas, heat exchangers, district heating and room heating, bakery ovens, and many other applications. The concept of PMC-based CHP is an excellent idea in terms of both energy efficiency and clean environment. Moreover, the recently introduced micro- and mesoscale applications provide a promising breakthrough in PMC technology, likely resulting in miniaturized combustion systems for MEMS (microelectromechanical system) applications. The need and wide scope for further research in all the aforementioned fields are quite obvious.

References

Akkutlu, I.Y., Yortsos, Y.C., 2003. The dynamics of in-situ combustion fronts in porous media. Combustion Flame 134, 229–247.

Al-Hamamre, Z., Diezinger, S., Talukdar, P., Issendorff, F.V., Trimis, D., 2006. Combustion of low calorific gases from landfills and waste pyrolysis using porous medium burner technology. Trans IChemE, Part B: Process Saf. Environ. Prot. 84 (B4), 297–308.

Bingue, J.P., Saveliev, A.V., Fridman, A.A., Kennedy, L.A., 2002. Hydrogen production in ultra-rich filtration combustion of methane and hydrogen sulphide. Int. J. Hydrogen Energy 27, 643–649.

Bingue, J.P., Saveliev, A.V., Kennedy, L.A., 2004. Optimization of hydrogen production by filtration combustion of methane by oxygen enrichment and depletion. Int. J. Hydrogen Energy 29, 1365–1370.

Dobrego, K.V., Gnezdilov, N.N., Kozlov, I.M., Bubnovich, V.I., Gonzalez, H.A., 2005. Numerical investigation of the new Regenerator-recuperator scheme of VOC oxidizer. Int. J. Heat Mass Transfer 48, 4695–4703.

Dobrego, K.V., Gnezdilov, N.N., Kozlov, I.M., Shmelev, E.S., 2006. Numerical study and optimization of the porous media VOC oxidizer with electric heating elements. Int. J. Heat Mass Transfer 6, 1–10.

Dobrego, K.V., Gnezdilov, N.N., Kozlov, I.M., 2007. Parametric study of recuperative VOC oxidation reactor with porous media. Int. J. Heat Mass Transfer 50, 2787–2794.

Durst, F., Weclas, M., 2001. A new type of internal combustion engine based on the porous-medium combustion technique. J. Automobile Eng. IMechE, Part D 215, 63–81.

Echigo, R., Ichimiya, K., Kurusu, M., Yoshizawa, Y., 1983. Combustion augmentation of extremely low calorific gases: application of the effective energy conversion method from gas enthalpy to thermal radiation. In: Lee, J.S. (Ed.), Proc. ASME-JSME Thermal Eng. Jt. Conference, Honolulu, 4, pp. 99–104.

Echigo, R., Yoshida, H., Tawata, H., Tada, S., 1993. 12th International Conference Thermoelectrics. Yokohama, Japan. VI-5.

Echigo, R., Yoshida, H., Tawata, K., Koda, M., Hanamura, K., 1994. 10th International Heat Transfer Conference. Brighton, UK, 1994, Vol. 3, p. 173.

Echigo, R., Yoshida, H., Tawata, K., Tada, S., 1995. Proceedings 4th ASME/JSME Thermal Engineering Joint Conference. Lahaina, HI. Vol. 4, p. 389.

Ellzey, J.L., William Jr., M., 2003. Porous burner for gas turbine applications. U.S. Patent No. 2003024655.

Fateev, G.A., Rabinovich, O.S., 1997. Metal hydride heat conversion on the basis of superadiabatic combustion waves in porous med. Int. J. Hydrogen Energy 22, 915–924.

Fateev, G.A., Rabinovich, O.S., 1998. Interaction of superadiabatic combustion and heat conversion waves in a porous medium with incorporated metal hydride element. Twenty-Seventh Symposium (International) on Combustion/The Combustion Institute, 2451–2458.

Fuse, T., Araki, Y., Kobayashi, N., Hasatani, M., 2003. Combustion characteristics in oil-vaporizing sustained by radiant heat reflux enhanced with higher porous ceramics. Fuel 82, 1411–1417.

Futko, S.I., Dobrego, K.V., Shmelev, E.S., Suvorov, A.V., Zhdanok, S.A., 2007. Thermal recovery of sorbents by filtration combustion. Combustion Sci. Technol. 179, 883–903.

Hanamura, K., Bohda, K., Miyairp, Y., 1997. A study of super-adiabatic combustion engine. Energy Convers. Mgmt 38, 1259–1266.

Hanamura, K., Kumano, T., 2003. Electric power generation by super-adiabatic combustion in thermoelectric porous element, 1st International Energy Conversion Engineering Conference, 17-21 August 2003. Portsmouth, Virginia, AIAA, 2003–6098.

Hanamura, K., Kumano, T., Iida, L., 2005. Electric power generation by super-adiabatic combustion in thermoelectric porous element. Energy 30, 347–357.

Howell, J.R., Hall, M.J., Ellzey, J.L., 1996. Combustion of hydrocarbon fuels within porous inert media. Prog. Energy Combust Sci. 122, 121–145.

Hunt, T.K., Ivanenok, J.F., Sievers, R.K., 1994. AMTEC auxiliary power unit for hybrid electric vehicles. In: Proceedings of the 29th Intersociety Energy Conversion Engineering Conference 1994.

Hunt, T.K., Sievers, R.K., Ivanenok, J.F., 1995. Low Emission AMTEC automotive power system. In: Proceedings of the 30th Intersociety Energy Conversion Engineering Conference, pp. 145–150.

Jugjai, S., Rungsimuntuchart, N., 2002. High efficiency heat-recirculating domestic gas burners. Exp. Thermal Fluid Sci. 26, 581–592.

Kamal, M.M., Mohamad, A.A., 2006. Combustion in Porous Media, a Review. J. Power Energy 220 (5), 487–508.

Kamal, M.M., Mohamad, A.A., 2006. Effect of Swirl on Performance of Foam Porous Medium Burners. Combustion Sci. Technol. 178 (4), 729–761.

Kamijo, T., Suzuki, Y., Kasagi, N., Okamasa, T., 2009. High-temperature micro catalytic combustor with Pd/nano-porous alumina. Proc. Combust. Inst. 32 (2), 3019–3026.

Li, Y.-H., Chao, Y.-C., Dunn-Rankin, D., 2008. Combustion in a meso-scale liquid-fuel-film combustor with central-porous fuel inlet. Combustion Sci. Technol. 180, 1900–1919.

Marbach, T.L., Agrawal, A.K., 2006. Heat-recirculating combustor using porous inert media for mesoscale applications. J. propulsion power 22, 145.

Marbach, T.L., Sadasivuni, V., Agrawal, A.K., 2007. Investigation of a miniature combustor using porous media surface stabilized flame. Combustion Sci. Technol. 179, 901–1922.

Mohamad, A.A., Ramadhyani, S., Viskanta, R., 1994. Modelling of Combustion and Heat Transfer in a Packed Bed with Embedded Coolant Tubes. Int. J. Heat Mass Transfer 37, 1181–1197.

Mohamad, A.A., Viskanta, R., Ramadhyani, S., 1994. Numerical Predictions of Combustion and Heat Transfer in Packed Bed with Embedded Coolant Tubes. Combustion Sci. Technol. J. 96, 387–407.

Mohamad, A.A., 2005. Combustion in porous media: Fundamentals and applications. Transport Phenomena Porous Media III, 287–304.

Mößbauer, S., Pickenäcker, O., Pickenäcker, K., Trimis, D., 1999. Application of the porous burner technology in energy and heat engineering. Fifth international conference on technologies and combustion for a clean environment (Clean Air V), Libson (Portugal), 12–15 July 1999, Volume 1. Lecture 20 (2), 519–523.

Mößbauer, S., Grüber, W., Trimis, D., 2001. Exhaust gas recirculation in porous burners for the target application zero emission steam engines. In: Orlandini, L. (Ed.), Proceedings of the Sixth International Conference on Technologies and combustion for a clean environment, vol. 2. Porto, Portugal, 9-12 July 2001, pp. 213–218.

Mujeebu, M.A., Abdullah, M.Z., Abu Bakar, M.Z., Mohamad, A.A., Muhad, R.M.N., Khalil, M., 2009a. Combustion in Porous media and its applications- A comprehensive survey. J. Environ. Manag. 90, 2287–2312.

Mujeebu, M.A., Abdullah, M.Z., Abu Bakar, M.Z., Mohamad, A.A., Abdullah, M.K., 2009b. Applications of porous media combustion technology—A review. Appl. Energy 86 (9), 1365–1375.

Mujeebu, M.A., Abdullah, M.Z., Abu Bakar, M.Z., Mohamad, A.A., Khalil, M., 2009c. A review of investigations on liquid fuel combustion in porous inert media. Prog. Energy Combustion Sci. 35, 216–230.

Noordally, E., Przybylski, J.M., Witton, J.J., 2004. Porous media combustors for clean gas turbine engines, Technical report, No: A318924, December 2004. Cranfield University, United Kingdom.

Park, C.W., Kaviany, M., 2002. Evaporation-combustion affected by in-cylinder, reciprocating porous regenerator. ASME J. Heat Transfer 124, 184–194.

Pedersen-Mjaanes, H., Chan, L., Mastorakos, E., 2005. Hydrogen production from rich combustion in porous media. Int. J. Hydrogen Energy 30, 579–592.

Periasamy, C., Chinthamony, S.K.S., Gollahalli, S.R., 2004. Numerical modelling of evaporation process in porous media for gas turbine applications, 42nd AIAA Aerospace Sciences Meeting and Exhibit 5–8 January 2004. Reno, Nevada, No. AIAA, 2004–2139.

Polasek, M., Macek, J., 2003. Homogenization of combustion in cylinder of CI engine using porous medium. SAE Int.. NO. 2003–01-1085.

Qiu, K., Hayden, A.C.S., 2006. Premixed gas combustion stabilized in fiber felt and its application to a novel radiant burner. Fuel 85, 1094–1100.

Qiu, K., Hayden, A.C.S., 2007. Thermophotovoltaic power generation systems using natural gas-fired radiant burners. Solar Energy Mater. & Solar Cell. 91, 588–596.

Qiu, K., Hayden, A.C.S., 2009. Increasing the efficiency of radiant burners by using polymer membranes. Appl. Energy 86 (3), 349–354.

Raviraj, S.D., Ellzey, J.L., 2006. Numerical and experimental study of the conversion of methane to hydrogen in a porous media reactor. Combustion Flame 144, 698–709.

Sadasivuni, V., Agrawal, A.K., 2009. A novel meso-scale combustion system for operation with liquid fuels. Proc. Combust. Inst. 32 (2), 3155–3162.

Sanmiguel, J.E., Mehta, S.A., Moore, R.G., 2003. An experimental study of controlled gas-phase combustion in porous media for enhanced recovery of oil and gas. ASME Trans. 125.

Slimane, R.B., Lau, F.S., Khinkis, M., Bingue, J.P., Saveliev, A.V., Kennedy, L.A., 2004. Conversion of hydrogen sulfide to hydrogen by super adiabatic partial oxidation: Thermodynamic consideration. Int. J. Hydrogen Energy 29, 1471–1477.

Shabunya, S.I., Martynenko, V.V., Yadrevskaya, N.L., Yakimovich, A.D., 2001. Modeling of the nonstationary process of conversion of methane to hydrogen in a filtration-combustion wave. J. Eng. Phys. Thermophysics 74 (5), 1059–1067.

Sozer, E., Shyy, W., Thakur, S., 2006. Multi-scale porous media modelling for liquid rocket injector applications, 42nd AIAA/ASME/SAE/ASEE Joint Propulsion Conference & Exhibit, 9-12 July. Sacramento, California, AIAA, 2006–5044.

Weclas, M., 2001. New strategies for homogeneous combustion in I.C. engines based on the porous medium (PM)-technology. ILASS Europe. June 2001.

Weclas, M., 2004. Strategy for intelligent Internal Combustion engine with homogeneous combustion in cylinder. Sonderdruck Schriftenr. Univ. Appl. Sci. Nuernberg No 26, 1–14.

Weclas, M., 2005. Porous media in internal combustion engines. In: Scheffler, M. (Ed.), Cellular Ceramics-Structure, Manufacturing, Properties and Applications. Wiley-VCH-Publ.2005, Colombo, P.

CHAPTER

Colloids in the Environmental Protection—Current and Future Trends

25

Qaisar Mahmood, (TI)*, Ather Farooq Khan†, Afsar Khan**

*Department of Environmental Sciences, COMSATS Institute of Information Technology, Abbottabad 22060, Pakistan, †Interdisciplinary Research Centre in Biomedical Materials, COMSATS Institute of Information Technology, Lahore-54000, Pakistan, **Department of Chemistry, COMSATS Institute of Information Technology, Abbottabad 22060, Pakistan

CHAPTER OUTLINE

The Role of Colloidal Systems in Environmental Protection. http://dx.doi.org/10.1016/B978-0-444-63283-8.00025-9

1 Introduction

Colloids are everywhere that we look, so why is it that most people know so little about them? Although Michael Faraday, one of the founders of colloid science, had beaten us to it by more than a century, as students we saw for ourselves the fascination of gold sols, nanometer-sized particles of gold dispersed in water as a stable colloid with a red coloration that changes to blue upon adding salt. Gold sols have currently gained renewed interest as a building block for nanotechnology.

Colloid means glue-like, originating from the Greek, koλλα. The term *colloidal* refers to a state of subdivision, implying that the molecules or polymolecular particles dispersed in a medium have at least in one direction a dimension roughly between 1 nm and 1 μm, or that in system discontinuities is found at distances of that order. It is not necessary for all three dimensions to be in the colloidal range: fibers in which only two dimensions are in this range and thin films, in which one dimension is in this range, may also be classified as colloidal. Nor is it necessary for the units of a *colloidal system* to be discrete: Continuous network structures, the basic units of which are of colloidal dimensions also fall in this class (e.g. porous solids, gels and foams). A *colloidal dispersion* is a system in which particles of colloidal size of any nature (e.g., solid, liquid, or gas) are dispersed in a *continuous phase* of a different composition (or state).

Colloids are materials that are predominantly liquid but that have other properties: either optical, giving rise to turbidity such as milk, or viscous, with characteristics of mucus, gelatin, or wet clay. These effects arise from the presence of macromolecules dissolved in liquid and/or by mixing two or more solid, liquid, or gas phases. Colloid science can therefore be described on the one hand as the study of solutions of macromolecules, for example, proteins in water or solutions of synthetic polymers, such as the clear glues for model construction kits. On the other hand, it is the study of dispersions of one phase in another, for example, emulsions (oil in water or water in oil), solid in liquid, foams, and the complex lyotropic liquid crystal dispersions of soap or synthetic detergents. Some readers will remember the old problem of mushy soap bars when left in contact with water, which arises from the ingress of water expanding the once hard compacted soap.

Most researchers agree that the term *colloid* is applied to suspended material in the size range of 1 μm to 1000 nm that may include inorganic materials, mineral fragments and mineral precipitates, biocolloids, as well as natural organic matter and other organic compounds and degradation products associated with low- and intermediate-level waste stream. Various workers have mentioned different size ranges; however, the upper limit is 1 μm. The term may be used to denote either the particles or the entire system. Colloidal systems (also called colloidal solutions or colloidal suspensions) are the subject of interface and colloid science. This field of study was introduced in 1861 by Scottish scientist Thomas Graham. Following are some explanations of various kinds of colloidal systems.

The name *dispersed phase* for the particles should be used only if they have essentially the properties of a bulk phase of the same composition.

The term *colloid* may be used as a short synonym for colloidal system. The size limits given above are not rigid since they will depend to some extent on the properties under consideration. This nomenclature can be applied to coarser systems, especially when a gradual transition of properties is considered.

The description of colloidal systems often requires numbering the components or constituents. It is felt that a fixed rule of numbering is unnecessarily restrictive. However, the author should make clear in all cases how he is numbering and in particular whether he is numbering by independent thermodynamic components (all neutral) or by species or constituents, of which some may be ionic, and which may be related by equilibrium conditions or by the condition of electroneutrality. A fluid colloidal system composed of two or more components may be called a *sol*, for example, a protein sol, a gold sol, an emulsion, a surfactant solution above the critical micelle concentration, or an aerosol.

In a *suspension,* solid particles are dispersed in a liquid; a *colloidal suspension* is one in which the size of the particles lies in the colloidal range. In an *emulsion* liquid, droplets and/or liquid crystals are dispersed in a liquid. In emulsions, the droplets often exceed the usual limits for colloids in size. An emulsion is denoted by the symbol O/W if the continuous phase is an aqueous solution and by W/O if the continuous phase is organic liquid (oil). More complicated emulsions such as O/W/O (i.e., oil droplets contained within aqueous droplets dispersed in a continuous oil phase) are also possible. Photographic emulsions, though colloidal systems, are not emulsions in the sense of this nomenclature.

A *latex* (plural = latices or latexes) is an emulsion or sol in which each colloidal particle contains a number of macromolecules.

A *foam* is a dispersion in which a large proportion of gas by volume in the form of gas bubbles is dispersed in a liquid, solid, or gel. The diameter of the bubbles is usually larger than 1 mm, but the thickness of the lamellae between the bubbles is often in the usual colloidal size range.

The term *froth* has been used interchangeably with foam. In particular, cases of froth may be distinguished from foam by the fact that the former is stabilized by solid particles (as in *froth-flotation*) and the latter by soluble substances.

Aerosols are dispersions in gases. In aerosols, the particles often exceed the usual size limits for colloids. If the dispersed particles are solid, one speaks of *aerosols of solid particles*; if they are liquid, one speaks of *aerosols of liquid particles*. Use of the terms *solid aerosol* and *liquid aerosol* is discouraged. An aerosol is neither "solid" nor "liquid" but, if anything, gaseous.

A great variety of terms such as dust, haze, fog, mist, drizzle, smoke, and smog are used to describe aerosols according to their properties, origin, and so on. Of these, only the terms *fog* and *smoke* are included in this nomenclature.

A *fog* is an aerosol of liquid particles, in particular a low cloud.

A *smoke* is an aerosol originating from combustion, thermal decomposition, or thermal evaporation. Its particles may be solid (magnesium oxide smoke) or liquid (tobacco smoke).

A *gel* is a colloidal system with a finite, usually rather small, yield stress. Materials such as silica gel which have passed a gel stage during preparation are improperly called gels.

The term *xerogel* is used for such dried-out open structures; and also for dried-out compact macromolecular gels such as gelatin or rubber.

The term *aerogel* is used when the openness of the structure is largely maintained.

Colloidal dispersions may be *lyophobic* (*hydrophobic*, if the dispersion medium is an aqueous solution) or *lyophilic* (*hydrophilic*). *Lyophilic sols* are formed spontaneously when the dry coherent material (e.g., gelatin, rubber, soap) is brought in contact with the dispersion medium; hence they are thermodynamically more stable than in the initial state of dry colloid material plus dispersion medium. *Lyophobic sols* (e.g., gold sol) cannot be formed by spontaneous dispersion in the medium. They are thermodynamically unstable with respect to separation into macroscopic phases, but they may remain for long times in a metastable state.

Lyophilic sols comprise both *association colloids* in which aggregates of small molecules are formed reversibly and *macromolecules* in which the molecules themselves are of colloidal size.

Mixtures of lyophobic and lyophilic colloids may form *protected lyophobic colloids*.

The terms *lyophilic* (*hydrophilic, lipophilic, oleophilic*, etc.) and *lyophobic* (*lipophobic*, etc.) may also be used to describe the character of interaction of a particular atomic group with the medium. In this usage, the terms have the relative qualitative meaning of "solvent preferring" (water-preferring, fat-preferring etc.) and "solvent rejecting" (water-rejecting, fat-rejecting, etc.), respectively.

The terms *solvent preferring* or *solvent rejecting* always refer to a differential process usually in the sense of preferring the solvent above itself or preferring itself above the solvent, but sometimes preferring one solvent (e.g., water) above another (e.g., oil).

A *colloidal electrolyte* is an electrolyte that gives ions, of which at least one is of colloidal size. This term, therefore, includes hydrophobic sols, ionic association colloids, and polyelectrolytes.

Ions of low relative molecular mass, with a charge opposite to that of the colloidal ion, are called *counter-ions*; if their charge has the same sign as that of the colloidal ion, they are called *co-ions*.

A *polyelectrolyte* is a macromolecular substance which, on dissolving in water or another ionizing solvent, dissociates to give *polyions* (*polycations* or *polyanions*)–multiply charged ions–together with an equivalent amount of ions of small charge and opposite sign. Polyelectrolytes dissociating into polycations and polyanions, with no ions of small charge, are also conceivable. A polyelectrolyte can be a *polyacid*, a *polybase*, a *polysalt,* or a *polyampholyte*.

If all particles in a colloidal system are of (nearly) the same size, the system is called *monodisperse*; in the opposite cases, the systems are *heterodisperse*.

If only a few particle sizes occur in a colloidal system, the system is *paucidisperse*, and if many particle-sizes occur it is *polydisperse*.

Because of their size, colloidal particles can pass through ordinary filters, but not through the extremely fine openings in a semipermeable membrane, such as parchment. A liquid cannot flow through a semipermeable membrane, but will diffuse through it slowly if liquid is on the other side. Although a colloidal dispersion cannot be purified by filtration, it can be dialyzed by placing it in a semipermeable bag with pure water on the outside. Dissolved impurities then gradually diffuse through the bag, while the colloidal particles remain imprisoned within it. If the process of dialysis is carried to completion, the suspension will often break down, or settle, because the stability of colloidal systems frequently depends on the electrical charges on the individual particles, and these are in turn generally dependent on the presence of dissolved electrolytes.

Although individual colloidal particles are too small to be seen with an ordinary microscope, they can be made visible by means of an ultra microscope, or dark-field microscope. If a colloidal dispersion is placed under a microscope and a beam of light is directed through from one side, the path of the beam becomes visible by scattering from the colloidal particles. This same phenomenon makes the path of a beam of light visible in a darkened room, but under the microscope separate flashes of light are observed. The particles are seen to be in random motion as the result of Brownian motion, and their speed is exactly that calculated for molecules the size of the colloidal particles. The particles are directly visible in an electron microscope. Some colloids are translucent because of the Tyndall effect, which is the scattering of light by particles in the colloid. Other colloids may be opaque or have a slight color. A colloidal dispersion of sub-μm particles may be stable or unstable to aggregation. Brownian motion ensures that the particles are in continual motion, giving rise to collisions at a rate determined by diffusion theory. Owing to the high interfacial free energy, lyophobic colloids are thermodynamically unstable and tend to aggregate. This is generally undesirable, and colloid scientists aim to prevent it from occurring.

In a stable dispersion, the particle collisions do not lead to aggregation because interparticle repulsion forces dominate. It will remain dispersed indefinitely, although particles bigger than about 0.1μm will sediment depending on their density. In an unstable dispersion, the collisions lead to aggregate formation; larger aggregates either sediment or cream depending on their relative density.

The repulsive forces in a stable dispersion were long ago identified as being electrical in origin. A surface potential exists at the interface between the solid particle and the surrounding liquid due to the presence of a surface charge. To maintain electrical neutrality, ions of opposite charge present in the medium are attracted closer to the particle surface, resulting in a diffuse layer of highly concentrated counter-ions. The concentration of counter-ions in this layer decays exponentially from the surface over a distance of tens of nanometers. The resulting ionic cloud is called the diffuse region of the electric double layer. On particle–particle collision, overlap of the ionic clouds gives rise to an osmotic repulsion that pushes the particles apart.

The DLVO theory of colloid stability, developed by Derjaguin and Landau and Verwey and Overbeek during the 1940s, proposes a balance of the repulsive electric double-layer forces (positive by convention) and the attractive van der Waals forces (negative by convention) that exist between all matters. These two forces were found to be of similar range and magnitude. The electrical forces increase exponentially as particles approach one another and the attractive forces increase as an inverse power of separation. As a consequence, these additive forces may be expressed as a potential energy versus separation curve. A positive resultant corresponds to an energy barrier and repulsion, while a negative resultant corresponds to attraction and hence aggregation. It is generally considered that the basic theory and its subsequent modifications provide a sound basis for understanding colloid stability.

The adsorption of lyophilic colloids—macromolecules—by the surface of lyophobic colloids gives rise to an additional repulsive force. Macromolecules attach to the surface to form a loop-like configuration of trains of segments attached to the surface, and loops and tails of segments extending out into the liquid phase. Research, mainly during the 1960s and 1970s, identified the nature of the repulsive forces arising from such adsorption. These are a combination of entropic repulsion, arising from the restricted configurational freedom of the adsorbed molecules when two particles collide; and osmotic repulsion, arising from the increased concentration of segments in the overlap region of the adsorbed layers on particle–particle contact. Except under special conditions, the presence of a saturated adsorbed layer always leads to a total stabilization of the dispersion to coagulation. Earlier publications referred to this effect as colloid protection, but it is now termed *steric stabilization*.

1.1 Environmental importance of colloids

Colloids are of environmental importance because their interaction with trace organic and inorganic pollutants, such as metals and persistent organic pollutants (POPs), play an important role in bio-uptake and biogeochemical cycling of the pollutants. In addition, colloids are important, often dominant, ligands affecting the speciation of trace metals (Geckeis et al., 2002) and other pollutants in environmental systems and affect the behavior of microbial pathogens. The importance of these colloids in pollutant transport in estuaries, lakes, and rivers has long been recognized (Tessier et al., 1994), with potential ecological and human health effects. Increasingly, it is recognized that bio-uptake and bioeffects of pollutants to many organisms such as microorganisms, filter and detritus feeders, fish, and even mammals may be altered in complex ways by association with colloids, which may be biologically or nonbiologically produced (Wilkinson and Buffle, 2004). A great deal of recent work has focused on bio-uptake of free metals and colloid-bound metals (e.g., Pan and Wang, 2003; Wang and Guo, 2000; Carvallho et al, 1999), although substantial progress is still required.

Despite the obvious importance, there is a lack of quantitative understanding of the structure of aquatic colloids and how this relates to their environmental "function" in trace pollutants and pathogen regulation (Muirhead and Lead, 2003). In addition, there has been an enormous growth in the number of techniques that can be used for the nonperturbing and quantitative separation and analysis of colloids and colloid-pollutant complexes, such as types of force and electron microscopy (Muirhead and Lead, 2003; Mondi et al., 2002; Balnois and Wilkinson, 2002), fluorescence correlation spectroscopy (FCS, Lead et al., 2003), field flow fractionation (FFF; Lyven et al., 2003), and X-ray spectroscopy (Gaillard et al., 2001). However, most of these methods are not currently being used for study of nonperturbed environmental colloidal systems.

2 Current and future trends in colloidal use for environmental protection

2.1 Superabsorbent materials

Special hydrogels as superabsorbent materials are widely employed in hygienic uses, particularly disposable diapers and female napkins where they can capture secreted fluids, for example, urine and blood. Agricultural grade of such hydrogels are used as granules for holding soil moisture in arid areas (Zohuriaan-Mehr and Kabiri, 2008). Superabsorbent polymer (SAP) materials are hydrophilic networks that can absorb and retain huge amounts of water or aqueous solutions. They can uptake water as high as 100,000%. Common SAPs are generally white sugar-like hygroscopic materials, which are mainly used in disposable diapers and other applications, including agricultural use. Due to variability of the possible monomers and macromolecular structure, many SAP types can be made. SAPs are originally divided into two main classes: synthetic (petrochemical-based) and natural (e.g., polysaccharide-and polypeptide-based). Most of the current superabsorbents, however, are frequently produced from acrylic acid (AA), its salts, and acrylamide (AM) via solution or inverse-suspension polymerization techniques. The main synthetic (internal) and environmental (external) factors affecting the acrylic anionic SAP characteristics are described briefly. The methods for quantifying the SAP practical features, that is, absorption capacity (both load-free and under load), swelling rate, swollen gel strength, wicking capacity, sol fraction, residual monomer, and ionic sensitivity were discussed. The SAP applications and the related research works, particularly the hygienic and agricultural areas, are reviewed. Meanwhile, the research findings on the effects of SAP in soil and agricultural achievements in Iran, as an arid country, are treated as well. Finally, the safety and environmental issues concerning SAP practical applications are discussed (Zohuriaan-Mehr and Kabiri, 2008).

Environmental sensitive hydrogels have the ability to sense environmental stimuli, such as changes of pH, temperature, or the concentration of metabolite and then release their load as a result of such a change. Hydrogels that are responsive to specific molecules, such as glucose or antigens, can be used as biosensors as well as in drug delivery systems (DDS). These kinds of hydrogels are also used as controlled-release delivery devices for bioactive agents and agrochemicals. Contact lenses are also based on hydrogels.

The hygroscopic materials are usually categorized into two main classes based on the major mechanism of water absorption: chemical and physical absorptions. Chemical absorbers (e.g., metal hydrides) catch water via chemical reaction, converting their entire nature. Physical absorbers imbibe water via four main mechanisms [8]:

1. Reversible changes of their crystal structure (e.g., silica gel and anhydrous inorganic salts).
2. Physical entrapment of water via capillary forces in their macroporous structure (e.g., soft polyurethane sponge).
3. A combination of the mechanism (2) and hydration of functional groups (e.g., tissue paper).
4. The mechanism that may be anticipated by combination of mechanisms of (2) and (3) and essentially dissolution and thermodynamically favored expansion of the macromolecular chains limited by cross linkages.

Superabsorbent (SAP) materials fit in the latter category, yet they are organic materials with enormous capability of water absorption. SAPs as hydrogels relative to their own mass can absorb and retain extraordinary large amounts of water or aqueous solution (Buchholz and Graham, 1998; Brannon-Peppas and Harland, 1990). These ultrahigh absorbing materials can imbibe deionized water as high as 1000–100,000% (10–1000 g/g), whereas the absorption capacity of common hydrogels is not more than 100% (1 g/g).

Commercial SAP hydrogels are generally sugar-like hygroscopic materials with white-light yellow color. The SAP particle shape (granule, fiber, film, etc.) has to be basically preserved after water absorption and swelling, that is, the swollen gel strength should be high enough to prevent a loosening, mushy, or slimy state. This is a major practical feature that distinguishes SAPs from other hydrogels.

Several papers have been published to review SAP hydrogel materials, each with its own individual outlook. As a general framework, synthetic methods and properties of hydrogel networks were reviewed (Mathur et al., 1996). Synthetic, semisynthetic, and biopolymeric hydrogels were also briefly reviewed (Kulicke and Nottelmann, 1989). The chemistry and physics of agricultural hydrogels were reviewed by Kazanskii and Dubrovskii (1992). Bouranis et al. (1995) have reviewed the synthetic polymers as soil conditioners.

Superabsorbents obtained from shellfish waste have also been reviewed (Dutkiewicz, 2002). Ichikawa and Nakajima have discussed the superabsorptive materials based on the polysaccharides and proteins (Ichikawa and Nakajima, 1996). A review profile of water-absorbing resins based on graft copolymers of acrylic acid and gelatinized starch was presented by Athawale et al. (2001). Buchholz has elaborated the uses of superabsorbents based on cross-linked, partially neutralized poly(acrylic acid) and graft copolymers of starch and acrylic acid (Buchholz, 1994). In another review, the synthesis of cross-linked acrylic acid-*co*-sodium/potassium acrylate has been described. The solution and suspension polymerization techniques used for preparing the acrylate superabsorbents have been discussed in detail (Dayal et al., 1999).

Two valuable books on the synthetic SAP materials were published in 1990–1998 (Buchholz and Graham, 1998; Brannon-Peppas and Harland, 1990), and the fundamental phenomena dealing with the synthetic hydrogels were reflected very clearly (Brannon-Peppas and Harland, 1990). In 2002, another valuable book was published, focused mainly on the fibers and textiles with high water absorbency characteristics (Chatterjee, 2002).

2.1.1 Characteristics of SAPs

The functional features of an ideal SAP material can be listed as follows (Zohuriaan-Mehr, 2006):

- The highest absorption capacity (maximum equilibrium swelling) in saline
- Desired rate of absorption (preferred particle size and porosity) depending on the application requirement
- The highest absorbency under load (AUL)
- The lowest soluble content and residual monomer
- The lowest price
- The highest durability and stability in the swelling environment and during the storage
- The highest biodegradability without formation of toxic species following the degradation
- pH-neutrality after swelling in water
- Colorlessness, odorlessness, and absolute nontoxicity
- Photostability
- Re-wetting capability (if required) (Zohuriaan-Mehr, 2006)

The SAP has to be able to give back the imbibed solution or to maintain it, depending on the application requirement (e.g., in agricultural or hygienic applications).

Obviously, it is impossible that a SAP sample would simultaneously fulfill all the above-mentioned required features. In fact, the synthetic components for achieving the maximum level of some of these features will lead to the inefficiency of the rest.

Therefore, in practice, the production reaction variables must be optimized such that an appropriate balance between the properties is achieved.

2.1.2 Applications of SAPs

Various applications and active fields of applied research works on SAPs are well reviewed by Po (1994). In addition to the hygienic and agricultural areas, SAP materials are (or can potentially be) used in many other fields (e.g., artificial snow, ornamental [colored] products, entertaining/educational toys and tools, building internal decoration, fire extinguishing/retarding gels, cryogenic gels, food/meat packaging, etc.) (Po, 1994). Concrete strengthening (Gao et al., 1997), reduction of the ground resistance in the electrical industry (Yamane et al., 1994), and controlled release of pesticides and agrochemicals (Bowman et al., 1990; Gao et al., 1997; Wu and Liu, 2008; Kenawy, 1998; Rudzinski et al., 2002; Guo et al., 2005, 2006; Liu et al., 2006, 2007; Liang et al., 2007; Levy et al., 1995, are other instances of SAP-applied research. In the field of food processing, for instance, yogurt dewatering was recently investigated using permeable membrane and acrylic SAP (Ahmadpour et al., 2007). Most recently, photochromic SAPs with excellent water absorption (2800 g/g) were synthesized using an azobenzene surface cross-linker (Mudiyanselage and Neckers, 2008). Under irradiation at 350 nm, water expulsion from the SAP is observed. The SAP preparation and characterization has been investigated in details (Mudiyanselage and Neckers, 2008a, 2008b). These photoactive hydrogels may be candidates to design new photochemically controlled systems for pharmaceutical, biomedical, or optical switching applications.

A surprising application of SAP materials was examined by Peter Cordani for modifying weather conditions (Cordani, 2001). Thus, a hurricane was seeded with almost 30,000 pounds of an SAP by means of a transport plane flying through the leading edge of the storm. Within 20 seconds, the SAP obtained over 70% of its absorption capacity, or nearly 300 times its weight. The winds of the storm would continue to disperse the materials, causing a form of internal flocculation and disrupting the feeding nature of the storm. When seeded close to land, the storm did not have sufficient time to reform to its previous destructive strength.

Research has shown little or no consistent adverse effects on soil microbial populations (Stahl et al., 2000). The environmental fate of SAPs and their microbial degradation were investigated by many researchers (Stahl et al., 2000; Wolter et al., 2002; Larson et al., 1997; Cutie et al., 1990; Barvenik and Polyacrylamide, 1994; Grula et al., 1994). The researchers at the University of California, Los Angles (UCLA), found that no toxic species remained in soil after several years of consuming SAP (Wallace et al., 1986).

2.1.3 Future trends

The discussion presented thus far provides ample evidence that colloids are very important chemical agents that play a pivotal role in environmental protection. There are many kinds of environmental stresses such as heavy metals in drinking water, pesticides, and food contaminants. Colloids are successfully applied in many areas of environmental protection. A number of environmental and energy technologies have already benefited substantially from colloidal technology.

1. Colloidal microgels have recently received attention as environmentally responsive systems and now are increasingly used in applications as carriers for therapeutic drugs and diagnostic agents. Synthetic microgels consist of a cross-linked polymer network that provides a depot for loaded drugs, protection against environmental hazards, and a template for postsynthetic modification or vectorization of the drug carriers. Vinogradov and Serguei (2006) has reviewed recent attempts to develop new microgel formulations for oral drug delivery. The synthesis of biologically active microgels will be of great beneficial value for protection against many pathogenic diseases in human and other domesticated animals.
2. Emulsions are successfully applied in many fields of human activity. When used as liquid colloidal carriers, the stability of emulsion droplets against coalescence often requires improvement. Additional protection against colloidal degradation or environmental stresses is almost an unavoidable precondition for employment of emulsion formulations in the food industry, pharmaceutics, cosmetics, and medicine (Grigoriev and Reinhard, 2009).
3. Another important application of colloids is their employment in the treatment of wastewater and contaminated soils. The world's current population is expected to grow to about 9 billion by the year 2025. This population increase will have direct influence on demand for water for domestic, industrial, and agricultural use. The use of colloids for water and soil purification from phenol, oil and oil products, and metal ions will greatly benefit humanity.

2.2 Nanoparticles in the environment

The increasing use of engineered nanoparticles (NP) in industrial and household applications will likely lead to the release of such materials into the environment (Nowackand Bucheli, 2007). Assessing the risks of these NP in the environment requires an understanding of their mobility, reactivity, ecotoxicity, and persistency. To date, only a few quantitative analytical techniques for measuring NP in natural systems are available, which results in a serious lack of information about their occurrence in the environment. Results from ecotoxicological studies show that certain NPs have effects on organisms under environmental conditions, though mostly at elevated concentrations. The next step toward assessing the risks of NP in the environment should therefore be to estimate exposure to the different NPs. It is also important to notice that most NP in technical applications are functionalized, and therefore studies using pristine NP may not be relevant for assessing the behavior of the NP actually used (Nowackand Bucheli, 2007).

Data on the current use and production of NP are sparse and often conflicting. One estimate for the production of engineered nanomaterials was 2000 tons in 2004, which was expected to increase to 58,000 tons in 2011–2020 (Maynard, 2006). Because of the potential of this technology, there has been a worldwide increase in investment in nanotechnology research and development (Guzman et al., 2006). The forecasted huge increase in the manufacture and use of NP makes it likely that increasing human and environmental exposure to NP will occur. Consequently, NP are beginning to come under scrutiny, and the discussion about the potential adverse effects of NP has increased steadily in recent years. In fact it has become a top priority for governments, the private sector, and the public all over the world (Roco, 2005; Helland et al., 2006; Siegrist et al., 2007). Most attention has thus far been devoted to the toxicology and health implications of NP (e.g., Oberdörster et al., 2005; Kreyling et al., 2006; Lam et al., 2006; Nel et al., 2006; Helland et al., 2007), while the behavior of NP in the environment (Biswas and Wu, 2005; Wiesner et al., 2006; Helland et al., 2007) and their ecotoxicology (Colvin, 2003; Moore, 2006; Oberdörster et al., 2006a) have been less often reviewed.

2.2.1 Engineered nanoparticles

Nanotechnology is defined as the understanding and control of matter at dimensions of roughly 1–100 nm, where unique physical properties make novel applications possible (EPA, 2007). NPs are therefore considered substances that are less than 100 nm in size in more than one dimension. There are various shapes of nanoparticles like spherical, tubular, or irregularly shaped, and they can exist in fused, aggregated, or agglomerated forms. NP can be divided into natural and anthropogenic particles (Nowack and Bucheli, 2007). The particles can be further separated based on their chemical composition into carbon-containing and inorganic NP. The C-containing natural NPs are divided into biogenic, geogenic, atmospheric, and pyrogenic NP (Table 25.1). Examples of natural NP are fullerenes and carbon nano-tubes (CNT) of geogenic or pyrogenic origin, biogenic magnetite, or atmospheric aerosols (both organic such as organic acids and inorganic such as sea salt). Anthropogenic NP can either be inadvertently formed as a by-product, mostly during combustion, or produced intentionally due to their particular characteristics (Nowack and Bucheli, 2004). In the latter case, they are often referred to as engineered or manufactured NP. Examples of engineered NP are fullerenes and CNT, both pristine and functionalized, and metals and metal oxides such as TiO_2 and Ag. Engineered NPs are the main focus of the current research on NP in the environment, but some of them also occur naturally, for example, as inorganic oxides or fullerenes.

Table 25.1 Various Kinds of Nanoparticles

		Formation		Examples
Natural	C-containing	Biogenic	Organic colloids Organisms	Humic, fulvic acids, viruses
		Geogenic	Soot	Fullerenes
		Atmospheric	Aerosols	Organic acids
		Pyrogenic	Soot	CNT Fullerenes Nanoglobules, onion-shaped nanospheres
	Inorganic	Biogenic	Oxides Metals	Magnetite Ag, Au
		Geogenic	Oxides Clays	Fe-oxides Allophane
		Atmospheric	Aerosols	Sea salt
Anthropogenic (engineered NP)	C-containing	By-products	Combustion by-products	CNT Nanoglobules, onion shaped nanospheres
		Engineered	Soot	Carbon black Fullerenes Functionalized CNT, fullerenes
			Polymeric NP	Polyethylene glycol (PEG) NP
	Inorganic	By-product	Combustion by-products	Platinum group metals
		Engineered	Oxides	TiO_2, SiO_2
			Metals	Ag, Iron
			Salts	Metal-phosphates
			Aluminosilicates	Zeolites, clays, ceramics

Source: Adapted from Nowack and Bucheli, 2007.

Environmental colloids include three major types of compounds: inorganic colloids, humic substances, and large biopolymers such as polysaccharides and peptidoglycans. Colloids must be seen as an essential building block of the abiotic medium supporting life in general (Buffle, 2006). Although knowledge of the structure and the environmental impact of natural colloids has significantly increased in recent years, their precise function and composition are still poorly defined.

Carbon Black (CB) is an industrial form of soot used in various applications such as filler in rubber compounds, primarily in automobile tires. The particle size of CB is partially in the nanometer range with average values between 20 and 300 nm for different materials (Blackford and Simons, 1987; Sirisinha and Prayoonchatphan, 2001; Tscharnuter et al., 2001). Although fullerenes and CNT are considered as engineered NP, they are also natural particles (fullerenes) or have close relatives in the environment (CNT). Whereas some of these fullerenes are of interstellar origin that have been brought to earth by comets or asteroids (Becker et al., 1996, 2001), the majority is believed to have formed from polycylic aromatic hydrocarbons (PAH) derived from algal matter during metamorphosis at temperatures between 300 and 500°C in the presence of elemental sulfur (Heymann et al., 2003) or during natural combustion processes.

A special class of unintentionally produced NP is composed of platinum (Pt) and rhodium (Rh) containing particles produced from automotive catalytic converters. Although most Pt and Rh are attached to coarser particles, about 17% was found to be associated with the finest aerosol fraction (< 0.43 mm) (Zereini et al., 2001).

Of the large family of fullerenes, the buckminsterfullerene C_{60} is by far the most widely investigated. Fullerenes are mainly proposed to be used in fullerene-polymer combinations, as thin films, in electro-optical devices, and in biological applications (Prato, 1999; Bosi et al., 2003). Due to the low water solubility of fullerenes, a lot of research is devoted to functionalization and a myriad of derivatives of C_{60} have been synthesized, which all have their peculiar characteristics and properties (Wudl, 2002). Carbon nanotubes (CNT) are considered the hottest topic in physics (Giles, 2006). Depending on the synthesis method, the technique used for separation from the amorphous by-products, subsequent cleaning steps, and finally different functionalizations, a variety of different CNTs are obtained that have very different properties (Dai, 2002; Niyogi et al., 2002).

Biological and medical applications in particular explore the potential of modifying the properties of CNT (Bianco and Prato, 2003). The NPs synthesized from organic polymers have gained widespread interest in medicine as carriers for drugs. The possibility of controlling size, surface charge, morphology, and composition make polymers especially well suited for designing NP with tailored properties. These NPs are taken up by a wide variety of cells and are studied for their ability to cross the blood-brain barrier (Koziara et al., 2003).

Several types of polymeric NPs have also been developed and proposed for soil and groundwater remediation. Micelle-like amphiphilic polyurethane particles have a hydrophilic outer side and a hydrophobic inner core and are therefore very well suited for the removal of hydrophobic pollutants (e.g., phenanthrene) from soils (Kim et al., 2000, 2003a,b, 2004a,b; Tungittiplakorn et al., 2004, 2005). The NP are able to extract the PAH in a similar manner to surfactant micelles, but unlike the micelles they do not sorb to soil particles.

Another polymeric nanoscale material is dendrimers that function as water-soluble chelators (Xu and Zhao, 2005, 2006). Engineered inorganic NP cover a broad range of substances, including elemental metals, metal oxides, and metal salts. Elemental silver is used in many products as bactericides (Morones et al., 2005), whereas elemental gold is explored for many possible applications and its catalytic activity (Brust and Liely, 2002). The use of nanoscale zero-valent iron (nZVI) for groundwater remediation ranks as the most widely investigated environmental nanotechnological application. Metallic iron is very effective in degrading a wide variety of common contaminants such as chlorinated methanes, brominated methanes, trihalomethanes, chlorinated ethenes, chlorinates benzenes, other polychlorinated hydrocarbons, pesticides, and dyes (Zhang, 2003). Successful results from field demonstrations using nZVI have been published, with injection of 1.7 to 400 kg of NP into the groundwater (Elliott and Zhang, 2001; Quinn et al., 2005).

To date, approximately 30 projects are under way where nZVI is used for site remediation (Li et al., 2006a). Nanoparticulate metal oxides are among the most used NP (Aitken et al., 2006). Bulk materials of TiO_2, SiO_2, and aluminum and iron oxides have been produced for many years. However, recently they have also been manufactured in nanosized form and have already entered the consumer market, for example, ZnO in sunscreens (Rittner, 2002). TiO_2 NP are widely used for applications such as photocatalysis, pigments, and cosmetic additives (Aitken et al., 2006).

A wide variety of other nanomaterials is under vigorous investigation by materials scientists. Nanosized zeolites (Larlus et al., 2006), clays (Yaron-Marcovich et al., 2005), and ceramics (Cain and Morrell, 2001) are other NPs that have been proposed for various applications. Several noncarbon nanotubes have also been synthesized (Pokropivnyi, 2001), for example, TiO_2 (Zhang et al., 2006). Quantum dots made from semiconductor materials such as CdSe, CdTe, or ZnS have attracted wide interest in areas such as information technology, molecular biology, and medicine (Gao et al., 2004).

2.2.2 Metabolism of NP in the environment

2.2.3.1 Aggregation of organic colloids and NP in water

In the environment, natural colloids or NP interact among themselves and with other natural NP or larger particles (Figures 25.1 and 25.2). The formation of aggregates in natural systems can be understood by considering physical processes, that is, Brownian diffusion, fluid motion, and gravity.

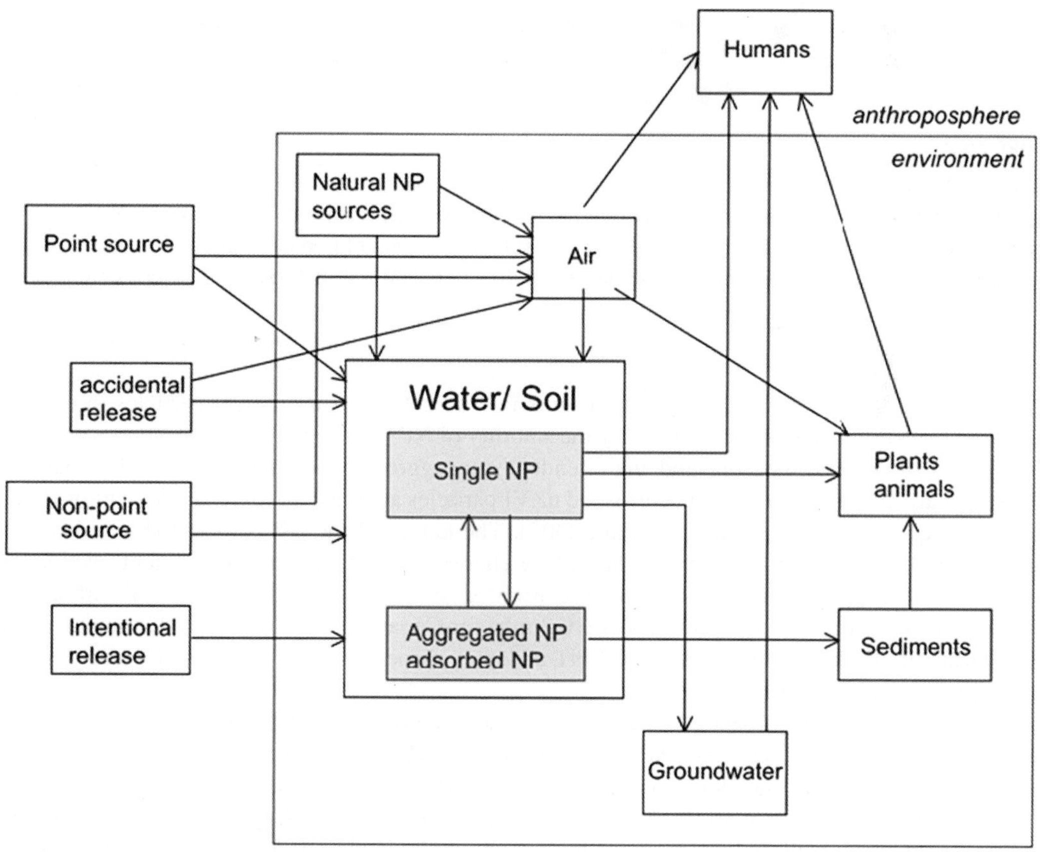

FIGURE 25.1

Nanoparticle pathways from the anthroposphere into the environment, reactions in the environment, and exposure of humans.

(adapted from Nowack and Bucheli, 2007).

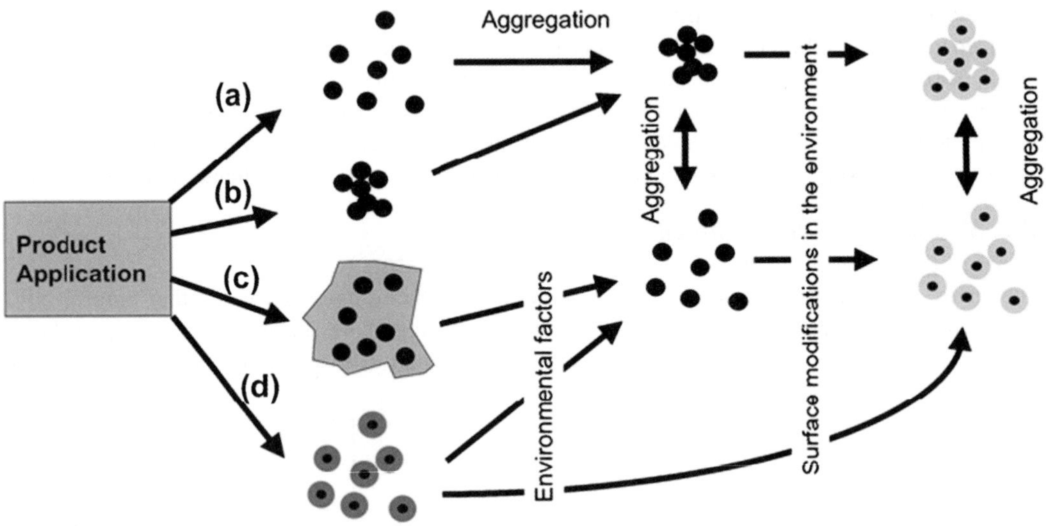

FIGURE 25.2

Release of NP from products and (intended or unintended) applications: (a) release of free NP, (b) release of aggregates of NP, (c) release of NP embedded in a matrix and (d) release of functionalized NP. Environmental factors (e.g., light, microorganisms) result in formation of free NP that can undergo aggregation reactions. Moreover, surface modifications (e.g., coating with natural compounds) can affect the aggregation behavior of the NP.

(adapted from Nowack and Bucheli, 2007).

Aggregation is particle-size dependent and results in efficient removal of small particles in environmental systems (Omelia, 1980). To quantify the stability of NP in the environment, we have to predict the stability of their suspension and their tendency to aggregate or interact with other particles (Mackay et al., 2006). Twenty nanometer-sized nZVI particles aggregated, for example, within 10 min to micrometer-sized clusters (Phenrat et al., 2007). The nature of the NP is modified by adsorption processes (Fukushi and Sato, 2005), and the surface charge in particular plays a dominant role (Kallay and Zalac, 2001, 2002). Cations, for example, are able to coagulate acid-treated CNT with critical coagulation concentrations of 37 mM for Na, 0.2 mM for Ca, and 0.05 mM for trivalent metals (e.g., La3þ) (Sano et al., 2001). Aggregation of CNT added as suspension to filtered pond water has been reported (Zhu et al., 2006c).

NPs are not necessarily released as a single NP (see Figure 25.2). In many applications, NPs are embedded in a matrix, and release of NP will occur through release of matrix-bound NP (Koehler et al., 2008). As many NPs are functionalized, release of functionalized NP is also possible. In the environment the released NPs are affected by environmental factors such as light, oxidants, or microorganisms. This can result in chemical or biological modification or degradation of the surface functionalization or the embedding matrix and may result in free NP. The surface of pristine NP can also be modified by environmental factors (e.g., coating by organic matter) or functionalized by chemical or biological processes. The effect of humic and fulvic acids to inhibit the aggregation of CNT has been recently shown (Hyung et al., 2007) and also nZVI is efficiently coated with humic acids (Giasuddin et al., 2007). Nanoparticulate ZnO, which was coated with the surfactant sodium dodecyl sulfate, was stable in soil suspension for 14 days without changes in particle-size distribution (Gimbert et al., 2007).

2.2.4 Behavior of NP in porous media

The transport of colloids in porous media and the colloid-facilitated transport of contaminants have received a lot of attention in the past (McGechan and Lewis, 2002; Sen and Khilar, 2006). The movement of colloids and therefore also of NP in porous media is impeded by two processes: straining or physical filtration where the particle is larger than the pore and is trapped; and true filtration where the particle is removed from the solution by interception, diffusion, and sedimentation.

Particles removed from solution by such processes can readily become resuspended upon changes in the chemical or physical conditions (e.g., changes in pH, ionic strength, and flow rate) (Sen and Khilar, 2006). Several studies have investigated the transport of a wide range of engineered NP through porous media (Lecoanet et al., 2004; Lecoanet and Wiesner, 2004; Dunphy Guzman et al., 2006). Particles smaller than 100 nm are predicted to have very high efficiencies of transport to collector surfaces due to Brownian diffusion. If all particle-collector contacts were to result in particle attachment to the collector, these small particles would be retained to a large extent by the porous medium. However, nanosized silica particles were not appreciably removed, and anatase NP were only removed between 55 and 70 %, depending on the flow velocity (Lecoanet and Wiesner, 2004). Fullerol (hydroxylated fullerene) and surfactant-stabilized CNT were almost completely mobile and only removed to a very low percentage (Lecoanet et al., 2004). The most efficient removal was observed for an iron oxide and for fullerene clusters (nC_{60}). The deposition efficiency for nC_{60} increased with time, and after 60 pore volumes virtually no nC_{60} were detected in the effluent. This effect was ascribed to "filter ripening," the increased filter efficiency by deposition of particles (Cheng et al., 2005a). These studies show that the collector efficiency for NP can be very different and that especially the surface-modified NP displayed high mobilities. Also, the environmental conditions are important, and efficient removal of titania NP was observed close to the pH at the point of zero charge (Dunphy Guzman et al., 2006). Because the nZVI particles used for groundwater remediation have a strong tendency to aggregate and adsorb to surfaces of minerals, a lot of effort has been directed toward methods to disperse the particles in water and render them mobile.

In one approach water-soluble starch was used as stabilizer (He and Zhao, 2005); in another, hydrophilic carbon or polyacrylic acid delivery vehicles were used (Schrick et al., 2004). Modified cellulose, sodium carboxymethyl cellulose (CMC), was found to form highly dispersed nZVI (He et al., 2007) and also several polymers have been tested and found to be very effective (Saleh et al., 2007). These modified NPs were found to be mobile under natural conditions, indicating the importance of knowing the exact surface properties of NP for a prediction of their potential mobility in the environment.

2.2.5 NP as adsorbents

The unique structure and electronic properties of some NPs can make them especially powerful adsorbents. Dissolved organic carbon and organic colloids in the submicron-size range have been recognized as a distinct nonaqueous organic phase to which organic pollutants are partitioning (Burkhard, 2000), which leads to their reduced bioavailability. Most data on dissolved organic carbon partition coefficients (KDOC) are available for PAH. In general, KDOC values for any individual compound vary widely over several orders of magnitude, depending on the characteristics of the sorbents such as size, conformation, and chemical composition, and probably also the experimental method applied. Moreover, they are significantly lower than predictions from octanol water partition coefficients, and partitioning relationships developed primarily from natural organic matter of sediments and soils (Burkhard, 2000). Still, by their shear abundance, such sorbents may significantly attenuate the truly dissolved exposures of organic pollutants. Note also that the partitioning of organic pollutants into colloidal organic carbon is not restricted to hydrophobic organic pollutants such as PAH, but also is of relevance for more polar compounds such as steroid hormones (Zhou et al., 2007a) or modern herbicides (Irace-Guigand and Aaron, 2003). The distribution of metal ions between solution and colloids strongly influences metal speciation and therefore metal bioavailability (Lead and Wilkinson, 2006). Colloids act, therefore, as a metal buffer that keep free metal ions, for example, of Cu^{2+}, within a range that is beneficial for life. Also, soot is a powerful adsorbent for organic compounds.

The nonlinear adsorption of organic compounds onto CB can completely dominate total sorption at low aqueous concentrations in soils and sediments (Cornelissen et al., 2005). The extremely efficient sorption to CB pulls highly toxic polycyclic aromatic hydrocarbons, polychlorinated biphenyls, dioxins, polybrominated diphenylethers, and pesticides into sediments and soils (Koelmans et al., 2006). The increased sorption is general but strongest for planar and most toxic compounds at environmentally relevant, low aqueous concentrations. The presence of CB can explain that the sorption of organic compounds into soils and sediments is much higher than expected based on absorption into organic matter alone (Cornelissen et al., 2005).

CNTs have received a lot of attention as very powerful adsorbents for a wide variety of organic compounds from water. Examples include dioxin (Long and Yang, 2001), PAH (Gotovac et al., 2006; Yang et al., 2006b,c), DDT and its metabolites (Zhou et al., 2006c), PBDEs (Wang et al., 2006), chlorobenzenes and chlorophenols (Peng et al., 2003; Cai et al., 2005), trihalomethanes (Lu et al., 2005, 2006), bisphenol and nonylphenol (Cai et al., 2003b), phthalate esters (Cai et al., 2003a), dyes (Fugetsu et al., 2004), the pesticides thiamethoxam, imidacloprid, and acetamiprid (Zhou et al., 2006a), and the herbicides nicosulfuron, thifensulfuron, metsulfuron, triasulfuron (Zhou et al., 2006b, 2007b), atrazine and simazine (Zhou et al., 2006d), and dicamba (Biesaga and Pyrzynska, 2006). It was found that purification (removal of amorphous carbon) of the CNT improved adsorption (Gotovac et al., 2006). The available adsorption space was found to be the cylindrical external surface; neither the inner cavity nor the interwall space of multiwalled CNT contributed to adsorption (Yang and Xing, 2007). Unlike fullerenes, no adsorption/desorption hysteresis was observed, indicating reversible adsorption (Yang and Xing, 2007).

Oxidized and hydroxylated CNT are also good adsorbers for metals. This has been found for various metals such as Cu (Liang et al., 2005a), Ni (Chen and Wang, 2006; Lu and Liu, 2006), Cd (Li et al., 2003; Liang et al., 2004), Pb (Li et al., 2002, 2006b), Ag (Ding et al., 2006), Am (III) (Wang et al., 2005), and rare earth metals (Liang et al., 2005b).

In most cases adsorption is highly pH-dependent, with increasing sorption and increasing pH as expected for adsorption of metals onto hydroxyl groups. Adsorption of organometallic compounds on pristine multiwalled CNT was found to be stronger than for CB (Munoz et al., 2005). Chemically modified NPs have been proposed for environmental cleanup and may therefore be released into the environment (Obare and Meyer, 2004). TiO_2 functionalized with ethylenediamine was tested for the ability to remove anionic metals from groundwater (Mattigod et al., 2005).

Fullerenes have also been tested for adsorption of organic compounds. Adsorption depends to a great extent on the dispersion state of the C_{60} (Cheng et al., 2004). Because C_{60} forms clusters in water, there are closed interstitial spaces within the aggregates where the compounds can diffuse into, which leads to significant adsorption/desorption hysteresis (Cheng et al., 2005b; Yang and Xing, 2007). Fullerenes were found to be not very good sorbents for a wide variety of organic compounds (e.g., phenols, PAH, amines), while they are very efficient for removal of organometallic compounds (e.g., organolead) (Ballesteros et al., 2000).

Many materials have properties that are dependent on size (Hochella, 2002). Hematite particles with a diameter of 7 nm, for example, adsorbed Cu ions at lower pH values than particles with 25 or 88 nm diameter, indicating the uniqueness of surface reactivity for iron oxide particles with decreasing diameter (Madden et al., 2006). However, an investigation of Pb adsorption onto TiO_2 NP showed that the bulk material exhibited stronger adsorption and higher adsorption capacity (Giammar et al., 2007).

2.2.6 Future trends

There are many types of nano materials (NMs), and the scientific community is making observations on NP ecotoxicity to inform the wider debate about the risks and benefits of these materials. Natural NPs have existed in the environment since the beginning of Earth's history, and natural sources can be found in volcanic dust, most natural waters, soils, and sediments. Natural NPs are generated by a wide variety of geological and biological processes, and while there is evidence that some natural NPs can be toxic, organisms have also evolved in an environment containing natural NPs. There are concerns that the natural nanoscale process could be influenced by the presence of pollution. Manufactured NPs show some complex colloid and aggregation chemistry, which is likely to be affected by particle shape, size, surface area, and surface charge, as well as the adsorption properties of the material. Abiotic factors such as pH, ionic strength, water hardness, and the presence of organic matter will alter aggregation chemistry and are expected to influence toxicity. The physicochemistry is essential to understanding the fate and behavior of NPs in the environment, as well as uptake and distribution within organisms, and the interactions of NPs with other pollutants.

Data on biological effects show that NPs can be toxic to bacteria, algae, invertebrates, and fish species, as well as to mammals. However, much of the ecotoxicological data is limited to species used in regulatory testing and freshwater organisms. Data on bacteria, terrestrial species, marine species, and higher plants is particularly lacking. Detailed investigations of absorption, distribution, metabolism, and excretion (ADME) remain to be performed on species from the major phyla, although there are some data on fish. The environmental risk assessment of NMs could be performed using the existing tiered approach and regulatory framework, but with modifications to methodology, including chemical characterization of the materials being used. Many challenges lie ahead, and controversies will arise (e.g., reference substances for ecotoxicology), but knowledge transfer from mammalian toxicology, colloid chemistry, as well as material and geological sciences, will enable ecotoxicology studies to move forward in this new multidisciplinary field.

Further work is required to develop new methods, ideally in situ, to fully eliminate sampling problems (Lead and Wilkinson, 2006). More importantly, available methods need to be systematically deployed, both correctly and routinely. This goal will require substantial collaboration, given the need for extensive, sophisticated, and rapid analysis involving minimal sample preparation and transport. To some extent, the methodological developments have been the only major advance in the last

5–10 years. Our knowledge of colloidal structure, trace elements, and pathogen binding, as well as the effects of binding on the bioavailability of pollutants and pathogens, has substantially broadened and deepened in the last few years and further important developments are underway. Among other objectives, the following will need to be addressed (Lead and Wilkinson, 2006):

1. The standardization and application to unperturbed samples of sampling, fractionation, and analytical methodologies. Further methodological developments are required but are perhaps secondary given the current state of knowledge.
2. The collection of good quality data on colloidal structure and colloidal interactions with pollutants, pathogens, and nutrients. Data needs to be collected on whole (unperturbed) samples, samples that have been (size or chemically) fractionated, and on reference or standard materials. Subsequently, current models will need to be tested and improved, to include, for instance, an appreciation of colloidal structure and its effects on pollutant binding.
3. Similarly, the effects of colloid–pollutant interactions on transport and, particularly, bioavailability, need to be explored more fully in both experimental and modeling studies.
4. Given that most water quality guidelines employ a 0.45-μm membrane filtration, some effort is required to transfer state-of-the-art scientific principles to the regulatory and policy-making bodies but also to correctly interpret the significant quantities of data that are available and based on such nonideal separations. Future requirements to consider colloids, at least implicitly, in regulations (such as the EU Water Framework Directive) should help to greatly push forward research in this area. Specific areas of interest include the structure and role(s) of the nanoparticles and the interaction between Fe and organic carbon, especially in seawater, where there is an impact on climate change issues (Lead and Wilkinson, 2006).

2.3 Industrial applications of colloids

Common stabilizers are the polymeric dispersants used in formulating printing inks to ensure that the dispersed phase remains as discrete units. Whole industries are built around the design and selection of dispersants to optimize product performance. Invariably, together with other macromolecules or liquid crystalline surface-active agents, these agents modify the flow characteristics of the product. The formulator's skill lies in achieving the necessary product characteristics such as mouth-feel (organoleptic properties) with synthetic foodstuffs, nondrip properties of paints, and prevention of fine mist with liquid aerosols. The list of uses is endless; Table 25.2 lists just a few of the industrial applications.

Table 25.2 Some Industrial Applications of Colloids

Effluent treatment	Precipitation and/or flocculation for clarification
Paint industry	Achieve homogeneous films, toughness and "hiding" power
Food industry	Stable creams and gels
Cosmetics and toiletries	Emulsions, toothpaste
Detergent industry	Stabilization of suspended soil, liquid abrasives
Pharmaceutical industry	Stable dispersions to ensure uniform dose of active drug
Agricultural industry	Pesticides formulated as dispersions

Although colloids have many applications in the industrial sector; we will describe their role in the environmental protection perspective only.

2.3.1 Colloids in water/wastewater treatment

Untreated surface water or groundwater is often contaminated with pathogenic organisms of fecal origin. If not, it can become contaminated during transport and storage (Esry et al., 1991; Mintz et al., 1995; Luby et al., 2001; Quick et al., 2002). Even water treated with a disinfectant often becomes contaminated when collected from a public standpipe and stored in the home. A recent review of the literature sponsored by the WHO concludes that simple, socially acceptable, and low-cost interventions at the household (point-of-use) and community level have the potential to significantly improve the microbial quality of household water and reduce the risk of diarrheal disease, dehydration, and death, particularly among children (Sobsey, 2002; Clasen et al., 2004).

Beginning in the late 1980s and early 1990s, ceramic filters for point-of-use water treatment began appearing in Third World marketplaces, and their performance has been evaluated in a small number of published studies (Kulkarni et al., 1980; Chaudhuri et al., 1994; Clasen et al., 2004). These studies have evaluated filters that are typically produced through an industrial design and manufacturing process (that does not use local labor) with high-purity ingredients. This often results in a filter price point that is beyond the reach of many residents of developing global communities.

The filters are typically made with local labor and materials (clay, water, and a combustible organic material such as sawdust, flour, or rice husks). The filter is formed using a filter press, air-dried, and fired in a flat-top kiln, increasing the temperature gradually to about 900°C during an 8-h period. This forms the ceramic material and combusts the sawdust, flour, or rice husk in the filters, making it porous and permeable to water. After firing, the filters are cooled and impregnated with colloidal silver by painting with, or dipping in, a colloidal-silver solution. The colloidal silver is hypothesized to act as a microbial disinfectant.

Each filter is tested by measuring the water-flow rate to ensure that it is between 1 and 2 L/hr. This test is the sole design criterion (other than the physical dimensions of the filter). A typical filter lifetime is two to three years; it must be periodically cleaned with a brush to maintain the design flow rate. Since 1998, Potters for Peace has aided NGOs in establishing filter factories throughout the world. They estimate that about 100,000 filters have been manufactured and distributed for use.

The Potters for Peace ceramic filter has many potential advantages as a point-of-use water treatment technology. It can be manufactured with mostly local materials and labor.

Since clay pots are often used as storage containers for water, it is a socially acceptable technology that can work year round in different climates. It does not impart an objectionable taste to the treated water. It is designed to remove both turbidity and pathogens, and its retail cost is low. No refereed journal publication has critically evaluated the performance of this filter. Lantagne (2001) has published the most comprehensive report on the filter to date.

There are several critical knowledge gaps related to filter design and performance. No information is available on how the manufacturing process or the composition and relative amounts of the raw materials used to make the filter affect its physical pore structure and treatment performance. The role of colloidal silver in the deactivation of pathogenic bacteria as they pass through the filter is poorly understood. It is uncertain if bacterial removal is caused by filtration/sorption, deactivation by colloidal silver, or some combination of these processes.

Cylindrical colloidal-silver-impregnated ceramic filters for household (point-of-use) water treatment were manufactured and tested by Oyanedel-Craver and Smith (2008) for performance in the laboratory with respect to flow rate and bacteria transport. Filters were manufactured by combining clay-rich soil with water, grog (previously fired clay), and flour, pressing them into cylinders and firing them at 900°C for 8 h. The pore-size distribution of the resulting ceramic filters was quantified by mercury porosimetry. Colloidal silver was applied to filters in different quantities and ways (dipping and painting). Filters were also tested without any colloidal-silver application. Hydraulic conductivity of the filters was quantified using changing-head permeability tests. $[^3H]H_2O$ water was used as a conservative tracer to quantify advection velocities and the coefficient of hydrodynamic dispersion. *Escherichia coli* (*E. coli*) was used to quantify bacterial transport through the filters. Hydraulic conductivity and pore-size distribution varied with filter composition; hydraulic conductivities were on the order of 10–5 cm/s, and more than 50% of the pores for each filter had diameters ranging from 0.02 to 15 μm. The filters removed between 97.8% and 100% of the applied bacteria; colloidal-silver treatments improved filter performance, presumably by deactivation of bacteria. The quantity of colloidal silver applied per filter was more important to bacteria removal than the method of application. Silver concentrations in effluent filter water were initially greater than 0.1 mg/L but dropped below this value after 200 min of continuous operation. These results indicate that colloidal-silver-impregnated ceramic filters, which can be made using primarily local materials and labor, show promise as an effective and sustainable point-of-use water treatment technology for the world's poorest communities.

2.3.2 Coagulants for water/wastewater treatment

Coagulation and flocculation are an essential part of drinking water treatment as well as wastewater treatment. Coagulants neutralize the repulsive electrical charges (typically negative) surrounding particles, allowing them to "stick together" creating clumps or flocks. Flocculants facilitate the agglomeration or aggregation of the coagulated particles to form larger floccules and thereby hasten gravitational settling. Some coagulants serve a dual purpose of both coagulation and flocculation in that they create large flocks that readily settle (Bratby, 2006).

Coagulation and flocculation are essential processes in various disciplines. In potable water treatment, clarification of water using coagulating agents has been practiced from ancient times. As early as 2000 BC, the Egyptians used almonds smeared around vessels to clarify river water. The use of alum as a coagulant by the Romans was mentioned around AD 77. By 1757, alum was being used for coagulation in municipal water treatment in England.

In modern water treatment, coagulation and flocculation are still essential components of the overall suite of treatment processes—understandably so, because since 1989 the regulatory limit in the United States for treated water turbidity has progressively decreased from 1.0 NTU in 1989 to 0.3 NTU today. Many water utilities are committed to consistently producing treated water turbidities of less than 0.1 NTU to guard against pathogen contamination (Bratby, 2006).

Coagulation is also important in several wastewater treatment operations. A common example is chemical phosphorus removal, and another, in overloaded wastewater treatment plants, is the practice of chemically enhancing primary treatment to reduce suspended solids and organic loads from primary clarifiers.

Wastewater is generated, usually from rinsing or cleaning manufactured products. For nonmetal-bearing waters, these include TSS (total suspended solids), BOD (biological oxygen demand), COD (chemical oxygen demand), and pH. For wastewater containing metals (either in solution or insoluble) certain metal limits must be met, along with all the other parameters. This is usually accomplished using pH adjustment, along with a coagulant (aluminum based or iron based). A coagulant is required to help give body to the water. Typically, you have a colloidal suspension that requires help in getting the particles to come together and form a larger particle. A coagulant will help neutralize similarly charged particles, allowing them to form small to midsize particles (sometimes called a pin-flock). This usually occurs in a pH environment between 6 and 9. Once the pin-flock has formed, a second chemical called a flocculent is required to make even larger particles. Addition of a flocculent occurs at a pH between 8 and 10 (depending on the characteristics of the wastewaters). Flocculent is added and acts as a net where it gathers up the smaller coagulated particles, making a larger particle. This larger particle will slowly drop to the bottom of the container (vessel), forming a sludge.

Many wastewater streams contain metals from the manufacturing processes. These metals can be either insoluble or soluble. Insoluble means that there are actual small particles floating in the water that are too light to settle out without aid. Soluble means that the metals have gone into solution and are actually part of the water. Many times, using hydroxide precipitation, along with the aid of a coagulant, most of these metals can be separated from the water (copper is most insoluble at a pH of 8, whereas nickel is most insoluble at a pH or 10.2, etc.) (Bratby, 2006).

Many cleaners contain chelators. These chelators have strong cleaning abilities (citric acid, edta, phosphates, etc). The chelators in the water combine with the metals in the water to form a very strong bond that in many instances cannot be broken with pH adjustment. An additional chemical must be added to cleave the metal/ chelator bond. These products are called precipitants. Precipitants have a stronger attraction to the metal than the chelator; therefore, they form a bond and can be separated from the water (i.e.; copper will form copper carbonate when poly thio carbonate is added). Coagulants are required to help these particles combine, along with a flocculent. Typically, the precipitant and coagulant can be added at the same pH range of 6–9. The pH needs to be between 8 and 10 for final flocculation. Table 25.3 shows various kinds of flocculants used in wastewater treatment.

2.3.3 The coagulants

The commonly used metal coagulants fall into two general categories: those based on aluminum and those based on iron. The aluminum coagulants include aluminum sulfate, aluminum chloride and sodium aluminate. The iron coagulants include ferric sulfate, ferrous sulfate, ferric chloride, and ferric chloride sulfate. Other chemicals used as coagulants include hydrated lime and magnesium carbonate.

The effectiveness of aluminum and iron coagulants arises principally from their ability to form multicharged polynuclear complexes with enhanced adsorption characteristics. The nature of the complexes formed may be controlled by the pH of the system.

When metal coagulants are added to water, the metal ions (Al and Fe) hydrolyze rapidly but in a somewhat uncontrolled manner, forming a series of metal hydrolysis species. The efficiency of rapid mixing, the pH, and the coagulant dosage determine which hydrolysis species is effective for treatment (Bratby, 2006).

There has been considerable development of pre-hydrolyzed inorganic coagulants, based on both aluminum and iron to produce the correct hydrolysis species regardless of the process conditions during treatment. These include aluminum chlorohydrate, polyaluminum chloride, polyaluminum sulfate chloride, polyaluminum silicate chloride and forms of polyaluminum chloride with organic

Table 25.3 Various Coagulants used in Wastewater Treatment

	Coagulants
CIAS	A liquid inorganic coagulant (aluminum sulfate) with a cationic charge. It is formulated to promote the coagulation of precipitated particles and assist in their rapid settling during wastewater treatment. CIAS is used as a coagulant for water clarification, wastewater treatment and related applications in water and wastewater treatment programs.
CIFCB	Liquid coagulant with a cationic charge. This product is excellent for breaking emulsions formed with oils, inks, surfactants, and so on. It is easy to handle and feed, economical to use, effective at low dosages, and performs well over a wide pH range. It also precipitates phosphates from wastewater streams.
CIFSB	A ferric sulfatebased liquid coagulant with a cationic charge. It is a blended coagulant formulated with an inorganic polymer and cationic polymer.
CIACH	An inorganic polymer (aluminum chlorohydrate) formulated for use as a coagulant. CIACH has been very successful in the replacement of alum, ferric chloride, ferric sulfate, and other inorganic salts, as well as organic polymers. The floc formed by CIACH is characterized as being small, dense particles and very sheer resistant. Using CIACH results in a reduction of overall treatment costs by lowering or eliminating the need for alkali and flocculent aids, reducing residual generation and handling and disposal costs, extending filter runs and water production, and reducing chemical handling and storage requirements.
CIOB-5	A blended coagulant (sluminum chloro hydrate) formulated with an inorganic polymer and organic cationic polymer (amine). Blend of above product.
CIASB-1	A liquid inorganic blended coagulant with a cationic charge. It is formulated with an inorganic polymer and cationic polymer (amine). Blend of CIASB.
CIOB-50	A liquid inorganic coagulant blended with organic polymer to accelerate the settling of suspended particles.
COHW-25	A concentrated liquid, organic, highly cationic, high-molecular-weight polyquaternaryamine.
CO-25	A concentrated liquid, organic, water-soluble, low cationic quaternary ammonium polyelectrolyte.
Flocculants	
FL-neg	A liquid organic acrylamide copolymer with a medium anionic charge.
FL-2	Ionic polyacrylamide (liquid). High-molecular-weight medium-charge flocculant. Used as an all-purpose flocculent particularly effective on inorganic streams.
FL-pos	Cationic flocculant (liquid). Medium molecular weight, medium charge used in many biological applications.
Specialty Chemicals	
NF-10 ANTIFOAM	Stable, water-based silicone defoamer recommended for wastewater treatment applications.
CM-C	An inorganic liquid multipurpose reagent for modifying components in wastewater requiring treatment. CM-C substitutes for heavy metals under chelated conditions to permit their precipitation; acts as a co-precipitant in hydroxide precipitation of zinc and other metals; aids in the clarification and dewatering operations of the treatment process; and deactivates sequestering agents and phosphates in spent cleaners

Table 25.3 Various Coagulants used in Wastewater Treatment—cont'd

Coagulants	
CM-M	An inorganic liquid reagent for modifying components in wastewater requiring treatment. CM-M substitutes for the heavy metals in chelated conditions to permit their precipitation, coagulation, and flocculation; demulsifies aqueous-based coolants and cutting oils; and neutralizes the dispersing characteristics of surfactants in spent cleaner baths.
CR-20	A liquid-stabilized reducing reagent for the reduction of hexavalent chromium (and other high valence metals) to its trivalent state as a preparatory stage to precipitation. CR-20 is also utilized as a reducing reagent for persulfate and permanganate solutions.
CitruClean	New green solvent. For use in industrial parts cleaning and paint cleanup application. Meets the toughest health, safety, and environmental regulations in existence today.

Source: www.ecologixsystems.com/product-specialty-chemicals-coag-floc.php.

polymers. Iron forms include polyferric sulfate and ferric salts with polymers. There are also polymerized aluminum-iron blends.

The principal advantages of pre-polymerized inorganic coagulants are that they are able to function efficiently over wide ranges of pH and raw water temperatures. They are less sensitive to low water temperatures; lower dosages are required to achieve water treatment goals; less chemical residuals are produced; and lower chloride or sulfate residuals are produced, resulting in lower final water TDS. They also produce lower metal residuals.

Pre-polymerized inorganic coagulants are prepared with varying basicity ratios, base concentrations, base addition rates, initial metal concentrations, ageing time, and ageing temperature. Because of the highly specific nature of these products, the best formulation for particular water is case specific and needs to be determined by Jar testing. For example, in some applications alum may outperform some of the polyaluminum chloride formulations (Bratby, 2006).

Polymers are a large range of natural or synthetic, water soluble, macromolecular compounds that have the ability to destabilize or enhance flocculation of the constituents of a body of water.

Natural polymers have long been used as flocculants. For example, Sanskrit literature from around 2000 BC mentions the use of crushed nuts from the Nirmali tree (*Strychnos potatorum*) for clarifying water – a practice still alive today in parts of Tamil Nadu, where the plant is known as Therran and cultivated also for its medicinal properties. In general, the advantages of natural polymers are that they are virtually free of toxins, biodegradable in the environment and the raw products are often locally available. However, the use of synthetic polymers is more widespread. They are, in general, more effective as flocculants because of the level of control made possible during manufacture.

Important mechanisms relating to polymers during treatment include electrostatic and bridging effects. Figure 25.3 shows schematic stages in the bridging mechanism. Polymers are available in various forms, including solutions, powders or beads, oil or water-based emulsions, and the Mannich types. The polymer charge density influences the configuration in solution: For a given molecular weight, increasing charge density stretches the polymer chains through increasing electrostatic repulsion between charged units, thereby increasing the viscosity of the polymer solution.

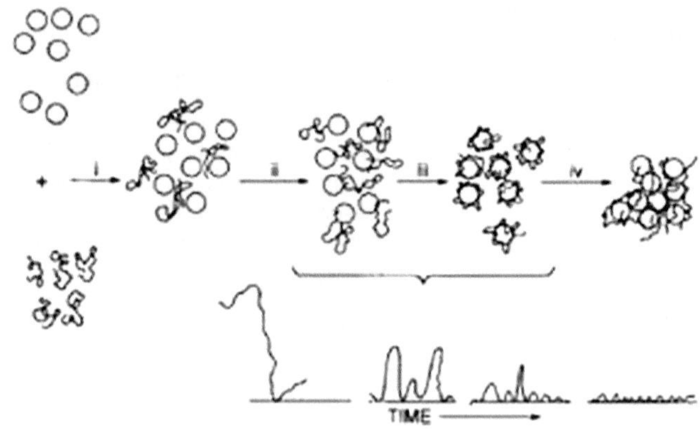

FIGURE 25.3

Stages in the bridging mechanism: **(i)** Dispersion; (ii) Adsorption; (iii) Compression or settling down (see inset); (iv) Collision (Akers, 1972).

One concern associated with synthetic polymers relates to potential toxicity issues, generally arising from residual unreacted monomers. However, the proportion of unreacted monomers can be controlled during manufacture, and the quantities present in treated waters are generally low (Bratby, 2006).

2.3.4 Removal of natural organic matter

Natural organic material (NOM) is usually associated with humic substances arising from the aqueous extraction of living woody substances, the solution of degradation products in decaying wood, and the solution of soil organic matter. These substances are objectionable for a number of reasons: They tend to impart color to waters; they act as a vehicle for transporting toxic substances and micropollutants, including heavy metals and organic pollutants; and they react with chlorine to form potentially carcinogenic by-products.

The degree to which coagulation can remove organic material depends on the type of material present. The specific ultraviolet absorption (SUVA) is related to the concentration and type of dissolved organic carbon (DOC) present, as follows:

$$\mathrm{SUVA} = \mathrm{UV}_{254}/\mathrm{DOC}\ (1/\mathrm{mg\ m})$$

where: UV_{254} is the ultraviolet absorbance measure at a wavelength of 253.7 nm, after filtration through 0.45-μm filters (m^{-1}); DOC is the dissolved organic carbon measured after filtration through 0.45-μm filters (mg/L).

In general, lower-molecular-weight species such as the fulvic acids are more difficult to remove by coagulation. Higher-molecular-weight humic acids tend to be easier to remove.

The United States Environmental Protection Agency (U.S. EPA) introduced enhanced coagulation for the removal of NOM. Enhanced coagulation is an elaboration of long-practiced techniques for removing organic color by coagulation. It requires the removal of NOM material, while still achieving good turbidity removal. These dual objectives can be met by selecting the best coagulant type, applying the best coagulant dosage and adjusting the pH to a value where best (or adequate) overall coagulation conditions are achieved (Bratby, 2006).

The enhanced coagulation approach recognizes that the constituents of any given water govern the practical degree of treatment achievable. Therefore, a water-specific point of diminishing returns

(PODR) is identified, at which a coagulant increment (10 mg/L for alum) results in a TOC removal increment of less than 0.3 mg/L.

Organics removal and enhanced coagulation are effective with traditional coagulants like aluminum sulfate, ferric chloride, and ferric sulfate, as well as formulations like polyaluminum chloride (PACl) and acid alum. Acid alum formulations are aluminum sulfate with 1 to 15% free sulfuric acid. Their effectiveness with TOC removal applications is due to the enhanced depression of pH.

TOC or NOM reductions depend on the type and dosage of coagulant, the pH, temperature, raw water quality, and NOM characteristics. In general, the optimum pH for ferric salts is in the range 3.7 to 4.2, and for aluminum sulfate in the range 5.0 to 5.5.

In some cases, the removal of lower weight organics has been improved by supplementing treatment with metal coagulants with powdered activated carbon (PAC). In one case with raw water TOC of 2.4 mg/L, a combination of an alum-polymer blend coagulant at 25 mg/L with PAC at 10 mg/L was optimal to achieve a 39% TOC reduction (Bratby, 2006).

In another case, water with a low humic content and low SUVA (1.43 L/mg.m) was treated with 65 mg/L $FeCl_3$ and 23 mg/L PAC. It was found that 56% of the TOC was nonhumic and 46% of the TOC had molecular weights less than 1000.

2.3.5 Pathogen Removal

The U.S. EPA surface water treatment rule requires 99.9% (3-log) *Giardia* removal or inactivation, and at least 99% (2-log) removal of *Cryptosporidium*. Adequately designed and operated water treatment plants, with coagulation, flocculation, sedimentation, and filtration, are assigned a 2.5-log removal credit for *Giardia*, leaving only 0.5-log inactivation to be achieved by disinfection.

Coagulation and flocculation, with dissolved air flotation (DAF) for clarification, has achieved average log removals of *Giardia* and *Cryptosporidium* of 2.4 and 2.1, respectively. Optimum coagulation conditions were governed by turbidity and NOM removal requirements, rather than by pathogen removals. Overall *Giardia* and *Cryptosporidium* removals, including the filtration step were approximately 5-log (Bratby, 2006).

Cryptosporidium oocyst surfaces are believed to consist of polysaccharide layers. The negative charge carried by the oocysts is believed to arise from carboxylic acid groups in surface proteins. Removal of *Cryptosporidium* using alum coagulation appears to be by a sweep floc mechanism. Zeta potential measurements suggest that removal does not appear to be by a charge neutralization mechanism at lower DOC concentrations. At higher DOC, it appears that the mechanism is mediated by a NOM-assisted bridging between aluminum hydroxide and oocyst particles.

Significant virus removals have been reported using metal coagulants and polyelectrolytes. Removals of up to 99.9% have been reported for both aluminum and ferric salts. Various polyelectrolytes (cationic) have effected removals of greater than 99% but have the disadvantage that if other material is present in the form of color, turbidity, and COD, removal of such material is poor. Using metal coagulants and poyelectrolytes conjointly has the advantage that better floc characteristics are produced. If a variety of substances are present in water, it is possible that the use of both metal coagulants and polyelectrolytes will effect a higher overall removal. However, this very much depends on the conditions pertaining for each case. When using polyelectrolytes as flocculant aids, floc formation improves but does not appear to improve virus removals beyond those achieved using metal coagulants alone.

Viruses are essentially DNA (deoxyribonucleic acid) or RNA (ribonucleic acid) units contained within a protein coat. The destabilization mechanism involves coordination reactions between metal coagulant species and carboxyl groups of the virus coat protein. Because of the similarity of the destabilization mechanisms for organic color and viruses, optimum removals tend to occur at similar pH values. The optimum pH for virus removal with aluminum sulfate has been found to be in the region of 5.0 with percentage virus removals in the range 97.7 to 99.8%. Using a cationic polyelectrolyte as flocculant aid, virus and turbidity removals were increased to 99.9 and 98.5%, respectively.

Metal coagulants or polyelectrolytes do not fully inactivate viruses. Therefore, a potential health hazard exists with the ultimate disposal of water treatment plant sludges. Furthermore, complete virus removal by destabilization with metal coagulants has not been reported. For a safe drinking water, disinfection of the water before distribution is required (Bratby, 2006).

However, some inactivation accompanies virus removal by coagulation. Some reports have shown that the infectious virus concentration only recovers partially after re-dissolution of aluminum hydroxide precipitates. This phenomenon has been interpreted as virucidal activity of the aluminum. PACl coagulants appeared to have a higher virucidal activity compared with alum. The presence of NOM in waters appears to inhibit the virucidal activity of the aluminum.

2.3.6 Removal of inorganics

In some cases coagulation operations can be useful for the removal of inorganics. Examples of successful applications are copper and mercury reductions from wastewater plant effluents. Two applications discussed in more detail in the next subsection are arsenic and fluoride removals in potable water treatment:

2.3.7 Arsenic removal

Arsenic is a commonly occurring toxic element, and long-term exposure to arsenic is injurious to health. The World Health Organization in 1993 reduced the arsenic limit for drinking water from 50 mg/L to 10 mg/l. In the United States, since passage of the Safe Drinking Water act in 1976, the maximum allowable arsenic concentration in drinking water was 50 mg/l. In 2002 this limit was lowered to 10 mg/L. Some states may adopt arsenic limits below the federal limit. For example, in 2005 the state of New Jersey announced a plan to adopt a 5 mg/L limit, and the state of California appeared to be considering an arsenic limit of approximately 4 mg/L.

Arsenic is stable in several oxidation states, under different redox conditions in water. However, when present in groundwater, arsenic occurs mostly in the forms of arsenite, As(III) and arsenate, As(V). As(III) is usually the predominant form in many groundwaters since it is more likely to be found in oxygen-free (anaerobic) conditions. As(V) is more common in aerobic waters. In general, As(V) is more readily removed than As(III).

Various technologies are used to remove arsenic from drinking water. These include adsorption on granular iron-based media; adsorption on ion-exchange resins; adsorption on activated alumina; co-precipitation in iron removal plants; coagulation with alum or ferric followed by conventional filtration; and coagulation with ferric followed by membrane filtration. Some studies have shown that the removal of arsenic by coagulation is more economical than other treatment alternatives.

With coagulation for arsenic removal, iron-based coagulants are generally more effective than aluminum coagulants. Iron coagulants added to water hydrolyze to form ferric hydroxide with a net positive charge. This net positive charge is a function of pH. As the pH decreases, the number of positively charged sites on the ferric hydroxide particles increases. Arsenate, As(V), is an anion and since it is negatively charged, it will adsorb to the positively charged ferric hydroxide particles by surface complexation. Arsenic removal is generally optimized at pH values of less than approximately 7.

Of the aluminum coagulants, the efficiencies of arsenic(V) removals are generally in the order polyaluminum chloride > polyaluminum sulfate > aluminum chloride > aluminum sulfate. Best results are obtained at a pH of 5.5. Arsenic removals have ranged from approximately 59 to 99% at dosages of 0.8 to 1.9 mg/l as Al, with sedimentation followed by filtration.

Using ferric coagulants efficiencies of arsenic(V), removals are in the order polyferric chloride > polyferric sulfate > feric chloride > ferric sulfate. Best results are obtained at a pH of 5.5. As(V) removals ranged from approximately 70 to 99.6% at dosages of 1.7 to 3.8 mg/L as Fe, with sedimentation followed by filtration (Bratby, 2006).

2.3.8 Fluoride removal

In 1975, the EPA named fluoride as a contaminant in the National Interim Primary Drinking Water Regulations. A Maximum Contaminant Level (MCL) was set at 1.4–2.4 mg/l to prevent dental fluorosis and more serious effects. To balance the benefits of fluoride for dental health, the deleterious effects of ingesting too much fluoride, and the costs of removing high concentrations of naturally occurring fluoride, the EPA in 1985 issued a new MCL of 4 mg/L for fluoride, with a secondary MCL of 2 mg/L. Systems with fluoride levels between 2 mg/L and 4 mg/L must provide the public with information about possible tooth discoloration.

The best available technologies for fluoride removal from water are generally considered to be activated alumina adsorption and reverse osmosis. However, in some cases, fluoride removal by aluminum coagulation has been shown to be cost effective. It appears that several aluminum-based coagulants are equally effective, based on the aluminum content added for treatment. Fluoride removal using aluminum-based coagulants is strongly affected by pH and aluminum dosage. Optimum pH varies from 6.0 to 7.5. A further factor is the residual aluminum remaining after treatment. Higher aluminum dosages often produce lower residual aluminum due to adsorption of fluoride to aluminum hydroxide flocks, rather than producing aluminum-fluoride complexes that remain in solution.

Aluminum dosages are generally high for appreciable fluoride removals. For example, to reduce fluoride from 3.6 mg/L to 1.8 mg/L, the aluminum dosage was 18 mg/L as Al, or 10 mg Al per mg fluoride removed, at an optimum pH of 6.5. To further reduce the fluoride to 1.0 mg/L, a dosage of 12 mg Al per mg fluoride removed was required (Bratby, 2006).

2.3.9 Chemical phosphorus removal

In many sensitive catchment areas, chemical phosphorus removal is also required for wastewater treatment. Inordinate dosages, beyond stoichiometric, are required to achieve very low effluent concentrations.

Within the stoichiometric range of phosphorus removal, there is a tightening of the optimal pH range as the metal coagulant dosage increases. However, beyond the stoichiometric range, when final phosphorus concentrations are progressively lower, the pH range widens again, toward the side of higher pH. For example, with alum the optimum pH range for effluent P concentrations down to approximately 0.2 is 5.5 to 6.0. However, as the Al ratio is increased for lower P concentrations, the required pH range widens to 6.0 to 7.0 (Bratby, 2006).

Within the stoichiometric P removal range, a precipitation model describes the interactions between metal and phosphorus. However, at very low P concentrations, more complex models that include precipitation, adsorption and floc- specific surface are required.

The benefits of sequential chemical addition for coagulation operations have been shown on many occasions. This is also the case with phosphorus precipitation. For very low final concentrations, overall coagulant dosages can be significantly reduced.

The degree of phosphorus removal depends not only on the coagulant added, but also on the mode of solid–liquid separation employed. This is particularly important for those cases where very low final phosphorus concentrations are achieved. Effluent suspended solids contribute significantly to effluent total phosphorus concentrations. For very low phosphorus residuals and high metal coagulant dosages, the phosphorus content of effluent suspended solids is significantly reduced. The reason is that at very high metal dosages, a larger proportion of the precipitates formed are metal hydroxides (Bratby, 2006).

2.3.10 Wastewater treatment

Physical-chemical treatment of wastewater was widely practiced until the late 19th century, up until the advent of the trickling filter for biological treatment. The early 1970s saw a partial revival of interest that has continued to the present day, particularly for treatment plants that are overloaded during peak flow events. The addition of coagulant chemicals to primary clarifiers, or to other dedicated physical separation processes, is an effective way of reducing the load to downstream biological processes, or in some cases for direct discharge. This practice is generally referred to as chemically enhanced primary treatment, or CEPT.

Principal disadvantages that might preclude a wholly physical-chemical solution to wastewater treatment are the problems associated with the highly putrescible sludge produced, and the high operating costs of chemical addition. However, much of the current interests in physical-chemical treatment stems from its suitability for treatment under emergency measures; for seasonal applications, to avoid excess wastewater discharges during storm events; and for primary treatment before biological treatment, where the above disadvantages become of lesser impact.

CEPT can also be an effective first step for pollution control in developing countries—particularly in large urban areas that have evolved with sewage systems but without centralized wastewater treatment, and have limited financial resources for more complete, but capital-intensive biological treatment options such as activated sludge systems. Such urban developments also may not have the areas available for appropriate technology options such as stabilization pond processes (Bratby, 2006).

The efficiency of CEPT, in terms of BOD or COD removal, depends on wastewater characteristics. With CEPT, one can expect to remove particulate components, together with some portion of the colloidal components. Therefore, with such a wastewater, it is feasible to achieve removals of more than: 95% TSS; 65% COD; 50-% BOD; 20% nitrogen; and 95% phosphorus. In practice, removals may be lower or higher: For example, in warmer climates, with larger collection systems, and relatively flat sewers, one would expect a higher degree of hydrolysis of particulate matter resulting in higher soluble fractions, and lower overall removals with CEPT. On the other hand, if the collection system is relatively small, the climate is cold, and wastewater is relatively fresh, there may be a higher proportion of particulate material, and CEPT removals could be higher.

Staged coagulation-flocculation can enhance CEPT performance. For example, at primary clarifier overflow rates of over 6 m/h (3600 gpd/ft^2) during peak flow treatment, TSS and BOD removals of 80 to 95%, and 58 to 68% were achieved, respectively, using 60 mg/L ferric chloride, followed by 15 mg/L polyaluminum chloride, followed by 0.5 mg/L anionic polymer. The total reaction time from the point of ferric chloride addition to entering the primary clarifiers was approximately 8 minutes at peak flow (Bratby, 2006).

2.4 Colloidal risk assessment

All scientific personnel working in colloidal group laboratories are required to carry out a mandatory risk assessment for all procedures carried out in the laboratory. No work goes on in the lab until a risk assessment has been done. Such assessments are for the benefit of workers as well as their associated coworkers in the laboratory and anyone else who may have to enter the laboratories routinely (cleaners, building supervisors) or in an emergency situation (porters, security, fire brigade). In preparing for an experiment, it must be remembered that coworkers may not fully understand the risks of the reactions, let alone other nonchemists who may be involved.

Normally, a Chemical Hygiene Plan (CHP) is devised whose purpose is to define work practices and procedures to help ensure that laboratory workers are protected from health and safety hazards associated with the hazardous chemicals with which they work. The Chemical Hygiene Plan is part of the compliance with the regulations promulgated on January 31, 1990 by the U.S. Department of Labor Occupational Safety and Health Administration (OSHA) and adopted by Kentucky OSH. This standard, titled "Occupational Exposures to Hazardous Chemicals in Laboratories," is hereafter referred to as the Lab Standard.

A Chemical Hygiene Plan is required to include the following.

- Standard operating procedures
- Criteria to determine and implement specific control measures, such as engineering controls and personal protective equipment
- An ongoing program to ensure that fume hoods and other engineering controls are functioning properly
- Information and training requirements
- Circumstances under which a particular laboratory function will require "prior approval"
- Provisions for medical consultation and medical exams
- Designation of the Principal Investigator/Laboratory Supervisor as the Chemical Hygiene Officer
- Additional precautions for work with select carcinogens, reproductive toxins, and extremely toxic substance

In general, colloidal chemical laboratories adhere to the following classification system for the risks of a procedure.

A	Those in which work may not be undertaken without close senior supervision, that is, the presence of the Supervisor or of a Senior Postdoctoral nominated by the Supervisor
Bp	Those in which work may not be started without the Supervisor's advice.
Bu	Those in which work may not be started without the Supervisor's advice.
C	Those with some risks (other than A and B) where care must be observed but it is considered that workers are adequately trained and competent in the procedures involved.
D	General laboratory practice.
E	Those which, even without training, have very low levels of risk.

2.4.1 Determining who is covered by the lab standard

OSHA defines a hazardous chemical as "a chemical for which there is statistically significant evidence based on at least one study conducted in accordance with established scientific principles that acute or chronic health effects may occur in exposed employees." In addition, OHSA defines a laboratory as "a workplace where relatively small quantities of hazardous chemicals are used on a non-production basis." Finally, the Laboratory Workers referred to in the Lab Standard are employees. OSHA defines an employee as "an individual employed in a laboratory workplace who may be exposed to hazardous chemicals in the course of his or her assignments." Students in an academic laboratory would not be considered laboratory workers by OSHA. Ideally, the principles outlined in this Chemical Hygiene Plan are to be followed:

All Laboratory Workers prior to the commencement of lab duties must read this Chemical Hygiene Plan. In addition to the Plan, the Laboratory Workers must be familiar with and adhere to prudent laboratory safety guidelines.

Training records should be kept by the Principal Investigator (PI) or the lab supervisor. After training, a written record stating that each Laboratory Worker has reviewed the Chemical Hygiene Plan and specific health and safety policies and guidelines for the individual lab must be kept by the person in charge of the lab. Each laboratory's CHP must be **reviewed annually** by the laboratory's Chemical Hygiene Officer and the "revised date" must be listed on the identification page.

There should be **Environmental Health and Safety Department** in each organization dealing with hazardous chemicals that have following responsibilities:

Appoint an Institutional Chemical Hygiene Officer who will routinely review the model Chemical Hygiene Plan and suggest modifications as needed

Provide technical assistance to Laboratory Supervisors and workers concerning appropriate storage, handling and disposal of hazardous chemicals

Provide general laboratory safety training upon request

Conduct exposure assessments and laboratory inspections upon request and on a routine basis

Provide technical assistance concerning personal protective equipment and laboratory safety equipment

Remain current on rules and regulations concerning chemicals used.

Laboratory Worker responsibilities regarding implementation of the Chemical Hygiene Plan:

Follow all health and safety standards and rules

Report all hazardous conditions to the laboratory supervisor

Wear or use prescribed protective equipment

Report any suspected job-related injuries or illnesses to the laboratory supervisor and seek treatment immediately

Refrain from the operation of any equipment or instrumentation without proper instruction and authorization

Remain aware of the hazards of the chemicals in the lab and how to handle hazardous chemicals safely, and

Request information and training when unsure how to handle a hazardous chemical or procedure

2.4.2 Combined risk assessment/experiment in progress

To minimize the amount of paperwork associated with undertaking routine risk assessments and preparing Experiment in Progress signs, the combined form shown below can be used. The following principles apply:

- Use of these forms as an "Experiment in Progress" form is mandatory.
- These forms are also a record of the risk assessment; paste the form into your lab books after use. This combined with the general risk assessments (updated at least yearly through DLM) and special risk assessments (as needed) should be sufficient to undertake good assessments of the risks involved in day-to-day work as well as complying with the safety regulations. As each of these forms is tailored to the specific reaction being performed, they will not have out-of-date information on them.
- Any procedure that requires boxes in the Hazard List section to be ticked is probably in category C or higher on the above list and will require the reverse side to be filled in to provide more details of the procedure being undertaken.
- Any category B or C procedures will require a supervisor's signature prior to laboratory work commencing, unless the worker is "licensed" for the procedure as detailed below. In such cases, the researcher can sign the "approved" section themselves and does not need a supervisor's signature on the form.
- Both electronic and paper versions of these forms will be available, and the forms themselves will be reusable in some circumstances.

2.4.3 Controlling chemical exposures

The Lab Standard requires the employer to determine and implement control measures to reduce employee exposure to hazardous chemicals; particular attention must be given to the selection of control measures for chemicals known to be extremely hazardous. There are three major routes for a chemical to enter the body: inhalation, absorption, and ingestion. Three types of controls for prevention of these various routes of entry include engineering controls, administrative or work practice controls, and personal protective equipment. Each route of entry a chemical can take to enter the body can be controlled in a number of ways, as explained below.

Evaluating the risk is the first step. A good place to start researching health and safety information is from the Material Safety Data Sheets (MSDS) that are shipped with the chemical. Further information on MSDS can be viewed at following website: http://www.ilpi.com/msds/index.html#Further.

2.4.4 Inhalation hazards

Inhalation of chemicals is the most common route a chemical can take to enter the body. One good way to reduce any type of exposure is substitution. Try to find a chemical that works just as well but is less volatile or toxic. It could also be changing to a liquid or a solid chemical from a gas. If substitution is not practical, engineering controls such as ventilation should be used to lessen the chance of exposure. The use of well-functioning local exhaust ventilation such as fume hoods, biological safety cabinets, vented glove boxes, and other local exhaust systems is often required to minimize exposure to hazardous chemicals. Dilution ventilation may be used to reduce exposure to nonhazardous nuisance odors. For extremely toxic chemicals such as those classified as poisonous gases by state or federal agencies (e.g., arsine, phosgene), the use of closed systems, vented gas cabinets, fail-safe scrubbing, detection, or other stricter controls may be required.

If both substitution and engineering controls are unavailable, the use of personal protective equipment may be required to reduce inhalation exposures.

Respiratory protection from dust masks to self-contained breathing apparatus may be utilized to this end. If laboratory employees wear respirators, requirements of the OSHA Respirator Standard (1910.134) must be met, and a written respirator program must be implemented. This Standard requires training on the proper use of respirators, medical surveillance to ensure the user is capable of wearing a respirator, and fit testing to ensure that the respirator fits properly. A lab worker or his/her supervisor should contact the Occupational Health and Safety Department (257-3827) in the event that respiratory protection is to be utilized to control exposures to hazardous chemicals.

In addition, the following principles should be utilized to reduce the risk of exposure to hazardous chemicals:

- Minimization of exposure time for individual employees
- Restricted access to an area where a hazardous chemical is used
- Proper signage on lab doors to indicate special hazards within 2.4.5. Skin/Eye Contact Hazards

To reduce the risk of a chemical entering the body via skin and eye contact, engineering controls include substitution and appropriate ventilation as described above in Inhalation Hazards. Also, the fume hood sash provides a good physical barrier. The more obvious means of preventing skin and eye contact is to wear personal protective equipment such as eye protection, face shields, gloves, appropriate shoes, lab aprons, lab coats, and other protective equipment as appropriate to the hazard. Chemical resistivity of the different types of protective equipment varies significantly. Safety showers/eye wash equipment is required where corrosive chemicals are used. Such equipment should be prominently labeled and not obstructed.

2.4.6 Ingestion hazards

Ingestion of chemicals is the least common route of entry into the body. However, a laboratory worker can easily ingest chemicals into the body via contaminated hands if they are not washed prior to eating, smoking, or sticking part of the hand or a writing tool that has been contaminated into the mouth. OSHA is strict on some activity in the lab to prevent this type of exposure. The Lab Standard forbids eating, drinking, and applying makeup and lip balm in areas where hazardous chemicals are used. Other examples of administrative controls are forbidding mouth pippeting and encouraging good personal hygiene. An engineering control for this type of hazard would be the use of a glove box. Finally, wearing the appropriate glove for personal protective equipment is necessary.

At the request of concerned laboratory workers, Occupational Health may conduct exposure evaluations for any suspected overexposure to substances regulated by OSHA. Records of exposure evaluations should also be kept in the Occupational Health and Safety Department and provided to the department and affected employees.

3 Conclusions

There is a growing public interest in developing risk assessment framework, environment regulations, as well as remedial strategies for protecting ecosystems and human from environmental hazards. A "colloidal" attitude means that systems are made of tiny building elements, and, as a consequence have a high specific surface area and unique properties that are different from those exhibited by ordinary homogeneous or heterogeneous systems. There are many kinds of environmental stresses such as heavy metals in drinking water, pesticides, and food contaminants. Colloids are successfully applied in many areas of environmental protection. A number of environmental and energy technologies have already benefited substantially from colloidal technology. The use of colloids for water and soil purification from phenol, oil and oil products, and metal ions will greatly benefit humanity.

Colloids are involved in environmental protection in areas such as drinking water, wastewater treatment, heavy metal remediation, treatment of contaminated soils, xenobiotic removal, abatement of pesticide contamination in food and soil, protection against radioactive materials, corrosion, used as drug carriers, emulsions, microgels, reduced-waste and improved energy efficiency; environmentally friendly composite structures; waste remediation; energy conversion, and so on.

Emulsions are successfully applied in many fields of human activity. When used as liquid colloidal carriers, the stability of emulsion droplets against coalescence often requires improvement. Additional protection against colloidal degradation or environmental stresses is almost an unavoidable precondition for employment of emulsion formulations in the food industry, pharmaceutics, cosmetics, and medicine. The synthesis of biologically active microgels will be beneficial as protection against many pathogenic diseases in human and domesticated animals.

There are concerns that the natural nanoscale process could be influenced by the presence of pollution. Manufactured NPs show some complex colloid and aggregation chemistry, which is likely to be affected by particle shape, size, surface area, and surface charge, as well as the adsorption properties of the material. Abiotic factors such as pH, ionic strength, water hardness, and the presence of organic matter will alter aggregation chemistry and are expected to influence toxicity. Data on biological effects show that NPs can be toxic to bacteria, algae, invertebrates, and fish species, as well as mammals. However, much of the ecotoxicological data is limited to species used in regulatory testing and freshwater organisms. Data on bacteria, terrestrial species, marine species and higher plants are particularly lacking. Detailed investigations of absorption, distribution, metabolism, and excretion (ADME) remain to be performed on species from the major phyla, although there are some data on fish. The environmental risk assessment of NMs could be performed using the existing tiered approach and regulatory framework, but with modifications to methodology including chemical characterization of the materials being used. Many challenges lie ahead, as well as controversies (e.g., reference substances for ecotoxicology), but knowledge transfer from mammalian toxicology, colloid chemistry, together with material and geological sciences, will enable ecotoxicology studies to move forward in this new multidisciplinary field.

Whole industries are built around the design and selection of dispersants to optimize product performance. Invariably, together with other macromolecules or liquid crystalline surface-active agents, these agents modify the flow characteristics of the product. The formulator's skill lies in achieving the necessary product characteristics such as mouth-feel (organoleptic properties) with synthetic foodstuffs, nondrip properties of paints, and prevention of fine mist with liquid aerosols. There are endless uses of colloids in the industry; however, from the environmental protection perspective, colloids have been used extensively in water and wastewater treatment.

All the scientific personnel working in colloidal group laboratories are required to carry out a mandatory risk assessment for all procedures carried out in the laboratory. No work goes on in the lab until a risk assessment has been done. Such assessment is for the benefit of workers as well as their associated coworkers in the laboratory and anyone else who may have to enter the laboratory routinely (cleaners, building supervisors) or in an emergency situation (porters, security, fire brigade). In preparing for an experiment, it must be remembered that coworkers may not understand fully the risks of the reactions, let alone other nonchemists who may be involved.

SYMBOLS AND ABBREVIATIONS

AA acrylic acid
ADME absorption, distribution, metabolism, and excretion
Ag silver
AM acrylamide
Am americium
Au gold
AUL absorbency under load
CB carbon black
Cd cadmium
CdSe cadmium selenide
CdSe cadmium telludride
CMC carboxymethyl cellulose
CNT carbon nanotubes
Cu copper
DDS drug delivery systems
DDT dichlorodiphenyltrichloroethane
EPA Environmental Protection Agency
Fe iron
g/g gram/gram
KDOC coefficient of dissolved organic carbon partition
mM millimole
nC_{60} fullerene
Ni nickel
nm nanometer
NMs nano naterials
NP nanoparticle
nZVI nanoscale zero-valent iron
O organic
PAH polycylic aromatic hydrocarbons
Pb lead
PBDEs polybrominated diphenyl ethers
PEG polyethylene glycol
Pt platinum
Rh rhodium
SAP superabsorbent polymer
SiO_2 silicon oxide
TiO_2 titanium oxide
W water

ZnO zinc oxide
ZnS zinc sulfide

Glossary

Aerogel a manufactured material with the lowest bulk density of any known porous solid
Aerosol a gaseous suspension of fine solid or liquid particles
Association colloids the systems in which the dispersed phase consists of clusters of molecules that have lyophobic and lyophilic parts
Co-ions ions of low relative molecular mass with the same charge as colloidal ion Continuous phase: the liquid in a disperse system in which solids are suspended or droplets of another liquid are dispersed. Also known as dispersion medium, external phase.
Colloid a system in which finely divided particles, which are approximately 10 to 10,000 angstroms in size, are dispersed within a continuous medium in a manner that prevents them from being filtered easily or settled rapidly
Colloidal a type of chemical mixture in which one substance is dispersed evenly throughout another
Colloidal dispersion See Colloidal mixture
Colloidal electrolyte an electrolyte that gives ions of which at least one is of colloidal size.
Colloidal mixture an intimate mixture of two substances, one of which, called the dispersed phase (or colloid), is uniformly distributed in a finely divided state through the second substance, called the dispersion medium (or dispersing medium); the dispersion medium or dispersed phase may be a gas, liquid, or solid. Also known as colloidal dispersion; colloidal suspension
Counter-ions ions of low relative molecular mass, with a charge opposite to that of the colloidal ion
Electroneutrality when there is no measurable charge excess in any side of the membrane
Emulsion when liquid droplets and/or liquid crystals are dispersed in a liquid
Foam a substance that is formed by trapping many gas bubbles in a liquid or solid
Fog an aerosol of liquid particles, in particular a low cloud
Froth a used interchangeably with foam
Froth-flotation a process for recovery of particles of ore or other material, in which the particles adhere to bubbles and can be removed as part of the froth
Gel a colloidal system with a finite, usually rather small, yield stress
Heterodisperse a colloidal system whose particle size are not uniform
Latex an emulsion or sol in which each colloidal particle contains a number of macromolecules
Lyophilic sols similarto true solutions in which the solute molecules are large and have an affinity for the solvent
Lyophilic having an affinity for a solvent ("solvent-loving"; if the solvent is water, the term *hydrophilic* is used)
Lyophobic sols in which there is no affinity between the dispersed phase and the liquid
Lyophobic lacking any affinity for a solvent (hydrophobic, if the dispersion medium is an aqueous solution)
Macromolecules a very large molecule, such as a polymer or protein, consisting of many smaller structural units linked together
Monodisperse a colloidal system containing the particles of nearly the same size
Paucidisperse when only a few particle sizes occur in a colloidal system
Polydisperse when many particle sizes occur in a colloidal system
Polyelectrolyte an electrolyte, such as a protein or polysaccharide, having a high molecular weight
Smoke an aerosol originating from combustion, thermal decomposition, or thermal evaporation. Its particles may be solid (magnesium oxide smoke) or liquid (tobacco smoke).
Suspension when solid particles are dispersed in a liquid
Xerogel a gel whose final form contains little or none of the dispersion medium used

References

Ahmadpour, A., Maskoki, A., Rezaie, M., 2007. Dewatering of yogurt using a permeable membrane and acrylic superabsorbent hydrogel. Iran. J. Polym. Sci. Technol. 20, 551–559.

Aitken, R.J., Chaudhry, M.Q., Boxall, A.B.A., Hull, M., 2006. Manufacture and use of nanomaterials: current status in the UK and global trends. Occup. Med. 56, 300–306.

Akers, R.J., 1972. Flocculation. Institute of Chemical Engineers, London.

Athawale, V.D., Lele, V., 2001. Recent trends in hydrogels based on starch-graft-acrylic acid: A review. Starch/Starke 3, 7–13.

Ballesteros, E., Gallego, M., Valcarcel, M., 2000. Analytical potential of fullerene as adsorbent for organic and organometallic compounds from aqueous solutions. J. Chromatogr. A. 869, 101–110.

Balnois, E., Wilkinson, K.J., 2002. Sample preparation techniques for the observation of environmental biopolymers by atomic force microscopy. Colloids Surfaces A. Physicochemical Eng. Aspects 207, 229–242.

Barvenik, F.W., 1994. Polyacrylamide characteristics related to soil applications. Soil Sci. 158, 235–243.

Becker, L., Poreda, R.J., Bada, J.L., 1996. Extraterrestrial helium trapped in fullerenes in the Sudbury impact structure. Science 272, 249–252.

Becker, L., Poreda, R.J., Hunt, A.G., Bunch, T.E., Rampino, M., 2001. Impact event at the PermianeTriassic boundary: evidence from extraterrestrial noble gases in fullerenes. Science 291, 1530–1533.

Bianco, A., Prato, M., 2003. Can carbon nanotubes be considered useful tools for biological applications? Adv. Mater. 15, 1765–1768.

Biesaga, M., Pyrzynska, K., 2006. The evaluation of carbon nanotubes as a sorbent for dicamba herbicide. J. Sep. Sci. 29, 2241–2244.

Biswas, P., Wu, C.Y., 2005. Nanoparticles and the environment. J. Air. Waste. Manag. Assoc. 55, 708–746.

Blackford, D.B., Simons, G.R., 1987. Particle-size analysis of carbon-black. Part. Charact. 4, 112–117.

Bratby, J., 2006. Coagulation and flocculation in water and wastewater treatment. IWA Publishing, London, Seattle.

Bosi, S., Da Ros, T., Spalluto, G., Prato, M., 2003. Fullerene derivatives: an attractive tool for biological applications. Eur. J. Med. Chem. 38, 913–923.

Bouranis, D.L., Theodoropoulos, A.G., Drossopoulos, J.B., 1995. Designing synthetic polymers as soil conditioners. Commun. Soil Sci. Plant Anal. 26, 1455–1480.

Bowman, D.C., Evans, R.Y., Paul, J.L., 1990. Fertilizer salts reduce hydration of polyacryamide gels and affect physical properties of gel-amended container media. J. Am. Soc. Hortic. Sci. 115, 382–386.

Brannon-Peppas, L., Harland, R.S., 1990. Absorbent polymer technology. Elsevier, Amsterdam. Ch 1–4.

Brust, M., Liely, C.J., 2002. Some recent advances in nanostructure preparation from gold and silver particles: A short topical review. Colloids Surf. A. 202, 175–186.

Bucheli, T.D., Blum, F., Desaules, A., Gustafsson, O., 2004. Polycyclic aromatic hydrocarbons, black carbon, and molecular markers in soils of Switzerland. Chemosphere 56, 1061–1076.

Buchholz, F.L., 1994. Recent advances in superabsorbent polyacrylates. Trend Polym. Sci. 2, 277–281.

Buchholz, F.L., Graham, A.T., 1998. Modern superabsorbent polymer technology. Wiley-VCH, New York. Ch. 1–7.

Buffle, J., 2006. The key role of environmental colloids/nanoparticles for the sustainability of life. Environ. Chem. 3, 155–158.

Burkhard, L.P., 2000. Estimating dissolved organic carbon partition coefficients for nonionic organic chemicals. Environ. Sci. Technol. 34, 4663–4668.

Cai, Y.Q., Cai, Y.E., Mou, S.F., Lu, Y.Q., 2005. Multi-walled carbon nanotubes as a solid-phase extraction adsorbent for the determination of chlorophenols in environmental water samples. J. Chromatogr. A. 1081, 245–247.

Cai, Y.Q., Jiang, G.B., Liu, J.F., Zhou, Q.X., 2003a. Multi-walled carbon nanotubes packed cartridge for the solid-phase extraction of several phthalate esters from water samples and their determination by high performance liquid chromatography. Anal. Chim. Acta. 494, 149–156.

Cai, Y.Q., Jiang, G.B., Liu, J.F., Zhou, Q.X., 2003b. Multiwalled carbon nanotubes as a solid-phase extraction adsorbent for the determination of bisphenol a, 4-n-nonylphenol, and 4-tert-octylphenol. Anal. Chem. 75, 2517–2521.

Cain, M., Morrell, R., 2001. Nanostructured ceramics: A review of their potential. Appl. Organomet. Chem. 15, 321–330.

Chatterjee, P.K., Gupta, B.S. (Eds.), 2002. Absorbent technology. Elsevier, Amsterdam. Ch. 1–2.

Chaudhuri, M., Verma, S.R., Gupta, A., 1994. Performance evaluation of ceramic filter candles. J. Environ. Eng. 120, 1646–1651.

Chen, C.L., Wang, X.K., 2006. Adsorption of Ni(II) from aqueous solution using oxidized multiwall carbon nanotubes. Ind. Eng. Chem. Res. 45, 9144–9149.

Cheng, X., Kan, A.T., Tomson, M.B., 2004. Naphthalene adsorption and desorption from aqueous C60 fullerene. J. Chem. Eng. Data 49, 675–683.

Cheng, X., Kan, A.T., Tomson, M.B., 2005a. Study of C60 transport in porous media and the effect of sorbed C60 on naphthalene transport. J. Mater. Res. 20, 3244–3254.

Cheng, X., Kan, A.T., Tomson, M.B., 2005. Uptake and sequestration of naphthalene and 1,2-dichlorobenzene by C60. J. Nanopart. Res. 7, 555–567.

Clasen, T.F., Brown, J., Collin, S., Suntura, O., Cairncross, S., 2004. Reducing diarrhea through the use of household-based ceramic water filters: A randomized, controlled trial in rural Bolivia. Am. J. Trop. Med. Hyg. 70, 651–657.

Colvin, V.L., 2003. The potential environmental impact of engineered nanomaterials. Nat. Biotechnol. 21, 1166–1170.

Cordani, P., 2001. Method of modifying weather. U.S. Patent 6,315,213.

Cornelissen, G., Gustafsson, O., Bucheli, T.D., Jonker, M.T.O., Koelmans, A.A., Van Noort, P.C.M., 2005. Extensive sorption of organic compounds to black carbon, coal, and kerogen in sediments and soils: Mechanisms and consequences for distribution, bioaccumulation, and biodegradation. Environ. Sci. Technol. 39, 6881–6895.

Cutie, S.S., Buzanowski, W.C., Berdasco, J.A., 1990. Fate of superabsorbents in the environment (analytical techniques). J. Chromatogr. A. 513, 93–105.

Dai, H.J., 2002. Carbon nanotubes: synthesis, integration, and properties. Acc. Chem. Res. 35, 1035–1044.

Dayal, U., Mehta, S.K., Choudhari, M.S., Jain, R., 1999. Synthesis of acrylic superabsorbents. J. Macromol. Sci. Rev. Macromol. Chem. Phys. C39, 507–525.

Ding, Q., Liang, P., Song, F., Xiang, A., 2006. Separation and preconcentration of silver ion using multiwalled carbon nanotubes as solid phase extraction sorbent. Sep. Sci. Technol. 41, 2723–2732.

Dunphy Guzman, K.A., Finnegan, D.L., Banfield, J.F., 2006. Influence of surface potential on aggregation and transport of titania nanoparticles. Environ. Sci. Technol. 40, 7688–7693.

Dutkiewicz, J.K., 2002. Superabsorbent materials from shellfish waste-A review. J Biomed. Mater. Res.(Appl. Biomater.) 63, 373–381.

Elliott, D.W., Zhang, W.X., 2001. Field assessment of nanoscale biometallic particles for groundwater treatment. Environ. Sci. Technol. 35, 4922–4926.

EPA, 2007. Nanotechnology White Paper. U.S. Environmental Protection Agency Report EPA 100/B-07/001, Washington, DC 20460.

Esrey, S.A., Potash, J.B., Roberts, L., Shiff, C., 1991. Effects of improved water-supply and sanitation on ascariasis, diarrhea, dracunculiasis, hookworm infection, schistosomiasis, and trachoma. Bull. WHO 69, 609–621.

Fugetsu, B., Satoh, S., Shiba, T., Mizutani, T., Lin, Y.B., Terui, N., Nodasaka, Y., Sasa, K., Shimizu, K., Akasaka, T., Shindoh, M., Shibata, K.I., Yokoyama, A., Mori, M., Tanaka, K., Sato, Y., Tohji, K., Tanaka, S., Nishi, N., Watari, F., 2004. Caged multiwalled carbon nanotubes as the adsorbents for affinity-based elimination of ionic dyes. Environ. Sci. Technol. 38, 6890–6896.

Fukushi, K., Sato, T., 2005. Using a surface complexation model to predict the nature and stability of nanoparticles. Environ. Sci. Technol. 39, 1250–1256.

Gao, D., Heimann, R.B., Alexander, S.D.B., 1997. Box-Behnken design applied to study the strengthening of aluminate concrete modified by a superabsorbent polymer/clay composite. Adv Chem. Res 9, 93–97.

Gao, X.H., Cui, Y.Y., Levenson, R.M., Chung, L.W.K., Nie, S.M., 2004. In vivo cancer targeting and imaging with semiconductor quantum dots. Nat. Biotechnol. 22, 969–976.

Geckeis, H., Rabung, T., Manh, T.N., et al., 2002. Humic colloid-borne natural polyvalent metal ions: Dissociation experiment. Environ. Sci. Technol. 36, 2946–2952.

Giammar, D.E., Maus, C.J., Xie, L.Y., 2007. Effects of particle size and crystalline phase on lead adsorption to titanium dioxide nanoparticles. Environ. Eng. Sci. 24, 85–95.

Giasuddin, A.B.M., Kanel, S.R., Choi, H., 2007. Adsorption of humic acid onto nanoscale zerovalent iron and its effect on arsenic removal. Environ. Sci. Technol. 41, 2022–2027.

Giles, J., 2006. Top five in physics. Nature 441, 265.

Gimbert, L.J., Hamon, R.E., Casey, P.S., Worsfold, P.J., 2007. Partitioning and stability of engineered ZnO nanoparticles in soil suspensions using flow field-flow fractionation. Environ. Chem. 4, 8–10.

Gotovac, S., Hattori, Y., Noguchi, D., Miyamoto, J., Kanamaru, M., Utsumi, S., Kanoh, H., Kaneko, K., 2006. Phenanthrene adsorption from solution on single wall carbon nanotubes. J. Phys. Chem. B. 110, 16219–16224.

Grigoriev, Dmitry, O., Miller, R., 2009. Mono- and multilayer covered drops as carriers. Curr. Opin. Colloid Interface Sci. 14, 48–59.

Grula, M.M., Huang, M.-L., Sewell, G., 1994. Interactions of certain polyacrylamides with soil bacteria. Soil. Sci. 158, 291–300.

Guo, M., Liu, M., Liang, R., Niu, A., 2006. Granular ureaformaldehyde slow-release fertilizer with superabsorbent and moisture preservation. J. Appl. Polym. Sci. 99, 3230–3235.

Guo, M., Liu, M., Zhan, F., Wu, L., 2005. Preparation and properties of a slow release membrane-encapsulated urea fertilizer with superabsorbent and moisture preservation". Ind. Eng. Chem. Res. 44, 4206–4211.

Guzman, K.A.D., Taylor, M.R., Banfield, J.F., 2006. Environmental risks of nanotechnology: national nanotechnology initiative funding, 2000–2004. Environ. Sci. Technol. 40, 1401–1407.

He, F., Zhao, D.Y., 2005. Preparation and characterization of a new class of starch-stabilized bimetallic nanoparticles for degradation of chlorinated hydrocarbons in water." Environ. Sci. Technol. 39, 3314–3320.

Helland, A., Kastenholz, H., Thidell, A., Arnfalk, P., Deppert, K., 2006. Nanoparticulate materials and regulatory policy in Europe: An analysis of stakeholder perspectives. J. Nanopart. Res. 8, 709–719.

Helland, A., Wick, P., Koehler, A., Schmid, K., Som, C., 2007. Reviewing the environmental and human health knowledge base of carbon nanotubes. Environ. Health. Perspect. 115, 1125–1131.

Heymann, D., Jenneskens, L.W., Jehlicka, J., Koper, C., Vlietstra, E., 2003. Terrestrial and extraterrestrial fullerenes. Fuller. Nanotub. Carbon Nanostruct. 11, 333–370.

Hochella, M.F., 2002. There's plenty of room at the bottom: nanoscience in geochemistry. Geochim. Cosmochim. Acta. 66, 735–743.

Hyung, H., Fortner, J.D., Hughes, J.B., Kim, J.H., 2007. Natural organic matter stabilizes carbon nanotubes in the aqueous phase. Environ. Sci. Technol. 41, 179–184.

Ichikawa, T., Nakajima, T., 1996. Superabsorptive Polymers (from natural polysaceharides and polypeptides). In: Salamone (Ed.), Polymeric MaterialsEncyclopedia. CRC Press, Boca Raton, FL, pp. 8051–8059.

Irace-Guigand, S., Aaron, J.J., 2003. The role of organic colloids in herbicide transfer to rivers: a quantitative study of triazine and phenylurea interactions with colloids. Anal. Bioanal. Chem. 376, 431–435.

Kallay, N., Zalac, S., 2001. Introduction of the surface complexation model into the theory of colloid stability. Croat. Chem. Acta. 74, 479–497.

Kallay, N., Zalac, S., 2002. Stability of nanodispersions: A model for kinetics of aggregation of nanoparticles. J. Colloid Interface Sci. 253, 70–76.

Kazanskii, K.S., Dubrovskii, S.A., 1992. Chemistry and physics of "agricultural" hydrogels. Adv. Polym. Sci. 104, 97–140.

Kenawy, E.-R., 1998. Recent advances in controlled release of agrochemical. J. Macromol. Sci. Rev. Macromol. Chem. Phys. C38, 365–390.

Kim, J.Y., Cohen, C., Shuler, M.L., 2000. Use of amphiphilic polymer particles for in situ extraction of sorbed phenanthrene from a contaminated aquifer material. Environ. Sci. Technol. 34, 4133–4139.

Kim, J.Y., Shim, S.B., Shim, J.K., 2004a. Comparison of amphiphilic polyurethane nanoparticles to nonionic surfactants for flushing phenanthrene from soil. J. Hazard. Mater. 116, 205–212.

Kim, J.Y., Shim, S.B., Shim, J.K., 2003a. Effect of amphiphilic polyurethane nanoparticles on sorptionedesorption of phenanthrene in aquifer material. J. Hazard. Mater. 98, 145–160.

Kim, J.Y., Shim, S.B., Shim, J.K., 2004b. Enhanced desorption of phenanthrene from aquifer sand using amphiphilic anionic polyurethane nanoparticles. J. Ind. Eng. Chem. 10, 1043–1051.

Kim, J.Y., Shim, S.B., Shim, J.K., 2003b. Synthesis of amphiphilic polyurethane nanonetwork particles and their application for the soil-washing process. J. Appl. Polym. Sci. 87, 1666–1677.

Koehler, A., Som, C., Helland, A., Gottschalk, F., 2008. Studying the potential release of carbon nanotubes throughout the application life cycle. J. Clean. Prod. 16, 927–937.

Koelmans, A.A., Jonker, M.T.O., Cornelissen, G., Bucheli, T.D., Van Noort, P.C.M., Gustafsson, O., 2006. Black carbon: The reverse of its dark side. Chemosphere 63, 365–377.

Koziara, J.M., Lockman, P.R., Allen, D.D., Mumper, R.J., 2003. In situ bloodebrain barrier transport of nanoparticles. Pharm. Res. 20, 1772–1778.

Kreyling, W.G., Semmler-Behnke, M., Moller, W., 2006. Health implications of nanoparticles". J. Nanopart. Res. 8, 543–562.

Kulicke, W.-M., Nottelmann, H., 1989. Structure and swelling of some synthetic, semisynthetic, and biopolymer hydrogels. Adv. Chem. Ser. 223, 15–44.

Kulkarni, D.N., Tajne, D.S., Parhad, N.M., 1980. Performance of domestic water filters. Indian J. Environ. Health 22, 30–41.

Lam, C.W., James, J.T., McCluskey, R., Arepalli, S., Hunter, R.L., 2006. A review of carbon nanotube toxicity and assessment of potential occupational and environmental health risks. Crit. Rev. Toxicol. 36, 189–217.

Lantagne, D.S., 2001. Investigation of the potters for peace colloidal silver impregnated ceramic filter. USAID, 79.

Larlus, O., Mintova, S., Bein, T., 2006. Environmental syntheses of nanosized zeolites with high yield and monomodal particle size distribution. Micropor. Mesopor. Mater. 96, 405–412.

Larson, R.J., Bookland, E.A., Williams, R.T., Yokom, K.M., Saucy, D.A., Freeman, M.B., Swift, G., 1997. Biodegradation of acrylic acid polymers and oligomers by mixed microbial communities in activated sludge. J. Environ. Polym. Degrad. 5, 41–48.

Lead, J.R., Wilkinson, K.J., 2006. Aquatic colloids and manoparticles: Current knowledge and future trends. Environ. Chem. 3, 159–171.

Lecoanet, H.F., Bottero, J.Y., Wiesner, M.R., 2004. Laboratory assessment of the mobility of nanomaterials in porous media. Environ. Sci. Technol. 38, 5164–5169.

Lecoanet, H.F., Wiesner, M.R., 2004. Velocity effects on fullerene and oxide nanoparticle deposition in porous media. Environ. Sci. Technol. 38, 4377–4382.

Levy, R., Nichols, M.A., Miller, T.W., 1995. Evaluation of superabsorbent polymer-pesticide formulations for prolonged insect control, ASTM Special Technical No. 1234, 330–339. In: Proceedings of the 14th Symposium on Pesticide Formulations and Application Systems. Dallas, TX. October 12-13.

Li, X.Q., Elliott, D.W., Zhang, W.X., 2006a. Zero-valent iron nanoparticles for abatement of environmental pollutants: materials and engineering aspects. Crit. Rev. Solid State Mater. Sci. 31, 111–122.

Li, Y.H., Wang, S., Wei, J., Zhang, X., Xu, C., Luan, Z., Wu, D., Wei, B., 2002. Lead adsorption on carbon nanotubes. Chem. Phys. Lett. 357, 263–266.

Li, Y.H., Wang, S.G., Luan, Z.K., Ding, J., Xu, C.L., Wu, D.H., 2003. Adsorption of cadmium(II) from aqueous solution by surface oxidized carbon nanotubes. Carbon 41, 1057–1062.

Li, Y.H., Zhu, Y.Q., Zhao, Y.M., Wu, D.H., Luan, Z.K., 2006b. Different morphologies of carbon nanotubes effect on the lead removal from aqueous solution. Diam. Relat. Mater. 15, 90–94.

Liang, P., Ding, Q., Song, F., 2005a. Application of multiwalled carbon nanotubes as solid phase extraction sorbent for preconcentration of trace copper in water samples. J. Sep. Sci. 28, 2339–2343.

Liang, P., Liu, Y., Guo, L., 2005b. Determination of trace rare earth elements by inductively coupled plasma atomic emission spectrometry after preconcentration with multiwalled carbon nanotubes". Spectrochim. Acta. B. 60, 125–129.

Liang, P., Liu, Y., Guo, L., Zeng, J., Lu, H.B., 2004. Multiwalled carbon nanotubes as solid-phase extraction adsorbent for the preconcentration of trace metal ions and their determination by inductively coupled plasma atomic emission spectrometry." J. Anal. At. Spectrom. 19, 1489–1492.

Liang, R., Liu, M., Wu, L., 2007. Controlled release NPK compound fertilizer with the function of water retention. React. Func. Polym. 67, 769–779.

Liu, M., Liang, R., Zhan, F., Liu, Z., Niu, A., 2007. Preparation of superabsorbent slow release nitrogen fertilizer by inverse suspension polymerization. Polym. Int. 56, 729–737.

Liu, M., Liang, R., Zhan, F., Liu, Z., Niu, A., 2006. Synthesis of a slow-release and superabsorbent nitrogen fertilizer and its properties. Polym. Adv. Technol. 17, 430–438.

Long, R.Q., Yang, R.T., 2001. Carbon nanotubes as superior sorbent for dioxin removal. J. Am. Chem. Soc. 123, 2058–2059.

Lu, C., Liu, C., 2006. Removal of nickel(II) from aqueous solution by carbon nanotubes. J. Chem. Technol. Biotechnol. 81, 1932–1940.

Luby, S., Agboatwalla, M., Raza, A., Sobel, J., Mintz, E., Baier, K., Rahbar, M., Qureshi, S., Hassan, R., Ghouri, F., Hoekstra, R.M., Gangarosa, E., 2001. Alow-cost intervention for cleaner drinking water in Karachi, Pakistan. Int. J. Infect. Dis. 5, 144–150.

Lyven, B., Hassellov, M., Turner, D.R., et al., 2003. Competition between iron- and carbon-based colloidal carriers for trace metals in a freshwater assessed using flow field-flow fractionation coupled to ICPMS. Geochim. Cosmochim. Acta. 67, 3791–3802.

Mackay, C.E., Johns, M., Salatas, J.H., Bessinger, B., Perri, M., 2006. Stochastic probability modeling to predict the environmental stability of nanoparticles in aqueous suspension. Integr. Environ. Assess. Manag. 2, 293–298.

Madden, A.S., Hochella, M.F., Luxton, T.P., 2006. Insights for size-dependent reactivity of hematite nanomineral surfaces through Cu2þ sorption. Geochim. Cosmochim. Acta. 70, 4095–4104.

Mathur, A.M., Moorjani, S.K., Scranton, A.B., 1996. Methods for synthesis of hydrogel networks: A review. J. Macromol. Sci. Rev. Macromol. Chem. Phys. C36, 405–430.

Mattigod, S.V., Fryxell, G.E., Alford, K., Gilmore, T., Parker, K., Serne, J., Engelhard, M., 2005. Functionalized TiO_2 nanoparticles for use for in situ anion immobilization. Environ. Sci. Technol. 39, 7306–7310.

Maynard, A.D., 2006. Nanotechnology: A Research Strategy for Addressing Risk. Woodrow Wilson International Center for Scholars, Washington, DC.

McGechan, M.B., Lewis, D.R., 2002. Transport of particulate and colloidsorbed contaminants through soil, part 1: General principles. Biosyst. Eng. 83, 255–273.

Mintz, E.D., Reiff, F.M., Tauxe, R.V., 1995. Safe water treatment and storage in the home. J. Am. Med. Assoc. 273, 948–953.

Mondi, C., Leifer, K., Mavrocordatos, D., et al., 2002. Analytical electron microscopy as a tool for accessing colloid formation process in natural waters. J. Microsc. Oxford 207, 180–190.

Moore, M.N., 2006. Do nanoparticles present ecotoxicologiocal risks for the health of the aquatic environment? Environ. Int. 32, 967–976.

Morones, J.R., Elechiguerra, J.L., Camacho, A., Holt, K., Kouri, J.B., Ramirez, J.T., Yacaman, M.J., 2005. The bactericidal effect of silver nanoparticles. Nanotechnology 16, 2346–2353.

Mudiyanselage, T.K., Neckers, D.C., 2008b. Highly absorbing superabsorbent polymer. J. Polym. Sci. A. Polym. Chem. 46, 1357–1364.

Mudiyanselage, T.K., Neckers, D.C., 2008a. Photochromic superabsorbent polymers. Soft Matter. 4, 768–774.

Muirhead, D., Lead, J.R., 2003. Measurement of the size and structure of natural aquatic colloids in an urbanised watershed by atomic force microscopy. Hydrobiologia 494, 65–69.

Munoz, J., Gallego, M., Valcarcel, M., 2005. Speciation of organometallic compounds in environmetal samples by gas chromatography after flow preconcentration on fullerenes and nanotubes. Anal. Chem. 77, 5389–5395.

Nel, A., Xia, T., Mädler, L., Li, N., 2006. Toxic potential of materials at the nanolevel. Science 311, 622–627.

Niyogi, S., Hamon, M.A., Hu, H., Zhao, B., Bhowmik, P., Sen, R., Itkis, M.E., Haddon, R.C., 2002. Chemistry of single-walled carbon nanotubes. Acc. Chem. Res. 35, 1105–1113.

Nowack, B., Bucheli, T.D., 2007. Occurrence, behavior and effects of nanoparticles in the environment. Environ. Pollut. 150, 5–22.

Obare, S.O., Meyer, G.J., 2004. Nanostructured materials for environmental remediation of organic contaminants in water. J. Environ. Sci. Health Part A. 39, 2549–2582.

Oberdörster, E., McClellan-Green, P., Haasch, M.L., 2006a. Ecotoxicity of engineered nanomaterials. In: Kumar (Ed.), Nanomaterials—Toxicity, Health and Environmental Issues. Wiley-VCH, Weinheim.

Oberdörster, G., Oberdörster, E., Oberdörster, J., 2005. Nanotoxicology: An emerging discipline evolving from studies of ultrafine particles. Environ. Health Perspect. 113, 823–839.

Omelia, C.R., 1980. Aquasols e the behavior of small particles in aquatic systems. Environ. Sci. Technol. 14, 1052–1060.

Pan, F., Lee, K., Wang, P., et al., 2003. Surface micropatterning of biomolecules through photolithography on block copolymer polymer brushes. Abstracts Pap. Am. Chem. Soc. 225. U618–U618.

Peng, X.J., Li, Y.H., Luan, Z.K., Di, Z.C., Wang, H.Y., Tian, B.H., Jia, Z.P., 2003. 2003. Adsorption of 1,2-dichlorobenzene from water to carbon nanotubes. Chem. Phys. Lett. 376, 154–158.

Phenrat, T., Saleh, N., Sirk, K., Tilton, R.D., Lowry, G.V., 2007. Aggregation and sedimentation of aqueous nanoscale zerovalent iron dispersions. Environ. Sci. Technol. 41, 284–290.

Po, R., 1994. Water-absorbent polymers: A patent survey. J. Macromol. Sci. Rev. Macromol. Chem. Phys. 34, 607–662.

Pokropivnyi, V.V., 2001. Non-carbon nanotubes (review). II. Types and structure. Powder Metall. Met. Ceram. 40, 582–594.

Prato, M., 1999. Fullerene Materials. Top. Curr. Chem. 199, 173–187.

Quick, R.E., Kimura, A., Thevos, A., Tembo, M., Shamputa, I., Hutwagner, L., Mintz, E., 2002. Diarrhea prevention through household-level water disinfection and safe storage in Zambia. Am. J. Trop. Med. Hyg. 66, 584–589.

Quinn, J., Geiger, C., Clausen, C., Brooks, K., Coon, C., O'Hara, S., Krug, T., Major, D., Yoon, W.S., Gavaskar, A., Holdsworth, T., 2005. Field demonstration of DNAPL dehalogenation using emulsified zero-valent iron. Environ. Sci. Technol. 39, 1309–1318.

Rittner, M.N., 2002. Market analysis of nanostructured materials. Am. Ceram. Soc. Bull 81, 33–36.

Roco, M.C., 2005. Environmentally responsible development of nanotechnology. Environ. Sci. Technol. 39, 106A–112A.

Rudzinski, W.E., Dave, A.M., Vaishnav, U.H., Kumbar, S.G., Kulkarni, A.R., Aminabhavi, T.M., 2002. Hydrogels as conclolled release devices in agriculture. Designed Monomers Polym. 5, 39–65.

Saleh, N., Sirk, K., Liu, Y.Q., Phenrat, T., Dufour, B., Matyjaszewski, K., Tilton, R.D., Lowry, G.V., 2007. Surface modifications enhance nanoiron transport and NAPL targeting in saturated porous media. Environ. Eng. Sci. 24, 45–57.

Sano, M., Okamura, J., Shinkai, S., 2001. Colloidal nature of single-walled carbon nanotubes in electrolyte solution: The Schulzee Hardy rule. Langmuir 17, 7172–7173.

Schrick, B., Hydutsky, B.W., Blough, J.L., Mallouk, T.E., 2004. Delivery vehicles for zerovalent metal nanoparticles in soil and groundwater. Chem. Mater. 16, 2187–2193.

Sen, T.K., Khilar, K.C., 2006. Review on subsurface colloids and colloid-associated contaminant transport in saturated porous media. Adv. Colloid Interface Sci. 119, 71–96.

Siegrist, M., Wiek, A., Helland, A., Kastenholz, H., 2007. Risks and nanotechnology: the public is more concerned than experts and industry. Nat. Nanotechnol. 2, 67.

Sirisinha, C., Prayoonchatphan, N., 2001. Study of carbon black distribution in BR/NBR blends based on damping properties: influences of carbon black particle size, filler, and rubber polarity. J. Appl. Polymer Sci. 81, 3198–3203.

Sobsey, M.D., 2002. Managing water in the home: Accelerated health gains from improved water supply. World Health Organization. http://www.who.int/water_sanitation_health/dwq/wsh0207/en/.

Stahl, J.D., Cameron, M.D., Haselbach, J., Aust, S.D., 2000. Biodegradation of superabsorbent polymers in soil. Environ. Sci. Pollut. Res. 7, 83–88.

Tessier, B., Terwind, T.J.H.J., 1994. An example of soft-sediment deformations in an Intertidal Enviornnment—The effect of a tidal bore. Comptes Rendus de l'academie des Sci. Serie II 319, 217–233.

Tscharnuter, W.W., Zu, L., Weiner, B.B., Bernt, W., 2001. ASTM Carbon black reference materials: particle sizing using a Brookhaven Instruments BI-DCP, Disc Centrifuge Photodensitometer. In: Proceedings of the Conference Carbon Black in the New Millenium. Gold Coast, Queensland, Australia. November 5–8.

Tungittiplakorn, W., Cohen, C., Lion, L.W., 2005. Engineered polymeric nanoparticles for bioremediation of hydrophobic contaminants. Environ. Sci. Technol. 39, 1354–1358.

Tungittiplakorn, W., Lion, L.W., Cohen, C., Kim, J.Y., 2004. Engineered polymeric nanoparticles for soil remediation. Environ. Sci. Technol. 38, 1605–1610.

Vinogradov, Serguei, V., 2006. Colloidal microgels in drug delivery applications. Curr. Pharm. Des. 12, 4703–4712.

Wallace, A., Wallace, G.A., Abuzamzam, A.M., 1986. Effects of a polymer as soil conditioner on yields and mineral nutrition of plants. Soil Sci. 143, 377–380.

Wang, J., Guo, P., Li, X., et al., 2000. Source identification of lead pollution in the atmosphere of Shanghai City by analyzing single aerosol particles (SAP). Environ. Sci. Technol. 34, 1900–1905.

Wang, J.X., Jiang, D.Q., Gu, Z.Y., Yan, X.P., 2006. Multiwalled carbon nanotubes coated fibers for solid-phase microextraction of polybrominated diphenyl ethers in water and milk samples before gas chromatography with electron-capture detection. J. Chromatogr. A. 1137, 8–14.

Wang, X., Chen, C., Hu, W., Ding, A., Xu, D., Zhou, X., 2005. Sorption of 243Am(III) to multiwall carbon nanotubes. Environ. Sci. Technol. 39, 2856–2860.

Wiesner, M.R., Lowry, G.V., Alvarez, P., Dionysiou, D., Biswas, P., 2006. Assessing the risks of manufactured nanomaterials. Environ. Sci. Technol. 40, 4336–4345.

Wolter, M., Wiesche, C., Zadrazil, F., Hey, S., Haselbach, J., Schnug, E., 2002. Biological degradability of synthetic superabsorbent soil conditioners. Landbauforschung Volkenrode 1, 43–52.

Wu, L., Liu, M., 2008. Preparation and characterization of cellulose acetate-coated compound fertilizer with controlled-release and water-retention. Polym. Adv. Technol. http://dx.doi.org/10.1002/pat.1034 in press.

Wu, L., Liu, M., 2008. Preparation and properties of chitosan-coated NPK compound fertilizer with controlled-release and water-retention. Carbohydr. Polym. 72, 240–247.

Wu, L., Liu, M., Liang, R., 2008. Preparation and properties of a double-coated slow-release NPK compound fertilizer with superabsorbent and waterretention. Bioresour. Technol. 99, 547–554.

Wudl, F., 2002. Fullerene materials. J. Mater. Chem. 12, 1959–1963.

Xu, Y.H., Zhao, D.Y., 2005. Removal of copper from contaminated soil by use of poly(amidoamine) dendrimers. Environ. Sci. Technol. 39, 2369–2375.

Xu, Y.H., Zhao, D.Y., 2006. Removal of lead from contaminated soils using poly(amidoamine) dendrimers. Ind. Eng. Chem. Res. 45, 1758–1765. www.ecologixsystems.com/product-specialty-chemicals-coag-floc.php.

Yamane, H., Ideguchi, T., Tokuda, M., Koga, H., 1994. A new ground-reducing material based on waterabsorbent polymer. Electron. Commun. Japan Part 1 77, 68–77.

Yang, K., Wang, X.L., Zhu, L.Z., Xing, B.S., 2006b. Competitive sorption of pyrene, phenanthrene, and naphthalene on multiwalled carbon nanotubes. Environ. Sci. Technol. 40, 5804–5810.

Yang, K., Xing, B., 2007. Desorption of polycyclic aromatic hydrocarbons from carbon nanomaterials in water. Environ. Pollut. 145, 529–537.

Yang, K., Zhu, L., Xing, B., 2006c. Adsorption of polycyclic aromatic hydrocarbons by carbon nanomaterials. bEnviron. Sci. Technol. 40, 1855–1861.

Yaron-Marcovich, D., Chen, Y., Nir, S., Prost, R., 2005. High resolution electron microscopy structural studies of organo-clay nanocomposites. Environ. Sci. Technol. 39, 1231–1238.

Zereini, F., Wiseman, C., Alt, F., Messerschmidt, J., Muller, J., Urban, H., 2001. Platinum and rhodium concentrations in airborne particulate matter in Germany from 1988 to 1998. Environ. Sci. Technol. 35, 1996–2000.

Zhang, H.M., Quan, X., Chen, S., Zhao, H.M., 2006. Fabrication and characterization of silica/titania nanotubes composite membrane with photocatalytic capability." Environ. Sci. Technol. 40, 6104–6109.

Zhang, W.X., 2003. Nanoscale iron particles for environmental remediation: an overview." J. Nanopart. Res. 5, 323–332.

Zhou, Q.X., Ding, Y.J., Xiao, J.P., 2006a. Sensitive determination of thiamethoxam, imidacloprid and acetamiprid in environmental water samples with solid-phase extraction packed with multiwalled carbon nanotubes prior to high-performance liquid chromatography. Anal. Bioanal. Chem. 385, 1520–1525.

Zhou, Q.X., Wang, W.D., Xiao, J.P., 2006b. Preconcentration and determination of nicosulfuron, thifensulfuron-methyl and metsulfuron-methyl in water samples using carbon nanotubes packed cartridge in combination with high performance liquid chromatography. Anal. Chim. Acta. 559, 200–206.

Zhou, Q.X., Xiao, J.P., Wang, W.D., 2006c. Using multi-walled carbon nanotubes as solid phase extraction adsorbents to determine dichlorodiphenyltrichloroethane and its metabolites at trace level in water samples by high performance liquid chromatography with UV detection. J. Chromatogr. A. 1125, 152–158.

Zhou, Q.X., Xiao, J.P., Wang, W.D., Liu, G.G., Shi, Q.Z., Wang, J.H., 2006d. Determination of atrazine and simazine in environmental water samples using multiwalled carbon nanotubes as the adsorbents for preconcentration prior to high performance liquid chromatography with diode array detector. Talanta 68, 1309–1315.

Zhu, Y., Zhao, Q., Li, Y., Cai, X., Li, W., 2006c. The interaction and toxicity of multi-walled carbon nanotubes with Stylonychia mytilus. J. Nanosci. Nanotechnol. 6, 1357–1364.

Zohuriaan-Mehr, M.J., 2006. Super-Absorbents (in Persian). Iran Polymer Society, Tehran, 2–4.

Index

Note: Page numbers followed by "f" denote figures; "t" tables.

A

B

C

D

E

F

G

H

I

L

M

N

O

P

Q